Proceedings of the 17th International Meshing Roundtable

Rao V. Garimella (Ed.)

Proceedings of the 17th International Meshing Roundtable

Springer

Editor

Rao V. Garimella
T-7, MS-B284
Los Alamos National Laboratory
Los Alamos, NM 87545
USA
Email: rao@lanl.gov

ISBN 978-3-540-87920-6 e-ISBN 978-3-540-87921-3

Library of Congress Control Number: 2008935489

Typesetting: Scientific Publishing Services Pvt. Ltd., Chennai, India.
Cover Design: WMX Design, Heidelberg

Printed in acid-free paper

12049370 5 4 3 2 1 0

springer.com

Reviewers

Alliez, Pierre	Inria, Sophia-Antipolis, France
Azernikov, Sergei	Siemens Corporate Research, Inc., USA
Benzley, Steven	Brigham Young Univ., USA
Blacker, Ted	Sandia Nat. Labs, USA
Blades, Eric	Mississippi State U., USA
Borden, Micheal	Sandia Nat. Labs, USA
Brewer, Michael	Sandia Nat. Labs, USA
Burgreen, Greg	Mississippi State U., USA
Chawner, John	Pointwise, Inc., USA
Chernikov, Andrey	C. of William and Mary, VA, USA
Clark, Brett	Sandia Nat. Labs, USA
Cuilliere, Jean-Christophe	Universit du Quebec a Trois-Rivieres, Canada
Dannenhoffer, John	Syracuse University, NY, USA
Dey, Tamal	Ohio State Univ., USA
Ebeida, Mohamed	U. of California, Davis, USA
Erickson, Jeff	U. of Illionois, Urbana-Champaign, USA
Federov, Andriy	C. of William and Mary, VA, USA
Frey, Pascal	Laboratorie Jacques Louis Lions. UPMC, France
Garimella, Rao	Los Alamos Nat. Labs, USA
Gillette, Andrew	U. of Texas, Austin, USA
Gosselin, Serge	U. of British Columbia, Canada
Guibault, Francois	Ecole Polytechnique de Montreal, Canada
Guoy, Damron	U. of Illionois, Urbana-Champaign, USA
Hudson, Benoit	Toyota Technilogical Institute at Chicago, USA
Ito, Yasushi	U. of Alabama, Birmingham, USA
Jiao, Xiangmin	State U. of New York, Stony Brook, USA
Kanai, Takashi	U. of Tokoyo, Japan
Karman, Steve	U. of Tennessee, Chattanooga, USA
Knupp, Patrick	Sandia Nat. Labs, USA
Kraftcheck, Jason	U. of Wisconsin, USA
Ledoux, Franck	CEA/DAM, USA
Lipnikov, Konstantin	Los Alamos Nat. Labs, USA
Luo, Xiaojuan	Rensselaer Polytechnic Institute, NY, USA
Maman, Nathan	Distene S.A.S, France
Marcum, David	Mississippi State U., USA
Mesri, Youssef	Inria, Sophia-Antipolis, France
Montenegro, Rafael	U. of Las Palmas De Gran Canaria, Spain
Mukherjee, Nilanjan	U.G.S., USA
Owen, Steve	Sandia Nat. Labs, USA

Patel, Paresh	MSC Software, USA
Pav, Steven	U. of California, San Diego, USA
Pebay, Philippe	Sandia Nat. Labs, USA
Quadros, William	Algor, USA
Rajagopalan, Krishnakumar	PDC, USA
Ray, Tathagata	Rensselaer Polytechnic Institute, NY, USA
Rivara, Maria-Cecilia	U. of Chile, Santiago, Chile
Sadowski, Torsten	Berlin Institute of Technology, Germany
Sarrate, Jose	Universitat Politecnica de Catalunya, Spain
Schneiders, Robert	MAGMA Giessereitechnologie GmbH, Germany
Shephard, Mark	Rensselaer Polytechnic Institute, NY, USA
Shepherd, Jason	Sandia Nat. Labs, USA
Shi, Xinwei	U. of California, Davis, USA
Shih, Alan	U. of Alabama, Birmingham, USA
Shimada, Kenji	Carnegie Mellon University, PA, USA
Shontz, Suzanne	Pennsylvania State Univ., USA
Si, Hang	Weierstrass Institute of Applied Analysis and Stochastics, Germany
Simpson, Bruce	U. of Waterloo, CA
Staten, Matthew	Sandia Nat. Labs, USA
Stewart, James	Queens University, Belfast, Ireland
Taghavi, Reza	Simulation Works, MN, USA
Tautges, Tim	Argonne Nationals Lab., WI, USA
Vassilevski, Yuri	Russian Academy of Sciences, Russia
Walsh, Joe	Simmetrix, Inc., NY, USA
Watanabe, Taichi	Tokyo Institute of Technology, Japan
Weill, Jean-Christophe	CEA, France
Xu, Dianna	Bryn Mawr College, PA, USA
Yamakawa, Soji	Carnegie Mellon U., PA, USA
Yvinec, Mariette	Inria-Prisme, France
Zhang, Yongjie (Jessica)	U. of Texas, Austin, USA

Preface

The papers in this volume were selected for presentation at the 17th International Meshing Roundtable (IMR), held October 12–15, 2008 in Pittsburgh, Pennsylvania, USA. The conference was started by Sandia National Laboratories in 1992 as a small meeting of organizations striving to establish a common focus for research and development in the field of mesh generation. Now after 17 consecutive years, the International Meshing Roundtable has become recognized as an international focal point annually attended by researchers and developers from dozens of countries around the world.

The 17th International Meshing Roundtable consists of technical presentations from contributed papers, research notes, keynote and invited talks, short course presentations, and a poster session and competition. The Program Committee would like to express its appreciation to all who participate to make the IMR a successful and enriching experience.

The papers in these proceedings were selected from more than 50 paper submissions. Based on input from peer reviews, the committee selected these papers for their perceived quality, originality, and appropriateness to the theme of the International Meshing Roundtable. We would like to thank all who submitted papers. We would also like to thank the colleagues who provided reviews of the submitted papers. The names of the reviewers are acknowledged in the following pages.

We extend special thanks to Jacqueline Hunter and Bernadette Watts for their time and effort to make the 17th IMR another outstanding conference.

August 2008 — 17th IMR Program Committee

Organization

Committee

Rao Garimella (Committee Chairman)
Los Alamos National Laboratory, Los Alamos, NM
rao@lanl.gov

Edward Ascoli
Pratt & Whitney Rocketdyne, Canoga Park, CA
edward.ascoli@pwr.utc.com

Bruce Simpson
UW, Dept. of Computer Science, Waterloo, Canada
rbsimpson@uwaterloo.ca

Joseph Walsh
Simmetrix, Inc., Clifton Park, NY
jwalsh@simmetrix.com

Soji Yamakawa
Carnegie Mellon University, Pittsburgh, PA
soji@andrew.cmu.edu

Brett Clark
Sandia National Laboratories, Albuquerque, NM
bwclark@sandia.gov

Steve Karman
University of Tennessee, Chattanooga, TN
Steve-Karman@utc.edu

Paresh Patel
MSC Software, Santa Ana, CA
Paresh.Patael@mscsoftware.com

Coordinators

Jacqueline A. Hunter
Sandia National Laboratories, Albuquerque, NM
jafinle@sandia.gov

Web Designer

Jacqueline A Hunter
Sandia National Laboratories, Albuquerque, NM
jafinle@sandia.gov

Web Site

http://www.imr.sandia.gov

Contents

Session 3A: Surface Meshing

Session 3B: Applications

Session 4A: Hexahedral Meshing

Sphere Packing and Applications to Granular Structure Modeling

Azeddine Benabbou[1], Houman Borouchaki[2], Patrick Laug[3], and Jian Lu[4]

[1] INRIA, BP 105, 75153 Le Chesnay cedex, France
azeddine.benabbou@inria.fr
[2] University of Technology of Troyes, 12 rue Marie Curie BP2060, 10010 Troyes Cedex, France
houman.borouchaki@utt.fr
[3] INRIA, BP 105, 75153 Le Chesnay cedex, France
patrick.laug@inria.fr
[4] Hong Kong Polytechnic University, Kowloon, Hong Kong
jian.lu@inet.polyu.edu.hk

Summary. This paper presents a new sphere packing algorithm for generating granular structures in either two or three dimensions. Such a structure is often modeled by a parallelepiped containing spherical balls in three dimensions or by a rectangle filled with disks in two dimensions. The grains (spherical balls or disks) are separated by interfaces with specific thicknesses, called grain boundaries, and their size correspond to a size distribution experimentally obtained. The geometrical modeling of such a structure, which consists in determining the repartition of the set of disjoint grains according to the above specifications, can then be considered as the classic sphere packing problem. The proposed method is a constructive algorithm based on an advancing-front approach, which is well known in a different context, namely mesh generation. Since the use of the advancing-front approach leads to empty areas near front collisions, a point relocation algorithm, using weighted Delaunay triangulation, is then introduced to balance the local density on the whole structure. Moreover, we propose a method to transform spherical balls (disks) into polyhedral (polygonal) cells similar to the real grain shape. Numerical 2D and 3D examples are provided to illustrate the capability and the efficiency of our approach. The algorithms and techniques presented here can find applications to generate aggregates in all fields concerned by the granular structures such as metallurgy, ceramics, soil science, cements, biomechanics, etc.

Keywords: sphere packing, advancing-front approach, granular, nanostructures, Laguerre diagram, quality meshes.

1 Introduction

The geometrical modeling of granular structures represents a key step in the simulations and studies of their behavior (mechanical, electrical, thermal, ...). This modeling is often based on experimental data provided by microscopic analysis in the form of a parallelepipedic (rectangular) sample and the distribution of the grain sizes within the considered structure. The physical results of simulations on these structures, by the molecular dynamics models or the finite element method or other methods of resolution, strongly depend on the geometric models provided. Hence, the geometrical modeling of these structures is a crucial task requiring special attention. Usually, a sample of a granular structure is modeled by a parallelepiped (rectangle) containing grains modeled by spherical balls (disks). Beyond the large simplifications in the geometrical calculations, this modeling provides an acceptable model in many cases where the computing with the real grain geometries (complex shapes) is very difficult to achieve [1]. Using this approximation, the geometrical modeling of granular structures is then very similar to the classic sphere (circle) packing which consists in filling a domain with spheres (circles) whose radii follow a size distribution. In our concern, we are particularly interested in the modeling and the meshing of nanomaterials considered as granular structures. This is due to their considerable scientific and industrial potentialities. Indeed, it is experimentally proved that the physical properties of nanomaterials are very different, and often better, than those of ordinary materials. This quality improvement of materials is due to changes of their structures on small scales, in other words to their nanostructuring. The structure of an ordinary material is constituted by a multitude of small polyhedral volumes called grains whose size usually ranges from 2 to 20 micrometers, separated by grain boundaries with specific thicknesses. Compared to the configuration of an ordinary material, the grain size in a nanomaterial is 1000 times smaller and the proportion of grain boundaries is more important. These two features, the grain size and the proportion of grain boundaries, are mostly responsible for changes in the material behavior when its structure changes from the micrometer to the nanometer scale [2, 3].

In our concern, the nanostructures studied are issued from the SMAT (Surface Mechanical Attrition Treatment) process [2], considered as granular structures.

2 Sphere Packing and Granular Structure Modeling

2.1 Sphere Packing

The term of sphere (circle) packing is generally used to evoke the mathematical study of the arrangements of non-overlapping spheres (circles) which fill a given domain. A typical sphere packing problem is to determine the repartition of the particles (spheres or circles) which maximizes the density in the domain. Indeed, the density is the proportion of space filled by the particles having this repartition. This density depends on the particle size distribution, the way these particles are packed and the volume of the domain to fill. In practice, to determine the efficiency of a filling algorithm to generate structures with high densities, we consider it in the particular case where the particles have an identical size and where the domain has a sufficiently large volume. Indeed, in this particular case, the density of the generated structure can be compared with $\frac{\pi}{\sqrt{18}} \simeq 0.74048$, the density of the cubic close packing arrangement in the three-dimensional Euclidean space (Kepler conjecture), and $\frac{\pi}{\sqrt{12}} \simeq 0.9069$, the density of the hexagonal packing arrangement in the two-dimensional Euclidean space. Note that the maximal density value in 3D concerns regular arrangements only. For irregular arrangements in 3D (which is usually the case in the sphere packing algorithms), the highest density (in the particular case where the particles have the same size) is generally about 0.64, which corresponds to the *jammed* configuration.

Several approaches have been proposed to fill a domain with particles in two and three dimensions. These methods can generally be classified into two major families: dynamic techniques and constructive techniques. The dynamic methods are based on the motion and/or the resizing of the particles. In these techniques, two approaches are essentially used. A purely geometric approach, where calculations are based only on the particles positions and sizes [4, 5, 6, 7], and a second approach in which physical properties are introduced allowing each particle to find an equilibrium position depending on its interactions with the other particles [8, 9, 10, 11]. Usually, the dynamic methods are very costly in terms of computing time, because the position and/or the size of each particle are/is modified during the whole filling process. The second family of methods is the constructive techniques, in which calculations are purely geometric: The position and often the size of each particle are kept throughout the filling process. In this kind of techniques, we can distinguish the methods where the filling in a given step depends on the situation of the system in the previous steps (like the advancing-front approach) [12, 1, 13, 14, 15], and the other pseudo-constructive methods where the repartition of the particles is randomly generated [16, 17, 18] or based on a pre-built triangulation [19] for example. In many contexts, the constructive techniques are considered as more advantageous than the dynamic ones.

2.2 Granular Structure Case

A granular domain, in particular a nanostructure, can be described by a parallelepiped (rectangle) containing spherical balls (disks) whose sizes are defined by a given distribution. The spherical balls (disks) represent the grains and the spaces between them the grain boundaries. The geometrical modeling of these structures can be considered as an application of the sphere (circle) packing when spheres (circles) are modeling grains. In addition to the conformity with grain size distribution, in the nanostructure case, the grains must be separated by grain boundaries with specific thicknesses and the generated aggregate must be irregular (no periodicity).

In terms of algorithmic aspects, further constraints must be considered. In fact, the algorithm for generating models of granular structures must also be fast and robust, especially in the three dimensions case.

Modeling grains by spherical balls (disks) obviously creates empty areas in the structure, *a fortiori* if the filling method is an advancing-front type. Since these empty areas, as well as their non homogeneous repartition, do not exist in the real structure, fatal errors may occur during the physical simulations. Thus, in most cases, an optimization algorithm, to equally distribute these empty areas on the whole structure, must be applied after the filling process. The resulting structure may be transformed, using the weighted Delaunay triangulation techniques, into an aggregate of grains in the form of Laguerre cells closer to reality.

3 Geometrical Modeling Scheme

3.1 Filling Algorithm (Sphere Packing Algorithm)

The algorithm we propose to generate granular structures, which is a new sphere packing algorithm, is a constructive method based on an advancing-front approach. The term "constructive" means that the construction at a given step depends and uses the result of the previous steps, and that particles keep their size and position during the filling process. The advancing-front approach used in our algorithm is a new variant of a classical approach in the context of meshing. Indeed, the idea of the advancing-front method is to mesh a domain by an iterative inwards spreading of a front initialized on its boundary until its complete recovery [20, 21]. The front is a set of elements (segments in 2D or triangles in 3D) connecting the vertices of the mesh. The advancing-front approach consists in using the front elements (a previous construction) to create and insert the new points in the mesh. At each step, an element of the front is selected to propose the position of the new point according to a given criterion. If the new point is valid, then the selected front element is deactivated and new front elements are created, connecting this new point to the points of the deactivated front element and possibly to other points of the mesh. A new point is said valid (and then inserted in the mesh) if it satisfies the specified criterion and if the possible created front elements do not intersect the current front elements. In the event that one or more points of the current mesh better fit the specified criterion using the selected front element (compared to the new point), then the best of these points is chosen and connected to the points of the selected front element. The convergence of the advancing-front method strongly depends on how the elements are selected from the front, the identification of the optimal points and the validation of the created front elements. The advancing-front method remains highly empirical, unlike Delaunay-type methods for example [22, 23], which are based on mathematical formulations. Thus, the advancing-front methods are often intuitive and their convergence, especially in the 3D case, cannot be guaranteed by a rigorous formalism. The use of the advancing-front approach, in our context, requires to introduce some new concepts in order to ensure its convergence.

In the analogy between this meshing advancing-front approach and our sphere packing algorithm, a node of the mesh becomes a particle in our geometrical modeling. The sizes of these particles are defined using a given grain size distribution. In our case, a front element is a virtual triangle (resp. segment) connecting the centers of 3 spherical balls (resp. 2 disks). A spherical ball (resp. disk) is the association of a center position and a radius defined according to the specified size distribution. The location of each spherical ball (resp. disk) is determined, using an element of the front, to maximize the local density with the particles (spherical balls or disks) constituting this front element. This position is valid if this particle does not overlap any of the existent particles and if it is completely contained in the domain. A similar sphere packing algorithm have been proposed by many authors including Yamakawa et al. [24], the proposed method defines the metric using a sphere (bubble) packing possibly overlapping approach. Our algorithm can be summarized by the following scheme:

1. Create an initial front
2. Do while the front still contains active elements
   a) Generate a random radius according to the grain size distribution
   b) Determine the level l of the front (= the first active element level),
   c) Do the following for the front of level l
      i. Select an active element if its level l_e is less than or equal to l
      ii. Place the new particle P_{new} using the selected front element e_s
      iii. Control the validity of P_{new}
      iv. If P_{new} is valid then
         - Deactivate e_s
         - Connect P_{new} to e_s particles [1] and go to (a)
      v. If the position of P_{new} is not valid and all the active elements with a level $\leq l$ have been tested to place P_{new} without success then
         - Deactivate the first active element e_f in the front
         - Determine the existent particle maximizing the local density with e_f particles
         - Add possible new elements to the front [2] and go to (b)
   d) End do
3. End while

The first step of our algorithm is to create an initial front by connecting the particles which are placed on the interior side of the domain boundary. Then the filling is made by an inwards spreading of the front. In figure 1, an illustration of a filling example in 2D is shown. Figure 2 shows a filling example using our algorithm in 3D by illustrating the increasing particle level towards the domain center (each color corresponds to a level in figure 2).

The convergence of the new variant of the advancing-front method is assumed by the front level concept. Indeed, the level of a front is defined by the level of its first active element. A front element is said active if it has not been deleted (deactivated) from the front, either because it has been used to insert a new particle or because it has been the first active element of a front in the case where no element of this front could insert the new particle. The level of an element is defined by the sum of the levels of its particles (spherical balls or disks). The particle level ranges from 0 (the level of the initial front particles) to a level n corresponding to the level of the last front particles. The level of

[1] This connection will create new front elements (two in 2D and three in 3D) with a level $\geq l_e$.

[2] For each set s containing $d-1$ particles of e_f, (d is the space dimension), do: if the determined particle is not connected to the particles of s, then connect it if the level of the resulting new front element is greater than l in the 3D case (this is necessary for the convergence of the method). In the 2D case, the new front element is created even if its level is less than or equal to l.

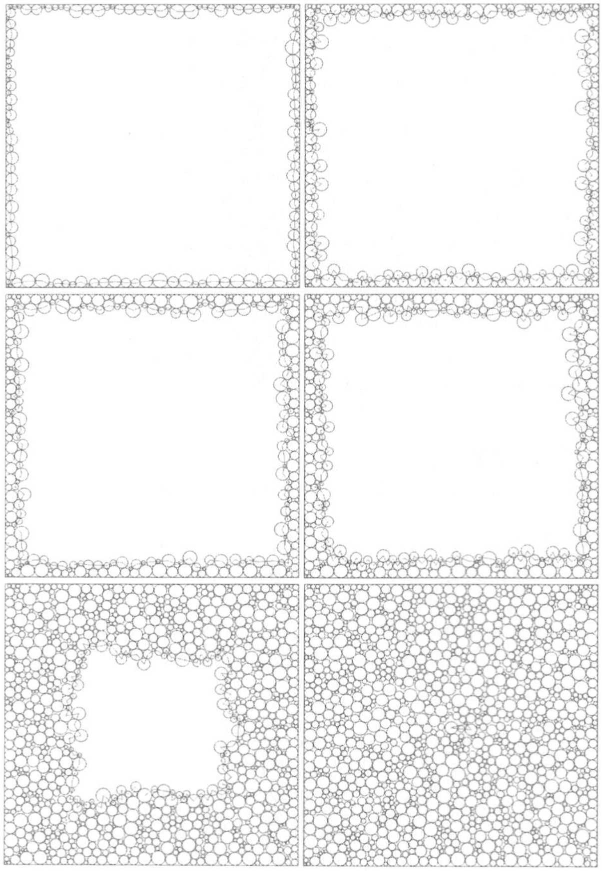

Fig. 1. A 2D illustration of a filling example using our algorithm

a new particle is a function of the levels of the particles of the front element used to insert this new particle. Indeed, the level l_{new} of the new particle P_{new} is calculated by:

$$l_{new} = \min_i(l_i) + max(1, \max_i(l_i) - \min_i(l_i)) \qquad (1)$$

where l_i are the levels of the particles of the front element used to insert the new particle. For the convergence of the method, the following two conditions must be satisfied: (a) a new particle level is equal or greater than to the minimum of the levels of the particles of the front element used to insert this new particle, (b) the filling using the elements of a front with a level l is initiated only when the fronts with levels less than l are completely saturated. The detection and analysis of conflicts (intersections) between the particles constitutes a crucial task which generally takes the half of the computing time of the whole filling

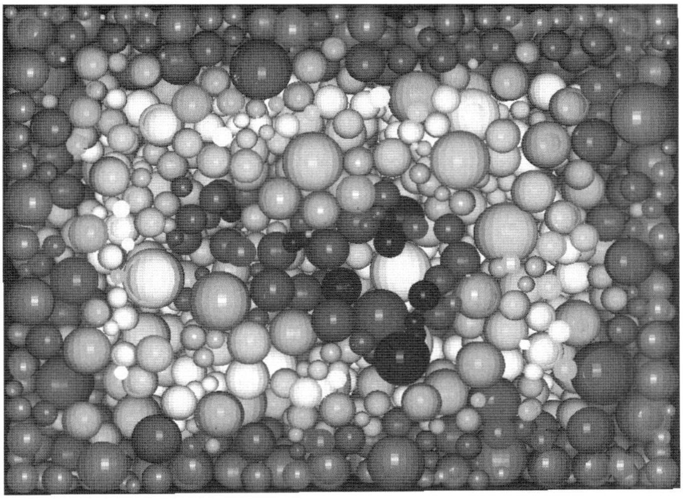

Fig. 2. Illustration of a filling example in 3D

process. To make this task faster, this detection is localized using a grid. In addition to the speed of the method, the use of this grid makes our algorithm linear. For more details of this algorithm see [1, 26, 27].

3.2 The Algorithm Efficiency

In this section, we give some criteria to characterize the sphere packing algorithm. In addition to the computing time and the density of the generated structure, another criterion is introduced. Indeed, this criterion characterizes the elements of the front by giving their quality. The quality q_e of a front element (P_1, P_2, P_3) (respectively (P_1, P_2) in the 2D case) is given by:

$$q_e = \frac{\delta_{12} + \delta_{23} + \delta_{31}}{3} \quad \text{(respectively, } q_e = \delta_{12}) \tag{2}$$

with $\delta_{ij} = d(P_i, P_j) - R_i - Rj$, where $d(P_i, P_j)$ denotes the Euclidean distance between the centers of the particles P_i and P_j with radii R_i and R_j respectively. The efficiency Q of the algorithm is presented by an histogram giving the frequency corresponding to each quality interval. Indeed, we define the minimum quality q_{min}, the maximum quality q_{max} and a quality step defined by $q_s = \frac{q_{max} - q_{min}}{10}$ for example. The frequency f_i corresponding to the ith quality interval $[q_{min} + (i-1)q_s, q_{min} + iq_s[$ is $\frac{n_i}{n_e}$, where n_i is the number of front elements whose quality is belonging to the quality interval i, and n_e is the total number of the front elements. It is said that a filling has a good quality if the histogram, representing Q, is a Gaussian centered in q_{min}.

3.3 Application

In this section, we consider a nanostructure of pure iron [25]. The domain to fill is a parallelepiped with dimensions $240 \times 300 \times 240$ nm (a square with dimensions 240×300 nm in the 2D case), the grain size distribution has radii ranging from 2 to 8 nm as shown in table 1, and the grain boundary thickness is a function of the grain radius given by $Gb(R) = 2/R$. In the filling process, the radius of each grain will be calculated as $R + Gb(R)$, where R is a radius randomly chosen in the grain size distribution (for more details see [1, 26, 27]).

Table 1. The grain size distribution of the pure iron nanostructured

radii (nm)	frequency (%)
[2 - 2.5[	15.8
[2.5 - 3[	21
[3 - 3.5[	20.2
[3.5 - 4[	16.7
[4.5 - 5[	6.8
[5 - 5.5[	4
[5.5 - 6[	3
[6 - 6.5[	0.8
[6.5 - 7[	0.7
[7 - 7.5[	0.6
[7.5 - 8[	0.4

In these conditions, using a machine with a 1.4 GHz processor, our algorithm generates a structure containing 33,738 spherical balls (respectively 1,330 disks in 2D) with a density of 0.494 (0.80) and a computing time less than 3 s (0.02 s). We can note that the filling is quite fast and that the density (in the 2D case) is quite satisfactory. The obtained density in the 3D case cannot be compared to a theoretical value because the distribution contains different grain sizes. To characterize our algorithm concerning the densities that can be obtained in the 3D case, we apply it in the particular case of a fixed radius (4 nm), keeping the same domain size and the same grain boundary expression. Indeed, the idea is to compare the density of the generated structure with the value 0.64 corresponding to the highest possible density in a irregular assembly of constant size particles. In this particular case, the generated structure contains 39,061 spherical balls with a density of about 0.606. This density is much better than the value 0.494 previously obtained with the specified grain size distribution.

The generated structures in 2D and 3D are illustrated by figure 3. We can observe that there are many empty areas in these structures (more visible in the 2D case). As mentioned above, this is due to the modeling of the grains by spherical balls (disks) and to the advancing-front method intrinsic characteristic of generating empty areas when two fronts are meeting. In addition to these factors, the constraints of generating random irregular structures (no periodicity) and the conformity with the grain size distribution, in our case, accentuate these

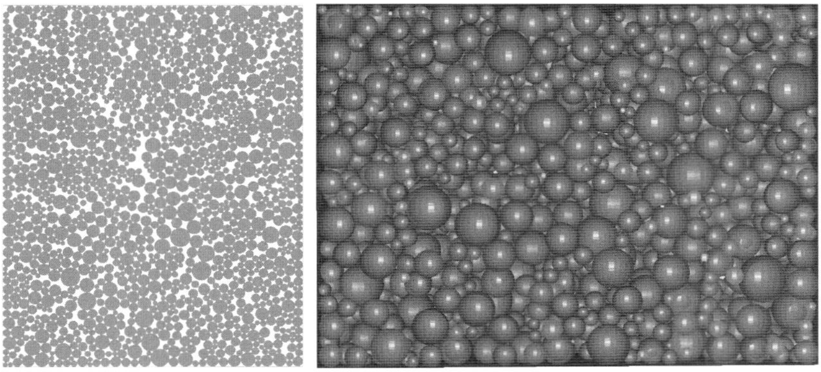

Fig. 3. Illustrations in 2D and 3D of the pure iron nanostructured model

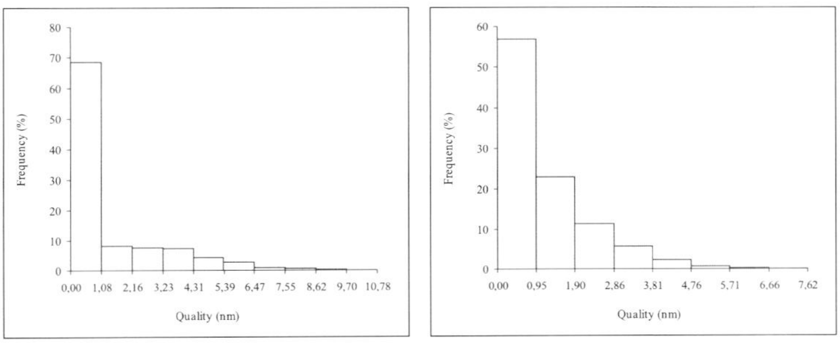

Fig. 4. The quality histogram for the cases 2D (left) and 3D (right)

empty areas. Since these empty areas, introduced by the filling algorithm, are residual, *i.e.* they do not physically exist in the real structure, then they must be equally distributed on the whole structure. This is the main purpose of the next section.

The filling quality histograms (of both 2D and 3D cases) for the pure iron example are shown in figure 4. We can observe that the filling quality in the 2D case is a Gaussian more pronounced (almost a Dirac) than in the 3D case. Indeed, the filling quality in the 3D case is calculated, on each triangle (a front element), by the average edges (of this triangle) quality, which does not show the gap between the qualities of these edges as in the 2D case.

4 Optimization of the Generated Structure and Cell Construction

The advancing-front method usually leads to an heterogeneity of the local density in the structure (some regions are denser than others, see figure 3). This feature can have a direct impact on the results of the physical simulations, particulary in the nanomaterials case. Indeed, in a nanostructure, an empty area is considered as an additional grain boundary. However, the proportion and the repartition of grain boundaries in such a structure are the major features of its behavior (obstacles to dislocations propagation for example [3]). Hence, a point relocation algorithm is necessary to equally distribute these additional empty areas on the whole structure, and then to reduce as much as possible their impact on the physical simulations.

The main idea of the point relocation algorithm that we propose is the following principle: "*in order to equally distribute the empty areas on the whole structure, one just has to equally distribute them locally around each particle*". Thus, we need to know the local configuration around each point (particle center), *i.e.* the positions and radii of its nearest particles. For this purpose, we use the weighted Delaunay triangulation (dual of Laguerre diagram) built on the point cloud constituted by the particle centers (where the weights are the radii of these particles) and the points of the domain boundary discretization (considered as particles with radius zero). The domain boundary discretization must be as fine as possible so that all the Laguerre diagram cells (dual of this triangulation) are completely contained in the domain.

4.1 Domain Boundary Discretization

Several approaches can be used to obtain the discretization of the domain boundary. As explained before, this discretization must allow all the Laguerre diagram vertices to be completely contained in the domain, but it must also be as least costly as possible (minimum number of points). To obtain this discretization, we are proceeding in an iterative scheme. At a first time, the weighted Delaunay triangulation, using the set of points constituted by the particle centers and the 8 points of the parallelepiped (4 points of the rectangle) representing the domain, is built. Then, for each tetrahedron (triangle) whose vertices are including at least one of the particle centers, we check if the center of the weighted "circumscribed" sphere (disk) is outside (or on) the domain boundary. If it is the case (outside or on the domain boundary), the centers of these particles are orthogonally projected onto the domain boundary side where the center of this "circumscribed" sphere (disk) is outside (or on it). These projected points are then added to the point cloud and a new weighted Delaunay triangulation is built. These second and third steps iterate until all the Laguerre diagram vertices (the tetrahedra (triangles) "circumscribed" spheres (disks) centers) are completely contained in the domain (including if they are on the domain boundary). In the 3D case, from two points associated with the same particle center (projection on two perpendicular planes), a third point (corresponding to the orthogonal projection of these two points on the segment, intersection of these two planes) is added to the point cloud. In figure 5, an example of the weighted Delaunay triangulation and its dual (Laguerre diagram) of a point cloud in 2D is given.

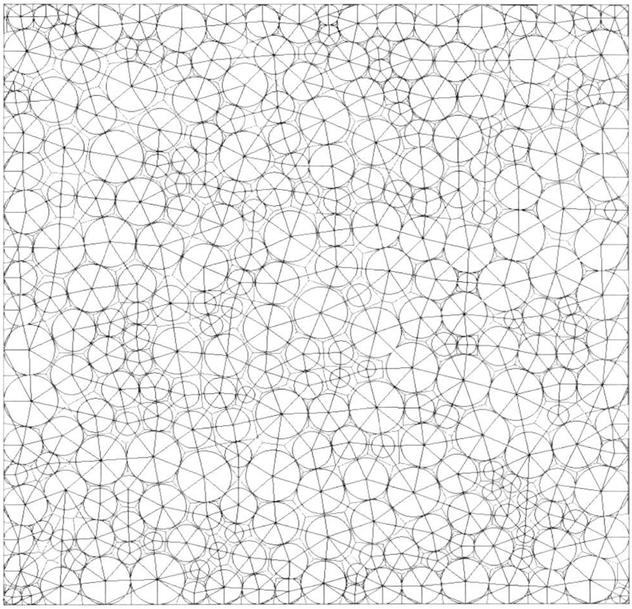

Fig. 5. The weighted Delaunay triangulation and its dual Laguerre diagram (red color) of a point cloud in 2D

4.2 The Point Relocation Algorithm

The algorithm we propose here to balance the density on the whole structure is an algorithm of point relocation under constraints. Indeed, the optimization produced by this algorithm must not lead to intersections between the particles, and thereby preserves the validity of the generated structure. The idea of this algorithm is to use the weighted Delaunay triangulation, as described above, to move locally each particle by considering its nearest neighbors (its topological ball). Indeed, each particle center is repositioned using the front defined by all the sides of its ball boundary. Let F_i (respectively S_i) be the ith face or triangle (segment) of the front defined by the sides of the point P ball boundary, we can assume that there is a unique position P_i of P calculated using F_i (S_i). This point relocation method is to move, step by step, the point P towards the barycenter P^* of the points P_i. At each step, the validity of the point P (no intersection with the other particles, and the particle centered on P is completely contained in the domain) is checked. The general scheme of our point relocation algorithm can be described by the few following lines:

- Do while there are relocated points
  - For each point P (particle center)
    - Define the ball of P using the weighted Delaunay triangulation
    - Calculate the positions P_i associated with the faces F_i (segments S_i) of the front defined by the sides of P ball boundary

  - Calculate the barycenter P^* of the points P_i
  - Move the point P, step by step, towards P^* (do $\alpha = 0.3\,\text{to}\,0.05\,\text{by}\,-0.05$)
    - $P_\alpha = (1-\alpha)P + \alpha P^*$
    - Control the validity of P_α
    - If the point P_α is valid then the position of the point P is replaced by P_α, exit from the loop on α
  - End of the loop on α
- End of the loop on the points P
- Reconstruct the weighted Delaunay triangulation

• End while

One may speculate about the use of this new point relocation algorithm, whereas other algorithms exist, especially the smoothing algorithm, usually used to smooth the edge lengthes in a mesh. Indeed, we have tested two variants of the smoothing algorithm in our context. The first variant is the classical smoothing scheme, which leads to intersections between the particles (and then do not preserve the validity of the generated structure). In the second variant, where the constraint of no overlapping between the particles is introduced, the optimization is very weak and the generated structure keeps a big heterogeneity in the local density.

4.3 Characterization of the Point Relocation Algorithm

In this section, we introduce a criterion to characterize the effectiveness of the point relocation algorithm. This criterion supposes that the weighted Delaunay triangulation *WDT*, based on the point cloud constituted by the generated particle centers and the points of the domain boundary discretization, is given. Indeed, with each point P of the *WDT* (except the centers of the border particles and the points of the domain boundary discretization) is associated a function $F(P)$ defined by:

$$F(P) = \max_j(\delta_j) - \min_j(\delta_j) \tag{3}$$

with $\delta_j = d(P, P_j) - w - w_j$, where $d(P, P_j)$ is the Euclidean distance between point P and point P_j (one of the points of the topological ball of point P given by the *WDT*), w if the weight of point P defined by $w = R + gb(R)$, where R is the radius of the particle centered on P and $gb(R)$ is the grain boundary around this particle, and w_j is the weight (in the same manner as w) of point P_j. Note that in our case, as shown in figure 6, we have necessarily $\delta_j \geq 0$.

By minimizing the function $F(P)$, the maximum and the minimum of δ_j tend to an average value δ_{moy} and then the optimization is realized. To present the results of the application of this criterion, we draw the graph of the function F between F_{min} (the minimum of $F(P)$) and F_{max} (the maximum of $F(P)$) by a frequency histogram in 10 intervals:

$$[F_{min} + i\frac{F_{max} - F_{min}}{10},\ F_{min} + (i+1)\frac{F_{max} - F_{min}}{10}[, \quad \text{for i=0,...,9}$$

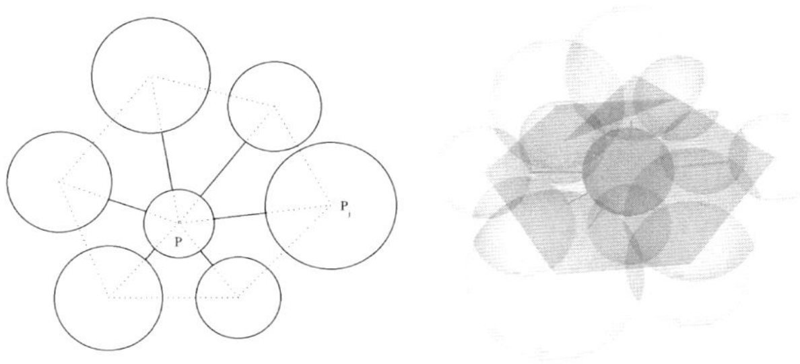

Fig. 6. The quantities δ_j are presented by solid lines for the 2D case (left) and by green lines for the case 3D (right)

4.4 Construction of Cells

The generated particles with spherical (circular) shape can be transformed, using the dual of the weighted Delaunay triangulation obtained after the point relocation procedure, into cells more similar to the grain shape observed in reality. This transformation procedure is optional, depending on the grain shape required by the user. To construct the cells representing the grains, we consider the dual of the weighted Delaunay triangulation (Laguerre diagram) obtained after the structure have been optimized (concerning the repartition of empty areas). As shown in figure 5, the cells entirely contain the particles, which is a principal feature of the duality between the weighted Delaunay triangulation and the Laguerre diagram. In our case, the weight of each particle, in the weighted Delaunay triangulation, is $w = R + gb(R)$, meaning that the grain boundaries are included in the weights of the particles. To separate the grain boundaries, these cells must be isotropically shrunk. Thus, from the vertices o_i of a cell and point P center of its correspondent particle, we define the vertices $\tilde{o}_i$ of the shrunk cell by:

$$\tilde{o}_i = o_i + (1 - \frac{R}{w})\overrightarrow{o_i P} \quad (4)$$

5 Numerical Applications

In this section, we consider two numerical examples in both 2D and 3D cases. The first example is a model where particles have the same radius, whereas in the second one, the particles have different radii. The machine used in the two examples has a Pentium M processor with 1.4 GHz.

5.1 Example 1

The 2D model consists in a domain with dimensions 100×100 nm and particles of radius 0.9 nm with a constant grain boundary of 0.1 nm. In the 3D model, the particles radius and the grain boundary are the same as in the 2D model and

Table 2. Modeling results in 2D and 3D cases

Results	2D model	3D model
Nb particles	2,789	17,787
Density	0.88	0.60
Time(s)	0.03	1

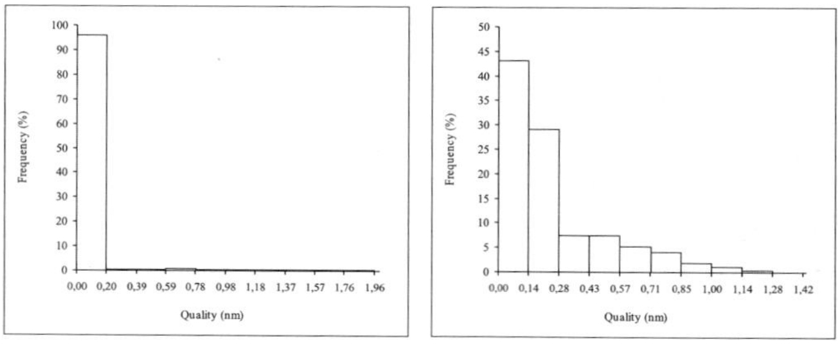

Fig. 7. The quality histograms for the cases 2D (left) and 3D (right)

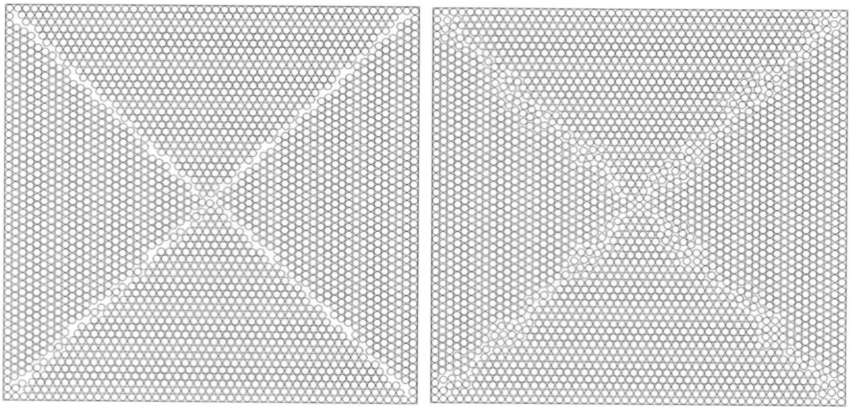

Fig. 8. The generated structure in the 2D model, before (left) and after (right) the application of the optimization algorithm

the domain is with dimensions $50 \times 50 \times 50$ nm. The results of the geometrical modeling are given in table 2.

We can observe that the filling is quite fast and gives satisfactory densities. The filling qualities, in both 2D and 3D are illustrated in figure 7. In figure 8 we illustrate the generated structure in the 2D model before and after the application of the point relocation algorithm. The histogram of function F, giving the repartition of the empty areas on the structure, before and after the application of the optimization algorithm is given in figure 9. We can observe that the two

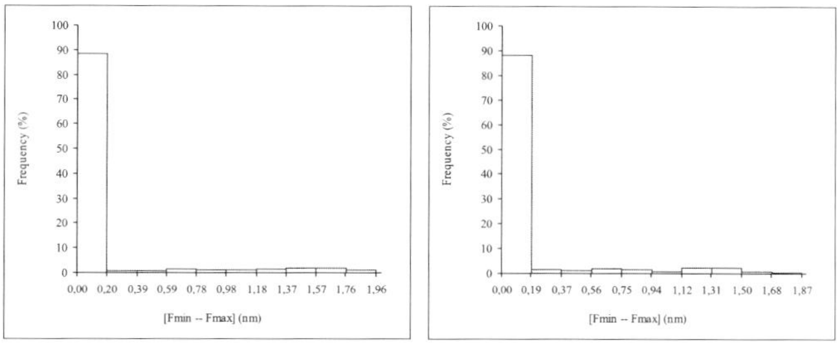

Fig. 9. The function F, in the 2D model, before (left) and after (right) the application of the optimization algorithm

histograms are quite similar: this is because, before the application of this algorithm, the function $F(P)$ was optimal (near 0) for most of points P (as we can see it in figure 8), and because for some points, the constraint of no intersection between the particles leads to a locked state.

5.2 Example 2

The 2D model consists in a domain with dimensions 200×200 nm and a uniform grain size distribution with radii in the interval $[0.8\,\text{nm}, 2.8\,\text{nm}]$ and a constant grain boundary of 0.2 nm. In the 3D model, the grain size distribution and the grain boundary are the same as in the 2D model and the domain has dimensions $50 \times 50 \times 50$ nm. The results of the modeling are given in table 3.

Table 3. Modeling results in 2D and 3D cases

Results	2D model	3D model
Nb particles	2324	1482
Density	0.81	0.51
Time(s)	0.02	0.05

In Figure 10, the structures with cells and particles in both 2D and 3D are presented. Note that each cell entirely contains a particle, and thus the volume (area) of the cell is always greater than that of its correspondent particle. Indeed, in the filling process, the thickness of grain boundaries is included in the radii of the particles, and despite of this, empty areas (intrinsic to the advancing-front method) still exist in the structure. These empty areas, which have no physical sense as nonexistent in the real structure (model obtained experimentally), are equally distributed on the whole structure leading to two kinds of models: either the model with spherical (circular) grains where the grain sizes are maintained and the thickness of grain boundaries is slightly increased, or the model with cells where the grain sizes are slightly increased and the thickness of grain boundaries is kept.

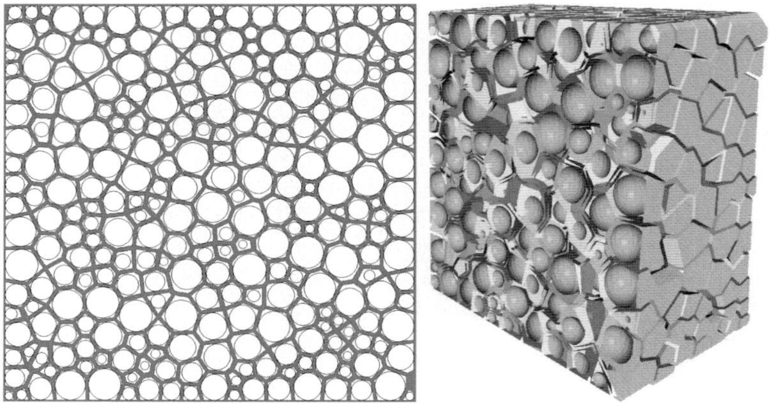

Fig. 10. Illustrations in 2D and 3D of the structures with cells and particles

6 Conclusions and Outlook

In this paper we have considered the problem of granular structure geometrical modeling under constraints in both two and three dimensions. We have considered this modeling as the classic sphere (circle) packing problem, and introduced a new method based on an advancing-front approach generally used in the meshing context.

Applying this new method in many cases, especially for nanostructures, we have shown that it can generate structures with a large number of grains (over 30,000) in a short computing time (about 2 s) with a satisfactory density and an ensured convergence.

The advancing-front approach used often generates heterogeneities in the repartition of empty areas in the structure. Thus, we have also proposed an optimization algorithm for equally distributing these areas on the whole structure. The proposed method is based on techniques of point relocation under constraints.

Finally, we have introduced a simple method to transform the structure with spherical (circular) grains into an assembly of cells closer to the grain shape observed in reality. This method is based on the weighted Delaunay triangulation, built on the point cloud constituted by particle centers and the points of the domain boundary discretization, by considering its dual Laguerre diagram.

To ensure the convergence in the 3D case, some faces (front elements) have been neglected in the filling process. An improvement in the algorithm on this point can be investigated, which may increase the density. Also, this improvement can be achieved by considering a densification procedure by a compromise between the density and the conformity with the grain size distribution.

In the point relocation algorithm, the optimization is based on the minimization of a functional calculated, for each particle, using the length of the empty areas (introduced by the advancing-front method) along some directions. Another approach is to consider this empty space in all directions by calculating the functional using the volume (the area) of this space around each particle. This method is more difficult to implement and more costly in terms of computing time, but it may give better results in the repartition of the empty areas on the structure.

This geometrical modeling of granular structures (nanostructures in particular) is a pre-step of their simulation by the finite element method. Another key task, which constitutes one of our principal activities, is the generation of quality meshes of these structures (in 2D and 3D) required by this simulation.

References

1. Benabbou, A., Borouchaki, H., Laug, P., Lu, J.: International Journal for Numerical Methods in Engineering (to appear, 2008)
2. Lu, K., Lu, J.: Materials Science and Engineering A 375, 38–45 (2004)
3. Kumar, K.S., Van Swygenhoven, H., Suresh, S.: Acta Materialia 51(19), 5743–5774 (2003)
4. Jodrey, W.S., Tory, E.M.: Phys Rev A 32, 4 (1985)
5. Lubachevsky, B.D., Stillinger, F.H.: Journal of Statistical Physics 60(5-6), 561–583 (1990)
6. Lin, X., Ng, T.: Geotechnique 47(2), 319–329 (1997)
7. Jiang, M.J., Konrad, J.M., Leroueil, S.: Computers and Geotechnics 30(7), 579–597 (2003)
8. Stillinger, F.H., Dimarzio, E.A., Kornegay, R.L.: J. Chem. Phys 40, 1564 (1964)
9. Visscher, W.M., Bolsterli, M.: Nature 239 (October 1972)
10. Bagi, K.: Mechanics of Materials 16(1-2), 101–110 (1993)
11. Han, K., Feng, Y.T., Owen, D.R.J.: Powder Technology 155, 33–41 (2005)
12. Bennett, C.H.: J Appl Phys 43, 6 (1972)
13. Feng, Y.T., Han, K., Owen, D.R.J.: International Journal for Numerical Methods in Engineering 56(5), 699–713 (2002)
14. Löhner, R., Onate, E.: International Journal for Numerical Methods in Engineering 61, 1977–1991 (2004)
15. Bagi, K.: Granular Matter 7, 31–43 (2005)
16. Stoyan, D.: Models of random systems of non-intersecting spheres. In: Prague Stochastics, JCMF 1998, pp. 543–547 (1998)
17. Häggström, O., Meester, R.: Random Struct. Algor. 9, 295–315 (1996)
18. Evans, J.W.: Rev Mod Phys 65, 1281–1304 (1993)
19. Cui, L., O'Sullivan, C.: Granular Matter 5(3), 135–145 (2003)
20. George, J.A.: Computer implementation of the finite element method, PhD thesis, Stanford university (1971)
21. Lo, S.H.: International Journal for Numerical Methods in Engineering 21, 1403–1426 (1985)
22. Delaunay, B.: Otdelenie Matematicheskikh i Estestvennykh Nauk 7, 793–800 (1934)
23. George, P.L., Borouchaki, H.: Delaunay Triangulation and Meshing, Hermes, Paris (1998)
24. Yamakawa, S., Shimada, K.: International Journal for Numerical Methods in Engineering 57(13), 1923–1942 (2003)

25. Tao, N.R., Wang, Z.B., Tong, W.P., Sui, M.L., Lu, J., Lu, K.: Acta Materialia 50(18), 4603–4616 (2002)
26. Benabbou, A., Borouchaki, H., Laug, P., Lu, J.: Rapport de recherche Inria (6414) (2008), http://hal.inria.fr/inria-00203375/fr/
27. Laug, P., Borouchaki, H., Benabbou, A., Lu, J.: C. R. Mecanique 336, 506–511 (2008)

Triangulation of Simple 3D Shapes with Well-Centered Tetrahedra

Evan VanderZee[1], Anil N. Hirani[2], and Damrong Guoy[3]

[1] Department of Mathematics
vanderze@illinois.edu
[2] Department of Computer Science
hirani@cs.uiuc.edu
[3] Center for Simulation of Advanced Rockets University of Illinois at Urbana-Champaign
guoy@uiuc.edu

Summary. A completely well-centered tetrahedral mesh is a triangulation of a three dimensional domain in which every tetrahedron and every triangle contains its circumcenter in its interior. Such meshes have applications in scientific computing and other fields. We show how to triangulate simple domains using completely well-centered tetrahedra. The domains we consider here are space, infinite slab, infinite rectangular prism, cube, and regular tetrahedron. We also demonstrate single tetrahedra with various combinations of the properties of dihedral acuteness, 2-well-centeredness, and 3-well-centeredness.

1 Introduction

In this paper we demonstrate well-centered triangulation of simple domains in $\mathbb{R}^3$. A well-centered simplex is one for which the circumcenter lies in the interior of the simplex (8). This definition is further refined to that of a k-well-centered simplex which is one whose k-dimensional faces have the well-centeredness property. An n-dimensional simplex which is k-well-centered for all $1 \leq k \leq n$ is called completely well-centered (15). These properties extend to simplicial complexes, i.e. to meshes. Thus a mesh can be completely well-centered or k-well-centered if all its simplices have that property. For triangles, being well-centered is the same as being acute-angled. But a tetrahedron can be dihedral acute without being 3-well-centered as we show by example in Sect. 2. We also note that while every well-centered triangulation is Delaunay, the converse is not true.

In (14) we described an optimization-based approach to transform a given planar triangle mesh into a well-centered one by moving the internal vertices while keeping the boundary vertices fixed. In (15) we generalized this approach to arbitrary dimensions and in addition to developing some theoretical properties of our method showed some complex examples of our method at work in the plane and some simple examples in $\mathbb{R}^3$. In this paper the domains we consider are space, slabs, infinite rectangular prisms, cubes and tetrahedra.

Well-centered triangulations find applications in some areas of scientific computing and other fields. Their limited use so far may be partly because until recently there were no methods known for constructing such meshes. One motivation for constructing well-centered meshes comes from Discrete Exterior Calculus, which is a framework for constructing numerical methods for partial differential equations (4, 8). The availability of well-centered meshes permits one to discretize an important operator called the Hodge star as a diagonal matrix, leading to efficiencies in numerical solution procedures. Other potential applications are the covolume method (10, 11), space-time meshing (13), and computations of geodesic paths on manifolds (9).

2 Well-Centeredness and Dihedral Acuteness for a Single Tetrahedron

An equatorial ball of a simplex is a ball for which the circumsphere of the simplex is an equator. Stated more precisely, if σ^k is a k-dimensional simplex, then the equatorial ball of σ^k is a $(k+1)$-dimensional ball whose center is $c(\sigma^k)$, the circumcenter of σ^k, and whose radius is $R(\sigma^k)$, the circumradius of σ^k. In (15) we showed that a simplex σ^n is n-well-centered if and only if for each vertex v in σ^n, v lies outside the equatorial ball of the facet τ_v^{n-1} opposite v. Figure 1 illustrates this characterization of well-centeredness; the vertex v at the top of the tetrahedron in Fig. 1 lies outside of the equatorial ball of the facet opposite v.

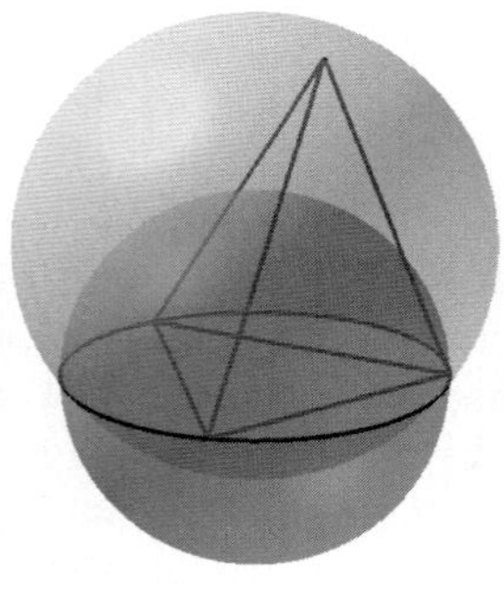

Fig. 1. One characterization of well-centeredness is that for each vertex v, the vertex lies outside of the equatorial ball of the facet opposite v. The larger sphere shown here is the circumsphere of the tetrahedron and the smaller sphere is the boundary of the equatorial ball of the bottom triangle. The top vertex of the tetrahedron is outside the equatorial ball.

We now have a definition of k-well-centeredness and an alternate characterization of well-centeredness, but these provide limited intuition for what it means to be well-centered. In this section we discuss a variety of tetrahedra, showing that in $\mathbb{R}^3$, a simplex that is 2-well-centered may or may not be 3-well-centered, and vice versa. We also discuss how being dihedral acute relates to being well-centered.

Figures 2 through 7 are pictures of six different tetrahedra that illustrate the possible combinations of the qualities 2-well-centered, 3-well-centered, and dihedral acute. Each picture shows a tetrahedron inside of its circumsphere. The center of each circumsphere is marked by a small, unlabeled axes indicator. In each case, the circumcenter of the tetrahedron lies at the origin, and the circumradius is 1. The coordinates given are exact, and the quality statistics are rounded to the nearest value of the precision shown.

The quality statistics displayed include the minimum and maximum face angles and dihedral angles of the tetrahedron. These familiar quality measures need no further explanation, and it is easy to determine from them whether a tetrahedron is 2-well-centered (having all face angles acute) or dihedral acute. The R/ℓ statistic shows the ratio of the circumradius R to the shortest edge of the tetrahedron, which has length ℓ. The range of R/ℓ is $[\sqrt{3/8}, \infty]$, with $\sqrt{3/8} \approx 0.612$. A single tetrahedron has a particular R/ℓ ratio, so the minimum R/ℓ equals the maximum R/ℓ in each of Figs. 2–7. Later, however, we will show similar statistics for tetrahedral meshes, and it is convenient to use the same format to summarize mesh quality in both cases. The R/ℓ ratio is a familiar measurement of the quality of a tetrahedron, especially in the context of Delaunay refinement.

The quality statistic h/R is less familiar. The R in this ratio is the circumradius. The h stands for height. For a given facet of the tetrahedron, h measures the signed height of the circumcenter of the tetrahedron above the plane containing that facet. The direction above the facet means the direction towards the remaining vertex of the tetrahedron, and h is positive when the circumcenter lies above the facet. In (15) the quantity h/R and its relationship to well-centeredness in any dimension is discussed at more length. For our purposes it should suffice to note that the range of h/R for tetrahedra is $(-1, 1)$, that a tetrahedron is 3-well-centered if and only if the minimum h/R is positive, and that $h/R = 1/3$ relative to every facet of the regular tetrahedron.

The regular tetrahedron is completely well-centered and dihedral acute. Our first example, shown in Fig. 2, is another tetrahedron that shares those properties with the regular tetrahedron. Not every completely well-centered tetrahedron is dihedral acute, and the second example, shown in Fig. 3, is a tetrahedron that is completely well-centered but not dihedral acute. Some sliver tetrahedra are completely well-centered and have dihedral angles approaching $180°$.

As one might expect, there are tetrahedra that have none of these nice properties. The tetrahedron shown in Fig. 4 is an example of such a tetrahedron. Most polar caps also have none of these nice properties. Some tetrahedra that are neither dihedral acute nor 2-well-centered nor 3-well-centered have much worse quality than the example in Fig. 4. This particular example is near the boundary dividing 3-well-centered tetrahedra from tetrahedra that are not 3-well-centered.

The tetrahedra we have considered so far have been either completely well-centered or neither 2-well-centered nor 3-well-centered. Our last three examples show that tetrahedra can be 2-well-centered without being 3-well-centered and vice-versa. The tetrahedron shown in Fig. 5 is 2-well-centered and dihedral acute, but not 3-well-centered. The example tetrahedron shown in Fig. 6 is similar but has been modified to no longer be dihedral acute. Our final example, shown in Fig. 7, is a tetrahedron that is 3-well-centered, but is neither 2-well-centered nor dihedral acute.

Table 1 summarizes the six examples presented in this section, indicating whether each example is 2-well-centered, 3-well-centered, and/or dihedral acute. Of the eight possible binary sequences, the two missing examples are the sequences N, N, Y and Y, N, Y, in which a tetrahedron would be dihedral acute but not 2-well-centered. These

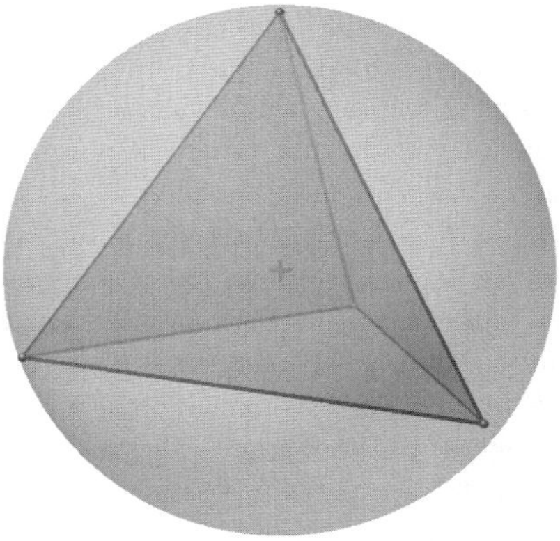

Vertex Coordinates		
x	y	z
0.6	−0.64	−0.48
0.48	0.8	−0.36
−0.96	0	−0.28
0	0	1

Quality Statistics		
Quantity	Min	Max
h/R	0.254	0.371
Face Angle	50.92°	67.08°
Dihedral Angle	58.76°	76.98°
R/ℓ	0.690	0.690

Fig. 2. A tetrahedron that is completely well-centered and dihedral acute

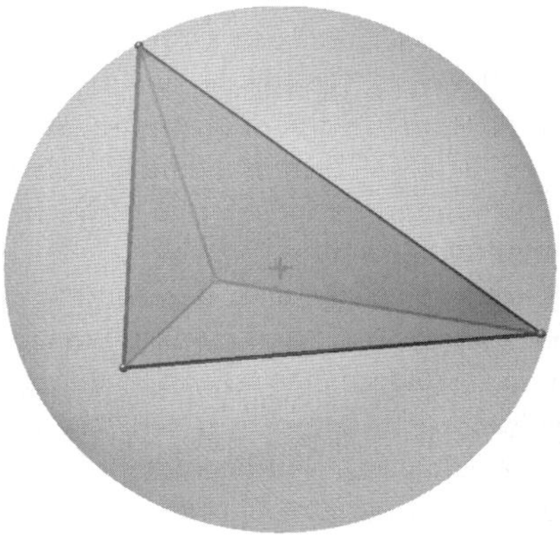

Vertex Coordinates		
x	y	z
0	0.96	−0.28
−0.744	−0.64	−0.192
0.856	−0.48	−0.192
−0.48	0.192	0.856

Quality Statistics		
Quantity	Min	Max
h/R	0.224	0.427
Face Angle	46.26°	77.62°
Dihedral Angle	52.71°	94.15°
R/ℓ	0.733	0.733

Fig. 3. A completely well-centered tetrahedron that is not dihedral acute

Vertex Coordinates		
x	y	z
0.224	−0.768	−0.6
0.8	0	−0.6
0.224	0.768	−0.6
−0.28	0	0.96

Quality Statistics		
Quantity	Min	Max
h/R	−0.029	0.600
Face Angle	29.89°	106.26°
Dihedral Angle	35.42°	116.68°
R/ℓ	1.042	1.042

Fig. 4. A tetrahedron that is not dihedral acute, 2-well-centered, or 3-well-centered

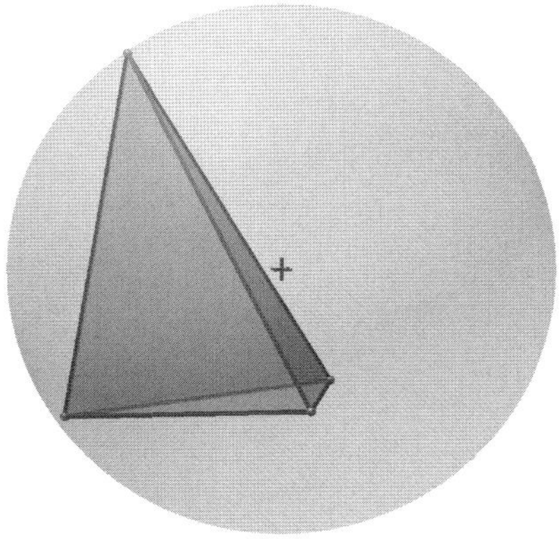

Vertex Coordinates		
x	y	z
0.36	−0.8	−0.48
0.768	0.28	−0.576
−0.6	0.64	−0.48
0.576	0.168	0.8

Quality Statistics		
Quantity	Min	Max
h/R	−0.109	0.562
Face Angle	41.71°	83.76°
Dihedral Angle	53.33°	85.72°
R/ℓ	0.863	0.863

Fig. 5. A tetrahedron that is dihedral acute and 2-well-centered, but not 3-well-centered

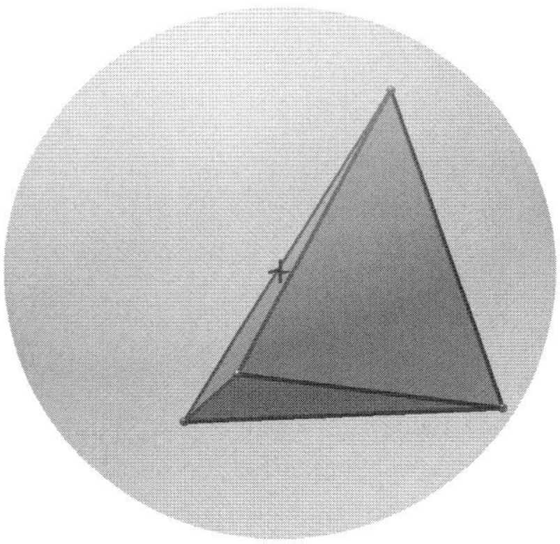

Vertex Coordinates		
x	y	z
−0.152	0.864	−0.48
−0.64	−0.6	−0.48
0.6	−0.64	−0.48
−0.192	−0.64	0.744

Quality Statistics		
Quantity	Min	Max
h/R	−0.024	0.630
Face Angle	42.08°	85.44°
Dihedral Angle	59.94°	91.20°
R/ℓ	0.806	0.806

Fig. 6. A tetrahedron that is not dihedral acute or 3-well-centered, but is 2-well-centered

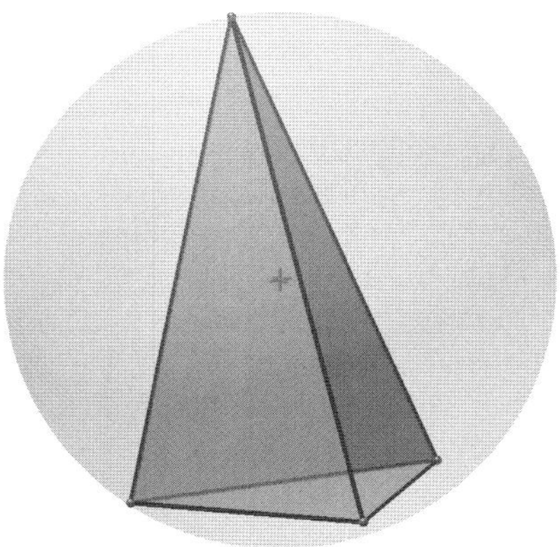

Vertex Coordinates		
x	y	z
0	−0.6	−0.8
0.64	−0.024	−0.768
−0.64	−0.024	−0.768
0	0.352	0.936

Quality Statistics		
Quantity	Min	Max
h/R	0.112	0.765
Face Angle	25.69°	95.94°
Dihedral Angle	40.33°	105.62°
R/ℓ	1.161	1.161

Fig. 7. A tetrahedron that is not dihedral acute or 2-well-centered, but is 3-well-centered

Table 1. A tetrahedron may have any of six different possible combinations of the qualities 2-well-centered, 3-well-centered, and dihedral acute

Figure	3-WC	2-WC	Acute
2	Y	Y	Y
3	Y	Y	N
4	N	N	N
5	N	Y	Y
6	N	Y	N
7	Y	N	N

examples are missing because they do not exist; every tetrahedron that is dihedral acute is also 2-well-centered. Eppstein, Sullivan, and Üngör provide a proof of this in Lemma 2 of (6), which states, among other things, that “an acute tetrahedron has acute facets.”

3 Tiling Space, Slabs, and Infinite Rectangular Prisms with Completely Well-Centered Tetrahedra

We have mentioned above that there are applications that could make good use of well-centered triangulations and have considered examples of single well-centered tetrahedra. In what follows, we give some examples of well-centered triangulations of simple domains in dimension 3.

In (6), Eppstein, Sullivan, and Üngör show that one can tile space, $\mathbb{R}^3$, and infinite slabs, $\mathbb{R}^2 \times [0, a]$, with dihedral acute tetrahedra. They also briefly discuss how high-quality tilings of space have been used to design meshing algorithms. The acute triangulations of space given in (6) all make use of copies of at least two different tetrahedra, and the authors suggest it is unlikely that there is a tiling of space with copies of a single acute tetrahedron. Their acute triangulation of the slab appears to use copies of seven distinct tetrahedra. The problem of triangulating an infinite rectangular prism, $\mathbb{R} \times [0, a] \times [0, b]$, or a cube, $[0, 1]^3$, with acute tetrahedra is still an open problem as far as the authors know.

In contrast to the complexity of tiling space with acute tetrahedra, there are fairly simple completely well-centered triangulations of space. Barnes and Sloane proved that the optimal lattice for quantizing uniformly distributed data in $\mathbb{R}^3$ is the body-centered cubic (BCC) lattice (2). Since this is related to centroidal Voronoi tesselations (CVTs) (5), and CVTs have been used for high-quality meshing of 3-dimensional domains (see (1)), it is not surprising that a Delaunay triangulation of the vertices of the BCC lattice gives rise to a high quality triangulation of space. The triangulation consists of congruent copies of a single completely well-centered tetrahedron, shown in Fig. 8. The tiling is one of four spatial tilings discovered by Sommerville (6, 12). The other three tilings Sommerville found are neither 3-well-centered nor 2-well-centered, though one of them has a maximum face angle of 90° and is dihedral nonobtuse. It is interesting to note that Fuchs algorithm for meshing spatial domains based on high-quality spatial tilings had “good performance ... when he used the second Sommerville construction,” S(6, 7) which is the completely well-centered tetrahedron shown in Fig. 8.

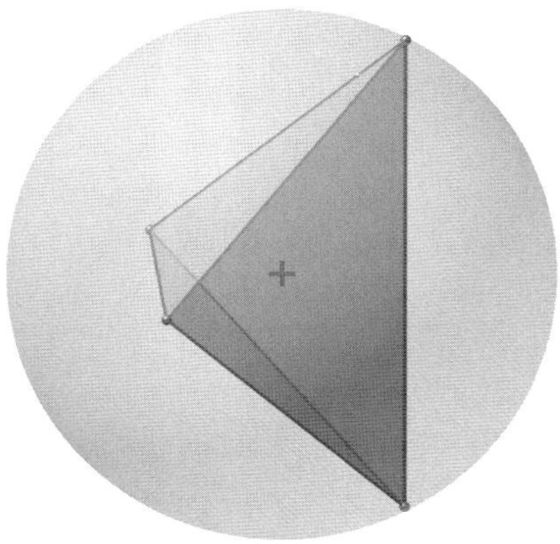

Vertex Coordinates		
x	y	z
-1	-2	0
1	0	-2
-1	2	0
1	0	2

Quality Statistics		
Quantity	Min	Max
h/R	0.316	0.316
Face Angle	54.74°	70.53°
Dihedral Angle	60.00°	90.00°
R/ℓ	0.645	0.645

Fig. 8. The second Sommerville tetrahedron – a completely well-centered tetrahedron that tiles space

Next we describe the space tiling that uses the Sommerville tetrahedron described above. The view of the tetrahedron shown in Fig. 8 has an elevation angle between 10 and 11 degrees, so it is not difficult to identify the horizontal edge and the vertical edge of the tetrahedron. The horizontal edge connects a pair of vertices from one of the cubic lattices, and the vertical edge connects a pair of vertices from the other cubic lattice. It is natural to think of this spatial tiling in terms of the interleaved cubic lattices, but we can also look at the tiling from a different perspective. Consider the tilted plane P that contains the bottom face of the tetrahedron shown in Fig. 8. Consider a single cube of one of the cubic lattices. The plane P is the one defined by two opposing parallel edges of the cube. At the center of the cube there is a vertex from the other cube lattice; this vertex also lies in P. By making translated copies of the vertices in P we can obtain all of the vertices of the BCC lattice. Each vertex in P has six adjacent vertices in P and four adjacent vertices in each of the plane above and below P. There are two types of tetrahedra in the tiling, but both types are copies of the same tetrahedron in this case. The first type is the convex hull of three vertices in a copy of P and one vertex from the plane above or below that copy of P. The second type of tetrahedron is the convex hull of an edge in a copy of P and a corresponding edge from the plane above or below that copy of P.

From this understanding of the structure of the BCC-based spatial tiling, we can generalize to an entire family of triangulations of space using copies of two different tetrahedra. We consider first the set of vertices $\{(i, 0, 0) : i \in \mathbb{Z}\}$. We will make translated copies of this line in one direction to make a plane P. To do this we choose a parameter $a > 0$ and make infinitely many copies of each vertex translating by the vector $(1/2, a, 0)$. Lastly we choose a parameter $b > 0$ and make translated copies of the plane P using the translation vector $(1/2, 0, b)$. Thus our set of vertices is

$$\{(i + j/2 + k/2, aj, bk) : i, j, k \in \mathbb{Z}\},$$

Fig. 9. Starting with a set of vertices equally spaced along a line, we make translated copies of the line in a plane and triangulate the plane with each vertex having six neighbors

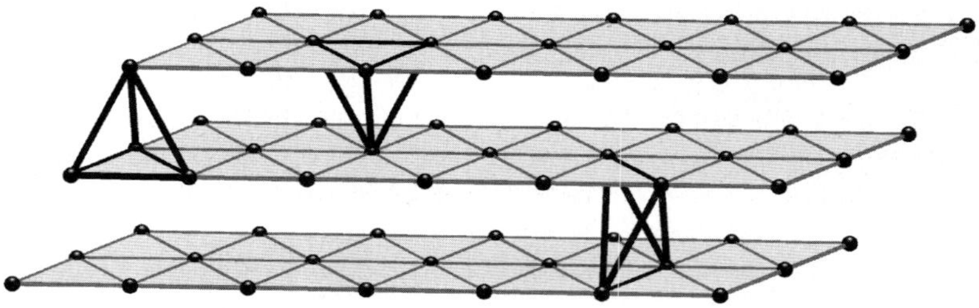

Fig. 10. We make translated copies of the plane into space and form tetrahedra by connecting vertices of adjacent planes as shown

i.e., the lattice $\Lambda = \{\sum_{i=1}^{3} c_i u_i : c_i \in \mathbb{Z}\}$ with basis vectors

$$u_1 = (1, 0, 0), u_2 = (1/2, a, 0), \text{ and } u_3 = (1/2, 0, b).$$

To turn this into a triangulation of space, we start by triangulating each copy of the plane P as shown in Fig. 9. Each vertex v in a copy of P is connected to the six vertices which lie at positions $v \pm (1, 0, 0)$, $v \pm (1/2, a, 0)$, and $v \pm (1/2, -a, 0)$. This yields the standard tiling of the plane with equilateral triangles if $a = \sqrt{3}/2$. Now for each triangle in a copy of P there is exactly one edge – the edge in the direction $(1, 0, 0)$ – for which a vertex lies directly above and directly below the midpoint of that edge. The first type of tetrahedron, then, is the convex hull of a triangle T in P and the vertex directly above or below the midpoint of one of the edges of T. If we add all possible tetrahedra of the first type, the gaps that remain are all tetrahedra of the second type. This second type of tetrahedron is the convex hull of an edge in P in the direction $\pm(1/2, a, 0)$ and the edge in the direction $\pm(1/2, -a, 0)$ whose midpoint lies directly above or below the midpoint of the given edge. Figure 10 shows three copies of the plane and highlights three of the tetrahedra that appear in the spatial tiling. The two tetrahedra on the left side of Fig. 10 are both copies of the first type of tetrahedron. The other tetrahedron is of the second type. For the BCC-based spatial triangulation, in which both types of tetrahedra are the same, the parameters are $a = b = \sqrt{2}/2$.

This family of triangulations of space is interesting for at least two reasons. First, the triangulation is completely well-centered if and only if both $a > 1/2$ and $b > 1/2$. Second, the family provides an elegant solution to the problems of tiling an infinite slab in $\mathbb{R}^3$ and tiling infinite rectangular prisms in $\mathbb{R}^3$. To tile the slab, one uses a finite number of translates of plane P. We see that any slab can be triangulated using copies of a single tetrahedron; for the parameters $a = b = \sqrt{2}/2$, the two types of tetrahedra are the same, and we can scale the result as needed to get a slab of the desired thickness. Triangulating rectangular prisms is also easy; it suffices to use a finite number of translates both of the initial line and of the resulting infinite strip. Again, this can be done using copies of a single tetrahedron, provided that the ratio of side lengths of the rectangle is a rational number. If the ratio of side lengths is p/q, one can use parameters $a = b = \sqrt{2}/2$ and take p copies of the initial line with q copies of the infinite strip. The result has the correct ratio of side lengths and can be scaled to get the desired rectangle.

4 Meshing the Cube With Well-Centered Tetrahedra

Meshing the cube with well-centered tetrahedra is significantly more difficult than meshing an infinite square prism, but it can be done. In this section we discuss several well-centered meshes of the cube that we constructed. In each case, the mesh was built by first designing the mesh connectivity, then moving the internal vertices using the optimization algorithm described in (15).

The first completely well-centered mesh of the cube that we discovered has 224 tetrahedra. We do not discuss the construction of this mesh in detail here, but it is worth mentioning the mesh because it has higher quality than the other well-centered meshes of the cube we will discuss. The quality statistics for the mesh are shown in the columns to the left in Table 2. The faces of this triangulation of the cube match up with each other, so it is relatively easy to make a well-centered mesh of any figure that can be tiled with unit cubes. One can use a copy of this well-centered mesh of the cube in each cube tile, using rotations and reflections as needed to make all the faces match.

Cassidy and Lord showed that the smallest acute triangulation of the square consists of eight triangles (3). Knowing that a completely well-centered mesh of the cube exists, it is natural to ask the analogous question for the cube. What is the smallest well-centered mesh of the cube? One can ask this question for the three different types of well-centered tetrahedral meshes – 2-well-centered, 3-well-centered, and completely well-centered. It is also conceivable that the well-centered mesh of the cube with the fewest tetrahedra is different from the mesh of the cube with the fewest vertices or edges, but we restrict our attention to the smallest well-centered mesh in the sense of fewest tetrahedra.

The answer to the question is not known for any of the three types of well-centered triangulations, but we can give some upper and lower bounds. The best known upper bound for 2-well-centered and completely well-centered triangulations of the cube is a completely well-centered mesh of the cube with 194 tetrahedra. Figure 11 shows a picture of this mesh. The quality statistics for the mesh are recorded in the middle columns of Table 2.

It is possible to improve the quality slightly by optimization of the location of the surface vertices, but for the version of the mesh shown in Fig. 11, the surface triangulation has some desirable symmetries, with every surface vertex at a cube corner, at the midpoint of an edge of the cube, or on a diagonal of a face of the cube. All of the faces match each other up to rotation and reflection, so this triangulation of the cube can also be used to create a well-centered mesh of any figure that can be tiled with unit cubes. Although the surface triangulation of this mesh is combinatorially the same as the triangulation with 224 tetrahedra, the vertices on the surfaces are not in the same location,

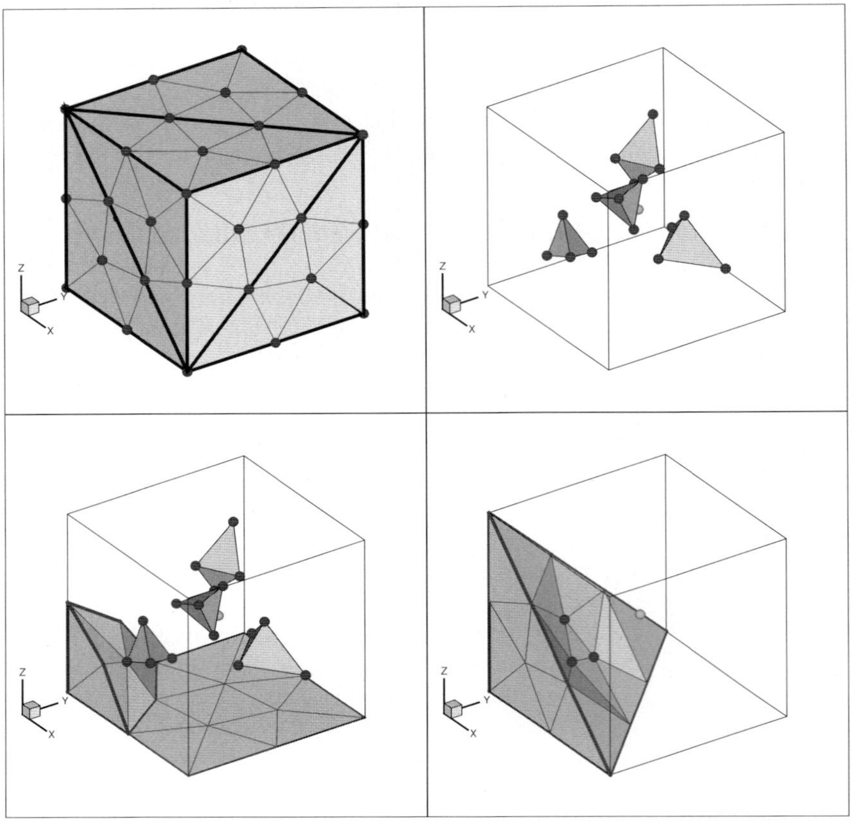

Fig. 11. A completely well-centered mesh of the cube with 194 tetrahedra

so it is not trivial to mix these two triangulations in meshing a cube-tiled shape. It is worth noting that the vertices of the surface triangulation that lie on the cube diagonals do not have coordinates of $1/3$ or $2/3$. Instead the coordinates are 0.35 or 0.65 for the vertices adjacent to an edge through the center of the cube face, and the coordinates are 0.295 or 0.705 for vertices not adjacent to such an edge.

Figure 11 is designed to make it possible to discern the structure of this triangulation of the cube. In the bottom right we see a triangulation of a region that fits into the corner of the cube and extends along the cube surface to diagonals of three of the faces of the cube. The triangulation basically has two layers of tetrahedra. The first layer is shown at bottom left. It consists of six tetrahedra that are incident to the corner of the cube. The two-layer triangulation is imprecisely replicated in four different corners of the cube. The thicker diagonal lines in the picture at top left help show which cube corners contain this type of triangulation. To complete the triangulation of the cube, a vertex at the center of the cube is added to the triangulations of the cube corners. The picture at

Table 2. The quality of our meshes of the cube decreases with the number of tetrahedra in the mesh of the cube

Cube Mesh Quality Statistics						
Quantity	224 Tets		194 Tets		146 Tets	
	Min	Max	Min	Max	Min	Max
h/R	0.041	0.850	0.005	0.790	0.016	0.854
Face Angle	21.01°	87.49°	26.93°	89.61°	17.09°	112.60°
Dihedral Angle	24.91°	105.61°	28.26°	126.64°	10.73°	163.17°
R/ℓ	0.618	1.569	0.612	1.134	0.711	1.835

top right of Fig. 11 shows all the interior vertices of the cube, combining the vertex at the center of the cube with one tetrahedron from each triangulation of a cube corner. The Delaunay triangulation defines the connectivity table, since any 3-well-centered triangulation is Delaunay.

This mesh establishes upper bounds for the 2-well-centered and completely well-centered cases. There is an even smaller triangulation of the cube that is known to be 3-well-centered. We do not describe it here except to say that the mesh consists of 146 tetrahedra and has a surface triangulation with fewer triangles than the two previously mentioned triangulations of the cube. The quality of that 3-well-centered mesh of the cube is shown in the rightmost columns of Table 2.

The upper bounds for this problem are simple to present, since they consist of constructive examples. The analytical lower bounds are rather more complicated to explain, and we discuss them here only briefly. One can show that in a 3-well-centered triangulation of the cube, no face of the cube is triangulated with two right triangles meeting along the hypotenuse (16). Thus each face of the cube must contain at least five vertices and at least three tetrahedral facets. Since there are six cube faces, this leads to a count of 18 tetrahedra, one adjacent to each facet. Some of these tetrahedra may have been counted twice, though, since a single tetrahedron may have a facet in each of two different cube faces. It can be shown that no tetrahedron having a facet in each of three different cube faces is a 3-well-centered tetrahedron (16). Thus no tetrahedron is triple-counted. A lower bound of 9 tetrahedra follows. It is possible to prove that any triangulation of the cube with all vertices lying on the edges of the cube is not 3-well-centered, but it is not immediately clear whether this can be used to improve the lower bound.

The lower bound for 2-well-centered and completely well-centered triangulations of the cube is slightly better. In this case, each face of the cube must be triangulated with an acute triangulation, so each cube face contains at least 8 tetrahedral facets. This gives a count of 48 tetrahedra, and once again we cannot have counted any tetrahedron three times (16). (For the 2-well-centered case this is not exactly the same reason as the 3-well-centered case.) A lower bound of 24 tetrahedra follows. This lower bound can be improved a little by paying attention to which triangular facets can and cannot lead to double-counted tetrahedra, and a lower bound of 30 tetrahedra can be obtained.

A careful analysis along these lines might improve the lower bound even more, since there is no way to conformally triangulate all the surfaces of the cube with an 8-triangle acute triangulation of each face. In any case, the authors suspect that the actual answer to these questions is close to 100 tetrahedra, if not greater, so the lower bounds mentioned here should be considered preliminary.

5 Some Subdivisions of Tetrahedra into Well-Centered Tetrahedra

Having seen that the cube can be subdivided into well-centered tetrahedra, one might also ask whether a tetrahedron can be subdivided into well-centered tetrahedra. Subdivisions of tetrahedra into well-centered tetrahedra could be used to refine an existing mesh. Subdividing a tetrahedron that tiles space into well-centered tetrahedra would provide new well-centered meshes of space. In general, well-centered subdivisions of tetrahedra might be used to design high-quality meshing algorithms.

One might suppose that subdividing the regular tetrahedron into smaller well-centered tetrahedra is relatively simple, but the problem is not so easy as it might seem. In two dimensions, the Loop subdivision, which refines a triangle by connecting the midpoints of each edge of the triangle, produces four smaller triangles. Each of these triangles is similar to the original triangle, so the Loop subdivision of an acute triangle is an acute triangulation of the triangle. In three dimensions, however, there is no obvious analog of the Loop subdivision.

Connecting the midpoints of the edges of a tetrahedron cuts out four corner tetrahedra that are similar to the original tetrahedron. The shape that remains in the center after removing these four tetrahedra is an octahedron. In the case of the regular tetrahedron, it is a regular octahedron, and it can be subdivided into four tetrahedra by adding an edge between opposite vertices of the octahedron. The result is not well-centered; the center of the octahedron is the circumcenter of all four tetrahedra, so the tetrahedra are not 3-well-centered. In addition, the facets incident to the new edge are right triangles, so the tetrahedra are not 2-well-centered.

We can turn this subdivision of the regular tetrahedron into a well-centered subdivision, however. By sliding some of the new vertices along the edges of the regular

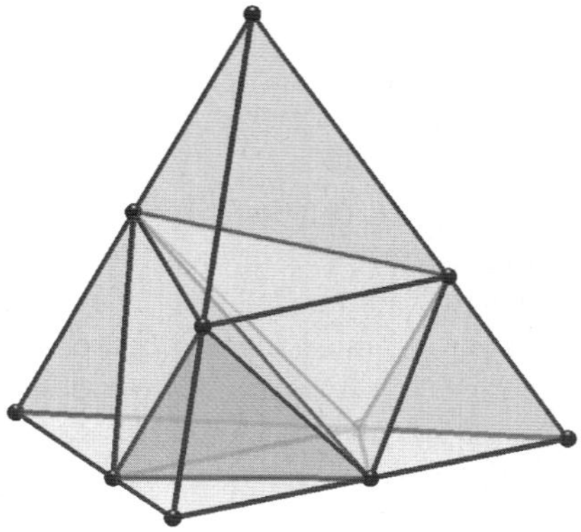

Quality Statistics		
Quantity	Min	Max
h/R	0.0345	0.712
Face Angle	38.87°	86.76°
Dihedral Angle	38.44°	121.37°
R/ℓ	0.702	0.934

Fig. 12. A simple subdivision of the regular tetrahedron into eight completely well-centered tetrahedra

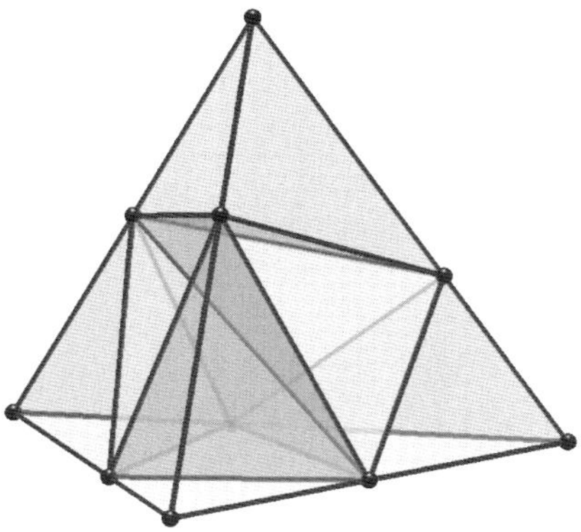

Quality Statistics		
Quantity	Min	Max
h/R	0.0448	0.584
Face Angle	$39.63°$	$87.43°$
Dihedral Angle	$46.72°$	$105.95°$
R/ℓ	0.777	0.826

Fig. 13. Another simple subdividision of the regular tetrahedron into eight completely well-centered tetrahedra

tetrahedron, moving them away from the edge midpoints, we can make the four center tetrahedra well-centered without degrading the four corner tetrahedra to the point of not being well-centered. Figures 12 and 13 illustrate two different successful ways we can slide the vertices along edges of the tetrahedra. In both cases, the midpoint vertices that are adjacent to the central edge remain stationary, to keep the central edge as short as possible. In Fig. 12, the four free midpoints all slide towards the same edge of the regular tetrahedron. In Fig. 13, the free midpoints slide along a directed four-cycle through the vertices of the regular tetrahedron.

There are also more complicated ways to divide the regular tetrahedron into smaller well-centered tetrahedra. Figure 14 shows the basic structure of a subdivision of the regular tetrahedron into 49 tetrahedra. A smaller regular tetrahedron is placed in the center of the large tetrahedron with the same orientation as the original. Each face of the smaller tetrahedron is connected to the center of a face of the larger tetrahedron. At each corner of the large tetrahedron, a small regular tetrahedron is cut off of the corner, and the resulting face is connected to a vertex of the central regular tetrahedron. After filling in a few more tetrahedral faces, six more edges need to be added to subdivide octahedral gaps into tetrahedra. The mesh shown in Fig. 15 is the completely well-centered mesh that results from optimizing the mesh shown in Fig. 14. This subdivision of the regular tetrahedron is interesting partly because all of the surface triangulations match and have three-fold radial symmetry. It is also possible that this type of subdivision will be easier to use in mesh refinement than the other two subdivisions.

It is not clear whether these constructions can be extended in some way to create a well-centered subdivision of any well-centered tetrahedron. The constructions cannot be extended to create well-centered subdivisions of all tetrahedra, since both constructions cut off the corners of the tetrahedron to create smaller tetrahedra that are nearly similar to the original tetrahedron. In particular, the cube corner tetrahedron, i.e., some scaled, rotated, translated version of the tetrahedron with vertices $(0, 0, 0)$, $(1, 0, 0)$, $(0, 1, 0)$, and $(0, 0, 1)$, cannot be subdivided into well-centered tetrahedra in this fashion; one can show that no tetrahedron with three mutually orthogonal faces is

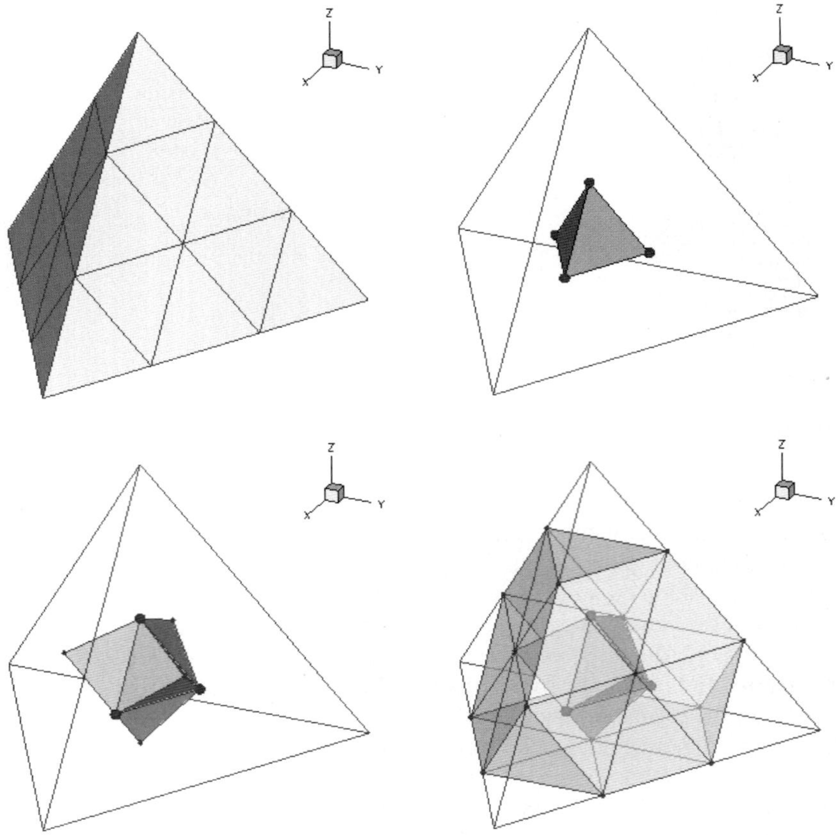

Fig. 14. A subdivision of the regular tetrahedron that can be made completely well-centered through optimization

3-well-centered (16). Subdividing the cube corner tetrahedron is particularly interesting, though, because it provides some guidance regarding what is needed to mesh a cube or, for that matter, any object having three mutually orthogonal faces that meet at a point.

In fact, the smallest known subdivision of the cube into well-centered tetrahedra is based on a subdivision of the cube corner into well-centered tetrahedra. The picture in the bottom right corner of Fig. 11 shows a triangulation of a region that fits into the corner of a cube. The three visible interior vertices in Fig. 11 are not coplanar with the cube diagonals, but they are nearly so, and there is a completely well-centered mesh of the cube corner that is combinatorially the same as that mesh. We actually obtained the mesh of the cube corner tetrahedron first, and the well-centered mesh of the cube in

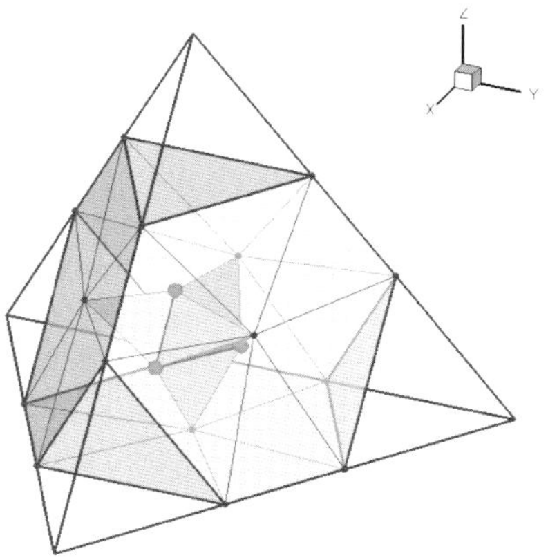

Quality Statistics		
Quantity	Min	Max
h/R	0.0146	0.845
Face Angle	23.36°	89.07°
Dihedral Angle	29.93°	107.73°
R/ℓ	0.612	1.305

Fig. 15. The completely well-centered subdivision of the regular tetrahedron that results from optimizing the mesh shown in Fig. 14

Fig. 11 was obtained by replicating the cube corner mesh as described earlier, adding a vertex at the cube center, computing the Delaunay triangulation, and optimizing with the software discussed in (15).

6 Conclusions and Questions

We have discussed some of the properties of well-centered tetrahedra and seen that it is possible to triangulate a variety of basic three-dimensional shapes with completely well-centered tetrahedra. The triangulations discussed suggest that it might soon be practical to mesh simple domains in $\mathbb{R}^3$ with well-centered tetrahedra. They also show that there are a rich variety of well-centered tetrahedra.

The authors hope that it will be possible to build robust software for meshing three-dimensional domains with this variety of well-centered tetrahedra, but there is still significant work to be done before that goal can be reached. Part of this work is to determine what properties the neighborhood of a vertex in a 3-dimensional triangulation must have in order to permit a well-centered triangulation. The question is partially answered in (16), but a complete answer is lacking. The ability to subdivide any tetrahedron (or even just any well-centered tetrahedon) into smaller well-centered tetrahedra would also be a significant advance toward this goal.

This work also raises some questions that are of more theoretical interest, though not without practical application. It would be interesting to construct the smallest possible well-centered mesh of the cube. How might one improve the lower bounds on the number of tetrahedra needed in a well-centered mesh of the cube? Also it is still an open question whether there are other tetrahedra for which copies of a single tetrahedron meet face to face and fill space. Could there be other well-centered tetrahedra that tile space? Are there other families of high-quality tetrahedra that tile space?

Acknowledgment

The authors thank Edgar Ramos and Vadim Zharnitsky for useful discussions. The work of Evan VanderZee was supported by a fellowship from the Computational Science and Engineering Program, and the Applied Mathematics Program of the University of Illinois at Urbana-Champaign. The work of Anil N. Hirani was supported in part by an NSF CAREER Award (Grant No. DMS-0645604) and the work of Damrong Guoy was supported by CSAR (Center for Simulation of Advanced Rockets). The CSAR research program is supported by the US Department of Energy through the University of California under subcontract B523819.

References

[1] Alliez, P., Cohen-Steiner, D., Yvinec, M., Desbrun, M.: Variational tetrahedral meshing. ACM Transactions on Graphics 24(3), 617–625 (2005), doi:10.1145/1073204.1073238

[2] Barnes, E.S., Sloane, N.J.A.: The optimal lattice quantizer in three dimensions. SIAM Journal on Algebraic and Discrete Methods 4(1), 30–41 (1983), doi:10.1137/0604005

[3] Cassidy, C., Lord, G.: A square acutely triangulated. J. Recreational Math. 13, 263–268 (1980)

[4] Desbrun, M., Hirani, A.N., Leok, M., Marsden, J.E.: Discrete exterior calculus. arXiv:math.DG/0508341 (2005), http://arxiv.org/abs/math.DG/0508341

[5] Du, Q., Faber, V., Gunzburger, M.: Centroidal Voronoi tessellations: applications and algorithms. SIAM Review 41(4), 637–676 (1999), doi:10.1137/S0036144599352836

[6] Eppstein, D., Sullivan, J.M., Üngör, A.: Tiling space and slabs with acute tetrahedra. Computational Geometry: Theory and Applications 27(3), 237–255 (2004), doi:10.1016/j.comgeo.2003.11.003

[7] Fuchs, A.: Automatic grid generation with almost regular Delaunay tetrahedra. In: Proceedings of the 7th International Meshing Roundtable, Dearborn, Michigan, October 26–28, 1998, pp. 133–147. Sandia National Laboratories (1998)

[8] Hirani, A.N.: Discrete Exterior Calculus. PhD thesis, California Institute of Technology (May 2003), http://resolver.caltech.edu/CaltechETD:etd-05202003-095403

[9] Kimmel, R., Sethian, J.: Computing geodesic paths on manifolds. Proc. Nat. Acad. Sci. 95, 8341–8435 (1998), doi:10.1073/pnas.95.15.8431

[10] Nicolaides, R.A.: Direct discretization of planar div-curl problems. SIAM Journal on Numerical Analysis 29(1), 32–56 (1992), doi:10.1137/0729003

[11] Sazonov, I., Hassan, O., Morgan, K., Weatherill, N.P.: Smooth Delaunay – Voronoi dual meshes for co-volume integration schemes. In: Proceedings of the 15th International Meshing Roundtable, Birmingham, Alabama, September 17–20, 2006, Sandia National Laboratories (2006), doi:10.1007/978-3-540-34958-7_30

[12] Sommerville, D.M.Y.: Space-filling tetrahedra in euclidean space. Proceedings of the Edinburgh Mathematical Society 41, 49–57 (1923)

[13] Üngör, A., Sheffer, A.: Pitching tents in space-time: Mesh generation for discontinuous Galerkin method. International Journal of Foundations of Computer Science 13(2), 201–221 (2002), doi:10.1142/S0129054102001059

[14] VanderZee, E., Hirani, A.N., Guoy, D., Ramos, E.: Well-centered planar triangulation – an iterative approach. In: Brewer, M.L., Marcum, D. (eds.) Proceedings of the 16th International Meshing Roundtable, Seattle, Washington, October 14–17, 2007, pp. 121–138. Springer, Heidelberg (2007), http://www.cs.uiuc.edu/hirani/papers/VaHiGuRa2007_IMR.pdf, doi:10.1007/978-3-540-75103-8_7

[15] VanderZee, E., Hirani, A. N., Guoy, D., Ramos, E.: Well-centered triangulation. Tech. Rep. UIUCDCS-R-2008-2936, Department of Computer Science, University of Illinois at Urbana-Champaign, Also available as a preprint at arXiv as arXiv:0802.2108v1 [cs.CG] (February 2008), http://arxiv.org/abs/0802.2108

[16] Vander Zee, E., Hirani, A.N., Guoy, D., Zharnitsky, V.: Conditions for well-centeredness. Tech. Rep. UIUCDCS-R-2008-2971, Department of Computer Science, University of Illinois at Urbana-Champaign (2008)

3D Delaunay Refinement of Sharp Domains without a Local Feature Size Oracle

Alexander Rand and Noel Walkington

Carnegie Mellon University

Summary. A practical incremental refinement algorithm for computing a quality, conforming Delaunay mesh of an arbitrary 3D piecewise linear complex is given. The algorithm allows small input angles and does not require the local feature size of the input to be computed before performing the refinement. Small input angles are protected with a new type of collar which is simpler to implement and analyze than previous approaches. The algorithm has been implemented and several computational examples are given.

1 Introduction

Ruppert's algorithm [9] is an elegant method for computing a quality conforming Delaunay mesh of a two-dimensional piecewise linear complex (PLC) which contain no acute angles. Based on Ruppert's original techniques of corner lopping and concentric shell splitting, the algorithm has been rigorously extended to 2D PLCs containing sharp input angles. Ruppert's algorithm has also been extended to handle three-dimensional PLCs without acute input angles [4]. Handling sharp angles in 3D leads to a variety of issues not seen in 2D. Subsequent algorithms have been complicated (in some cases to the point of prohibiting implementation) and lacked certain desirable properties seen in the original method.

Early algorithms for computing conforming Delaunay tetrahedralizations (without considering radius-edge quality) in 3D [6, 3] involved computing local feature size on all segments of the input through a brute force search before inserting Steiner points. Several extensions of these methods to the problem of quality mesh generation [2, 1] have also required the computation of local feature size as a preprocess. An algorithm of Pav and Walkington [7] gives an alternative view: Delaunay refinement should be considered an algorithm for computing local feature size and thus local feature size should not be a necessary input for the algorithm. They gave a 3D Delaunay refinement algorithm for which all steps require only "local" information about the mesh in the Delaunay tetrehedralization. Still, this algorithm is not local in the same way as Ruppert's algorithm: steps in the algorithm require searching the Delaunay triangulation over a prescribed distance rather than only querying the immediate Delaunay neighbors. This issue of finding a local algorithm for computing constrained Delaunay tetrahedralizations has been resolved [12, 11], but has not been addressed in the case of (usual) Delaunay tetrahedralization.

We present a Delaunay refinement algorithm for general PLCs which attempts to preserve as many features of Ruppert's algorithm as possible. The algorithm does not require local feature size information to be precomputed. Each operation of the algorithm is local to the current Delaunay tetrahedralization in the sense that encroachment can be determined considering only a point's immediate Delaunay neighbors. The result of the algorithm is a conforming Delaunay tetrahedralization consisting of tetrahedra with bounded radius-edge ratio away from sharp input angles. The strategy for protecting sharp angles is designed for simplicity both in implementation and analysis. The authors are unaware of any algorithm for quality, conforming Delaunay refinement in 3D which has been implemented. Existing software packages which handle small input angles are based on constrained Delaunay tetrahedralization [12, 11] and weighted Delaunay tetrehedralization [1].

To highlight the differences between the algorithm and other extensions of Ruppert's algorithm to three dimensions, an analogous two dimensional version is developed and analyzed. In both cases, a Delaunay refinement algorithm is used to prove estimates on the local feature size of the mesh at input points (in 2D) or on input segments (in 3D). These are the reverse inequalities of those usually seen in the analysis of Delaunay refinement algorithms; while typically local feature size is shown to be a lower bound for the distance between points in the mesh, we show that local feature size bounds this distance (in the form of segment lengths in 3D) from above. With these estimates, a "collar" can be inserted near these sharp angles which ensures that the resulting mesh conforms to the input at termination. An example of the collar used to protect sharp input features can be seen in Figure 6 in Section 4.

In Section 2, we define the problem and a number of feature size functions used in the analysis. Sections 3 and 4 outline the algorithms in two and three dimensions, respectively, and justify termination and correctness. Several resulting meshes from the 3D implementation are given in Section 5.

2 Problem Description and Notation

We describe algorithms for meshing a general PLC given by sets of input points and input segments in 2D, $\mathcal{C} = (\mathcal{P}, \mathcal{S})$, or sets of input points, input segments and input faces in 3D, $\mathcal{C} = (\mathcal{P}, \mathcal{S}, \mathcal{F})$. In 2D, a PLC $\mathcal{C}' = (\mathcal{P}', \mathcal{S}')$ is a refinement of the PLC $(\mathcal{P}, \mathcal{S})$ if $\mathcal{P}' \subset \mathcal{P}$ and each segment in $\mathcal{S}$ is the union of segments in $\mathcal{S}'$. In 3D, a PLC $(\mathcal{P}', \mathcal{S}', \mathcal{F}')$ refining $(\mathcal{P}, \mathcal{S}, \mathcal{F})$ must satisfy the additional condition that each face in $\mathcal{F}$ must be the union of faces in $\mathcal{F}'$.

An important parameter to the algorithm is a raduis-edge ratio threshold, κ, which must be greater than 2. The algorithm eliminates all tetrahedra which have radius-edge ratio greater than κ except for some which are near the sharp angles in the input.

The algorithms described involve incrementally refining the input PLC until the Delaunay triangulation/tetrahedralization of the point set $\mathcal{P}'$ conforms to the input PLC and satisfies the radius-edge quality criteria. Throughout this refinement, it is necessary to identify some important simplices based on their relation to the input complex.

Definition 1. *Consider a refinement* $(\mathcal{P}', \mathcal{S}', \mathcal{F}')$ *of an input PLC* $(\mathcal{P}, \mathcal{S}, \mathcal{F})$.

- *An* **end segment** *is a segment in* $\mathcal{S}'$ *for which at least one endpoint is an input point in* $\mathcal{P}$.
- *An* **end facet** *is a facet in* $\mathcal{F}'$ *for which at least one vertex lies on an input segment in* $\mathcal{S}$.

The smallest angle in the input PLC will be denoted by α. In 2D, this is the smallest angle between adjacent input segments. In 3D, this is the smallest angle between any pair of adjacent input features: faces or segments. Let $\mathbf{C}_{d,\alpha}$ denote the set of d-dimensional PLCs for which no two features meet at an angle of less than α.

An appropriate notion of feature size is essential to the analysis of Delaunay refinement algorithms. The standard definition of local feature size is given below as well as another sizing function (mesh feature size defined below) which is used throughout the arguments.

Definition 2. *Let* $\mathcal{C}$ *be a PLC.*

- *The* i-**local feature size** *at point* x *with respect to* $\mathcal{C}$, $\text{lfs}_i(x, \mathcal{C})$, *is the radius of the smallest circle centered at* x *which touches two* disjoint *features of* $\mathcal{C}$ *of dimension no greater than* i.
- *The* i-**mesh feature size** *at point* x *with respect to* $\mathcal{C}$, $\text{mfs}_i(x, \mathcal{C})$, *is the radius of the smallest circle centered at* x *which touches two features of* $\mathcal{C}$ *of dimension no greater than* i.
- *The* **current feature size** *with respect to* $\mathcal{C}'$, $N(x, \mathcal{C}') := \text{lfs}_0(x, \mathcal{C}')$, *is the distance from* x *to its second nearest neighbor in* $\mathcal{P}'$.

Each of these feature size functions is Lipschitz (with constant 1). For a fixed PLC, local and current feature size each have a strictly positive minimum value while mesh feature size can equal zero.

If the argument supplied to any of the above feature size functions is a set of points, rather than a point, then the result is defined to the be infimum of the function over the set. The dimension argument will be omitted when considering the $d-1$ dimensional feature sizes. For instance $\text{lfs}_i(s, \mathcal{C}) := \inf_{x \in s} \text{lfs}_i(x, \mathcal{C})$ and, in 3D, $\text{lfs}(x, \mathcal{C}) := \text{lfs}_2(x, \mathcal{C})$.

The distinction in notation between local/mesh feature size and current feature size is deliberate: these functions will be used differently.

Convention: Throughout the analysis, local/mesh feature size will always be evaluated with respect to the input PLC, while current feature size will always involve the current refined PLC. By adhering to this usage, the PLC argument

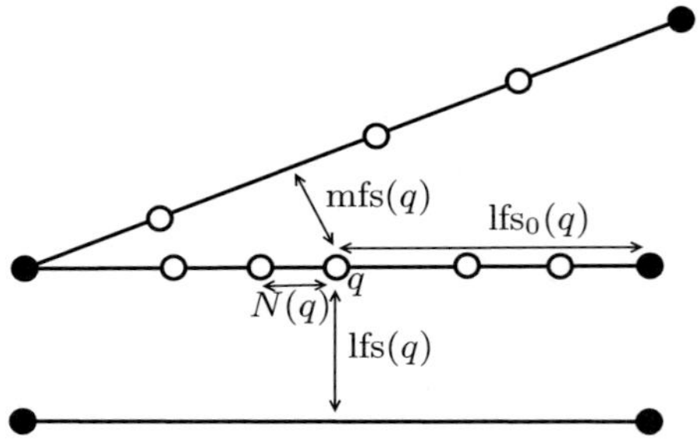

Fig. 1. Example of sizing functions in Definition 2 for a 2D PLC. The black points represent input points while the white points represent points added by the algorithm.

can be omitted. See Figure 1 for an example the different feature size functions for a simple PLC.

Finally, much of the analysis involves giving identical estimates for the local feature size of end segments and the mesh feature size of non-end segments. It is useful to refer to these two cases with the same notation.

Definition 3. *The i-**feature size** of segment s is defined as follows.*

$$\mathrm{fs}_i(s) = \begin{cases} \mathrm{lfs}_i(s) & \text{if } s \text{ is an end segment,} \\ \mathrm{mfs}_i(s) & \text{if } s \text{ is a non-end segment} \end{cases}$$

*Given a segment or triangle s in $\mathcal{C}'$, point x is called an i-**feature size witness** for s if x is contained in a feature of $\mathcal{C}$ of dimension at most i which is disjoint from s. For segment or triangle s in $\mathcal{C}'$, point x is called a **local feature size witness** for s if x is contained in a feature of $\mathcal{C}$ which is disjoint from another feature of $\mathcal{C}$ containing s.*

This definition of feature size is similar to the local gap size used by Chen and Poon [2]. The notion of i-feature size witness is used by recognizing that if x is an i-feature size witness of segment s then

$$\mathrm{fs}_i(s) \leq \mathrm{dist}(x, s).$$

3 Delaunay Refinement Algorithm in 2D

The following algorithm can be used to mesh any 2D PLC and extends many of the properties of Ruppert's Delaunay refinement algorithm for PLCs in $\mathbf{C}_{2,90}$. Several effective approaches to protecting acute angles from cascading encroachment sequences have already been developed for this problem [9, 10, 5, 8]. However, we present a different corner-lopping algorithm which mirrors the 3D algorithm given in Section 4. In what follows, small angles are protected by splitting all acutely adjacent segments at an equal distance. Then, these segments are protected by rejecting any circumcenters proposed for insertion which encroach the diametral disk of an end segment. This algorithm differs from previous versions as we compute an acceptable length for the protected end segments in the output mesh before refining any triangles for quality.

Algorithm Overview

The algorithm consists of the following steps.

- (Step 0) Compute the Delaunay triangulation of the input point set.
- (Step 1a) Estimate local feature size at all input points.
- (Step 1b) Protect small input angles with a "collar".
- (Step 2) Perform standard Delaunay refinement away from the collar.

Step 0 is a standard procedure and common in many Deluanay refinement algorithms. Steps 1a and 2 are Delaunay refinement algorithms. Step 1b involves a single pass over the set of input points to insert points. Properties of the mesh generated by this algorithm are summarized in two theorems.

Theorem 1 (Termination/Grading). *The algorithm terminates. Moreover, there exists $c_\alpha > 0$ such that for each point p added by the algorithm,*

$$N(p) \geq c_\alpha \text{lfs}(p)$$

Verifying Theorem 1 holds for each step in the algorithm uses the standard arguments; the insertion radius of each point is inductively bounded below by feature size.

Ruppert's analysis of the standard Delaunay refinement algorithm for non-acute domains showed that this condition is sufficient to ensure a size-competitive output mesh. In our case, $c_\alpha = O(\frac{1}{\sin\alpha})$ and thus, this condition does not guarantee a competitive output size over the class of all 2D PLCs but does ensure that the output is size competitive on $\mathbf{C}_{2,\alpha}$ for any fixed α.

When considering the standard Delaunay refinement algorithm for non-acute domains, termination implies that the output conforms to the input PLC and that no poor quality triangles remain. A guarantee of this type requires more attention when dealing with acute angles.

Theorem 2 (Conformity/Quality). *At termination, the resulting Deluanay triangulation conforms to the input PLC. Moreover, any remaining poor quality triangles are near the collar region.*

Ensuring that the conformity/quality theorem holds requires a series of lemmas related to each step. With the lemmas, the notion that a triangle is near the collar will be made precise.

Step 1a

This step is a Delaunay refinement algorithm. Segments are queued to be split based on the following encroachment criteria and no triangles are split for quality during this step.

- (Segment Encroachment) Segment s is encroached if it has an endpoint q with a Delaunay neighbor p which lies in the diametral disk of s and either p is a feature size witness for s or s is an end segment and is more than twice as long as its shortest adjacent end segment.

This is the usual definition of encroachment with one modification: an end segment cannot be encroached by a point on a sufficiently long adjacent end segment. This ensures that adjacent end segments do not alternately encroach each other and prevent termination of the algorithm. Thus, this definition of encroachment is strictly weaker than the standard diametral disk notion, yet shares the property that encroachment of a segment can be checked by only considering the Delaunay neighbors of its endpoints.

The triangulation resulting from this step does not necessarily conform to the input PLC and may contain poor quality elements. However, using this triangulation, it is possible to bound the local feature size at input points in terms of a local quantity (in the Delaunay triangulation): the distance from the input point to its nearest neighbor.

Lemma 1. *After Step 1a terminates, the following inequality holds for all input points $q_0 \in \mathcal{P}$.*

$$N(q_0) \leq \sqrt{2}\mathrm{lfs}(q_0).$$

This inequality is in the opposite direction of those in the standard analysis of Delaunay refinement algorithms and Theorem 1. Rather than guaranteeing points are far enough apart, it ensures that the distance from each input point to its nearest neighbor is not much greater than the local feature size.

While the constant in Theorem 1 depends on α and can be large, the constant in Lemma 1 is sharp and independent of α. This estimate gives an appropriate length at which segments can be split to protect acutely adjacent segments.

While the bound in Lemma 1 can be shown for any point in the mesh, $q \in \mathcal{P}'$, the statement is deliberately only made for input points. In order to protect sharp angles, it is necessary to estimate the feature size on all $d-2$ dimensional input features. The same principle will be applied in the 3D version.

Step 1b

End segments are split in this step to form a "collar" around each input point which ensures end segments conform to the input, even near skinny input angles. For each input point q_0, all segments containing q_0 are split at a distance of $\frac{N(q_0)}{3}$ away from q_0.

Any point added during this step will be called a **collar point**. All end segments are considered to be **collar simplices**. (This distinction is not particularly useful in the 2D algorithm but it will be necessary to consider the analogous simplices in the 3D version.) The **collar region** is defined to be the set of collar simplices.

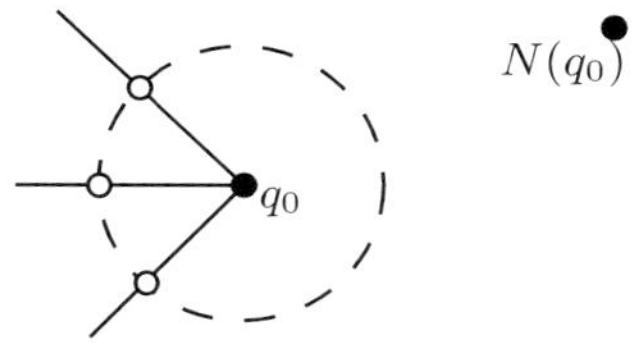

Fig. 2. Adjacent segments are split at equal length in Step 1b

Lemma 2. *Following Step 1b, the following properties hold.*

- *The diametral disk of each collar simplex contains no points of* $\mathcal{P}'$.
- *The diametral disk of each collar simplex does not intersect any non-collar segment.*

Lemma 1 is used to show that the ball of radius $\frac{N(q_0)}{3}$ centered at q_0 does not intersect any segments which are disjoint from q_0. The results of Lemma 2 will ensure that these segments will not need to be split again in the following Delaunay refinement step.

Step 2

With the collar in place, it is now safe to perform a slightly modified version of Ruppert's algorithm. In this step, segments are queued for conformity based on the standard diametral disk encroachment criteria, and triangles are queued based on the radius edge-quality criteria. As the input may contain small angles, an additional rule is needed to ensure the algorithm terminates despite some poor quality triangles near these small input angles. When processing poor quality triangles from the queue, the following policy is considered before inserting the circumcenter.

- Let c be a circumcenter proposed for insertion to split a poor quality triangle. If c lies in the diametral disk of a collar simplex, reject c.

Lemma 3. *No point is inserted in the diametral disk of a collar segment.*

With this lemma, standard techniques can be used to verify Theorem 1 and Theorem 2. Any remaining poor quality triangles must have circumcenters which lie in the diametral disk of some collar simplex (i.e. end segment). This makes precise what it means in Theorem 2 for a triangle to be "near the collar."

4 Delaunay Refinement Algorithm in 3D

A similar approach to that given in 2D can be applied to 3D PLCs. It is important to note that *input points* in the two dimensional problem play the role of *segments* in the 3D case. In 2D, we used the distance to the nearest neighbor to estimate the local feature size at an input point, while, in 3D, each segment's length will be used to estimate its feature size.

Algorithm Overview

The first steps of the algorithm provide an estimate of the feature size on each segment of the input. Then a collar is inserted around the 1-skeleton to ensure the mesh conforms to the input near skinny input angles. Finally, the usual Delaunay refinement algorithm is performed with a rejection policy to ensure both that the algorithm terminates and that the resulting mesh conforms to the input PLC.

- (Step 0) Compute the Delaunay triangulation of the set of input points.
- (Step 1a) Protect small angles between segments by splitting based on 0-local feature size.
- (Step 1b) Estimate the 1-feature size on all segments.
- (Step 2a) Split segments to improve the 1-feature size estimate.
- (Step 2b) Estimate the local feature size on all segments.
- (Step 2c) Protect the 1-skeleton with a collar.
- (Step 3) Perform standard Delaunay refinement away from the collar.

Steps 1a, 2a, and 2c occur in one pass over the mesh: they loop through all the segments and add an appropriate number of points for each segment. Steps 1b, 2b and 3 are Delaunay refinement algorithms. Each of these steps has its own notion of encroachment by which simplices are queued and split according to some priority. Only in Step 3 are tetrahedra split based on radius-edge quality. Figure 9 in Section 5 shows a face of an example mesh produced at each step of the algorithm.

The output guarantees of the algorithm are split into two theorems.

Theorem 3 (Termination/Grading). *The algorithm terminates. Moreover, there exists $c_\alpha > 0$ such that for any point p in the output mesh,*

$$N(p) \geq c_\alpha \text{lfs}(p).$$

This theorem is shown via induction using similar techniques to algorithms for large angle input. In the case of inputs with small angles, the next theorem is also essential.

Theorem 4 (Conformity/Quality). *The resulting mesh is a conforming Delaunay refinement of the input PLC. Moreover, all resulting tetrahedra with radius-edge ratio greater than κ are near the collar region.*

Throughout the refinement, independent Delaunay meshes for each of the input features (segments and faces) are maintained in additional to the complete tetrahedralization. (This also occurs in 2D case, but the meshes of segments are simple and no hierarchy needs to be considered.) To ensure correctness of the algorithm, simplex splits and point insertions must be processed following a partial order given in [4]. Segments and triangles in the refinement refer to simplices in the separate meshes of the input features. However, when discussing Delaunay neighbors of the endpoint of a segment or vertex of a triangle, we are referring to the neighbors in the 3D mesh. To ensure that this makes sense, we further restrict the required partial order to always handle forced point insertions before handling simplex splits. For example, if a segment is split, then its midpoint is immediately inserted into all meshes containing the segment before another simplex is split in any mesh. This ensures that for any point in a lower dimensional mesh, its Delaunay neighbors in the 3D mesh can still be considered.

In order to prove Theorem 4, conditions on the mesh following each step are shown.

Step 1a

Once the Delaunay triangulation of the point set has been computed, each segment is split at one-third the distance from an input point to its nearest neighbor. This is a typical "grooming" step in two dimensional algorithms [5]. This ensures that each end segment in the mesh conforms to the input.

Step 1b

This is the first Delaunay refinement step. In this case, segments are queued based on the following encroachment criteria.

- (Segment Encroachment) Segment s is encroached if it has an endpoint q with a Delaunay neighbor p such that p is a 1-feature size witness for s and $|p - q| < |s|$.
- (End Segment Balance) Non-end segment s_n is encroached if it is adjacent to an end segment s_e and $|s_e| < |s_n|$.

The encroachment criteria only depend on the Delaunay neighbors of the endpoints of the segment in question in the current Delaunay triangulation.

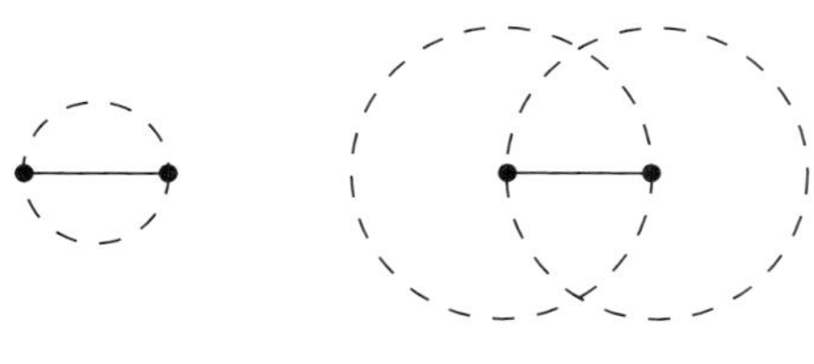

Fig. 3. (Left) Encroachment region for a segment in Ruppert's algorithm. (Right) Encroachment region for a segment in Step 1b.

Also, observe that the end segment balance criterion is necessary in order to ensure that the feature size bound in the next lemma holds. This ensures that a long non-end segment segment will be split even if a nearby feature is hidden from the non-end segment by an input point. See Figure 4.

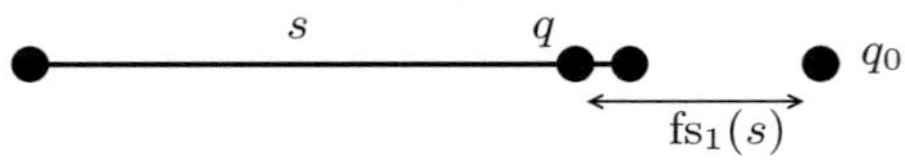

Fig. 4. End segment balance is essential. Without this, segment s may be much longer than the local feature size on the segment.

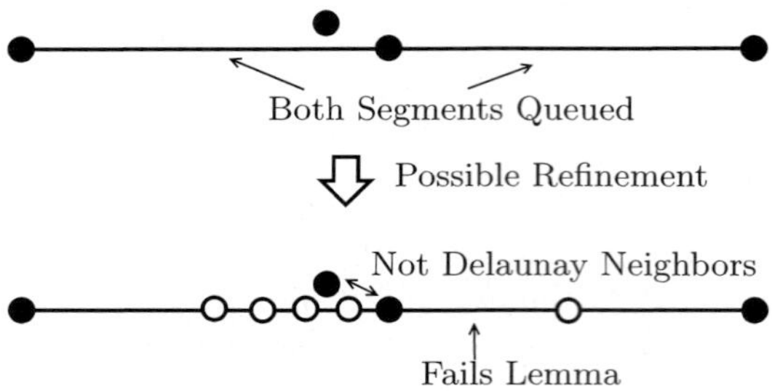

Fig. 5. Poor refinement order can lead to a failure of Lemma 4

Lemma 4. *After Step 1b, for any segment s in the mesh, $|s| \le \sqrt{2}\,\mathrm{fs}_1(s)$.*

Ideally, this lemma would follow from the algorithm without further restriction. However, for the algorithm as specified this may not be true! Consider a situation as seen in Figure 5. If the algorithm always processes segments on the left first, once the segment on the right splits for the first time, it may no longer have a 1-feature size witness which is on its Delaunay cavity. At this point, though, the segment will not have been split enough to satisfy Lemma 4.

To prevent this issue we prioritize the queue of segments to be split according to the following rule.

- Segments are prioritized by length. Longer segments are always split first.

We choose this ordering to get a sharp constant in the lemma. This can be seen from the following example. Consider the segment between $(-1, 0, 0)$ and $(1, 0, 0)$ and the segment between $(0, -1, \sqrt{2})$ and $(0, 1, \sqrt{2})$. Each segment has length two and the distance between the two segments is $\sqrt{2}$. The distance between any pair of endpoints is 2, and thus neither segment is encroached according to this step of the algorithm. The following proposition shows that this example gives the nearest two segments which are not encroached.

Proposition 1. *Let s and $\bar{s}$ be segments in $\mathbb{R}^3$. If $|s| \ge |\bar{s}|$ and $\mathrm{dist}(s, \bar{s}) < \frac{|s|}{\sqrt{2}}$, then there are endpoints q and $\bar{q}$ of s and $\bar{s}$, respectively, such that $|q - \bar{q}| < |s|$.*

Step 2a

The constant in the bound of Step 1b is not strong enough for the upcoming argument in Step 2b. Thus, in this step, all segments are split into fourths, giving the needed bound.

Lemma 5. *Following Step 2a, for any segment in the mesh,*

$$|s| \leq \frac{1}{2\sqrt{2}} \mathrm{lfs}_1(s).$$

The estimate $|s| \leq \frac{1}{2\sqrt{2}} \mathrm{fs}_1(s)$ does not hold after this step. This is due to the fact that when an end segment is split, the resulting non-end segment can have a 1-mesh feature size which is much smaller than its 1-local feature size.

Step 2b

This step is similar to Step 1b with an additional encroachment rule for triangles in the current Delaunay triangulation of the faces. Segments and faces are queued for encroachment according to the following rules.

- (Segment Encroachment) Segment s is encroached if it has an endpoint q with a Delaunay neighbor p such that p is a 1-feature size witness for s or a local feature size witness for s and $|p - q| < |s|$.
- (End Segment Balance) Non-end segment s_n is encroached if it is adjacent to an end segment s_e and $|s_e| < |s_n|$.
- (Face Encroachment) A triangle t is encroached if it has a vertex q with Delaunay neighbor p such that p is not in the face containing q and $|p - q| < 2r_t$. (The circumradius of t is denoted r_t.)

Without a rejection policy, the above encroachment rules can (and do) lead to non-termination. This can occur when two acutely adjacent faces alternately cause each other to split. This is prevented by the following rule.

- Suppose a circumcenter of a triangle, c_t, is proposed for insertion into a face F and has a prospective Delaunay neighbor, p, which is the end point of a segment, s, contained in F, and $|c_t - p| < |s|$. Then c_t is not inserted into the mesh.

As in Step 1b, an ordering of the queue is necessary to get the proper bounds on segment lengths. The queue priority is as follows.

1. Triangles.
2. Segments by length. Longer segments are split first.

Using this algorithm, it is now possible to extend the result in Lemma 4 to include a bound based on local feature size.

Lemma 6. *After Step 2b, for any segment s in the mesh, $|s| \leq \frac{5}{3}\mathrm{lfs}(s)$ and $|s| \leq \sqrt{2}\mathrm{fs}_1(s)$.*

Step 2c

Lemma 6 allows for the insertion of points to form a collar region around the segments in each face. This will ensure both that the mesh conforms to the input

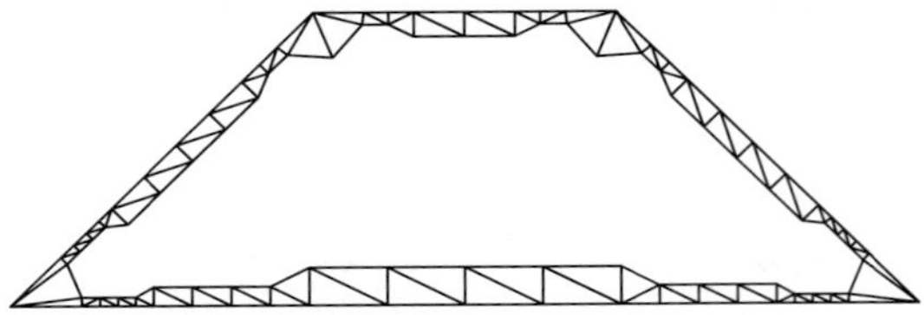

Fig. 6. Collar simplices in a face

near skinny angles and that Ruppert's algorithm will terminate when applied away from these regions.

The collar ensures conformity near the boundary by placing points in each face based only on the segments on the boundary of the face. This ensures that end triangles in adjacent faces "line up" so they do not encroach each other.

In order to prevent the collar from tangling, we need a slightly stronger bound on the feature size of segments: $|s| < \text{fs}_1(s)$. This can be attained by splitting all non-end segments, as all end segments satisfy this bound following Step 2a.

The collar is formed by inserting points according to the following rules.

- If s and s' are adjacent non-end segments containing point q, then a point p is inserted at distance $\frac{\max(|s|,|s'|)}{2}$ from q, in the direction into the face perpendicular to s.
- If s is an end segment and s' is an adjacent non-end segment, both containing q, then insert p at the intersection of the line parallel to s in the face at distance $\frac{|s'|}{2}$ away from s and on the circle of radius $|s|$ around the input point on s.
- Given any input point q_0 in the face, insert a point p in the face such that the circle of radius $N(q_0)$ around q_0 has no arcs of angle greater than 90 degrees.

Let each point added during this step be called a **collar point**. For each segment, let the segment between the two collar points corresponding to the end points of the segment be called a **collar segment**. Similarly, the segments which connect collar points on the disk in a face around an input point are also considered collar segments. **Collar simplices** are both subsegments of input segments and triangles which are between a segment or input point and its associated collar segment. While some collar segments may be split in Step 3 of the algorithm, the collar region never changes. This is analogous to the 2D version: once adjacent end segments are split at an equal distance, they are never split again. See Figure 6 for an example of the points added to a face to form the collar.

Lemma 7. *After Step 2c, the following properties hold.*

- *All adjacent collar segments meet at non-acute angles.*
- *In each face, the diametral disk of each collar segment contains no points of $\mathcal{P}'$.*
- *The circumball of any collar simplex contains no points of $\mathcal{P}'$.*

- *The circumball of any collar simplex does not intersect any disjoint faces or segments.*

Each of the above properties is important. Since the circumball of each collar simplex is empty, this ensures that the collar simplices conform to the input. Collar segments meeting non-acutely ensures that the complement of the collar region in each face is well-suited for Ruppert's algorithm. The final property is a result of the feature size bounds in the Step 2b, and is needed to guarantee that subsequent points inserted in the mesh for conformity will not encroach disjoint collar simplices.

Step 3

In the final step, the volume mesh is refined based on both quality and conformity criteria using Ruppert's algorithm. Similar to the non-acute case, any maximum radius-edge threshold $\kappa > 2$ can be selected for determining poor quality tetrahedra. As seen in the 2D case, a rejection policy must be followed to ensure that adjacent faces do not lead to cascading encroachment.

- Let c be a circumcenter proposed for insertion to split a skinny tetrahedron. If c lies in the circumball of a collar simplex, reject c.

During the algorithm, it is important to ensure that the properties of the collar in Lemma 7 continue to hold while allowing refinement of the non-collar region of each face to create a conforming mesh. As in 2D, the "collar region" does not change, however the set of collar simplices does change. This occurs when the standard Delaunay refinement algorithm seeks to insert a point in a face that encroaches upon a collar segment. Instead of adding this encroaching point, this collar segment is split. This new point is considered a collar point and the collar segment is broken into two new collar segments. The collar region has not changed but there are new collar simplices. This new point may encroach upon the circumball of another collar element in an adjacent face. In this face, the collar segment associated with this encroached circumball is also split so that the collar simplices on adjacent faces again "line up."

Proposition 2. *Let A be the set of vertices on the boundary of some face F and let A' be a set of points inside F. Suppose that for each boundary segment of F there is a circle through the end points of the segment which does not contain any points of $A \cup A'$ in its interior. Then the Delaunay triangulation of $A \cup A'$ conforms to F. Moreover, for any Delaunay triangle t in the interior of F, the circumcenter of t either lies inside F or inside the empty disk associated with a boundary segment.*

A face with the collar simplices removed satisfies the hypothesis of the lemma (as the circumdisk of each collar segment is empty). This ensures that whenever a circumcenter of a triangle in a face is proposed for insertion, it either lies in the face or encroaches a collar segment. Thus whenever a triangle is processed to be split, some action is taken and encroached faces are never forgotten. At termination, all faces must conform to the input.

Another proposition used in this step comes from [4].

Proposition 3. *[4, Lemma 4.5] Let $\mathcal{T}$ be the Delaunay triangulation of a planar face F such that the circumball of each bounding segment is empty. Let $\mathcal{B}_1$ be the union of the circumballs of all bounding segments of F and $\mathcal{B}_2$ be the union of the circumballs of all triangles in F.*

Let $p \in F \setminus \mathcal{B}_1$. If $\mathcal{T}'$ is the Delaunay triangulation of the face resulting from the addition of p, define $\mathcal{B}_1'$ and $\mathcal{B}_2'$ to be the union of the circumballs of bounding segments and triangles, respectively. Then,

$$\mathcal{B}_1' \cup \mathcal{B}_2' \subset \mathcal{B}_1 \cup \mathcal{B}_2$$

During this step of the algorithm, when a circumcenter is inserted it does not encroach any collar simplices. The above proposition ensures that as the collar is split the new collar simplices are not encroached by previously added circumcenters. Observe that once the collar is added the circumball of each segment always remains empty. Thus the conditions of Proposition 3 hold for every face in the mesh.

Lemma 8. *Whenever the queue of collar segments to be split is empty, the following properties hold.*

- *The circumball of any collar element contains no points in $\mathcal{P}'$.*
- *Adjacent collar segments meet at non-acute angles.*
- *In each face, the protecting disk of each collar segment contains no points in $\mathcal{P}'$.*

In the final property, the term protecting disk is used, rather than diametral disk. For the initial collar segments this protecting disk is the diametral disk. When a collar segment is split, the protecting disk for a new collar segment is the smallest disk containing the new collar segments and contained in the old protecting disk. This disk is not the diametral disk when collar segments corresponding to input points are split. In this case, the split point is added on the circle around the associated input point, and the diametral disks of the new collar segments are not contained in the old protecting disk. See Figure 7.

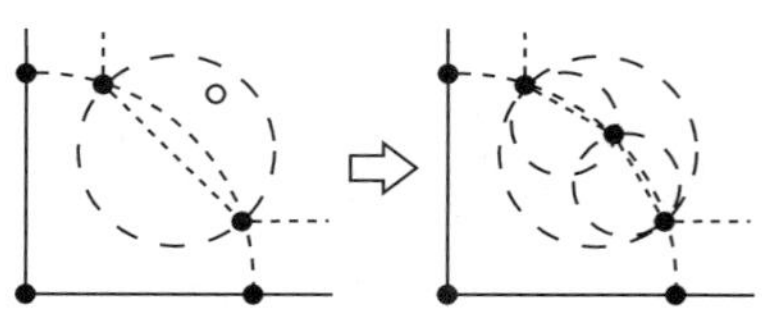

Fig. 7. A collar segment associated with an input point is split. When a proposed point (denotd by the empty dot) encroaches a collar segment, the segment is split by adding the midpoint of the arc of a circle around the input point. The new collar segments are protected with disks which are contained in the old protecting disk.

Also, Lemma 8 only applies when the queue of collar segments to be split is empty. Collar segments can be queued when a collar point on an adjacent face encroaches some collar simplex. When this occurs, the circumball of some collar simplices are not empty. However, once all collar segments have been split on adjacent faces, the circumball of each collar simplex is empty and thus the collar simplices conform to the input.

5 Implementation Details and Examples

When implementing the algorithm, several minor enhancements have been made which greatly reduce the output mesh size. Since the collar is needed to protect sharp input angles, it only needs to be inserted near segments and input points which violate the standard requirements for the termination of Ruppert's algorithm. (Any acute angle between adjacent faces is protected with a collar. For angles involving segments, this bound can be weakened as stated in [4].) In some examples, this greatly reduces the number of points which must be added to the mesh in Step 2c and thus the size of the final result.

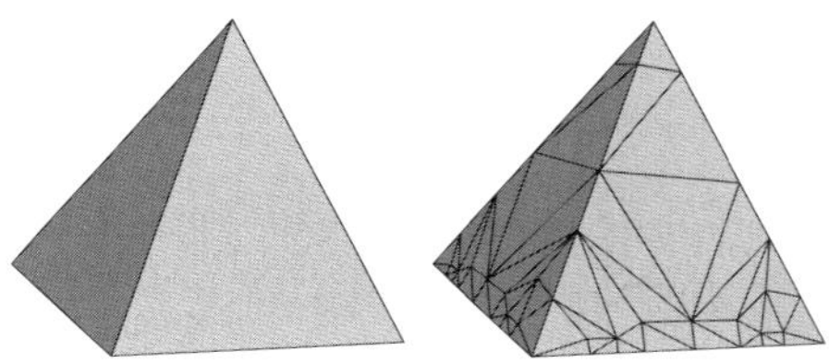

Fig. 8. Initial PLC input and final refined mesh for the pyramid example

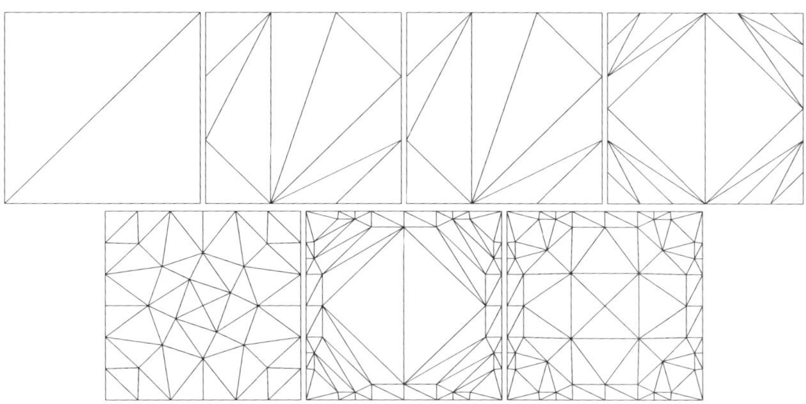

Fig. 9. Refinement of the base of the pyramid after each step, showing the mesh of the face after Steps 0, 1a, 1b, 2a, 2b, 2c, 3

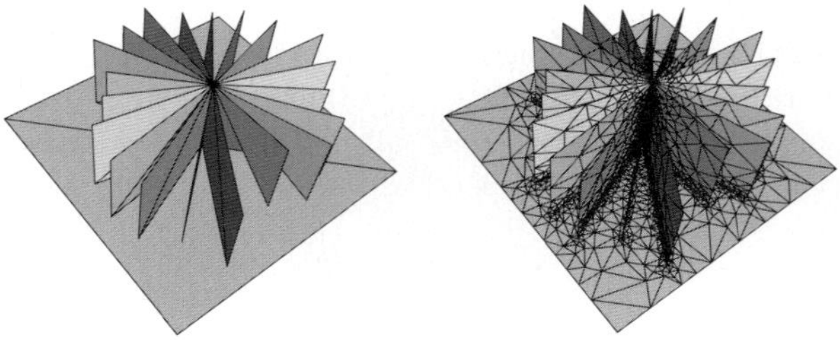

Fig. 10. Initial PLC input and final refined mesh for the wheel example. The base face is not connected to the other faces.

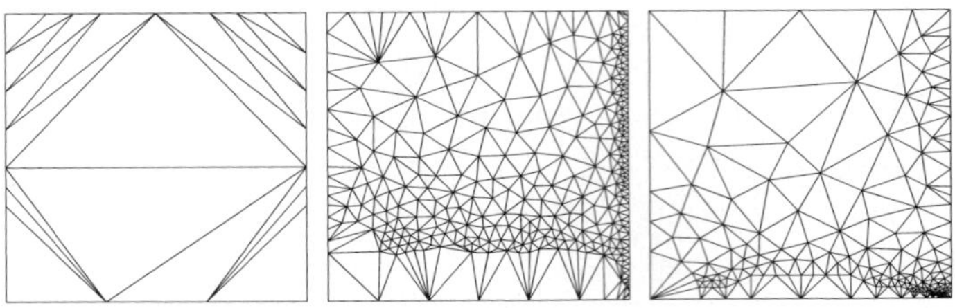

Fig. 11. One spoke in the wheel example is shown following Step 1b (left), Step 2b (center), and Step 3 (right). The center of the wheel is at the bottom of the face and the disjoint, nearby plane is very close to the right side of the face.

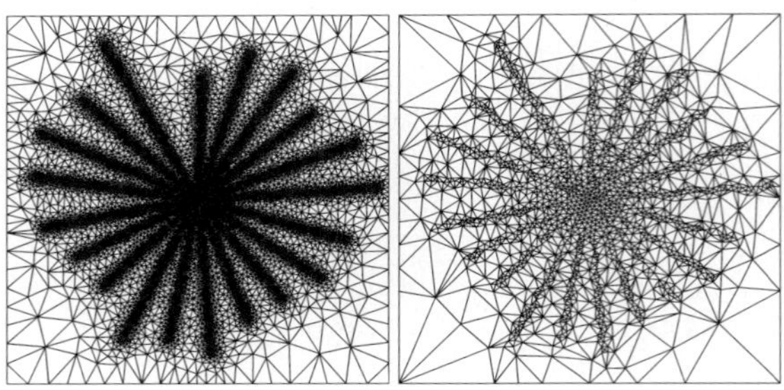

Fig. 12. Base plane in the wheel example is shown following Step 2b (left) and Step 3 (right)

Next, instead of adding the collar to the mesh produced by Step 2b, it is instead added to a blank copy of the input PLC. Since the encroachment property in Step 2b is stronger than that in Step 3, this often results in a mesh which is less refined on the faces than the mesh produced in Step 2b. This is especially effective when using a large radius-edge threshold.

Finally, while the Lemma 5 requires each segment in the mesh to be divided in to fourths, the implementation has been seen to terminate when splitting only in

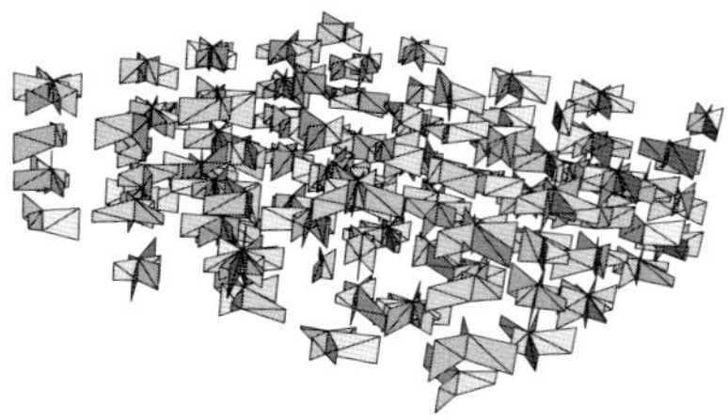

Fig. 13. A PLC with many small angles

Table 1. Number of vertices in the mesh following each step. Following Step 0, the mesh contains the 1556 input points plus 26 points for the initial bounding box. Step 3 was performed with several maximum radius-edge parameters and the final mesh size is given for each of these cases with the parameter in parentheses.

Step	0	1a	1b	2a	2b	2c	3 (2)	3 (4)	3 (8)
Vertices	1,582	5,850	13,590	26,628	105,781	13,695	76,522	54,392	46,776

half. We continue to seek a example causing non-termination or stronger proof which justifies this practical modification.

The first example is a mesh of a pyramid with a square base. The angles between the base face and each of the side faces is less than 90°, so each of the edges around the base is protected with a collar. See Figures 8 and 9.

Another example involves a "wheel" of twenty rectangular faces which meet at a single segment. This wheel of faces lies very close to an additional disjoint face. In this example, the collar is only added around the segment in the center of the wheel. Away from this center segment all of the faces and segments meet at non-acute angles and do not need to be protected. See Figures 10, 11, and 12.

In the final example, depicted in Figure 13, 100 "wheel" complexes similar to the previous example are meshed together. Each wheel contains in this complex contains between 3 and 16 spokes. The input contains 1556 vertices, 2134 segments and 678 faces. Table 1 contains a list of mesh sizes after each step of the algorithm. Note that the final mesh is substantially smaller than the mesh created in Step 2b. Removing unneeded points before the encroachment criteria is relaxed in Step 3 greatly reduces the output size.

References

1. Cheng, S.-W., Dey, T.K., Levine, J.A.: A practical Delaunay meshing algorithm for a large class of domains. In: Proceedings of the 16th International Meshing Roundtable (October 2007)
2. Cheng, S.-W., Poon, S.-H.: Three-dimensional Delaunay mesh generation. Discrete Comput. Geom. 36(3), 419–456 (2006)
3. Cohen-Steiner, D., de Verdière, E.C., Yvinec, M.: Conforming Delaunay triangulations in 3D. Comput. Geom. Theory Appl. 28(2-3), 217–233 (2004)

4. Miller, G.L., Pav, S.E., Walkington, N.J.: An incremental Delaunay meshing algorithm. Technical Report 02-CNA-023, Center for Nonlinear Analysis, Carnegie Mellon University, Pittsburgh, Pennsylvania (2002)
5. Miller, G.L., Pav, S.E., Walkington, N.J.: When and why Ruppert's algorithm works. In: Proceedings of the 12th International Meshing Roundtable, pp. 91–102 (September 2003)
6. Murphy, M., Mount, D.M., Gable, C.W.: A point-placement strategy for conforming Delaunay tetrahedralization. International Journal of Computational Geometry and Applications 11(6), 669–682 (2001)
7. Pav, S.E., Walkington, N.J.: Robust three dimensional Delaunay refinement. In: Proceedings of the 13th International Meshing Roundtable (September 2004)
8. Pav, S.E., Walkington, N.J.: Delaunay refinement by corner lopping. In: Proceedings of the 14th International Meshing Roundtable (September 2005)
9. Ruppert, J.: A Delaunay refinement algorithm for quality 2-dimensional mesh generation. J. Algorithms 18(3), 548–585 (1995)
10. Shewchuk, J.R.: Mesh generation for domains with small angles. In: Symposium on Computational Geometry, pp. 1–10 (2000)
11. Si, H.: On refinement of constrained Delaunay tetrahedralizations. In: Proceedings of the 15th International Meshing Roundtable (September 2006)
12. Si, H., Gartner, K.: Meshing piecewise linear complexes by constrained Delaunay tetrahedralizations. In: Proceedings of the 14th International Meshing Roundtable (September 2005)

Tetrahedral Mesh Generation for Non-rigid Registration of Brain MRI: Analysis of the Requirements and Evaluation of Solutions

Andriy Fedorov and Nikos Chrisochoides

Center for Real-Time Computing, College of William and Mary, Williamsburg, VA
{fedorov,nikos}@cs.wm.edu

Summary. The application we target in this paper is the registration of pre-operative Magnetic Resonance Imaging with the scans acquired intra-operatively during image-guided neurosurgery. The objective of this application is improved tracking of tumor boundaries and surrounding brain structures during open skull tumor resection. We focus on a validated, physics-based non-rigid registration approach, which has been used in clinical studies for the last three years. This approach requires tetrahedral tessellation of the brain volume for biomechanical model construction. The analysis of the requirements and available methods to construct such a discretization is the objective of our paper.

The paper presents a number of practical contributions. First, we survey the proposed approaches to tetrahedral mesh generation from medical image data. Second, we analyze the application-specific requirements to mesh generation. Third, we describe an end-to-end procedure of tetrahedral meshing for this application using off-the-shelf non-commercial software. Finally, we compare the performance of the considered mesh generation tools in the application context using generic and application-specific quantitative measures.

1 Introduction

Mesh generation for medical imaging applications has been attracting a lot of attention in the mesh generation community. Finite Element Method (FEM) is essential in modeling tissue deformation for these applications, therefore necessitating research and development of new mesh generation methods and tools. The existing imaging modalities, such as Magnetic Resonance Imaging (MRI), provide only limited knowledge about the internal organs. The lack of precise geometric models and the need to construct volume tessellations from the image data is a very practical limitation, which complicates the use of existing off-the-shelf meshing tools. Application-specific requirements make the problem even more challenging. In practice, there is no single widely accepted method to address the mesh generation needs of all applications.

In this paper we focus on tetrahedral mesh generation for physics-based Non-Rigid Registration (NRR) of brain MRI during Image-Guided Neurosurgery (IGNS). Specifically, we focus on the NRR approach developed by Clatz et al. [1], and subsequently validated in the clinical setting by Archip et al. [2]. The method has been studied at Brigham and Women's Hospital (Harvard Medical School, Boston) for the last three years.

The main goal in neurosurgery is maximum removal of the tumor tissue with the minimum damage to the healthy brain structures. It is very difficult to identify boundaries of certain tumors, e.g., gliomas, with the naked eye. Instead, pre-operative imaging is typically used to precisely locate the tumor, and the neighboring life-critical structures. Brain shift is often unavoidable during the tumor resection, making the pre-operative images invalid. Intra-operative images acquired periodically during the tumor resection have lower quality, and usually cannot substitute the pre-operative data. NRR is used to align the pre-operative MRI with the intra-operative scans. The specific NRR method we study uses a patient-specific bio-mechanical model to facilitate the brain shift estimation. Tetrahedral tessellation of the brain volume (more specifically, skull Intra-Cranial Cavity (ICC)) is essential for this application.

Our paper is the continuation of an earlier study [3]. We consider the complete process of mesh generation from the segmented image, and discuss specific application requirements derived from the established registration method. Our contributions are the summary of the state of the art methods for constructing tetrahedral meshes from images, and a practical evaluation of the existing off-the-shelf meshing tools. Our evaluation is based on the number of qualitative and quantitative metrics, which allow to compare the studied methods in the context of the FEM computations that facilitate IGNS.

2 Physics-Based Non-rigid Registration of Brain MRI

2.1 Formulation

The objective of image registration is to determine the transformation that aligns features in one (floating) image with the features in another (target) image. Image registration is a fundamental problem in medical image processing. The reader is referred to the survey by Hill et al. [4] for a thorough review on this topic. Non-rigid image registration is used when the imaged structure is subject to a non-affine transformation. Image registration methods are usually tailored to a specific clinical application. The clinical application we target is IGNS facilitated by a physics-based NRR approach.

The NRR method in [1] consists of the following steps. First, a sparse set of mathematical landmarks, which we call *registration points*, is identified within the pre-operative image of the brain volume. Once the intra-operative image is available, the time-critical part of the computation is initiated. The deformation is estimated at each registration point using block matching between the pre- and intra-operative images (in the context of registration, these are the floating and target images, respectively). Block matching results inherently contain incorrect matches (outliers). Because of outliers, the deformation field cannot be derived by interpolating displacements at the registration points. Mechanical energy of the deforming mesh is used to regularize the displacements, i.e., smooth the impact of outliers on the solution:

$$W = (\mathbf{HU} - \mathbf{D})^T \mathbf{S} (\mathbf{HU} - \mathbf{D}) + \mathbf{U}^T \mathbf{KU}.$$

Here $\mathbf{U}$ is the vector of displacements at the mesh nodes, $\mathbf{K}$ is the stiffness matrix [5], $\mathbf{H}$ is the interpolation matrix from the tetrahedra vertices to the registration points inside a tetrahedron, $\mathbf{D}$ is the vector of displacements recovered at the registration points, and $\mathbf{S}$ is the block-diagonal positive semi-definite matrix which captures confidence in the corresponding entry of $\mathbf{D}$ [1].

The outliers are found by iteratively comparing the displacements at the registration points with those interpolated from the mesh vertices, and discarding the matches with the largest error magnitude. The displacements at the mesh vertices are estimated using the following formulation:

$$\mathbf{F}_0 = 0,\ \mathbf{U}_i = [\mathbf{H}^T \mathbf{S}_i \mathbf{H} + \mathbf{K}]^{-1} [\mathbf{H}^T \mathbf{S}_i \mathbf{D} + \mathbf{F}_{i-1}],\ \mathbf{F}_i = \mathbf{KU}_i. \tag{1}$$

During each iteration, the registration points are ordered by the absolute value of the error $\|\mathbf{D}_i - \mathbf{HU}_i\|$ weighted by the 3×3 image structure tensor $\mathbf{T}_i$ at the registration point. The weight is used because block matching can only recover displacements in the directions orthogonal to the edges in the image [1]. Outlier registration points are selected as a pre-defined, e.g., 5%, number of registration points with the largest error magnitude.

2.2 Finite Element Mesh Generation

Tetrahedral mesh has a dual role in the formulation. First, it is used to find the stiffness matrix in the mechanical energy component. Second, it allows to regularize, or smooth, the displacements recovered by block matching locally within the mesh vertex neighborhood. The displacement recovered at each of the registration points within the mesh vertex cell complex affects the displacement at the corresponding mesh vertex. Therefore, it is important to maintain the empirically obtained ratio between the number of mesh vertices and the number of registration points under 0.1 (at least 10 registration points per mesh vertex) [1]. Ideally, this ratio should be maintained within each mesh vertex cell complex. We define the cell complex as a set of mesh tetrahedra incident on a mesh k-cell, e.g., mesh vertex is a 0-cell, and mesh edge 1-cell. The matrix $\mathbf{H}^T \mathbf{SH}$ has a non-zero 3×3 entry for each mesh vertex and edge with the cell complexes containing registration points. The corresponding sub-matrices can be expressed as the sum over the registration points in a cell. For example, the diagonal 3×3 sub-matrix that corresponds to the mesh vertex $\mathbf{v}_i$ can be calculated as the following summation over the registration points in the cell complex $\mathcal{S}$ of $\mathbf{v}_i$:

$$[\mathbf{H}^T \mathbf{SH}]_i = \sum_{\forall \mathcal{T} \in \mathcal{S}(\mathbf{v}_i)} \sum_{\forall k \in \mathcal{T}} h^{\mathcal{T}}_{\mathbf{v}_i}(k)^2 \frac{\mathrm{tr}(\mathbf{K})}{np} c(k) \mathbf{T}(k). \tag{2}$$

Here, $h^{\mathcal{T}}_{\mathbf{v}_i}(k)$ is the barycentric coordinate of the kth registration point with respect to $\mathbf{v}_i$ in the containing tetrahedron $\mathcal{T}$, n is the number of mesh vertices,

p is the number of the registration points, $c(k)$ is the correlation coefficient from block matching, and $\mathbf{T}(k)$ is the image structure tensor at k.

The result of non-rigid registration is a deformation field, which defines deformation vector at each point in the image. That vector is computed by interpolating the displacements at the mesh vertices after the last iteration [5]:

$$\forall \mathbf{x} \in \mathcal{T} : \mathbf{u}(\mathbf{x}) = \sum_{j=0}^{3} h_{\mathbf{v}_j}^{\mathcal{T}}(\mathbf{x})\mathbf{u}(\mathbf{v}_i). \tag{3}$$

Note that the iterative procedure in Equation 1 requires solving the linear system of equations at each iteration within the time-critical part of the computation. The size of this system depends on the number of degrees of freedom in the mesh model. Therefore, it is important to minimize the size of the mesh as much as possible without sacrificing the accuracy of the solution.

2.3 Application-Specific Requirements to Mesh Generation

Based on the registration formulation, we can derive the following application-specific requirements to mesh generation:

- R1 *Equi-distribution of the registration points w.r.t. mesh vertex cells*: small number of registration points (e.g., less than 10, but greater than 0) within the vertex cell complex makes the formulation more sensitive to outliers and introduces additional displacement error [1].
- R2 *Minimization of the approximation error at registration points*: error of the displacements recovered at registration points can be reduced locally by using smaller mesh elements [6].
- R3 *Prevention of tetrahedron inversion during mesh deformation*: while the interpolation error shown in Equation 3 does not depend on the tetrahedron shape, inversion or collapse of a tetrahedron will result in an unrealistic deformation field, e.g., points inside the different tetrahedra can map to the same image location. We can attempt to remedy this problem by adjusting element size locally according to the expected deformation.

Adaptive refinement of the mesh following the simulation, if necessary, is outside the scope of this paper.

3 Image-to-Mesh Conversion

3.1 Formulation and Generic Requirements to Mesh Generation

Following the notation of Hill et al. [4], we define the image domain Ω as the overlap between the bounded continuous set $\tilde{\Omega}$ (image field of view) and the infinite discrete sampling grid Γ_ς, characterized by the anisotropic sample spacing $\varsigma = (\varsigma^x, \varsigma^y, \varsigma^z)$, $\Omega = \tilde{\Omega} \cap \Gamma_\varsigma$. Voxel is an orthogonal parallelepiped-shaped volume element centered at the sampling grid point. Its dimensions are defined by the grid spacing ς. For the considered application, image A is a mapping of points in the image domain $\mathbb{R}^3$ to $\mathbb{R}$, $A : \mathbf{x} \in \Omega \mapsto A(\mathbf{x})$. In this paper we are mostly concerned with the medical applications that provide a *segmentation* of the object of interest. The output of segmentation is a binary image, i.e., $A(\mathbf{x}) \in \{0, 1\}$, with the subset $\{\mathbf{x} \in \Omega | A(\mathbf{x}) = 1\}$ corresponding to the voxels located inside the object. Let Σ be the surface that separates zero and non-zero voxels of this binary image. The surface Σ is defined implicitly, as we only know whether a given voxel is inside or outside the object.

The objective of mesh generation for FEM computations from the binary image data is to construct a conforming tetrahedral mesh $\mathcal{M} = (\mathcal{V}, \mathcal{T})$, which satisfies the following *generic requirements*:

R4 The mesh boundary (triangulation) should be close to Σ.
R5 Mesh size should be minimized to reduce the computational costs.
R6 Mesh elements should not have small angles [6].

These generic requirements have been in the focus of the mesh generation community for decades. The application-specific requirements are usually addressed by constructing a customized mesh sizing function, or developing customized mesh generation methods. In this paper we explore the first approach. Therefore, the ability to accept a user-defined sizing function is an essential feature for a mesh generation method to be considered for our application.

3.2 Related Work

An intrinsic difficulty of generating meshes from the binary image data is the processing and recovery of the object geometry. General-purpose mesh generators (for solid and geometric modeling applications) expect that the object boundary is parametrized, i.e., it is defined by means of constructive solid geometry primitives, or explicitly (e.g., through the boundary discretization, as a collection of patches). Therefore, in order to convert the binary image into a tetrahedral mesh, one can either (1) recover the parametrized object surface followed by a conventional mesh generation technique, or (2) use a mesh generation method, which operates directly on the binary image. The reader is referred to the survey by Owen [7] for a discussion of the classical approaches to volume meshing. Next, we overview some of the methods designed to construct tetrahedral discretizations directly from the image data.

A step which precedes geometry processing is segmentation of the structure of interest from the multivalued image. Segmentation is a fundamental problem in medical image processing [8], and is outside the scope of this paper. We assume that the segmented ICC is provided by the application.

The objective of the surface recovery step is to construct an explicit representation of Σ. The methods which recover piecewise-linear surface approximation and provide surface triangulation are most practical, as this is the input most mesh generation methods expect. It is desired for the triangulated surface (1) to have the same topology as Σ, (2) to be sufficiently close to Σ, and (3) to have guaranteed quality of the triangles in the surface discretization, as defined by

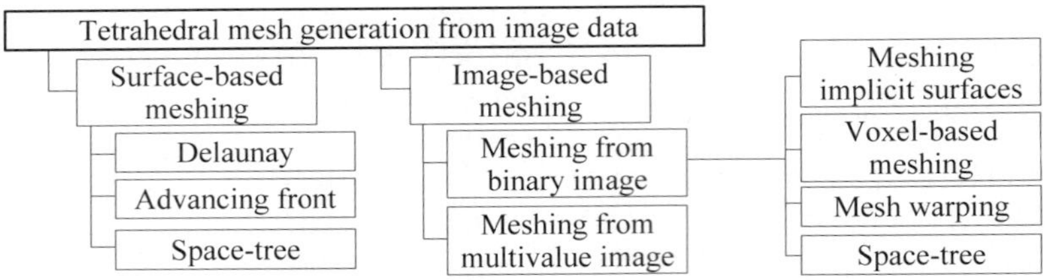

Fig. 1. A hierarchy of the studied approaches to mesh generation

the triangle aspect ratio, while (4) minimizing the number of triangles. The first two requirements are important for the accurate representation of the object, while the other two are essential in order to satisfy the generic requirements to volume mesh generation.

A straightforward approach to recover iso-surface is by means of the Marching Cubes (MC) algorithm [9]. However, the original version of this algorithm may produce a triangulation with topological problems. Another practical problem is the inability of the MC algorithm to generate adaptive surface triangulations. The surface produced by MC has "jagged" artifacts because of voxel sampling, which may create subsequent problems with the simulation [10]. Surface mesh simplification, or remeshing, is often a necessary post-processing step when MC is used [11, 12].

Parametrized surface representation can be recovered by more advanced techniques, which provide theoretical guarantees about the recovered surface. Since the seminal work of Amenta and Bern [13], a number of algorithms have been proposed [14, 15, 16]. These methods differ in their performance, robustness to noise and theoretical guarantees.

A number of approaches to volume mesh construction were suggested, that are capable to operate on image date directly. We separate such methods into the following two categories.

The methods from the first category do not require segmentation and create meshes from the multi-value image data [17, 18]. The assumption is usually made that the pixels which correspond to the same tissue have similar intensity, and the object boundary can be defined by some isosurface value. Therefore, these methods attempt to minimize the error of approximating this isosurface, while maintaining good quality of the mesh tetrahedra. Such approach to mesh generation is very practical for volume rendering and certain FEM applications. However, brain segmentation is one of the very challenging problems in image segmentation, which cannot be solved by thresholding only [19, 8]. Direct isosurface-based meshing of the brain volume from the multivalued image may lead to large errors in the surface recovery.

The second category of the image-based mesh generation methods operate on binary images, produced by specialized image segmentation algorithms. Among the methods in this category we separate four groups.

The methods in the first group attempt to recover the surface of the object at the level of voxel resolution [20, 21, 22]. Therefore, we call those techniques *voxel-based meshing*. The sizes of surface triangles in the meshes constructed using this approach are comparable with the size of voxels, i.e., the surface triangulation is not adaptive. The use of such techniques is problematic for our application, because the control over the element size is limited due to the fixed high resolution of the surface discretization.

A large number of methods that are based on space-tree decompositions have been proposed recently for meshing binary images. Conceptually, this approach has a long history in classical mesh generation. Yerry and Shephard [23] were some of the first to present an octree-based approach to 3-d mesh generation. Mitchell and Vavasis [24] describe a quadtree-based algorithm with theoretical bounds on the mesh size. These ideas have been adopted to construct tetrahedral meshes from binary images [25, 26, 27]. Such methods recover the surface by finding the points of intersection of the adaptive space-tree with the surface of the object defined as a binary image. Mesh quality near the surface can be compromised, as the newly inserted mesh nodes can be arbitrarily close to the existing nodes. Mesh optimization [28] is commonly used as a post-processing step. In practice, the methods from this group are well-suited for meshing binary images, and were shown to be quite effective for a number of medical applications. Some of these methods were designed and evaluated on the segmented brain MRI data [25, 26, 27], which is the geometry used in our application. However, the control over the customized element sizing is usually very limited, and has not been evaluated previously.

The methods based on surface matching use a template volume mesh, which is warped to match the surface of the modeled object [29]. While the advantages of this approach are good surface fidelity, control over the mesh size and high speed, the quality of the elements undergoing deformation during warping can be compromised. Mesh optimization is a commonly used post-processing step for the methods in this group. This concept is most suitable for meshing objects that have very similar geometries. The geometry of ICC is quite similar between different subjects. However, the mesh element sizing depends on criteria R1-3, which are case-specific. Therefore, it is not feasible to construct a single template mesh for our application to satisfy case-dependent distributions.

Representation of the object as a binary image conveniently lends itself to the construction of an implicit function describing the object surface. Implicit function is a mapping $\phi : \mathbb{R}^3 \mapsto \mathbb{R}$, and the object surface is defined as the kth level set of this function, $\phi(\mathbf{x}) = k$. An approximation of the implicit function representing the object surface can be easily obtained by computing the distance transform on the binary image [27], and using the zero level set as a surface definition. A number of volume mesh generation methods have been introduced recently to mesh implicitly defined surfaces [30, 31].

Overall, we observe that a great variety of methods for tetrahedral meshing of binary images have been developed. Most of these methods were proposed and evaluated in the context of their fitness to a specific application. Little or no attention is usually paid to the comparison of the newly proposed techniques with the existing methods, and few implementations are available to conduct such an evaluation by external groups. Also, most of the effort is usually directed to developing a method that delivers good practical results, and not on establishing theoretical guarantees about the produced meshes. The lack of such guarantees makes it even more difficult to select the most appropriate algorithm from the range of seemingly similar methods.

With the specific application we target, our goal is first to derive the precise requirements to mesh generation. These requirements can then be used to customize and evaluate readily available, established methods to address the problem of mesh generation for the NRR application. The results of such evaluation can be used next to identify problems within the existing approaches and justify the development of new mesh generation methods for this application. However, before such necessity is justified, we believe the possibility of using existing off-the-shelf tools must be carefully examined.

4 Methodology

4.1 Mesh Generation Tools

We evaluate three conceptually different approaches to constructing tetrahedral meshes from binary images. Common to all these methods is their ability to adjust the mesh element size locally according to the value of sizing function, or local refinement rules.

Tetgen is an implementation of the Delaunay mesh generation and refinement algorithm by Si [32]. This method is accompanied by the proof of termination and bounds on the circumradius to shortest edge ratio. The implementation has a number of very practical features: it accepts user-defined sizing function, the exterior boundaries are refined simultaneously with the volume and "... are never over-refined" [32]. The size of the mesh can be controlled by the *alpha* parameters, which are not part of the basic Delaunay refinement [32]. We used *Tetgen* version 1.4.2. The implementation works with the input surface defined as a piecewise linear complex (PLC).

NETGEN is an advancing front mesh generator developed by Shöberl [33]. We used version 4.5rc2 of the code. The implementation can be used to construct both surface triangulations of the parametrized surfaces, and adaptive volume tetrahedralizations. It gives the user some control over the mesh grading, its implementation is accompanied by a GUI environment, and the acceptable inputs include triangulated surface of the domain.

RGM is a space-tree based mesher we presented earlier [27] that works directly on binary image data. Our implementation is based on the algorithm of Molino et al. [30], which builds a mesh from the implicit definition of the domain. *RGM* is designed to work directly with the binary image, and we use the *Mesquite* [34] mesh optimization library to improve the mesh quality following the surface recovery. We used slightly modified version compared to the code available online. Specifically, we use *Mesquite* instead of *GRUMMP* for mesh optimization, and implement a custom subdivision rule, as described in the next section.

Both *Tetgen* and *NETGEN* require parametrized surface representation. We use implicit surface meshing method by Boissonnat and Oudot [16] implemented in the Surface mesh generation package of CGAL [35]. The implementation generates adaptive triangular surface directly from the binary image. The guarantees of approximation accuracy and surface quality are provided. All the interaction with the image data is done via the Insight Toolkit (ITK) [36][1].

4.2 Adaptive Mesh Generation

A sizing function $H(p) > 0$ specifies the desired length of edges at point p [32]. It can be defined analytically, or, more often, its values are prescribed at the vertices of a background mesh. The value of sizing at the non-vertex locations can then be derived by interpolation. Both *Tetgen* and *NETGEN* accept background mesh to control local mesh size. We use the sizing function defined at the vertices of a background mesh to address the application-specific requirements to mesh generation.

We use the same CGAL-recovered surface mesh for both background mesh and the mesh used in FEM calculations. The background mesh is built using *Tetgen*, with the small uniform bound on the tetrahedron volume. The sizing value at each mesh vertex was initialized with the distance to the kth registration point closest to it, to reflect the density of the registration point distribution. The idea here is that the shape of a perfect mesh vertex cell complex is close to a ball with the radius prescribed by the background mesh. We use CGAL [35] k-neighbor search to find the k closest registration points and the distance to the furthest point for each background mesh vertex. The process of background mesh initialization is parametrized by the number of closest registration points k. Our goal was to have around 30 registration points in the cell complex of each vertex. However, based on the experimental results, the mean value in distribution of the registration points both for *Tetgen* and *NETGEN* was not approaching the desired bound when we set $k = 30$. Experimentally, we arrived at a result that the best distribution is obtained by using larger values for k (we used $k = 100$), and adjusting the *Tetgen* mesh by reducing the *alpha* parameters of the implementation, see [32]. The *NETGEN* mesh was constructed using the same background mesh, but the sizing values were scaled down by constant to have similar number of nodes compared to the adaptive *Tetgen* mesh.

The sizing of the adaptive mesh constructed with *RGM* was controlled by a custom subdivision function. This function is called for each tetrahedron during the refinement at each mesh resolution, and returns *true* if the tetrahedron

---

[1] Off-the-shelf software tools we used (links valid as of July 15, 2008): *Tetgen*: `http://tetgen.berlios.de/`, *NETGEN*: `http://www.hpfem.jku.at/netgen/`, *RGM*: `http://www.na-mic.org/svn/NAMICSandBox/trunk/TetrahedralMeshGeneration/`, CGAL: `http://cgal.org/`, ITK: `http://itk.org/`, VTK: `http://vtk.org/`, Paraview: `http://paraview.org`.

requires subdivision. We calculate the number of registration points inside the cells of the four tetrahedron vertices. The tetrahedron is refined if the number of the registration points in each vertex cell exceeds parameter n. We used two subdivision resolutions in all cases. The process of adaptive mesh construction with *RGM* is parametrized by the value of constant n, and the spacing of the initial lattice, see [27]. These parameters were experimentally chosen so that the size of the adaptive mesh (the number of mesh vertices) is approximately the same as the size of the adaptive *Tetgen* mesh.

4.3 Evaluation

We compare two sets of meshes constructed with each of the evaluated methods, i.e., with and without using custom mesh sizing. The meshes are used to register real MRI subject to synthetic deformations. We use synthetically deformed images, because the true deformation field (the ground truth deformation) cannot be recovered in the real IGNS cases, and the registration accuracy cannot be evaluated at an arbitrary image location. The ability to assess the registration error is essential for our evaluation. We provide the details on the generation of the synthetic deformation field in [37].

We try to create meshes with the similar number of vertices within each group (uniform and adaptive), and compare them using a set of quantitative metrics. There are two groups of metrics we use. The first group includes the mesh properties, which can be directly optimized during the process of mesh construction. These include element shape and surface approximation accuracy. We assess the element shape by the minimum dihedral angle for each tetrahedron of the mesh. Surface approximation accuracy is evaluated as the percentage of the registration points covered by the mesh, which is a practical measure for the NRR application.

The second group includes quantitative metrics, which cannot be directly optimized by the existing mesh generation methods. Let $\bar{\mathbf{D}}_i$ be the ground truth displacement at the registration point i, which aligns given point in the floating image with the corresponding point in the target image. This value is known to us, because the true deformation field is synthetic. We define the following application-specific quantitative metrics:

1. *Approximation error* at a registration point is defined as $\|\mathbf{D}_i - \mathbf{HU}_i\|$. We assess the accuracy of approximation by the percentage of the registration points, where the magnitude of this error exceeds 1.0. We call those registration points “error points”, while reporting results (errors below this threshold are in the sub-voxel range).
2. *Outlier detection sensitivity*, defined as the ratio of the true outliers within the discarded registration points to the total number of the discarded registration points. The true outlier is defined as a registration point, where $\|\mathbf{T}_i\mathbf{D}_i - \bar{\mathbf{D}}_i\| > 1.0$.

3. *RMS of the absolute error* at the registration points, absolute error being defined as $\|\mathbf{HU}_i - \bar{\mathbf{D}}_i\|$.
4. *Distribution of the registration points* with respect to mesh vertices.

Note that the goal of the study was not to tune the NRR parameters to get the optimal registration results. We attempt to perform a controlled study of the impact of mesh generation on the NRR performance, while keeping fixed the other parameters that can influence registration accuracy.

5 Results

We constructed synthetic deformation fields and performed NRR on the MRI scans 1 through 3 in the set of 18 images available from the Internet Brain Segmentation Repository (IBSR)[2]. The synthetic deformations were generated with 20 "knots" in the deformation grid on average [37]. The deformation magnitude at each "knot" was under 5% of the brain size to maintain the validity of the linear elastic physical model [5, 1].

The parameters used for CGAL surface mesh generation [35] were: angular bound 30°, surface radius and distance bounds 10.0, surface precision bound 0.001. Each method was used to construct two meshes for each registration case. The prescribed element size was uniform throughout the volume of the first mesh. The second mesh was constructed to adapt the element size according to the sizing function designed in Section 4.2. We adjusted the implementation-specific parameters to have the uniform and adaptive meshes with approximately 1.5k and 6k vertices, respectively.

Figure 2 shows cross-sections of the adaptive meshes. The adaptive meshes generated with *NETGEN* have a layer of relatively large elements near the surface of the mesh. This is explained by the nature of the Advancing Front algorithm, which does not insert new points on the triangulated surface. We explored the option to use the CGAL triangulation as the support surface, and instructed *NETGEN* to construct a new triangulation to respect the prescribed element sizing. However, the re-triangulated surfaces contained small triangles, which did not obey the prescribed edge sizing.

None of the meshes contained sliver elements. We observed, that the minimum dihedral angle was the largest, 14°, in the *NETGEN*-generated meshes. The values of this metric for *Tetgen* and *RGM* were 8° and 5° respectively.

The synthetic registration cases differ in the number of registration points, and in the number of true outliers. Moreover, because of the differences in mesh surfaces recovered by CGAL and *RGM* respectively, different percentage of those points are located inside the mesh domain, as summarized in Table 1. This is an important observation, because the percentage of outliers impacts the registration error, which is also used in the evaluation. Also, due to the lower precision

[2] The MR brain data sets were provided by the Center for Morphometric Analysis at Massachusetts General Hospital and are available at `http://www.cma.mgh.harvard.edu/ibsr/`.

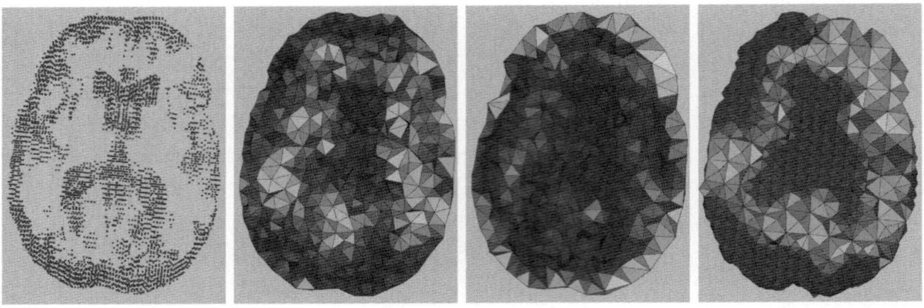

Fig. 2. Left to right: selected registration points, and adaptive tetrahedral mesh cuts (same slice) generated with *Tetgen*, *NETGEN*, and *RGM*. Tetrahedra are colored according to their volume, from blue (smallest) to red (largest).

Table 1. Registration points, true outliers, and their coverage by the mesh surfaces

case id	reg. points	outliers,%	reg. points inside,%		outliers inside,%	
			CGAL	RGM	CGAL	RGM
IBSR01	56447	7.8%	95.6%	91.1%	6.9%	5.5%
IBSR02	57526	16.2%	94.9%	90.8%	14.3%	12.4%
IBSR03	46525	18.8%	95.3%	90.5%	16.0%	13.1%

Table 2. Min/average/max of registration points per mesh vertex cell complex

case id	uniform-graded meshes			adaptive meshes		
	Tetgen	NETGEN	RGM	Tetgen	NETGEN	RGM
IBSR01	0/135/460	0/137/715	0/130/390	0/36/111	0/36/432	0/33/277
IBSR02	0/131/479	0/136/667	0/124/374	0/35/118	0/35/444	0/30/168
IBSR03	0/127/408	0/126/611	0/121/345	0/27/87	0/27/320	0/28/254

of surface approximation by *RGM*, about 4% fewer registration points and 1% to 3% fewer outliers are located within the *RGM*-generated meshes. The distribution of outliers is non-uniform, and many of them are located close to the surface.

Based on the results presented in Table 2 and Figure 3, the use of custom mesh sizing can significantly improve the distribution of registration points compared to uniform-sized meshes. In conjunction with the sizing function, *Tetgen* achieves the best distribution results overall. Although the average values for the distributions are similar for all meshes, *Tetgen* meshes have lower maximum values and better distribution: the distribution curve approaches normal distribution with the mean close to the desired number ($k = 30$). The advantage of *Tetgen* over *NETGEN* is that mesh points can be inserted at arbitrary locations on the surface during refinement. *RGM* is limited even more than *NETGEN*, as

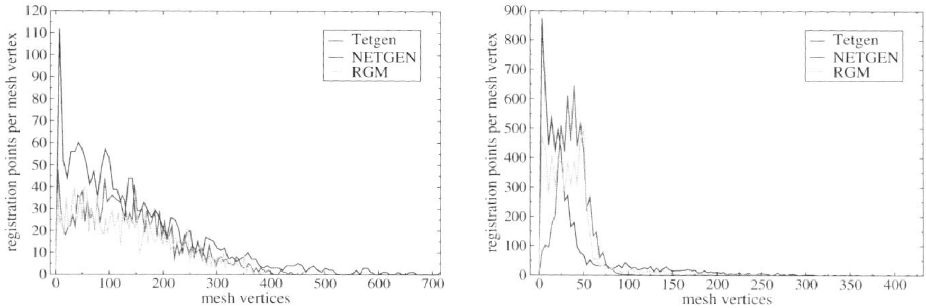

Fig. 3. Distribution of the number of registration points per mesh vertex: uniform-graded meshes (left) and adaptive refined meshes (right); IBSR01

Table 3. Application-specific metrics for the evaluated meshes

case id	metric	uniform size meshes			adaptive meshes		
		Tetgen	NETGEN	RGM	Tetgen	NETGEN	RGM
IBSR01	mesh points	1617	1596	1607	6044	6020	6209
	error points	2.4%	2.6%	1.9%	1.6%	1.8%	1.1%
	sensitivity	53.5%	52.6%	43.8%	52.3%	52.5%	42.8%
	RMS error	1.60	1.62	1.61	1.69	1.65	1.60
IBSR02	mesh points	1682	1617	1696	6166	6255	6993
	error points	6.7%	7.1%	5.9%	4.9%	5.8%	4.3%
	sensitivity	64.7%	64.4%	59.4%	63%	64.7%	60.5%
	RMS error	1.98	1.92	1.82	2.55	2.27	2.23
IBSR03	mesh points	1410	1413	1404	6631	6503	6033
	error points	7.7%	8.1%	6.3%	4.8%	6.4%	4.5%
	sensitivity	72.9%	71.2%	65.7%	70.7%	74.6%	69.3%
	RMS error	2.61	2.53	2.25	3.52	3.08	3.05

new points can be inserted only at the periodic predefined locations, based on the initial lattice structure. Nevertheless, the distributions in *RGM* meshes are consistently better compared with *NETGEN*. This might be caused by large elements near the surface of the *NETGEN* meshes.

Note that empty vertex cells do not pose a problem. The corresponding mesh nodes will move following the neighboring vertices during registration. Problems can be caused by few registration points (the contribution of outliers is not smoothed by the correctly recovered displacements), or by very large number of registration points in the cell (increased approximation error).

The non-rigid registration was performed with the default parameters suggested by Clatz et al. [1]. The quantitative metrics that are not directly optimized by mesh generation are summarized in Table 3. The approximation accuracy is consistently improved for all mesh generation methods when the refined meshes are used. However, this is the only metric that is clearly connected with the size

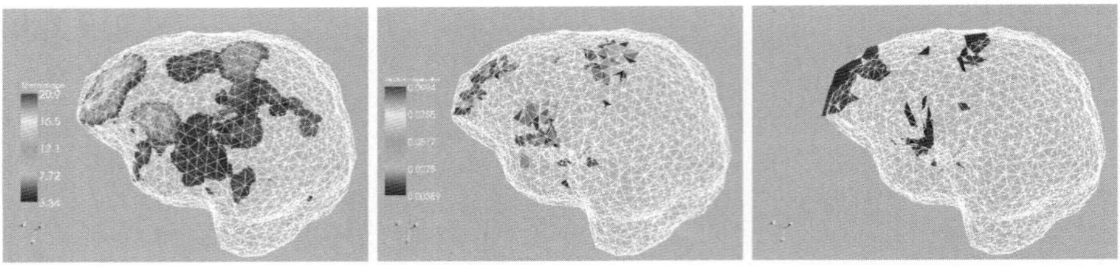

Fig. 4. Left: Image voxels with the largest values of the error with respect to ground truth. Center: mesh elements with the minimum eigenvalue of the dilation matrix below 0.1. Right: inverted mesh elements after NRR. IBSR02, *Tetgen* adaptive mesh, wireframe shows the undeformed mesh.

of the elements of the mesh. We do not observe improvement in either outlier detection sensitivity or RMS of the registration error. On the contrary, RMS error is increasing in the refined meshes. We suggest that there are two major reasons why this may be the case.

First, the adaptively refined meshes are more susceptible to the element inversion during NRR. The areas of the high error in the deformation field recovered by NRR are spatially co-located with the areas of the mesh, where tetrahedra invert or become highly skewed. We use the tetrahedron measure proposed by Baker [38] and identify highly skewed mesh elements by the minimum eigenvalue of the element dilation matrix. Figure 4 shows spatial correlation of the locations of skewed and inverted elements with the areas of image with the largest registration error.

Second, refined meshes decrease the error of approximating displacements recovered by block matching both at the outlier and non-outlier registration points. Ideally, the mesh should be constructed in such a way that in each mesh vertex cell the number of outliers is less than the number of correct displacements. In general outliers are distributed non-uniformly. As we decrease the size of the mesh elements, it becomes more likely that the registration points inside some cells will be dominated by the outliers, causing higher error with respect to the true deformation.

We tried to resolve the first problem by modifying the sizing function to reflect the deformation magnitude averaged over the k closest registration points. The parameters for CGAL surface mesh recovery were chosen according to the maximum of the averaged deformation magnitude near the object surface. None of the evaluated meshers was able to follow the prescribed sizing distribution closely. Both *Tetgen* and *NETGEN* created large tetrahedra near the mesh surface. We cannot attempt to improve the fitness of the mesh to the sizing function by reducing the default values of the *alpha* parameters, as we have done for the meshes evaluated previously. The *alpha* parameters control the bound on the shortest edge length at a mesh point, see Lemma 1 by Si [32]. Their reduction introduces small volume elements. The construction of meshes that adapt to the degree of deformation requires further study.

6 Discussion

In this section we summarize our quantitative and qualitative analysis for each of the six major requirements identified for the NRR application.

R1: Equi-distribution of the registration points. We were not able to achieve equi-distribution, which may be an NP-hard problem. However, in all cases, the use of custom sizing function significantly improved that distribution. The mean values for the number of registration points were close for all meshes. In all test cases, meshes constructed with *Tetgen* had lower values of the maximum number of registration points and better distribution overall.

R2: Reduction of the approximation error at registration points. In the general case, approximation error can be reduced by refining the mesh. This is observed in Table 3 when comparing uniform and adaptive meshes. Further refinement is problematic, as it would violate the requirement R1. When comparing the meshes with respect to the number of error points, we observe that this number is consistently lower for *Tetgen* vs. *NETGEN*. Because *Tetgen* better follows the values of sizing, it creates smaller elements in the areas with high registration point density. Therefore, the approximation error will also be reduced for the points in those areas. *RGM* has consistently the lowest number of error points. This can be explained, because there are fewer registration points located inside the mesh, while the mesh sizes are comparable with the *Tetgen* and *NETGEN* counterpart meshes.

R3: Prevention of tetrahedra inversion. Tetrahedra inversion can be prevented by increasing the size of the mesh elements. This goes contrary to what is required by the requirements R1 and R2. Moreover, it is not clear what should be the optimum element size to avoid inversion. The balancing of this requirements with R1 and R2 is the subject of future work.

R4: Object surface approximation accuracy. CGAL implicit surface recovery procedure was used to construct surface triangulations for *Tetgen* and *NETGEN*. This allows to control over both the surface approximation accuracy and the angles in surface triangulation. The distribution of registration points is highly non-uniform, and is more dense in the areas of the prominent features of the brain, cortex being one of those areas. Poor surface approximation accuracy discards registration points located in the cortex area.

R5: Mesh size optimality. The number of vertices in the meshes we constructed were adjusted using the custom parameters for mesh generation in each particular case. However, given the similar number of mesh vertices, the evaluated meshes perform differently with respect to the quantitative metrics we compared. An important practical concern is the ability to control the total size of the mesh. Based on our experience, *Tetgen* is the most flexible in this respect. In order to control the size of the mesh with *NETGEN*, the sizing values at the background mesh vertices should be scaled. Moreover, the size of the elements near the surface is very hard to control. The size of the *RGM* mesh can be changed by varying the size of the initial lattice, and by parametrizing the refinement rules, which is not straightforward in practice (e.g., changing the number of the maximum registration points for the tetrahedron vertex cells, which can only be done in integer increments).

R6: Control over minimum dihedral angle. Based on our experimental data, the meshes produced by all three methods are very similar when judged by the distribution of minimum dihedral angle. *NETGEN* produces meshes with the relatively largest values of the minimum dihedral angle. However, all of the compared meshes had minimum angle larger than $5°$, which is an acceptable value for the stiffness matrix calculations.

In summary, we were able to address all of the application-specific requirements except R3 using the off-the-shelf mesh generation tools. Custom sizing function was essential to meet requirements R1 and R2. Among the evaluated meshing tools, the Delaunay-based approach provides best theoretical guarantees, best practical results, and is the most flexible in the mesh size control. At the same time, the RMS error values, which are the metrics of the most practical relevance, are very similar for all the evaluated methods.

The open questions related to this study are the following: (1) construction of the sizing function, which balances the conflicting requirements, and gives the ability to assign weights according to their importance; (2) improved fitness of the generated meshes to the desired mesh sizing, and studying the guarantees of such fitness; (3) further evaluation of the impact of the mesh on the error of registration with respect to the ground truth.

Acknowledgement. We thank Olivier Clatz for insightful discussions and for contributing the NRR code for this study, Simon K. Warfield and Ron Kikinis for providing the opportunity of evaluating our work in the clinical setting. We thank Hang Si and Joachim Shöberl for the development and continuing support of extraordinary mesh generation software.

This research was supported in part by NSF grants CSI-0719929 and CNS-0312980, and by John Simon Guggenheim Memorial Foundation.

References

1. Clatz, O., Delingette, H., Talos, I.-F., Golby, A.J., Kikinis, R., Jolesz, F.A., Ayache, N., Warfield, S.K.: Robust non-rigid registration to capture brain shift from intra-operative MRI. IEEE Trans. Med. Imag. 24, 1417–1427 (2005)
2. Archip, N., Clatz, O., Whalen, S., Kacher, D., Fedorov, A., Kot, A., Chrisochoides, N., Jolesz, F., Golby, A., Black, P.M., Warfield, S.K.: Non-rigid alignment of pre-operative MRI, fMRI, and DT-MRI with intra-operative MRI for enhanced visualization and navigation in image-guided neurosurgery. NeuroImage 35, 609–624 (2007)
3. Fedorov, A., Chrisochoides, N., Kikinis, R., Warfield, S.K.: An evaluation of three approaches to tetrahedral mesh generation for deformable registration of MR images. In: Proc. of IEEE ISBI 2006, pp. 658–661 (2006)

4. Hill, D.L.G., Batchelor, P.G., Holden, M., Hawkes, D.J.: Medical image registration. Physics in Medicine and Biology 46, R1–R45 (2001)
5. Delingette, H., Ayache, N.: Soft tissue modeling for surgery simulation. In: Computational models for the human body, 1st edn. Handbook of Numerical Analysis, vol. XII, pp. 453–550. Elsevier, Amsterdam (2004)
6. Shewchuk, J.R.: What is a good linear element? Interpolation, conditioning, and quality measures. In: Proc. of 11th IMR, pp. 115–126 (2002)
7. Owen, S.J.: A survey of unstructured mesh generation technology. In: Proc. of 7th IMR, pp. 239–267 (1998)
8. Yoo, T.S.: Insight into images. A K Peters (2004)
9. Lorensen, W.E., Cline, H.E.: Marching cubes: A high resolution 3d surface construction algorithm. In: Proc. of ACM SIGGRAPH, pp. 163–169 (1987)
10. Boyd, S.K., Müller, R.: Smooth surface meshing for automated finite element model generation from 3d image data. J. of Biomechanics 39, 1287–1295 (2006)
11. Dey, T.K., Li, G., Ray, T.: Polygonal surface remeshing with Delaunay refinement. In: Proc. of 14th IMR, pp. 343–361 (2005)
12. Cebral, J., Castro, M.A., Löhner, R., Burgess, J.E., Pergolizzi, R., Putman, C.M.: Recent developments in patient-specific image-based modeling of hemodynamics. Mecanica Computacional XXIII, 1471–1482 (2004)
13. Amenta, N., Bern, M.: Surface reconstruction by Voronoi filtering. In: Proc. of 14th SCG, pp. 39–48. ACM Press, New York (1998)
14. Kolluri, R., Shewchuk, J.R., O'Brien, J.F.: Spectral surface reconstruction from noisy point clouds. In: Proc. of SGP 2004, pp. 11–21 (2004)
15. Dey, T.K., Levine, J.A.: Delaunay meshing of isosurfaces. In: Proc. of IEEE Int. Conf. on Shape Modeling and Applications (SMI 2007), pp. 241–250 (2007)
16. Boissonnat, J.-D., Oudot, S.: Provably good sampling and meshing of surfaces. Graphical Models 67, 405–451 (2005)
17. Frey, P., Sarter, B., Gautherie, M.: Fully automatic mesh generation for 3-d domains based upon voxel sets. Int. J. for Num. Meth. in Engng. 37, 2735–2753 (1994)
18. Zhang, Y., Bajaj, C., Sohn, B.-S.: 3d finite element meshing from imaging data. Comp. Meth. in App. Mech. and Engng. 194, 5083–5106 (2005)
19. Cline, H.E., Lorensen, W.E., Kikinis, R., Jolesz, F.: Three-dimensional segmentation of MR images of the head using probability and connectivity. J. of Comp. Assisted Tomography 14, 1037–1045 (1990)
20. Cebral, J.R., Löhner, R.: From medical images to CFD meshes. In: Proc. of 8th IMR, pp. 321–331 (1999)
21. Hartmann, U., Kruggel, F.: A fast algorithm for generating large tetrahedral 3d finite element meshes from magnetic resonance tomograms. In: IEEE Workshop on Biomedical Image Analysis, p. 184 (1998)
22. Udeshi, T.: Tetrahedral mesh generation from segmented voxel data. In: Proc. of 12th IMR, pp. 425–436 (2003)
23. Yerry, M.A., Shephard, M.S.: Automatic three-dimensional mesh generation by the modified-octree technique. Int. J. Num. Meth. Engng 20, 1965–1990 (1984)
24. Mitchell, S.A., Vavasis, S.A.: Quality mesh generation in higher dimensions. SIAM J. Comp. 29, 1334–1370 (2000)
25. Ferrant, M.: Physics-based Deformable Modeling of Volumes and Surfaces for Medical Image Registration, Segmentation and Visualization. PhD thesis, Universite' Catholique de Louvain (2001)
26. Mohamed, A., Davatzikos, C.: Finite element mesh generation and remeshing from segmented medical images. In: Proc. of IEEE ISBI, pp. 420–423 (2004)

27. Fedorov, A., Chrisochoides, N., Kikinis, R., Warfield, S.K.: Tetrahedral mesh generation for medical imaging. ISC Insight Journal, 2005 MICCAI Open Source Workshop (2005), http://hdl.handle.net/1926/35
28. Freitag, L.A., Knupp, P.M., Munson, T.S., Shontz, S.M.: A comparison of optimization software for mesh shape-quality improvement problems. In: Proc. of 11th IMR, pp. 29–40 (2002)
29. Castellano-Smith, A.D., Hartkens, T., Schnabel, J., Hose, D.R., Liu, H., Hall, W.A., Truwit, C.L., Hawkes, D.J., Hill, D.L.G.: Constructing patient specific models for correcting intraoperative brain deformation. In: Niessen, W.J., Viergever, M.A. (eds.) MICCAI 2001. LNCS, vol. 2208, pp. 1091–1098. Springer, Heidelberg (2001)
30. Molino, N.P., Bridson, R., Teran, J., Fedkiw, R.: A crystalline, red green strategy for meshing highly deformable objects with tetrahedra. In: Proc. of 12th IMR, pp. 103–114 (2003)
31. Persson, P.O., Strang, G.: A simple mesh generator in MATLAB. SIAM Review 46, 329–345 (2004)
32. Si, H.: On refinement of constrained Delaunay tetrahedralizations. In: Proc. of 15th IMR, pp. 509–528 (2006)
33. Schöberl, J.: NETGEN: An advancing front 2d/3d-mesh generator based on abstract rules. Computing and Visualization in Science 1, 41–52 (1997)
34. Brewer, M., Diachin, L.F., Knupp, P.M., Leurent, T., Melander, D.J.: The Mesquite mesh quality improvement toolkit. In: Proc. of 12th IMR, pp. 239–250 (2003)
35. CGAL: CGAL, Computational Geometry Algorithms Library (2008), http://www.cgal.org
36. Ibanez, L., Schroeder, W.J.: The ITK Software Guide. Kitware Inc. (2003)
37. Billet, E., Fedorov, A., Chrisochoides, N.: The use of robust local Hausdorff distances in accuracy assessment for image alignment of brain MRI. ISC Insight Journal (January-June 2008), http://hdl.handle.net/1926/1354
38. Baker, T.J.: Mesh movement and metamorphosis. Engng with Comp. 18, 188–198 (2002)

Quadrilateral Meshes with Bounded Minimum Angle

F. Betul Atalay[1], Suneeta Ramaswami[2,⋆], and Dianna Xu[3]

[1] St. Joseph's University, Philadelphia, PA
fatalay@sju.edu
[2] Rutgers University, Camden, NJ
rsuneeta@camden.rutgers.edu
[3] Bryn Mawr College, Bryn Mawr, PA
dxu@cs.brynmawr.edu

Summary. This paper presents an algorithm that utilizes a quadtree to construct a strictly convex quadrilateral mesh for a simple polygonal region in which no newly created angle is smaller than $18.43^\circ (= \arctan(\frac{1}{3}))$. This is the first known result, to the best of our knowledge, on quadrilateral mesh generation with a provable guarantee on the minimum angle.

1 Introduction

The generation of quadrilateral meshes with provable guarantees on mesh quality poses several interesting open questions. While theoretical properties of triangle meshes are well understood [4, 8, 9, 7, 11, 12, 15, 14], much less is known about algorithms for *provably good* quadrilateral meshes. Analysts, however, prefer quadrilateral and hexahedral meshes for better solution quality in numerous applications [1, 2, 6, 10, 16]. This is because they have better convergence properties, and hence lower approximation errors, in finite element methods for solutions to systems of partial differential equations. Quadrilateral meshes also offer lower mesh complexity, and better directionality control for anisotropic meshing. For stable analytical results, however, it is critical to construct meshes with certain quality guarantees. Specifically, algorithms that construct well-shaped elements by providing bounds on minimum and maximum angles have much practical value. Techniques such as paving [5] work well in practice, but do not give provable angle guarantees. Circle-packing techniques have been used to construct quadrangulations with no angles larger than 120° for polygon interiors [3], but with no bound on smallest angle. An algorithm to construct linear-sized strictly convex quadrilateral meshes for arbitrary planar straight line graphs is given in [13].

Our contribution. In this paper, we present a new algorithm to generate quadrilateral meshes for simple polygonal regions, possibly with holes, with a provable guarantee on the minimum angle. We use quadtrees to show that no newly created angle in the quadrilateral mesh is smaller than 18.43°. The quadrilaterals are strictly convex, i.e., the maximum angle is strictly less than 180°. This is the first known quadrilateral mesh generation algorithm with a provable bound on the *minimum* angle. (Quadtrees have been used to give triangular meshes without small angles for point sets and polygons

⋆ Research partially supported by NSF Grant 0204293.

in 2D [4], and octrees have been utilized to construct tetrahedral meshes with bounded aspect-ratio elements for polyhedra [12].)

In Section 2, we use quadtrees to construct a quadrilateral mesh for a point set in which the minimum angle is bounded below by $45° - \arctan(\frac{1}{3}) = 26.57°$. We then describe in Section 3 an algorithm that adapts the guaranteed-quality mesh of polygon vertices to polygon edges to construct a quadrilateral mesh for the interior of a simple polygon (possibly with holes) in which new angles (angles other than those determined by the input) are bounded below by $\arctan(\frac{1}{3}) = 18.43°$.

Throughout this paper, we use the shorter terms "quadrangulate" and "quadrangulation" instead of "quadrilateralize" and "quadrilateralization". We also sometimes use the word "quad" for quadrilateral. *Steiner points* are additional points, other than those provided by the input, inserted during the mesh generation process.

2 Point Set Mesh with Bounded Minimum Angle

We first describe an algorithm to construct a quadrilateral mesh with a minimum angle bound of $26.57°$ for a given point set X.

2.1 Construction of the Quadtree

Given a point set X, we construct a quadtree for X with the following separation and balancing conditions. These conditions are similar to those in [4], but adapted to particular requirements for quadrilateral (rather than triangle) meshing.

A. Split a cell C (with side length l) containing at least one point if it is crowded. A cell is crowded if one or more of the following conditions hold:
   1. it contains more than one point from X.
   2. one of the extended neighbors is split (an extended neighbor is a cell of same size sharing either a side or corner of C).
   3. it contains a point with a nearest neighbor less than $2\sqrt{2}l$ units away.

B. When a crowded cell C is split, split those extended neighbors of C that share an edge or corner with a child of C containing an original point in X.

C. The final quadtree is balanced so that the edge lengths of two adjacent cells differ at most by a factor of 2 (neighbors of C with side length l have length $l/2$ or $2l$).

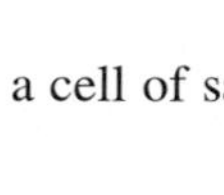

Observe that in a quadtree with the above separation and balancing conditions, a cell containing a point from X is guaranteed to be surrounded by 8 empty cells of the same size. We refine the quadtree decomposition further to do the following: Split each of these eight empty quadtree cells into 2×2 cells and rebalance the quadtree. This converts the original 3×3 grid around every point $p \in X$ into a 6×6 grid. Furthermore, now p lies at the center of a 5×5 equal-sized grid (outlined in bold in figure), and is surrounded by twenty-four empty quadtree cells of the same size. There are two reasons for this refinement step:

1. The final step of our algorithm to construct a quadrilateral mesh for X consists of warping a Steiner point in the mesh to an original point $p \in X$ (Section 2.4). This step is simplified considerably due to the refinement.

2. The algorithm to construct a quadrilateral mesh for non-acute polygons (Section 3.1) uses the 5×5 grid to mesh the region around polygon vertices.

We construct a quadrilateral mesh with bounded minimum angle for X by placing Steiner points in the interior of the quadtree cells. The placement of the Steiner points is determined by identifying and applying templates to the quadtree decomposition. A leaf of the quadtree is an unsplit cell and we refer to these as *1-cells* in our discussion. A template is applied to each internal node of the quadtree.

2.2 The Templates

A template is labeled by the number of children of a quadtree node that are 1-cells. Hence we have 6 template configurations, for nodes with zero ($\mathcal{T}^{(0)}$), one ($\mathcal{T}^{(1)}$), two, three ($\mathcal{T}^{(3)}$) or four ($\mathcal{T}^{(4)}$) 1-cell children. Nodes with two 1-cell children have two layouts, $\mathcal{T}^{(2a)}$ and $\mathcal{T}^{(2b)}$.

Templates at the deepest level of subdivision. The templates at the deepest level of subdivision are shown in Fig. 1. Note that, all other possible configurations are symmetric to the depicted ones. In order to quadrangulate a template, first, a Steiner point is placed at the center of each quadtree cell. These points are denoted with full circles. We then place *extra* Steiner points, which are denoted by empty circles in the figure, for one of two reasons: (i) In $\mathcal{T}^{(1)}$, the top-left extra point and in $\mathcal{T}^{(2b)}$ the middle extra point are added to be able to quadrangulate *properly* within the template. (ii) The remaining extra points are added in the 1-cells, halfway on the diagonal between the center Steiner point and the outer cell corner. The reason for adding the second type of Steiner points is that after an internal node is quadrangulated, it will provide a polygonal chain with an even number of points (we will call them *even-connector chains*) to which its neighbors can connect.

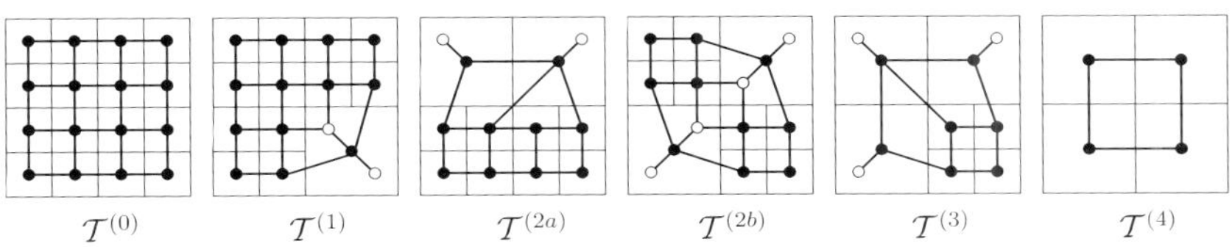

Fig. 1. Templates at the deepest level of the subdivision

General Templates. Our recursive algorithm applies templates to all internal nodes starting with the deepest ones. We generalize the templates to apply to an arbitrarily deep internal node as shown in Fig. 2. In general, when a template is applied to an internal node, its children which are not 1-cells have already had templates applied to them. Each such child has been quadrangulated internally and provides even-connector chains on all four sides. The corresponding endpoints of two neighboring chains are then connected to construct a polygon with guaranteed even number of vertices which can therefore be quadrangulated. We name this process "stitching", illustrated by the cross-hatched regions in Fig. 2. The processed internal nodes are

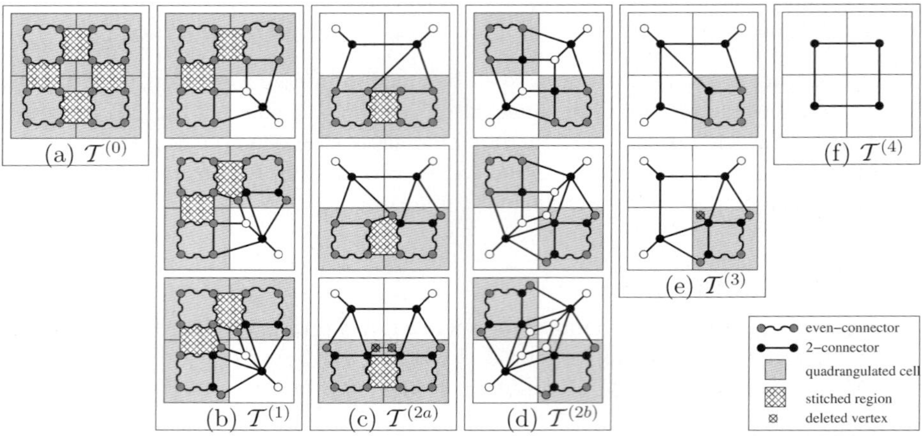

Fig. 2. General templates at arbitrary level of subdivision

depicted as blackboxes with even-connector chains at each side. Note that the placement of a chain's endpoint does not necessarily correspond to the exact location of the endpoint within the actual cell, due to the existence of type (ii) Steiner points.

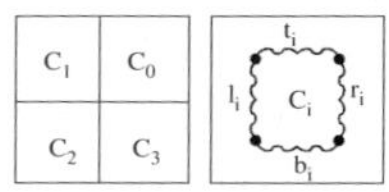

Labeling the chains. Children quadrants of a cell are labeled C_0, C_1, C_2, and C_3 in counterclockwise order starting from the northeast quadrant. The four chains surrounding a processed quadrant C_i are labeled l_i, r_i, t_i and b_i (see figure).

2.3 The Algorithm

The recursive procedure *applyTemplate* that applies a template to an internal node is presented in the code block given in Fig. 3. It is initially called with the root node of the quadtree. Note that the algorithm is presented only with respect to the depicted configurations of the templates. Symmetric configurations are handled similarly.

Stitching Chains. Procedure *stitchChains* connects the four endpoints of two neighboring even-connector chains and quadrangulates the resulting polygon. Note that such a polygon is guaranteed to have an even number of vertices on the boundary. The algorithm is illustrated in Fig 4. Procedure *stitchChains* is only called if the current template is of type $\mathcal{T}^{(0)}$, $\mathcal{T}^{(1)}$ or $\mathcal{T}^{(2a)}$. The action of this procedure is also illustrated by the crosshatched areas in Fig. 2(a), (b) and (c).

The quadrangulation process divides the chains into half chains, each of which spans the corresponding edge of a child quadrant. These half chains are then recursively

```
applyTemplate(QuadtreeNode N)
  templateType ← whichTemplate(N)
  for C_i ∈ children(N)
    if C_i is not a 1-cell
      (l_i, r_i, t_i, b_i) ← applyTemplate(C_i)
    else
      construct (l_i, r_i, t_i, b_i) for C_i.
  switch (templateType)
    case T^(0):
      stitchChains(l_0, r_1), stitchChains(b_1, t_2)
      stitchChains(r_2, l_3), stitchChains(t_3, b_0)
    case T^(1):
      stitchChains(l_0, r_1), stitchChains(b_1, t_2)
      Place Steiner points and quadrangulate per Fig. 2(b).
    case T^(2a):
      stitchChains(r_2, l_3)
      Place Steiner points and quadrangulate per Fig. 2(c).
    case T^(2b):
      Place Steiner points and quadrangulate per Fig. 2(d).
    case T^(3):
      Place Steiner points and quadrangulate per Fig. 2(e).
    case T^(4):
      Place Steiner points and quadrangulate per Fig. 2(f).
  return (l_1 + l_2, r_0 + r_3, t_1 + t_0, b_2 + b_3)
```

Fig. 3. *applyTemplate* quadrangulates node N

```
stitchChains(Chain ch_1, Chain ch_2)
  switch (length(ch_1), length(ch_2))
    case (2 − 2), (2 − 4), (4 − 2):
      Apply appropriate base case from Fig. 5.
    case (2 − 6), (2 − 8)):
      Apply appropriate base case from Fig. 6.
    default:
      (f_1, s_1) ← getHalfChains(ch_1)
      (f_2, s_2) ← getHalfChains(ch_2)
      stitchChains(f_1, f_2)
      stitchChains(s_1, s_2)
```

Fig. 4. *stitchChains* stitches two even-connector chains, one from each of the two neighbor cells sharing an edge

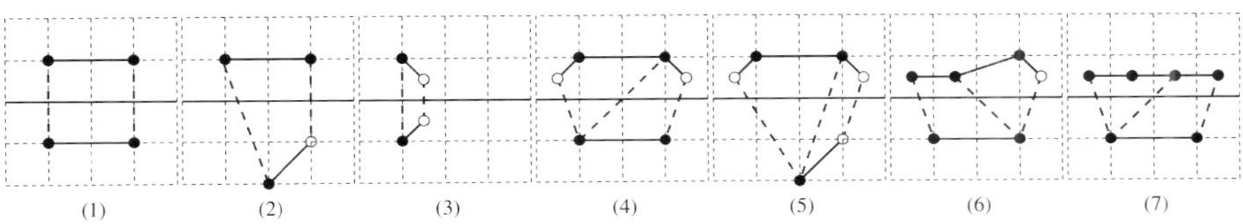

Fig. 5. Stitching 2-2 and 2-4 or 4-2 connector chains

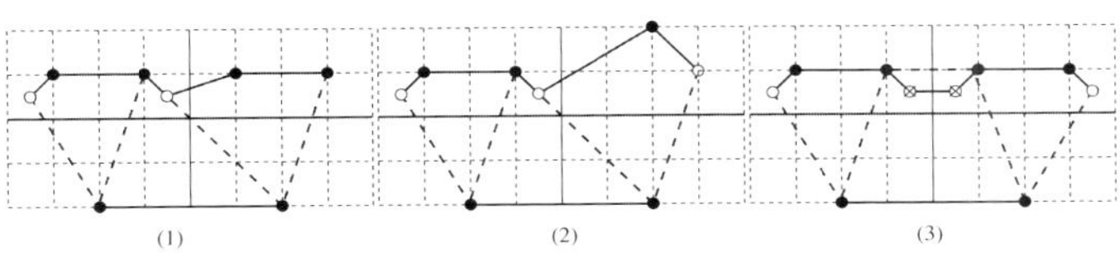

Fig. 6. Stitching 2-6 and 2-8 connector chains

stitched. Although the even-connector chains can be arbitrarily long, at the base case there are only four types of chains: chains with 2, 4, 6 or 8 connectors. Figs. 5-6 illustrate how the base-case chains are stitched (the stitching edges are dotted). Symmetric cases are not listed in the illustrations.

2.4 Angle Bounds

Minimum Angle. We analyze the minimum angle bound in each of *applyTemplate*, the base case of *stitchChains*, and the recursive step of *stitchChains*.

General templates: By construction, the minimum angle appears in templates $\mathcal{T}^{(1)}$ and $\mathcal{T}^{(2b)}$ and equals $45° - \arctan(\frac{1}{3}) = 26.57°$ (illustrated in Fig. 7).

Stitching base case: The base cases of stitching generate the same minimum angle of $45° - \arctan(\frac{1}{3}) = 26.57°$ which can be found in in Fig. 5-(5).

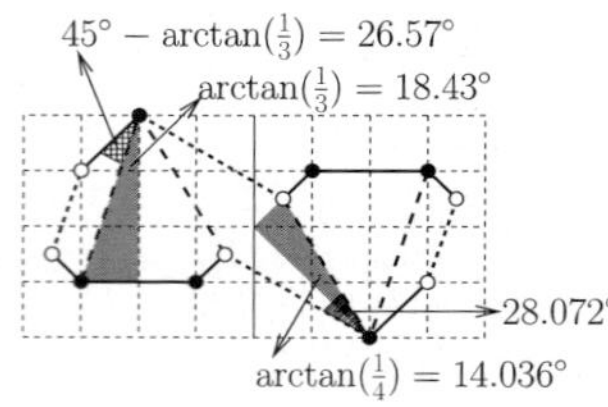

Fig. 7. Minimum angle bounds

Stitching merging step: After the corresponding half-chains are stitched in the recursive step of *stitchChains*, a middle quad is formed by the four end points of the stitched half-chains. This middle quad gives a minimum angle of $2 \times \arctan(\frac{1}{4}) = 28.07°$. See Fig. 7. Recall that these four points are by construction on the two diagonals that cross at the center of four quadtree quadrants. Furthermore, they are either at the center of the quadtree quadrant, or halfway down the diagonal from the center. The worst-case configuration is illustrated in Fig. 7. This results from connecting any Fig. 5-(5) connector chain with an inverted version of itself.

Degenerate quads. In the stitching cases illustrated by Fig. 5-(7), Fig. 6-(1), Fig. 6-(2) as well as template $\mathcal{T}^{(2a)}$ (Fig. 1), there are degenerate quads with two edges on a straight line. In all cases, the $180°$ vertex is connected to a third vertex on the other side of the degenerate quad, by construction. This allows perturbation of the degenerate vertex along the third edge, which reduces the $180°$ angle and increases the other two, thus eliminating the degenerate quad.

Maximum Angle. The quadrilaterals generated by our algorithms (including the one in Section 3) are *strictly* convex; i.e., the maximum angle is bounded away from $180°$. The perturbation used to handle degenerate quads (above) implies the maximum angle is less than $(180 - \epsilon)$ for some $\epsilon > 0$. We conjecture the value of ϵ can be bounded from below, but do not explicitly address that question in this paper.

Warping to Original Points. After the construction of the quadrilateral mesh using quadtree cell centers and extra points as Steiner points, we warp certain mesh vertices to the original points from the input point set X. Recall that the quadtree splitting rules of Section 2.1 ensure that the quadtree cell containing an original point $p \in X$ is surrounded by twenty-four empty quadtree cells of the same size. Moreover, the eight empty cells immediately surrounding p do not contain any extra points. Therefore, the warping step simply consists of translating the Steiner point in p's cell to p, along with all the incident edges.

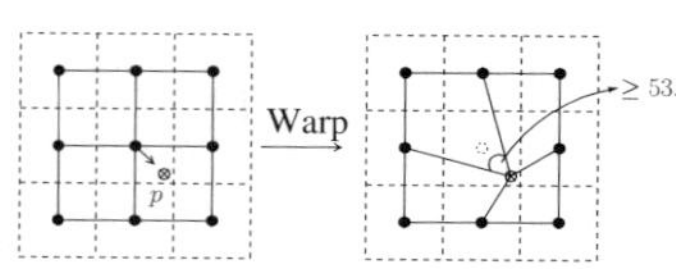

The worst-case minimum angle arising from these nine cells after the warping step is $2 \times 26.57 = 53.14°$. In summary, we have shown the following result:

Theorem 1. *Given a quadtree decomposition with N quadtree cells satisfying the point set separation conditions for a point set X, applyTemplate constructs a mesh for X with at most $3N$ quadrilaterals in which every angle is at least 26.57 degrees.*

Observe that the value of N in the above theorem depends on the geometry of the point set as well as the size of the point set. Due to the point set separation conditions, which are derived from [4] and as was shown there, the size of the quadtree decomposition increases as the distance between the closest pair of points decreases.

3 Polygon Mesh with Bounded Minimum Angle

Given a simple polygon P, possibly with holes, with vertex set X, we give an algorithm to construct a quadrilateral mesh for P and its interior in which no new angle is larger than $18.43°$. The basic idea behind the algorithm is to first construct a guaranteed-quality mesh for X as described in the previous section, and then adapt this mesh to incorporate the edges of P. We use δP to refer to the polygon boundary, and P to refer to the union of the boundary as well as interior.

We describe in Section 3.1 a provably good algorithm to construct a quadrilateral mesh with bounded minimum angle for a simple polygon P in which all interior angles are non-acute (i.e., greater than or equal to $90°$). In Section 3.2, we describe how to handle acute angles.

3.1 Non-acute Simple Polygons

Let P be a non-acute polygon with vertex set X and edges oriented counter-clockwise about the boundary. Let $\mathcal{QT}$ be a quadtree decomposition of X satisfying the point set separation conditions of Section 2.1. Let $\mathcal{Q}$ be a quadrilateral mesh for X with minimum angle $26.57°$, as guaranteed by Theorem 1. In this section, we describe a method to adapt $\mathcal{Q}$ to δP to create a *constrained* quadrilateral mesh for P. In a constrained quadrilateral mesh, we allow Steiner points to be inserted on δP as well, so that the union of the finite elements of the mesh is equal to P.

We start by describing an algorithm to adapt $\mathcal{Q}$ to include a single edge of P. In order to use this algorithm on all edges of P, $\mathcal{QT}$ must satisfy certain polygon edge separation conditions, which are discussed towards the end of the section. We conclude the section by describing how to construct the final constrained mesh for P by adapting to the regions around the vertices.

Inserting an edge into $\mathcal{Q}$

Consider an edge $\vec{e} = (a, b)$ of P oriented from a to b, where $a, b \in X$. Assume that $\vec{e}$ makes an angle between $-45°$ and $45°$ with the positive x axis (if not, orient the x axis so that this is the case). We say that a point lies "above" $\vec{e}$ if it lies in the open halfspace to the left of the oriented line through $\vec{e}$. We use $\vec{e}$ to define two chains of edges from $\mathcal{Q}$ and $\mathcal{QT}$, as described below:

(i) $\vec{e}$ intersects quadrilaterals of $\mathcal{Q}$. Edges of these quadrilaterals are used to define a chain of edges called the *quadrangulation chain* α associated with $\vec{e}$.
(ii) $\vec{e}$ intersects quadtree cells of $\mathcal{QT}$. The centers of these cells are used to define a chain of edges called the *quadtree chain* β associated with $\vec{e}$.

Quadrangulation Chain. Let $q_1, q_2, \ldots, q_k$ be the quadrilaterals of $\mathcal{Q}$ intersected by $\vec{e}$ in left to right order as traversed from a to b (since the quadrilaterals are convex, each q_i is unique). Let E_i be the edges of q_i that lie entirely above $\vec{e}$. E_i may have 0, 1, or 2 edges. If E_i has two edges, they are listed in clockwise order about q_i. Then the *quadrangulation chain* α is defined as $\alpha = E_1 \cdot E_2 \ldots \cdot E_k$, where $\cdot$ represents edge concatenation. See Fig. 8 for an example of a quadrangulation chain, in which E_1 has 1 edge, E_2 has 2 edges, and E_3 has 0 edges. Note that the same edge may repeat twice in α (the repetitions always appear consecutively) and such an edge is incident to q_i that has $|E_i| = 0$. For example, the quadrangulation chain in Fig. 8 has three repeating edges, which are incident to the quadrilaterals q_3, q_5 and q_{10}. If we drop the repetitions from α, then α is a weakly simple polygonal chain.

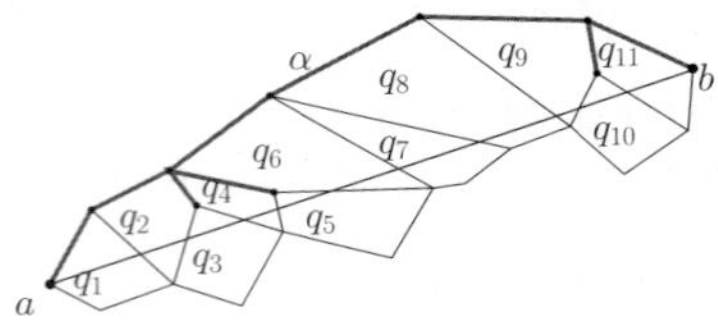

Fig. 8. Quadrangulation chain α

A vertex belongs to an edge if it is one of the endpoints of the edge. We say that $v \in \alpha$ if v is a vertex of $\mathcal{Q}$ and belongs to one of the edges of α. If we quadrangulate the region bounded by α and $\vec{e}$ by adding Steiner points either in the interior of the region or on $\vec{e}$ itself, the resulting quadrangulation is compatible with $\mathcal{Q}$ (since edges of α are edges in $\mathcal{Q}$). However, in order to quadrangulate the region with the desired angle bounds, we need to know more about the geometry of α. The quadtree chain, described below, allows us to establish the required geometric properties for α.

Quadtree Chain. In the remainder of the paper, we use the same symbol to refer to a quadtree cell as well as its center whenever the meaning is clear from the context. Given a cell c, $N(c), W(c)$, and $E(c)$ denote, respectively, the set of north, west, and east neighbor cells of c (note that each set has at most two elements in it because of the balancing conditions for $\mathcal{QT}$).

Let C be the set of cell centers of quadtree cells in $\mathcal{QT}$ that are intersected by $\vec{e}$. C does not include the starting or ending cells (i.e., the cells containing a and b, respectively). Let θ be the angle (in degrees) that $\vec{e}$ makes with the positive x axis. The *quadtree chain* β is defined as follows:

1. If $c \in C$ and c lies above $\vec{e}$, then c belongs to β.
2. If $c \in C$ and c lies below $\vec{e}$, then $N(c) \subset \beta$. Note that the centers of cells in $N(c)$ must lie above $\vec{e}$ under our assumption that $-45 \leq \theta \leq 45$.
3. If $c \in C$, c lies below $\vec{e}$, and $0 \leq \theta \leq 45$ (resp., $-45 \leq \theta < 0$), then a cell center in $W(c)$ (resp., $E(c)$) belongs to β if it lies above $\vec{e}$.

Let $\{c_1, c_2, \ldots, c_m\}$ be the cell centers in β in lexicographically sorted (by x, then y) order. Recall that $\mathcal{Q}$ is constructed from the quadtree decomposition $\mathcal{QT}$ of X. The overall approach to incorporating edge $\vec{e}$ into $\mathcal{Q}$ is summarized below:

(A) We first show that quadtree chain β is a subset of quadrangulation chain α.
(B) This fact allows us, in turn, to exploit the structure provided by $\mathcal{QT}$ and our algorithm from Section 2 to identify a small number of possible ways in which two consecutive points c_i and c_{i+1} of β can be connected along the chain α. We use α_i, $1 \leq i \leq m-1$, to refer to the subchain of α starting at c_i and ending at c_{i+1}. α_i may lie under $\overrightarrow{c_i c_{i+1}}$. In this case, we choose instead a chain of edges in $\mathcal{Q}$ lying above $\overrightarrow{c_i c_{i+1}}$ in order to simplify the final quadrangulation step in part (C) below.
(C) Finally, we quadrangulate the region bounded by $\vec{e}$ and α by breaking it into smaller sub-regions defined by perpendicular projections from c_i and c_{i+1} onto $\vec{e}$. The case analysis form part (B) is then used to prove a minimum angle guarantee of $18.43°$ for the quadrangulation of each subregion.

Lemmas 1-3 are required for steps (A)-(C) and stated here without proof.

Lemma 1. *For $1 \leq i \leq m, c_i \in \alpha$. That is, every cell center in the quadtree chain belongs to the quadrangulation chain.*

Lemma 2. *For $1 \leq i \leq m-1$, c_i and c_{i+1} are edge or corner neighbors in $\mathcal{QT}$.*

Lemma 3. *Let v_i be the vertical projection of c_i on $\vec{e}$. For $1 \leq i \leq m$, the segment $\overline{c_i v_i}$ does not intersect α.*

Lemmas 1 and 3 imply that the edge sequence $(v_i, c_i) \cdot \alpha_i \cdot (c_{i+1}, v_{i+1}) \cdot (v_{i+1}, v_i)$ defines a simple polygon for all $1 \leq i \leq m-1$. Call this polygon A_i. We now use Lemma 2 to prove that α_i is composed of at most four edges. This is done via a case analysis on the ways in which c_i and c_{i+1} are connected in $\mathcal{Q}$.

Lemma 4. *The number of edges in α_i is at most four.*

Proof: We know from Lemma 2 that c_i and c_{i+1} are either edge or corner neighbors in $\mathcal{QT}$. We consider each case separately. Our case analysis only depicts α_i with two or more edges (i.e., when c_i and c_{i+1} are not directly connected). Let s_i, $1 \leq i \leq m$ refer to the size of c_i's cell (by "size", we mean "side length").

Case 1: c_i and c_{i+1} are edge neighbors. In this case, the connectivity between c_i and c_{i+1} in $\mathcal{Q}$ may come from either the application of a template (*applyTemplate*) at some level of recursion, or the application of the stitching step (*stitchChains*) at some level of recursion. We consider different possibilities based on the ratio $s_i : s_{i+1}$, which may be $1:1, 1:2$, or $2:1$. Configurations for these cases are shown in Figs. 9, 10, and 11, respectively. Each of these figures indicates the minimum internal angle in A_i along α_i. Note that each of them is well above $18.43°$. We depict only distinct α_i that differ in either the number of edges, or the angles at the vertices (that is, we do not show other, symmetric configurations that lead to the same α_i).

Case 2: c_i and c_{i+1} are corner neighbors. In this case, the connectivity between c_i and c_{i+1} in $\mathcal{Q}$ may come from the application of *applyTemplate*, the application of *stitchChains*, or through a *center quad*. The center quad is the quadrilateral formed at the center, i.e. the meeting point of the four quadrants, after a general template (ref. Fig. 3) is applied during the recursive step. Since c_i and c_{i+1} are corner neighbors, the

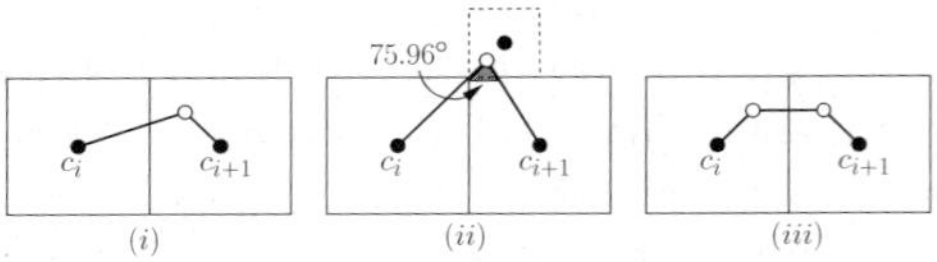

Fig. 9. Configurations for α_i when $s_i : s_{i+1}$ is $1 : 1$. (i) and (ii) come from stitching base cases, and (iii) from stitching merge steps.

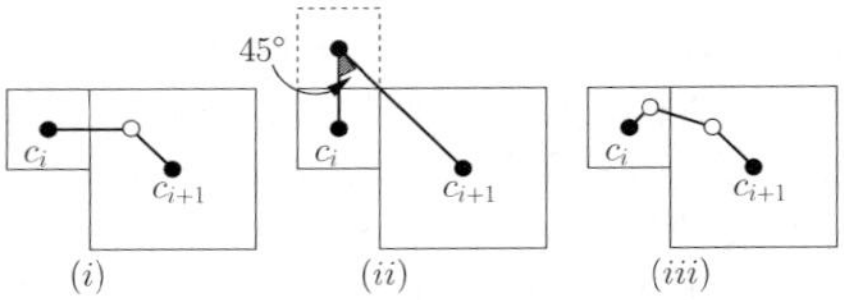

Fig. 10. Configs. for α_i when $s_i : s_{i+1}$ is $1 : 2$. (i) and (ii) from *applyTemplate* and stitching base cases. (iii) from stitching merge steps.

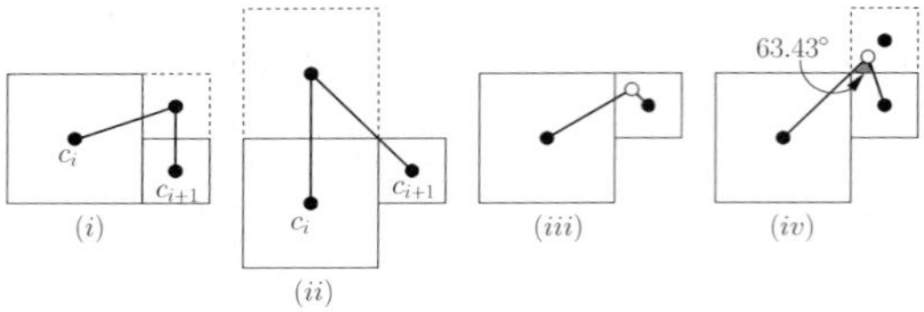

Fig. 11. Configurations for α_i when $s_i : s_{i+1}$ is $2 : 1$. (i) and (ii) come from *applyTemplate* and stitching base cases. (iii) and (iv) occur only in the stitching base cases.

ratio $s_i : s_{i+1}$ can be $1 : 1, 1 : 2, 2 : 1, 1 : 4$, or $4 : 1$. We consider the case of center quads first, and then consider templates and stitchings.

Case 2.1: α_i contains center quad edges. Let s be the size of the cell adjacent to c_i as well as c_{i+1} and lying above $\overrightarrow{c_i c_{i+1}}$. Possible configurations for α_i when $s_i : s : s_{i+1} \equiv 1 : 1 : 1$ are shown in Fig. 12. In 12(i), 12(iv), and 12(vi), the point in the cell adjacent to c_i and c_{i+1} may be either a cell center or an extra point of a larger cell. When $s_i : s : s_{i+1} \equiv 1 : 2 : 1$ or $s_i : s : s_{i+1} \equiv 1 : \frac{1}{2} : 1$, possible configurations of α_i are shown in Figs. 13 and 14, respectively.

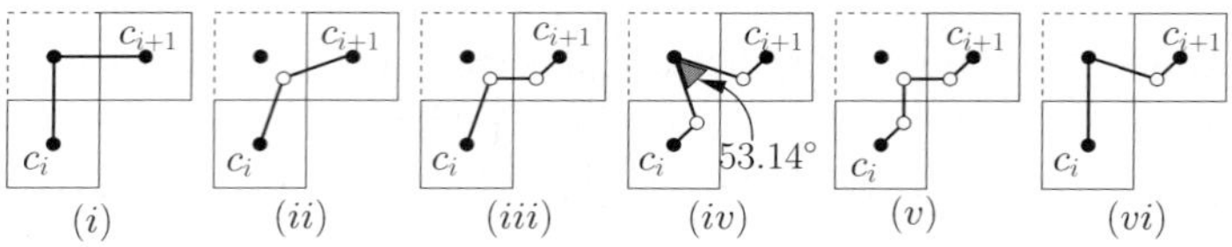

Fig. 12. $s_i : s : s_{i+1} \equiv 1 : 1 : 1$

When $s_i : s : s_{i+1} \equiv 1 : 1 : 2$, possible configurations of α_i are shown in Fig. 15. For the case when $s_i : s : s_{i+1} \equiv 2 : 1 : 1$, the α_i are obtained by reflections about the line $y = x$ of those in Case 2.1.4. Hence the minimum internal angle shown in Fig. 15 holds here as well. Similarly, Fig. 16 depicts α_i when $s_i : s : s_{i+1} \equiv 1 : 2 : 2$ and the chains in this figure are reflections about the line $y = x$ of possible α_i when $s_i : s : s_{i+1} \equiv 2 : 2 : 1$

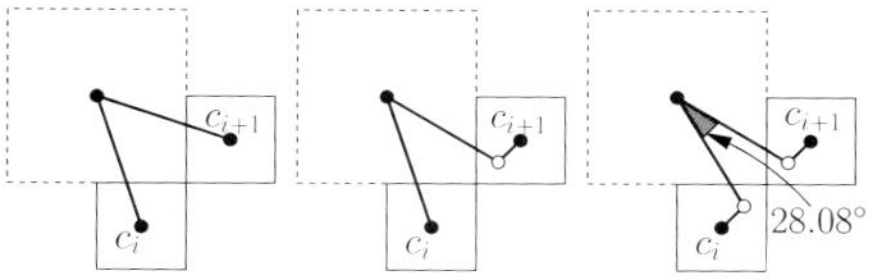

Fig. 13. $s_i : s : s_{i+1} \equiv 1 : 2 : 1$

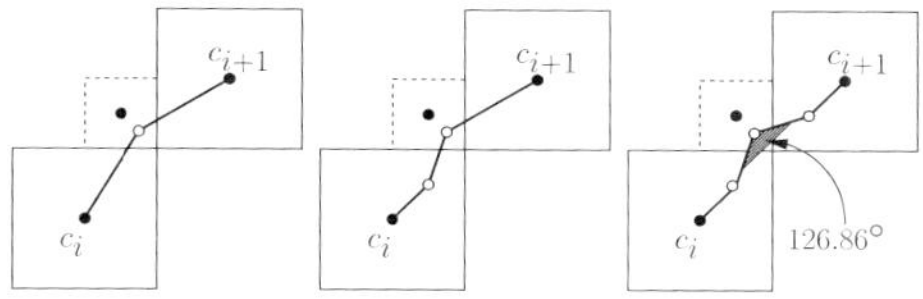

Fig. 14. $s_i : s : s_{i+1} \equiv 1 : \frac{1}{2} : 1$

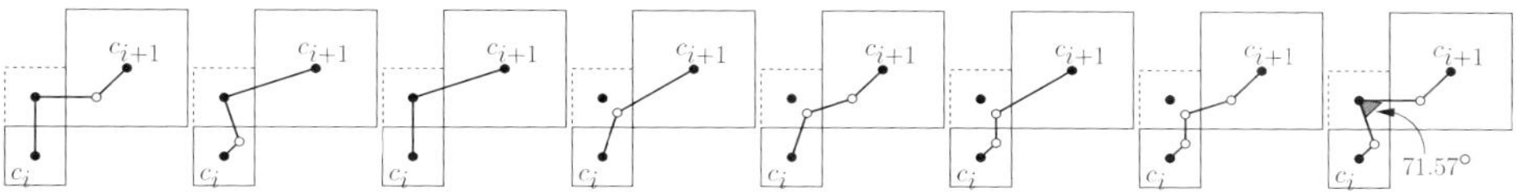

Fig. 15. $s_i : s : s_{i+1} \equiv 1 : 1 : 2$

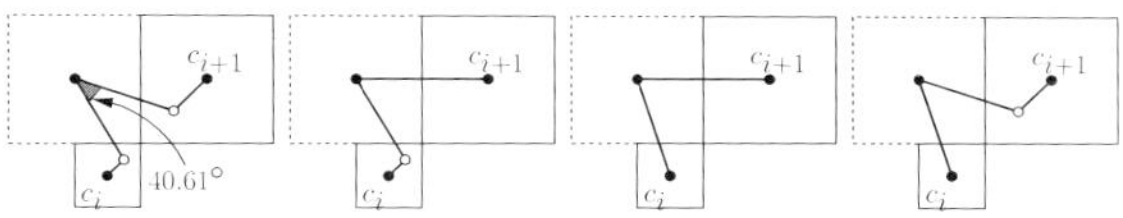

Fig. 16. $s_i : s : s_{i+1} \equiv 1 : 2 : 2$

Finally, Fig. 17 shows possible configurations of α_i when $s_i : s : s_{i+1} \equiv 1 : 2 : 4$. For the case when $s_i : s : s_{i+1} \equiv 4 : 2 : 1$, the α_i are obtained by 180° rotations of those in Fig. 17.

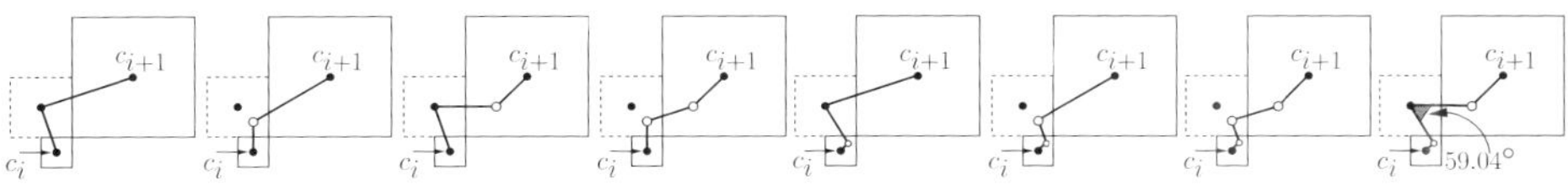

Fig. 17. $s_i : s : s_{i+1} \equiv 1 : 2 : 4$

Case 2.2: α_i constructed by application of applyTemplate or stitchChains. All new configurations of α_i that occur by a template application, or a stitching step at some level of recursion are listed. By "new", we mean configurations that do not appear in Figs. 12-17. Note that when c_i and c_{i+1} are connected via templates or stitchings, $s_i : s_{i+1}$ is $1 : 1$ (see Fig. 18), $1 : 2$, (see Fig. 19) or $2 : 1$ (reflections about the line of $y = x$ of the α_i in Fig. 19), but not $1 : 4$ or $4 : 1$.

While Figs. 12-19 all depict c_i and c_{i+1} in the southwest and northeast quadrants respectively, note that each of the α_i in these figures has a 90° rotational symmetry corresponding to c_i and c_{i+1} in the northwest and southeast quadrants, which does not change the minimum internal angles indicated in those figures.

It follows from the above case analysis that α_i has at most four edges. □

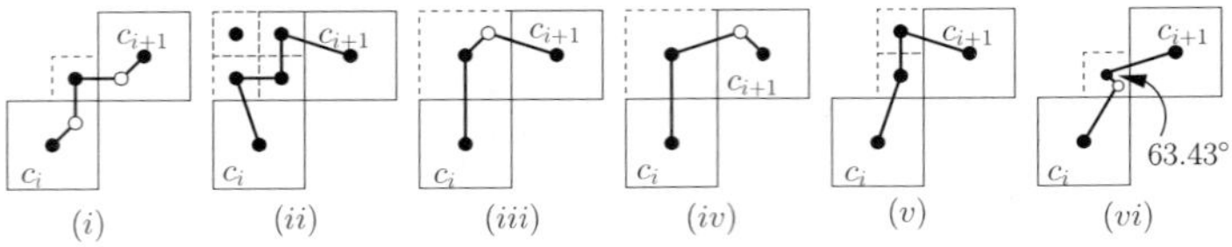

Fig. 18. $s_i : s_{i+1} \equiv 1 : 1$

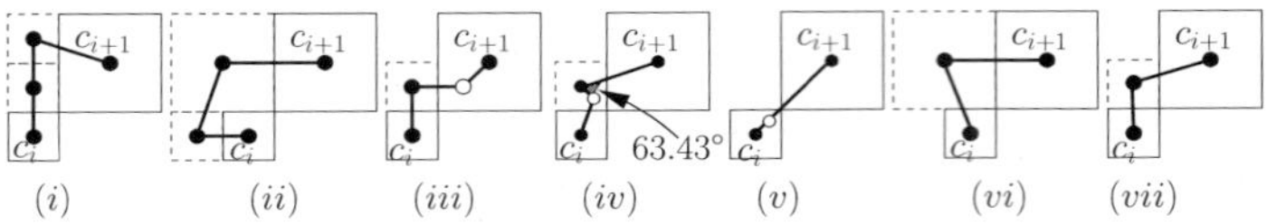

Fig. 19. $s_i : s_{i+1} \equiv 1 : 2$

We now describe how to quadrangulate each polygonal region $A_i = (v_i, c_i) \cdot \alpha_i \cdot (c_{i+1}, v_{i+1}) \cdot (v_{i+1}, v_i)$ independently for $1 \leq i \leq m$. Before doing this, we first show that rather than using the vertical projections v_i and v_{i+1}, we may instead use perpendicular projections of c_i and c_{i+1} onto edge $\vec{e}$ (Lemma 6). This allows us to prove angle bounds for quadrangulating A_i that are independent of the angle that $\vec{e}$ makes with the horizontal (recall that this is between $-45°$ and $45°$).

Lemma 5. *Let δ_i be the signed angle (in degrees) between $\overrightarrow{c_i c_{i+1}}$ and the positive x-axis. Then abs$(\delta_i) \in \{0, 18.43, 45, 71.57, 90\}$.*

Proof: Since c_i and c_{i+1} are both cell centers in $\mathcal{QT}$, and we know from Lemma 2 that they are edge or corner neighbors, it follows that there are a constant number of possibilities for δ_i: If c_i and c_{i+1} are edge neighbors with $s_i = s_{i+1}$, then δ_i is either 0 or 90°. If c_i and c_{i+1} are edge neighbors with $s_i \neq s_{i+1}$, then $\tan(\delta_i) = \frac{1}{3}$, i.e., abs$(\delta_i) = 18.43°$, or $\tan(\delta_i) = 3$, i.e., abs$(\delta_i) = 71.57°$. If c_i and c_{i+1} are corner neighbors, then abs$(\delta_i) = 45°$. □

Let θ be the signed angle made by $\vec{e}$ with the positive x-axis. The value of θ determines the range of possibilities for δ_i. This is because of our definition of the quadtree chain, which specifies that either cell c_i (resp. c_{i+1}) has center above $\vec{e}$ and is intersected by $\vec{e}$, or it is the north/west neighbor of a cell intersected by $\vec{e}$ whose center lies below $\vec{e}$. Table 1 summarizes the possible values of δ_i for given ranges of θ. This relationship also allows us to prove Lemma 6, stated here without proof.

Lemma 6. *Let p_i be the perpendicular projection of c_i on $\vec{e}$, and v_i the vertical projection of c_i on $\vec{e}$. Assume $\vec{e}$ makes an angle between $-45°$ and $45°$ with the positive x axis. Then for all $1 \leq i < m$, c_{i+1} lies outside the triangle $\Delta(p_i c_i v_i)$.*

We know from Lemma 3 that for all $1 \leq i \leq m$, α_i does not intersect $\overline{c_i v_i}$ or $\overline{c_{i+1} v_{i+1}}$. Furthermore, we know from the proof of Lemma 4 that α_i lies above $\overrightarrow{c_i c_{i+1}}$ (that is, it does not intersect the region bounded by $c_i v_i v_{i+1} c_{i+1}$). Therefore, Lemma 6 implies

Table 1. Range of values for θ and δ_i

Range of θ	**Values of δ_i**
$18.43° \leq \theta < 45°$	$-18.43° \leq \delta_i \leq 90°$
$0 < \theta < 18.43°$	$-45° \leq \delta_i < 90°$
$-18.43° < \theta \leq 0$	$-71.57° \leq \delta_i \leq 45°$
$-45° < \theta \leq -18.43°$	$-71.57° \leq \delta_i \leq 18.43°$

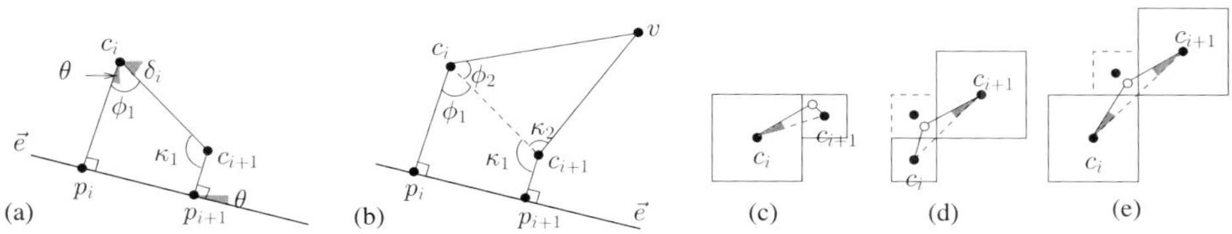

Fig. 20. (a) $\min\{\phi_1, \kappa_1\} \geq 18.43°$. (b) $\min\{\phi_1 + \phi_2, \kappa_1 + \kappa_2\} \geq 2 \times 18.43°$. (c) Shaded angle is $12.54°$. (d)-(e) Shaded angle is $14.04°$.

that α_i does not intersect $\overline{c_i p_i}$ or $\overline{c_{i+1} p_{i+1}}$ either. We redefine polygon A_i to be $(p_i, c_i) \cdot \alpha_i \cdot (c_{i+1}, p_{i+1}) \cdot (p_{i+1}, p_i)$ (that is, it is defined by the perpendicular projections rather than the vertical ones). Lemmas 7 and 8 establish bounds on some angles in A_i when α_i has one and two edges, respectively.

Lemma 7. *Let* $\phi_1 = \angle p_i c_i c_{i+1}$ *and* $\kappa_1 = \angle c_i c_{i+1} p_{i+1}$. *Then* $\min\{\phi_1, \kappa_1\} \geq 18.43°$.

Proof: Refer to Fig. 20(a). Since $\phi_1 = 90 - \theta + \delta_i$ and $\kappa_1 = 90 + \theta - \delta_i$ (recall θ and δ_i are signed angles), and the fact $-71.57° \leq (\theta - \delta_i) \leq 71.57°$ (refer to Table 1), it follows that $\phi_1 \geq 18.43°$ and $\kappa_1 \geq 18.43°$. □

Lemma 8. *Suppose* α_i *has two edges,* $c_i v$ *and* $v c_{i+1}$. *Let* $\phi_1 = \angle p_i c_i c_{i+1}$, $\kappa_1 = \angle c_i c_{i+1} p_{i+1}$, $\phi_2 = \angle c_{i+1} c_i v$, *and* $\kappa_2 = \angle c_i c_{i+1} v$. *Then (i)* $\min\{\phi_1, \kappa_1\} > 18.43°$ *and (ii)* $\min\{\phi_1 + \phi_2, \kappa_1 + \kappa_2\} \geq 2 \times 18.43°$.

Proof: (i) From Lemma 7 we know that $\min\{\phi_1, \kappa_1\} \geq 18.43°$. To see that it must be *strictly* greater, note that if $\min\{\phi_1, \kappa_1\} = 18.43°$ then $\text{abs}(\theta - \delta_i) = 71.57°$. From Lemma 5 and Table 1, it can be seen that $\text{abs}(\theta - \delta_i) = 71.57°$ when *(a)* $\theta = 18.43°$ and $\delta_i = 90°$, which is impossible because c_i and c_{i+1} are directly connected whenever $\delta_i = 90°$, or *(b)* $\theta = 0$ and $\delta_i = 71.57°$, which is also impossible because θ must be strictly greater than 0 whenever $\delta_i = 71.57°$.

(ii) Refer to Fig. 20. First observe that if $\min\{\phi_2, \kappa_2\} \geq 18.43°$, then part (i) implies the claim. Hence assume that $\min\{\phi_2, \kappa_2\} < 18.43°$. The only configurations of α_i for which $\min\{\phi_2, \kappa_2\} < 18.43°$ are shown in Fig. 20(c)-(e). We use the angle dependency in Table 1 to prove that these configurations imply $\min\{\phi_1 + \phi_2, \kappa_1 + \kappa_2\} \geq 2 \times 18.43°$. Further details are omitted here. □

Lemma 9. *For* $1 \leq i \leq m - 1$, *the simple polygon* $A_i = (p_i, c_i) \cdot \alpha_i \cdot (c_{i+1}, p_{i+1}) \cdot (p_{i+1}, p_i)$ *can be quadrangulated with at most five quadrilaterals with a minimum angle of* $18.43°$.

Proof: Since α_i has at most four edges (Lemma 4), we have four cases.

Case 1: α_i has one edge. In this case, A_i is already a quadrilateral. The fact that all angles of A_i are at least $18.43°$ follows from Lemma 7.

Case 2: α_i has two edges. Let $c_i v$ and $v c_{i+1}$ be the two edges of α_i. Let ϕ_1, κ_1, ϕ_2, and κ_2 be as in Fig. 20(b). Let $\gamma = \angle c_i v c_{i+1}$. Observe that $\gamma \geq 26.57°$ because the edges of α_i come from $\mathcal{Q}$. We quadrangulate A_i according to the angles ϕ_2 and κ_2.

- If $\min\{\phi_2, \kappa_2\} \geq 18.43°$, place a Steiner point s on $\overline{c_i c_{i+1}}$ and at the perpendicular projection p of s onto $\vec{e}$. Connect s to c_i, c_{i+1}, and p to obtain a quadrangulation of A_i. Perturb s towards p to obtain strictly convex quadrilaterals. We know from Lemma 8(i) that there is always a small perturbation of s that maintains all angles in the resulting quadrangulation at or above $18.43°$. See Fig. 21(a).

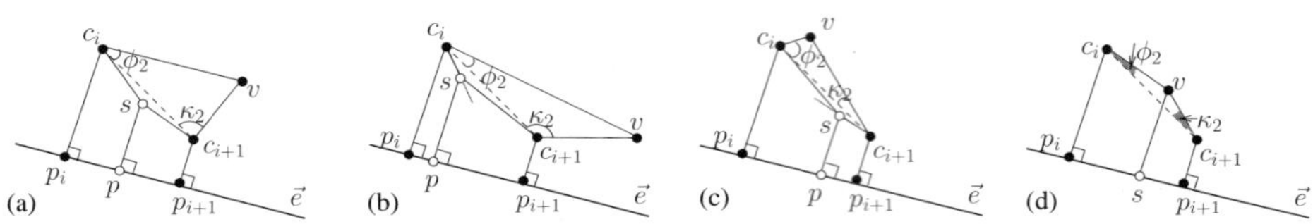

Fig. 21. α_i has two edges. (a) $\min\{\phi_2, \kappa_2\} \geq 18.43°$. (b) $\phi_2 < 18.43°$ and $\kappa_2 \geq 18.43°$. (c) $\phi_2 \geq 18.43°$ and $\kappa_2 < 18.43°$. (d) $\phi_2 < 18.43°$ and $\kappa_2 < 18.43°$.

- If $\min\{\phi_2, \kappa_2\} < 18.43°$, the placement of Steiner points depends on which of ϕ_2 and κ_2 are smaller than $18.43°$ (assume wlog that $\phi_1 + \phi_2 \leq \kappa_1 + \kappa_2$). If exactly one of ϕ_2 or κ_2 is less than $18.43°$, then Lemma 8 guarantees a quadrangulation of A_i with the required angle bounds. We omit details and refer to Fig. 21(b)-(c). If both ϕ_2 and κ_2 are less than $18.43°$, the only possible configuration for α_i is shown in Fig. 20(e). Observe that in this case, v can see $\vec{e}$. Let s be the perpendicular projection of v onto $\vec{e}$, unless $0 < \theta < -18.43°$, in which case let s be the vertical projection of v onto $\vec{e}$. Connect v to s to obtain a quadrangulation of A_i in which all angles satisfy the lower bound .

Case 3: α_i has three edges. The method used to quadrangulate A_i depends on the number of reflex internal vertices of α_i, which is zero or one (note that since α_i lies above $\overrightarrow{c_i c_{i+1}}$, it is not possible for both internal angles to be reflex):

- If the two internal angles along α_i are both convex, draw an edge between c_i and c_{i+1}, which quadrangulates A_i with two quadrilaterals. Some examples of such α_i can be seen in Fig. 10(iii) and 19(ii). In all such cases, Lemma 7 guarantees that all angles in the quad below $\overrightarrow{c_i c_{i+1}}$ is at least $18.43°$. The quad above $\overrightarrow{c_i c_{i+1}}$ has a minimum angle of $26.57°$. See Fig. 22(a).
- If one of the internal angles along α_i is reflex, c_i and c_{i+1} must be corner neighbors. Let r be the reflex vertex. r either lies on the segment $\overline{c_i c_{i+1}}$ (Fig. 22(b)), or belongs to the quadtree cell $N(c_i)$ adjacent to c_i and c_{i+1} and lying above $\overrightarrow{c_i c_{i+1}}$ (Fig. 22(c)). Several examples of the former appear in Figs. 12-17. For the latter, see Fig. 18(v)-(vi) and Fig. 19(iv). We can show that in both cases,

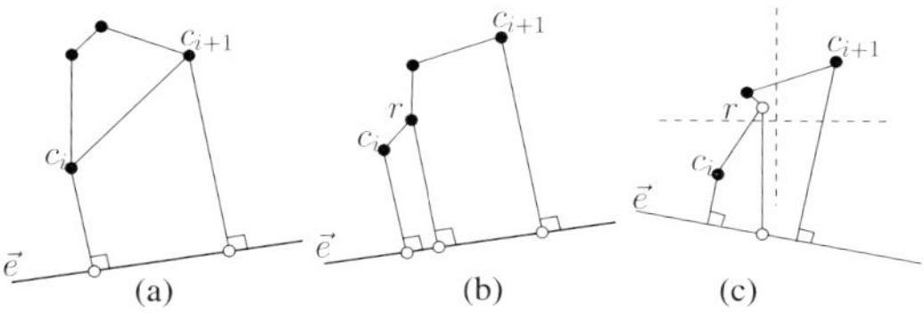

Fig. 22. α_i has three edges

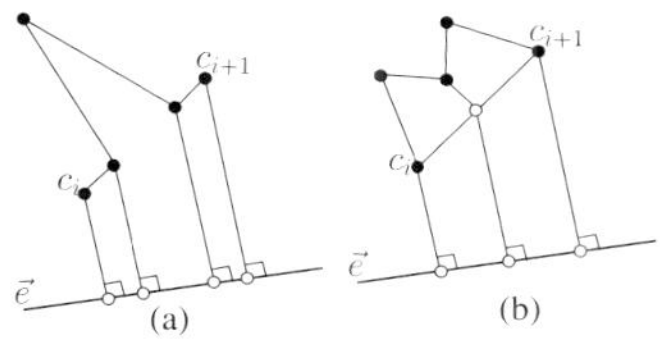

Fig. 23. α_i has four edges

A_i can be decomposed into a quadrilateral and a pentagon that is further decomposed into three quads with the required minimum angle bounds. Details are omitted here.

Case 4: α_i has four edges. α_i is classified according to the three internal vertices:

- If the three internal vertices consist of two reflex vertices separated by a convex vertex (e.g., Fig. 18(i)), the reflex vertices always lie on $\overline{c_i c_{i+1}}$. Insert edges from each reflex vertex to its perpendicular projection onto $\vec{e}$. This decomposes A_i into two quads and a pentagon. Lemmas 7 and 8 provide the required minimum angle bounds. See Fig. 23(a).
- If the three internal vertices consist of two convex vertices separated by a reflex vertex (Fig. 18(ii)), decompose A_i into four quads as shown in Fig. 23(b). Note that this decomposition has the required minimum angle bounds regardless of the value of θ.

This completes the proof that A_i can be quadrangulated with at most five quadrilaterals with a minimum angle of $18.43°$. □

Edge separation conditions for quadtree

Every edge $\vec{e}$ of the polygon P defines a chain of edges given by $\cup_{1 \le i \le m} \alpha_i$. From this chain, we obtain the polygons A_i, each of which is then quadrangulated as described above. In order to conduct this process independently for every edge of the polygon, we impose an *edge separation condition* on $\mathcal{QT}$. The edge separation condition requires that all quadrangulation chains $\cup_{1 \le i \le m} \alpha_i$ defined by the edges of the polygon be disjoint from each other. Recall that these chains do not start in the cell containing the segment endpoint, but rather in one adjacent to it. This allows quadrangulation chains to be separated completely, except in the 5×5 grid of cells around each polygon vertex. In the worst case, the edge separation condition requires that every cell intersected by a polygon edge be surrounded by a 3×3 grid of empty cells, but in practice, this requirement does not apply uniformly across the entire segment.

Connecting quadtree chains around polygon vertices

For every edge of the polygon P, the quadtree chain starts and ends at a cell center within the 3×3 grid of quadtree cells that is guaranteed to exist around each of its endpoints. Let v be a vertex of P and let e and f be the two oriented edges incident on v (the interior of P lies to their left). Let u be the last quadtree chain vertex for edge e and let w be the first quadtree chain vertex for edge f. Note that u and w are both cell centers in the 3×3 grid around v. Let $\bar{u}$ and $\bar{w}$ be the perpendicular projections of u and w onto e and f, respectively. Let E be a sequence of edges connecting u to w in the 3×3 grid. The region around vertex v is meshed by quadrangulating the polygon P_v defined by the edges $v\bar{u}, \bar{u}u, E, w\bar{w}, \bar{w}v$. The method used to quadrangulate P_v depends on the number of edges in E, which is between one and seven (inclusive). Refer to Fig. 24 for an illustration of some cases. The underlying 3×3 grid is used to prove the following lemma, stated here without proof.

Lemma 10. *P_v can be decomposed into at most seven quadrilaterals with a minimum angle of* $18.43°$.

Figure 25 shows quadrangulation chains $\cup_{1 \le i \le m} \alpha_i$ for some edges of a polygon (the entire polygon is shown in Fig. 28). Quadtree chain vertices are highlighted.

Summary of algorithm. We summarize in Figure 26 the algorithm to quadrangulate the interior of a non-acute simple polygon P of n edges $e_1, e_2, \ldots, e_n$ and vertices $v_0, v_1, \ldots, v_{n-1}$, where $e_i = (v_{i-1}, v_i)$ (where $v_n = v_0$). The resulting quadrilaterals have a minimum angle bound of $18.43°$.

Theorem 2. *Given a quadtree decomposition with N quadtree cells satisfying the edge separation condition for a simple polygon P, Quadrangulate(P, n) constructs a mesh for P with at most $5N$ quadrilaterals in which every angle is at least* $18.43°$.

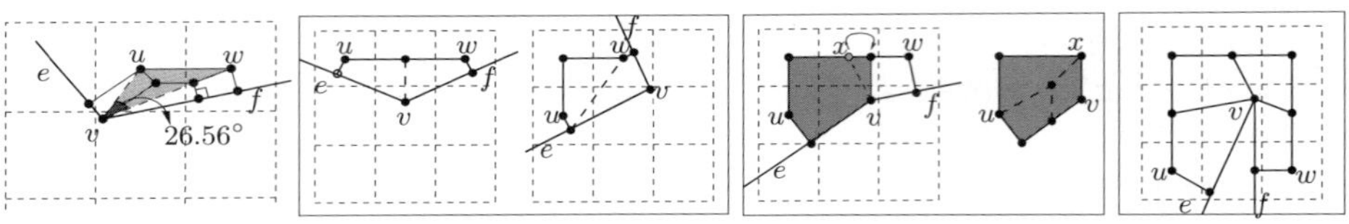

(a) (b) (c) (d)

Fig. 24. Connecting at corners. Number of edges in E is (a) one, (b) two, (c)-(d) three or more.

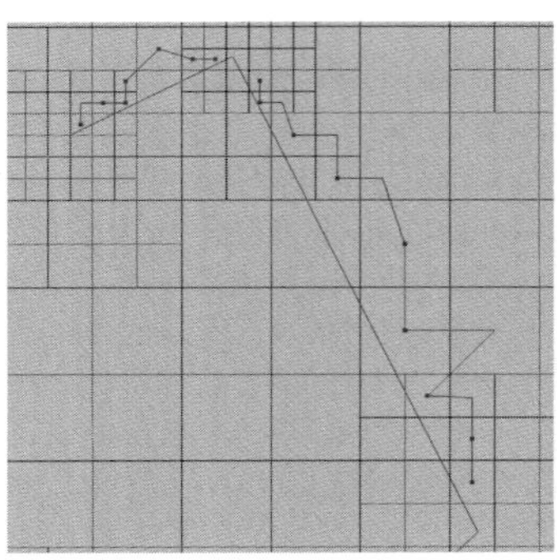

Fig. 25. Quadrangulation chains

Quadrangulate(P, n):
  $\mathcal{QT} \leftarrow$ Quadtree decomposition satisfying edge separation conditions for vertices of P
  $\mathcal{Q} \leftarrow$ Quadrangulation resulting from *applyTemplate* on $\mathcal{QT}$.
  for $e_i \in \{e_1, e_2, \ldots, e_n\}$
    $\alpha(e_i) \leftarrow$ Quadrangulation chain for e_i
    $\mathcal{Q}(e_i) \leftarrow$ Quadrangulation of region bounded by e_i and $\alpha(e_i)$, as given by Lemma 9.
    $\mathcal{Q}(v_i) \leftarrow$ Quadrangulation of corner polygon P_{v_i}, as given by Lemma 10.
  $\mathcal{Q}' \leftarrow \bigcup_{1 \le i \le n} \mathcal{Q}(e_i) \cup \bigcup_{1 \le i \le n} \mathcal{Q}(v_i)$

  return $\mathcal{Q}' \cup (\mathcal{Q} \cap (P - \mathcal{Q}'))$

Fig. 26. Summary of algorithm

3.2 General Simple Polygons

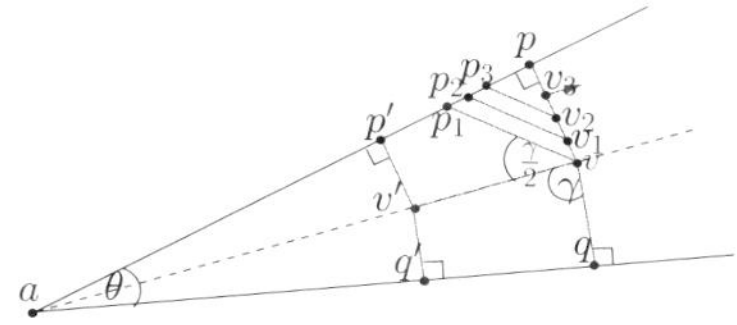

Fig. 27. Handling acute angles in P

Let P be a general simple polygon containing acute angles. We first convert P into a polygon that contains only obtuse angles by "cutting off" the acute angle vertices. Let a be an acute angle vertex of P. Let $\theta, 0 \leq \theta < 90$, be the angle at that vertex. Let v be a point on the angle bisector of a, and let p and q be the perpendicular projections of v onto the two edges incident at a. v is chosen so that the quadrangular region $apvq$ does not contain any other vertices of the polygon P. Cut all such regions $apvq$ from P. Let B be the polygon resulting from this procedure. Construct a quadrilateral mesh for B using the algorithm in Section 3.1. Observe that now there might be Steiner points on pv (e.g., v_1, v_2, v_3 in Fig. 27) and vq. These are used to quadrangulate the region $apvq$ with the required angle bounds. We omit details and refer to Fig. 27 for an illustration of the quadrangulation.

4 Conclusion

Sample meshes generated by our algorithm are shown in Figures 28 and 29. Observe that in these examples the ratio of the number of quadrilaterals to the number of quadtree cells is less than one. The image on the right shows a zoomed-in portion of the mesh on the left.

This paper presents the first known result on the generation of a quadrilateral mesh for the interior of a simple polygon (possibly with holes) in which every new angle in

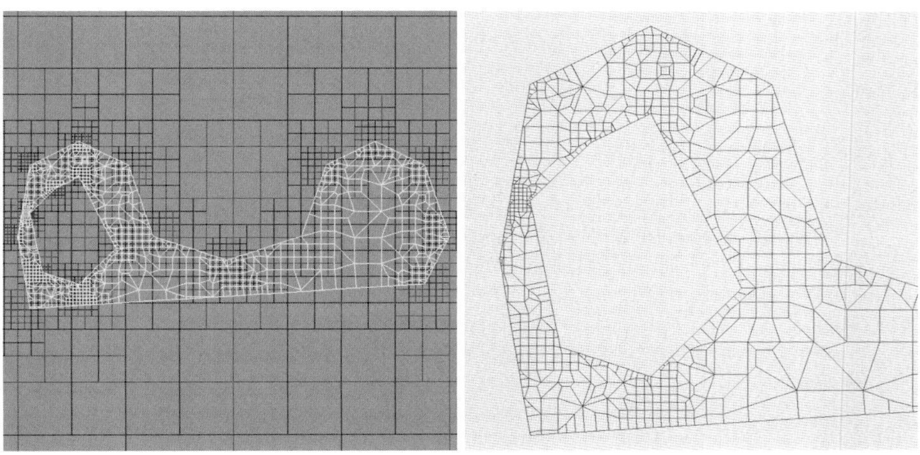

Fig. 28. Number of polygon edges: 19. Minimum mesh angle: $24.26°$. Number of quadtree cells: 1399. Number of mesh vertices: 989. Number of mesh faces (quads): 841.

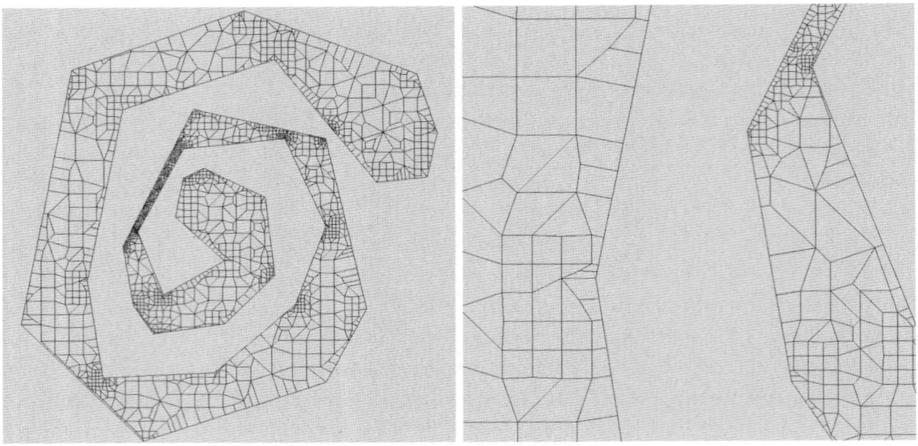

Fig. 29. Number of polygon edges: 33. Minimum mesh angle: 20.67°. Number of quadtree cells: 2623. Number of mesh vertices: 2213. Number of mesh faces (quads): 1859.

the mesh is bounded from below. The main open question resulting from this work is its extension to polygon interior as well as exterior. While our algorithm itself is applicable to the interior or the exterior of the polygon, the difficulty of adapting it to both lies in resolving mesh compatibility at the boundary without propagating the changes throughout the mesh. We are currently investigating alternative strategies to mesh the region bounded by quadtree chains on both sides of each polygon edge.

Acknowledgments. The authors would like to thank anonymous reviewers for helpful comments that served to improve the presentation in the paper.

References

1. Allman, D.J.: A quadrilateral finite element including vertex rotations for plane elasticity analysis. Int. J. Numer. Meth. in Engineering 26, 717–730 (1988)
2. Benzley, S., Perry, E., Merkley, K., Clark, B., Sjaardema, G.: A Comparison of All Hexahedral and All Tetrahedral Finite Element Meshes for Elastic and Elastic-Plastic Analysis. In: Proc. of the 4th International Meshing Roundtable, pp. 179–192 (1995)
3. Bern, M., Eppstein, D.: Quadrilateral meshing by circle packing. In: 6th IMR, pp. 7–19 (1997)
4. Bern, M., Eppstein, D., Gilbert, J.: Provably good mesh generation. J. Comp. Sys. Sci. 48, 384–409 (1994)
5. Blacker, T., Stephenson, M.: Paving: A new approach to automated quadrilateral mesh generation. Int. J. Numer. Meth. in Engineering 32(4), 811–847 (1991)
6. Brauer, J.R. (ed.): What every engineer should know about finite element analysis, 2nd edn. Marcel-Dekker, New York (1993)
7. Cheng, S.-W., Dey, T., Ramos, E., Ray, T.: Quality meshing for polyhedra with small angles. In: Proc. 20th Annual Symposium on Computational Geometry (2004)

8. Chew, L.P.: Guaranteed-quality mesh generation for curved surfaces. In: Proc. of the 9th ACM Symposium on Computational Geometry, pp. 274–280 (1993)
9. Chew, L.P.: Guaranteed-quality delaunay meshing in 3d. In: Proc. of the 13th ACM Symposium on Computational Geometry, pp. 391–393 (1997)
10. Johnston, B.P., Sullivan, J.M., Kwasnik, A.: Automatic conversion of triangular finite meshes to quadrilateral elements. Int. J. Numer. Meth. in Engineering 31(1), 67–84 (1991)
11. Miller, G.L., Talmor, D., Teng, S.-H., Walkington, N.: A delaunay based numerical method for three dimensions: Generation, formulation, and partition. In: Proc. of the 27th ACM Symposium on the Theory of Computing, pp. 683–692 (1995)
12. Mitchell, S., Vavasis, S.: Quality Mesh generation in Three Dimensions. In: Proceedings of the 8th Annual Symposium on Computational Geometry, pp. 212–221 (1992)
13. Ramaswami, S., Siqueira, M., Sundaram, T., Gallier, J., Gee, J.: Constrained quadrilateral meshes of bounded size. Int. J. Comp. Geom. and Appl. 15(1), 55–98 (2005)
14. Shewchuk, J.: Constrained Delaunay Tetrahedralizations and Provably Good Boundary Recovery. In: Proc. of the 11th International Meshing Roundtable, pp. 193–204 (2002)
15. Shewchuk, J.R.: Triangle: Engineering a 2D quality mesh generator and Delaunay triangulator. In: Lin, M.C., Manocha, D. (eds.) FCRC-WS 1996 and WACG 1996. LNCS, vol. 1148. Springer, Heidelberg (1996)
16. Zienkiewicz, O.C., Taylor, R.L.: The finite element method. McGraw-Hill, New York (1989)

Automated Quadrilateral Coarsening by Ring Collapse

Mark W. Dewey[1], Steven E. Benzley[2], Jason F. Shepherd[3], and Matthew L. Staten[4,5]

[1] Elemental Technologies Inc.
mildewey@elemtech.com
[2] Brigham Young University
seb@byu.edu
[3] Sandia National Labs*
jfsheph@sandia.gov
[4] Sandia National Labs*
mlstate@sandia.gov
[5] Carnegie Mellon University
mstaten@cmu.edu

Summary. A uniform finite element mesh rarely provides the best discretization of a domain to accommodate a solution with both optimal efficiency and minimal error. Mesh adaptation can approach a more optimal solution by accommodating regions of the mesh with higher or lower element density. Extensive attention has been given to mesh adaptation in both computational mechanics and computer graphics to provide or improve methods for increasing the model resolution or solution accuracy. The algorithm developed in this paper, entitled Automated Quadrilateral Coarsening by Ring Collapse (AQCRC), provides a unique solution to allow mesh coarsening of both structured and unstructured quadrilateral meshes. The algorithm is based on modification and removal operations utilizing the dual description of the quadrilateral mesh. The AQCRC algorithm iterates on five steps: 1) input of a coarsening region and a coarsening factor, 2) selection of coarsening rings, 3) mesh quality improvement, 4) removal of coarsening rings, and 5) mesh clean-up. Examples are presented showing the application of the algorithm.

1 Introduction

Finite element analysis (FEA) continues to push the limits of computing power in terms of the size of models being analyzed. The computation time required even by the most powerful machines can be hours or days on complex problems, with computational time increasing proportional to the cube of the number of nodes in the finite element mesh [1]. However, the accuracy of the finite element analysis is also proportional to the number of nodes in the mesh, increasing with finer resolution meshes. In many analysis situations, there may be specific areas in the mesh where accuracy, and thus mesh resolution, is more important than in other locations where lower resolution may be tolerated. For example, if a solution is shown to have high gradient in a particular location, the errors produced by a low density mesh may be significant and

* Sandia is a multiprogram laboratory operated by Sandia Corporation, a Lockheed Martin Company, for the United States Department of Energy under Contract DE-AC04-94AL85000.

prevent proper engineering conclusions from the analysis process. Therefore, adapting the mesh to allow higher nodal density in the areas of interest can be critical to accurate analysis, while lower density in low gradient areas may allow for reductions in the solution time. If a uniform mesh is used, analysis time may be much greater than necessary while overall accuracy in the analysis may not improve significantly. The competing objectives of accuracy and computation time have prompted investigation into the field of mesh adaptation for the purpose of optimizing meshes.

Most research in mesh adaptation for computational mechanics centers on refinement algorithms that increase element density locally [2]. A complementary algorithm for decreasing local element density by removing elements (i.e., coarsening) could be a powerful companion tool to refinement algorithms, potentially allowing more flexible mesh adaptation. Unfortunately, coarsening is an area of research which has received limited attention. The algorithm presented in this paper, Automated Quadrilateral Coarsening by Ring Collapse (AQCRC), provides a fully automated conformal coarsening algorithm suitable for use with generalized unstructured quadrilateral finite element meshes.

Tools for mesh manipulation by both refinement and coarsening operations increase the ability to adapt a mesh. For example given a uniform mesh, the mesh density in an area of interest may be increased by established refinement techniques [3] and decreased away from the areas of interest using the coarsening technique described in this paper. An initial analysis on a base mesh may be used to indicate locations where high density meshes and low density meshes are appropriate based on gradients of the initial solution. Rather than remeshing the model, the base mesh may be modified using refinement and coarsening tools. This would allow increased resolution and accuracy in the results while maintaining a similar computation time for the entire model. Furthermore, a given model may require adaptation in different locations depending on different load cases, adaptation by both refinement and coarsening from a single base mesh may allow more efficient and robust generation of meshes appropriate for varied circumstances.

This paper describes the development and implementation of the AQCRC algorithm for use in quadrilateral finite element analysis. The background of the problem and discusses the limitations of previously developed coarsening and simplification algorithms is presented first, followed by a description of the implementation of the algorithm. A brief case study is then presented to show the merit of the algorithm and, finally, conclusions are drawn with recommendations of further research.

2 Background

Mesh adaptation is a field which has received extensive study among both computational mechanics and computer graphics researchers. Generally these two fields have not collaborated due to the many additional restrictions that apply to computational mechanics but are not necessary in computer graphics. The adaptation algorithms developed for computer graphics are, therefore, rarely suited to computational mechanics. One example of these additional restrictions in computational mechanics involves the requirement that a mesh must accurately represent the geometry of the model by insuring that the nodes representing a curve or surface of the model do not move off the geometry, whereas in graphics a sufficiently low level of detail might justify combining surfaces and/or curves. In the next section, we provide some background on efforts in mesh adaptation and give some additional motivation for our work.

2.1 Triangle Mesh Simplification and Coarsening

The use of triangular meshes in computer graphics and computational mechanics is common due to the relative simplicity of generating the meshes from these simplex elements. Triangle meshing algorithms are well-established and on-going efforts in the research community continue to advance the quality of these meshes.

One of the foremost algorithms of triangle mesh simplification was developed by Garland, et al. [4]. His approach is fast, reliable, and is also generally applicable to any polygon mesh. The algorithm assumes that the mesh is composed entirely of triangles or can be broken into a mesh entirely composed of triangles. It is designed to combine surfaces and curves that are indistinguishable when rendered at a low level of detail. Hoppe, et al. [5], demonstrate mesh adaptation respecting geometric curves and surfaces in order to preserve sharp corners and edges in the mesh representation.

A survey of triangle mesh coarsening algorithms is documented by Cignoni, et al. [6]. Cignoni et al. identify the following major simplification methodologies: coplanar facets merging, controlled vertex/edge/face decimation, retiling, energy function optimization, vertex clustering, wavelet based approaches, and simplification via intermediate hierarchical representation. Additional reviews that compare smaller sets of algorithms are also found in [7, 8].

While triangle meshes have widespread use, quadrilateral meshes are sometimes preferred in computational analysis due to beneficial mathematical properties of the quadrilateral element that result in reduced solution error with fewer elements than triangle meshes [1]. Unfortunately, despite the wide availability of triangle mesh adaptation algorithms, most of the algorithms developed for triangle meshes cannot be adapted for use on quadrilateral meshes.

2.2 Quadrilateral Mesh Simplification and Coarsening

Various efforts have been made to develop all-quadrilateral coarsening algorithms; unfortunately, all of these algorithms have significant restrictions which prevent use with unstructured meshes. Takeuchi, et al. [9], modified the approach developed by Garland, et al. [4], to simplify quadrilateral meshes; however, the process is designed for full-model simplification and may produce degenerate elements (i.e. quadrilaterals which are inverted or concave). Cheng, et al., developed a method of coarsening a structured, all-quadrilateral mesh specifically for use on auto-body parts [10]; however, this method has not been adapted for use in unstructured meshes. Kwak, et al., performs simplification using remeshing algorithms [11]; however, this global approach can be slow when only local adaptation is needed. Choi describes an algorithm which can be used to undo previous refinement on both quadrilateral and hexahedral meshes [12]; however, the reliance on knowledge of previous refinement restricts the algorithm from being used on a base mesh that has not been refined. Nikishkov developed a quadtree method for mesh adaptation that allows both refinement and coarsening [13]; however, his method requires the use of special elements or else it produces non-conforming elements. The left panel of **Fig. 1** shows an initial uniform mesh with the region to be coarsened highlighted in grey. The center panel shows the mesh once Nikishikov's coarsening method has been applied. The nodes marked A, B, C, and D are locations where the mesh is not conforming. We note that conforming quadrilateral meshes, where the elements are simply connected and there are no gaps, are required by most FEA solvers.

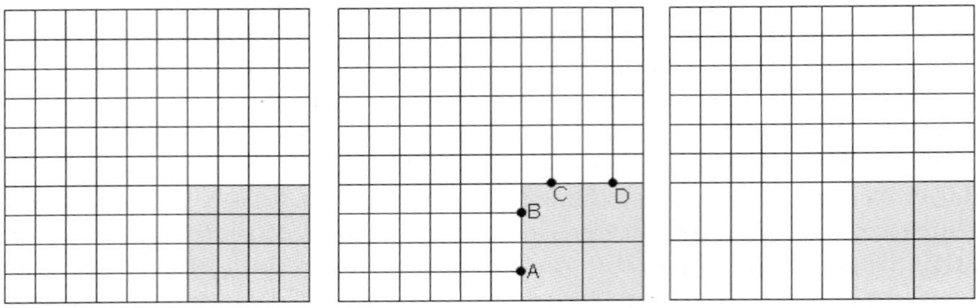

Fig. 1. Quadtree and chord removal coarsening

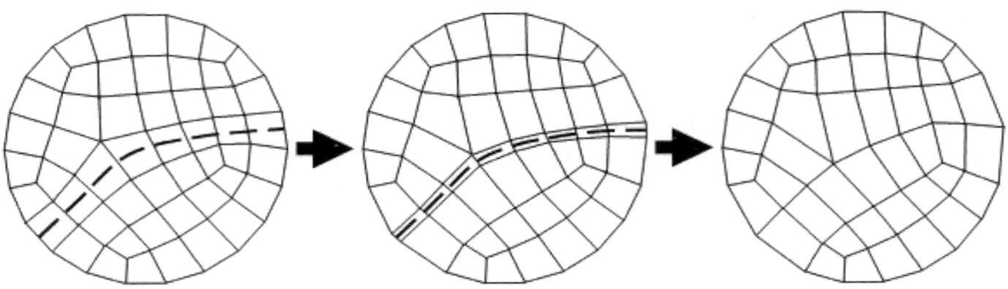

Fig. 2. Removing a chord from a mesh

The basis of the AQCRC algorithm derives from the dual chord representation of a quadrilateral/hexahedral mesh [14,21]. A dual chord is a set of quadrilaterals connected through pairs of opposite edges extending through the mesh or connecting back on the original starting edge. In **Fig. 2** a dashed line is shown highlighting one chord of the mesh.

Borden, et al. [15], recognized that it is possible to remove an entire chord from a quadrilateral mesh, maintaining conformal connectivity, by simply collapsing the defining edges of the chord as shown in **Fig. 2.** The removal of a chord reduces the number of quadrilaterals in the mesh and coarsens the quadrilaterals adjacent to the chord. The right panel of **Fig. 1.** shows this coarsening applied to the selected coarsening region. Unfortunately, while the local region, highlighted in gray, is significantly coarsened, the effect of the coarsening may extend well beyond the boundaries of the coarsening region. The research of Borden, et al., was continued in the paper by Benzley, et al. [16], where initial steps were made to localize the coarsening region.

Staten, et al. [17], speculated that if a circular chord (i.e. a chord which forms a closed loop) could be created in a specific region, then the removal of that chord would produce localized coarsening. Staten, et al., showed that if the portions of chords bounding a region can be established, then simple chord operations (i.e. alterations to the mesh which change the connectivity of chords) can be performed at the intersections of these bounding partial chords to combine them into a single continuous chord surrounding the region to be coarsened. The operations used to combine the partial chords into a single circular chord included the edge swap, face close and doublet insertion operations that are described in the following paragraph. The faces within the region to be coarsened in these partial chords form a ring of quadrilaterals. A ring of quadrilaterals within the coarsening region will be known as a 'coarsening ring' throughout the remainder of this paper.

Fig. 3 shows an example of the coarsening process developed by Staten, et al. [17]. The left panel shows the initial mesh. The dashed lines mark chords which bound the region to be coarsened. The quadrilaterals highlighted in grey are the bounding partial chords and define a coarsening ring. In the second panel, the quadrilaterals where two bounding partial chords intersect have been modified with doublet insertion, face close, and edges swap operations. The top left intersection is modified using a doublet insertion operation. The doublet insertion operation causes the quadrilateral at the intersection of two bounding chords to be divided into two degenerate quadrilaterals by inserting two edges and a node between opposite nodes on a single quadrilateral as shown in detail in the bottom left corner of **Fig. 4**. In both figures, the doublet node is circled for clarity. A doublet node is an internal node in a quadrilateral mesh connected to only two edges. The bottom two chord intersections in **Fig. 3** are modified by a face close operation. The face close operation results in the quadrilateral at the intersection of the bounding chords to be deleted by merging two nodes opposite each other as shown in detail in the top right corner of **Fig. 4**. The top right chord intersection in **Fig. 3** is modified by an edge swap operation. The edge swap operation results in an edge between the quadrilateral at the intersection of the bounding chords and one of the adjacent quadrilaterals within one of the bounding chords to change node connectivity as shown in detail in the bottom right corner of **Fig. 4**. The combination of these operations result in the mesh shown in **Fig. 3**, which contains a single circular chord that now bounds the coarsening region as shown in the center panel of the same figure. This chord is removed by an extraction operation (see **Fig. 2.**), resulting in the mesh shown in the right panel of **Fig. 3**.

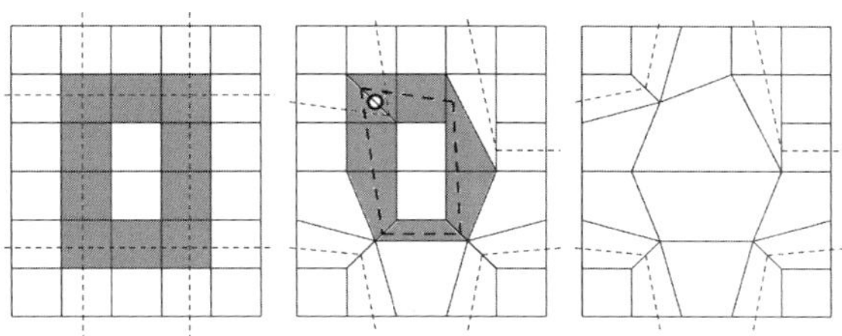

Fig. 3. Chord operations and removal

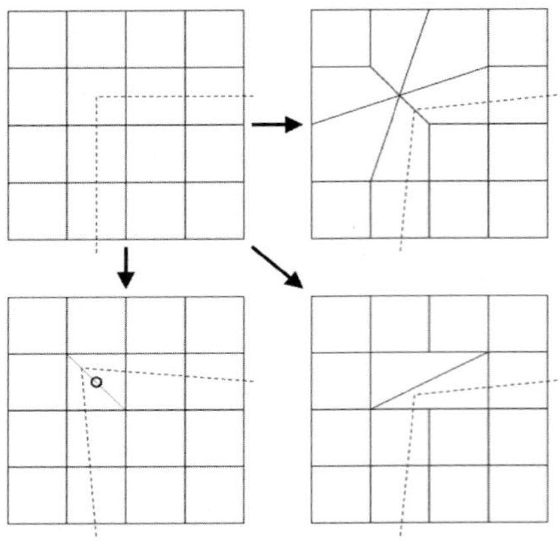

Fig. 4. Chord operations: top right, Face Close; bottom left, Doublet Insertion; bottom right, Edge Swap

3 Automated Quadrilateral Coarsening by Ring Collapse

The algorithm developed by Staten, et al., is a coarsening algorithm which can locally coarsen an unstructured, conforming, all-quadrilateral mesh [17]. The AQCRC algorithm, presented in this Section, generalizes, further develops and automates the work of Staten, et al. The AQCRC algorithm advances the basic methodology by extracting multiple coarsening rings simultaneously, introducing logic to maximize element quality, and eliminating the step of combining the bounding partial chords into a single circular chord.

One of the key developments of the AQCRC algorithm is the use of coarsening rings, i.e. a closed ring of quadrilaterals contained within the coarsening region, rather than circular chords. The AQCRC algorithm does not take the step of modifying the bounding partial chords with chord operations to create a single chord (see **Fig. 3.**). Instead the coarsening ring is removed directly.

The AQCRC algorithm iterates over five steps until sufficient coarsening has been accomplished. These five steps are outlined below and will be examined in detail in the following sections.

1. A contiguous coarsening region and the final mesh coarseness are specified.
2. One or more coarsening rings are selected within the coarsening region containing a number of quads less than or equal to a goal number of quads.
3. The bounding partial chord intersections may be altered with chord operations to increase the final quality of the mesh or to prevent merging nodes illegally.
4. The identified coarsening rings are collapsed from the mesh and the mesh reconnected in a manner that retains its conformal properties.
5. The mesh is checked, cleaned up, and smoothed to ensure that elements have acceptable quality.

3.1 Defining the Coarsening Region and Removal Parameter

The first step in the AQCRC algorithm is to specify a coarsening region. The left panel of **Fig. 5** shows an example of a coarsening region highlighted in dark gray. An additional layer of elements marked with light gray surrounds the coarsening region and is used as a transition layer. Elements within the coarsening region may be moved and/or deleted during the coarsening process; the elements in the transition region may be moved slightly, but not deleted. If the coarsening region includes a boundary of the initial mesh (i.e. a curve which only bounds one meshed surface), the elements along that boundary are also considered transition elements.

In addition to specifying the coarsening region, the number of quadrilaterals to be removed must be defined. A target size or coarsening factor may be given to establish the goal number of quadrilaterals which are to be removed from the mesh. A coarsening factor corresponds to the multiplicative increase in average area that should occur within the coarsening region during the execution of the AQCRC algorithm. A target size corresponds to the average length of the edges in the coarsening region after the execution of the AQCRC algorithm.

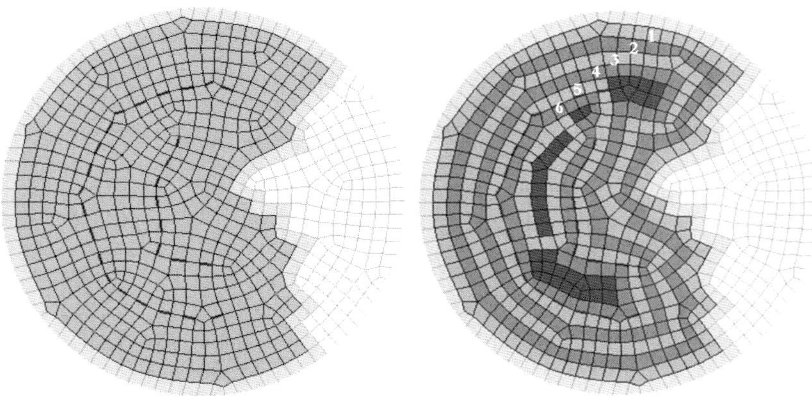

Fig. 5. Coarsening region selection and coarsening rings

Equation 1 shows how a coarsening factor is used to determine the number of quadrilaterals that should be removed.

$$N_{e-r} = E_t - \frac{E_t}{F} \tag{1}$$

where N_{e-r} = the number of elements to be removed
E_t = the number of elements in the coarsening region
F = coarsening factor

Equation 2 converts the target size into an equivalent coarsening factor which is then converted into the number of elements to be removed by Equation 1.

$$F = \frac{l_f^2}{l_0^2} \qquad (2)$$

where F = coarsening factor
l_f = the final average edge length specified by the user
l_0 = the initial average edge length in the coarsening region

Once the number of quadrilaterals to be removed is determined, a 10% tolerance factor (t_f) is also calculated such that the number of quadrilaterals actually removed is within ± t_f of the calculated goal. This tolerance factor is limited to being a minimum of 3 quadrilaterals or a maximum of 50 quadrilaterals. The values of 10%, 3 and 50 are heuristic values shown to give reasonable results during our experimentation.

3.2 Selecting Coarsening Rings

Once a coarsening region has been defined, a set of concentric coarsening rings is developed for removal. Each of the coarsening rings is analyzed to determine which coarsening ring removals will preserve a high quality mesh and the goal number of elements to be removed.

Ring Identification
A coarsening ring is identified by locating an element in the coarsening region which is adjacent to or shares a node with an element marked as a boundary element. Because the coarsening region is always bounded, the set of elements adjacent to the boundary forms a coarsening ring. The elements in the new coarsening ring are set aside forming a new, but smaller, coarsening region. This process is iterated, until all coarsening rings in the coarsening region are identified. During the process of identifying coarsening rings there are a few cases where the ring identified could not be collapsed or there are not enough elements to form a closed loop. These invalid cases are handled by marking some of the elements as boundary elements. Eventually all of the elements in the coarsening region will be marked as boundary elements, which effectively ends the ring identification portion of the algorithm. In most cases, the coarsening region is large enough that several rings are created concentrically. The right panel of **Fig. 5** shows the rings developed within the coarsening region. The alternating numbered regions of darker and lighter shaded grey elements show the set of rings. The dark regions not numbered are locations of elements which are not included as rings because they were part of an invalid ring case. Further details on handling invalid cases can be found in [18].

Ring Selection
Once the set of coarsening rings has been created, a subset is chosen for removal. To facilitate choosing an optimum set of rings for removal, each node connected to a ring is assigned to a node group. A node group is the set of nodes that will be merged into a single node when the coarsening ring is collapsed. The node groups are used to facilitate choosing the best set of rings to collapse by providing a rough estimate of the mesh connectivity and element shapes that will be created upon collapse. Each node group will be assigned a projected location. Generally, the projected location is the

centroid of all the nodes in the group. This location may be modified so that the mesh continues to adhere to the geometry of the model.

The projected location of the node groups is calculated to enable quality metric calculations for each of the coarsening rings. The rings are then ordered based on prospective quality. Coarsening rings are removed based on the ordered list until the coarsening goal is satisfied (within the tolerance factor).

3.3 Improving Quality

In some cases, modifying the connectivity of a ring can improve the quality of the final mesh. As each ring is chosen for collapse, it is examined more closely to see if there is room for quality improvements. Quality is a major consideration in coarsening because ring removal generally reduces mesh quality. Quality improvement operations typically prevent the creation of high-valence nodes, (e.g., nodes with more than 5 edges), as well as to prevent merging nodes from different curves (see [17, 18] for details).

3.4 Collapsing Coarsening Rings

Once the coarsening rings have been selected and the various quality improvement operations have been applied, the mesh is ready to be coarsened. Each of the selected coarsening rings is collapsed in succession using the following procedure. The nodes in each node group are moved to the projected location and the quadrilaterals that are part of the ring are deleted. As the quadrilaterals are deleted, any edge that is no longer associated with a quadrilateral (i.e. the quadrilaterals on either side of it have been removed) is deleted and the nodes on either end of the deleted edge are merged together. At corners, a simple collapse as shown in **Fig. 6** creates a conformal mesh (compare to the operation shown in **Fig. 3.**). In the left and center panels of **Fig. 6**, the node groups are circled with darkened edges connecting each group. The dashed line indicates the coarsening ring of quadrilaterals being collapsed. In the right panel the circled nodes are the locations of the merged nodes in the final mesh.

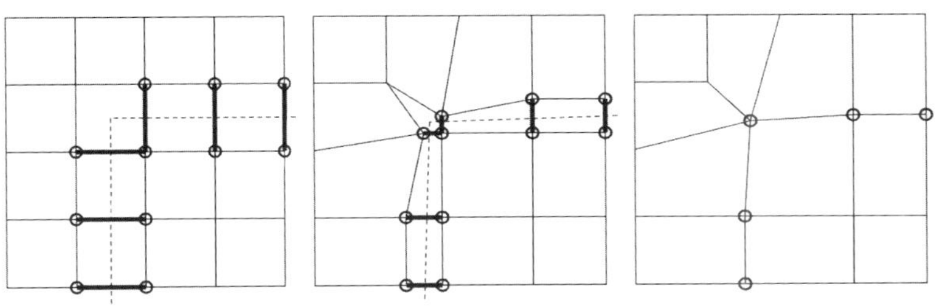

Fig. 6. Collapsing node groups

3.5 Mesh Clean-Up

Despite efforts to minimize high-valence nodes during quality improvement, a few cases remain where high-valence nodes are formed. Furthermore, the collapse of two rings separated by a single layer of elements may reduce element quality due to multiple node projections for quadrilaterals in the non-removed coarsening ring. Additionally, collapsing quadrilaterals along geometric curves may result in elements with low quality which cannot be smoothed. To improve mesh quality, clean-up operations like those described in [18, 20], which change mesh connectivity to create a more structured mesh, and smoothing operations, like those described in [23], which improve the mesh quality by simply moving node coordinates, are applied.

3.6 Coarsening Iterations

At the end of each coarsening cycle the net number of quadrilaterals removed is determined. This net number includes any quadrilaterals added or removed by the clean-up procedures. If too few quadrilaterals have been removed, the algorithm is executed again. If a sufficient number of quadrilaterals have been removed, or if no quadrilaterals have been removed, the new, coarsened mesh is integrated into the original mesh. The algorithm provides a message if insufficient coarsening has taken place to reach the goal number of elements to be removed. **Fig. 7** shows the example given in **Fig. 5** after it has gone through several coarsening iterations. The region exterior to the coarsening region is identical in both meshes.

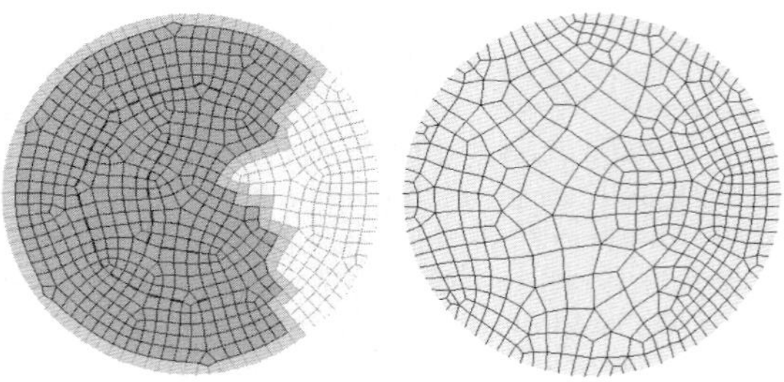

Fig. 7. Original mesh compared to coarsened region

4 Results and Example

One example of mesh coarsening using the AQCRC algorithm is given in this section. The resulting mesh and quality of the initial mesh and of a coarsened mesh will be compared. The quality metric used is the scaled Jacobian metric [19] which ranges from-1.0 to 1.0, where a value of 1.0 represents a perfect square while anything below 0.0 is an inverted (non-convex) element (0.0 typically being a triangle-shaped element). A scaled Jacobian value greater than 0.2 is generally considered acceptable for analysis accuracy. A scaled Jacobian value below 0.2 are considered marginal [21].

4.1 Lever Mesh Example

Fig. 8. shows a lever model that has been meshed with shell elements. An initial mesh and a mesh that has had most of the inside area coarsened to a factor of four are shown. Previous analysis showed that high stress gradients occurred around the outside boundaries. In this example the mesh has had an analysis run on it using Calculix, an open source finite element analysis package. Both analyses had nearly identical results; the error introduced by the coarsening is less than 0.2% while the calculation speed increased 347%. Table 1 shows the element count of the original and coarsened meshes, the minimum scaled jacobian of the elements in the mesh, the analysis time and peak Von Mises Stress of the model. The accuracy of the analysis has not been compromised, but the speed of analysis has increased by more than three.

Table 1. Lever mesh results

Mesh	Element Count	Min. Scaled Jacobian	Analysis Time	Peak Von Mises Stress
Initial	11113	0.71	3m 11s	5.34 E+4 psi
Coarsened (factor 4)	3261	0.39	55s	5.33 E+4 psi

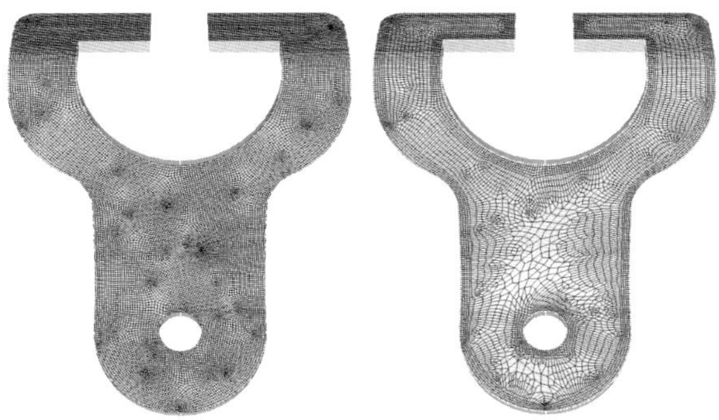

Fig. 8. Lever shell mesh

5 Conclusions and Recommendations

The level of coarsening to be achieved is limited primarily by element quality considerations. Generally speaking, the more the mesh is coarsened, the greater the reduction in mesh quality. This can largely be attributed to the quality of the transition elements between the fine and coarse regions of the mesh. At least one layer of elements must take on a trapezoidal shape to enable the sizing transition. As the aspect ratio increases, the quality of the mesh is reduced. This should not deter from the use of the coarsened mesh so long as the quality remains within acceptable ranges.

The availability of a fast, robust coarsening algorithm for unstructured, all-quadrilateral, conformal meshes expands the tools available for computational modeling. The use of coarsening coupled with refinement provides and effective means to adapt a mesh more effectively, increasing model accuracy while reducing computation times.

This paper presented the Automated Quadrilateral Coarsening by Ring Collapse (AQCRC) algorithm. The algorithm conformally coarsens localized regions of quadrilateral meshes by creating, modifying and removing coarsening rings. The automation procedure optimizes final element quality while attempting to heavily coarsen the defined region. Application of the AQCRC algorithm demonstrates that it is capable of removing enough nodes from the mesh to increase speeds by many times while maintaining a mesh quality sufficient for accurate analysis. The potential speed benefits of coarsening algorithms could significantly reduce the time and cost of computational analysis of large scale problems.

5.1 Further Research and Development

Further research and development of this algorithm continues. Clean-up operations are already a focus of significant effort to address problems coarsening around complex surfaces. An alternative method of coarsening, partial chord removal, is also being explored.

One of the methods of coarsening discussed in the Section 2 was undoing refinement. The AQCRC algorithm could be developed into a reversible coarsening procedure, allowing refinement of regions previously coarsened. This could be accomplished by storing the changes made to the mesh in the remaining mesh entities. As some nodes or edges may be merged and then merged again, a set of changes could be pushed onto a stack so that the reverse operations could be performed one at a time, restoring the mesh to its original state.

References

1. Bathe, K.: Finite Element Procedures. Prentice-Hall, Upper Saddle River (1996)
2. Schneiders, R.: Octree-based hexahedral mesh generation. Int. Journal of Comp. Geom. & Applications 10(4), 383–398 (2000)
3. Parrish, M.: A Selective Approach to Comformal Refinement of Unstructured Hexahedral Meshes, MS Thesis, Brigham Young Univ., Utah (2007)
4. Garland, M., Heckbert, P.: Surface Simplification Using Quadric Error Metrics, Carnegie Melon University (1997), `http://graphics.cs.uiuc.edu/%7Egarland/papers/quadrics.pdf`
5. Hoppe, H., DeRose, T., Duchamp, T., McDonald, J., Stuetzle, W.: Mesh Optimization, University of Washington (1993), `http://citeseer.ist.psu.edu/cache/papers/cs/797/ftp:zSzzSzftp.cs.washington.eduzSztrzSz1993zSz01zSzUW-CSE-93-01-01.pdf/hoppe93mesh.pdf`
6. Cignoni, P., Montani, C., Scopigno, R.: A Comparison of Mesh Simplification Algorithms. Computers & Graphics 22(1), 37–54 (1997)
7. Silva, S., Silva, F., Madiera, J., Santos, B.: Evaluation of Mesh Simplification Algorithms Using PolyMeCo: A Case Study. In: Proceedings of SPIE – The International Society for Optical Engineering, Proceedings of SPIE-IS and T Electronic Imaging - Visualization and Data Analysis 2007 vol. 6495, p. 64950D (2007)
8. Silva, S., Madiera, J., Ferreira, C., Santos, B.: Comparison of Methods for the Simplification of Mesh Models Using Quality Indices and an Observer Study. In: Proceedings of SPIE – The International Society for Optical Engineering, Proceedings of SPIE-IS and T Electronic Imaging - Visualization and Data Analysis XII, vol. 6492, p. 64921L (2007)

9. Takeuci, S., Kanai, T., Suzuki, H., Shimada, K., Kimura, F.: Subdivision Surface Fitting Using QEM-based Mesh Simplification and Reconstruction of Approximated B-spline Surfaces. In: Proceedings of The Eighth Pacific Conference on Computer Graphics and Applications, October 3-5, 2000, pp. 202–212 (2000)
10. Cheng, A., Zhong, Z.: Local Coarsening for Quadrilateral Meshes on Autobody Tool Surface. Journal of Computer-Aided Design and Computer Graphics 14(1), 50–55 (2002)
11. Kwak, D., Cheon, J., Im, Y.: Remeshing for metal forming simulating-Part 1: Two-dimensional Quadrilateral Remeshing. International Journal for Numerical Methods in Engineering 53(11), 2463–2500 (2002)
12. Choi, C.: Adaptive Mesh Refinement/Recovery Strategy for FEA. Structural Engineering and Mechanics 7(3-4), 379–391 (2004)
13. Nikishkov, G.: Finite Element Algorithm with Adaptive Qaudtree-Octree Mesh Refinement. Anziam J. 46(E), C15–C18 (2005)
14. Murdoch, P., Staten, S., Blacker, T., Mitchell, S.: The Spatial Twist Continuum: A Connectivity Based Method for Representing All Hexahedral Finite Element Meshes. Finite Element Analysis and Design 28(2), 137–149 (1997)
15. Borden, M., Staten, S., Shepherd, J.: Hexahedral Sheet Extraction. In: 11th International Meshing Roundtable, Sandia National Laboratories, September 2002, pp. 147–152 (2002)
16. Benzley, S., Harris, N., Scott, M., Borden, M., Owen, S.: Conformal Refinement and Coarsening of Unstructured Hexahedral Meshes. Journal of Computing and Information Science Engineering 5, 330–337 (2005)
17. Staten, M., Benzley, S., Scott, M.: A Methodology for Quadrilateral Finite Element Mesh Coarsening. To Engineering With Computers 24(3), 241–251 (2008)
18. Dewey, M.: Automated Quadrilateral Coarsening by Ring Collapse, MS Thesis, Brigham Young University, Utah (2008)
19. Knupp, P.: Achieving Finite Element Mesh Quality via Optimization of the Jacobian Matrix Norm and Associated Quantities. International Journal for Numerical Methods in Engineering 48(8), 1165–1185 (2000)
20. Kinney, P.: CleanUp: Improving Quadrilateral Finite Element Meshes. In: 6th International Meshing Roundtable, Sandia National Laboratories, October 1997, pp. 449–461 (1997)
21. Shepherd, J.: Topologic and Geometric Constraint Based Hexahedral Mesh Generation. PhD Thesis, University of Utah, Utah (2007)
22. Benzley, S., Dewey, M., Woodbury, A., Staten, M.: Enhanced Mesh Adaptation Using Localized Conformal Quadrilateral and Hexahedral Coarsening. In: Inaugural International Conference of Engineering Mechanics Institute (to be published, 2008)
23. Freitag, L., Plassmann, P.: Local Optimization-Based Simplical Mesh Untangling and Improvement, Technical report, Mathematics and Computer Science Division, Argonne National Laboratory (1997), http://citeseer.ist.psu.edu/277987.html

The Effect of Vertex Reordering on 2D Local Mesh Optimization Efficiency

Suzanne M. Shontz[1,⋆] and Patrick Knupp[2,⋆⋆]

[1] Department of Computer Science and Engineering, 343J Information Sciences and Technology Building, The Pennsylvania State University, University Park, PA 16802
shontz@cse.psu.edu

[2] Applied Mathematics and Applications, MS 1318, P.O. Box 5800, Sandia National Laboratories, Albuquerque, NM 87185-1318
pknupp@sandia.gov

1 Local Mesh Optimization

Many applications in computational science such as heat transfer, advection-diffusion, and fluid dynamics numerically solve partial differential equations. To numerically solve the equations, finite element, finite volume, and other PDE-discretization methods are commonly used, along with meshes to discretize the physical domain. It is well-known that the mesh and its quality can greatly impact the accuracy of simulations, as well as solver efficiency [1], [2]. Mesh quality can be improved by various methods including adaptivity [3], [4], smoothing [5], [6], and swapping [7], [8]. In mesh smoothing, one employs vertex-movement strategies to change the coordinates of mesh vertices, leaving initial mesh connectivity intact. Mathematically rigorous methods for formulating the smoothing problem entail the use of an objective function that measures quality; included in this category are the variational methods for structured meshes [9], [10], [11], and the direct optimization methods based on mesh entities such as lengths and angles [12], [13].

A commonly-used method for smoothing meshes is the *local patch* method wherein one loops over a set of patches containing one free vertex and the

⋆ This work was funded in part by NSF grant CNS 0720749 and a Grace Woodward grant from The Pennsylvania State University.

⋆⋆ This work was funded by the Department of Energy's Mathematics, Information, and Computational Science Program (SC-31) and was performed at Sandia National Laboratories. Sandia is a multiprogram laboratory operated by Sandia Corporation, a Lockheed Martin Company, for the U.S. Department of Energy under contract DE-ACO4-94AL85000. This work was performed by an employee of the U.S. Government or under U.S. Government contract. The U.S. Government retains a nonexclusive, royalty-free license to publish or reproduce the published form of this contribution, or allow others to do so, for U.S. Government purposes.

adjacent vertices [14], [15], [16]. On each patch, the coordinates of the free vertex are updated by using an update rule or by solving a local optimization problem. These solution methods are required to be very fast in order to keep pace with other parts of a PDE-based simulation. As such, one usually only performs a few sweeps to obtain an inexact solution, i.e., one that is closer to optimal than the initial mesh, but differs from the true optimum.

The local patch smoothing method strongly resembles local iterative methods such as Gauss-Seidel for solving systems of equations. From that perspective, it is natural to ask whether or not the knowledge of linear solvers can be applied to mesh smoothing methods. One solver topic that seems particularly relevant to local mesh smoothing is that of vertex ordering. It is well-known that, in solving systems of equations via local relaxation methods, the order in which the local sets of equations associated with a vertex are considered can make a difference in the time needed to solve the equations. For example, [17] describes natural and red-black orderings, also known as point partitionings. This must also be the case when smoothing meshes via local patches; however, it does not appear to have been investigated to any large extent. In the present study, we examine the effect that altering the vertex ordering has on the CPU time to solution, with the goal of accelerating the convergence.

2 Timing Considerations

Pseudocode for the local patch optimization scheme is shown below. The smoothing or optimization procedure used to update the free vertex coordinates may itself be iterative, but the details of this inner iteration procedure are not needed for the present purposes. For this study, only interior vertices are allowed to move; boundary vertices are treated as fixed and therefore do not appear in the smooth list. Note that if the termination criterion (for the outer 'while' loop) is not tight, then the final mesh will depend on which fixed vertex ordering is selected.[1] The outer termination criterion used in this study is a comparison of the value of the objective function at each iteration (representing the mesh quality) to a predetermined value. Therefore, we are interested in the ordering scheme that reaches the given level of quality in the least amount of time. The inner termination criteria varied as described later.

To assess the effectiveness of the reordering schemes in different situations, we recorded the Optimization Time per Iteration (OTI), which is the difference between Timer3 and Timer2. The cumulative optimization time (COT), T_j, is the sum of all the OTI's over the 'while' loop, with j denoting the reordering scheme $j = 1, 2, ..., J$.[2] Also recorded were the cumulative reordering time, CRT, (Timer4 - Timer3) and the total time, TT, (Timer 5 - Timer 1).

[1] In general, the final mesh can also depend on which fixed vertex ordering is selected even when the tolerance is tight due to the existence of non-unique minimae. We have attempted to avoid this possibility by using a convex local metric.

[2] The timings include a negligible amount of time due to overhead operations.

```
Pseudocode for Local Mesh Smoothing and Optimization

1. Initialization: vertex list, termination criterion,
     tolerances, objective function, control parameters.
   -- CPU Timer_1
2. While tolerance not satisfied and iteration count
     not exceeded
          -- CPU Timer_2
          Loop over list of free vertices
            - update free vertex coordinates (optimization)
          End loop over free vertices
          -- CPU Timer_3
          Reorder vertex list every k-th iteration
          -- CPU Timer_4
   End while loop
   -- CPU Timer_5
```

3 Study Design

There are many variables, parameters, and choices one might make in a study of the impact of vertex reordering on the overall efficiency of local patch-based mesh optimization. Among the possible choices are different quality metrics, objective function templates, optimization solvers, serial or parallel solvers, computer architecture, 2D/3D, element type, type of reordering scheme, mesh anisotropy/isotropy, mesh heterogeneity/homogeneity, mesh connectivity, tightness of the termination criterion of the outer 'while' loop, and the number of free vertices in the mesh. For this initial work we confine our focus by fixing the local quality metric ($||T - I||_F^2$), the template (linear averaging), the solver (Feasible Newton [18]), serial computation, and the dimension (2D). The quality metric is described in [19] where its utility is demonstrated. The quality metric takes on values from 0 to ∞ with 0 being the best value; a value of 0 is obtained when $A = W$, i.e., whenever the physical and reference elements agree. Further, the metric and objective function are proven convex in [20]. The machine employed for this study is equipped with two Intel (R) Xeon (TM) processors each having four cores; each processor has an 8MB cache shared between its cores. The 64-bit machine has 4GB of RAM and runs Linux. Optimizations and reorderings were performed using the Mesquite code (Version 1.99 Alpha) [21].

The remaining free variables that were studied in this research are: (i) several vertex reordering schemes to be described in the next section, (ii) the number of free vertices in the optimization problem, (iii) mesh type (element type and degree of heterogeneity/isotropy), and (iv) the frequency with which the vertex list reordering is performed.

The outer termination criterion for the runs was based on the desire to have them complete in roughly five minutes, because, in this amount of time, the resulting meshes are neither highly accurate nor grossly inaccurate (the inaccurate case, i.e., requiring about one minute, was also considered). The outer termination criterion value was determined in a set of baseline calculations by running the code for more than five minutes to find the value of the objective function at five minutes. This value was then used to terminate subsequent runs.

The optimization problem that is solved in this study used an objective function that was the average over the entire mesh of the values of the local metric at the element corners (quadrilateral case) or at the barycenter (triangle case). The matrix $T = AW^{-1}$ was determined from A and W, where A represents the quality of the physical mesh, and W represents the target quality, as determined from a reference mesh. See [22] and [19] for details.

4 Vertex Reordering Schemes

Vertex reordering can be done either statically or dynamically. In the static case, one takes as input the initial ordering of the vertex list provided to the optimization scheme and reorders using some criterion. The reordered list is then used and kept fixed during the optimization. In the dynamic case, the reordered list is updated at the end of each iteration (or at the end of every k^{th} iteration) within the optimization, thus creating a sequence of lists. Both strategies were considered in this study. The reordered lists contain the same number of vertices as the initial list. The challenge in all of these approaches is to devise a reordering scheme that accelerates convergence while keeping the computational cost of the reordering low so that the latter does not dominate the time to optimize.

The vertex reordering schemes chosen for this study were ones that were easy to implement and make sense from a geometric point of view. Admittedly, from a theoretical point of view, it is more appealing to choose reordering schemes based on prior knowledge and experience from linear solvers. However, since it is not necessary to derive the global linear systems in order to solve our optimization problem, this approach is reserved for future work. Specific vertex reordering schemes that were investigated are described next.

Scheme (N): Null Order. Do not reorder the vertex list.
Scheme (R): Random Order. Generate a random integer from 1 to N, with N being the total number of local patches in the mesh. Place the vertex assigned the value 1 first in the list and N last.
Scheme (WQP): Order by Worst Quality Patch. Evaluate the objective function on a local patch to measure local patch quality. Sort by putting the worst quality patch first in the list, and so on.
Scheme (GAVM): Order by Greatest Absolute Vertex Movement. Evaluate the position of the free vertex in the local patch before and after the optimization and calculate the absolute distance moved. Sort by putting the patch with the greatest absolute distance moved first in the list, and so on.
Scheme (GRVM): Order by Greatest Relative Vertex Movement. Evaluate the relative distance moved by the free vertex in the local patch before and

after the local optimization. Relative distance was measured by dividing the absolute distance moved by the initial absolute distance moved; the normalizing quantity thus varies from patch to patch. Sort by putting the patch with the greatest relative distance moved first in the list, and so on.
Scheme (LNG): Order by Largest Norm of Gradient. Evaluate the ℓ_2-norm of the local gradient of the objective function. Sort by putting the largest norm first in the list, and so on.

For the schemes WQP, GAVM, GRVM, and LNG, there were associated 'distance' schemes D-WQP, D-GAVM, D-GRVM, and D-LNG which are created by measuring the distance of the free vertex in each patch from the position of the free vertex which had the (i) worst patch quality, (ii) greatest average vertex movement, (iii) greatest relative vertex movement, or (iv) largest norm of the gradient, respectively. Then sort by putting the patch with the smallest distance first in the list, and so on.

For each of the schemes above, there is a corresponding ordering scheme produced by reversing the ordering (the abbreviation for these schemes is the same as the forward ordering except there is a 'minus' at the start of the abbreviation). For example, scheme -WQP orders the vertex patches from best quality first to worst quality last. There are a total of 20 ordering schemes.

A QuickSort algorithm was used in this study to do the reordering. The expected runtime of the sort is $O(N \log N)$, with the worst case being $O(N^2)$, where N is the number of patches to smooth per outer iteration. The larger the number of patches, the more time it takes to reorder them, so the key question is whether or not the time saved by the reordering (as a function of N) increases as fast as the time to do the reordering.

5 Mesh Selection Criteria

Five meshes were created, as described in Table 1. Two basic cases are the structured quadrilateral meshes and the unstructured triangle meshes. The meshes all have the same physical domain, i.e., the horseshoe. The five mesh types contain roughly 20K, 40K, or 80K free vertices each, for a total of 15 meshes included in the study.

Table 1. Mesh Identification Table

MeshID	Description
SQI	Structured Quad, 1:1 cell aspect ratio
SQII	Structured Quad, 1:5 cell aspect ratio
SQIV	Structured Quad, Biased 1:5 cell aspect ratio
UTI	Unstructured Triangle Mesh, Trapezoidal Logical Domain
UTIV	Unstructured Triangle Mesh, Biasing

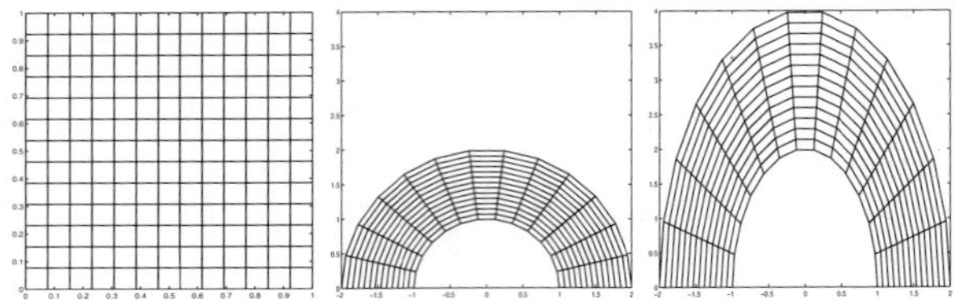

Fig. 1. Mesh Type SQI for the Unit Square, Reference Semi-Annulus, and Horseshoe

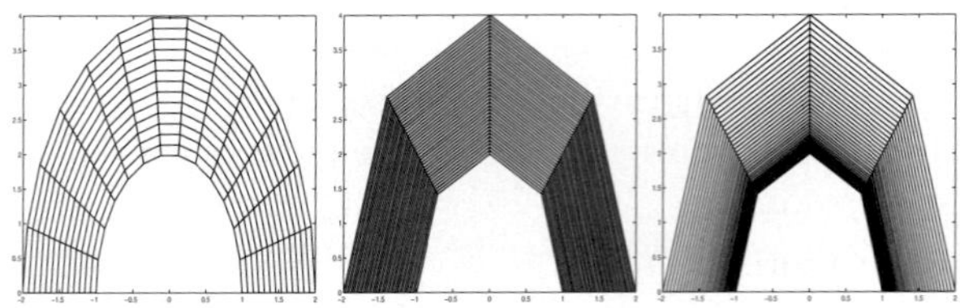

Fig. 2. The Three Quadrilateral Mesh Types. Left to Right: SQI, SQII, SQIV

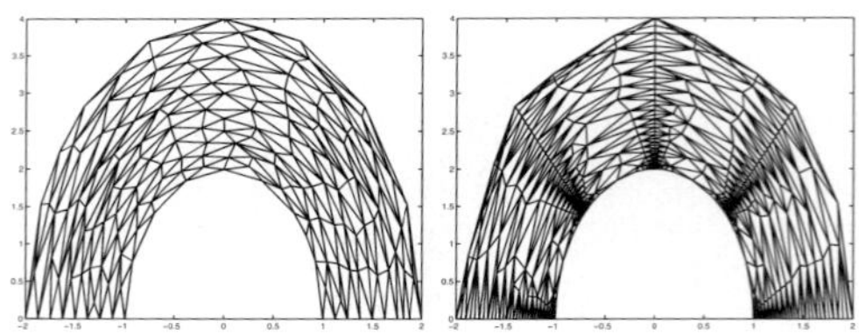

Fig. 3. Two Triangular Mesh Types. Left to Right: UTI, UTIV

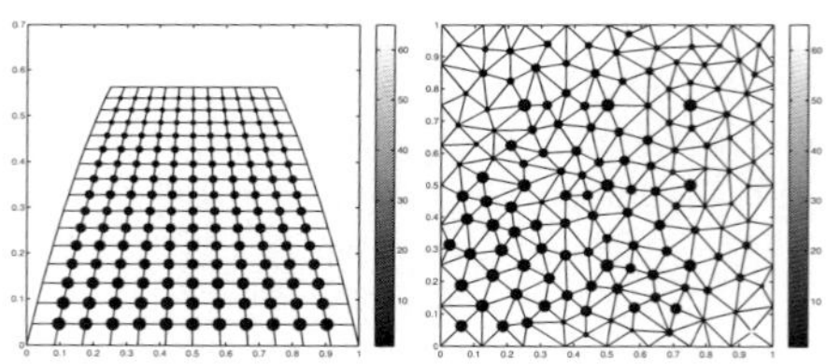

Fig. 4. Initial Vertex Ordering on Structured and Unstructured Meshes

The approach to generating these meshes was to first generate regular meshes on the unit square. For each such mesh, a reference mesh on the semi-annulus and an initial mesh on the horseshoe were created by analytically mapping each vertex in the unit mesh to give corresponding coordinates on the other two domains. An example of the mesh triples (logical, reference, and horseshoe domains) generated using the mappings for Mesh Type SQI is given in Figure 1. Optimization was always performed on the horseshoe domain using target matrices based on the corresponding semi-annulus mesh.

Figures 2 and 3 show coarse versions of the Horseshoe meshes in Table 1.

Figure 4 shows the initial ordering of the vertices as given by the mesh generator in the structured and unstructured meshes on the unit square. The mapping, of course, preserves the orderings. Large vertices occur early in the list, while smaller vertices occur towards the end. The initial (row-by-row) ordering for all of the structured meshes is the same as shown in the figure, while the ordering for the unstructured mesh in the figure is a representative ordering for both types of unstructured meshes used in this document.

6 Numerical Experiments and Results

Eight numerical experiments were run using the Mesquite code to investigate the behavior of the reordering schemes under various circumstances such as mesh type, mesh size, and reordering strategy. Table 2 summarizes the eight experimental set-ups.

Table 2. Table of Experiments

Exp. No.	Exp. Name	Mesh Types	Mesh Sizes	Schemes	S/D	Frequency	# Runs
6.1	Baseline	all 5	all sizes	N	D	N/A	30
6.2	Static Reordering	all 5	80K	20	S	N/A	100
6.3	Sensitivity to IM	SQI	80K	20	S/D[5]	N/A / 1[6]	40
6.4	Dynamic Reordering	all 5	all sizes	20	D	1	300
6.5	Frequency	all 5	80K	20	D	1,2,5,10	400
6.6	ITC	all 5	80K	20	D	1	100
6.7	Inaccurate	all 5	all sizes	20	D	1	300
6.8	Red-Black	all SQ	80K	20	D	1	60

[5] One static (S) and one dynamic (D) experiment were conducted.
[6] The frequency value is not applicable (N/A) for the static experiment and is set to one (1) for the dynamic experiment.

The output of these experiments took the following form: 1) input summary tables to record the key inputs, 2) overlay plots showing average mesh quality vs. total time (TT) for all twenty reordering schemes, and 3) ranking tables giving optimization (COT), reordering (RT), and total times (TT), along with rankings from 1 to 20 for each of these items (plus a ranking based on outer iteration count). Typical examples of the quality vs. time plots are shown in Figure 5. Additional post-processing of the results was also performed. The purpose of each experiment is described next, along with selected results. At the end of the section, observations that pertain to all of the experiments are given.

6.1 Baseline Experiment

The purpose of this 'experiment' was to determine the quality level that would result in 1 or 5 minutes of smoothing. One minute of smoothing gives a relatively inaccurate approximation to the optimal mesh, while five minutes gives moderately accurate results. Runs were terminated after one or five minutes, and the quality level was recorded, to be used in subsequent experiments.

Selected Results
By examining the quality vs. time plots, it was noted that the shape of these curves generally followed two basic patterns: either the curves were straight lines, or they were highly curved, similar to a rapidly decaying L-shaped function (see Figure 5). The pattern was found to continue over the seven other experiments that were performed. The explanation appears to be that if the initial mesh is relatively 'far' from the optimal mesh then rapid convergence occurs at early times and linear convergence at later times, giving the L-shaped curve. On the other hand, if the initial mesh is relatively 'close' to the optimal mesh, one only sees the tail of the L-shaped function, which appears linear. For the most part, the curves for the SQI and SQII meshes appear linear, while the SQIV, UTI, and UTIV meshes often showed L-shaped convergence behavior. In general, the curves for the finer meshes (40-80K) had a steeper slope than the coarser meshes (20-40K) and satisfied the convergence criterion more quickly.

6.2 Static Experiment

The purpose of this experiment was to observe the behavior of the reordering schemes using a 'static' reordering strategy in which the reordering is done only once, i.e., before optimization begins.

Selected Results
The average quality vs. time curves were observed to be monotonically decreasing in most cases; however, for a few ordering schemes (GRVM, R, WQP) on the SQII meshes, they actually increased initially, before decreasing and satisfying the convergence tolerance.[3] This placed these schemes in last place in terms of TT ranking. In very rare instances (and this applies to all experiments), quality vs. time curves were observed to cross one another, meaning that ordering schemes that were effective at early times, became relatively ineffective at later times. For the most part, however, quality vs. time curves do not cross one another. Moreover, in the static experiment, they tend to be parallel to one another, as seen in Figure 5.

The choice of reordering scheme had a noticeable effect on time to convergence (TT). The variation in total time-to-convergence relative to the average time-to-convergence for the given mesh type over the reordering schemes in the static

[3] The increase is attributed to the fact that it is theoretically possible for the local average quality to increase when using the Feasible Newton solver on a local patch; a poor choice of search direction is one way this can happen. If the local average quality can increase, then so can the global average quality.

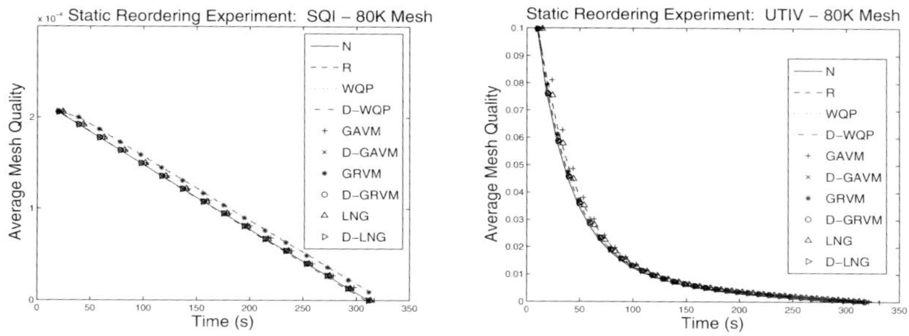

Fig. 5. Sample Quality vs. Time Plots: Linear (left) and L-shaped (right)

experiments was, for the 80K meshes, SQI: 312 ± 5 seconds, SQII: 366 ± 61 seconds, SQIV: 258 ± 62 seconds, UTI: 312 ± 10 seconds, and UTIV: 319 ± 13 seconds. That is, the variation in time-to-convergence due to the choice of reordering schemes was: 1.6% for SQI, 16.7% for SQII, 24.0% for SQIV, 3.2% for UTI, and 4.1% for UTIV. It appears that the choice of reordering scheme is more important for biased or anisotropic mesh types. Although the difference between using one reordering scheme versus another is noticeable, the maximum effect was less than 40% of the total run times.

Nearly all reorderings took less than 1.5 seconds. Smoothing times varied from 105 to 315 seconds, so reordering time was less than one percent of the smoothing time. In terms of total time, -N was the best at 196 seconds, while -GAVM was the worst at 320 seconds. Notably, Scheme N took about 314 seconds while scheme -N took 196 seconds, about a 30% difference in total time. Other mesh types had different reordering schemes giving the least TT.

6.3 Sensitivity of Results to Initial Mesh

The purpose of this experiment was to observe the behavior of the reordering schemes when a different initial mesh was used, with both static and dynamic (see Experiment 6.4) reordering strategies. The second initial mesh differed from the first initial mesh only in the vertex coordinates. The vertices in both of these initial mesh types were ordered in the same way (see Experiment 6.8 for an example of different vertex orderings in the initial guess). The vertex coordinates in the second initial mesh were obtained from the first initial mesh by randomly perturbing them by the largest possible amount without creating inverted mesh elements; the second mesh is farther from optimal.

Selected Results for the Static Case

The quality vs. time plots showed that the curves that were linear when using the first initial guess became L-shaped when using the second initial guess, as expected since the second initial guess is farther from optimal than the first.

The variation in total time-to-convergence relative to the average time-to-convergence for the 80K SQI mesh type over the reordering schemes in this experiment was 895 ± 51 seconds, or 5.6%, compared to 1.6% when the first initial guess was used. Thus, it appears that the farther the initial guess is from optimal, the larger the variation, and thus the more significant the choice of reordering scheme becomes.

6.4 Dynamic Experiment

The purpose of this experiment was to observe the behavior of the reordering schemes using a different reordering strategy. The 'dynamic' strategy is to perform the reordering once per outer iteration. This will cost more than the static strategy in terms of time spent reordering, so the question is whether or not the added expense is worth it in terms of the total CPU time expended.

Selected Results
The quality vs. time curves differed from those in the static case in that the linear curves that were parallel in the static case remained linear, but with differing slopes, depending on the reordering scheme. This reflects the fact that in the dynamic strategies, differences amongst ordering schemes become cumulative with time, whereas this does not happen in the static case. As a result, the quality vs. time curves in the dynamic case diverge from one another with increasing time. A similar effect was seen for the L-shaped curves.

The choice of reordering scheme did have a noticeable effect on the time to convergence. The variation in total time-to-convergence relative to the average time-to-convergence for the given mesh type over the reordering schemes in the dynamic experiments was, for the 80K meshes, SQI: 365 ± 55 seconds, SQII: 411 ± 107 seconds, SQIV: 345 ± 116 seconds, UTI: 488 ± 182 seconds, and UTIV: 421 ± 120 seconds. That is, the variation in time-to-convergence due to the choice of reordering schemes was: 15.1% for SQI, 26.0% for SQII, 33.6% for SQIV, 37.3% for UTI, and 28.5% for UTIV. Comparing these results to the static strategy, one sees that both the average total time-to-convergence and the variation is larger in the dynamic case.

Although, the *average* time-to-convergence in the dynamic case is larger than in the static case, it does not necessarily mean that there may not be particular reordering schemes that, when used in the dynamic strategy, are competitive with the static strategy. To investigate, we compared the best time-to-solution for the static and dynamic cases for each of the 80K mesh types (excluding the $\pm N$ and $\pm R$ schemes). It appears that the best dynamic schemes (in terms of least TT) are not quite as effective as the same schemes when used statically. The only exception was for SQII, and that is attributed to the fact that in the static runs the D-GAVM quality versus time curve initially increased, causing it to lag the others; this did not happen in the dynamic case. Evidently, updating the ordering once every iteration is too expensive compared to the gain in overall solution time. The next experiment investigates a modified dynamic strategy.

Table 3. Best Dynamic Scheme vs. Corresponding Static Result

Mesh Type	Best Scheme	Static Time	Dynamic Time
SQI	D-GRVM	312.74	314.81
SQII	D-GAVM	326.13	313.40
SQIV	-GRVM	203.35	228.77
UTI	-GRVM	321.21	346.49
UTIV	D-LNG	318.47	333.08

6.5 Frequency of Dynamic Reordering Experiment

The purpose of this experiment was to observe the behavior of the reordering schemes with a modified dynamic reordering strategy. Instead of reordering every iteration, in this experiment, reordering was performed every iteration, every other iteration, every fifth iteration, and every tenth iteration.

Selected Results
The case of the 80K SQIV mesh is instructive. Consider the behavior of scheme -GRVM vs. the reordering frequency (k); this scheme had the smallest TT over the 20 schemes when $k = 1, 2$.

Table 4. Scheme -GRVM vs. Reordering Frequency

Frequency	COT	COT Rank	RT	RT Rank	TT	TT Rank
1	198.07	1	33.48	12	231.55	1
2	224.76	6	9.55	10	234.31	1
5	261.32	6	0.05	5	261.37	6
10	307.20	7	0.02	6	307.22	7
Static	203.33		0.02		203.35	

The fraction of time spent reordering vs. the total time decreases as reordering becomes less frequent. The optimization time is smallest with frequency 1, as is total time. In this case, reordering every iteration is well worth it compared to the other reordering frequencies. However, the total time in the static case is even less. Figure 6 shows TT vs. reordering frequency for schemes -GRVM, -WQP, and N on the 80K, SQIV mesh. Note that the figure shows that it is best to apply the -WQP scheme as infrequently as possible!

This experiment shows that the ideal frequency of reordering depends on the reordering scheme in question. If the reordering scheme is an effective one for the given problem, it should be applied as frequently as possible.

Does the dynamic strategy ever result in a total time that is less than the corresponding static case? In general, the answer is yes, but not significantly. For each of the five mesh types one can find a reordering scheme whose total time when $k = 1$ or $k = 10$ is less than its own static time, but, even then,

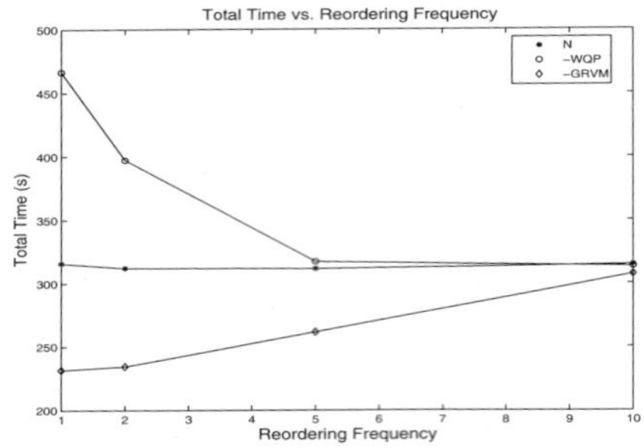

Fig. 6. Total Time vs. Frequency for Schemes -GRVM, -WQP, and N

the dynamic times are only a few percent less than the static times. Therefore, the dynamic strategy, even with infrequent reordering, seems to be mostly less efficient than the static strategy.

6.6 Inner Termination Criterion Experiment

The purpose of this experiment is to determine whether or not the number of inner iterations in the optimization had any significant impact on the results of the reordering experiments.

Selected Results

Increasing the number of inner iterations to 2 increases the time-to-convergence in many of these runs. Since the variation in total times over the twenty reordering schemes does not increase with the number of inner iterations, the impact of the choice of reordering scheme becomes less as the number of inner iterations increases. In addition, the rankings of the reordering schemes changes with the number of iterations.

6.7 Inaccurate Experiment

The purpose of this experiment was to determine how the results of the reordering experiments might change if only an inaccurate (1 minute) solution were sought when dynamic vertex reordering was performed.

Selected Results

The main result here was that, indeed, the rankings of the reordering schemes did change when comparing the 'inaccurate' case to the 'medium-accurate' case. The variations in total time-to-convergence relative to the average time-to-convergence for the given mesh type over the reordering schemes in the Inaccurate Experiment were, for the 80K meshes, SQI: 101 $\pm$ 25 seconds, SQII: 204 $\pm$ 70 seconds, SQIV: 100 $\pm$ 39 seconds, UTI: 113 $\pm$ 35 seconds, and UTIV: 107 $\pm$ 40 seconds. That is, the variation in time-to-convergence due to the choice of reordering schemes was: 24.8% for SQI, 34.3% for SQII, 39.0% for SQIV, 31.0%

for UTI, and 37.4% for UTIV. Comparing these results to the medium-accurate dynamic strategy, one sees that although both the total time-to-convergence and the variation is smaller in the inaccurate case, the choice of reordering scheme becomes more significant percentage-wise.

6.8 Red-Black Experiment

The purpose of this experiment was to investigate whether or not the ordering of the vertices in the initial mesh had a significant impact on the behavior of the ordering schemes. The first (original) initial meshes for the structured mesh types were modified to create a red-black (checkerboard) ordering of the vertices while keeping the vertex coordinates the same. Results were then compared to the original initial mesh.

For this experiment, the percent variation over the reordering schemes was much larger than when using the ordering in the first initial mesh: SQI - 144%, SQII - 209 %, and SQIV - 222 %. For example, the SQI curves diverge much more than in the original ordering, giving a larger variation and much longer run-times overall (300-800 sec vs. 300-450 in original). Scheme rankings are somewhat similar to original, but not too close. For the SQII curves, it was observed that the plots no longer show an initial increase in average quality for any scheme, whereas in the original some did. The scheme rankings are completely different. For SQIV, there was slightly more L-shaped convergence behavior, and overall run times are longer, and the scheme rankings change. The -N scheme is conspicuously bad in the red-black experiments.

The results in the scheme ranking tables showed that for SQI, the best red-black total time is 308.72 sec (for scheme N), while the best original total time was 309.51 sec (again for N), so red-black ordering per se is hardly worse than the original ordering. However, the worst red-black time is 752.21 sec (for GAVM), compared to the worst original (WQP, 420.21 sec). Table 5 directly compares results for three reordering schemes. The table shows that the ordering of the initial mesh itself can alter the final results, but it is hard to predict in advance what will happen.

Table 5. Comparison of Natural and Red-Black Orderings (SQI)

Scheme	COT	COT Rank	RT	RT Rank	TT	TT Rank
D-GRVM-or	309.07	4	5.74	9	314.81	3
D-GRVM-rb	314.57	12	6.48	8	321.05	4
GAVM-or	316.50	14	79.53	16	396.03	17
GAVM-rb	599.84	19	152.37	20	752.21	20
WQP-or	391.84	20	28.37	11	420.21	20
WQP-rb	392.21	15	28.25	10	420.46	16

6.9 Reordering Time

For each of the runs performed in this study, we measured the cumulative CPU time (RT) expended during the reordering step. To investigate which reordering schemes tended to take the least amount of CPU time, we computed a set of histograms. For each reordering scheme we counted the number of times it was ranked first (in terms of reordering time), second, third, and all the way to twentieth. The counting was performed over the entire set of runs performed in this study. Results are summarized in Table 6 and the schemes assigned an overall rank.

Table 6. A Rough Reordering Time Ranking Over All Runs

Rank	Scheme	Mode	2nd Mode	Best Rank	Worst Rank
1	N	1	-	1	1
2	-N	2	3	2	4
3	R	3	4	2	5
4	-R	4	3	2	5
5	GRVM	5	6	4	6
6	D-GAVM	7	8	6	12
7	-GRVM	6	12	4	16
8	D-LNG	7	8	6	11
9	D-GRVM	9	8	6	12
10	WQP	10	11	9	19
11	-WQP	11	14	10	20
12	D-WQP	12	10/13	9	17
13	-D-GRVM	14	11	10	19
14	-D-GAVM	14	13	7	19
15	-D-LNG	15	13	7	19
16	-D-WQP	16	15	13	20
17	-GAVM	17	18	12	19
18	LNG	19	18	10	20
19	GAVM	20	18	10	20
20	-LNG	20	19	11	20

Scheme N ranks first since it takes no CPU time at all, being a null operation. The next three schemes also take very little work since they only reverse a given ordering that was fast to create in the first place. It is difficult to explain the rankings for the remaining schemes, although perhaps a look at operation counts might help. 'Distance' scheme ranks are mixed with the schemes upon which they are based, a somewhat surprising result. As expected, the reverse orderings ranked below the forward orderings (except GAVM).

6.10 Overall Scheme Rankings

The selected results reported in the previous experiments are supplemented by taking a look at TT scheme ranking results over the entire set of runs that were performed over all the experiments. From the results obtained, it was found that the 'best' orderings varied from experiment to experiment, mesh type to mesh type, mesh size to mesh size, etc., with no discernible pattern. If one were going to do a lot of mesh optimization calculations with these variables fixed, one might benefit by determining which ordering scheme is best in that specific instance. For a general purpose tool such as Mesquite, it is reasonable to ask if there are reordering schemes which generally performed well over all the runs (and which generally did poorly). To investigate, we created a set of histograms, one for each reordering scheme, in which we counted the number of times the scheme was ranked first, second, third, and all the way up to twentieth. The counting was over the entire set of runs.

Table 7. A Rough Scheme Ranking Over All Runs

Rank	Scheme	Mode	2nd Mode	Best Rank	Worst Rank
1	N	1	2	1	20
2	-N	1	2	1	20
3	D-GAVM	3	4	1	15
4	D-LNG	4	3	1	15
5	D-GRVM	4	5	2	16
6	D-WQP	6	9	2	17
7	-D-GAVM	7	8	1	17
8	-D-LNG	9	7	1	17
9	-GRVM	2	15	1	18
10	-D-GRVM	8	9	2	17
11	GRVM	16	2	2	20
12	-D-WQP	10	15	1	19
13	-R	18	17	1	20
14	R	17	18	9	20
15	-GAVM	14	11	5	20
16	-LNG	15	12	2	19
17	LNG	13	18	3	19
18	GAVM	20	12	6	20
19	WQP	20	19	3	20
20	-WQP	20	19	4	20

To rank the schemes using these histograms, we looked at the mode for each scheme and the general distribution around the mode. Based on this, the overall rankings are shown in Table 7, with schemes N, -N, D-GAVM, D-LNG among the best and LNG, GAVM, WQP, and -WQP among the worst. The best 'worst rank' over all the schemes was 15th, so no scheme was immune from a circumstance in which it did poorly. The worst 'best rank' was 9th, so nearly all the schemes had at least one circumstance it which it did very well relative to the others. On the whole, though, the top five of these schemes stand out as being good general purpose reordering schemes. It is perhaps not too surprising to see schemes N and -N at the top of the rankings since they incur virtually no cost in terms of reordering time. Furthermore, there may be an advantage to leaving the original orderings alone since these reflect the order in which the mesh vertices were likely created when the mesh was generated. Not surprising perhaps is the relatively low rankings of the $\pm R$ schemes since they do not take advantage of problem structure. Nevertheless, they ranked higher than six other schemes. As a general rule, the 'distance'-based schemes were ranked higher than the schemes from which they were derived. For example, GAVM ranks 18th while D-GAVM ranks 3rd; this is because reordering for D-GAVM takes less time than that of GAVM, and often the optimization for D-GAVM is more effective than that of GAVM in that the number of outer iterations is reduced thus requiring less time spent on optimization. Counterintuitively, reordering by patch quality is not advantageous. Ordering of patches to maintain or create locality seems to be an important variable in creating an effective reordering scheme.

7 Conclusions and Future Work

The results of this study are somewhat preliminary but a number of important trends are suggested. First, it appears that the static strategy is superior to the dynamic ones because the cost of additional reordering does not result in enough of a speedup in the optimization process. Second, reordering appears most important when 'inaccurate' meshes are considered satisfactory or when the initial mesh is far from optimal. Third, it is difficult to predict in advance which reordering scheme will be best for a given problem. It was noted that optimization of meshes with strong biasing or heterogeneity will likely benefit most from a consideration of vertex ordering. Additionally, reordering may be beneficial if the ordering of the patches in the initial mesh lacks locality, as seen in the red-black experiment. Fourth, although total time to solution is impacted noticeably by vertex ordering, it is usually within a factor of two, as opposed to orders-of-magnitude. This could change if other mesh sizes and types were considered. Fifth, the natural ordering provided by the mesh generator resulted in scheme N being one of the more efficient orderings. Reversing the ordering of the list (forward to reverse) is inexpensive and often improved TT. Sixth, distance-based reordering schemes appear to be more effective than the schemes upon which they are based; once again, this is most likely due to an increase in locality. Seven, no discernible pattern of results was observed related to mesh size. Perhaps much smaller or larger meshes would have resulted in a pattern with respect to this parameter. Some of these observations can perhaps be explained theoretically, as has been done for the linear equation solver case. Nonetheless, it is useful to have empirical data to suggest possible theoretical results and to provide additional insight.

More data was generated in this study than could be presented due to space limitations. Additional post-processing and analysis of the data is planned in the future to better quantify the results, check for over-looked tendencies, and to determine if there are particular circumstances which can be predicted in advance as to which of the reordering schemes should be used. Future work on this topic might include the creation of better ordering schemes to provide more locality, further investigations of reordering efficiency vs. mesh size, other dynamic strategies, and the effect of using other optimization solvers.

Acknowledgements

The authors would like to thank Jason Kraftcheck for advice on implementing algorithms in Mesquite, Anirban Chatterjee for C++ consultation, and the reviewers for helpful comments.

References

1. Shewchuk, J.: What is a good linear finite element? (unpublished preprint, 2002)
2. Batdorf, M., et al.: Computational study of the effect of unstructured mesh quality on solution efficiency. In: Proceedings 13th Annual AIAA Computational Fluid Dynamics Conference. AIAA (1997)
3. Bank, R., Smith, R.: Mesh smoothing using a posteriori error estimates. SIAM J. Numer. Anal. 34, 979–997 (1997)
4. Lague, J., Hecht, F.: Optimal Mesh for P1 Interpolation in H1 Seminorm. In: Proceedings 15th Intl. Meshing Roundtable, pp. 259–270. Springer, Heidelberg (2006)
5. Montenegro, R., Escobar, J., Montero, G., Rodriguez, E.: Quality Improvement of Surface Triangulations. In: Proceedings 14th Intl. Meshing Roundtable, pp. 469–484. Springer, Heidelberg (2005)
6. Branets, L., Carey, G.: Smoothing and adaptive redistribution for grids with irregular valence and hanging nodes. In: Proceedings 13th Intl. Meshing Roundtable, pp. 333–344 (2004)
7. Freitag, L., Ollivier-Gooch, C.: Tetrahedral Mesh Improvement Using Swapping and Smoothing. Int. J. Num. Meth. Engr. 40, 3979–4002 (1997)
8. Klingner, B., Shewchuk, J.: Aggressive Tetrahedral Mesh Improvement. In: Proceedings 16th Intl. Meshing Roundtable, pp. 3–23. Springer, Heidelberg (2008)
9. Brackbill, J., Saltzman, J.: Adaptive Zoning for singular problems in two dimensions. J. Comp. Phys. 46, 342–368 (1982)
10. Knupp, P.: Jacobian-weighted Elliptic Grid Generation. SIAM J. Sci. Comp. 17(6), 1475–1490 (1996)
11. Liseikin, V.: A Computational Differential Geometry Approach to Grid Generation. Springer, Heidelberg (2004)
12. Castillo, J.: A discrete variational method. SIAM J. Sci. Stat. Comput. 12(2), 454–468 (1991)
13. Freitag, L., Knupp, P.: Tetrahedral mesh improvement via optimization of the element condition number. Int. J. Num. Meth. Engr. 53(6), 1377–1391 (2002)
14. Canann, S., Tristano, J., Staten, M.: An approach to combined Laplacian & Optimization-Based Smoothing for Triangular, Quadrilateral, & Quad-dominant Meshes. In: 7th Intl. Meshing Roundtable, Dearborn MI, pp. 479–494 (1998)
15. Freitag, L.: On Combining Laplacian and Optimization-Based Smoothing Techniques, Trends in Mesh Generation, AMD vol. 220, pp. 37–43. ASME (1997)
16. Hansbo, P.: Generalized Laplacian smoothing of unstructured grids. Com. Num. Meth. Engr. 11(5), 455–464 (1995)

17. Hageman, L., Young, D.: Applied Iterative Methods. Academic Press, London (1981)
18. Munson, T.: Mesh Shape-Quality Optimization Using the Inverse Mean-Ratio Metric. Math. Program. 110, 561–590 (2007)
19. Knupp, P.: Updating Meshes on Deforming Domains. Communications in Numerical Methods in Engineering 24, 467–476 (2008)
20. Vander Zee, E., Knupp, P.: Convexity of Mesh Optimization Metrics using a Target-Matrix Paradigm, Sandia Laboratories, SAND2006-4975J (2006)
21. Brewer, M., Freitag Diachin, L., Knupp, P., Leurent, T., Melander, D.: The Mesquite Mesh Quality Improvement Toolkit. In: Proceedings 12th Intl. Meshing Roundtable, Sandia Laboratories, Albuquerque, NM, pp. 239–250 (2003)
22. Knupp, P.: Formulation of a Target-Matrix Paradigm for Mesh Optimization, SAND2006-2730J, Sandia National Laboratories, Albuquerque NM (2006)

Flattening 3D Triangulations for Quality Surface Mesh Generation

Kirk Beatty and Nilanjan Mukherjee*

Meshing & Abstraction Group
Digital Simulation Solutions
SIEMENS
2000 Eastman Dr., Milford, Ohio 45244 USA
kirk.beatty@siemens.com, mukherjee.nilanjan@siemens.com

Abstract. A method of flattening 3D triangulations for use in surface meshing is presented. The flattening method supports multiple boundary loops and directly produces planar locations for the vertices of the triangulation. The general nonlinear least-square fit condition for the triangle vertices includes conformal (angle preserving) and authalic (area preserving) conditions as special cases. The method of Langrange multipliers is used to eliminate rotational and translation degrees of freedom and enforce periodic boundary conditions. Using matrix partitioning, several alternative sets of constraints can be efficiently tested to find which produces the best domain. A surface boundary term is introduced to improve domain quality and break the symmetry of indeterminate multi-loop problems. The nonlinear problems are solved using a scaled conformal result as the initial input. The resulting 2D domains are used to generate 3D surface meshes. Results indicate that best mesh quality is achieved with domains generated using an intermediate *altitude preserving* condition. Apart from an admirable robustness and overall efficiency, the 2D developed domains are particularly suited for structured transfinite/mapped meshes which often reveal wiggly irregularities with most conventional developed domains. Flattening and meshing (both free and transfinite/mapped) results are presented for several 3D triangulations.

Keywords: flattening, free-boundary, interpolation, Langrange multipliers, parameterization, periodic, transfinite, triangle.

1 Introduction

Although direct methods of generating meshes on 3D surfaces abound, a large majority of mesh generation programs used in both the industry and the research laboratories continue to use indirect method of generating the mesh on a flattened 2D parameter space first, transforming the mesh back to the original 3D surface later. Finite element analyses of structures, fluid flow and interaction problems have increased in complexity over the last few decades. The mesh to solve (especially quad-form meshes) often have pressing quality requirements. Crash analyses limit the variation of element size in anisotropic meshes. A large range of boundary-flow and structural problems require meshes to be boundary structured or partially or wholly transfinite.

* Corresponding author.

A general purpose mesh generation engine needs to be able to cater to most classes of problems. Such meshes of high quality and constraint are often virtually impossible to generate directly on real, complex 3D geometry efficiently.

Thus the problem of generating a low distortion 2D parameter space from the 3D surface continues to enthrall engineers and mathematicians. The goal is to create a 2D flattened domain of a crack-free triangulated (tessellated) 3D surface where the mesh could be generated. A discrete, continuous transformation function should be made available to finally transform the mesh nodes to the 3D surface. The same transform can be also used in 2D space to check the quality of the 3D elements being generated in 2D space.

1.1 Use of 2D Domains Created from 3D Triangulation

Two-dimensional domains created from 3D triangulations have a variety of applications. These include the fitting of parametric surfaces, the use and creation of texture maps, and as a parameterization to solve differential equations on 3D surfaces. In our case, the domains are used in the meshing of tessellated faces, which are based on a 3D triangulation. These faces can be derived by combining triangulations of several NURBS surfaces. The topology associated with these faces can include many boundary loops and/or repeated edges. To see how we use the 2D domains, we give a simplified version of our meshing process:

1.2 Simplified Meshing Method – Free or Transfinite Meshing

Before we start the meshing process, we abstract the tessellated faces. This geometry abstraction toolbox includes a host of basic discrete geometry operators which involve splitting and combining the facetted faces, cutting sharp corners that cause meshing issues. First mesh all edges at 3D locations. (For mapped meshes one must match the number of vertices on matched edges for the whole system.) For each surface:

1) Create 2D domain from 3D triangulation and topology. Split edge uses of same edge in domain (to create non-periodic surfaces from periodic) if needed for quality.
2) Transform nodes on 3D edges to 2D domain.
3) Mesh domain in 2D using boundary nodes.
4) Map 2D node positions to 3D using triangle coordinates.

In this paper we will focus on how the domains produced in step 1 affect our meshing result.

2 Past Research

There is an extensive literature for the parameterization of 3d triangulations [1]. Some techniques require a fixed convex boundary in UV space as an initial condition. Floater [2] has developed an improvements to this approach by considering harmonic function theory. Using his approach, the results vary smoothly with input data, are fast to solve, and guarantee no triangle flips in UV space. Though the technique is powerful, the requirements of a non-convex boundary (and the filling in of holes) can often cause significant distortion. Instead we focus on free boundary methods.

One of the first free-boundary approaches was developed by Hormann and Greiner [3]. They developed the Most Isometric Parameterization (MIPS) based on minimizing a metric of distortion. They realized that directly using the Dirichlet energy did not punish the collapsing of triangles, or a degenerate solution. Instead they found a nonlinear criterion that was more difficult to solve. They had good results on small models, but the overall approach was computationally expensive. Levy et al. [4] considered a quasi-conformal approach (Least square conformal mapping -LSCM) in the context of automatic atlas generation. For smaller regions this method was quite fast and could produce good results. They found they could avoid degeneracy by fixing two points of the system in UV space. They chose two far apart in geodesic distance along the surface. Desbrun et al [5] developed a similar least squares conformal approach independently. They found they could linearly combine the results with area preserving (authalic) solution that was not formulated for a free boundary. They developed a general approach of constraints using Lagrange multipliers. Even though that constraint approach was quite general, in practice they used a more simplistic approach to avoiding degeneracy than Levy et al. They fixed two points farthest apart in 3d space. An angle based flattening (ABF) approach was found by Sheffer and de Sturler [6]. They considered equations for angles that were consistent with the plane. When they found a solution to their nonlinear problem with constraints, they could guarantee a no triangle flip answer. Sheffer et al. [7] developed a hierarchical approach for large systems and pointed out their approach produced significantly less stretch than the LSCM method. We suggest that in the comparison the LSCM procedure suffered from a simplistic pinning condition and a "collapsing triangle" problem evident with large triangulations. Recently Zayer et al [8] have found a fast linear approach that gives very similar results to the ABF method. After the angles of the triangles are determined by the ABF (or derivative) procedure, it is still necessary to find a solution in UV space. Here at Siemens, we have used a flattening approach based on preserving the lengths of triangle sides. Chief architects of the technology included Evan Sherbrooke and David Gossard (formerly of New Technologies Inc.) Similarly to the ABF method, constraint equations are developed such that the result is consistent with the restriction of the plane. The constraints were based on the law of cosines for the wheel of triangles around all interior vertices. Once the lengths of the triangles are found, they must be rolled out into the 2D plane. (The work in this paper originated from improvements to the robustness of the roll-out procedure used in NTI's method.) A global parameterization based on gradient fields was developed by Gu and Yau [9]. In their method they found the underlying basis vectors of the linear space. The bounding conditions were based on periodic boundary conditions on cuts of closed surfaces or a double covering of the non-closed surface. By tying together the boundaries of the surface together, and avoiding pinning of points, very uniform conformal solutions were found.

3 Problem Statement

Our meshing method requires a procedure to create a 2D domain for meshing from a 3d triangulation of a face. Associated with the facetted face there is a topology which includes loops and edges. This topology is a important component of our meshing process, which in turn puts requirements on our flattening procedure. The facetted faces consist of triangles defined by facet vertices. The facetted faces to be flattened have a loop and edge topology. The edges are defined on edges of facets. The edges can be both on the exterior and interior of the 3d face. Edges on the interior of a face have two uses by the loops, which are in different orientations. Edges interior to a face can be used to make non-manifold connections between different faces or enforce meshing along an edge. They can also be added to help the creation of a low distortion domain in two dimensions. A cylinder can be represented with one loop of edges with a repeated edge use. If the repeated edge is allowed to split apart in a UV domain, then we can get a rectangle where the vertices along the repeated edge have different uses with different UV values. An edge so used is commonly referred to as a "seam". Originally, we distinguished between interior edges that were from cutting a cylinder, and edges that were added for other purposes. However, there were two problems with this approach. The first is that an edge could have different uses on different faces. (This problem could be rectified by considering edge uses.) The second problem was that a small modification of the topology of a surface could require reevaluating the status of all edges of the surface. These considerations lead us to the following requirements for our flattening process.

i) Produces a 2D domain with minimal global and local distortion. Ability to trade-off shape distortion, dimension distortion, and performance depending on application.
ii) Supports multiple loops and non-convex boundaries
iii) Automatically finds constraints between the repeated edges of a face's topology that lower distortion in the 2d domain.
iv) Overall procedure must be fast and robust.

4 Weighted Edge Flattening Method (WEFM)

The WEFM will be presented as both a local and a global solution. The local solution is a simpler framework, but can become unstable with large systems of badly shaped triangles. It also requires explicit constraints. A global solution that overcomes these problems is presented.

4.1 Rolling Out a 3D Triangulation into 2D Using Iteration

Consider creating a 2D domain from a 3D triangulation surface with the five triangles shown in Figure 1. All of the triangles of the surface are equilateral, and four of them are from the top of a square based pyramid. An additional (equilateral) triangle is added with one side attached to one side of the square at the base of the pyramid. In the case of this example, it is impossible to put the vertices into 2-d without distorting the lengths of the edges and the angles of the triangles. Notice the angles of the four triangles the meet the vertex of the pyramid are 60 degrees. However, as this point is interior to the surface the total angle must be 360 degrees. But only triangles with all angles the same have all sides of the same length, so the sides can not be all the same length in 2D. If we put a cut at the thick solid line in Figure 1a, and let these sides

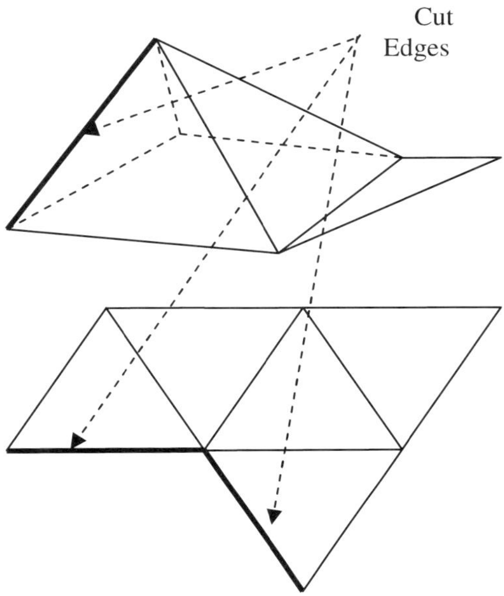

Fig. 1a. Simple 3D Triangulation

Fig. 1b. Flattened 2D domain with cut (5 equilateral triangles)

separate in the UV plane, then it is possible to find a UV domain with no area/length distortion except measurements that involve the cut. The result is given in Figure 1b.

When rolling the triangulation out to the plane requires deformation, averages/compromises are used for determining the positioning of vertices. The first step of the iterative procedure is to lay a long perimeter edge of the triangulation along the x-axis. The edge is given the same length that it has in 3D. For the given figure, the edge is placed from vertex 1 to vertex 2 on the x – axis. Then the vertex is placed opposite to the edge in the domain to keep the same shape and orientation. This process is continued, picking a vertex that neighbors the edges placed in 2D. In Fig. 2a, this is done with vertices 3 and 4. When vertex 5 comes, it is found that it is neighboring two different triangles, and the coordinates found for the vertex are different. The vertex is placed at weighted average of these results. Each triangle is given a weight which is determined by the length of the opposite side divided by the altitude to the vertex we are placing in 2D. Up until vertex 5 is placed in 2D, all of the edges have their original length. This makes it possible for each triangle to have a position for a neighboring vertex that does not change any lengths or angles. Once the results are averaged, vertex 5 is no longer at each desired position, and distortion results.

When we come to placing vertex 6 in the plane we have a new problem. The length from vertex 4 to vertex 5 in the plane is different then they are in 3D. Vertex 6 could be placed such that:

1) The triangle in 2D has the same angles as in 3D.
2) The altitude to the vertex has the same length in 2D and 3D.
3) The triangle in 2D has the same (absolute) area as in 3D.

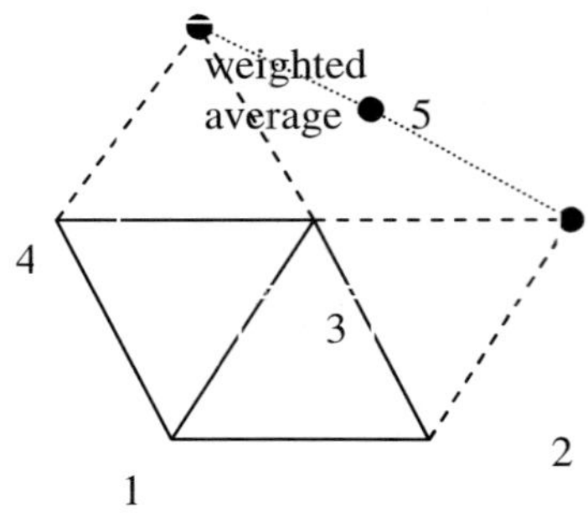

Fig. 2a. Simple triangulation roll-out where vertex 5 is weighted average from triangles 4-3-5 and 3-2-5

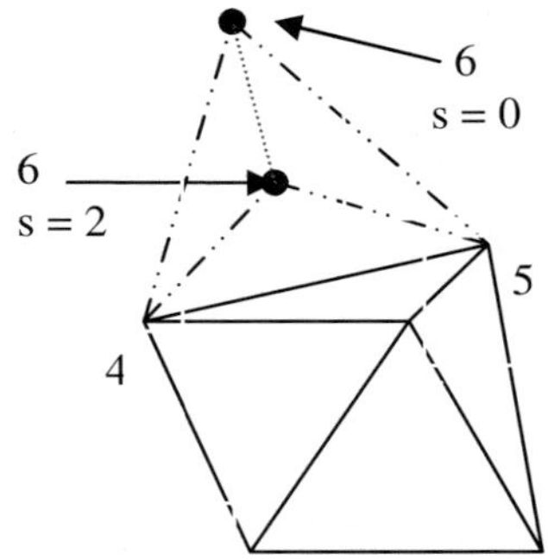

Fig. 2b. Simple triangulation roll-out where vertex 6 placement can vary from preserving angles to preserving area

To calculate the location of vertex 6 for these different cases a notation based on half-edges is used. (Choices 2 and 3 will be made unique with further restrictions.) A half-edge is associated with one and only one side of a triangle. There are one half-edge for each exterior edge of the triangulation, and two half-edges for each interior edge of the triangulation. A special notation is used to indicate the vertex indices associated with an half-edge with index h. The vertex indices along a half-edge with index h are designated in the counter-clockwise order of the triangle by $I(h,1)$ and $I(h,2)$. The vertex opposite the half-edge with index h is designated by $I(h,3)$. Consider this notation applied to Fig. 2b for the half-edge the goes from vertex 4 to vertex 5. This half-edge is in triangle 4-5-6, since is assumed the triangles are numbered consistently and counter-clockwise. If the half-edge from vertex 4 to vertex 5 has an index of 13, then $I(13,1) = 4$, $I(13,2) = 5$, and $I(13,3) = 6$.

Associated with each half-edge are two parameters that give the shape of the triangle. The first gives how far along the half-edge is the opposite vertex. Using the dot product to find the parameter (α_h) along the half-edge:

$$\alpha_h = \frac{\left(v_{I(h,2)} - v_{I(h,1)}\right) \cdot \left(v_{I(h,3)} - v_{I(h,1)}\right)}{[L_h]^2} \tag{1}$$

where L_h= length of the half edge with index h. The second parameter (β_h) is a ratio of the altitude to the opposite vertex to the length of the half-edge with index h. Using the cross product for the parameter perpendicular to the half-edge:

$$\beta_h = \left| \frac{\left(v_{I(h,2)} - v_{I(h,1)}\right) \times \left(v_{I(i,3)} - v_{I(h,1)}\right)}{[L_h]^2} \right| f\left(I(h,3)\right) \tag{2}$$

where $f(I(h,3))$ is an adjustment for Gaussian curvature around the opposite vertex. For a vertex on the boundary it is set to 1.0. For an interior vertex it is the sum of the angles around the vertex divided by 2π. Also associated with an half-edge is the ratio

(ρ_h) of the geodesic length of the half-edge with index h (which we take to be the 3D length) to the length of the edge in the UV domain. That is

$$\rho_h = \frac{L_h}{\left|z_{I(h,2)} - z_{I(h,1)}\right|} \tag{3}$$

where the each z refers to the UV position of the vertex with the same index. The formula to calculate a new UV location of a vertex from anopposite half-edge is given by eqn.(4) where the UV coordinates have been

$$z_{I(h,3)} = z_{I(h,1)} + \alpha_h(z_{I(h,2)}) - z_{(I(h,1))}) + \beta_h(z_{I(h,2)} - z_{I(h,1)})(\rho_h)^s i \tag{4}$$

represented as complex numbers, i is the imaginary unit vector, and s is a global parameter which gives different types of "roll outs". Choices 1,2, and 3 correspond to s values of 0, 1, and 2 respectively. These choices are referred to as conformal, altitude preserving, and authalic. Notice for the s =0 solution the value of ρ does not matter. When one vertex is across from many half-edges, then the weight for each half-edge is proportional to the reciprocal of its associated β. The weights are normalized to sum to 1.0. Taking half-edge 13 which goes from vertex 4 to vertex 5, and using an s value of 1.0 the equation (4) becomes:

$$z_6 = z_4 + \alpha_{13}(z_5 - z_4) + \beta_{13}(z_5 - z_4)(\rho_{13})i \tag{5}$$

The order in which points are selected for putting into 2d affects the solution that is found. Ordering the vertices by the ratio of the 3d lengths of the edges which are in 2d that ring a vertex to the perimeter of the ring in 3d provides a good heuristic for sorting the list of possible vertices. Once all the points are in the UV plane, we can continue to iterate through all vertices updating their position. Choice (1) can lead to degenerate solutions, and choice (3) is often unstable.Though choice (2) often works well for domain generation, it is not always stable and can produce domains that are of poor quality. It is also order dependent; heuristic choices of which vertex to calculate next can dramatically affect the result. Though it is straightforward to constrain points of the iterative solution, creating periodic boundary conditions is more problematic.

4.2 A Global Method for Triangle Roll-Out

To overcome the limitations of the iterative scheme, we move to a global solution using a least squares fit. The solution is directly in terms of planar coordinates, and the constraints can be any linear combination of vertex positions. The function to be minimized consists of four terms. The first term is a surface energy term based on the same formulation as the iterative method. The second term uses Lagrange multipliers to take out the translation degree of freedom by constraining a weighted average of the UV points to be at the origin. The third term is a more general constraint term used to enforce boundary conditions and eliminate a rotational degree for freedom. The rotational degree of freedom can be eliminated by enforcing the difference between the UV position of two vertex uses to be a vector. (At repeated edges the vertices can be split in the domain) The last term is based on angles in UV space between edges on the boundary loops. The angles are for a counterclockwise direction in the designated outer loop, and for a clockwise direction in all other loops. This convention breaks the symmetry of the problem for uncut cylinders. The upper limit n is for the number of vertex uses of the model. The upper limit m is for the number of constraints in the model. The upper limit d is for the loops of the model. l_i is the number of facet segments for loop i. The four term function to be minimized for the vector of unknown UV positions:

$$F(Z)=\sum_{i=1}^{n} q_i \overline{q_i} + \lambda_1 \sum_{i=1}^{n} \eta_i z_i + \sum_{k=2}^{m} \lambda_k \left(\left[\sum_{i=1}^{n} g_i^k z_i \right] - c_k \right) + \sum_{i=1}^{d} \mu_d \sum_{j=1}^{l_i} \frac{p_{i,j} \overline{p_{i,j}}}{\left| l_{i,j} \right|} \tag{6}$$

qi of the first term is a 2d vector (or complex number). Multiplying q_i by its complex conjugate gives a distance squared. The vector q_i starts at a weighted average of UV positions determined by all half-edges opposite the vertex use. (There is one such half-edge for each triangle that contains vertex use i.) It ends at the UV position of the vertex use times the sum of the weights of all half-edges for which it is an opposite.

$$q_i = \sum_{\substack{h \in H \\ \forall I(h,3)=i}} w_h \left[z_i - \left[z_{I(h,1)} + \sigma_h \left(z_{I(h,2)} - z_{I(h,1)} \right) \right] \right] \tag{7}$$

The index notation for vertices based on half-edges, $I(h,1)$ etc., was described in the previous section. The scaling-rotation vector, which depends on the type of "roll out, is:

$$\sigma_h = \alpha_h + \beta_h (\rho_h)^s \mathrm{i}. \tag{8}$$

In the special case s is set to zero, a value for ρ is not needed. The weight of each half-edge is given by

$$w_h = \frac{1}{\beta_h} \left(\sum_{\substack{h' \in H \\ \forall I(h',3)=I(h,3)}} \frac{1}{\beta_h'} \right)^{-k} \tag{9}$$

where β_h is defined in eqn. 2. The sum is over the set of all half-edge indices that have the same opposite vertex use. Setting k to 1 normalizes the sum of the weights around a vertex use to 1. Improved results are often found when k is set to ½. (In the iterative scheme k is always set to 1.) For the second term of eqn. (6), η_i is the area of all 3D triangles connected to vertex $\boldsymbol{i}$. For the second and third terms the λ_k are Lagrange multipliers. For the third term, g_i^k is a weight for a point of the constraint, and c_k is a fixed vector [5]. In our current usage, g_i^k is usually -1, 0, or 1 and c_k is zero. But these constants can be any complex value. The last term of the global equation involves loop segments of different loops. $p_{i,j}$ is a vector relation for loop i segment j. The term preserves boundary angle and length ratios:

$$p_{i,j} = \left(z_{i,j-1} - z_{i,j}\right) \cdot r_{l,-1} \cdot e^{i\varphi_{i,j}} + \left(z_{i,j+1} - z_{i,j}\right) \cdot r_{i,1} \cdot e^{-i\varphi_{i,j}} \tag{10}$$

φ is half the goal angle between two neighboring segments of a loop. $r_{l,+-l}$ is the ratio of segment length before/after the vertex to their average. In the last term, $l_{i,j}$ is the average length of two consecutive segments of the loop. μ_d is a weighting factor for the boundary. $z_{i,j}$ is the UV position for the point on loop i at the end of segment j. Solve the system by setting the partial derivatives for the points and Lagrange multipliers to zero.

4.3 General Solution Process

The solution of the least squares minimization problem is found by solving the equations resulting from setting the partials for the points and the Lagrange Multipliers to zero. Approaching the general nonlinear solution is done in three steps:

1) Solving the (linear) conformal system without repeated edge constraints. Return if the solution is of sufficient quality.
2) Compare solutions for repeated edges to find if each repeated edge should be periodic, free, or together (scar). Again, can return with solution if domain is of sufficient quality.
3) Repeat solve with increasing s value using the constraint decision found in step 2. Iterate to an s=1.0 solution. Use the boundary energy term in the nonlinear solution. When s is not 0.0, a value of ρ_h determined from equation 3 must be used in equation 8. The value is determined from the solution at the previous s.

During the solution process we often make use of a score to assess the quality of the domain. The score is from zero to one based on a weighted average of the square of the ratio of areas for the triangles in UV and 3d space. (The reciprocal of the ratio is taken if it is above 1.0). One is subtracted from the total weighted score for all triangles if there is any triangle flip. Two is subtracted for boundary intersection. (A flip of a triangle would result in a mapping such that a position in the UV domain would not correspond to one point in the 3-d triangulation. This could lead to an unusable mesh.).

To eliminate these degrees of freedom for step 1 of the process, Levy fixed two points that were far away in a 3D geodesic sense. This requires finding geodesic distances in 3D with holes. Instead of fixing two points *apriori*, we do most of the work of solving the linear system, and add the constraints later. To do this extensive use was made of the method of Lagrange Multipliers and matrix partitioning. To eliminate the translational degree of freedom, a weighted average of the vertex UV positions was constrained to be at the origin. To eliminate a rotational degree of freedom, we fix the vector between two points that are on the boundary. Once this solution is found, we can find two points on the boundary that are far apart in the UV domain. We eliminate the constraint fixing the facet edge, and add the constraint for the two points that are far apart in UV. We reuse the LU decomposition of the part of our matrix that involves points, so the second solution is found very quickly. One example where the two-step approach works well, while the methods of Levy and Desbrun do not, is in the flattening of the triangulation of a sphere with a small hole. A two level matrix partitioning procedure significantly speeded up step 2 of the process.

5 Surface Mesh Generation

Once a 2D triangulated domain is generated, the boundary of the domain is discretized with nodes (driven by a global element size) and a variety of 2D mesh generation algorithms are used to fill the area. The algorithms used can be broadly classified into three categories, namely

a. Unstructured free mesh generation using subdivision technique [10,11]
b. Unstructured free mesh generation using a combination of subdivision and advancing loop-front methods [12]
c. Structured mesh generation using boundary-blended transfinite interpolation techniques [13,14]

5.1 Mesh Quality Measure

For free, unstructured meshes, mesh distortion can be measured by an element quality metric, λ, where

$$\lambda_{mesh} = (\Pi_{i=0}^{n} \lambda_i)^{1/n} \qquad \text{denotes the quality of the mesh} \tag{11}$$

λ_i (mesh quality of the i-th triangle) where λ_i is as proposed by Lo[15].
All elements need to pass a critical threshold defined by $\lambda_{cr} = 0.6$.

Both λ_{mesh} and λ_i of the worst element in the mesh can be compared for different domains to decide which is better.

For transfinite meshes however, element or mesh quality, as determined by λ_{mesh} or λ_i is not necessarily a good measure. Bad domains can cause the transfinite nature (structured) of the mesh to be destroyed even for the same number of nodes/elements and element connectivity. The rail lines of the mesh can be wiggly, or unparallel – all of these could destroy the properties of a transfinite mesh. To measure the wiggle of the rail lines a "wiggle factor" is used. The wiggle factor, ε_{fi} for the i-th rail line is an error norm computed for each interior mesh node on the line summed over the length of the rail line in j direction made up of n nodes.

$$\varepsilon_{fi} = 1/n \sum_{j=0}^{n} (L_{ij} - L_{ij-1})^2 / \bar{L}_i^2 \tag{12}$$

Lij denotes the length vector at node ij (jth node on the i-th rail line). $\bar{L}_i$ denotes the average length vector of all nodes on the i-th rail line.

6 Examples and Discussion

Three different examples will be discussed to show the benefit of using an efficient flattening or domain development technique for high quality surface meshing. Certain specific aspects of mesh quality or pattern will be discussed in this connection. In all of the results presented a k value of 1 was used in eqn. (9), though it has subsequently found a k value of ½ slightly improved the results for the boundary structured free mesh.

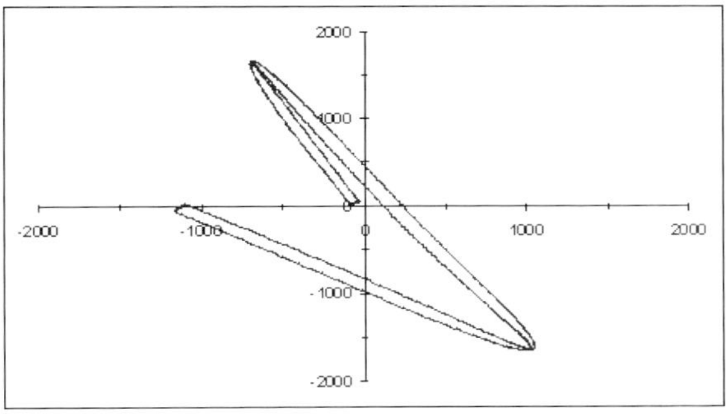

Fig. 3a. 2D domain of a helical surface developed by WFEM at s=0 (conformal) with no boundary constraint

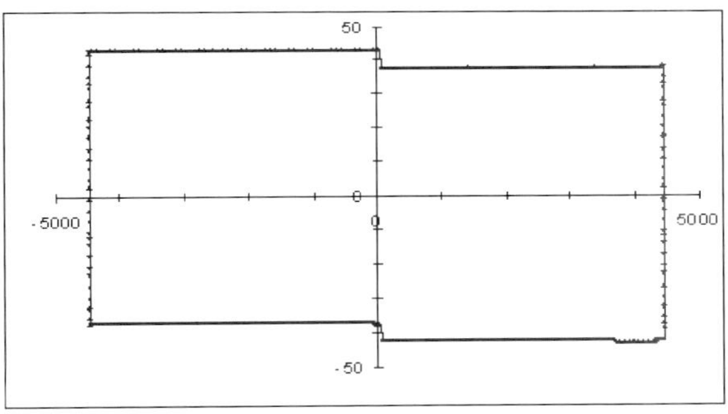

Fig. 3b. 2D domain of the same helical surface by WFEM at s=1 with boundary constraint

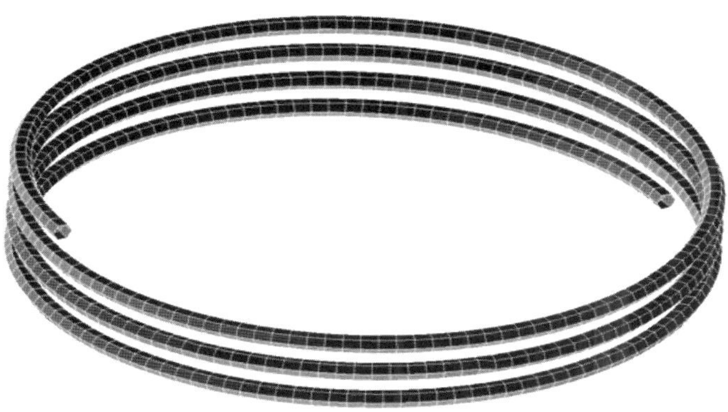

Fig. 3c. 3D final transfinite mesh on the helical surface

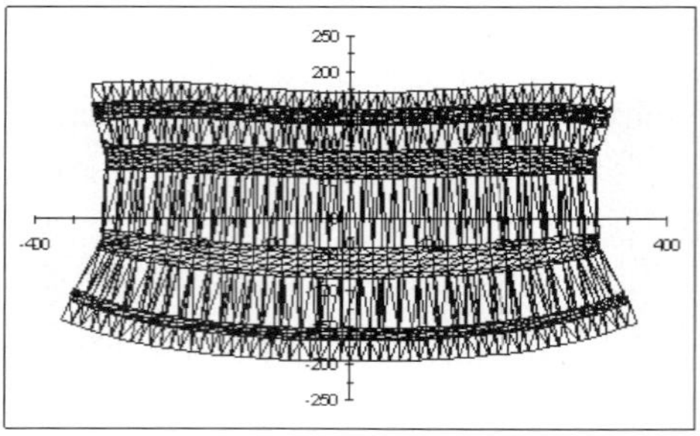

Fig. 4a. 2D domain of a pressure vessel wall surface by WEFM at s = 0 (conformal)

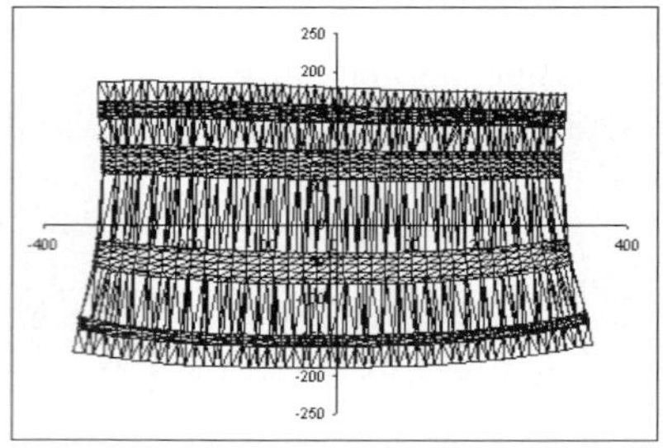

Fig. 5a. 2D domain of a pressure vessel wall surface by WEFM at s = 0.3 with boundary energy term

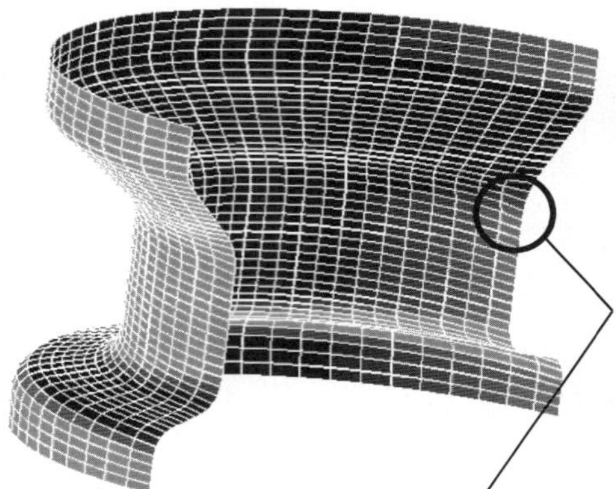

Fig. 4b. Corresponding final 3D transfinite mesh with wiggly ends

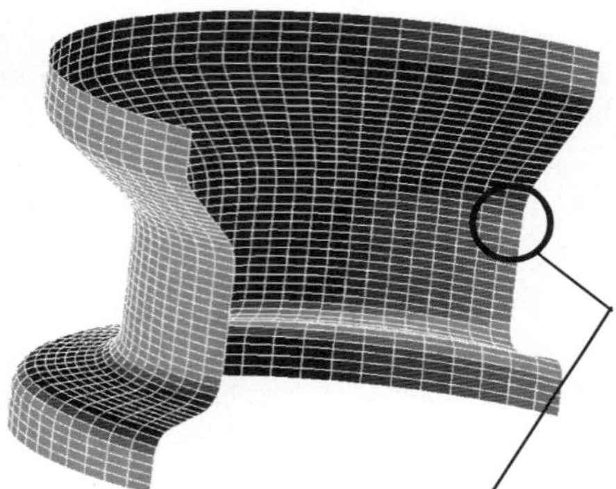

Fig. 5b. Corresponding final 3D transfinite mesh

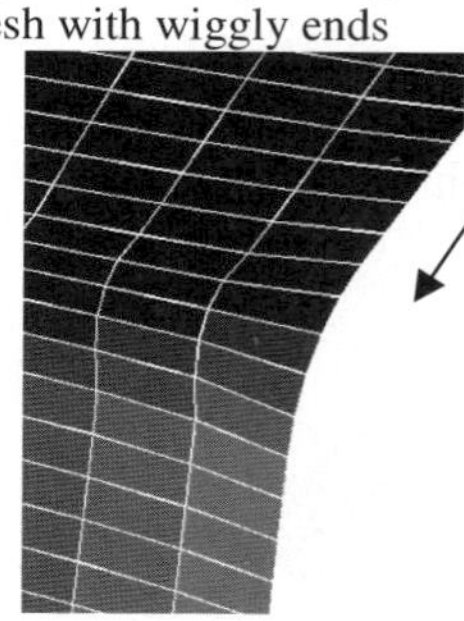

Fig. 4c. Rail lines at near face boundary showing wiggles

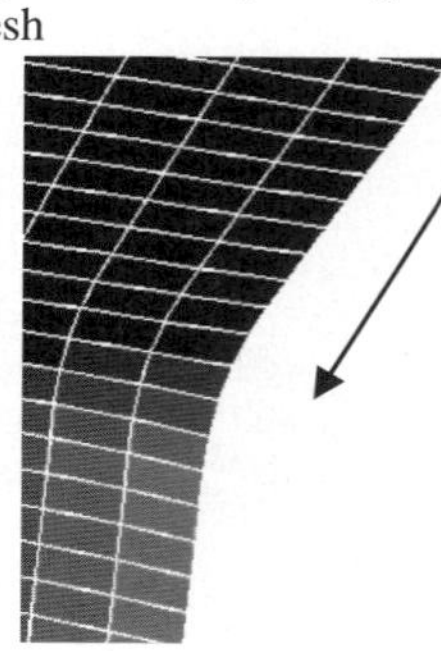

Fig. 5c. Smooth rails

6.1 Helical/Spiral Surface Meshing

Meshing is often challenged by merged tessellated surfaces that are helically/spirally coiled. Such surfaces abound in threaded mechanical enginering parts and are often encountered in finite element analysis. Fig. 3a shows a typical rolled out 2D domain

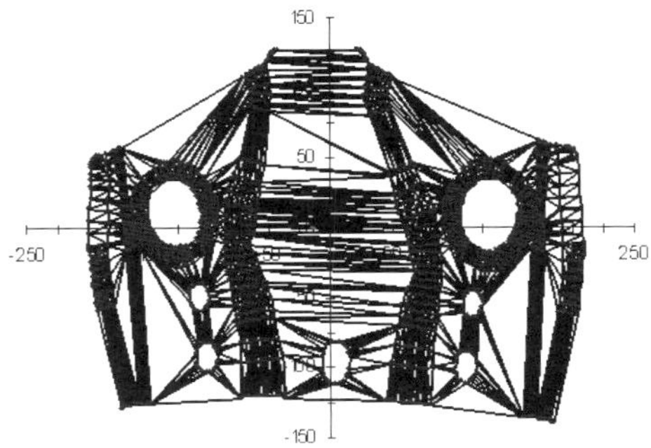

Fig. 6a. 2D domain of an automobile frame by WEFM at s=0.8 with boundary energy term

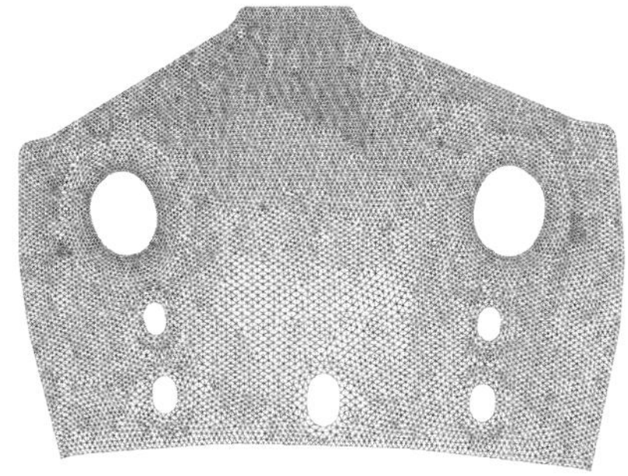

Fig. 6b. 2D triangular CSALF mesh on the 2D domain

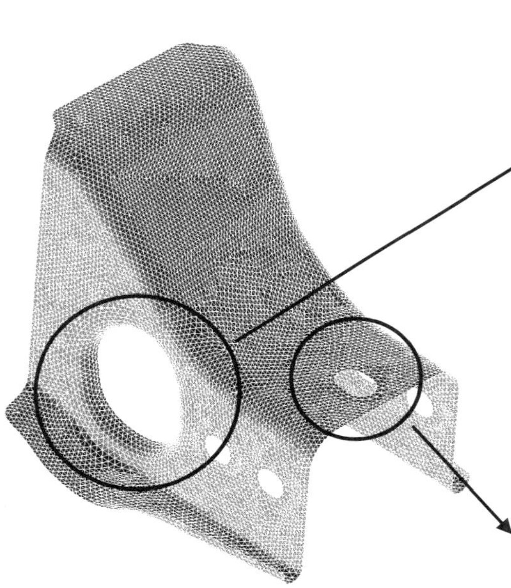

Fig. 6c. Corresponding final 3D mesh

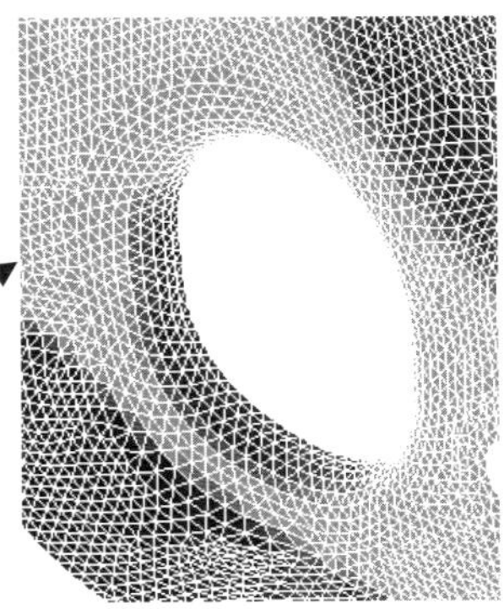

Fig. 6d. Layered mesh around frame cut-out

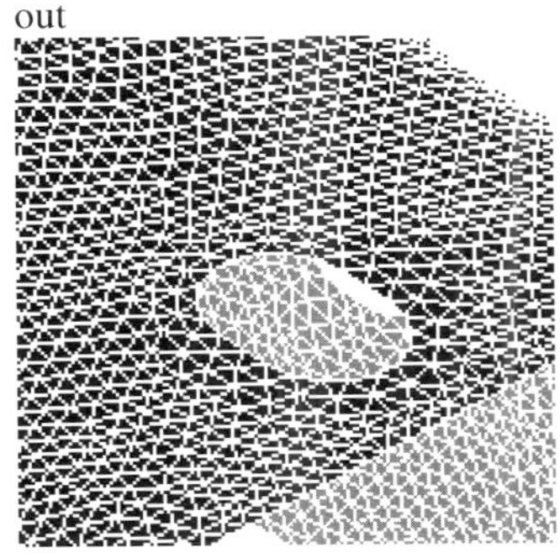

Fig. 6e. Boundary structured mesh around holes

which has appreciable distortion but is not suitable for map meshing. Fig. 3b presents a continuous section of a 2D domain developed by WFEM with s=1. This domain is regular and can be meshed by transfinite interpolation (mapped meshing) which is depicted in Fig. 3c. Note, the domain has a small kink resulting from an irregular cut line chosen along the facet edges. The 3D mapped mesh reflects a small local distortion at that location.

6.2 Accurate Transfinite Meshing Needs

Transfinite or mapped meshes are structured and mathematically accurate. The Coon's surface is not easily available on discrete geometry. Thus the efficiency and the accuracy of the flattening strategy is important for generating such meshes. Fig. 4a shows the 2D domain of a pressure vessel wall surface developed by WFEM (s=0.0) with no boundary energy term. The corresponding 3D transfinite mesh is shown in Fig. 4b. When we zoom in on the elements near the vertical edge, as is evident in Fig. 4c, we notice a lateral waviness in a certain region. This region corresponds to the two longitudinal kinks that exist on the vertical edges of the 2D domain. The wiggle factor of the transfinite distortion parameter for this mesh is 0.12 which is far above the limiting value of 0.01. With a domain developed by WFEM (s = 0.3) with a boundary energy term as pictured by Fig. 5a, these irregularities vanish and we get a perfect 3D transfinite mesh as shown in Fig. 5b-5c. The wiggle factor, described by eqn. (12) is 0.007.

6.3 Boundary-Structured Free Meshing on Complex Geometry

Good quality boundary-structured free meshes are important for a host of structural and fluid flow analysis problems. Generation of such surface and volume meshes on complex industrial geometry is key to the success of an engineering design process. Fig. 6a shows the 2D domain of such a frame structure developed by WFEM (s=0.8). A triangular CSALF mesh is shown on the 2D domain in Fig. 6b. These zones are of importance to the designer as they are usually susceptible to high stresses. The final 3D mesh shown in Fig. 6c reveals the same. It is important to note that the mesh, although unstructured in a global sense, is extremely layered/structured around all the inner circular/oval boundaries (see insets in Fig. 6d -6e).

7 Conclusion

It was found the conformal (s=0) solution for the WEFM can be used for meshing processes based on a 2d domain. However, as in all conformal based solutions, the dimensions for non-developable triangulations are changed when they are mapped to the UV domain. This can be a problem when high quality meshes are required. It was found an intermediate "altitude preserving" condition (s=1) of the WEFM, which compromises shape and dimension distortion, could give superior results when used with the presented map meshing process, and with forms of "mesh paving". It was found that automatically choosing from all the valid solutions gave us the best overall strategy. An automatically procedure for creating and evaluating constraints for repeated edges was developed. It was used to determine if each repeated edge should be left free, treated as a scar, or be part of a periodic boundary condition. A boundary energy term in the WEFM formulation enabled the solution of indeterminate multi-loop problems, and generally improved the quality of the resulting domain. The difference in quality was readily apparent in the resulting 3D mapped mesh produced from the different domains.

References

1. Floater, M.S., Hormann, K.: Surface parameterization: A tutorial and survey. In: Dodgson, M.S.E.N., Sabin, M. (eds.) Advances on Multiresolution in Geometric Modelling, Springer, Heidelberg (2004)
2. Floater, M.S.: Mean value coordinates. Computer Aided Geometric Design 14(3), 231–250 (1997)
3. Hormann, K., Greiner, G.: MIPS: An Efficient Global Parameterization Method. In: Laurent, J.-P., Sablonniere, P., Schumaker, L.L. (eds.) Proceedings, Curve and Surface Design, pp. 153–162. Vanderbilt University Press (2000)
4. Levy, B., Petitjean, S., Ray, N., Maillot, J.: Least Squares Conformal Maps for Automatic Texture Atlas Generation. In: Proceedings, ACM, SIGGRAPH (July 2002)
5. Desbrun, M., Meyer, M., Alliez, P.: Intrinsic Parameterization of Surface Meshes. Computer Graphics Forum (EUROGRAPHICS 2002) 21(3), 209–218 (2002)
6. Sheffer, A., de Sturler, E.: Parameterization of faceted surfaces for meshing using angle based flattening. Engineering with Computers 17, 326–337 (2001)
7. Sheffer, A., Levy, B., Mogilnitsky, M., Bogomyakov, A.: ABF++: Fast and Robust Angle Based Flattening. Proceedings, ACM Transactions on Graphics 24(2), 311–330 (2005)
8. Zayer, R., Levy, B. Seidel, H.-P.: Linear Angle Based Parameterization. In: Proceedings, Eurographics Symposium on Geometry Processing (2007)
9. Gu, X., Yau, S.-T.: Global Conformal Surface Parameterization. In: Proceedings, Eurographics Symposium on Geometry Processing, pp. 127–137 (2003)
10. Cabello, J.: Towards Quality Surface Meshing. In: Proceedings, 12th International Meshing Roundtable, pp. 201–213 (2003)
11. Schoof, A.J.G., Th, L.H., Van Beukering, M., Sluiter, M.L.C.: A general purpose two-dimensional mesh generator. Advances in Engineering Software 1(3), 131–136 (1979)
12. Mukherjee, N.: CSALF: A combined subdivision and advancing loop-front approach to generating controlled boundary structured free meshes. International Journal of CAD/CAM (submitted, 2007)
13. Perronnet, A.: Triangle, tetrahedron, pentahedron transfinite interpolations. Application to the generation of C0 or G1-continuous algebraic meshes. In: Proceedings. Internationa Conference of Numerical Grid Generation in Computational Field Simulations, Greenwich, England, July 6-9, 1998 vol. 7, pp. 467–476 (1998)
14. Mukherjee, N.: High Quality Bi-Linear Transfinite Meshing with Interior Point Constraints. In: Pebay, P.P. (ed.) Proceedings, 15th International Meshing Roundtable, pp. 309–323. Springer, Heidelberg (2006)
15. Lo, S.H.: A New Mesh Generation Scheme For Arbitrary Planar Domains. International Journal For Numerical Methods in Engineering (21), 1403–1426 (1985)

Delaunay-Based Anisotropic Mesh Adaptation

Doug Pagnutti and Carl Ollivier-Gooch

Advanced Numerical Simulation Laboratory
The University of British Columbia

Summary. Science and engineering applications often have anisotropic physics and therefore require anisotropic mesh adaptation. In common with previous researchers on this topic, we use metrics to specify the desired mesh. Where previous approaches are typically heuristic and sometimes require expensive optimization steps, our approach is an extension of isotropic Delaunay meshing methods and requires only occasional, relatively inexpensive optimization operations. We use a discrete metric formulation, with the metric defined at vertices. To map a local sub-mesh to the metric space, we compute metric lengths for edges, and use those lengths to construct a triangulation in the metric space. Based on the metric edge lengths, we define a quality measure in the metric space similar to the well-known shortest-edge to circumradius ratio for isotropic meshes. We extend the common mesh swapping, Delaunay insertion, and vertex removal primitives for use in the metric space. We give examples demonstrating our scheme's ability to produce a mesh consistent with a discontinuous, anisotropic mesh metric and the use of our scheme in solution adaptive refinement.

1 Introduction

Numerical solution of PDEs is an important analysis tool for scientists and engineers. Ideally, such numerical software should require only specifying the physics, geometry, and boundary conditions of the problem and then produce a solution of a certain accuracy. Unfortunately, assessing and minimizing discretization error is a non-trivial task, requiring time consuming mesh dependence analysis or computation of solutions on meshes that are much finer than the required solution accuracy dictates. Mesh adaptation seeks to automate this process by first computing the solution on a coarse mesh and then successively refining the mesh so that each mesh is more nearly optimal for the solution being computed. Unstructured meshes are well-suited for adaptation because they can be modified locally much more easily than structured meshes, which have a fixed topology.

Labelle and Shewchuck [7] have developed a theoretically sound extension of guaranteed-quality Delaunay meshing by constructing anisotropic Voronoi diagrams and then dualizing to a Delaunay triangulation. Unfortunately, these anisotropic Voronoi diagrams can be difficult to generate, and the dualization to a valid triangulation requires restrictions on metric smoothness. Alternatively, there are other more heuristic methods for generating anisotropic meshes that rely on local optimization algorithms. Dompierre et al. [5] suggest splitting edges with length[1] above a threshold, removing nodes with all edge lengths below a threshold, and swapping edges and moving nodes to equidistribute length. New edges are assigned lengths based on an interpolated error estimate from the background mesh. Buscaglia and Dari [3] take this a step further by testing a large set of potential operations on a sub-mesh and then selecting the one that most increases quality according to an error estimate on a background mesh; this approach can produce excellent meshes, but at a high cost.

Our algorithm combines the heuristic nature of these two approaches with some of the attributes of isotropic Delaunay meshing. This gives us two principal advantages over other heuristic methods. First, we do not rely on moving vertices and so metric information can be easily retained without storing and referencing the original mesh. Second, because of an improved choice of where to insert vertices, our algorithm produces quality anisotropic meshes without the expense of methods like Buscaglia and Dari's [3], or the challenges in updating the metric when moving nodes of the scheme of Dompierre et al [5]. In addition, our scheme does not require the use of a background mesh to maintain error / metric information.

To explain our algorithm we will examine, in Section 2 metric spaces and how they can be used for anisotropic meshing. Then, we will create a quality measure for our anisotropic mesh (Section 3) and modify several existing mesh operations to increase that quality measure (Section 4). Next, we turn our attention to treatment of boundaries during anisotropic adaptation (Section 5). Finally, we will demonstrate that our algorithm produces meshes well suited to the desired metric.

2 Metric Spaces

Before defining what a quality anisotropic mesh is, we must first define the desired cell sizes and shapes using a metric to measure distance between two points. For a vector space X, a metric is a function $d : X \times X \rightarrow \mathbb{R}$ for which the following conditions are satisfied for every $x, y, z \in X$.

$$\begin{aligned} d(x,y) = d(y,x) > 0 &\iff x \neq y \\ d(x,x) &= 0 \\ d(x,z) &\leq d(x,y) + d(y,z) \end{aligned}$$

For uniform isotropic meshing, the metric is the standard distance measure $\|x - y\|$. More complicated metrics allow for distance to be measured differently depending on orientation, thus introducing anisotropy. In particular, the metric

[1] The term error is used in the cited paper, however the metric length can be defined as equivalent to the error and so we will use the term length to be consistent with usage later in the paper.

most commonly used in anisotropic meshing can be found by linearly transforming the standard distance measure $d_{\text{iso}}(x, y)$ into one which is anisotropic $d_{\text{ani}}(x, y)$ as follows:

$$d_{\text{iso}}(x, y) = \sqrt{V^T V}$$
$$d_{\text{ani}}(x, y) = \sqrt{(TV)^T (TV)}$$
$$d_{\text{ani}}(x, y) = \sqrt{V^T (T^T T) V}$$

Where $V = x - y$ and T is an invertible linear transformation. Since a triangle is perfectly defined by the lengths of its sides, any triangle in an anisotropic mesh has an equivalent triangle when its edge lengths are measured by the metric. Quality measures can typically be rewritten to use only edge lengths as well. For example, if a triangle's edges have metric lengths (a, b, c) then the area A and circumradius R of a triangle can be calculated by the following formulas

$$A = \frac{1}{4}\sqrt{(a+b+c)(-a+b+c)(a-b+c)(a+b-c)}$$
$$R = \frac{abc}{4A} \tag{1}$$

and subsequently used to compute quality measures in the metric space. Although we use the linear transformation metric throughout this article, our algorithm is based purely on vertex spacing in the metric space and could be adapted to use any metric.

For most problems of interest, different metrics must be used in different areas of the domain. This is the principal difficulty faced by anisotropic meshing algorithms, because even if the metric function is valid at a specific point, the interpolated metric space between any two points might not be valid. We deal with this problem by defining the metric discretely at vertices. When refining the mesh, we measure the length of each triangle's edges according to the average metric of the three vertices. This provides a good estimate of whether the triangle is large enough that it requires insertion. If a vertex is inserted, the metric at that vertex is assigned based on a linear interpolation from the three vertices of the triangle in which it was inserted.[2] Alternatively, when coarsening the mesh we estimate the length of each edge by first calculating what that length is according to the two adjacent triangles and then averaging those lengths. These two definitions give unique triangle quality and edge length measures which can then be used for mesh improvement operations.

3 Quality Measures

There are a staggering number of triangle quality measures used to determine how good the elements of a mesh are. Since no quality measure is known to be

[2] Note that this is not necessarily the triangle it was inserted to split!

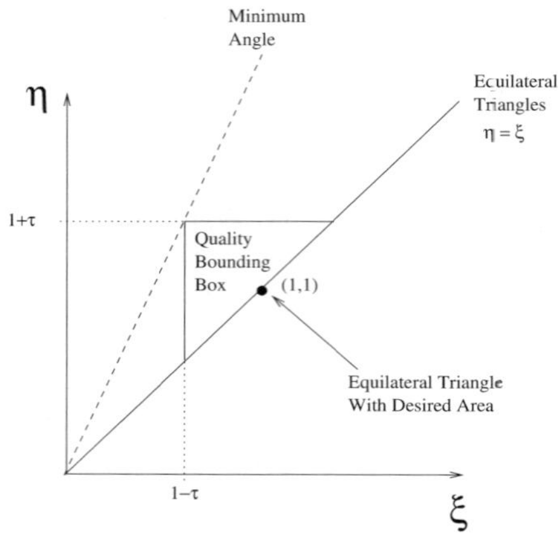

Fig. 1. The quality space used to evaluate triangles in the mesh

superior for every numerical solver, we have chosen to use the quality measure that best suits our meshing approach. Miller, Talmor and Teng [10] pointed out that the most natural quality measure for Delaunay refinement is the ratio of circumradius to shortest edge. We will therefore define a *quality space*

$$(\xi, \eta) = \left(\frac{3^{\frac{1}{4}}}{2\sqrt{A_{\text{des}}}} l_{\min}, \frac{3^{\frac{3}{4}}}{2\sqrt{A_{\text{des}}}} R \right) \doteq \left(\frac{0.658}{\sqrt{A_{\text{des}}}} l_{\min}, \frac{1.140}{\sqrt{A_{\text{des}}}} R \right) \tag{2}$$

where $l_{\min}$ is the length of the shortest edge, R is the circumradius, and A_{des} is the desired area. For a uniform isotropic mesh, the ideal triangle is an equilateral triangle with area $A_{\text{des}} = \frac{A_\Omega}{N_v}$ where A_Ω is the total area of the domain Ω and N_v is the number of triangles; this triangle has $(\xi, \eta) = (1, 1)$ for our chosen constants. We use this same concept to choose A_{des} in Equation 2. To estimate the area A_Ω of the domain Ω in the metric space, we sum the area of individual triangles in the metric space. Then, the ideal area A_{des} is found by dividing the total area A_Ω by the number of triangles N_v and a desired refinement ratio r, so that $A_{\text{des}} = \frac{A_\Omega}{r N_v}$. The equilateral triangle with this area would be located at $(\xi_{\text{des}}, \eta_{\text{des}}) = (1, 1)$ in the quality space.

Having defined the perfect triangle, we define the range of triangle qualities are considered acceptable based on a value τ such that triangles with $\xi < 1 - \tau$ or $\eta > 1 + \tau$ are considered bad. To ensure shape quality, we restrict τ to so that we have a lower bound on the minimum angle of the triangle, which can be written as $\frac{l_{\min}}{R} = 2 \sin \theta_{\min}$. Eliminating $l_{\min}$ and R in favor of ξ and η, we have

$$\frac{\xi\sqrt{3}}{\eta} = 2 \sin \theta_{\min}$$

and so for the corner point $(1 - \tau, 1 + \tau)$ in Figure 1,

$$\frac{1-\tau}{1+\tau} = \frac{2}{\sqrt{3}} \sin \theta_{\min} \tag{3}$$

$$\tau = \frac{3 - 2\sqrt{3} \sin \theta_{\min}}{3 + 2\sqrt{3} \sin \theta_{\min}} \tag{4}$$

So choosing θ_{min} is equivalent to choosing τ. Chew's uniform isotropic Delaunay refinement algorithm[4] produces triangles whose angles are all greater than thirty degrees so we have chosen to use this as the target angle giving $\tau = 2 - \sqrt{3} \approx 0.2679$. An illustration of the quality space is found in Figure 1.

4 Mesh Operations

There are three principal operations that our algorithm uses to improve the quality distribution of triangles in the mesh. The first two operations, edge swapping and point insertion, are extensions of isotropic Delaunay operations for use with metric spaces. The last operation, vertex removal, is done simply to remove the smallest edges in the mesh.

4.1 Swapping

The first and simplest operation is edge swapping. For isotropic Delaunay meshing algorithms, edge swapping is used to ensure all triangles have empty circumcircles. For our discrete representation of the metric, this is not necessarily possible. Furthermore, whenever an edge is swapped, the metric used to examine the new triangles can differ from the metric of the old triangles, possibly reversing the swapping decision. Both of these problems are solved using a heuristic approach. Instead of swapping for cells with non-empty circumcircles, we instead swap to maximize the minimum quality ratio ξ/η — after mapping according each triangle according to its own metric — for each pair of adjacent triangles. For an isotropic metric, this is equivalent to maximizing the minimum angle, which produces a Delaunay triangulation. To determine which configuration does this, we calculate (ξ, η) explicitly both for the current triangles and the triangles that would be formed if the edge were swapped. Note that our discrete metric definition — specifically, the metric function we use to compute metric edge lengths for a triangle — is dependent on vertex connectivity, so swapping an edge effectively changes the metric. Because of this, a Delaunay triangulation cannot be found by simply mapping the local vertex locations to the metric space, as one might do with a continuous metric.

4.2 Vertex Insertion

The second mesh modification operation that we use is vertex insertion. In isotropic Delaunay refinement, vertices are inserted for triangles with large circumradii or poor shape. After insertion, the resulting triangles have smaller circumradii, and no edge smaller than the original circumradius is created. In

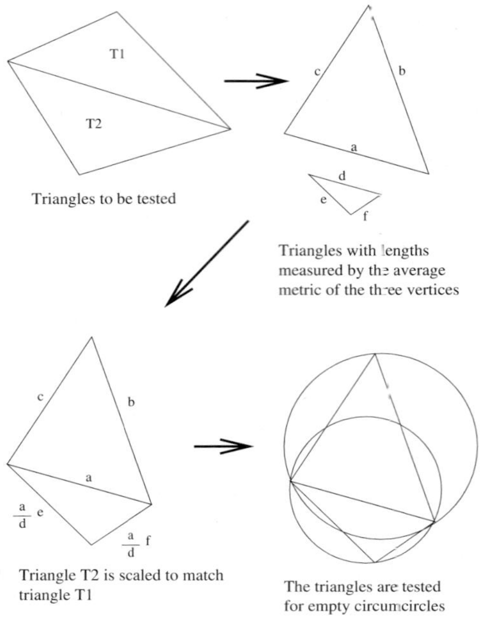

Fig. 2. Testing for Empty Circumcircles in the Metric Space

our anisotropic adaptation, we hope for similar behavior by inserting at an approximate circumcenter. We find this approximate circumcenter by creating a virtual sub-mesh around the triangle we want to refine. To construct this virtual sub-mesh, we first transform the initial triangle to one with lengths as described by the metric. Then, we calculate that triangle's circumcenter and build the virtual sub-mesh one triangle at a time until the circumcenter lies within it. Once this circumcenter is found, we transform it back to the actual mesh and insert a vertex at that location.

Due to our choice of metric, the measured length of a shared edge is not necessarily the same for both triangles. To address this when creating the virtual sub-mesh, we scale any added triangle so that the edges match. This does not deteriorate our insertion result since the Delaunay property is independent of size. In Figure 2, the triangles T1 and T2 are first transformed to have the lengths assigned by their respective metrics. Then T2 is scaled by a/d so that the edges can be aligned. Since the swapping algorithm discussed in Section 4.1 does not guarantee empty circumcircles, each time a triangle is added to the virtual sub-mesh it must be tested to determine whether it is Delaunay.

The construction of a virtual sub-mesh is illustrated in Figure 3. In this case, triangle $T1$ is chosen for insertion and the circumcenter X of the mapped triangle $T1'$ is calculated in the virtual space. Additional triangles are added to the virtual sub-mesh: their shape is determined by their relative edge lengths in the metric space, and they are scaled so that both copies of each edge have the same length

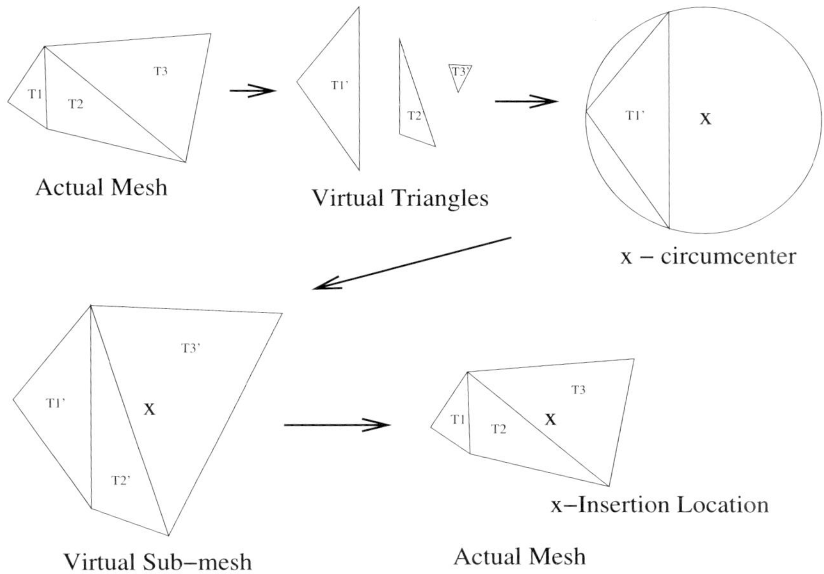

Fig. 3. Finding an approximate circumcenter of triangle T1 by constructing the virtual mesh

in the metric space. In this case, $T2$ and $T3$ are mapped, scaled, and stitched onto the virtual sub-mesh, at which point the sub-mesh contains the circumcenter of $T1'$ (in this case, in $T3'$). Once that insertion location is chosen in the virtual sub-mesh, it is linearly transformed back to the real mesh so that it has the same barycentric coordinates in both the real and the virtual triangle. The new vertex is inserted with the most simple connectivity and swapping is performed recursively. The desired effect of each insertion is to decrease circumradii without dramatically reducing the quality of the local triangles. Unfortunately, due to the discrete nature of the metric, there are some cases where this approach fails.

The first degenerate case occurs when the circumcircle of the triangle we wish to eliminate by insertion is not empty in the virtual mesh. When this occurs, we instead split the longest edge of that triangle. There is theoretically no limit on the minimum edge that this operation will create. To address this, we smooth the new vertex to maximize the minimum angle in the virtual mesh, resulting in a much better insertion location. A similar approach is used in other anisotropic refinement algorithms [1, 3]. In our algorithm, however, optimization is only required for degenerate cases, which are a small fraction of the total number of insertions.[3]

[3] The exact percentage of degenerate cases is entirely dependent on how the metric varies within the mesh. Experience has shown that degeneracy in 5-10% of insertions is typical for the metric we used.

The second degenerate case occurs when inserting the new vertex does not affect the triangle for which it was inserted. In other words, insertion did not reduce the largest circumradius in the mesh. This happens most often when there is large difference in metric size between triangles. Currently, insertion is simply repeated for the original triangle. Since that triangle will no longer have an empty circumcircle, refinement will split the longest edge and then locally optimize the location.

4.3 Vertex Removal

The final operation that we use to improve the mesh is coarsening. Coarsening is not a natural component of Delaunay meshing so there are no standard isotropic techniques to apply on the virtual mesh. Instead, we know that removing vertices increases edge lengths in the mesh. Thus, removing a vertex that is part of the shortest edge will always increase the minimum edge length. To remove a vertex, we first swap edges until there are only three edges incident on that vertex and combine the three incident triangles to form one triangle when the vertex is removed. The edges are then swapped recursively in the same manner described in Section 4.1. Coarsening is the only operation that can destroy metric information. Because of this, repeated refinement and coarsening tends to smooth the metric.

5 Boundary Protection

A key element of Delaunay meshing is boundary treatment. While the previous discussion of point insertion assumed that the circumcenter lies within the domain, that is not necessarily true. For cases when the circumcenter lies outside the domain or across a boundary, the virtual mesh cannot be completely constructed. When this happens, the boundary edge which terminated the virtual mesh is split. For cases where the circumcenter of a triangle lies inside the domain, but near the boundary, there are no constraints on the circumradius of the new triangle formed. To address this problem, we adapt the boundary protection proposed by Shewchuk [19]. That is, we disallow any insertion within the diametral lens of a boundary segment and instead split that boundary segment. A diametral lens can be found by taking the intersection of the two circles of radius $\frac{1}{\sqrt{3}}L$ with the boundary segment (of length L) as a chord. For an anisotropic mesh, the diametral lens is approximated by transforming any potential boundary triangles using the average metric of the two boundary vertices.

A further complication with boundaries is that any curvature of the boundary must be preserved throughout refinement. If the curvature of the boundary is not preserved, then the geometry will not converge to the modeled geometry, introducing unnecessary modeling error. This is also very important for high-order solutions because correct curved boundary representation is essential to achieving high order accuracy. Ollivier-Gooch and Boivin [2] solved this problem for isotropic meshes by limiting the curvature of the boundary discretization and

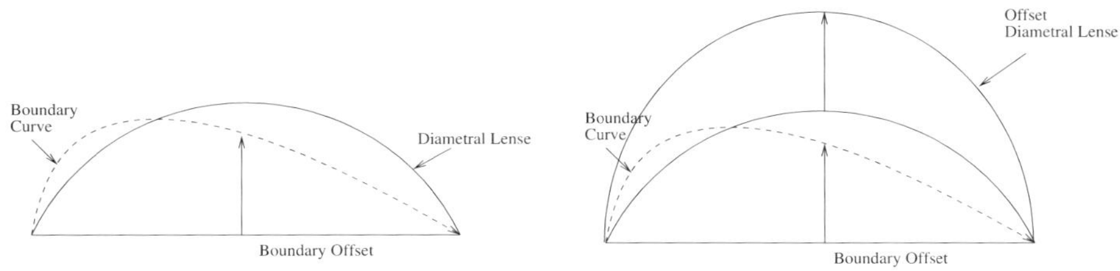

Fig. 4. Isotropic, modified diametral lens for curved boundary

thus enforcing that the diametral lens lie outside the boundary curve. This is a much more difficult task when the metric is allowed to vary along the boundary edge because the measured curvature changes depending on what metric is used. Instead we rely again on heuristics: expansion of the diametral lens and encroachment checking before boundary insertion.

First, we expand the diametral lens to better account for the curvature. To do this, we first calculate the vertical offset between the linear boundary segment and the actual curved boundary. Then, we scale the diametral lens of the segment to have a height that is the sum of the original lens height and the offset. An isotropic example of this is shown in Figure 4. For an anisotropic metric, we compute the offset in the metric space based on the average metric of the two boundary vertices.

As with similar approaches in isotropic meshing, making this adjustment still does not guarantee that new vertices will be inserted within the domain or that splitting a boundary curve will produce good quality triangles. To remedy this, whenever a boundary edge is split, any vertex that encroaches on the new edges in the virtual mesh must be removed. Together, these methods ensure that the refined meshes will respect the boundary discretization and that no overly-flat triangles will be created along the boundary.

6 Algorithm Summary

We apply the following algorithm to produce quality anisotropic meshes that match an arbitrary metric and have a prescribed maximum number of vertices.

1. Assign a metric to each vertex.
2. Sum the anisotropic area of each triangle and calculate A_{des} and thus (ξ, η) for every triangle.
3. Remove any interior vertices that encroach on a boundary segment according to the metric.
4. Recursively swap edges to minimize the ratio η/ξ. If this does not result in an anisotropic Delaunay triangulation, mark offending edges as being non-Delaunay.
5. Insert vertices for triangles with $\eta > (1 + \tau)$ until the maximum number of vertices is reached or there are no more triangles with large circumradii.

Assign metrics to new vertices based on a linear interpolation from the three vertices of the triangle in which the vertex was inserted. If a face used in constructing the virtual mesh is marked as non-Delaunay, then the treatment for non-empty circumcircles from Section 4.2 is applied.

6. Remove vertices for edges with $\xi < (1-\tau)$ until there are no more small edges or sufficiently many triangles with large η are created. Limiting how many large triangles can be created is an effective way of reducing the overlap between refinement and coarsening. If this limit is a fraction of the number of vertices inserted in the previous refinement step then it adds a guarantee of termination irrespective of how many iterations are allowed.
7. Repeat the process of inserting and removing vertices until all triangles are within the acceptable tolerance, no more vertices can be removed, or a fixed number of iterations are performed. For quality meshes, the final process must always be mesh refinement.

7 Results

To demonstrate our algorithm's ability to create anisotropic meshes from a metric, the metric $d(\mathbf{x}_1, \mathbf{x}_2)$ defined by Equation 5 is assigned at every vertex.

$$\begin{aligned} d(\mathbf{p}_1, \mathbf{p}_2) &= \sqrt{(\mathbf{p}_2 - \mathbf{p}_1)^T M (\mathbf{p}_2 - \mathbf{p}_1)} \\ M(x,y) &= \begin{bmatrix} a & 0 \\ 0 & b \end{bmatrix} \\ a &= \begin{cases} 1 \times 10^4 & 0.475 < x < 0.525 \\ 1 & \text{Elsewhere} \end{cases} \\ b &= \begin{cases} 1 \times 10^4 & 0 \leq x < 0.05 \\ 1 & \text{Elsewhere} \end{cases} \end{aligned} \tag{5}$$

This metric should produce a mesh with distinct anisotropic regions. We first assign the analytic metric to vertices of a very coarse mesh and then refine the mesh so that the metric-based area per triangle is approximately halved. Then, the vertices of the new mesh are again assigned the metric from Equation 5. We repeat this process four times to produce a mesh that is approximately sixteen times more refined then the original mesh and well representative of the analytic metric. The meshes produced from one iteration to the other are shown in Figure 6. To demonstrate that the mesh achieved the desired anisotropy we can view the mesh around the point (0.525, 0.1) with different axis scalings. It is obvious in Figure 7 that the mesh has achieved the desired anisotropy because the mesh appears isotropic when certain regions are scaled appropriately.

For a more practical example, and one involving curved boundaries, we examine laminar flow around the NACA-0012 airfoil. To compute the solution on each mesh, we use a second-order, vertex-centered finite-volume solver. This solver uses least-squares reconstruction [14], Roe's scheme [18], and Newton-GMRES for rapid convergence [9, 11]. Viscous terms are discretized as described in [14].

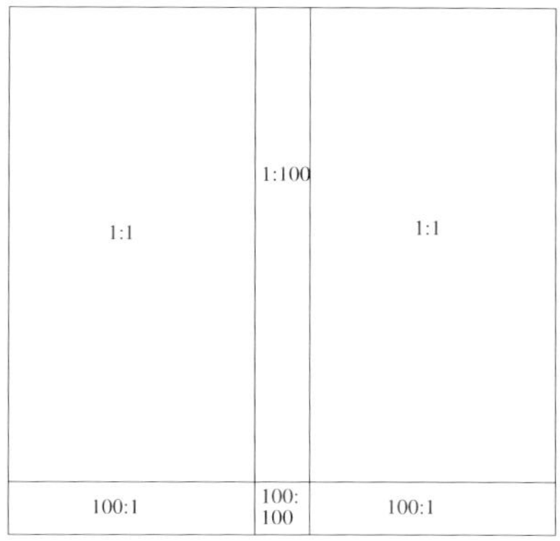

Fig. 5. Expected Triangle aspect ratios given the metric from Equation 5

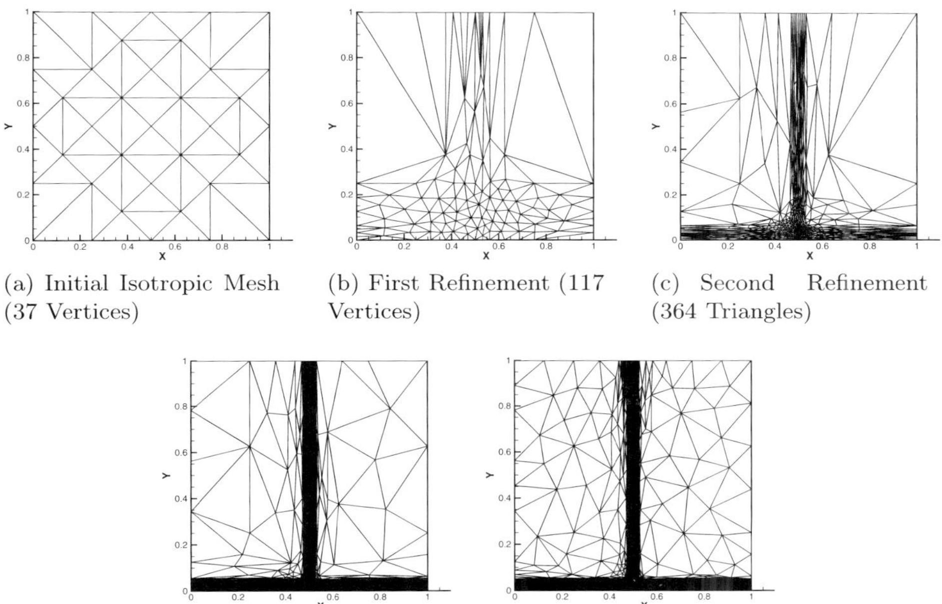

(a) Initial Isotropic Mesh (37 Vertices)

(b) First Refinement (117 Vertices)

(c) Second Refinement (364 Triangles)

(d) Third Refinement (985 Vertices)

(e) Fourth Refinement (2423 Vertices)

Fig. 6. A square mesh being recursively refined according to the metric defined by Equation 5

To calculate the metric between adaptation loops, we use the reconstruction error metric proposed in [15]. The initial mesh, shown in Figure 8a, is constructed using the isotropic mesh generator in GRUMMP [2, 13, 12]. This mesh is obviously ill-suited to computing viscous flow around the airfoil since refinement is

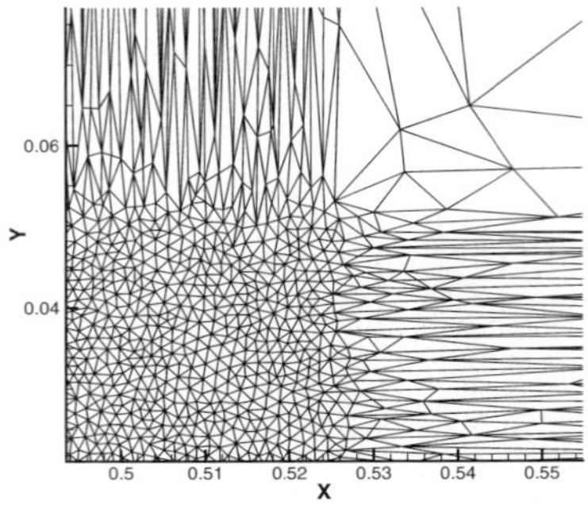

(a) 1:1 Axis Scaling

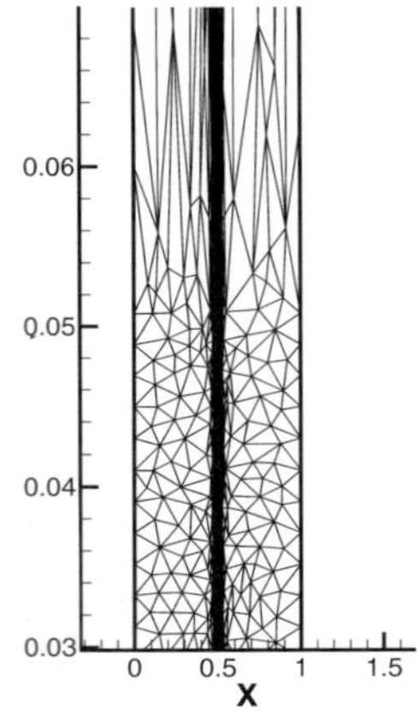

(b) 1:100 axis scaling

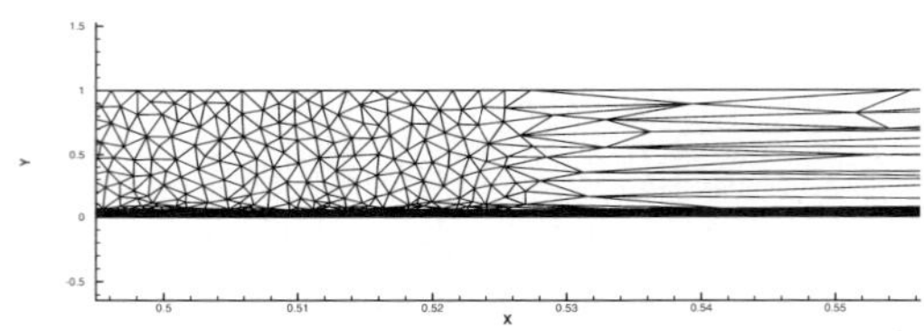

(c) 100:1 Axis Scaling

Fig. 7. Mesh from Figure 6e viewed with different axis scalings

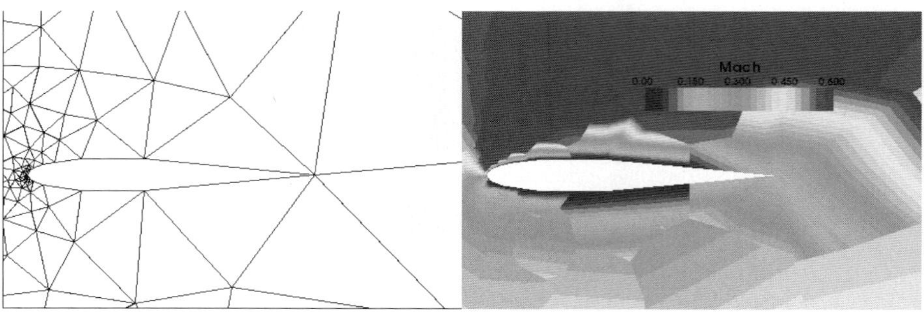

(a) 197 Vertices (b)

Fig. 8. NACA 0012, Mach=0.5, Re=5000, 197 Vertices

done based only on boundary curvature. The resulting solution in Figure 8b is evidence of how poor the mesh is.After choosing a metric based on this solution, we create a new mesh that should be better suited to the flow solution. We

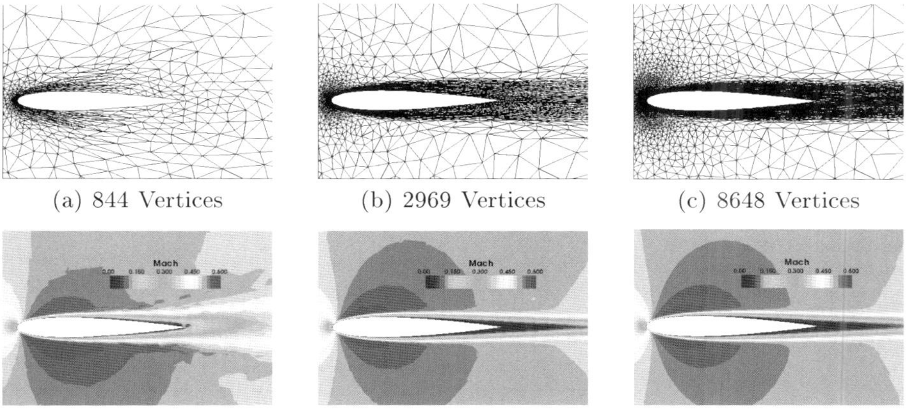

Fig. 9. Intermediate Meshes and Solutions: NACA 0012, Mach=0.5, Re=5000

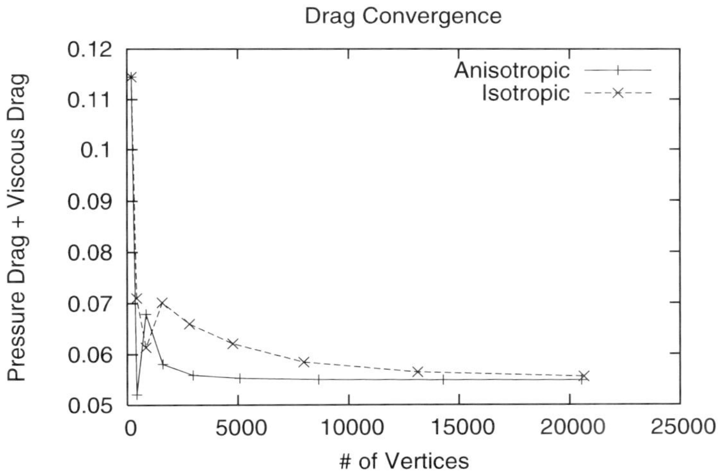

Fig. 10. NACA 0012, Mach=0.5, Re=5000: Drag Convergence

repeat this process and generate the intermediate meshes in Figure 9. The drag convergence plot of Figure 10 illustrates that the solution is converging to a single value and that this convergence occurs much more quickly with anisotropic refinement than isotropic refinement. This increase in accuracy is also visible in the improved resolution of the wake on the final mesh of Figure 9.In Table 1 we show that the fine mesh results for both anisotropic and isotropic refinement are comparable to other results in the literature. Close inspection of the stagnation point in Figure 11 shows that cells are small and isotropic around the stagnation point and quickly become highly anisotropic as the boundary layer develops.

Table 1. Comparison of drag, lift, and separation point for NACA 0012 airfoil: $Mach = 0.5$, $Re = 5000$, $\alpha = 0$

Method	Final Mesh Size	$C_{D,P}$	$C_{D,\nu}$	C_L	$\frac{x_{sep}}{x_{chord}}$
Anisotropic	20187	0.0226	0.0323	0.000114	0.795
Isotropic	20642	0.0224	0.0332	0.0095	0.834
ARC2D [16]	320 x 128 cells	0.0221	0.0321	–	0.824
Mavriplis [8]	320 x 64 cells	0.0229	0.0332	–	0.814
Radespiel [17]	512 x128 cells	0.0224	0.0330	–	0.814

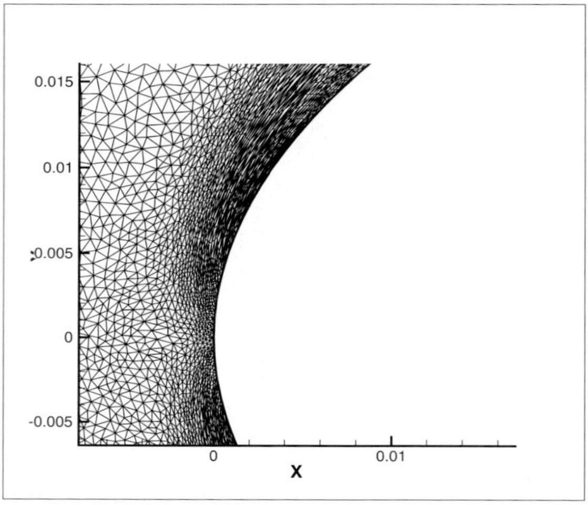

Fig. 11. Close up view of the stagnation point of the finest mesh in Figure 9 a

Since the metric in this case is based on the Hessian of the flow variables, this is exactly the type of mesh that one would expect.

For each mesh adaptation, there is an inner series of refinement and coarsening loops that aim to improve the final mesh. To illustrate the effect of these inner iterations on the quality of the mesh, we have shown the distribution of the triangles within the quality space in Figure 12 for the intermediate mesh created in Figure 9a. The change between distributions a and b in Figure 12 clearly shows that the initial refinement step is very effective at concentrating triangles within the the desired area. Unfortunately, it also produces many small edges and some poor angles. After coarsening is performed in Figure 12c, most of the triangles with small edges are removed but some triangles with very large circumradii are created. Those small edges that do remain are all located in regions of high boundary curvature. The next refinement again eliminates the large circumradii triangles without creating many small edges. The final two passes of coarsening and refinement have very little net impact on the quality of the mesh. While

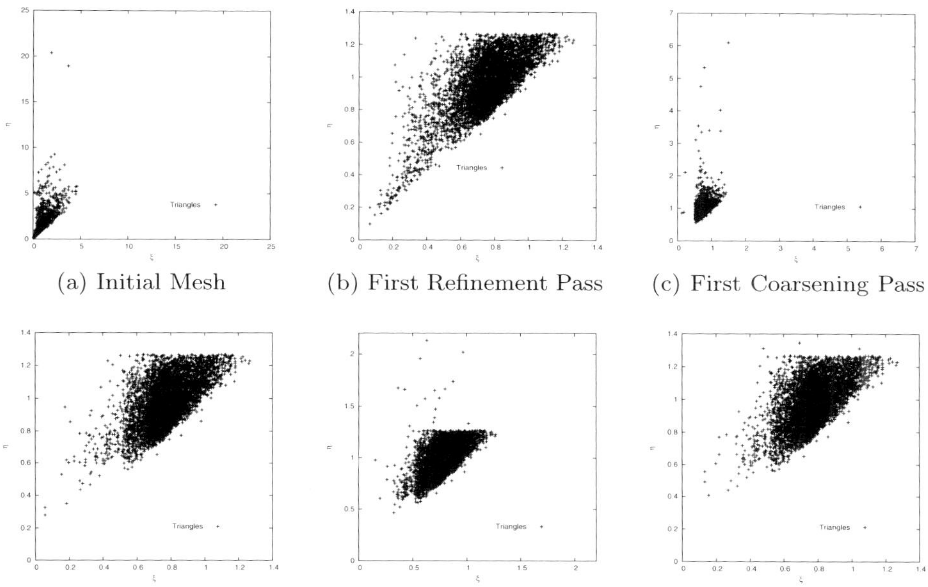

(a) Initial Mesh (b) First Refinement Pass (c) First Coarsening Pass

(d) Second Refinement Pass (e) Second Coarsening Pass (f) Final Refinement Pass

Fig. 12. Sample Quality Distribution During Refinement and Coarsening

the final mesh contains some triangles outside the desired bounds, this appears unavoidable for a general geometry and metric.

8 Conclusions and Future Improvements

We have presented a method for anisotropic mesh refinement that produces high quality meshes without expensive local optimization routines. Our methods assigns metrics to vertices and then defines triangle quality based on approximate metric lengths. The three mesh modification tools we use are swapping to maximize quality, inserting at approximate circumcenters to decrease circumradii, and removing vertices to eliminate small edges. Because there are no guarantees on the results of these modification tools, we use them iteratively to produce an optimal mesh. We have presented examples demonstrating that our algorithm produces meshes that are well adapted to the prescribed metric. When we combine this anisotropic adaptation with an effective metric estimator, we can recursively improve CFD solutions much more cheaply than with uniform refinement.

There are also several improvements to our algorithm that could be implemented in the future. For example, we have stated that local mesh optimization is generally an expensive procedure. However, experience with isotropic meshing in GRUMMP has shown that if the mesh is already of high quality, local optimization-based smoothing can be a fast and effective way of further improving the mesh [6]. It is therefore our belief that the most efficient way of creating high-quality anisotropic meshes would be to insert and remove vertices according to our algorithm and then perform a limited number of passes over the mesh using local optimization. Since the mesh is already of high quality, many fewer passes over the mesh would be required to maximize the quality then if local optimization was the principal tool for mesh improvement. The challenge here will be defining an appropriate objective function for the anisotropic mesh, including determining its gradient with respect to the location of the vertex being smoothed.

Another potential improvement to our algorithm is a modification of the desired quality bounds in areas near boundaries with small features. The isotropic approach is to assign length scales based on the boundary geometry [19]. A similar approach might work for anisotropic meshing by decreasing the area of the desired equilateral triangle used in the calculation for $(\xi_{\text{des}}, \eta_{\text{des}})$ in Figure 1. This would likely reduce the number of instances where vertex removal creates poor quality triangles and thus it should reduce the need for repeated vertex insertion and removal iterations.

Finally, our meshing algorithm produces some small angle triangles that could be improved by further insertion. This would result in smaller triangles than desired, but the improvement in triangle quality might be worth the extra computation cost associate with more triangles. A final pass of quality-based refinement seems a likely way of addressing this problem.

Although many two dimensional meshing algorithms proclaim a straightforward extension to three dimensions, experience in three-dimensional isotropic meshing shows that this is rarely the case. The main stumbling block in extending the current algorithm to three dimensions will be developing an analog to the triangle mapping procedure described in Section 4.1. Mapping tetrahedra and scaling faces will definitely fail because of shape differences between the two mapped copies of a single face; mapping faces and patching those together to form tetrahedra in the metric space seems the most promising approach.

Acknowledgments

This work has been funded by the Canadian Natural Sciences and Engineering Research Council under Special Research Opportunities Grant SRO-299160.

References

1. Ait-Ali-Yahia, D., Baruzzi, G., Habashi, W.G., Fortin, M., Dompierre, J., Vallet, M.G.: Anisotropic mesh adaptation: towards user-independent, mesh-independent and solver-independent CFD. Part II: Structured grids. Inter. J. Numer. Meth. Fluids 39(8), 657–673 (2002)

2. Boivin, C., Ollivier-Gooch, C.F.: Guaranteed-quality triangular mesh generation for domains with curved boundaries. Inter. J. Numer. Meth. Eng. 55(10), 1185–1213 (2002)
3. Buscaglia, G.C., Dari, E.A.: Anisotropic mesh optimization and its application in adaptivity. Inter. J. Numer. Meth. Eng. 40, 4119–4136 (1997)
4. Chew, L.P.: Guaranteed-quality triangular meshes. Technical Report TR-89-983, Dept. of Computer Science, Cornell University (1989)
5. Dompierre, J., Vallet, M.G., Bourgault, Y., Fortin, M., Habashi, W.G.: Anisotropic mesh adaptation: towards user-independent, mesh-independent and solver-independent CFD. Part III. Unstructured meshes. Inter. J. Numer. Meth. Fluids 39(8), 675–702 (2002)
6. Freitag, L.A., Ollivier-Gooch, C.F.: Tetrahedral mesh improvement using swapping and smoothing. Inter. J. Numer. Meth. Eng. 40(21), 3979–4002 (1997)
7. Labelle, F., Shewchuk, J.R.: Anisotropic voronoi diagrams and guaranteed-quality anisotropic mesh generation. In: Proceedings of the Nineteenth Annual Symposium on Computational Geometry, Association for Computing Machinery, pp. 191–200 (2003)
8. Mavriplis, D., Jameson, A.: Multigrid solution of the two-dimensional Euler equations on unstructured triangular meshes. AIAA paper 87-0353 (January 1987)
9. Michalak, K., Ollivier-Gooch, C.: Matrix-explicit GMRES for a higher-order accurate inviscid compressible flow solver. In: Proc. 18th AIAA CFD Conf. (2007)
10. Miller, G.L., Talmor, D., Teng, S.-H.: Optimal coarsening of unstructured meshes. J. Alg. 31(1), 29–65 (1999)
11. Nejat, A., Ollivier-Gooch, C.: A high-order accurate unstructured finite volume newton-krylov algorithm for inviscid compressible flows. J. Comp. Phys. 227(4), 2592–2609 (2008)
12. Ollivier-Gooch, C.F.: An unstructured mesh improvement toolkit with application to mesh improvement, generation and (de-)refinement. AIAA 98-0218 (January 1998)
13. Ollivier-Gooch, C.F.: GRUMMP — Generation and Refinement of Unstructured, Mixed-element Meshes in Parallel (1998–2005), `http://tetra.mech.ubc.ca/GRUMMP`
14. Ollivier-Gooch, C.F., Van Altena, M.: A high-order accurate unstructured mesh finite-volume scheme for the advection-diffusion equation. J. Comp. Phys. 181(2), 729–752 (2002)
15. Pagnutti, D., Ollivier-Gooch, C.: A generalized framework for high order anisotropic mesh adaptation. Comp. Struct. (submitted 2008)
16. Pulliam, T.H.: Efficient solution methods for the navier-stokes equations. In: Numerical Techniques for Viscous Flow Computation in Turbomachinery Bladings, von Kármán Institute for Fluid Dynamics Lecture Series. von Kármán Institute, Rhode-St-Genese, Belgium (1985)
17. Radespiel, R.: A cell-vertex multigrid method for the Navier-Stokes equations. NASA TM-101557 (1989)
18. Roe, P.L.: Approximate Riemann solvers, parameter vectors, and difference schemes. J. Comp. Phys. 43, 357–372 (1981)
19. Shewchuk, J.R.: Delaunay Refinement Mesh Generation. PhD thesis, School of Computer Science, Carnegie Mellon University (May 1997)

Anisotropic Level Set Adaptation for Accurate Interface Capturing*

V. Ducrot[1] and P. Frey[1,2]

[1] UPMC Univ Paris 06, UMR 7598, Laboratoire J.L. Lions, F-75005 Paris, France
ducrot@ann.jussieu.fr
[2] Universidad de Chile, UMI 2807, Centro de Modelamiento Matemático, Santiago, Chile
frey@ann.jussieu.fr

Summary. In fluid-structure interactions and fluid simulations, like incompressible two-phase flows involving high viscosity and density ratios, interface capturing and tracking is considered as a very challenging problem and has a strong impact on industrial applications. Usually, an adaptive grid is used to resolve the problem as well as to track the interface. In this paper, we describe an adaptive scheme based on the definition of an anisotropic metric tensor to control the generation of highly streched elements near an interface described with a level set function. In our approach, quasi-uniform anisotropic meshes are created with the objective of minimizing the interpolation errors by capturing the interface features using curvature-adapted anisotropic elements. The accuracy of the method is verified and numerical experiments are presented to show its efficiency.

1 Introduction

In this paper, we consider the very challenging problem that consists in capturing and tracking moving fronts and interfaces. This situation arises typically in bi-fluid flow simulations, free surface flows, non linear processes like dendritic growth, for example. Such numerical simulations are considered difficult to carry on because the quality of the numerical solution strongly depends on the accuracy of the interface approximation, which is in principle related to the minimal size of the discretisation. Hence, it seems often tedious to find a right compromise between the desired level of accuracy and the overall computationnal cost (memory and time ressources) of the simulation.

Since the pionnering works of Dervieux [11] and Osher [28], level set methods have been successfully used in numerous scientific computing applications where time-dependent equations and moving interfaces were involved. Until recently, most applications of level sets involved structured grids on which high-order schemes can be implemented and computed easily. On such grids and on related spatial decompositions (e.g., octree), the differential operators and the solutions

* Research supported by the INRIA ARC-LMN and a BQR grant from UPMC Scientific committee.

of time-dependent Hamilton-Jacobi equations can be efficiently computed [15]. The simplicity of Cartesian grids associated with the simplicity of level set formulations compared to Lagrangian or ALE methods seemed to have frozen the situation. For dealing with complex geometries (*i.e.*, domain boundaries and interfaces) however, geometry-adapted unstructured meshes have proved more versatile and accurate in finite element simulations, notably because the number of vertices for a given accuracy is lower than with (graded) Cartesian grids. Moreover, high-order non-oscillatory schemes have been efficiently introduced on triangulations[1]. The next step would be to generate anisotropic curvature-adapted meshes to represent the interface manifold. Then, the numerical schemes may be more expensive to solve, but the substantial reduction of the number of elements will largely compensate this drawback and favorably impacting the overall computational cost of the simulation.

Over the last decade, several anisotropic mesh adaptation techniques have been developed and efficient adaptive algorithms are now available, at least for steady-state problems. For time-dependent problems however, and especially dynamically evolving interfaces, the principal difficulty is to follow the arbitrary evolution of the phenomenon in space and time, including the possible topology modification. Indeed, in these simulations, the geometry is intrinsically a part of the solution of the simulation. Usually, the triangulation is adapted frequently, to maintain a valid representation of the interface with minimal approximation errors. This issue has been partially addressed by using a fixed point algorithm and a metric-based refinement procedure to locally adapt a triangulation within each time step with respect to the solution variations (first and second-order derivatives) [4]. In [25], the triangulation of the interface is directly adapted to the interface curvatures using local mesh modifications thus making it difficult to handle topology changes during the simulation. To overcome this problem, we suggest to use the level set formulation of a time-dependent problem that offers the advantage of embedding the manifold in a higher dimensional space and does not require to discretize it explicitly, and anisotropic mesh adaptation techniques to refine the triangulation in the vicinity of the manifold for approximation error control purposes.

The aim of this paper is to present an anisotropic mesh adaptation method, based on a geometric error estimate, for capturing an interface described by a level set function, with a high level of accuracy. This result is obtained by defining a Riemannian metric tensor based on the differential properties of the manifold of codimension one corresponding to the interface [14]. Mesh adaptation consists then in creating a quasi-uniform mesh with respect to the metric tensor field, using local mesh modifications: edge split, edge collapse, edge flip and vertex relocation. As the interface evolves in time, we rely on a fixed point algorithm to correctly locate the interface positions at the beginning and at the end of each time step (before and after each advection step of the level set function). By intersecting the metric tensors corresponding to these relative positions, we will show that it is possible to substantially reduce the diffusion problem. Moreover, we solve this advection step by a second order scheme based on the method of characteristics. This approach can be used with any implicitly-defined, scalar valued function. This allows us to envisage other categories of applications, for instance: in computational solid geometry modeling (where objects are commonly defined by an implicit function) and in scientific visualization for representing isovalue curves or surfaces of a numerical finite element solution accurately.

The remainder of this paper is organized as follows. In Section 2, we briefly review the level set formulation of the interface tracking problem and we outline the related numerical issues we have specifically addressed. In Section 3, we introduce the main notions of anisotropic mesh adaptation based on Riemannian metric tensor. A complete description of this method can be found in the general purpose book [18] and the references therein. In particular, we outline the definition and construction of a metric tensor and the fixed point algorithm for time dependent problems. In Section 4, we explain how to control and bound the geometric approximation of an interface and we introduce an anisotropic error estimate for this problem. Numerical examples of simulations are given in Section 5 to show the efficiency of this approach.

2 The Level Set Context

2.1 Level Set Formulation

We consider a level set formulation of an time dependent problem in a computational domain $\Omega \subset \mathbb{R}^d$ $(d = 2, 3)$ which is subdivided in two subdomains Ω^+ and Ω^- sharing a common interface Γ, such that $\Omega^+ \cup \Omega^- = \Omega$ and $\Omega^+ \cap \Omega^- = \emptyset$ and $\Gamma = \partial\Omega^+ \cap \partial\Omega^-$, where $\partial\Omega^*$ denotes the boundary of the domain Ω^*. Typically, we introduce a Lipschitz-continuous function $u : \mathbb{R}^+ \times \mathbb{R}^d \to \mathbb{R}$ and the interface Γ is chosen to coincide with the level set where $u(t, x) = 0$. As suggested by Osher [28], the level set function u is initialized with the distance function to the initial interface $u(0, x) = \pm d(x)$, the sign $\pm$ is negative in Ω^- and positive in Ω^+.

Usually, the motion of the interface is driven by a velocity field $v(t, x)$ related to the differential properties of the manifold. This leads to a first-order advection equation posed in $\mathbb{R}^d$, for all $x \in \Omega$, $t \in [0, T]$:

$$\partial_t u(t, x) + v_n(t, x)|\nabla u(t, x)| = 0 \tag{1}$$

where v_n denotes the normal component of the velocity to each level set. It is well known that singularities are often encountered in solving this equation and thus appropriate techniques must be developed to select the unique viscosity solution [10].

2.2 Numerical Issues

The computational domain Ω is covered by an anisotropic unstructured triangulation T_h and the distance value to the interface is associated with its vertices, thus defining the level set function u in a discrete manner. The numerical resolution of Equation (1) raises several issues that we emphasize in the next sections.

Level set advection. We use a finite element technique to solve the level set advection problem on anisotropic unstructured meshes. Suppose the vector field $v(t,x)$ is sufficiently smooth, we recall that a *trajectory* associated to the field v is a solution of the differential problem :

$$\frac{dX}{dt}(t,x_0) = v(t,X(t,x_0)), \qquad X(0,x_0) = x_0$$

where the point $X(t,x_0)$ represents the position at time $t \in [0,\infty]$ of a particule initially in x_0 at time $t = 0$. Such a trajectory is a parametrized curve $X(t,x)$. The method of characteristics solves this problem by converting the PDE into a system of ODE. Using a finite element scheme to find an interpolation of the desired solution u using piecewise affine continuous functions, the variational problems reads, with the basis functions φ_i:

$$\iint_\Omega u^{n+1}(x,y)\varphi_i(x,y)dxdy = \iint_\Omega u^n \circ X^n(x,y)\varphi_i(x,y)dxdy\,. \qquad (2)$$

The linear system $Au^{n+1} = f$ is solved using a second-order scheme thanks to a prediction-correction loop [26]. The main advantage of this formulation is that it accounts for large time stepping as compared with other explicit method.

Reinitialization procedure. An interesting property of the signed distance function is that we have $|\nabla u| = 1$, which greatly simplifies the calculation of the normal $\mathbf{n}$, the mean curvature H and the Gaussian curvature K. Hence, we have:

$$\begin{cases} \mathbf{n(x)} = \dfrac{\nabla u(x)}{|\nabla u(x)|} = \nabla u(x)\,, \\ H(x) = \mathrm{div}(\mathbf{n(x)}) = \Delta u(x) \\ K(x) = det(\Pi^t(x) D^2 u(x) \Pi(x))\,, \end{cases} \qquad (3)$$

where Π is the matrix of the projection on the tangent plane at x. Unfortunatly, almost all motion fields, but a constant velocity field v, do not preserve this property. Therefore, in order to recover it and the simple formulas, a *reinitialization* stage is introduced.

Two different approaches have been proposed in the literature. On the first hand, one can solve the non-linear equation at steady state [33]:

$$\partial_t \psi - S(\psi)(1 - |\nabla \psi|) = 0\,, \qquad (4)$$

where $S(\psi)$ is a sign function taken as 1 in Ω^+, -1 on Ω^- and 0 on the interface Γ, and $\psi(0,x)$ is taken as the current value of $u(t,x)$. At convergence, the function ψ is substituted back to the function u. A drawback of this methods is related to the regularity of the function ψ that may cause the interface Γ to move incorrectly from its starting position, if ψ is not smooth. Moreover, in the numerical schemes, the time step is usually bounded by a stability condition related to the minimal size h_{min} of the triangulation T_h. Since we are concerned with anisotropic triangulations with high aspect ratio elements and a minimal size at least three orders of magnitude smaller than the domain size to resolve small features, such technique would lead to a prohibitive computational cost.

On the other hand, a Fast-Marching method is used to evolve a front at constant speed initiated from the interface Γ [34, 31, 13]. It consists in solving the following equation in each element:

$$\sum_{i=1}^{d+1} |\nabla \varphi_i u_i|^2 = 1 \tag{5}$$

where the φ_i are the shape functions in the element and u_i is the value of the level set function at the vertices of the element. The problem is usually solved using a first-order scheme. However, this approach could be improved when considering that the function u is a *signed* distance function to an interface Γ. Since the sign is not modified by the reinitialization stage, it is sufficient to compute the Euclidian distance to Γ in order to determine function φ. Furthermore, we have a bound ε on the distance between the approximation Γ_h and the manifold Γ on T_h using the Riemannian metric tensor field. Hence, function φ can be determined with an accuracy of order ε by computing the distance of the mesh vertices to Γ_h. This can be achieved in a efficient by a parallel algorithm for computing point distance to polyline [21]. The complexity of our approach is then in $(m+n)\log m$, where m represents the number of entities in the reconstructed interface and n denotes the number of mesh vertices, $n \gg m$, in general.

3 Anisotropic Mesh Adaptation

In order to resolve the velocity field accurately, it is mandatory to have a sufficient number of grid points in the vicinity of the interface. For this reason, unstructured triangulations offer a better alternative to Cartesian grids in solving this equation [1]. The approximation error is then strongly related to the quality of the triangulation and to the mesh density. To this end, the definition of a Riemaniann metric tensor helps to control the generation of anisotropic elements close to the interface.

In this section, we briefly present some basic notions and definitions on computational triangulations, metric tensor fields and mesh adaptation. We introduce the following notations: Ω denotes a simply connected open bounded domain in $\mathbb{R}^d$, $\overline{\Omega}$ is a closure of Ω and $|\Omega|$ is the d-dimensional measure or the volume of Ω.

3.1 Uniform Triangulations

Suppose given a family of triangulations T_h on the domain Ω, h representing the characteristic element size. Each element $K \in T_h$ is a closed subdomain of Ω and we assume that $\overline{\Omega} \subset \bigcup_{K \in T_h} K$ and that the usual finite element requirements are verified (*i.e.* elements do not overlap). Then, a *uniform* mesh T_h of Ω is a mesh in which all element are equally sized and regular. In such case, if $|T_h|$ represents the number of mesh elements and $h_K = diam(K)$ the diameter of K, the size $h = \max_{K \in T_h} h_K$ is given by the relation:

$$h_K \approx \left(\frac{|\Omega|}{|T_h|} \right)^{1/d} \qquad \forall K \in T_h \,.$$

A *quasi-uniform* mesh is a mesh for which (i) there exists a constant τ such that

$$\frac{h_K}{\rho_K} \leq \tau \qquad \forall K \in \bigcup_h T_h \,,$$

where, for any open ball $B_i \subset K$, $\rho_K = \sup_i \{diam(B_i)\}$ is the in-diameter of K and (ii) the variation of h is bounded by a constant. This does not mean a constant mesh size over the domain.

3.2 Anisotropic Mesh Adaptation Scheme

Nowadays, mesh adaptation is widely used to improve the efficiency of the numerical schemes as well as the accuracy of numerical solutions in computational simulations. It consists in concentrating a maximum number of nodes in regions of large solution variations and a minimal number of nodes in other regions of the computational domain. Consequently, the overall number of nodes required to achieved a desired accuracy can be reduced by a substantial amount, thus impacting favorably the computational cost of the simulation. In many applications, solutions exhibit a large gradient variation in some regions, often combined with highly anisotropic features (shock waves or boundary layers in fluid dynamics, for instance), and elements with high aspect ratio are needed. In this case, isotropic triangulations would contain too many elements. It is thus desirable to adjust the element size as well as the element shape and orientation to better match the solution variations. In the past decade, the theory of error estimates as been largely investigated by various research groups [5, 8, 12, 16, 19] and has provided the concept of *optimal* triangles with respect to a so-called *metric tensor* field based on the gradient and Hessian of the solution [7].

3.3 Metric Tensor Field

From the geometric viewpoint, the metric defining the element size, shape and orientation is represented by an ellipsoid. Hence these notions are related to its volume, the lengths of its semi-axes and its principal axes vectors, respectively. The metric tensor denoted as $M(x)$ is used to generate a quasi-uniform mesh in the metric related to M. More precisely, the volume of an element $K \in T_h$ is unitary:

$$\int_K \sqrt{det(M(x))}\, dx = 1 \,, \qquad \forall K \in T_h$$

which corresponds to the discrete formulation

$$|K|\sqrt{det(M_K)} = 1\,, \qquad \forall K \in T_h$$

where M_K is an average of $M(x)$ on K. By extension, the length of a curve γ in a metric given by $M(x)$ for any $x \in \Omega$ is defined as:

$$|\gamma|_M = \int_0^1 \sqrt{\langle \gamma'(t), M(\gamma(t))\gamma'(t)\rangle}dt\,,$$

where $\gamma(t) : [0,1] \to \mathbb{R}^d$ is a parametrization of γ. Since the metric tensor $M(x)$ is a symmetric positive definite matrix, the spectral decomposition theorem allows to decompose M as:

$$M = P\,\Lambda\,P^t = \sum_{i=1}^{d} \lambda_i \mathbf{e_i}\mathbf{e_i}^t\,,$$

where the normalized eigenvectors of M are the columns of matrix $P = [\mathbf{e_1}, \ldots, \mathbf{e_d}]$ such that $P\,P^t = I_d$ and Λ is the diagonal matrix of the eigenvalues λ_i. Notice that the matrix P prescribes the orientation and the matrix Λ prescribes the size and shape of any element K.

3.4 Mesh Quality

As we have numerical simulations in mind, it is important to introduce quality measures in order to evaluate how close the metric prescriptions are satisfied at any x in the domain. Mesh quality assessement has been an active area of research for isotropic meshes (see [6] for a survey). For practical reasons, we will consider a single measure to evaluate the quality of an element K:

$$Q_{ani}(K) = \alpha_d \frac{\left(\displaystyle\sum_{i=1}^{k} \langle \mathbf{e_i}, M_K \mathbf{e_i}\rangle\right)^d}{|K|\sqrt{det(M_K)}}\,,$$

where $\mathbf{e_i}$ represents here any of the k egdes of K and α_d is a normalisation constant such that $Q_{ani}(K) = 1$ for a regular element. Notice that $Q_{ani} \geq 1$ for all $K \in T_h$ and thus the larger $\max_K Q_{ani}(K)$ is, the more the triangulation T_h deviates from the metric prescriptions.

3.5 Anisotropic Mesh Adaptation

Traditionally, mesh adaptation algorithm are based on either r-methods or h-methods. In the first approach, the number of mesh vertices is kept constant, only the vertex locations are changed throughout the simulation. Algorithms in the second category proceed by local mesh modifications like edge flipping, edge contraction, vertex addition and vertex relocation. In our approach, anisotropic meshes are created using a Delaunay-based method in which a metric tensor

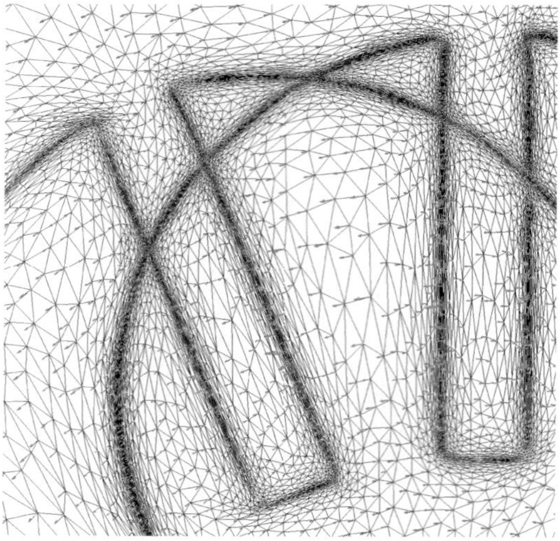

Fig. 1. Anisotropic mesh for the transient fixed-point problem. Adapted to the solutions u_n and u_{n+1}, and the corresponding velocity field at time t_n.

prescribing the size, stretching and orientation of the element is associated with the mesh vertices [18]. Moreover for unsteady physical problems, several metric tensor fields must be combined to overcome the numerical diffusion associated with the numerical scheme [4, 20].

Solution prediction. To deal with unsteady phenomena, we use a fixed-point mesh adaptation algorithm to predict the solution evolution in the domain Ω. This algorithm solves a transient fixed-point problem for the couple (mesh, solution) at each iteration of the classical adaptation scheme. Suppose that the triangulation T_h is adapted to the behavior of the solution $u_n = u(t_n, x)$, for all x in Ω, at the beginning of a time period $[t_n, t_{n+1}]$, the aim of this scheme is to predict the solution behavior at t_{n+1} and to refine the mesh accordingly. Therefore, within this time period at least two anisotropic metric tensors must combined together, producing a mesh that is refined in regions of large gradient variations of both u_n and u_{n+1}. This procedure can be seen as an internal loop within a classical adaptation scheme. In each internal iteration j, the solution u^j_{n+1} converges toward the solution u_{n+1}. This can be checked numerically using the L^1-norm of the difference between the solutions u^j_{n+1} and u^{j-1}_{n+1}.

When two solutions are considered their relative metric tensors must be reduced to a single one for mesh generation purpose. To this end, a metric intersection procedure is used, based on the simultaneous reduction of the quadratic forms associated with the tensors [20]. Figure 1 shows the anisotropic mesh used to resolve the internal fixed-point loop.

Local mesh adaptation. In our approach, we rely on local topological and geometrical mesh modifications (edge split, edge collapse, vertex relocation, edge flip). At each step, the current triangulation is modified to generate quasi-uniform triangulations with respect to the metric tensor field M. The method is based on the analysis of the mesh edges: all edges must belong to the interval $[l_{min}, l_{max}]$ and the mesh elements quality must be close to the optimal unit value. Long edges $l_M(\mathbf{e}) > l_{max}$ are splitted into unit subsegments with respect to M and the new vertices are introduced using a modification of the Deaunay kernel to account for anisotropic elements. Short edges $l_M(\mathbf{e}) < l_{min}$ are collapsed by merging their two endpoints at one of the extremity. A mesh optimization stage involving vertex relocation procedure is then applied to improve the overall mesh quality [18].

4 Geometric Approximation

In the kind of problem we consider, an interface is defined as a manifold of codimension one embedded in $\mathbb{R}^d$. We introduce an implicit level set interface representation, where the interface Γ is the zero isocontour of a Lipschitz-continuous function u. We choose to have a $V_h = \mathbb{P}_1(T_h)$ Lagrange approximation space, and we denote u_h the projection of u on V_h.

4.1 Error Bound

We aim at controlling the approximation error between the exact (unknown) curve Γ and an approximate curve Γ_h defined as a zero isocontour of the function u_h:

$$\Gamma_h = \{x \in \Omega\,,\; u_h(x) = 0\}\,.$$

This control can be obtained by introducing a bound on the Hausdorff distance between Γ and Γ_h, *i.e.*, by prescribing a value of ε sufficiently small such that:

$$\max(\max_{x \in \Gamma} \min_{y \in \Gamma_h} |x - y|, \max_{y \in \Gamma_h} \min_{x \in \Gamma} |x - y|) < \varepsilon\,.$$

Actually, we established the following result that provides an indirect yet practical control on the Hausdorff distance [14]. Suppose u is an implicit real-valued function defined on a domain Ω such that ∇u exists ($|\nabla u| = n$, $n > 0$) and such that its derivatives are k-lipschitz ($k \geq 1$) on an open set. Then, we have the theoretical bounds:

Theorem 1. *Under the previous hypothesis, if we denote the set of elements intersected by Γ as $\mathcal{E} = \{x \in K, K \in T_h$ and $K \cap \Gamma \neq \emptyset\}$, then for all $x \in \Gamma_h$:*

$$d(x, \Gamma) \leq \frac{1}{\min_{p \in \mathcal{E}} |\nabla u(p)|} \left| \sum_{j=1}^{d+1} \lambda_j \left({}^t(x - x_j)(D^2 u(x_j) - D^2 u(x))(x - x_j)\right) \right|,$$

and conversely, for all $x \in \Gamma$,

$$d(x,\Gamma_h) \leq \frac{1}{|\nabla u_h(K)|}\left|\sum_{j=1}^{d+1}\lambda_j\left({}^t(x-x_j)(D^2u(x_j)-D^2u(x))(x-x_j)\right)\right|.$$

where λ_j denotes here the barycentric coordinate of a point x associated with the vertex x_j of K and D^2u represents the Hessian of the function u (the gradient of u_h is piecewise constant per element).

The complete proof of this result can be found in [14] and relies on the analysis of a Taylor expansion with integral rest in any simplex intersected by the desired level set. In the next section, we will show how this result can be used to define a metric tensor field to prescribe appropriate element sizes and shapes to minimize the approximation error in a triangulation.

4.2 A Geometric Metric Tensor

Let us consider the set $\mathcal{E}$ of mesh simplices intersected by the zero level set manifold Γ, *i.e.* such that the sign of the function u is not constant on the element. The following results relate the size of the elements of the set $\mathcal{E}$ to the local curvature(s) of the manifold in two and three dimensions and provide a bound on the distance between the manifold and its piecewise affine approximation.

Corollary 1. *In two dimensions, we have:*

$$d(\Gamma_h,\Gamma) \leq \max_{K\in\mathcal{E}} \frac{l^2\kappa}{1-h\kappa-h^2\kappa^2}$$

where h (resp. l) represents the size of the element K in the normal (resp. tangential) direction to the curve Γ, κ is the maximum of the absolute value of the local curvature to Γ in K.

Corollary 2. *Similarly, in three dimensions, we have:*

$$d(\Gamma_h,\Gamma) \leq \max_{K\in\mathcal{E}} \frac{l_1^2\kappa_1}{1-h\kappa_1-h^2\kappa_1^2}+\frac{l_2^2\kappa_2}{1-h\kappa_2-h^2\kappa_2^2}$$

where h (resp. l_1, l_2) represents the size of the element K in the direction normal (resp. of the principal curvatures) to the manifold and κ_1, κ_2 are the minumum and maximum values of the two local principal curvatures of Γ in K.

Again, the full proof of these results has been given in [14]. These bounds on the approximation error can be conveniently used to define a suitable discrete metric tensor field at each mesh vertex of every element in the set $\mathcal{E}$. Indeed, setting the level of accuracy of the piecewise affine approximation to an arbitrary value ε is equivalent to imposing $d(\Gamma_h,\Gamma)<\varepsilon$. Hence, we propose to define the metric tensor at each vertex as follows:

$$M = \frac{\nabla u \nabla u^t}{h_{min}^2} + \frac{(D^2u)}{\varepsilon} \tag{6}$$

where h_{min} represents the smallest (user-defined) element size in T_h. With this definition, the elements will be aligned with the principal directions of curvatures and with the normal to Γ and their relative sizes will be related to the local curvatures by setting:

$$l_i = \sqrt{\frac{\varepsilon}{\kappa_i}}.$$

4.3 Numerical Issues

The geometric metric tensor definition involves the Hessian matrix of the level set function u. Practically, the solution u_h is piecewise affine and only known at the vertices of triangulation T_h. In order to compute the gradient and the Hessian matrices, we propose to reconstruct a high-order solution $\tilde{u} \in C^2(\Omega)$ from u_h in order to write:

$$|u - u_h|_X \approx |\tilde{u} - u_h|_X$$

where $|\cdot|_X$ is the norm related to the approximation space. Notice that in our approach, the solution $\tilde{u}$ does not need to be known, as only the Hessian matrix is required to define the metric tensor.

We propose a least-squares approximation of the second derivative of $\tilde{u}$ based on a Taylor expansion of the solution $\tilde{u}$ at vertex x, following [3]:

$$\tilde{u}(x_i) = \tilde{u}(x) + \mathbf{xx}_i \cdot \nabla\tilde{u}(x) + \frac{1}{2}\langle \mathbf{xx}_i, M(x)\mathbf{xx}_i\rangle$$

where $\tilde{u}(x_i)$ is the solution associated with the vertex x_i connected to the vertex x. This leads to a usually overdetermined linear system $Ay = b$, that is solved using a least square approximation $A^tAy = A^tb$, where y is a $\frac{d(d+1)}{2}$ vector of the Hessian coefficients. Solving it directly allows us to avoid all problems related with ill-conditioned matrices.

We have also investigated the Hessian reconstruction using a dual L^2 projection scheme of the numerical solution u [18]. This robust technique, did not provide better results for our purposes in the numerical simulations, thus we have used the least-square approximations in all following examples.

5 Numerical Examples

The authors and colleagues have implemented all the numerical techniques described in this paper (the error estimate, the adaptive mesh generation algorithm, the advection and the Stokes flow solvers) in two and three dimensions.

5.1 Numerical Examples in Two Dimensions

In this section, we investigate the efficiency and the accuracy of our interface tracking mesh adaptation method on numerical simulations of Zalesak rigid body rotation and on a viscous flow problem.

Zalesak rigid body

This test case has been originally proposed by Zalesak [36] and we consider here a slightly modified version suggested by Rudman [30] and commonly used for volume integrity checking. The computational domain Ω is here a $[-2,2]\times[-2,2]$ square, the circular body of unit radius is centered at position $(0,0.75)$ and the width of the slot is 0.12. We consider a steady advection problem where the disk is rotated with a constant angular velocity of 1 time unit, which constitutes a good test of how efficiently the interface is advected in case of high curvatures. The body should return to the initial position after a revolution of 2π, without any modification of the interface shape and volume (Figure 2). To overcome numerical dissipation problems when solving the advection equation, we have used a simple mass-correction procedure. Assuming that the error is constant over the domain, we use an optimization scheme to determine a constant value by which the level set is normalized so as to preserve the total mass.

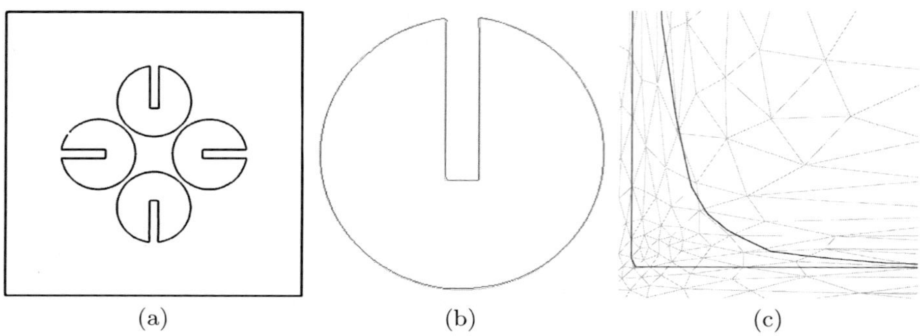

Fig. 2. Evolution of the rigid body Zalesak disk: (a) four positions of the rotated body, (b) superposition of the initial and final (after one complete revolution) discretized level set curves , (c) zoom near a singularity

Figure 2 shows the evolution of the rigid body in the velocity field corresponding to angles of rotation $\theta = k\pi/2,\ k \in \{1,\ldots,4\}$. At each time step, the solution is computed using a high-resolution adapted mesh, for which the smallest prescribed edge length is 0.001 and the maximal prescribed element stretching is given by the ratio between the maximal and the minimal edge length, $0.2/0.001 = 200$. Each anisotropic adapted mesh contains about 6 500 vertices whereas the corresponding isotropic graded mesh (*i.e.*, non uniform) would contain about 10 times more vertices. The generation of each anisotropic mesh corresponding to a rotation angle of $\pi/8$ is obtained using a fixed point algorithm involving 5 metric intersections in time and requires about 46 seconds while the advection time requires 4.3 seconds on a laptop.

The accuracy of the level set approximation can be evaluated by two measures. First, we evaluated the perimeter of the interface and the area of the domain

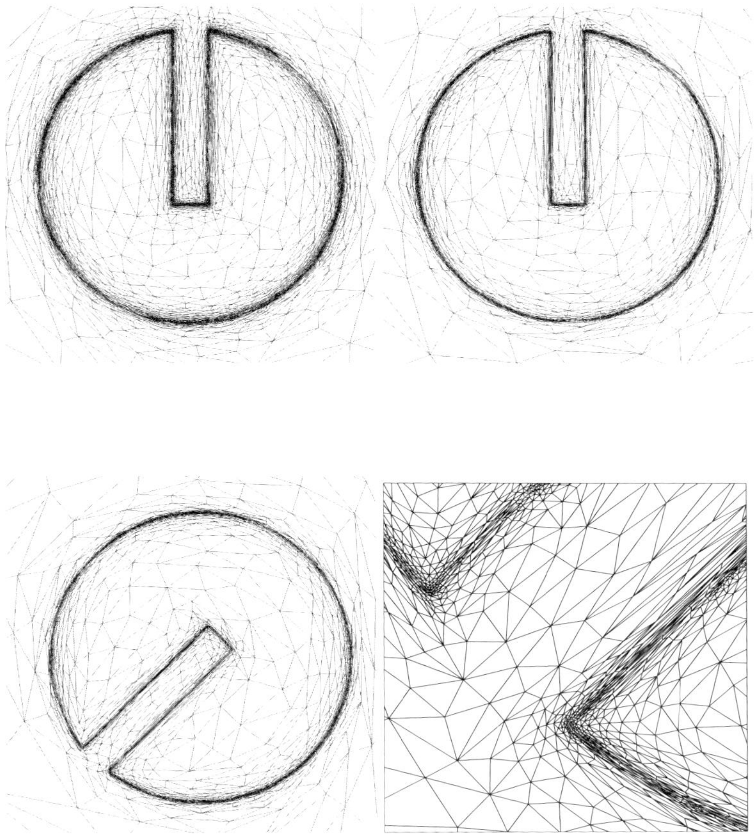

Fig. 3. Zalesak disk test. Advection of the level set corresponding to rotation angles $\theta = 0$, $\theta = 2\pi$ (top) and $\theta = \dfrac{3\pi}{4}$ with a local enlargement close to the singularities (bootom).

delineated by of the interface during the simulation without and with the mass correction scheme. Without mass correction, the perimeter values are ranging from 4.3275 (initial length) to 4.2985 (final length), corresponding to a shrinkage of less than one percent of the interface curve. The area of the internal domain is equal to 0.71316 (initial) and to 0.706 (final), the area (mass) loss is of the order of one percent. With the mass correction scheme, the final surface is 0.7131677, less than 0.01 percent difference wih the initial surface,but the final perimeter value is 4.450584 (Figure 3).

A viscous flow example

The next example concerns a viscous flow computation. Here, we attempt to solve the Stokes problem (7) describing an incompressible flow of two immiscible fluids with very different viscosities μ_i on a domain Ω subdivided in two subdomains Ω_i sharing a common interface:

$$\begin{cases} -\mu^i \Delta \mathbf{u}^i + \nabla p^i = \rho^i \mathbf{f}^i \\ \text{div } \mathbf{u}^i = 0 \end{cases} \tag{7}$$

where ρ_i and p_i represent the volumic mass and pressure, respectively and with the continuity condition of the velocity field and the equilibrium of the normal constraints with the surface tension at the interface Γ:

$$\begin{cases} \mathbf{u}^1 - \mathbf{u}^2 = \mathbf{0} \\ (\boldsymbol{\sigma}^1 - \boldsymbol{\sigma}^2) \cdot \mathbf{n}^1 = -K\, \mathbf{n}^1 \end{cases} \tag{8}$$

endowed with Dirichlet or Neumann boundary conditions:

$$\mathbf{u} = \mathbf{u}_D \text{ on } \Gamma_D, \quad \sigma \cdot \mathbf{n} = \mathbf{s}_N \text{ on } \Gamma_N, \quad \Gamma_D \cup \Gamma_N = \partial\Omega, \quad \Gamma_D \cap \Gamma_N = \emptyset \tag{9}$$

In the variational formulation, we consider as approximation space of the velocity $\mathbf{u}$ (resp. of the pressure p) the space of Lagrange finite elements $\mathbb{P}_1$-bubble (resp. $\mathbb{P}_1$). Uzawa's method is then used to solve the resulting linear system, in which several equations of the form $AX = G$, A being the stiffness matrix, are solved using a conjugate gradient method. In our approach, the condition number of A depends only on the mesh size h of the triangulation T_h covering the domain Ω. The Uzawa solution $(\mathbf{u}, p)$ of the Stokes problem is now obtained as the limit of a sequence of solutions $(\mathbf{u}_k, p_k)$ when $k \to \infty$ [9]. If this example, the critical part is related to the anisotropic discretization of the interface.

Here, we present the simulation of a Couette flow in a unit square $\Omega = [0,1] \times [0,1]$, a disk of radius 0.1 centered inside the domain corresponds to the domain Ω^1. We considered the following parameter values: viscosities $\mu^1 = 1$, $\mu^2 = 10^{-6}$

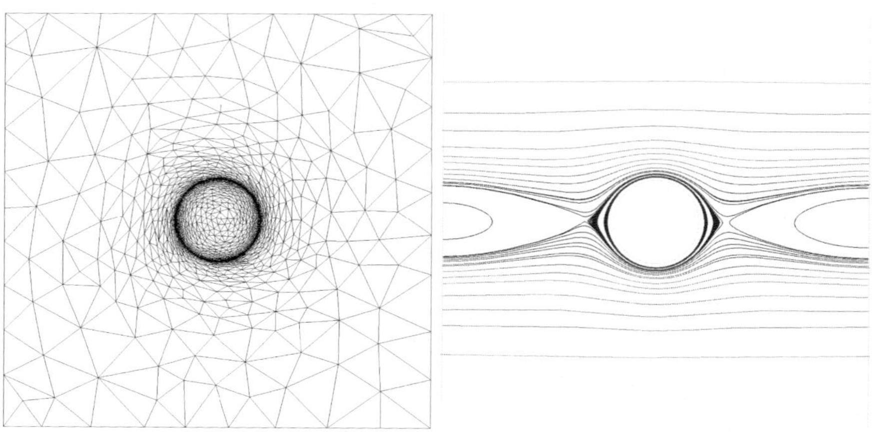

Fig. 4. Mesh of the computational domain (notice the anisotropy in the vicinity of the interface) and streamlines of the velocity field

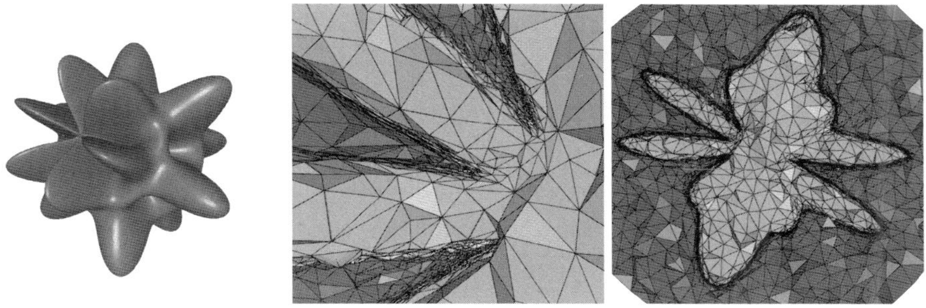

(a) (b) (c)

Fig. 5. Example of mesh adaptation of a surface embedded in the unit sphere. (a) the reconstructed piecewise affine surface, (b) cutting plane through the tetrahedral mesh and (c) local enlargement of the cutting plane.

and volumic mass $\rho^1 = \rho^2 = 1$. Dirichlet boundary conditions have been prescribed on the external boundaries of Ω: $\mathbf{u}_x = 2(y - 0.5)$, $\mathbf{u}_y = 0$. Figure 4 shows the anisotropic mesh used to perform the simulation and the streamlines of the velocity field.

5.2 A Numerical Example in Three Dimensions

This approach has been implemented in three dimensions for dealing with interface capturing problems. In order to illustrate this feature, Figure 5 shows an anisotropic mesh adapted to a slightly complex although regular analytical surface embbeded in a unit sphere defined in spherical coordinates as:

$$\rho = 0.45 + 0.3\cos(6\theta)\cos^2(3\phi) \qquad \theta \in [0, 2\pi]\,, \phi \in [0, 2\pi]\,.$$

Here, the accuracy level is controlled by the parameter $\varepsilon = 10^{-3}$, leading to a mesh containing $242\,887 < 3.10^5$ vertices. Notice that an isotropic mesh corresponding to the same level of accuracy would contain about 10^9 vertices, 3 to 4 orders of magnitude more vertices than with an anisotropic mesh. The apparent irregularity of the mesh Figure 5 (middle and right-hand side) is due to the cutting plane through tetrahedral elements. Actually, the piecewise affine discretization of the manifold is surprisingly smooth (Figure 5).

6 Conclusions and Perspectives

In this paper, we have presented an efficient method for obtaining very accurate piecewise affine approximations of manifolds of co-dimension one based on the definition of a metric tensor field used to govern the generation of anisotropic meshes. The results obtained so far in numerical simulations are more accurate than similar results obtained using isotropic meshes. The next stage will be to

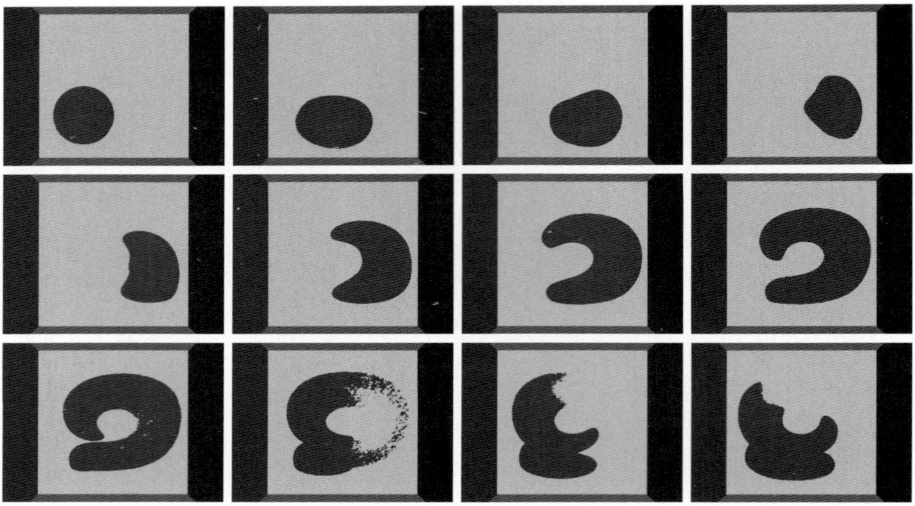

Fig. 6. Preliminary results for capturing accurately a dynamically evolving manifold in three dimensions

handle dynamically evolving interfaces in time-dependent simulations. Figure 6 shows the preliminary results obtained on the deformation of a unit sphere under a deformation field. This approach seems quite successfull in preserving the volume of the level set when the interface undergoes large amount of deformation and stretching induced by an incompressible flow field and can very well compete with methods like particle level sets [15]. By nature, this approach is well suited to maintain regions of high-curvature and the thin surface that developed during the deformation. Less than 10^5 vertices are required to resolve this test case, a number that can be compared with the 100^3 grid generally used in other approaches.

References

1. Abgrall, R.: Numerical discretisation of boundary conditions for first order Hamilton Jacobi equation n triangular meshes. Comm. Pure Appl. Maths. 49, 1339–1373 (1996)
2. Adalsteinsson, D., Sethian, J.A.: A fast level set method for propagating interfaces. Journal of Computational Physics 118(2), 269–277 (1995)
3. Alauzet, F., Frey, P.: Estimateur d'erreur géométrique et métriques anisotropes pour l'adaptation de maillage. Partie I: aspects théoriques, preprint INRIA, RR-4759 (2003)
4. Alauzet, F., Frey, P., George, P.L., Mohammadi, B.: 3D transient fixed-point mesh adaptation for time dependent problems: application to CFD simulations. J. Comp. Phys. 222, 592–623 (2007)

5. Apel, T.: Anisotropic finite elements: local estimates and applications, B.G. Teubner Stuttgart, Leipzig, Series of Advances in Numerical Mathematics (1999)
6. Apel, T., Berzins, M., Jimack, P.K., Kunert, G., Plaks, A., Tsukerman, I., Walkley, M.: Mesh shape and anisotropic elements: theory and practice. In: Whiteman, J.R. (ed.) The Mathematics of Finite Elements and Applications X, pp. 367–376. Elsevier, Oxford (2000)
7. DAzevedo, E.F., Simpson, R.B.: On optimal triangular meshes for minimizing the gradient error. Numer. Math. 59, 321–348 (1991)
8. Babuška, I., Aziz, A.K.: On the angle condition in the finite element method. SIAM J. Numer. Anal. 13, 214–226 (1976)
9. Bui, T.T.C., Frey, P., Maury, B.: Méthode du second membre modifié pour la gestion de rapports de viscosité importants dans le problme de Stokes bifluide. C.R. Acad. Sci. Série I, 524–529 (2008)
10. Crandall, M., Lions, J.L.: Two approximations of solutions of Hamilton-Jacobi equations. Math. Comp. 43, 1–19 (1984)
11. Dervieux, A., Thomasset, F.: Multifluid incompressible flows by a finite element method. Lecture Notes in Physics 11, 158–163 (1981)
12. Dompierre, J., Vallet, M.-G., Bourgault, Y., Fortin, M., Habashi, W.G.: Anisotropic mesh adaptation: towards user-independent, mesh-independent and solver-independent CFD. Part III: unstructured meshes. Int. J. Numer. Meth. Fluids 39, 675–702 (2002)
13. Dijkstra, N.: A note on two problems in connexion with graphs. Numerische Mathematik 1(1), 269–271 (1959)
14. Ducrot, V., Frey, P.: Contrôle de l'approximation géométrique d'une interface par une métrique anisotrope. C.R. Acad. Sci. 345, Série I, 537–542 (2007)
15. Enright, D., Fedkiw, R.: Robust treatment of interfaces for fluid flows and computer graphics. In: Hou, T., Tadmor, E. (eds.) Hyperbolic Problems: Theory, Numerics, Applications, pp. 153–164. Springer, New York (2003)
16. Formaggia, L., Perotto, S.: New anisotropic a priori error estimates. Numer. Math. 89, 641–667 (2001)
17. Fortin, M.: Etude numérique d'estimations d'erreur a posteriori. REEF 9, 467–486 (2000)
18. Frey, P.J., George, P.L.: Mesh generation. Application to finite elements, 2nd edn. Wiley, Chichester (2008)
19. Frey, P., Alauzet, F.: Anisotropic mesh adaptation for CFD computations. Comput. Methods Appl. Mech. Engrg. 194, 5068–5082 (2005)
20. Frey, P.J.: A differential approach to mesh generation. In: Ciarlet, P.G., Li, T. (eds.). Series in Contemporary Applied Mathematics. World Scientific, Singapore (2008)
21. Guèziec, A.: Meshsweeper: Dynamic Point-to-Polygonal-Mesh Distance and Applications. IEEE Trans. Visualization and Computer Graphics 7(1) (2001)
22. Hoch, P., Rascle, M.: Hamilton-Jacobi equations on a manifold and applications to grid generation or refinement. SIAM J. Sci. Comput. 23(6), 2055–2073 (2002)
23. Huang, W.: Metric tensors for anisotropic mesh generation. J. Comput. Phys. 204, 633–665 (2005)
24. Jiang, G.S., Peng, D.: Weighted ENO schemes for Hamilton-Jacobi equations. SIAM J. Sci. Comput. 21, 2126–2143 (2000)
25. Jiao, X., Colombi, A., Ni, X., Hart, J.C.: Anisotropic mesh adaptation for evolving triangulated surfaces. In: Proc. 15th Int. meshing Roundtable, pp. 173–190 (2006)

26. Kim, B., Liu, Y., Llamas, I., Rossignac, J.: Advections with significantly reduced dissipation and diffusion. IEEE Transactions on Visualization and Computer Graphics (2006)
27. Nagrath, S., Jansen, K.E., Lahey, R.T.: Computation of incompressible bubble dynamics with a stabilized finite element level set method. Comput. Methods Appl. Mech. Engrg. 194, 4565–4587 (2005)
28. Osher, S., Sethian, J.A.: Fronts propagating with curvature-dependent speed: Algorithms based on Hamilton-Jacobi formulations. J. Comp. Phys. 79, 12–49 (1988)
29. Osher, S., Shu, C.W.: High-order essentially nonoscillatory schemes for Hamilton-Jacobi equations. SIAM J. Numer. Anal. 28, 907–922 (1991)
30. Rudman, M.: Volume tracking methods for interfacial flow calculations. Int. Journal for Numerical Methods in Fluids 24(7), 671–691 (1997)
31. Sethian, J.: Level set methods and fast marching methods. Cambridge University Press, Cambridge (1999)
32. Smereka, P.: Semi-implicit level set method for curvature and surface diffusion motion. Journal of Scientific Computing 19, 439–456 (2003)
33. Sussman, M., Smereka, P., Osher, S.: A level set approach for computing solutions to incompressible two-phase flow. J. Comput. Phys. 114, 146–159 (1994)
34. Tsitsiklis, J.N.: Efficient algorithms for globally optimal trajectories. IEEE Trans. on Automatic Control 40(9), 1528–1538 (1995)
35. Yamakawa, S., Shimada, K.: High quality anisotropic tetrahedral mesh generation via ellipsoidal bubble packing. In: Proc. 9th Int. Meshing Roundtable, pp. 263–273 (2000)
36. Zalesak, S.T.: Fully multidimensional flux-corrected transport algorithms for fluids. Journal of Computational Physics 31(3), 335–362 (1979)
37. Zheng, X., Lowengrub, J., Anderson, A., Cristini, V.: Adaptive unstructured volume remshing: II: application to two and three dimensional level-set simulations of multiphase flow. J. Comput. Phys. 208, 626–650 (2005)
38. Zhu, J.Z., Zienkiewicz, O.C.: Superconvergence recovery technique and a posteriori error estimators. Inter. J. Numer. Meth. Engrg. 30, 1321–1339 (1990)

Anisotropic Delaunay Mesh Adaptation for Unsteady Simulations

C. Dobrzynski[1] and P. Frey[2,3]

[1] IMB, UMR 5251, Université Bordeaux 1, 351 cours de la Libération, 33405 Talence cedex, France and INRIA Futurs, Scalapplix team
cecile.dobrzynski@math.u-bordeaux1.fr

[2] UPMC Univ Paris 06, UMR 7598, Laboratoire J.L. Lions, F-75005 Paris, France
frey@ann.jussieu.fr

[3] Universidad de Chile, UMI 2807, Centro de Modelamiento Matemático, Santiago, Chile

Summary. Anisotropic mesh adaptation is a key feature in many numerical simulations to capture the physical behavior of a complex phenomenon at a reasonable computational cost. It is a challenging problem, especially when dealing with time dependent and interface capturing or tracking problems. In this paper, we describe a local mesh adaptation method based on an extension of the Delaunay kernel for creating anisotropic mesh elements with respect to adequate metric tensors. In addition, we show that this approach can be successfully applied to deal with fluid-structure interaction problems where parts of the domain boundaries undergo large displacements. The accuracy and efficiency of the method is assessed on various numerical examples of complex three-dimensional simulations.

1 Introduction

In the context of numerical simulations based on variational methods, adaptive and anisotropic triangulations have proven to be very effective for solving complex physical and biomedical problems described by a set of partial differential equations; see for instance [19, 30, 32]. Actually, many applications (e.g., in solid and fluid dynamics, combustion, heat transfer, etc.) require localized regions of the computational domain to have a larger mesh density, *i.e.*, closely-spaced vertexs, to capture the singular or nearly singular solutions that develop in such regions and to resolve large solution variations sufficiently accurately. Solving these equations with a uniform mesh would require a huge number of mesh vertexs, often out of reach of the current computer technology. Indeed, the aim of mesh adaptation is twofold: improving the efficiency of the method for better accuracy and stability at a lower computational expense.

On the other hand, dynamically evolving surfaces arise in numerous computational applications, such as free surfaces in multiphase flows or moving and deforming interfaces in fluid-structure interactions, biomedical surfaces, etc. These applications require or involve potentially large displacements or deformations of the domain geometry in time. Furthermore, this moving geometry is generally part of the solution of a set of partial differential equations and thus it is not known by or described by an analytical model. Such surfaces can be successfully handled by level set formulations [29] or implicit surfaces. In many cases, it is again more efficient to use anisotropic elements and to adapt the mesh to capture the interface or to follow the severe deformations of the geometry. The goal of this paper is to provide such mesh adaptation features for unstructured simplicial meshes in view of time-dependent and geometry evolving simulations.

In general, anisotropic mesh adaptation aims at equidistributing the approximation error by adjusting locally the mesh density according to a metric tensor field based on the Hessian of the numerical solution [16, 18]. It relies on the ability to control the *size*, the *shape* and the *orientation* of the mesh elements. In addition to improving the accuracy of the solution, anisotropy allows to preserve the order of convergence of the computational schemes [19]. It has already been largely shown that highly stretched mesh elements can interpolate a smooth function much more accurately than an isotropic mesh with regular elements [3, 33, 34]. As we will emphasize here, anisotropic elements also have the advantage of introducing regularity in the approximation of interfaces between physical domains.

The contributions of this paper are the following. First, it provides a general context for anisotropic Delaunay-based mesh adaptation in three dimensions, based on a local point insertion procedure. In this respect, it can be considered as an extension of previous works on anisotropic meshing for complex domains [7, 9, 12, 13, 22, 32]. It can also be related to computational geometry results on Delaunay insertion of Steiner points in a triangulation, sometimes called point placement methods [11, 26]. However, here we provide a straightforward and convergent algorithm to locally adapt the mesh elements to any anisotropic metric tensor. Furthermore, we show how a slight modification of the Delaunay kernel allows us to prevent the generation of badly-shaped elements. Since our approach is based on local mesh modifications, the termination of the algorithm is straightforward, although the resulting triangulation fulfill the Delaunay criterion only locally. Second, this paper explains how to build an anisotropic metric tensor for level set interface tracking, following the ideas of [14]. Then, we show how this method can be efficiently used to resolve fluid-structure interactions and moving mesh problems where the computational domain geometry undergoes large displacements. We demonstrate benchmark and simulation results on rigid-body and on fluid dynamics simulations. For the sake of simplicity however, we will assume in all these simulations that the initial surface mesh is not concerned by adaptation. The local adaptation of surface meshes has been largely discussed in other papers over the last years (surveys can be found in [17, 27], for instance).

The remainder of this paper is organized as follows. In Section 2, we review the main issues of anisotropic mesh adaptation based on Riemannian metric tensors. This notion of an anisotropic metric tensor has been described in a general purpose book [18] and in many research papers. In particular, we outline the definition of a metric tensor, the notion of metric intersection and interpolation. In Section 3, we show how the classical Delaunay mesh generation procedure can be extended to the anisotropic context. In particular, we show how to compute the Delaunay kernel used for point insertion in the Bowyer-Watson approach to mesh generation. In Section 4, we describe our local mesh adaptation algorithm. Numerical examples of simulations are given in Section 5 to show the efficiency of the approach.

2 Anisotropy and Metric Related Issues

As pointed out already [18, 23], anisotropic meshing is closely related to differential geometry concepts and numerical error estimates. Here, we will briefly review the theoretical material (curvature estimates, metric tensor field, mesh quality) for anisotropic meshing strictly required to understand our approach, as it has already been thoroughly described in various papers [1, 3, 7, 12, 13, 15, 22, 19].

2.1 Basic Definitions

We introduce the following notations: Ω denotes a simply connected open bounded domain in $\mathbb{R}^3$, $\overline{\Omega}$ is the closure of Ω and $|\Omega|$ is the d-dimensional measure or the volume of Ω. We suppose we are given a conforming triangulation T_h on Ω, h representing the characteristic element size. Each element $K \in T_h$ is a closed subdomain of Ω and we assume that $\overline{\Omega} = \bigcup_{K \in T_h} K$ and that the usual finite element requirements are satisfied (*i.e.*, non-overlapping and intersecting elements are disallowed).

A *uniform mesh* of Ω is then a mesh in which all elements are equally-sized and regular. In such case, if $|T_h|$ represents the number of mesh elements and $h_K = diam(K)$ the diameter of K, the size $h = \max_{K \in T_h} h_K$ is given by the relation:

$$h_K \approx \left(\frac{|\Omega|}{|T_h|}\right)^{1/d} \qquad \forall\, K \in T_h \,.$$

A *quasi-uniform* mesh is a mesh for which (i) there exists a constant τ such that

$$\frac{h_K}{\rho_K} \leq \tau \qquad \forall K \in \bigcup_h T_h \,,$$

where, for any open ball $B_i \subset K$, $\rho_K = \sup_i\{diam(B_i)\}$ is the in-diameter of K and (ii) the variation of h is bounded by a constant. Notice that this notion does not assume a constant mesh size over the domain.

2.2 Metric Tensors

Essential to mesh adaptation is the ability to control the *size*, the *shape* and the *orientation* of mesh elements. This specification is usually based on an error estimate or an error indicator. Typically, it uses a matrix-valued field for anisotropic mesh adaptation.

On the continuous level, it is suitable to consider that mesh elements are represented by ellipsoids. In this geometric representation, the size of the element is its volume, the shape is associated with the ratio of the lengths of its semi-axes and the orientation is provided by its principal axis vectors. Then, their control can be achieved by specifiying a metric tensor $M(x)$ to indicate the size, shape and orientation of mesh elements on the whole domain. Here, $M(x)$ is a 3×3 symmetric positive definite matrix, sometimes referred to as a *monitor function* [23]. The function $\rho(x) = \sqrt{\det(M(x))}$ is called the *adaptation function.* This metric tensor is used to generate a quasi-uniform mesh of Ω in the metric related to M. The volume of an element $K \in T_h$ is unitary:

$$\int_K \sqrt{det(M(x))}\, dx = \int_K \rho(x)\, dx = 1\,, \qquad \forall K \in T_h$$

and corresponds to the discrete formulation:

$$|K| \sqrt{det(M_K)} = 1\,, \qquad \forall K \in T_h\,,$$

where M_K is an average of $M(x)$ on K. By extension, the length of a curve γ in a metric M given by $M(x)$ for any $x \in \Omega$ is defined as:

$$|\gamma|_M = \int_0^1 \sqrt{\langle \gamma'(t), M(\gamma(t))\gamma'(t)\rangle} dt\,,$$

where $\gamma(t) : [0,1] \to \mathbb{R}^d$ is a parametrization of γ. By analogy, the length of a mesh edge $\mathbf{e}$ is defined as:

$$l_M(\mathbf{e}) = \int_0^1 \sqrt{\mathbf{e}^t M(t) \mathbf{e}}\, dt\,.$$

The edge length value represents also the number of subdivisions of the edge required to match the mesh size prescribed by the metric tensor. Since the metric tensor $M(x)$ is supposed a symmetric positive definite matrix, the spectral decomposition theorem allows to decompose M as:

$$M = P \Lambda P^t = \sum_{i=1}^{d} \lambda_i \mathbf{e_i} \mathbf{e_i}^t\,,$$

where the normalized eigenvectors of M are the columns of matrix $P = [\mathbf{e_1}, \ldots, \mathbf{e_d}]$ such that $P P^t = P^t P = I_d$ and Λ is the diagonal matrix of the eigenvalues $\lambda_i = {h_i}^{-1}$, where the h_i are the sizes in each eigendirection. It is obvious to see that the matrix P prescribes the orientation, and the matrix Λ prescribes the size and shape of any element K.

Metric intersection and interpolation

Metric intersection

Suppose that two metric tensors are specified at a vertex $p \in T_h$. For mesh generation purposes, we would like to deal with a single metric at the vertex. To this end, we define a metric intersection procedure. Geometrically speaking, it consists in defining the largest ellipsoid E included in the intersection of the two ellipsoids E_1 and E_2 associated with the two metric tensors. From the algebraic point of view, we use the simutaneous reduction of the two underlying quadratic forms to find a basis of the vector space in which the matrices M and N associated with E_1 and E_2 are represented respectively by I_3 and D, a diagonal matrix of $M_3(\mathbb{R})$.

Denoting by $P = \begin{pmatrix} e_1 & e_2 & e_3 \end{pmatrix}$ the invertible matrix of $GL_3(\mathbb{R})$ formed by the eigenvectors $e_i, i = 1,3$ of the matrix $N' = M^{-1}N$ leads to define the intersection matrix $M_\cap$ as: $M_\cap = (P^t)^{-1}\Lambda P^{-1}$, where Λ is the diagonal matrix with the coefficients $\max(\lambda_i, \mu_i)$.

Metric interpolation

Consider a parametrization of a mesh edge pq as $c : [0,1] \to \mathbb{R}^3$, $c(t) = (1-t)p+tq$ and two metric tensors M_p and M_q associated with the endpoints. We are looking for the metric tensor at t, hence for a matrix $M(t)$ defined along the segment $c(t)$ for any value of the parameter $t \in [0,1]$. The definition of this matrix $M(t)$ involves the interpolation of the eigenvalues of the matrices M_p and M_q. This procedure allows one to define a continuous metric field along the segment. To this end, we suggest the following linear interpolation scheme:

$$M(t) = \left((1-t)M_p^{-\frac{1}{2}} + tM_q^{-\frac{1}{2}}\right)^{-2}, \quad 0 \le t \le 1\,. \tag{1}$$

Finding the interpolated metric tensor M_t requires us to express the two matrices in a basis $\{e_i\}$ in which both are congruent to a diagonal matrix and then to deduce the metric tensor at point t. In other words, this scheme is similar to reducing simultaneously the two quadratic forms associated with the metrics.

2.3 Mesh Quality Measures

There are several reasons for assessing a mesh and controlling its quality. In particular, it is useful to know if the mesh elements are aligned with the physical solution, especially in the anisotropic context. Then, in the adaptive context, it is important to know how closely the equiditribution and alignment conditions are satisfied by the mesh. Finally, in three dimensions, a quasi-uniform mesh is not easy to produce if only controlling the lengths of mesh edges. It is well known that *slivers* (null volume) elements may occur and their creation cannot be prevented by simply checking the their edge lengths.

The topics of mesh quality and mesh assessment have been studied in the context of finite element methods in numerous papers. We refer the reader to [2] and the references therein for more details on the related issues. Classical quality measures include minimal angle condition [35] or maximal angle condition [4] as well as aspect ratio considerations. Other works include error estimates to account for the shape of the element and the solution behavior [6]. The adequacy between anisotropic elements and the anisotropy of the solution can be measured by a *matching function* [25].

Given a metric tensor field $M(x)$, it is natural to define critera to measure how closely the mesh elements are aligned and equidistributed with respect to $M(x)$. For practical reasons, we introduce a single measure to evaluate the quality of an element K (given here in dimension d):

$$Q_{ani}(K) = \alpha_d \frac{\left(\sum_{i=1}^{k} \langle \mathbf{e_i}, M_K \mathbf{e_i} \rangle\right)^d}{|K| \sqrt{\det(M_K)}} ,$$

where $\mathbf{e_i}$ represents any of the k egdes of K and α_d is a normalisation constant such that $Q_{ani}(K) = 1$ for a regular element. Notice that $Q_{ani} \geq 1$ for all $K \in T_h$ and thus the larger $\max_K Q_{ani}(K)$ is, the less the triangulation T_h matches the metric specifications. This quality measure is able to indicate how well a mesh element match the metric specification, both in terms of size (edge lengths) and of shape (aspect ratio).

In addition, we define an *efficiency index* τ that provides a single scalar value to evaluate how well a mesh complies with the metric requirements:

$$\tau = \exp\left(\frac{1}{ne} \sum_{1 \leq i \leq ne} \ell_M(\mathbf{e}_i)\right) , \quad \ell_M(\mathbf{e}_i) = \begin{cases} l_M(\mathbf{e}_i) - 1 \text{ if } l_M(\mathbf{e}_i) < 1 \\ l_M^{-1}(\mathbf{e}_i) - 1 \quad \text{else} \end{cases} \tag{2}$$

and ne denotes the total number of mesh edges. The objective is to generate meshes for which the efficiency index is close to the optimal value of one. In the simulations, a value $\tau \geq 0.85$ will be considered as an acceptable lower bound.

2.4 Error Estimates

A function $e(h, u_h, f)$ is called an *a posteriori error* if it provides an upper bound on the approximation error: $\|u - u_h\|_W \leq e(h, u_h, f)$. In the context of numerical simulations, the aim is to obtain an anisotropic bound where the physical derivatives are related to the size, the shape and the orientation of mesh elements. Research has been very active these recent years to develop mathematically-based error estimates and several references are provided in the bibliography section of this paper. From the numerical point of view, it is interesting to obtain estimates for the classical L^1 and L^2 norms or the H^1 seminorm. Such estimates have been provided for the interpolation error on linear Lagrange finite elements and involve the eigenvalues and eigenvectors of the Jacobian matrix of the affine mapping between the reference element and a mesh element [15, 31] or are based on the Hessian matrix of the solution [18, 1]. A local error model can be defined at a mesh vertex p as follows:

$$e_M(p) = \sum_{i=1}^{d} h_i^2 \left| \frac{\partial^2 u}{\partial \alpha_i^2} \right| ,$$

where the α_i are the coefficients of the diagonal matrix Λ (see above) and the h_i indicate the local sizes in the directions of the eigenvectors of the Hessian matrix. For our purposes, we define a local error indicator the L_∞ norm as:

$$e_K = \|u - P_h u\|_{\infty,K} \leq C \max_{y\in K} \max_{\mathbf{c}\subset K} \langle \mathbf{v}, |D^2 u(y)| \mathbf{v} \rangle \,,$$

where D^2u represents the Hessian matrix of the function u and C is a constant independent of h.

3 Anisotropic Delaunay Mesh Adaptation

Our approach suggests a modification of the classical point insertion procedure in a Delaunay triangulation to account for the anisotropic metric specifications. At first, we briefly recall the Delaunay kernel in the isotropic context, and then we introduce the modifications in the anisotropic context and for local mesh adaptation.

3.1 The Delaunay Kernel

In the classical isotropic context, mesh points are inserted in an existing Delaunay triangulation using the so-called *Delaunay kernel* [18]. This incremental yet practical method provides an efficient means for constructing a Delaunay triangulation.

Theorem 1. (Delaunay kernel) *Let T_i be the Delaunay triangulation of the convex hull of a set of points $\mathcal{S} \subset \mathbb{R}^d$, and let $p \notin \mathcal{S}$ be a vertex enclosed in T_i. The* Delaunay kernel *procedure can be written as:*

$$T_{i+1} = T_i \setminus \mathcal{C}(p) + \mathcal{B}(p) \,, \tag{3}$$

and provides T_{i+1}, a Delaunay triangulation of the convex hull of $\mathcal{S} \cup \{p\}$.

The proof is obtained by duality with the Voronoï diagram and can be found in [21, 18]. In this fundamental result, $\mathcal{C}(p)$ stands for the *cavity* of point p: the set of simplices in T_i such that their circumspheres contain point p and $\mathcal{B}(p)$ denotes the *ball* of point p: the set of simplices formed by joining p to the external faces of $\mathcal{C}(p)$. Practically, this result ensures that the cavity $\mathcal{C}(p)$ is a star-shaped polytope with respect to point p.

However, in numerical simulations, at least two specific problems arise, related to the necessity of:

i) inserting specific entities in the triangulations (a given set of edges defining the domain boundary, for instance) and
ii) creating additional vertices in the triangulation that are not part of the initial set $\mathcal{S}$ (for instance internal vertices during mesh adaptation).

Regarding the last requirement, the following result provides the existence of a triangulation:

Lemma 1. *Let T_i be an arbitrary triangulation and let $p \in \mathbb{R}^d$ be a point enclosed in T_i, p not being a vertex of T_i. Then, a valid conforming triangulation T_{i+1} having p as vertex can be created using the Delaunay kernel, Theorem 1.*

The proof relies on the star-shaped character of the region $\mathcal{C}(p)$ with respect to point p [18]. This is only possible if the cavity is created incrementally from a *seed* tetrahedron (see above). Notice that due to this restriction, the resulting triangulation T_{i+1} may not be a Delaunay triangulation, although it is conforming to finite elements requirements [10].

Point insertion issues

Defining the cavity $\mathcal{C}(p)$ of a point p requires identifying all simplices having a circumsphere that contains point p. The radius of the sphere circumscribed to a simplex K corresponds to the distance between any vertex p_i of K and the circumcenter O of K: $r_K = d(p_i, O) = \|Op_i\|$. The circumcenter O can be computed as the solution of a linear system of the form:

$$\|Op_1\| = \|Op_i\|\,, \ \forall p_i \in K\,,\ p_i \neq p_1,\ i = 2, \ldots, d\,.$$

Hence, given a point p, an element K belongs to the cavity $\mathcal{C}(p)$ iff the following relation holds:

$$\alpha(p, K) = \frac{\|Op\|}{r_K} < 1\,,$$

where $\alpha(p, K)$ is the *Delaunay measure* of point p with respect to the simplex K. First, we have to find the simplex or, in some peculiar cases, the set $\{K \in T_h\,, p \in K\}$ of all simplices containing p the point to be inserted. Then, by incorporating all adjacent simplices K_i such that $\alpha(p, K_i) < 1$, the set $\mathcal{C}(p)$ is obtained. Each time a new simplex K_i is added to $\mathcal{C}(p)$, the star-shapedness of this set is checked and the simplex is eventually removed if $\mathcal{C}(p)$ loses its property.

In addition, to avoid the generation of badly-shaped elements like *slivers*, we introduced a minimal volume requirement: all new simplices obtained by connecting p to the external faces of $\mathcal{C}(p)$ must have a measure larger than a given lower bound: $|K| > \varepsilon$, with ε set to 1.10^{-5} in the numerical experiments. This simple check has revealed especially and quite surprinsingly efficient in preventing the generation of most of the slivers, thus impacting favorably the optimization stage and the mesh quality histograms (cf. Section 4).

3.2 Anisotropic Delaunay Kernel

As expected, the extension to the anisotropic case consists of introducing a metric tensor, hence a symmetric positive definite matrix M_p at each point $p \in \mathbb{R}^d$. This will allow us to consider the Euclidean norm of any vector in $\mathbb{R}^d$ given the inner product $\langle \cdot, \cdot \rangle_M$. Hence, all distance checks involved in the Delaunay measure will be replaced by length checks according to the given matrix M_p, namely:

$$\alpha_{M_p}(p, K) = \frac{\|Op\|_{M_P}}{(r_K)_{M_p}} < 1\,.$$

The Delaunay kernel (Theorem 1) relies on distance evaluations to define the cavity of the point p to be inserted in an existing Delaunay triangulation. Using the Delaunay measure α_{M_p}, it is possible to define a valid the cavity *i.e.*, star-shaped with respect to p. In [21, 18], we have the following result:

Lemma 2 (anisotropic Delaunay kernel). *Given a metric tensor M_p at a point $p \in \mathbb{R}^d$, the anisotropic Delaunay measure, $\alpha_{M_p}(p, K) < 1$, provides a valid anisotropic Delaunay kernel.*

Again, the proof of the validity relies on the definition of a cavity, star-shaped with respect to p. Notice however that the resulting triangulation T_{i+1} may not be a Delaunay triangulation [26], but it is still conforming to finite elements requirements. In this respect, this result provides a practical way of inserting a point in an anisotropic triangulation.

Numerical issues

Obviously, taking into account only the metric tensor at the given point p is not sufficient and accurate in many applications as the mesh element size may change rapidly from one point to another in space. Therefore, we advocate considering all metric tensors related to the vertices of element K [20]. But doing so however, leads to solving a nonlinear system of equations. The center O_K of the topological sphere circumscribed to tetrahedron $K = \{p_1, \ldots, p_4\}$ is the solution of a set of equations:

$$l_{M_p}(O_K, p_1) = l_{M_p}(O_K, p_i) \qquad i = 2, \ldots, 4\,,$$

where the length of the edge $O_K p_i$ in the metric M_p is given by:

$$\begin{aligned} l_{M_p}(O_K, p_i) &= 2\, m_{12}\, (x_i - O_x)(y_i - O_y) + m_{11}\, (x_i - O_x)^2 \\ &+ 2\, m_{13}\, (x_i - O_x)(z_i - O_z) + m_{22}\, (y_i - O_y)^2 \\ &+ 2\, m_{23}\, (y_i - O_y)(z_i - O_z) + m_{33}\, (z_i - O_z)^2, \end{aligned}$$

with $p_i = (x_i, y_i, z_i)^t$, $O_K = (O_x, O_y, O_z)^t$, and $M_p = (m_{ij})_{1 \le i,j \le 3}$. Finding the center O_K simply leads to solving the following linear system:

$$\begin{pmatrix} a_2\; b_2\; c_2 \\ a_3\; b_3\; c_3 \\ a_4\; b_4\; c_4 \end{pmatrix} \begin{pmatrix} O_x \\ O_y \\ O_z \end{pmatrix} = \begin{pmatrix} d_1 \\ d_2 \\ d_3 \end{pmatrix},$$

with the coefficients:

$$\begin{aligned} a_i &= 2\, (m_{11}\, (x_i - x_1) + m_{12}\, (y_i - y_1) + m_{13}\, (z_i - z_1)), \\ b_i &= 2\, (m_{22}\, (y_i - y_1) + m_{12}\, (x_i - x_1) + m_{23}\, (z_i - z_1)), \\ c_i &= 2\, (m_{33}\, (z_i - z_1) + m_{13}\, (x_i - x_1) + m_{23}\, (y_i - y_1)), \\ d_i &= m_{11}\, x_i^2 + 2\, m_{12}\, x_i\, y_i + 2\, m_{13}\, x_i\, z_i + m_{22}\, y_i^2 + 2\, m_{23}\, y_i\, z_i + m_{33}\, z_i^2 \\ &\quad -(m_{11}\, x_1^2 + 2\, m_{12}\, x_1\, y_1 + 2\, m_{13}\, x_1\, z_1 + m_{22}\, y_1^2 + 2\, m_{23}\, y_1\, z_1 + m_{33}\, z_1^2). \end{aligned}$$

Furthermore, numerical experiments revealed that it is interesting to account for all metric tensors at the vertices p_i in order to define the cavity. We found efficient to assess the cavity with two inequalities:

$$\begin{cases} \alpha_{M_p}(p, K) < 1, \\ \displaystyle\sum_{i=1}^{4} \alpha_{M_{p_i}}(p, K) + \alpha_{M_p}(p, K) < 5. \end{cases} \tag{4}$$

Like in the generic case, we avoided the creation of most *slivers* by introducing a minimal volume requirement $|K| > \varepsilon$, the measure of the element being computed in the relevant metric.

4 Local Anisotropic Mesh Adaptation

Adaptive meshing methods belong to one of two categories depending on whether they proceed by global or local remeshing of the computational domain at each iteration. Global remeshing techniques consist in constructing a new mesh of the domain at each iteration, to ensure that the elements are in good agreement with the anisotropic metric-related prescriptions. The latter are supplied at the vertices of the previous mesh that is then acting as a control space. Obviously, the order of complexity of the meshing method remains the same throughout the whole adaptation scheme. In steady-state adaptative simulations, the number of modifications usually decreases with the iterations since a fixed point of the pair (mesh,solution) is targeted. In other words, once mesh features have been identified and captured by adjusting the local vertex density, numerical accuracy is only a matter of introducing a few more vertexs in critical regions while most of the mesh is kept unchanged. Hence, remeshing at each iteration the whole domain results in a loss of efficiency (even if improvements have been proposed [9, 19]). It seems more advantageous that fewer vertices are inserted in critical regions over time, as this will help to minimize the run time of the algorithm while at the same time converging to a solution. Hence, in our approach, we perform local mesh modifications.

4.1 Mesh Modification Operations

Mesh modification are either geometrical (edge split, edge collapse, vertex relocation) or topological (edge flip). In the anisotropic context, we assume that a metric tensor field is provided at the mesh vertices of a given triangulation. The objective is then to modify this triangulation iteratively by local operations in order to obtain a quasi-uniform triangulation wih respect to this field. An important feature consists in modifying the metric specifications to account for a desirable (*i.e.*, user-specified) mesh gradation. This procedure is fully described in [8] and is used here as such.

Our approach is based on the analysis of the edge lengths. Given a metric field prescription, all mesh edges have to belong to the interval $[l_{min}, l_{max}]$ and the mesh element quality has to be close to the optimal unit value. Theoretically, a quasi-uniform triangulation is characterized by the fact that all edges have a unit length, *i.e.*, $l_{min} = l_{max} = 1$. However, it is easy to understand that such restriction is highly improbable and we suggest to set the lower and upper bounds of the edge lengths to the values $l_{min} = \sqrt{2}^{-1}$ and $l_{max} = \sqrt{2}$. All local mesh modifications can be simulated in order to check whether the resulting configuration is better than the previous one. A configuration is considered as improved if the mesh quality of all the elements involved is better (or almost identical in the case of an edge collapse) than the original configuration.

Short edges $l_M(\mathbf{e}) < l_{min}$ are simply collapsed by merging the two endpoints of such edge at one of the endpoints. All mesh elements surrounding this edge are then removed. If a short edge cannot be collapsed, an edge flip operation attempts to solve the problem. The edge swap consists in remeshing the set of elements surrounding the edge by changing the vertex connectivity. This is an expensive modification as it requires the computation of a large (combinatorial) number of quality measures. For such reason, no edge flip will be checked if the set of elements contains more than 7 elements.

Long edges $l_M(\mathbf{e}) > l_{max}$ are normally removed by splitting them into unit subsegments with respect to the metric specification. This consists in introducing the new vertices using the anisotropic Delaunay kernel. The main advantage of this approach is that the set of elements concerned by the vertex insertion (the cavity) is usually much larger than the set of elements strictly surrounding the edge (the initial cavity). Computationally, it revealed more efficient to proceed like this and the quality of the resulting mesh is better than with the classical edge split operation. Furthermore, the creation of *slivers* is explicitely checked and prevented during this stage. Sometimes, in peculiar cases, an edge flip operation can be used to remove a long edge.

In addition, a mesh optimization stage involves a vertex relocation procedure. This procedure aims at equidistributing the edge lengths, with the objective of improving the efficiency index. Given a mesh vertex p, the objective is to improve the lengths of all edges connected to p. To this end, the vertex is moved in the direction of an optimal vertex location corresponding to the average optimal positions along each edge. This operation does not affect the mesh topology, nonetheless optimal vertex positions have to be checked to prevent invalid configurations.

All mesh modification operations are described in details in the general purpose book [18].

4.2 Mesh Adaptation Scheme

Our mesh adaptation algorithm can then be described as follows: we assume given a triangulation T and a metric tensor field M defined at the mesh vertices,

```
Enforce the desired h-gradation,
do
   1. loop over mesh edges /* edge analysis */
      compute l_M(e) for every edge e
      if (l_M(e) < l_min) then
         collapse e
      else if (l_M(e) > l_max) then
```

```
            split e and insert new vertex (Delaunay kernel)
         endif
      while ( T is modified )
      2. loop over mesh elements /* quality optimization */
         perform edge flips
         perform vertex relocation
      end for
   while ( triangulation T is modified ).
```

Notice that this algorithm (composed of two levels of imbricated loops) always terminates in a finite number of iterations. Indeed, mesh modifications are only applied if they result in the mesh quality improvement. An infinite loop could potentially occurs between several configurations if the analysis is based on the edge lengths only. For instance, if the values of l_{min} and l_{max} are too close, the edge split and the edge collapse operations could possibly lead to a previous configuration. However, since we apply the mesh modification if the quality is strictly improved, this case cannot happen in our scheme. In addition, for efficiency purposes, we have restricted the overall number of iterations to a maximum value.

The complexity of the algorithm is related to the number of loops that are performed. The inner loops are in $O(nt)$, where nt is the number of mesh elements. Since we limit the nmber of outer loop to a maximum value c, the overall complexity is $c\,O(nt)$, *i.e.*, linear in number of mesh elements.

4.3 Application to Large Displacements Problems

As pointed out in the introduction, our local anisotropic mesh adaptation approach allows us to handle rigid-body displacement problems without difficulty. Fluid-structure interactions involves a moving structure, rigid or deformable, and a fluid in flow around a part or the whole structure. The domain boundary is moving, however moving only boundary vertices would quickly result in an invalid mesh. Therefore, internal vertices are also relocated using a linear elasticity analogy [5]. Actually, the whole procedure is straightforward and can be decomposed in three successive stages: given a field of displacements prescribed at the domain boundaries, *e.g.*, resulting from a fluid calculation, (i) solve a linear elasticity equation as suggested by [5] to define a discrete displacement field at all mesh vertices and (ii) move the mesh vertices to the positions prescribed by this field and (iii) optimize the resulting mesh.

This boundary value problem assumes that part of the boundary remains fixed as another part is moving. It reads:

$$\begin{cases} -div\sigma(u) = f \text{ in } \Omega \\ u = 0 \text{ on } \Gamma_{fixed} \\ \sigma(u)\cdot n = 0 \text{ on } \partial\Omega\backslash\Gamma_{fixed} \end{cases}$$

and the Hook's law gives the relation between the stress tensor σ and the linearized strain tensor e as: $\sigma = 2\mu e(u) + \lambda tr(e(u)) I_3$, with $e(u) = \frac{1}{2}(\nabla u + \nabla u^t)$, λ and μ are the Lamé coefficients describing the mechanical properties of the material. The variational formulation is solved using affine P_1 Lagrange finite elements. The Lamé coefficients have been set experimentally in our simulations. The linear system is solved using a conjugate gradient routine with a LDL^t preconditioning step. The convergence of the linear system is not affected too much by the anisotropy of the mesh, indicating that the condition number does not degrade too much with the anisotropy.

Notice that our approach preserves the number of mesh vertices during this stage and involves vertex relocation and edge flips only. However, edge splits and edge colapses can also be used to achieve a better efficiency in large displacement problems.

5 Application Examples

Our approach has been tested on numerous test cases and is currently used by several research groups. We provide here a set of numerical examples to assess this method.

5.1 An Analytical Metric

This first test case is provided to illustrate our mesh adaptation scheme on a steady-state problem where the size, the shape and the orientation of the mesh elements are prescribed using the following analytical metric:

$$M = \begin{pmatrix} h_1^{-2} & 0 & 0 \\ 0 & h_{max}^{-2} & 0 \\ 0 & 0 & h_{max}^{-2} \end{pmatrix} \quad \text{with} \quad \begin{cases} h_1 & = h_{max}|1 - e^{-|x-0.5|}| + 0.003 \\ h_{max} & = 0.2 \end{cases} \tag{5}$$

This metric simulates the capture of a planar shock in the computational domain: $\Omega = [-1\ ,\ 1]^3$. We started with an initial surface-adapted mesh T_0 containing no internal vertex containing 4,399 vertices and 13,978 tetrahedra (Figure 1). The final adapted mesh after 5 iterations contains 51,440 tetrahedra and its efficiency index is 0.87. Regarding the mesh quality, 99.96 percents of the elements have a quality measure between 1 and 3.

5.2 Interface Capturing

Our approach has also proven useful in accurately tracking and approximating a dynamically evolving interface [14]. In this approach, the metric tensor field is related to the intrinsic properties of the manifold of codimension one that correspond to the interface. We consider an implicitly-defined, scalar valued function u on a domain $\Omega \subset \mathbb{R}^3$ and we denote by Γ the surface associated to the isovalue $u = 0$. The objective is to produce a mesh where the density

Fig. 1. Initial surface-adapted mesh (left), cut though the final adapted mesh (middle) and local enlargment (right)

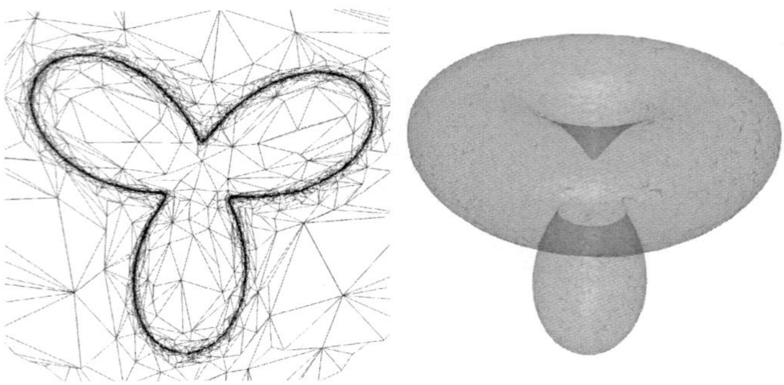

Fig. 2. Example of mesh adaptation to capture an implicitly defined surface: cutting plane through tetrahedral elements (left) and isosurface reconstruction (right)

is high in the vicinity of the isosurface so as to minimize the piecewise affine approximation of this interface. To this end, we defined the following metric tensor at the vertices of all mesh elements intersected by the manifold:

$$M = R \begin{pmatrix} 1/\varepsilon^2 & 0 & 0 \\ 0 & |\lambda_2|/\varepsilon & 0 \\ 0 & 0 & |\lambda_3|/\varepsilon \end{pmatrix} {}^tR,$$

with $R = \begin{pmatrix} \nabla u \; v1 \; v2 \end{pmatrix}$, where (v_1, v_2) is a basis of the tangent plane to the surface and λ_i are the eigenvalues of the Hessian of u. At all other vertices, we define the metric αI_3 with $\alpha \in \mathbb{R}^+$. We consider the analytical surface defined in spherical coordinates as: $r = 0.45 + 0.3 \sin(3\phi)$, with $\theta \in [0; 2\pi]$ and $\phi \in [-\frac{\pi}{2}; \frac{\pi}{2}]$. We started from an initial uniform mesh of size $h = 0.2$ (Figure 2). The final mesh after 8 iterations contains 10^5 vertices for a minimal element size $h = 10^{-3}$. The approximation error in the L_∞ norm between the surface and its piecewise affine discretization is lesser than 10^{-4}.

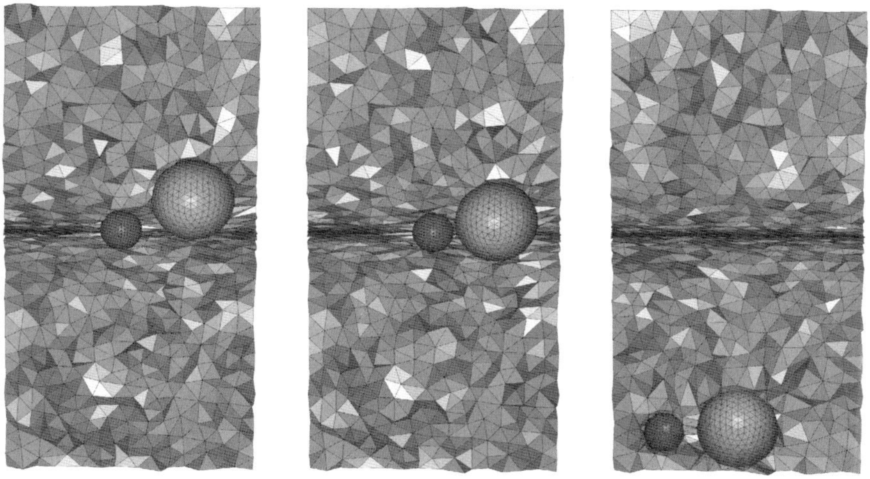

Fig. 3. Two moving spheres are crossing an anisotropic field

5.3 Moving Spheres in Anisotropic Field

We consider two spheres of different radii moving in a computational domain $\Omega = [-2\,,\,2] \times [-2\,,\,6] \times [-10\,,\,1.5]$ (Figure 3). The initial mesh is adapted to a planar shock corresponding to a metric specification very similar to that given by Equation (5). The main difficulty is to preserve the anisotropic refinement throughout the simulation. To this end, we have allowed edge split and edge collapse operations in addition to vertex relocation and edge flips. Nevertheless, the number of vertices remains almost constant: $12,200$ for he inital mesh and $11,282$ at the 34th time step. The efficiency index is also constant $\tau = 0.86$.

5.4 Airflow around a Rotating Helicopter Propeller

The flow is governed by the classical compressible Euler equations of the fluid dynamics and the numerical resolution uses the software `Fluidbox` based on finite volume and Arbitrary Lagrangian Eulerian (ALE) method [28] that imposes a constant number of vertices and the connectivity to remains identical to that of the initial mesh. The solver uses implicit time stepping scheme. Furthermore, we consider that the propeller turns with a constant angular velocity $\theta = 3.6deg$, *i.e.*, 10 full rotations per second. One mesh is generated at each time step $dt = 1/6,000$ seconds and 100 meshes are needed to achieve a complete revolution. However, the constant topology and number of vertices constraints have been relaxed: a few edge flips were introduced to remove badly-shaped elements and the solution had then to be interpolated on the optimized mesh (Figure 4). All meshes contain about 3.10^5 elements for an efficiency index close to 0.85 (more than 98 percent of the elements have a quality better than 3).

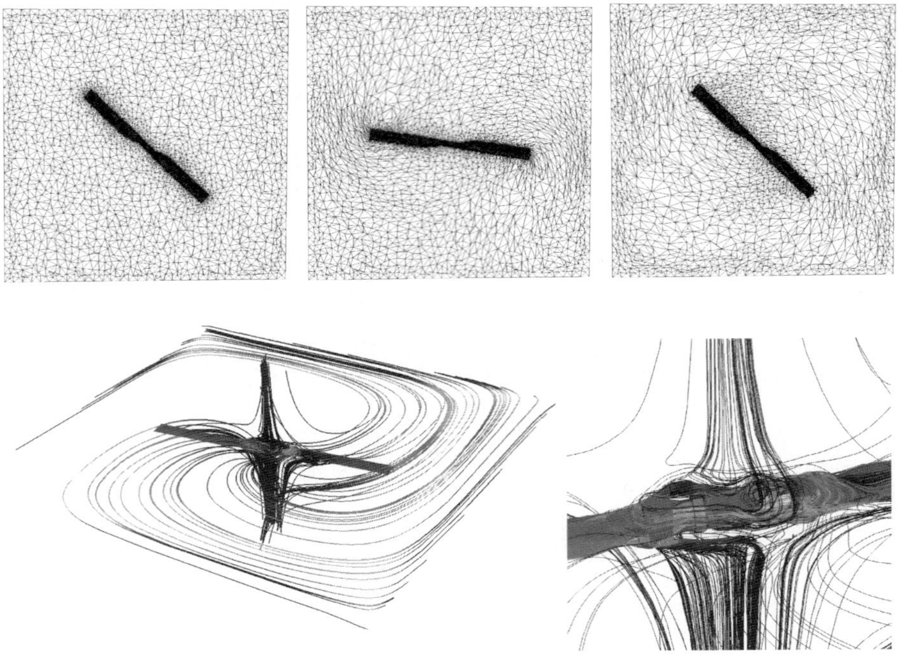

Fig. 4. Airflow simulation: adapted meshes at iterations 1, 99 and 500 (top); streamlines and velocity modulus (bottom)

6 Conclusions and Perspectives

In this paper, we have presented an efficient method for obtaining anisotropic adapted meshes based on Riemannian metric specifications. This approach is based on a modification of the classical Delaunay kernel and involve local mesh modification operations. The results obtained so far in the numerical simulations are promising and confirm the cogency of the local modification strategy. The next stage will be to handle dynamically evolving domains where both the geometry and the topology of the domains change in time.

References

1. Alauzet, F., Loseille, A., Dervieux, A., Frey, P.: Multi-dimensional continuous metric for mesh adaptation. In: Proc. 15th Int. Meshing Roundtable, pp. 191–214 (2006)
2. Apel, T., Berzins, M., Jimack, P.K., Kunert, G., Plaks, A., Tsukerman, I., Walkley, M.: Mesh shape and anisotropic elements: Theory and practice. In: Whiteman, J.R. (ed.) The Mathematics of Finite Elements and Applications X, pp. 367–376. Elsevier, Oxford (2000)

3. DAzevedo, E.F., Simpson, R.B.: On optimal triangular meshes for minimizing the gradient error. Numer. Math. 59, 321–348 (1991)
4. Babuška, I., Aziz, A.K.: On the angle condition in the finite element method. SIAM J. Numer. Anal. 13, 214–226 (1976)
5. Baker, T.J.: Mesh Movement and Metamorphosis. Eng. Comput. 18(3), 188–198 (2002)
6. Berzins, M.: A solution-based triangular and tetrahedral mesh quality indicator. SIAM J. Sci. Comput. 19, 2051–2060 (1998)
7. Bossen, F.J., Heckbert, P.S.: A pliant method for anisotropic mesh generation. In: Proc. 5th Int. Meshing Roundtable, pp. 63–74 (1996)
8. Borouchaki, H., Hecht, F., Frey, P.: Mesh gradation control. Int. J. Numer. Methods Engng. 43(6), 1143–1165 (1998)
9. Castro-Diaz, M.J., Hecht, F., Mohammadi, B., Pironneau, O.: Anisotropic unstructured mesh adaption for flow simulations. Int. J. Numer. Meth. Fluids 25, 475–491 (1997)
10. Ciarlet, P.G.: The Finite Element Method for Elliptic Problems. North-Holland, Amsterdam (1978)
11. Chernikov, A., Crisochoides, N.: Three-dimensional semi-generalized point placement method for Delaunay mesh refinement. In: Proc. 16th Int. Meshing Roundtable, Seattle, pp. 25–44 (2007)
12. Dolejsi, V.: Anisotropic mesh adaptation for finite volume and finite element methods on triangular meshes. Computing and Visualisation in Science 1, 165–178 (1998)
13. Dompierre, J., Vallet, M.-G., Bourgault, Y., Fortin, M., Habashi, W.G.: Anisotropic mesh adaptation: Towards user-indepedent, mesh-independent and solver-independent CFD. Part III: Unstructured meshes. Int. J. Numer. Meth. Fluids 39, 675–702 (2002)
14. Ducrot, V., Frey, P.: Contrôle de l'approximation géométrique d'une interface par une métrique anisotrope. C.R. Acad. Sci. 345, Série I, 537–542 (2007)
15. Formaggia, L., Perotto, S.: New anisotropic a priori error estimates. Numer. Math. 89, 641–667 (2001)
16. Fortin, M.: Etude numérique d'estimations d'erreur a posteriori. REEF 9, 467–486 (2000)
17. Frey, P.: About surface remeshing. In: Proc. 9th Int. Meshing Roundtable, pp. 123–136 (2000)
18. Frey, P.J., George, P.L.: Mesh generation. Application to finite elements, 2nd edn. Wiley, Chichester (2008)
19. Frey, P.J., Alauzet, F.: Anisotropic mesh adaptation for CFD simulations. Comput. Methods Appl. Mech. Engrg. 194(48-49), 5068–5082 (2005)
20. George, P.L.: Premières expériences de maillage automatique par une méthode de Delaunay anisotrope en trois dimensions, RT-0272, Inria (2002)
21. George, P.L., Borouchaki, H.: Delaunay triangulation and meshing. Applications to finite elements, Hermès Science, Paris (1998)
22. Gruau, C., Coupez, T.: 3d unstructured and anisotropic mesh generation with adaptation to natural and multidomain metric. Comput. Methods Appl. Mech. Engrg. 194(48-49), 4951–4976 (2005)
23. Huang, W.: Metric tensors for anisotropic mesh generation. J. Comput. Phys. 204, 633–665 (2005)
24. Jiao, X., Colombi, A., Ni, X., Hart, J.: Anisotropic mesh adaptation for evolving triangulated surfaces. In: Proc. 15th Int. Meshing Roundtable, pp. 173–190 (2006)

25. Kunert, G.: A local problem error estimator for anisotropic tetrahedral finite element meshes. SIAM J. Numer. Anal. 39, 668–689 (2001)
26. Labelle, F., Shewchuk, J.: Anisotropic Voronoi Diagrams and Guaranteed-Quality Anisotropic Mesh Generation. In: Proc. 19th Annual Symposium on Computational Geometry, pp. 191–200. ACM, San Diego (2003)
27. Löhner, R.: Regridding surface triangulations. J. Comput. Physics 26(1), 1–10 (1996)
28. Nkonga, B.: On the conservative and accurate CFD approximations for moving meshes and moving boundaries. Comput. Methods Appl. Mech. Engrg. 190, 1801–1825 (2000)
29. Osher, S., Fedkiw, R.: Level Set Methods and Dynamic Implicit Surfaces, vol. 153. Springer, New York (2002)
30. Peraire, J., Vahdati, M., Morgan, K., Zienkiewicz, O.C.: Adaptive remeshing for compressible flow computations. J. Comput. Phys. 72, 449–466 (1997)
31. Picasso, M.: An anisotropic error indicator based on Zienkiewicz-Zhu error estimator: Application to elliptic and parabolic problems. SIAM J. Sci. Comput. 24, 1328–1355 (2003)
32. Remacle, J., Li, X., Shephard, M.S., Flaherty, J.E.: Anisotropic adaptive simulation of transient flows using discontinuous Galerkin methods. Int. J. Numer. Meth. Engr. 62(7), 899–923 (2003)
33. Rippa, S.: Long and thin triangles can be good for linear interpolation. SIAM J. Numer. Anal. 29, 257–270 (1992)
34. Shewchuk, J.R.: What Is a Good Linear Element? Interpolation, Conditioning, and Quality Measures. In: 11th Int. Meshing Roundtable, pp. 115–126 (2002)
35. Zlámal, M.: On the finite element method. Numer. Math. 12, 394–409 (1968)

Dynamic Parallel Adaption for Three Dimensional Unstructured Meshes: Application to Interface Tracking

Youssef Mesri, Walid Zerguine, Hugues Digonnet,
Luisa Silva, and Thierry Coupez

CIM Group, Center for Material Forming/ MINES ParisTech, Rue Claude Daunesse
BP 207 06904 Sophia Antipolis cedex France
youssef.mesri@ensmp.fr

1 Introduction

The anisotropic mesh adaption techniques in the last decade have dramatically improved the numerical simulations *accuracy* of complex problems. An optimal anisotropic mesh adaption consists in refining and coarsening the mesh, by using a metric to specify stretching directions, in order to accurately capture physical anisotropy such as shock waves, contact discontinuities, vortexes, boundary layers and free surfaces. Thus, we propose in this paper, an anisotropic *a posteriori* error estimator that controls the error due to mesh discretization in all space directions. From the *a posteriori* error analysis, we obtain an optimal metric (optimal mesh) as a minimum of an error indicator function and for a given number of elements. The optimal metric obtained is used to build an optimal mesh for the given number of elements. Furthermore, solutions for the physical problems illustrated here are often more accurate on adapted meshes than those obtained on globally-refined meshes and at a much lower cost.

The mesh adaption procedure consists in improving iteratively the couple mesh/solution until convergence. This iterative process becomes more expensive when we handle the complex industrial three-dimensional problems.

In this paper, we propose a dynamic parallel mesh adaption on unstructured mesh that is powerful for computing unsteady three-dimensional problems. In this context, we decompose the original 3-dimensional mesh adaption problem into N_s smaller subproblems which are solved (i.e., meshed) concurrently using P processors.

Unfortunately, the adaptive solution of unsteady problems causes load imbalance among processors on a parallel machine. This is due to the fact that the computational intensity is not only time dependent, but also varies spatially over the problem domain. However, balancing dynamically the computational load is very difficult. It requires reliable measurements of processor workload and the amount of data transfer, as well as the minimization of interprocessor communication. For this reason, we have built a parallel mesh partitioning/re-partitioning procedure that is extended and customized for FEM computations [7]. The partitioning/

re-partitioning scheme allows also to dynamically load balance the workload over heterogeneous architectures [8].

Figure 1 depicts our framework for parallel adaptive flow computation. It consists of a CFD solver and mesh adapter, with a partitioner that redistributes the computational mesh when necessary. The mesh is first partitioned and mapped among the available processors. The flow solver runs for several iterations, updating solution variables. Once a solution is obtained, the mesh adaption procedure is invoked. It targets a re-meshing based on the *a posteriori* error estimator computed from the flow solution. The old mesh is then locally adapted, generating a new one and followed by the repartitioning procedure to load balance the new workload.

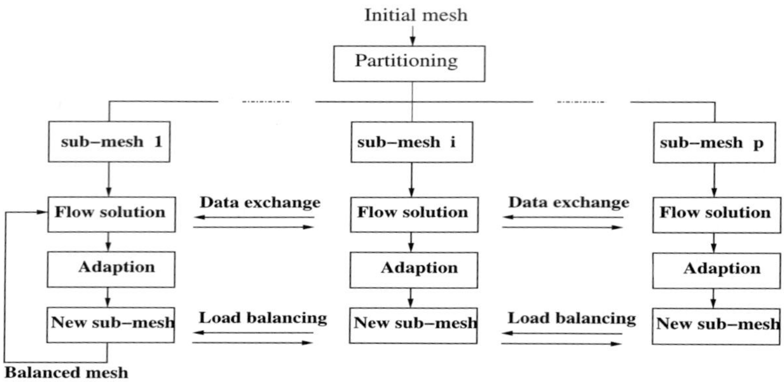

Fig. 1. Overview of our framework for parallel adaptive flow computation

This paper is organized as follows: the first section details the notations and notions that will be used in the sequel. The second section is devoted to the main contribution of this paper, the generalization of Almeida's anisotropic estimator [1] to multi-dimensional unstructured meshes ($n \geq 3$). The second major contribution consists in the parallel implementation of the mesh adaption procedure, presented in Section 3. In Section 4 we depict numerical results which exhibit the advantage of the combined parallel mesh adaption algorithm and anisotropic error estimator to handle accurately complex CFD problems.

2 The Optimal Adaptive Mesh Procedure

2.1 Notations and Notions

Given a polygon $\Omega \in R^d$, we consider a set of triangulations $\{\mathcal{T}_h\}$. We use the standard subspace of approximation

$$V_h = \{v \in H_0^1(\Omega) : v|_T \in \mathcal{P}_k(T)\} \quad (1)$$

where $\mathcal{P}_k(T)$ denotes the space of polynomials of degree k. We associate, for each node n_i, $1 \leq i \leq \mathcal{N}_{\mathcal{T}_h}$ of the triangulation, a basis function $\varphi_i \in V_h$. For each i, we set $S_i = \text{supp } \varphi_i$.

An *a posteriori* error estimator for the difference between a given function $u \in W^{2,p}(\Omega)$ and a discrete function u_h which is an approximation of $u \in \Omega$ is presented as,

$$\|u - u_h\|_{L^p(\Omega)} \leq C\|u - \Pi u\|_{L^p(\Omega)} \leq C\Big(\sum_{T \in \mathcal{T}_h} \|H(u)(x)\|^p_{L^p(T)}\Big)^{\frac{1}{p}}, \tag{2}$$

where $\Pi : W^{2,p}(\Omega) \to V_h$ is the Clément interpolation operator, $H(u)(x) = D_2u(x)(x - x_T)(x - x_T)$, $D_2u(.)$ is a Hessian operator and x_T is the barycenter of the element T.

The *a posteriori* error estimator is based on processing the function $u_h \in V_h$ in order to obtain a better approximation to the Hessian of u. We use an approximation instead of the exact Hessian matrix to estimate the L^p-norm of the error $e = u - u_h$.

The process to obtain a *recovered* Hessian matrix from the function u_h is based on a technique to recover the gradient. Zienkiewicz and Zhu [9] used a recovered gradient to estimate the energy norm of the error of the finite element approximation (for more information on recovering first derivatives, see also [10]). Furthermore, Almeida et al. [1] presented an upper bound of the error $\|u - u_h\|_{L^p(\Omega)}$ that depends on the recovered Hessian and on the number of elements,

$$\begin{aligned}\|u - u_h\|_{L^p(\Omega)} \leq C\|u - \Pi u\|_{L^p(\Omega)} &\leq C\Big(\sum_{T \in \mathcal{T}_h} \|H(u)(x)\|^p_{L^p(T)}\Big)^{\frac{1}{p}} \\ &\leq C' N_{\mathcal{T}_h}^{-\alpha} \|H_R(u_h)(x)\|_{L^p(\Omega)}\end{aligned} \tag{3}$$

where $\alpha \geq 0$ and $N_{\mathcal{T}_h}$ denotes the number of elements of the mesh.

In our work, we use the L^p-norm of the recovered Hessian as an *a posteriori* error estimator. To simplify the notation, the recovered Hessian H_R will be noted only H.

In what concerns re-meshing, our algorithm implies that around an arbitrary point P of the mesh, we try to build equilateral tetrahedrons in the metric defined by the local metric field $\mathcal{M}$, according to a local topological technique. This metric is defined in R^d by:

$$\mathcal{M}(P) = \frac{1}{h_1(P)} e_1 \otimes e_1 + \cdots + \frac{1}{h_d(P)} e_d \otimes e_d \tag{4}$$

where $(e_i)_{i=1,d}$ are the eigenvectors of the recovered hessian $H(u_h(P))$ and $h_i(P)$ are the mesh sizes in the e_i directions.

2.2 Anisotropic *A Posteriori Error* Estimator

We assume that the function $u_h \in V_h$ is a good approximation of the function u. Hence, we get

$$\|u - u_h\|_{L^p(\Omega)} \approx C\|H(u_h(x))(x - x_0) \cdot (x - x_0)\|_{L^p(\Omega)} \tag{5}$$

where $H(u_h)$ is the recovered Hessian. This shows that the interpolation error in one point x such that $|x - x_0|$ is small enough, is governed by the behavior of the second order derivative in such point. Thus, the interpolation error is not distributed in an isotropic way around point x_0, i.e., the error depends on direction $x - x_0$ and the Hessian matrix value in this point, $H(u_h(x))$. Therefore, we suggest as in [15, 1, 16, 14, 13, 4] the use of the expression (5) as a *directional local estimator*. However, the Hessian matrix is not a metric: it is not positive definite. Therefore, the following tensor is introduced

$$\mathcal{H} = R\Lambda R^T \tag{6}$$

where R is the orthonormal matrix which corresponds to the eigenvectors $(e_i)_{i=1,d}$ of the Hessian matrix, while $\Lambda = diag(|\lambda_1|, ..., |\lambda_d|)$ is the diagonal matrix of the absolute values of the eigenvalues of $H(u_h(x))$. $\mathcal{H}$ can be also written as follows:

$$\mathcal{H} = R\Lambda R^T = |\lambda_1| e_1 \otimes e_1 + \cdots + |\lambda_d| e_d \otimes e_d \tag{7}$$

Given a mesh discretization $\mathcal{T}_h$ of Ω, we define the anisotropic error estimator of the element $T \in \mathcal{T}_h$ by

$$\eta_T = \Big(\int_T (\mathcal{H}(u_h(x_0))(x - x_0) \cdot (x - x_0))^p \, dT \Big)^{\frac{1}{p}} \tag{8}$$

and the global estimator η by

$$\eta = \big(\sum_{T \in \mathcal{T}_h} \eta_T^p \big)^{\frac{1}{p}} \tag{9}$$

where x_0 is the barycenter of T.

The definition of η_T can be used to get the following upper bound.

$$\eta_T^p = \int_T (\mathcal{H}(u_h(x_0))(x - x_0) \cdot (x - x_0))^p dT \tag{10}$$

Substituting (7) in (10) we get,

$$\begin{aligned}
\eta_T^p &= \int_T \big(\sum_{i=1,d} |\lambda_i(x_0)| [e_i(x_0) \otimes e_i(x_0)](x - x_0) \cdot (x - x_0) \big)^p dT \\
&= \int_T \big(\sum_{i=1,d} |\lambda_i(x_0)| [e_i(x_0) \cdot (x - x_0)]^2 \big)^p dT
\end{aligned}$$

If we write $x - x_0$ in the eigenvectors basis: $x - x_0 = \sum_{i=1,d} x_i e_i$ where $|x_i| \leq h_i$, the projection of $x - x_0$ on e_i direction is $[e_i \cdot (x - x_0)]^2 = x_i^2 \leq h_i^2$ and then,

$$\eta_T^p \leq \int_T \Big(\sum_{i=1,d} |\lambda_i(x_0)| h_i^2 \Big)^p dT \tag{11}$$

We notice that an optimal mesh regarding a solution field u is the one that is aligned with this solution. It means that the *shape* of each element in the mesh is such that the local error in any direction attains the same value. This is equivalent to the fact that the local error in the principal directions of curvature is constant per element i.e.

$$|\lambda_1| h_1 = \cdots = |\lambda_d| h_d^2 = cte. \tag{12}$$

Then, the stretching of the element T is defined as

$$s_{1T} := \frac{h_1}{h_2} = \Big(\frac{|\lambda_2|}{|\lambda_1|}\Big)^{1/2}, s_{iT} := \frac{h_i}{h_{i+1}} = \Big(\frac{|\lambda_{i+1}|}{|\lambda_i|}\Big)^{1/2}, s_{(d-1)T} := \frac{h_{d-1}}{h_d} = \Big(\frac{|\lambda_d|}{|\lambda_{d-1}|}\Big)^{1/2} \tag{13}$$

where h_i are sorted in the decreasing order, i.e., $h_1 \geq h_2 \geq \cdots \geq h_d$. Combining (11) and (12) we get

$$\eta_T^p \leq |T| \big(d |\lambda_d(x_0)| h_d^2\big)^p \tag{14}$$

where $|T|$ is the volume of element T.

Finally, we obtain the following upper bound of the local error estimator

$$\eta_T \leq d |T|^{\frac{1}{p}} |\lambda_d(x_0)| h_d^2 \tag{15}$$

This shows that the local estimator is bounded by the maximum of the second order derivative in the barycenter of the element times the square of the longitude in this direction.

We may introduce as the local estimator the above upper bound and is given by

$$\eta_T \;=\; d|T|^{\frac{1}{p}} |\lambda_d(x_0)| h_d^2 \tag{16}$$

In the next section, we define a minimization problem where the functional is the error indicator. The solution of this problem is the metric used to construct an optimal adapted mesh.

2.3 Optimal Mesh as a Solution of an Optimization Problem

Let $\mathcal{T}_h$ denote the current finite element discretization of the domain Ω, u_h denotes the approximate solution associated to the mesh $\mathcal{T}_h$, $h_{old}(P)$ describes the local mesh size at node P in the direction of the maximum value of the directional second order derivative at the point P. Then, given the number of desired elements $N_{\mathcal{T}_h'}$ in the new adapted mesh, the optimal mesh adaptive procedure generates a *new* mesh, $\mathcal{T}_h'$, such that the new distribution $h_{new}(P)$, for all $P \in \mathcal{T}_h$ minimizes the global estimator error. Hence, the optimal mesh adaptive procedure looks for an optimal mesh as a solution of the following constrained optimization problem

$$\begin{cases} \text{Find } h_T = \{h_{1T}, \cdots, h_{dT}\}, T \in \mathcal{T}_h \text{ that minimizes the cost function} \\ F(h_T) = \sum\limits_{T\in\mathcal{T}_h} (\eta_T)^p \\ \text{under the constraint} \quad N_{\mathcal{T}_h'} = C_0^{-1} \sum\limits_{T\in\mathcal{T}_h} \int_T \prod\limits_{i=1}^{d} \frac{1}{h_{iT}}\, dT \end{cases} \tag{17}$$

where C_0 is the volume of a regular tetrahedron. In the constraint equation, the number of desired elements is obtained as follows:

$$|\Omega| = \sum_{T\in\mathcal{T}_h} |T|_{\mathcal{M}} = N_{\mathcal{T}_h'} C_0 = \sum_{T\in\mathcal{T}_h} \int_T det(\mathcal{M})\, dx = \sum_{T\in\mathcal{T}_h} \int_T \prod_{i=1}^{d} \frac{1}{h_{iT}}\, dx. \tag{18}$$

The optimization problem (17) was studied in the case of two dimensional meshes ($d = 2$) in [1]. Here, we generalize the study to three dimensional unstructured meshes. The main difficulty of the 3D analysis is, that there are different directions of stretching, in opposition to 2D case where there is only one direction. The following theorem generalizes the anisotropic error estimator proposed in [1] to three dimensional unstructured meshes.

Theorem 1. *For $d = 3$, the optimization problem (17) has a unique solution and is given by*

$$\begin{cases} h_{3T} = \left[\dfrac{\beta}{\frac{(2p+3)}{3}\, C_{1T}} \int_T C_{2T}\, dT \right]^{\frac{1}{2(p+3)}} \\ h_{2T} = s_{2T} h_{3T} \\ h_{1T} = s_{1T} s_{2T} h_{3T} \end{cases} \tag{19}$$

where

$$C_{1T} = 3^p C_0 s_{1T} s_{2T}^2 |\lambda_3|^p, \qquad C_{2T} = C_0^{-1} \frac{1}{s_{1T} s_{2T}^2} \qquad \textit{and}$$

$$\beta^{\frac{1}{\frac{2}{3}(p+3)}} = N_{\mathcal{T}_h'}^{-1} \sum_{T\in\mathcal{T}_h} \left\{ \left(\frac{1}{\int_T C_{2T}\, dT} \right)^{\frac{1}{\frac{2}{3}(p+3)}} \int_T C_{2T} \left[\frac{2p+3}{3} C_{1T} \right]^{\frac{1}{\frac{2}{3}(p+3)}} dT \right\}.$$

Proof. Let us consider the following optimization problem:

$$\begin{cases} \text{Find } h_T = \{h_{1T}, h_{2T}, h_{3T}\}, T \in \mathcal{T}_n \text{ such that minimizes the cost function} \\ F(h_T) = \sum\limits_{T\in\mathcal{T}_h} |T| \left[3|\lambda_3| h_{3T}^2 \right]^p \\ \text{under the constraint} \quad N_{\mathcal{T}_h'} = C_0^{-1} \sum\limits_{T\in\mathcal{T}_h} \int_T \prod\limits_{i=1}^{3} \frac{1}{h_{iT}}\, dT \end{cases} \tag{20}$$

Let $s_{1T} = \frac{h_{1T}}{h_{2T}}$ and $s_{2T} = \frac{h_{2T}}{h_{3T}}$ be the two stretching directions of the element T in R^3. Then, we write h_{1T} and h_{2T} as a function of s_{1T}, s_{2T} and h_{3T}:

$$h_{1T} = s_{1T} s_{2T} h_{3T}, \quad h_{2T} = s_{2T} h_{3T} \tag{21}$$

The volume of a tetrahedron T on the metric space can be written as:

$$|T|_{\mathcal{M}} \;=\; |T|\; det(\mathcal{M}) \rightarrow |T| \;=\; |T|_{\mathcal{M}}\; det(\mathcal{M})^{-1} = C_0\; h_{1T} h_{2T} h_{3T}. \tag{22}$$

where $C_0 = \frac{\sqrt{2}}{12}$ is the volume of a regular tetrahedron. Thus, with the variable change (21), the 3D optimization problem becomes a 1D optimization problem such as the unique value to look for is h_{3T}

$$\begin{cases} \text{Find } h_{3T},\ T \in \mathcal{T}_n \text{ that minimizes the cost function} \\[2ex] F(h_{3T}) = C_0 \displaystyle\sum_{T \in \mathcal{T}_h} \; 3^p s_{1T} s_{2T}^2 |\lambda_3|^p h_{3T}^{2p+3}\; dT \\[2ex] \text{under the constraint} \qquad N_{\mathcal{T}'_h} = C_0^{-1} \displaystyle\sum_{T \in \mathcal{T}_h} \int_T \frac{1}{s_{1T} s_{2T}^2} \frac{1}{h_{3T}^3}\; dT \end{cases} \tag{23}$$

Introducing the following notations

$$C_{1T} \;=\; 3^p C_0 s_{1T} s_{2T}^2 |\lambda_3|^p, \qquad C_{2T} = C_0^{-1} \frac{1}{s_{1T} s_{2T}^2} \quad \text{and} \quad \gamma_T = \frac{1}{h_{3T}^3}. \tag{24}$$

The minimization problem can be then written as follows:

$$\begin{cases} \text{Find } \gamma_T,\ T \in \mathcal{T}_n \text{ such that minimizes the cost function} \\[2ex] F(\gamma_T) = \displaystyle\sum_{T \in \mathcal{T}_h} C_{1T} \gamma_T^{-\frac{2p+3}{3}}\; dT \\[2ex] \text{under the constraint} \qquad N_{\mathcal{T}'_h} = \displaystyle\sum_{T \in \mathcal{T}_h} \int_T C_{2T} \gamma_T\; dT \end{cases} \tag{25}$$

which is equivalent to the following min-max problem

$$\min_{\gamma_T} \max_{\beta}\; \mathcal{L}(\gamma_T, \beta) = F(\gamma_T) - \beta \Big[N_{\mathcal{T}'_h} - \sum_{T \in \mathcal{T}_h} \int_T C_{2T} \gamma_T\; dT \Big]. \tag{26}$$

The solution of this min-max problem is given by

$$\frac{\partial \mathcal{L}}{\partial \gamma_T} = 0 \Rightarrow -\frac{2p+3}{3} C_{1T} \gamma_T^{-\frac{2}{3}(p+3)} + \beta \int_T C_{2T}\; dT \;=\; 0 \tag{27}$$

$$\frac{\partial \mathcal{L}}{\partial \beta} = 0 \;\Rightarrow\; N_{\mathcal{T}'_h} \;-\; \sum_{T \in \mathcal{T}_h} \int_T C_{2T} \gamma_T\; dT \;=\; 0. \tag{28}$$

From (27) we get

$$\gamma_T = \left[\frac{\frac{2p+3}{3} C_{1T}}{\beta \int_T C_{2T}} \right]^{\frac{1}{\frac{2}{3}(p+3)}}. \tag{29}$$

and substituting the expression of γ_T in (28) we obtain

$$\beta^{\frac{1}{\frac{2}{3}(p+3)}} = N_{\mathcal{T}_h'}^{-1} \sum_{T \in \mathcal{T}_h} \left\{ \left(\frac{1}{\int_T C_{2T}} \right)^{\frac{1}{\frac{2}{3}(p+3)}} \int_T C_{2T} \left[\frac{2p+3}{3} C_{1T} \right]^{\frac{1}{\frac{2}{3}(p+3)}} dT \right\}. \tag{30}$$

From the definition of the third equation of (24)

$$h_{3T} = \left(\frac{1}{\gamma_T} \right)^{\frac{1}{3}} = \left[\frac{\beta}{\frac{(2p+3)}{3} C_{1T}} \int_T C_{2T} \, dT \right]^{\frac{1}{2(p+3)}} \tag{31}$$

We finally get the optimal value of h_{3T} and then h_{1T} and h_{2T} thanks to expression (21) . □

Remark 1. We can get the same proof for $d > 3$, and that by using a recurrence argument. The generalized form of the theorem can be written as follows:

$$\begin{cases} h_{dT} = \left[\frac{\beta}{\frac{(2p+d)}{d} C_{1T}} \int_T C_{2T} \, dT \right]^{\frac{1}{2(p+d)}} \\ h_{iT} = \left(\prod_{k=i}^{d-1} s_{kT} \right) h_{dT}, \quad 1 \le i \le d-1 \end{cases} \tag{32}$$

where

$$C_{1T} = C_0 d^p \left(\prod_{i=1}^{d-1} s_{iT}^i \right) |\lambda_d|^p, \quad C_{2T} = C_0^{-1} \frac{1}{\left(\prod_{i=1}^{d-1} s_{iT}^i \right)} \quad \text{and}$$

$$\beta^{\frac{1}{\frac{2}{d}(p+d)}} = N_{\mathcal{T}_h'}^{-1} \sum_{T \in \mathcal{T}_h} \left\{ \left(\frac{1}{\int_T C_{2T}} \right)^{\frac{1}{\frac{2}{d}(p+d)}} \int_T C_{2T} \left[\frac{2p+d}{d} C_{1T} \right]^{\frac{1}{\frac{2}{d}(p+d)}} dT \right\}. \ \square$$

We know now the optimal distribution of the element's shapes and the stretching directions that are given by the eigenvectors of the recovered Hessian as well as the optimal metric that will be used to modify the background mesh. Both are used as an input of the mesh generator tool in order to obtain a new (optimal) mesh. The mesh generator used here and its parallelization will be described in the following section.

3 Parallel Mesh Generation

In this section, we review a parallel mesh generation and adaption procedure based on a topological mesh generator. In the first part (3.1) we describe briefly the sequential mesh generator MTC and then (section 3.2) we illustrate the strategy adopted to parallelize this mesh generator.

3.1 Mesh Generation

MTC is a mesh generator developed by Thierry Coupez at the Ecole des Mines de Paris, Center for Material Forming, Sophia Antipolis. It is based on the idea to improve iteratively, an initial unsatisfactory mesh by local improvements.

MTC mesh generator re-meshes the initial mesh iteratively by a local mesh optimization technique. This technique consists in local re-meshing of cavities formed by small clusters of elements in order to increase the "quality" of the elements of the cluster. In the re-meshing process, two principles are enforced:

- **The minimal volume**, which assures the conformity of the mesh, with no element overlaps: let $\mathcal{T}_i(C)$ denote the i-th set of elements T filling the local cavity. Following the minimum volume principle we choose as an optimal (possibly not unique) re-triangulation of the cavity the one satisfying

$$\sum_{T \in \mathcal{T}_i(C)} |(Volume)(T))| \rightarrow \min, \tag{33}$$

  where the minimization is done over a small set $i = \{1, \ldots, I\}$ of possible triangulations $\mathcal{T}_i(C)$ of elements (Fig. 2 right) connecting the nodes on the border of the cavity, or other nodes like the cavity barycenter, with all boundary faces.

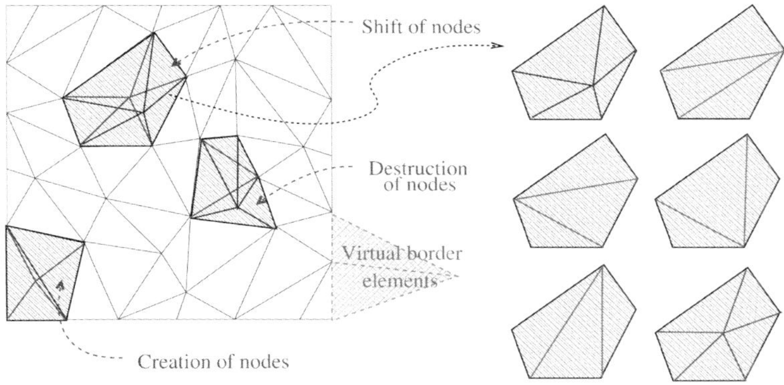

Fig. 2. Local mesh optimization process in MTC

- **The geometrical quality** $Q(T)$, which is evaluated for each element. If the minimum of (33) is not unique, this criterion picks among all admissible cavity re-triangulations the one improving the geometrical quality of the mesh by improving the quality of the worst element of the triangulation.

While the former criterion assures the conformity of the mesh, if the initial mesh was conforming, the latter handles improvements of element shape, size, connectivity, etc., depending on the quality function $Q(T)$. Usually, the quality function $Q(T)$ is a function of the geometry of the element T and the prescribed background metric, which give together a measure for the element size and the element form (aspect ratio). For further details see [2].

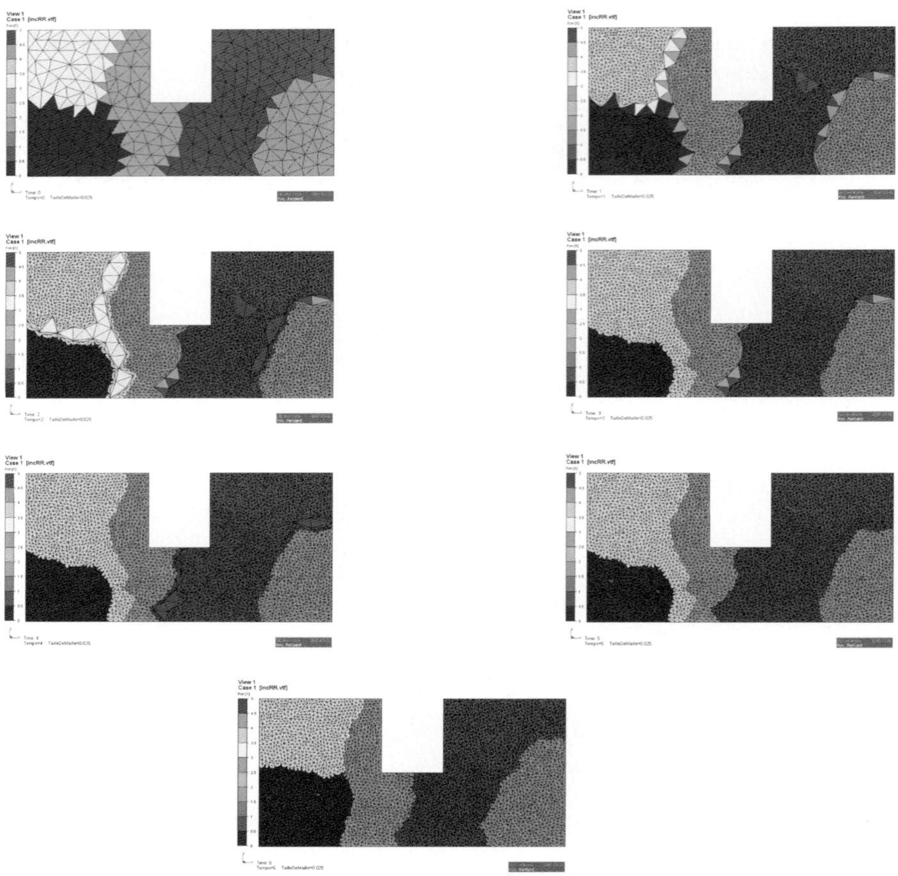

Fig. 3. Illustration of the strategy used to parallelize the mesh generator. From the top to the bottom, we show the successive steps of parallel repartitioning and parallel re-meshing by keeping interfaces unchanged.

3.2 Parallel Mesh Adaption Procedure

The parallelism of the mesh generator MTC is performed by partitioning/repartitioning the initial mesh into submeshes. The individual submeshes are refined/derefined or adapted to an error estimator by using sequential MTC techniques. In order to achieve mesh conformity and correctness, the interface faces between the submeshes should be subdivided the same way in all submeshes that share them to avoid the non conforming points throughout the process. For this reason, we have chosen to keep interfaces unchanged during the local (inside each subdomain) re-meshing and then to move the interface inside the domain in order to enable re-meshing in a next phase. Figure (3) shows this strategy applied to a simple 2D mesh with 6 submeshes partitioned onto 6 processors. The partitioning/re-partitioning of a mesh is performed in parallel using a generic graph partitioning that is extended and customized for FEM computations. This software was developed as part of the DRAMA project [7].

4 Numerical Results

Using the anisotropic adaption technique described in Section 2 and 3, we study the evolution of the error regarding the number of mesh elements on analytical functions, in two and three dimensions. The global *a posteriori* error estimator defined in (9) is used here to construct an optimal mesh with respect to the L^2 error estimator. The global error estimator becomes then:

$$\eta = \{ \sum_{T \in \Gamma_h} (\eta_T)^2 \}^{1/2} \tag{34}$$

4.1 2D Analytical Cases

Let us consider the following function

$$f(x,y) = tanh(100y - 50 + 20sin(-4x+4))), (x,y) \in \Omega = [0,1] \times [0,1] \tag{35}$$

We represent on Fig.4 (left) the analytical function f defined on a square of size $[0,1]\times[0,1]$. The Hessian of the solution is then built by means of a double $L2$ projection technique. With a number of elements equal to 2000 the minimization process leads to a new size distribution and the metric tensor generates a new adapted mesh described on Fig.4 (right). The final anisotropic mesh obtained is aligned with the analytical solution. We are now interested in the convergence of the L^2 global estimator regarding the number of elements in the mesh. For this function, a convergence of order 2 is reached Fig.5.

4.2 3D Analytical Case

We choose now a more complex function in 3D in order to check the efficiency of our adaption technique.

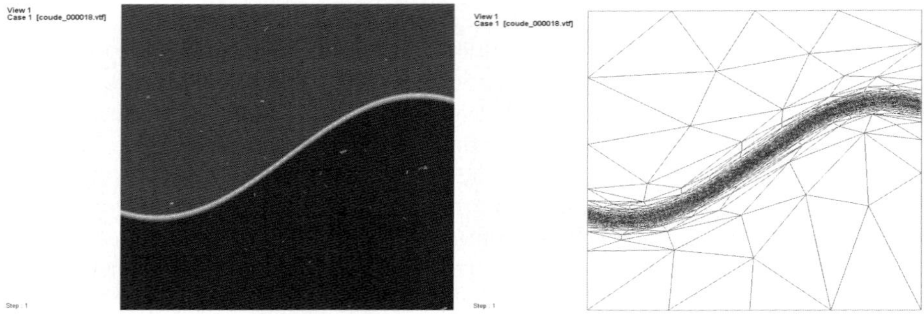

Fig. 4. Visualization of the function f (left) on the adapted mesh (right)

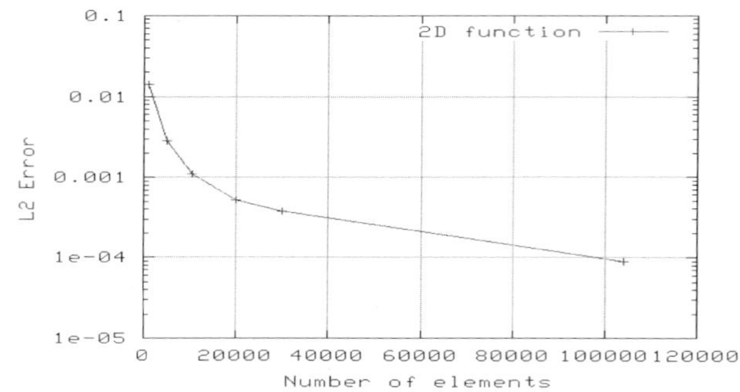

Fig. 5. Evolution of the L^2 error vs. the number of elements in the mesh for a 2D analytical test case

$$h(x, y, z) = \tanh((x + 1)^{20}(y - 0.5)^9 z), (x, y, z) \in \Omega = [0, 1] \times [0, 1] \times [0, 1] \quad (36)$$

We represent on Fig.6 (left) the analytical function h defined on a cube of size $[0, 1] \times [0, 1] \times [0, 1]$ and, after the mesh adaption strategy, the resulting mesh is depicted in Fig.6 (right). The number of elements targeted is of 100000. As in 2D case, the mesh emphasizes well with the solution of the analytical function. Moreover, we overcome with success the difficulty occurring in the corners. Indeed, the steep change of directions in the corner regions may be cause of element distortions. We can show that, with this more complex analytical solution and in 3D, we also reach a convergence of order 2 in what concerns the $L2$ error (Fig.7).

4.3 Dynamic Adaption Flows

In this paper, we apply our mesh adaption technique to simulate complex multiphase problems involved in manufacturing processes like Water-Assisted

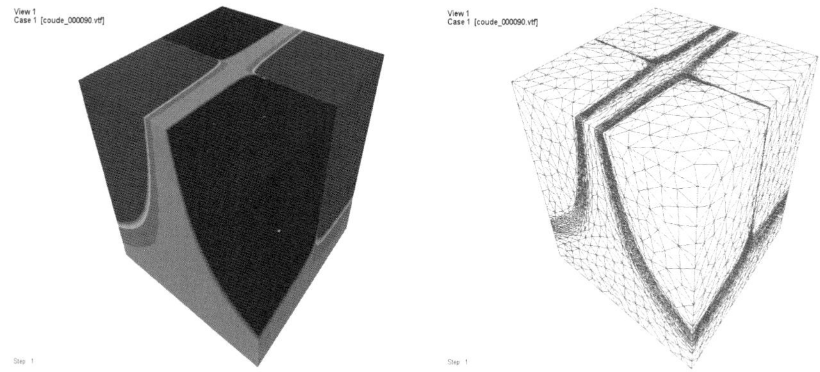

Fig. 6. Visualization of the function h (left) on the adapted mesh (right)

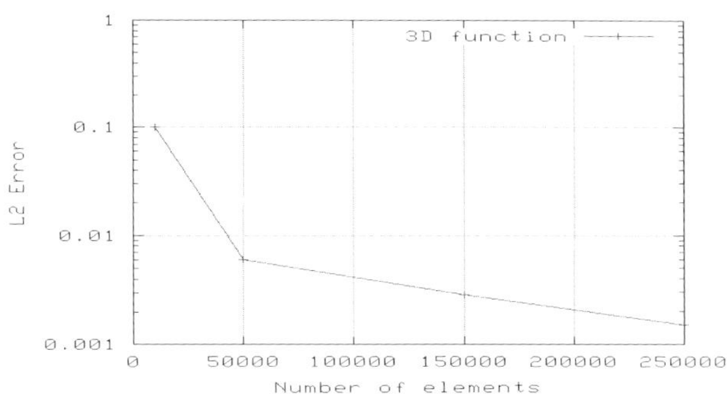

Fig. 7. Evolution of the L^2 error vs. the number of elements in the mesh for a 3D analytical test case

Injection Molding (WAIM). The simulation of the process needs an accurate description of the water evolution. The anisotropic strategy aims at improving multiphase flow computations by allowing a better description of physical parameters that characterize strongly heterogeneous flows. Flow equations are solved by considering heterogeneous incompressible Navier-Stokes equations coupled with the heat equation. The water is supposed a newtonian fluid whereas the polymer is considered non newtonian: its viscosity described by WLF and Carreau Yasuda laws, respectively, for the dependency on the temperature field and on the shear rate.

$$\begin{cases} \rho\frac{\partial v}{\partial t} + \rho\nabla v \cdot v - \nabla \cdot (2\eta\varepsilon(v)) + \nabla p = \rho g \\ \nabla \cdot v = 0 \\ \rho = 1_{\Omega water}\rho_{water} + (1 - 1_{\Omega water})\rho_{polymer} \\ \eta = 1_{\Omega water}\eta_{water} + (1 - 1_{\Omega water})\eta_{polymer} \end{cases} \tag{37}$$

where ρ,η,v,p,g represent respectively the density, viscosity, velocity, pressure and gravity fields. We denote by ρ_{water}, η_{water}, $\rho_{polymer}$, $\eta_{polymer}$ the density and viscosity of the water and the polymer. The Navier Stokes equations are solved with a mixed finite element formulation. We consider a linear and continuous velocity enriched spatial discretization (P1+/P1).

$$\begin{cases} \rho C(\frac{\partial T}{\partial t} + \overrightarrow{v} \cdot \overrightarrow{\nabla T}) - \nabla \cdot (k\overrightarrow{\nabla T}) = f \qquad \text{on } \Omega\times]0,t_n[\\ T(\overrightarrow{x},t) = T_{imposed} \qquad \text{on } \Omega\times]0,t_n[\\ T(\overrightarrow{x},t=t_0) = T_0(\overrightarrow{x}) \\ \eta(T,\dot{\gamma}) \ (CarreauYasuda/WLF) \text{ Computation of the thermodependent viscosity} \end{cases} \tag{38}$$

We follow the polymer/air and water/polymer interfaces with two Level Set functions [12]. The Navier Stokes equations will provide us the velocity and pressure fields and this same velocity will transport the Level Set functions whose zero iso-values represent the interfaces. Therefore, a good description of the interfaces depends on the accurate resolution of the coupled problem. The Level Set strategy applied here is based on a new approach that allows to keep an unitary gradient for the Level Set function without regularizing it periodically with an Hamilton-Jacobi Equation [12].

2D dynamic adaption flows

We apply our adaptive strategy on 2D dynamic flows. We aim to simulate the WAIM process on a 2D cavity. We impose at the entrance of the cavity a constant

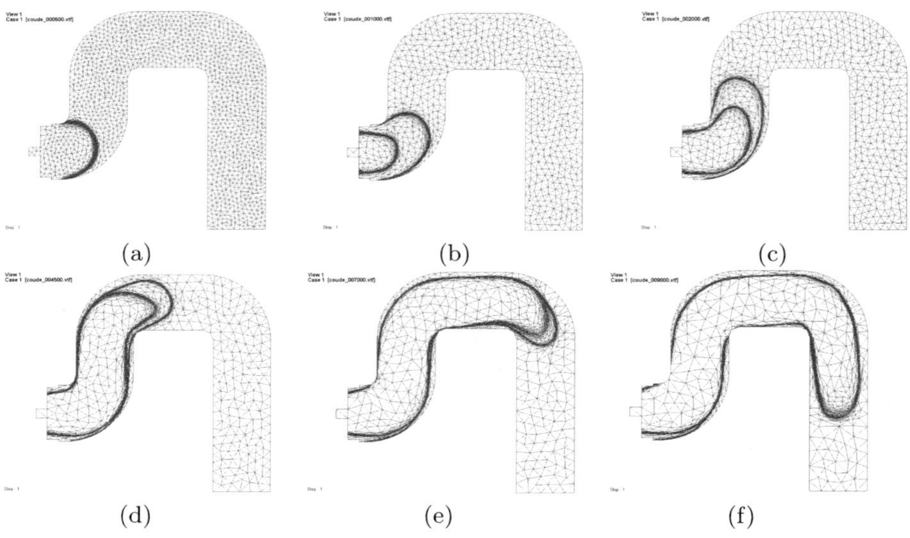

Fig. 8. Evolution of the polymer/air and water/polymer fronts during the water assisted injection process

(a) (b)

Fig. 9. Zoom in the mesh

velocity field of $0.1m.s^{-1}$ in order to be close to the industrial process and to fill the cavity in two seconds. During the first time of the simulation only the polymer is in the cavity Fig.8 (a). Then we change the boundary conditions for the Navier-Stokes equations and we impose a water velocity of $0.2m.s^{-1}$ at the same entrance. Thus, the water pushes the polymer outside the cavity (Fig.8 from (b) to (f)). The mesh adaption strategy is applied on a reconstructed Level Set function(combination of the polymer/air and water/polymer Level Set functions) and the number of elements aimed is of 20000. Notice that the interface is well captured thanks to anisotropic mesh refinement around the interface, whereas the isotropic region is derefined in order to satisfy the number of elements constraint.

A zoom in the mesh Fig.9 shows us the accuracy of our method.

3D industrial Water Assisted Injection Molding case

The aim of this section is to study the efficiency of our adaptive strategy on an industrial 3D WAIM process Fig.10. So as to achieve our simulation we extend to 3D space the coupled system described on section 4.3. Few seconds before the water injection the cavity is already filled with the polymer. For this specific simulation the adaption strategy is applied on the water/polymer Level Set function(and not as in the 2D dynamic case on the two Level Set functions).

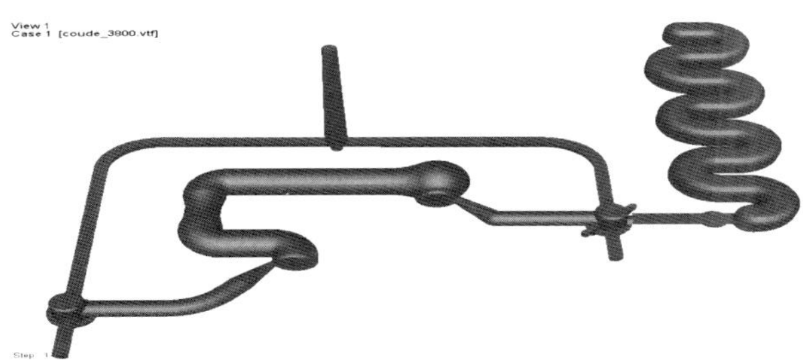

Fig. 10. Water Assisted Injection Molding Cavity

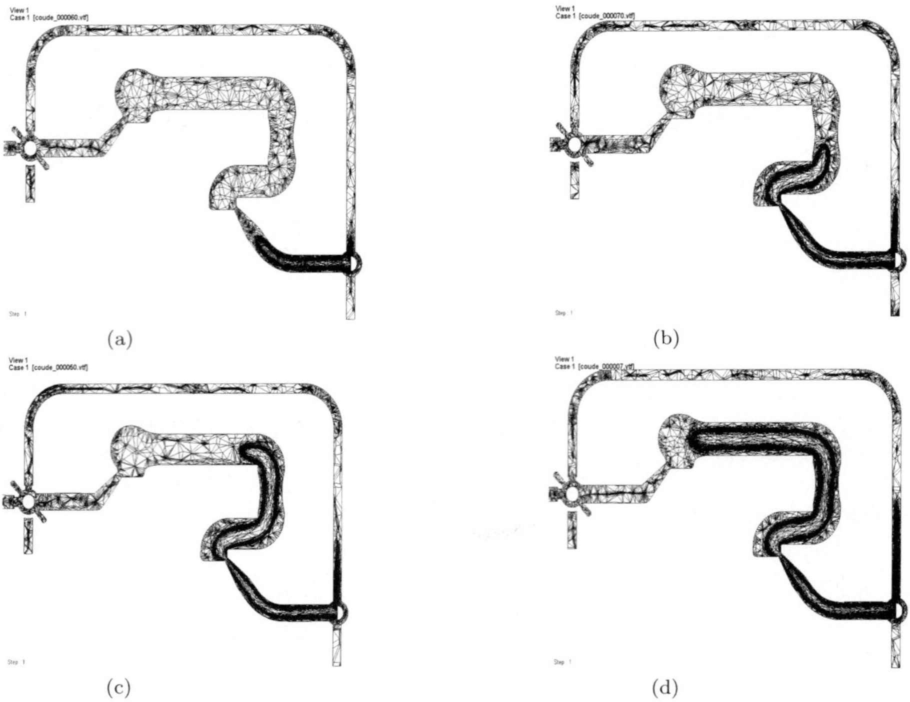

Fig. 11. Evolution of water/polymer fronts during the water assisted injection process

During the water injection and regarding a certain percentage of injected water a new mesh is generated dynamically.

The simulation is run onto 16 processors and we fix a constraint of 10 millions of elements. A cut of the mesh allows to depict the water tracking inside the cavity at different time of the simulation Fig.11. The mesh follows dynamically the evolution of the water. The adaption process allows an accurate description of the water vein and a considerable gain of CPU time, since, far from the water/polymer front, the size of the elements increases. Therefore the number of unknowns of our system is dramatically reduced.

5 Conclusions

We have presented a dynamic parallel mesh adaption procedure. It is based on the definition of an anisotropic *a posteriori* error estimator, the search of the optimal mesh (metric) that minimizes the error estimator and the use of the serial mesh generator (MTC) in a parallel context. The parallelization strategy consists in balancing dynamically the workflow by repartitioning the mesh after each re-meshing stage. The numerical 2D and 3D applications show that the proposed anisotropic error estimator gives an accurate representation of the exact error.

It was shown also, that the optimal adaptive mesh procedure provides a mesh refinement and element stretching which appropriately captures interfaces for industrial injection polymers problems. Finally, the anisotropic adapted meshes provide a highly accurate solutions that are often better than those obtained on globally-refined meshes and at a much lower cost.

References

1. Almeida, R., Feijoo, R., Galeao, A., Padra, C., Silva, R.: Adaptive finite element computational fluid dynamics using an anisotropic error estimator. Comput. Methods Appl. Mesh. Engrg. 182, 379–400 (2000)
2. Coupez, T.: Génération de maillage et adaptation de maillage par optimisation locale. In Revue européene des éléments finis 9(4), 403–423 (2002)
3. Coupez, T., Digonnet, H., Ducloux, R.: Parallel meshing and remshing. Appl. Math. Modelling 25(2), 83–98 (2000)
4. Mesri, Y., Alauzet, F., Loseille, A., Hascoet, L., Koobus, B., Dervieux, A.: Continuous metric for computational fluid dynamics. CFD Journal 16(4), 346–355 (2008)
5. Tam, A., Ait-Ali-Yahia, D., Robichaud, M.P., Moore, M., Kozel, V., Habashi, W.G.: Anisotropic mesh adaptation for 3D flows on structured and unstructured grids. Computer Methods in Applied Mechanics and Engineering 189(4), 1205–1230 (2000)
6. Huang, W.: Metric tensors for anisotropic mesh generation. J. Comput. Phys. 204, 633–665 (2005)
7. Basermann, A., Clinchemaillie, J., Coupez, T., Fingberg, J., Digonnet, H., Ducloux, R., Gratien, J.M., Lonsdale, G., Maerten, B., Roose, B., Walshaw, C.: Dynamic load balancing of finite element applications with the DRAMA library. Appl. Math. Modelling 25(2), 83–98 (2000)
8. Mesri, Y., Digonnet, H., Guillard, H.: Mesh Partitioning forParallel Computational Fluid Dynamics Applications On a Grid. In: Finite Volumes for complex applicationsIV, pp. 631–642. Hermes Science Publisher (2005)
9. Zienkiewicz, O.C., Zhu, J.Z.: The superconvergent patch recovery and a posteriori error estimator. Part 1. The recovery technique. Int. J. Numer. Meth. Eng. 33, 1331–1364 (1992)
10. Zienkiewicz, O.C., Zhu, J.Z.: The superconvergent patch recovery and a posteriori error estimator. Part 2. Error estimates and adaptivity. Int. J. Numer. Meth. Eng. 33, 1365–1382 (1992)
11. de Cougny, H., Shephard, M.: Parallel volume meshing using face removals and hierarchical repartitioning. Comp. Meth. Appl. Mech. Engng. 174(3-4), 275–298 (1999)
12. Coupez, T.: Réinitialisation convective et locale des fonctions Level Set pour le mouvement de surfaces et d' interfaces. Journées Activités Universitaires de Mécanique LaRochelle (2006)
13. Lipnikov, K., Vassilevski, Y.: Error bounds for controllable adaptive algorithms based on a Hessian recovery. Computational Mathematics and Mathematical Physics 45(8), 1374–1384
14. Alauzet, F., Loseille, A., Dervieux, A., Frey, P.: Multi-dimensional continuous metric for mesh adaptation. In: Pebay, P. (ed.) Proceedings of 15th International Meshing Roundtable, pp. 163–171. Springer, Heidelberg (2006)

15. Castro-Diaz, M., Hecht, F., Mohammadi, B., Pironneau, O.: Anisotropic unstructured mesh adaptation for flow simulations. Int. J. Numer. Meth. Fluids 25, 475–491 (1997)
16. Dervieux, A., Leservoisier, D., George, P.-L., Coudiere, Y.: About theoretical and practical impact of mesh adaptations on approximation of functions and of solution of PDE. Int. J. Numer. Meth. Fluids 43, 507–516 (2003)

Transient Mesh Adaptivity with Large Rigid-Body Displacements

G. Compère*, J.-F. Remacle, and E. Marchandise

Institute of Mechanical, Materials and Civil Engineering (iMMC)
Université catholique de Louvain (UCL)
Avenue Georges Lemaitre 4, 1348 Louvain-la-Neuve, Belgium
{gaetan.compere,jean-francois.remacle,emilie.marchandise}@uclouvain.be

Summary. This paper presents a procedure for computing fluid-structure interaction problems when the boundaries of the domain undergo large displacements. The algorithm is based on both mesh motion and mesh adaptation techniques. More specifically, we use a node repositioning algorithm based on an elastic analogy together with a mesh adaptation procedure based on local mesh modifications [1]. This paper also includes a new technique for eliminating sliver tetrahedra. Some computational results are finally presented that include statistics on mesh quality measures.

1 Introduction

This paper deals with the issue of computing fluid-structure interaction (FSI) problems with large displacements of the structure. The most common way of dealing with FSI is to adopt an Arbitrary Lagrangian Eulerian (ALE) formulation of the fluid equations. ALE formulations allow to take into account small motions of the nodes in the fluid caused by the displacement of the structure. This approach suffers from obvious limitations as node repositioning cannot always provide a valid mesh when significant displacements or deformations of the structure are considered.

One way of addressing this problem is to remesh the entire domain when the displacements of the structure are too large to be handled [2, 3]. In this work, we rather use local mesh modifications [1, 4, 5, 6] both to optimize the quality of the tetrahedra and to comply a mesh size field. This approach is globally advantageous compared to global remeshing:

- Local solution projection procedures can be built in a way that ensures local conservation [4],
- The mesh remains unchanged in large parts of the domain,
- Local mesh modifications can be performed in parallel, enabling transient adaptive simulation to run on parallel computers [5].

There exist a small number of other approaches to handle large mesh deformations, among which the sliding mesh techniques [7]. Compared to local mesh

* Fonds de la Recherche Scientifique (FRS-FNRS), Belgium.

modifications, they have the disadvantage of requiring a flux computation at the interface and working with non coinciding meshes. The overset grid methods [8] are also developed in the literature but require grid assemblies and donor cell search processes. The local mesh modification technique has a unique mesh and a single prescribed boundary motion, which leads to a conceptually simple method and robust mesh movements [5].

Transient adaptive computations using local mesh modifications have already been applied to transient multiphase flow simulations in [9]. In this previous work, the mesh was adapted in order to capture the interface between two fluids, described by a level set function. In this paper, we extend the method to FSI problems.

In this context, two new issues have to be addressed. The first issue is the formulation of the mesh motion problem. In FSI, only the motion of the boundaries of the domain is prescribed. Typically, some kind of elastic analogy, possibly with a variable stiffness [10], is used to extend this motion inside the domain. Here, we do not make the assumption that the boundary motion is small or that the volume of the domain remains unchanged. An elastic approach does not give any guarantee on the validity of the mesh for large displacements. Even when the motion is limited, ill-shaped elements are produced in the process. This is the second issue. A new efficient procedure is presented that enables to eliminate all ill-shaped elements that are inevitably produced during the mesh motion.

The remainder of this paper is organized as follows: the first section introduces the different edge size fields. Section 3 describes the set of local mesh modification operators while section 4 presents the ill shaped elements elimination algorithm. In section 5, we recall the principles of the elastic analogy for mesh motion and discuss the choice of local element stiffness. The global procedure is then described in section 6. Section 7 presents some numerical results.

2 Mesh Size Field

The aim of the mesh generation process is to build elements of controlled shape and size. Mesh generators are usually able to adapt to a so called mesh size field (see for instance [11, 1]). An isotropic mesh size field is a scalar function $\delta(\mathbf{x}, t)$ that defines the optimal length of an edge at position $\mathbf{x}$ of the domain and at time t.

We typically define the non-dimensional length L_e^{tr} of edge e as

$$L_e^{tr}(t) = \int_e \delta^{-1}(\mathbf{x}, t) dl. \quad (1)$$

The quantity L_e^{tr} represents the number of subdivisions of edge e that are necessary for having an edge that has exactly the right size with respect to the size field.

There are lots of ways to define a size field: some are based on rigorous error estimation procedures [12] but there are lots of heuristics. For example, it is often considered that mesh size should be smaller near boundaries: when dealing with viscous flows, a large part of the vorticity is created near the walls, when dealing with solid mechanics problems, stress concentration are usually located near the boundaries.

We define $d(\mathbf{x},t)$ as the distance to the closest boundary at time t. This distance can be computed *in place* using the Approximated Nearest Neighbor Algorithm [13]. A first mesh size field $\delta_1(\mathbf{x},t)$ is computed as follows:

$$\delta_1(\mathbf{x},t) = \delta_1^{\text{small}} + \alpha_1(\mathbf{x},t)(\delta_1^{\text{large}} - \delta_1^{\text{small}}),$$

where

$$\alpha_1(\mathbf{x},t) = \begin{cases} 0 & \text{if} \quad d(\mathbf{x},t) \leq d_1^{\min} \\ \frac{d(\mathbf{x},t)-d_1^{\min}}{d_1^{\max}-d_1^{\min}} & \text{if} \quad d_1^{\min} < d(\mathbf{x},t) < d_1^{\max} \\ \infty & \text{if} \quad d(\mathbf{x},t) \geq d_1^{\max} \end{cases}$$

with δ_1^{large} and δ_1^{small} a large and a small desired mesh sizes, $d_1^{\max}$ and $d_1^{\min}$ two field values that define the zone of refinement. An example of use of δ_1 is presented on figure 1.

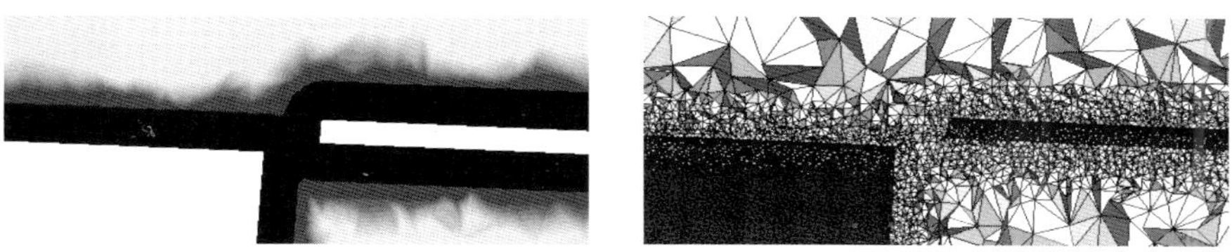

Fig. 1. Mesh adapted using the distance to boundaries

Other size fields $\delta_2(\mathbf{x},t), \delta_3(\mathbf{x},t), \ldots$ can be defined:

1. mesh sizes prescribed at model vertices and interpolated linearly on model edges;
2. prescribed mesh gradings on model edges (geometrical progressions, ...);
3. mesh sizes defined on another mesh (a background mesh) of the domain;
4. mesh sizes that adapt to the principal curvature of model entities.

The size field $\delta(\mathbf{x},t)$ is computed, at time t, as the minimum of all size fields. It is usually bounded by upper and lower values of mesh sizes.

In an ideal mesh, each edge has an adimensional length $L_e^{tr} = 1$. This ideal situation cannot be attained in practice. In the adaptation procedure one has to decide whether an edge is acceptable, i.e. find a range $[L_{low}, L_{up}]$ for which an edge is considered to have a good size. Then a long edge has a size $L_e^{tr} > L_{up}$ and a short edge has a size $L_e^{tr} < L_{low}$.

In [1], the authors show that choosing a too narrow range for acceptable edge sizes may lead to infinite loops between splits and collapses. In § 6 and § 7, we show the influence of this interval on the mesh quality and on the number of infinite loops.

3 Local Mesh Modifications

The three basic local mesh modifications are the edge split, the edge collapse and the edge swap. Other operators like face swap and some compound operators are also used in this work.

There are two ways of dealing with long edges. The first one consists in tagging every long edge of the mesh. Then, every tetrahedron of the mesh is subdivided using a template that is function of the number of edges that have been split. The second manner consists in splitting all tetrahedra surrounding a long edge, and then proceed to the next edge. The latter approach has been used because it is simpler (only one template has to be defined), more robust (no Steiner point has to be introduced) and (surprisingly) more efficient. Moreover, we have found out that both methods lead to meshes with elements of similar qualities. **The edge split** operator is depicted on figure 2.

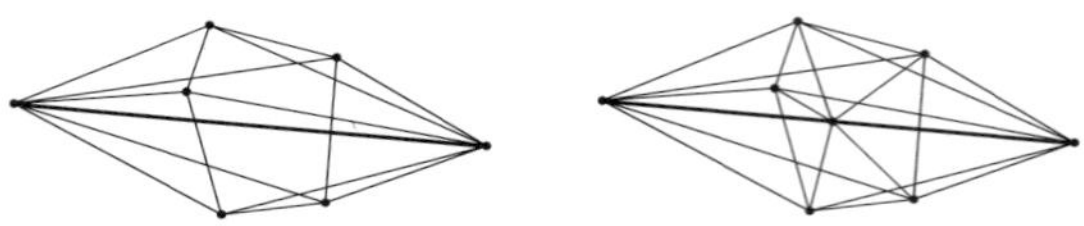

(a) (b)

Fig. 2. Edge split operation: (a) initial cavity, (b) cavity after the edge split

When a short edge is found, an **edge collapse** is applied. The edge collapse operator (see figure 3) removes an edge and all its bounding elements from the mesh by merging its two extremities at one of their locations.

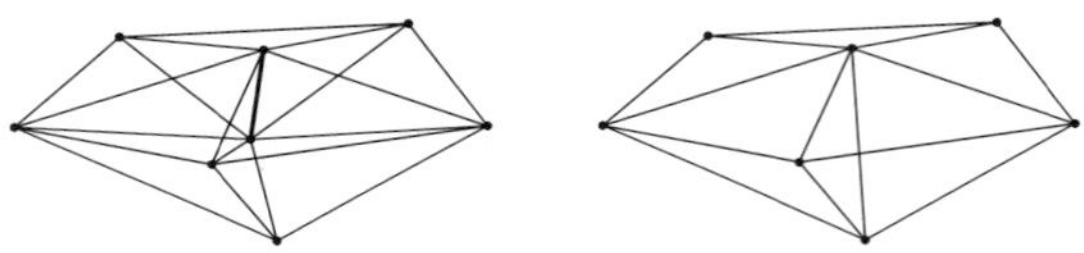

(a) (b)

Fig. 3. Edge collapse operation: (a) initial cavity, (b) cavity after the edge collapse

The edge swap consists in remeshing the cavity surrounding the edge with the aim of improving the worst element shape in that cavity. Figure 4(a) shows an edge to be swapped that is surrounded by five tetrahedra. Figures 4(c), (d) and (e) depict three of the five possible configurations after the swap. The retained configuration is the one that has the best minimal element quality in the cavity.

The face swap removes a face and replaces it by an edge, leading to the creation of three tetrahedra instead of two. This operation can be seen as the inverse edge swap with $n = 3$.

The face collapse operator can be seen as the compound of and edge split and an edge collapse. The operation is depicted on Figure 5. One of the edges of

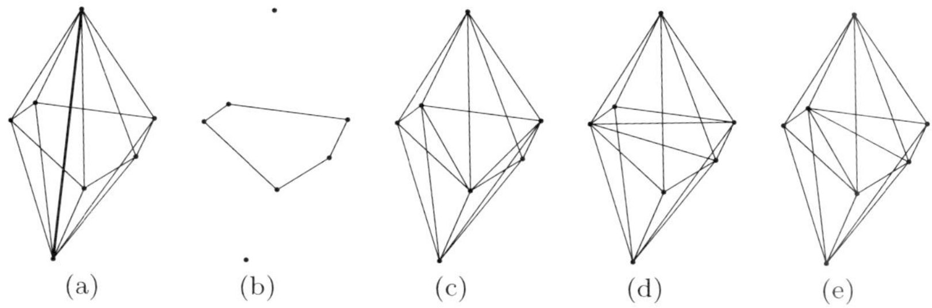

Fig. 4. Edge swap operation: (a) initial cavity with $n = 5$, (b) mean surface that is going to be triangulated, (c),(d) and (e) three possible configurations after the swap

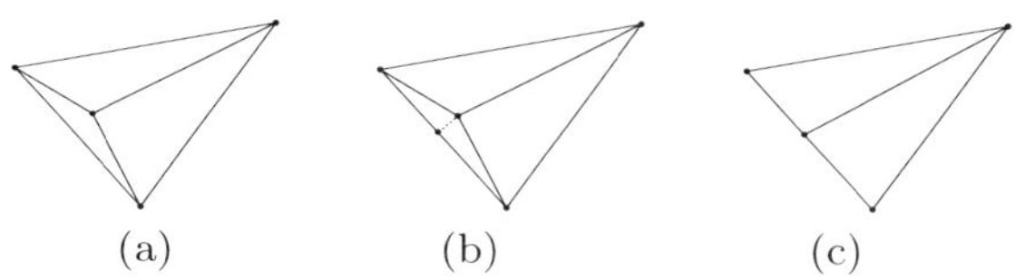

Fig. 5. Face collapse operation: (a) initial tetrahedron, (b) split of an edge of the concerned face, (c) collapse of the new edge on the new vertex

the face is split and the resulting new edge is collapsed on the new vertex. This operation is particularly well suited to eliminate an ill-shaped element that has a face with a small ratio area/edge lengths.

The double edge split collapse compound operator consists in splitting two edges of a tetrahedron and then collapsing the edge joining the two new vertices. This operator is used when a tetrahedron with a small volume has two edges nearly intersecting. The operation is depicted on Figure 6.

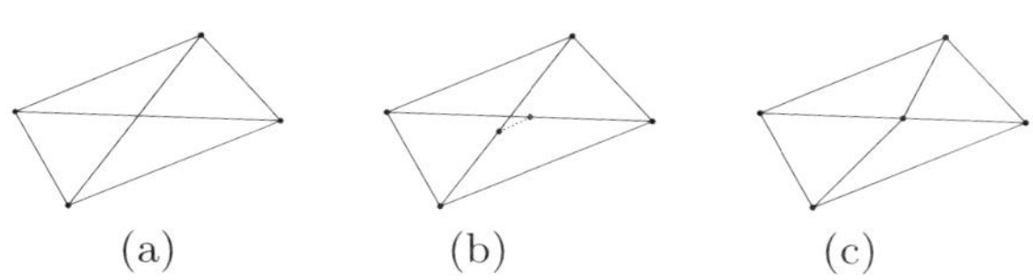

Fig. 6. Double edge split + edge collapse operation: (a) initial tetrahedron, (b) situation after the edge splits, (c) situation after the edge collapse

4 Sliver Tetrahedra Handling

A tetrahedron is said to be a sliver when it has a small volume and no short edge. Such a tetrahedron can be classified in one of the two categories [14] depicted

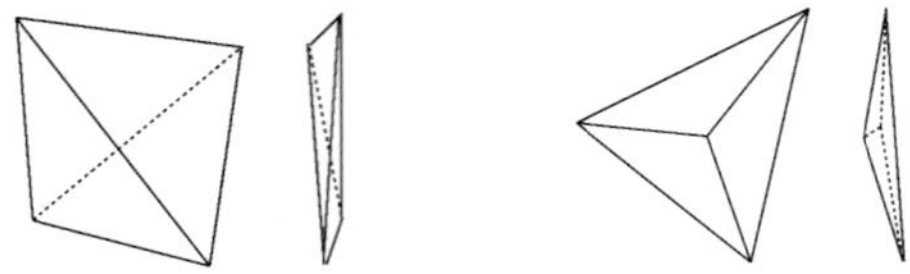

(a) (b)

Fig. 7. Classification of the sliver tetrahedra: (a) type I, (b) type II

on figure 7. Type I slivers are tetrahedra in which two edges almost intersect. In type II slivers, one vertex is very close to its opposite face. The determination of the sliver type is important because it gives a useful information about the best local mesh modification that can be applied in order to eliminate it. The projection algorithm presented in [1] is used in this work to determine the type of sliver.

In FSI, when the mesh motion solver is pushed to its limits, it inevitably generates slivers near the boundaries undergoing a motion. In [1], an algorithm selecting the local mesh modifications to be applied was proposed to eliminate most of them. In this work, we propose an extension to this algorithm that eliminates more slivers.

A new sequence of mesh modification operations is proposed in Table 1. The operations are sorted with respect to their average efficiency to eliminate slivers of type I and II. The efficiency measure that we have used takes into account both the rate of success of eliminating slivers and CPU time consumption of the given operator. The rates of success of the different operations have been deduced from various test cases. Some of them are presented in section 7. We notice that the vertex motion is the very last solution as it is highly probable that the tetrahedron becomes a sliver again after the next step of the boundaries motion. The target location for the motion is computed with the method proposed in [15].

Note that those sequences are designed to be part of the sliver elimination algorithm inserted in the global adaptation procedure presented in section 6. Other sequences could be more efficient if the global procedure was modified.

In this work, we consider that eliminating the slivers is more important than enforcing locally the length criterion, which means that an operation that can eliminate a sliver or at least improve the local quality is performed, even if it creates long or short edges. The long or short edges will be eliminated later on by the global procedure but only if they do not create a sliver tetrahedron (see section 6.2).

The construction of the list of sliver tetrahedra is made by computing the quality of every element of the mesh. The quality measure is chosen to be the cubic mean ratio [16], defined by

$$\eta^3 = 15552 \frac{V^2}{\left(\sum_{i=1}^{6} (L_i)^2\right)^3}, \tag{2}$$

Table 1. Sequence of local mesh modifications attempted to eliminate a sliver

Sliver type	Priority level	Local mesh modification
I	1	Split one of the two key edges
	2	Collapse one of the edges of the tetrahedron
	3	Split both key edges and collapse the new interior edge
	4	Split one of the key edges and collapse the new vertex
	5	Swap one of the two key edges
	6	Relocate a vertex
II	1	Collapse one of the edges of the tetrahedron
	2	Collapse the key face
	3	Swap any of the edges bounding the key face
	4	Swap the key face
	5	Relocate a vertex

with V the volume of the element and L_i the length of the ith edge of the element. The 15552 factor is set to scale η^3 such that it ranges from 0 to 1. Its value is 1 for an equilateral tetrahedron and 0 for a flat tetrahedron. An element is considered as a sliver if its quality is under a certain threshold T_{sliver}. In all simulations presented in this paper, T_{sliver} is set to 0.01.

5 Mesh Motion Solver

In FSI problems, the structure imposes its motion to some of the boundaries of the fluid domain. Moving only the boundary nodes leads to the generation of ill conditioned tetrahedra. A way to circumvent partially this issue is to relocate the nodes of the volume using a linear elasticity analogy [17, 6].

A linear elastic problem is solved every time the mesh is moved. Here, we use the approach of [10] in which a stiffness alteration based on the Jacobian of the elements is used to stiffen the smaller elements. The corresponding finite element formulation is written as follows. Consider the vector of displacements of the nodes $\mathbf{y}^h \in \mathcal{S}^h$, where $\mathcal{S}^h$ is the piecewise linear nodal finite element space. The modified energy of deformation of a tetrahedron e submitted to a displacement $\mathbf{y}^h$ (with the corresponding strain and stress tensors $\boldsymbol{\epsilon}$ and $\boldsymbol{\sigma}$) is written as:

$$\frac{1}{2}\int_e \left[\boldsymbol{\epsilon}(\mathbf{y}^h) : \boldsymbol{\sigma}(\mathbf{y}^h)\right]^e J^e \left(\frac{J^0}{J^e}\right)^\chi dv, \tag{3}$$

where χ is the stiffening parameter and J^0 a fictious volume constant over the mesh.

Figure 8 illustrates the influence of χ. With $\chi = 0$, we observe ill shaped elements near the walls. The elastic solver fails to achieve the mesh deformation with positive volumes of the elements if a motion of the cube of more than $0.1 \times h$ where h is the size of the cube is imposed. With $\chi = 1$ and $\chi = 2$, a motion of $1.0 \times h$ can be attained without any particular problem. In the $\chi = 2$ case, the

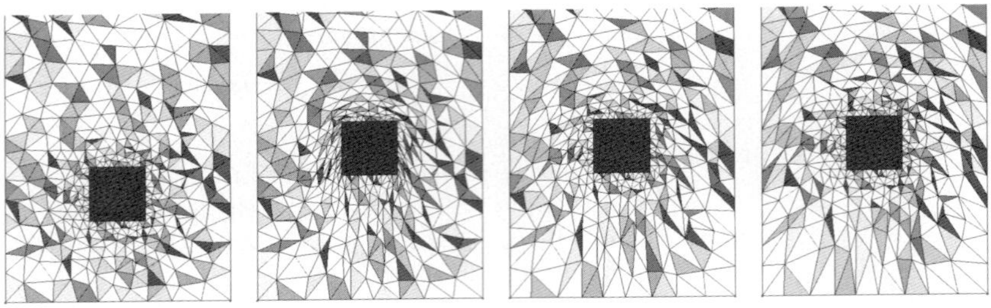

Initial mesh $\chi = 0.0$ $\chi = 1.0$ $\chi = 2.0$

Fig. 8. Mesh around a cube and its deformations for various values of χ

elements are very well preserved around the cube but the quality falls outside of the refined zone. The choice of $\chi = 1$ seems to be a good compromise and has been used in all the computations presented in this paper.

6 Global Mesh Modification Procedure

Our FSI procedure has two different stages: the node repositioning stage and the mesh adaptation stage. The node repositioning step does not imply any change in the mesh topology. As it does not need to reallocate the resources for the storage of the topology, the solution or any data, it can be performed often at a reasonable computational cost, that is, each time a displacement is prescribed at the interface between the fluid and the structure. Moreover, the node repositioning is explicitly taken into account in the ALE formulation of the governing equations and does not need any mesh to mesh interpolation. In the second step of the procedure, the mesh is adapted using the local mesh modifications presented in section 3 and the global size field presented in section 2.

6.1 The Mesh Adaptation Algorithm

The aim of the adaptation procedure is twofold. First, the adimensional size of all edges of the mesh have to lie in the interval $[L_{low}, L_{up}]$. Then, the quality of the resulting mesh has to be optimal. Minimal quality criterions can be expressed for instance in terms of a minimal quality for every element and/or a minimal mean quality of the elements.

The mesh adaptation procedure can be described as follows:

`Do {`

- Collapse short edges i.e. edges that have lengths $L_e^{tr} < L_{low}$.
- Do an edge swap loop: loop over all edges and compute the minimal quality of the elements surrounding the edge. The quality is given by the cubic mean ratio (equation 2). If the quality is lower than a given threshold T_{swap}, try to find a swap configuration. Apply the swap if it improves the minimal quality

in the cavity and if it does not create a long or a short edge. The threshold T_{swap} is fixed to 0.1 in all the simulations presented in this paper.
- Eliminate sliver tetrahedra with the algorithm described in section 4.
- Split long edges i.e. edges that have lengths $L_e^{tr} > L_{up}$.

} `While the mesh is modified.`

The collapse loop is performed first in order to avoid memory peaks. Indeed, the number of nodes decreases during this loop. As the edge swap loop is costly, it is performed just after the coarsening loop, with the minimum number of nodes. The sliver region elimination can then be performed without having to compute any element shape, as all element qualities are computed at the previous step.

6.2 Infinite Loops

There are a few ways in which the adaptation procedure can degenerate in an infinite loop between two or more configurations.

If the interval of tolerance for the length of the edges is not large enough, an infinite loop between split and collapse operations can appear. Due to the heuristic nature of the mesh adaptation, we cannot guarantee that such a loop will never appear, even for a large interval. For that reason, a maximum number of iterations is imposed in the global procedure. The edge length interval should then be carefully chosen in order to strongly limit the number of infinite loops without being too far from the unit mesh. More detailed results are presented in section 7.

Another possibility of having an infinite loop is illustrated in Figure 9, in which an edge split is followed by an edge collapse and an edge swap. This is possible because the goal of the edge split and collapse operators is to respect a criterion on edges length while the edge swaps tend to improve the quality of the tetrahedra. A way to eliminated such a scenario is to forbid an edge swap if it creates a long or short edge.

Finally, the operators used in the sliver regions handler can also create a set of complex infinite loops. This possibility can be avoided if any operation that creates a sliver tetrahedron is forbidden. Of course, as an exception, we allow the replacement of a sliver by a better sliver.

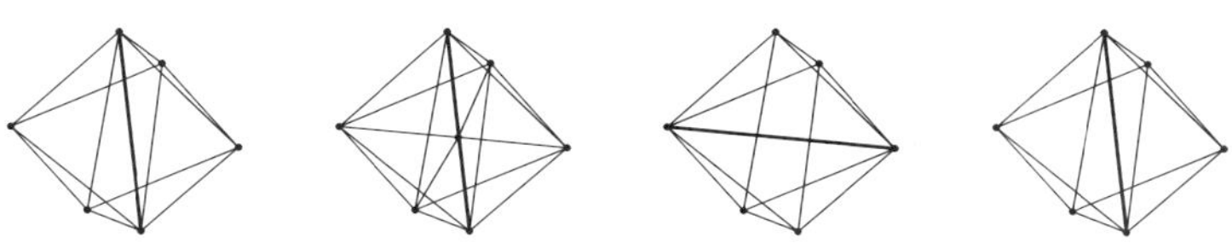

(a) (b) (c) (d)

Fig. 9. Infinite loop between an edge swap, an edge split and an edge collapse: (a) initial cavity, (b) cavity after the split, (c) cavity after the collapse, (d) cavity after the swap, identical to (a)

7 Computational Results

Our adaptation procedure has been evaluated on three test cases. The quality of the meshes, the elimination of the sliver tetrahedra and the evolution of the number of nodes are presented in the first and the second cases. The third test case is a fluid-structure interaction computation with a very large motion of one or two spheres in a fluid at rest.

7.1 Cylinder and Tube

This test case consists in the penetration of a cylinder in a tube, the diameters of the objects being quite close. The initial geometry and mesh are shown in figure 10. The cylinder has a radius of 0.9 and a length of 3.0. The tube has the same length with an internal radius of 1.0 and an external radius of 1.2. The initial distance between the two objects is 1.0. The two objects move with a velocity of 1.0 for the cylinder and −1.0 for the tube. The end of the test is fixed at time 4.0, when the objects are completely separated and the distance between them is 1.0.

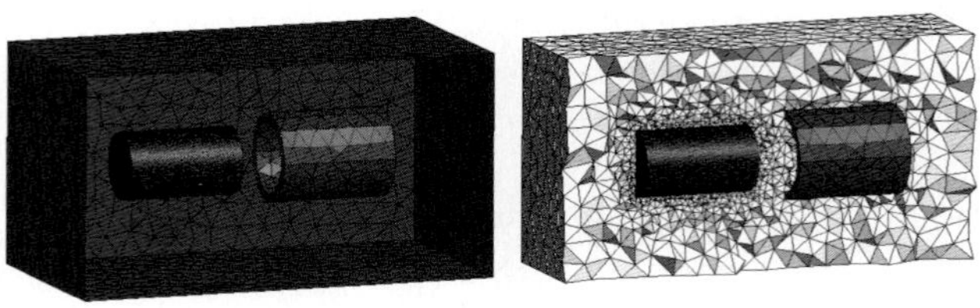

Fig. 10. Cylinder and tube test case: initial geometry and mesh

A global edge length of 0.6 is prescribed on the whole domain and both objects are equipped with a local size field. The parameters of the local size fields for the cylinder and the tube are

$$\begin{cases} d_{cyl}^{\min} = 0 \\ d_{cyl}^{\max} = 1.0 \end{cases} \quad \begin{cases} \delta_{cyl}^{\text{small}} = 0.1 \\ \delta_{cyl}^{\text{large}} = 0.6 \end{cases} \quad \begin{cases} d_{tube}^{\min} = 0 \\ d_{tube}^{\max} = 1.0 \end{cases} \quad \begin{cases} \delta_{tube}^{\text{small}} = 0.4 \\ \delta_{tube}^{\text{large}} = 0.6 \end{cases}$$

For an interval $[L_{low}, L_{up}]$ set at $[0.5, 1.4]$, the number of nodes ranges from about 23.500 to 33.000 during the adaptation. This is sufficient to get representative data and statistics about the mesh quality and the sliver elements handling.

A time step of 0.01 has been chosen for this test case. It corresponds to a relative motion of the objects of $0.2d_{cyl}^{\min}$ at each time step.

The mesh aspect obtained at different times with a time step of 0.01 and an interval $[L_{low}, L_{up}]$ set at $[0.5, 1.4]$ is shown in figure 11.

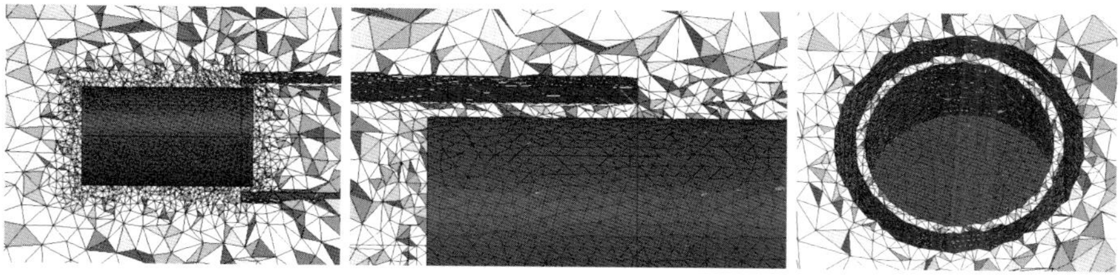

Fig. 11. Mesh aspect at different times with $\Delta t = 0.01$, $L_{low} = 0.5$ and $L_{up} = 1.4$

Table 2. Tube and cylinder test case: statistics about sliver tetrahedra handling

		Successes	New sliver	Failures	Success rate (%)
Type I (547)					
	All operators	541	1	5	98.9
	Edge collapse	244	3	300	44.6
	Edge split	346	20	181	63.3
	Double split+collapse	210	11	326	38.4
	Face collapse	279	13	255	51.0
	Edge swap	306	17	224	55.9
	Vertex relocation	504	10	33	92.1
Type II (308)					
	All operators	306	1	1	99.4
	Edge collapse	119	2	187	38.6
	Face collapse	155	1	152	50.3
	Edge swap	202	5	101	65.6
	Face swap	112	6	190	36.4
	Vertex motion	292	1	15	94.8

Sliver tetrahedra handling

In order to get more information about the efficiency of each mesh modification applied in the sliver region elimination, we have run the adaptation procedure in which the slivers handling algorithm is enhanced with a routine that tests all operations individually on each sliver before going in the normal algorithm of elimination. This routine has no influence on the results, but provides the statistics shown in table 2. The first columns indicates the number of sliver tetrahedra that the operators can eliminate without creating a new sliver. The second column shows the number of situations in which the operator can only eliminate the sliver by creating another sliver with a better quality, while the third column shows the number of slivers that could not be eliminated or for which a worse sliver would appear. The rate of success of the operator (not including the cases in which a new sliver is created) is indicated in the last column.

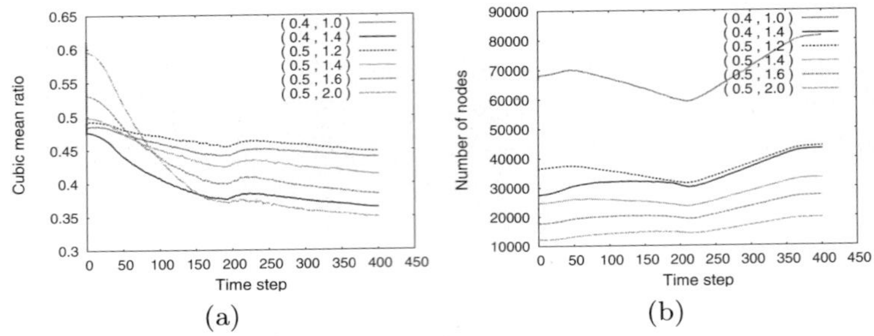

Fig. 12. Cylinder and tube test case: comparison between intervals $[L_{low}, L_{up}]$: (a) evolution of the mean quality of the mesh and (b) the number of nodes

Table 3. Tube and cylinder test case: comparison between intervals $[L_{low}, L_{up}]$: number of infinite loops

Interval	[0.4, 1.0]	[0.4, 1.4]	[0.5, 1.2]	[0.5, 1.4]	[0.5, 1.6]	[0.5, 2.0]
Infinite loops	400	0	400	81	0	0

We can see that combining all operators do not lead to a total elimination of the slivers but provides a very good rate of elimination. The vertex repositioning looks attractive but we observed that a lot of slivers eliminated by a repositioning become slivers again once a global node motion is applied. For type I slivers, the edge split is very efficient if we notice that it is a very fast operator compared to the edge swaps. Note that those results are influenced by the fact that the sliver elimination is done after the edge collapse and edge swap loops and before the edge split loop.

Mesh quality and infinite loops

A particular result that we observe in our simulations is that the mean quality of the mesh depends on the boundaries of the interval $[L_{low}, L_{up}]$. Figure 12 (a) shows the evolution of the quality for various intervals, figure 12 (b) shows the evolution of the number of nodes in each case, while table 3 shows the number of time steps in which an infinite loop was created.

We can see that the condition $L_{low} < 0.5\, L_{up}$ is not sufficient to avoid infinite loops. Indeed, there is an infinite loop at each time step with the intervals $[0.4, 1.0]$ and $[0.5, 1.2]$ while some infinite loops appear with $[0.5, 1.4]$. We would then recommend to use an interval with a similar range than $[0.4, 1.4]$ or $[0.5, 1.6]$.

We also observe that the quality is dependent on the boundaries of the interval. If we compare the intervals $[0.4, 1.4]$ and $[0.5, 1.4]$, the quality seems to be better with the largest L_{low}. If we look at the results for $[0.5, L_{up}]$, we can see that the quality decreases to a smaller value when L_{up} is bigger. These two observations lead to the conclusion that the quality tends to a bigger value if the interval is smaller.

In order to get the best mean quality without running into infinite loops, a compromise has to be found on the size of the interval. The smaller interval which do not lead to infinite loop should be used. From the set of intervals tested here above, our choice would turn towards [0.4, 1.4] or [0.5, 1.6].

7.2 Worm Screw

The second test case is a worm screw whirling inside a cylinder. It has been chosen to demonstrate the capabilities of the adaptation procedure on relatively complex geometries.

Figure 13 shows the geometry and the initial mesh of the domain at time 0. The domain is bounded by a worm screw of diameter 4.94 and length 15.6 turning with an angular speed of 1.0, and a fixed cylinder of diameter 5.2 and length 16.2. The computation ends when a complete turn has been operated, i.e. at $t = 2\pi$. A time step of 0.02 is chosen.

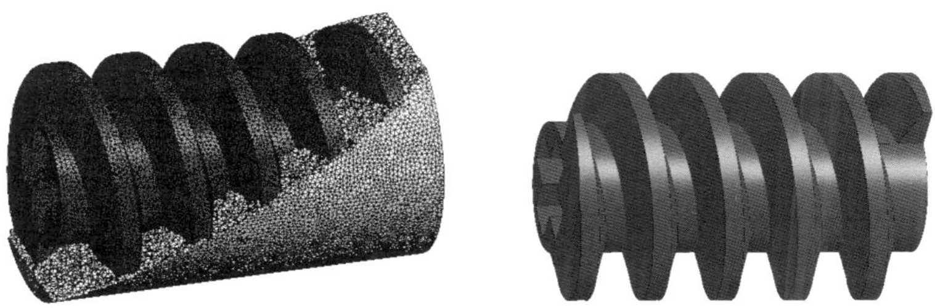

Fig. 13. Worm screw: initial geometry and mesh

The maximum size on the whole domain is set to 1.0, while a size field with the following parameters is prescribed at every wall:

$$d_1^{\text{min}} = 0 \qquad d_1^{\text{max}} = 0.4 \qquad \delta_1^{\text{small}} = 0.2 \qquad \delta_1^{\text{large}} = 0.4$$

The interval $[L_{low}, L_{up}]$ is set to $[0.5, 1.6]$. With these parameters, the number of nodes ranges from 125.000 to 155.000 during the computation.

Figure 14 shows the aspect of the mesh at times 4.0 and 6.0.

The evolution of the mean quality is shown in figure 15 (a). The cubic mean ratio tends to a value close to 0.40, which is quite similar to what was observed in the previous test case with the same interval $[L_{low}, L_{up}]$. The distribution of the quality in the final mesh is shown in figure 15 (b). The evolution of the number of nodes is drawn on figure 15 (c). It stabilizes around 155.000.

The statistics about the elimination of the sliver tetrahedra are shown in table 4. Every sliver is eliminated and most of them are eliminated at the first attempts.

(a) (b)

Fig. 14. Worm screw: cuts in the mesh at times 4.0 and 6.0

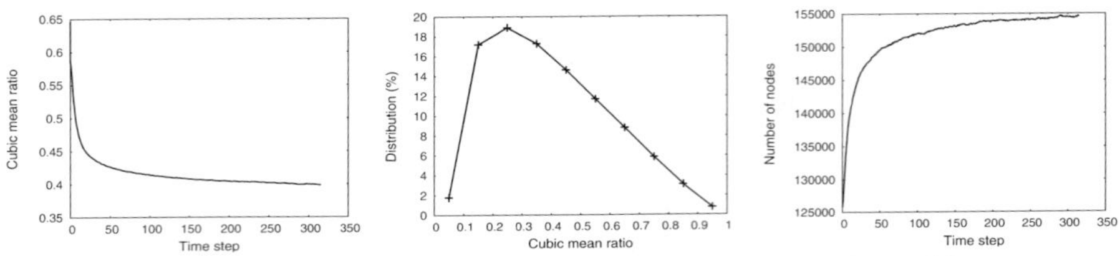

(a) (b) (c)

Fig. 15. Worm screw: (a) evolution of the mean cubic mean ratio, (b) distribution of the quality in the final mesh, (c) evolution of the number of nodes

Table 4. Worm screw: statistics about sliver tetrahedra handling

		Applied (no sliver)	Applied (new sliver)
Type I (1532)			
	Edge split	1182	0
	Edge collapse	239	0
	Double split+collapse	15	1
	Face collapse	18	0
	Edge swap	4	1
	Vertex relocation	72	0
	Remaining	0	
Type II (1521)			
	Edge collapse	1430	0
	Face collapse	33	1
	Edge swap	17	0
	Face swap	1	0
	Vertex motion	38	1
	Remaining	0	

7.3 Spheres Falling in a Fluid

In this test case, the mesh adaptation procedure is applied to a fluid-structure interaction solver. The problem solved implies the motion of one or two spheres falling in a fluid.

The fluid is governed by the incompressible Navier-Stokes equations. The equations are discretized in a dual finite elements / finite volumes formulation.

The motion of the sphere is computed with the Newmark method and the coupling is achieved with a CSS procedure [18] equipped with sub-iterations cycles.

Single sphere

For this computation, a single sphere is immersed in the fluid. We fix the parameters of the fluid so that the Reynolds number Re is equal to 1 at equilibrium. Due to the gravity, the sphere accelerates until the gravity is exactly balanced by the drag force and the Archimede's force. We fix the mass of the sphere such that the theoretical velocity at equilibrium is 1.

Figures 16 shows the evolution of the velocity and displacement of the sphere as well as the evolution of the mesh quality. We notice that the velocity tends to 1 as expected and that the quality decreases slowly so that the loss along the computation is not significant.

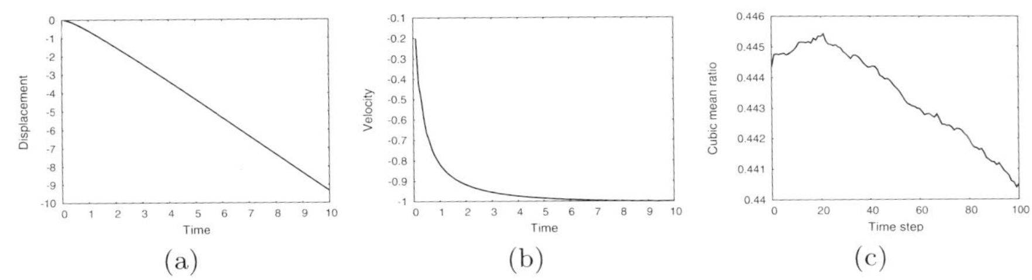

Fig. 16. Single sphere test case: (a) evolution of the displacement, (b) velocity and (c) mean quality of the mesh

Two spheres

In the next computation, the motion of two vertically aligned spheres are investigated.

The evolution of the displacements and the velocities of the spheres are shown in figures 17 (a) and (b). We observe that from the time 1.5, the upper sphere is aspirated by the flow around the first one. Eventually, the spheres touch and the computation stops.

The cut of the computational mesh is represented on figure 18 for different times as well as the pressure field.

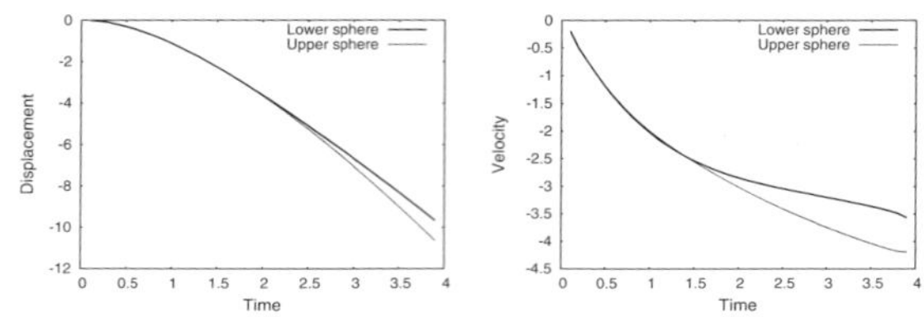

(a) (b)

Fig. 17. Two spheres test case: (a) evolution of the displacements and (b) velocities

Time: 0.0 2.0 3.4 3.9

Fig. 18. Two spheres test case: mesh and pressure field at different time steps

8 Conclusion

A procedure to handle large deformations of a mesh in FSI problems by nodes movement and local mesh adaptation was presented. The procedure is robust and can handle complex geometries with very large deformations of the domain.

For that purpose, a node repositioning technique with a selective treatment for the elements previously applied for two-dimensional meshes was applied for three-dimensional meshes.

A new adaptation procedure has been proposed in which an efficient handling of the sliver tetrahedra is included. The procedure achieves a good quality for the mesh and complies a mesh size field. Important parameters of the procedure like the boundaries of the interval $[L_{low}, L_{up}]$ were studied with a set of computations.

In particular, some clues are given to choose them by looking at the quality, number of nodes and production of infinite loops.

Finally, mesh deformation tests and fluid-structure computations have been performed in order to show the potential of the presented approach.

References

1. Li, X., Shephard, M.S., Beall, M.W.: 3d anisotropic mesh adaptation by mesh modification. Computer Methods in Applied Mechanics and Engineering 194, 4915–4950 (2005)
2. Almeida, R., Feijoo, P., Galeao, A., Padra, C., Silva, R.: Adaptive finite element computational fluid dynamics using an anisotropic error estimator. Computer Methods in Applied Mechanics and Engineering 182(3-4), 379–400 (2002)
3. Borouchaki, H., George, P., Hecht, F., Laug, P., Saltel,: Delaunay mesh generation governed by metric specifications - part i: Algorithms and part ii: Applications. Finite Elements in Analysis and Design 25, 61–83, 85–109 (1997)
4. Remacle, J.-F., Li, X., Shephard, M., Flaherty, J.: Anisotropic adaptive simulation of transient flows using Discontinuous Galerkin methods. International Journal for Numerical Methods in Engineering 62, 899–923 (2004)
5. Cavallo, P.A., Sinha, N., Feldman, G.M.: Parallel unstructured mesh adaptation method for moving body applications. AIAA Journal 43, 1937–1945 (2005)
6. Baker, T.J.: Mesh movement and metamorphosis. Engineering With Computers 18(3), 188–198 (2002)
7. Mathur, S.R.: Unsteady flow simulations using unstructured sliding meshes. AIAA Journal (June 1994)
8. Lijewski, L.E., Suhs, N.E.: Time-accurate computational fluid dynamics approach to transonic store separation trajectory prediction. Journal of Aircraft 31(4), 886–891 (1994)
9. Compère, G., Marchandise, E., Remacle, J.-F.: Transient adaptivity applied to free surface flows. Journal of Computational Physics 227, 1923–1942 (2008)
10. Stein, K., Tezduyar, T., Benney, R.: Mesh moving techniques for fluid-structure interactions with large displacements. Journal of Applied Mechanics 70, 58–63 (2003)
11. Peraire, J., Peiro, J., Morgan, K.: Adaptive remeshing for three dimensional compressible flow computation. Journal of Computational Physics 103, 269–285 (1992)
12. Ainsworth, M., Oden, J.T.: A Posteriori Error Estimation in Finite Element Analysis. Wiley, New York (2000)
13. Arya, S., Mount, D.M., Netanyahu, N.S., Silverman, R., Wu, A.Y.: An optimal algorithm for approximate nearest neighbor searching. Journal of the ACM 45, 891–923 (1998), `http://www.cs.umd.edu/~mount/ANN/`
14. de l'Isle, E.B., George, P.L.: Modeling, Mesh Generation, and Adaptive Numerical Methods for Partial Differential Equations, I. In: Babuska, I., et al. (eds.), vol. 75, pp. 97–128. Springer, Heidelberg (1993)
15. de Cougny, H.L., Shephard, M.S., Georges, M.K.: Explicit node point mesh smoothing within the octree mesh generator. Scorec report, Rensselaer Polytechnic Institute, 10 (1990)
16. Liu, A., Joe, B.: On the shape of tetrahedra from bisection. Mathematics of computation 63, 141–154 (1994)

17. Tezduyar, T.E., Behr, M., Mittal, S., Liou, J.: Computation of unsteady incompressible flows with the finite element methods–space-time formulations, iterative strategies and massively parallel implementations vol. 246, ASME Pressure Vessels Piping Div Publ PVP., ASME, NY, pp. 7–24 (1992)
18. Piperno, S., Farhat, C.: Partitioned procedures for the transient solution of coupled aeroelastic problems - part ii: energy transfer analysis and three-dimensional applications. Computer Methods in Applied Mechanics and Engineering 190, 3147–3170 (2001)

Filling Arbitrary Holes in Finite Element Models

Andreas Schilling[1], Katrin Bidmon[1], Ove Sommer[2], and Thomas Ertl[1]

[1] Visualization Research Centre (VISUS), Universität Stuttgart, Germany
{schilling,bidmon,ertl}@visus.uni-stuttgart.de
[2] science + computing ag, Tübingen, Germany
ove.sommer@science-computing.de

Summary. We present a new approach for filling arbitrary holes in Finite Element (FE) models composed of independent meshes which are commonly used in Computer Aided Engineering (CAE). During the preprocessing of an FE model for fluid structure simulation a closed structure hull is required in order to subsequently generate a volume mesh representing the fluid. Today's complex CAE models include thousands of holes that need to be effectively processed. The widespread algorithms mostly fail the special requirements in this context as they can only process holes within a single mesh or smooth surfaces without feature lines. Extending these features into the filling mesh is important for both the visual quality as well as the simulation results. The presented approach handles these cases and automatically detects ordinary holes, e.g. with topological defined boundary. Additionally, holes with more complex boundaries, e.g. defined by several independent meshes, can be specified interactively by the user. Our hole quantification enables the user to automatically close the vast majority of holes and substantially accelerates this preprocessing step. Furthermore, the presented meshing algorithm offers extensive control to gain satisfying results for the described use case.

1 Introduction

Growing standards for new industrial products require the close collaboration of different numerical simulation disciplines during virtual prototyping using multiphysics solvers. One example is the combination of computational fluid dynamics and structural dynamics simulation. In order to allow the interaction between structure and bounded or surrounding fluid the different properties of a mixed CAE model need to be represented by capable finite elements and their interface areas have to be coupled. Some application fields of such models are ship simulation, acoustic analysis of a power train or a whole car passenger compartment. To complete the CAE model for the latter case based on an existing structure model the fluid volume has to be meshed by 3D finite elements in a preprocessing step. A tetrahedral mesher requires a watertight bounding structure. Generally this is not given because today's car body models consist of hundreds of independently meshed car body parts (see Fig. 1) which are connected to each other by special finite elements.

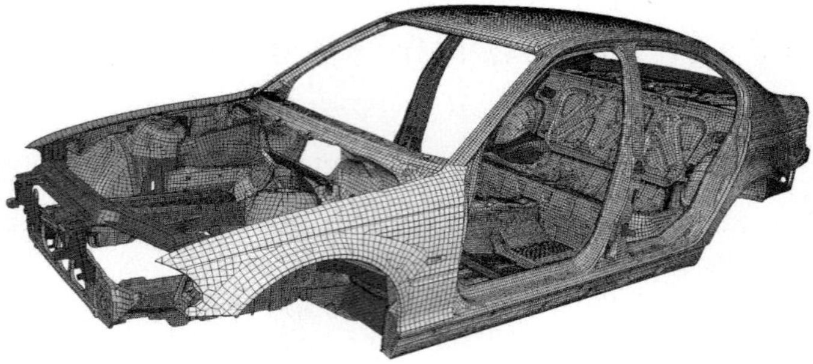

Fig. 1. FE models generally consist of many independently meshed parts that provide arbitrary holes due to unmodelled parts as well as constructive cut-outs

In the application we consider a fluid mesh is generated using a hexahedral volume growing approach. The user specifies a seed point inside the structure model that surrounds the fluid volume. The volume mesh extends with a specified minimal cell size in all directions until it will be stopped by any intersecting structure. Therefore, all holes in the surrounding structure that are at least as large as the minimal cell size need to be closed to prevent the generation of fluid cells outside the fluid volume. Until now those holes had to be found visually by the user or they were detected after the volume has grown to the outside. These preprocessing steps took several days. Our approach allows for cutting down on this processing time significantly by providing the potential for processing the majority of those holes automatically.

Aside from ordinary holes inside a structural mesh, such as constructive cut-outs, the CAE models also contain holes with complex boundaries, e.g. composed and defined by more than one continuous finite element mesh. These holes, for example, may result from missing model parts which are not available in the current structural model, e.g. the front shield or doors. Whereas the former type of holes can be detected automatically, the latter need to be defined interactively by the user. We describe a method to define holes effectively and an algorithm that allows the meshing of detected and defined holes in consideration of requirements included by the application area in CAE modelling. A hole classification facilitates the meshing automation to a large extent which leads to a significant speed-up of the overall preprocessing step.

2 Related Work

Holes in meshes are a very common problem in several parts of visualization. They often arise in datasets derived from scanned geometry where data is missing or as polygonization artefacts during isosurface extraction. Hence, there is a wide range of methods – some follow a very general approach, others are tailored to suit more special cases. All these methods are mainly developed with closed, oriented 2-manifold meshes in mind. In contrast to holes in scanned geometry or resulting from surface reconstruction, holes in CAE models usually are not erroneous parts in the model but intended constructive cut-outs or unmodelled parts and thus may spread over numerous independently meshed parts. Thus, the meshes involved are not necessarily oriented and 2-manifold and especially are not closed surfaces. However, some steps in the workflow require a watertight model in some parts. Therefore, these gaps and holes have to be closed.

The related algorithms can be split into two main categories: Volumetric approaches and geometry-based techniques. The volumetric methods split the region around a hole into inner part and outer part, usually based on the volumetric hull. Hence, the boundary layer defines the missing surface. In [5] the model is converted into volume representation by a signed distance function and a diffusion process is applied. Thus, the function is extended through the volume until it spans all holes which can be meshed subsequently. Additional constraints have to be added to avoid unintended merging of surfaces. Tailored to repair meshes from range scans, the quality of the resulting surface reconstruction is not sufficient for CAE models.

A very promising algorithm to fill complex holes was presented in [8]. So-called *atomic volumes* are generated by space decomposition and determined to be either completely inside or outside the model, based on a graph representation of the volume structure. Additionally, different topological filling styles can be chosen by the user. Even though this is a very powerful method, it does not fit the special needs of CAE models consisting of numerous independently meshed parts. The same holds for the method presented in [10]. There, the holes are filled using a context-based method: The characteristics of the surface are analysed and parts of the mesh determined to fit the region around the hole are copied into the hole and thus fill it iteratively. Although this gives impressive results for surfaces resulting from erroneous surface sampling, it is not applicable to our scenario, due to the reasons mentioned above.

Radial basis functions and the resulting implicit surface are used e.g. in [2] and [3]. This approach was extended in [4] to recover sharp feature lines in holey areas using an iterative and time-consuming error reduction step. Yoo [14] presents a method capable of handling large polygon models. Domain decomposition is used to solve the problems locally and an implicit surface is defined going through the hole. The final mesh results from a combination of smoothing and refinement scheme together with marching cubes.

Altogether, these methods do not fit too well the needs for filling holes in CAE models. In this context geometrical approaches are more promising, described in the following.

In [1] an early method for closing gaps in CAE models is presented. Nevertheless, only stitching and filling of small holes is performed. Newer methods are far more flexible. In [13] an algorithm based on moving least squares is presented, reconstructing a locally smooth surface. This limitation makes the approach unsuitable for the more general shapes of holes in CAE models. An updated version of this method was presented in [12] but the algorithm itself stayed mainly unchanged. The algorithm presented in [7] fills holes considering the curvature of the enclosing mesh.

Another geometrical approach for filling holes in meshes is [11]. The algorithm uses an advancing front method to close the holes incrementally. Usually, MLS projection, evaluating the vertices' neighbourhood to get a suitable projection plane, is used to insert additional triangles at the front. A pro of this method is that it is also applicable to noisy and unstructured data. By contrast, [6] uses minimal surfaces to fill arbitrary and not necessarily plane holes. Using dynamic programming an ideal solution for the entire problem is computed. This approach holds the advantage of being applicable to arbitrary holes as well as being independent of the surrounding mesh and its characteristics. On the other hand, minimal surfaces do not always meet the requirements of FE models.

Nevertheless, the methods in [6] and [11] turned up to be most promising to be adapted to our problem and thus were used as a base of our solution. We go into more detail about it in the following sections.

3 Preliminaries

In the first step of the complete hole filling process the present holes in a model have to be identified. Before we explain the detection process for holes we first want to clarify the meaning of a hole in the scope of this work. Two classes of holes can be distinguished exemplified in Fig. 2:

A *topological hole* is a hole that can be identified by topological properties, i.e. by searching a loop of boundary edges (see Fig. 2(a)). A boundary edge in a mesh is an edge that is adjacent to only one element.

A *semantic hole* is a hole that, in general, can not be identified by topological properties. Reasons for that may be, that the hole spans over several independent meshes or that it is not made up by boundary edges, e.g. a cut-out in a thick-walled part. An example of a semantic hole resulting from an unmodelled part is depicted in Fig. 2(b), the selection mechanism for semantic holes is described in Sec. 4.

Topological holes can easily be found, so the detection process can run automatically once it is started. They can be found by searching a boundary edge and tracing connected boundary edges until a closed loop is found. This process is repeated until all edges of the mesh have been processed. Alternatively, the user can interactively select a seed boundary edge to start the detection. In case the meshes may have T-joints or boundary vertices that are part of several boundary edges, the detection algorithm has to deal with these special cases additionally.

If the mesh is not closed the described algorithm will also detect the outline as a hole, as it applies to the above definition of a hole. Either a test has to be applied to sort out boundaries automatically or the user has to sort out these false holes by selection after the detection process has ended.

In most cases, semantic holes can not be identified automatically. Holes of this class are e.g. unmeshed parts or constructive openings (see Fig. 2). In the most complex cases these holes may appear by a certain configuration of several

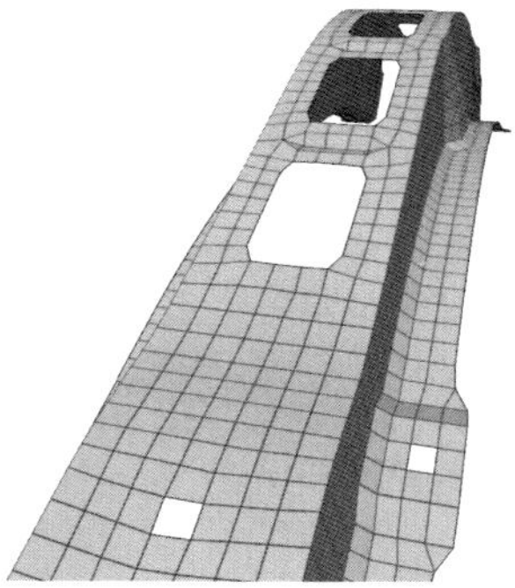

(a) Topological holes are enclosed by a continuous mesh.

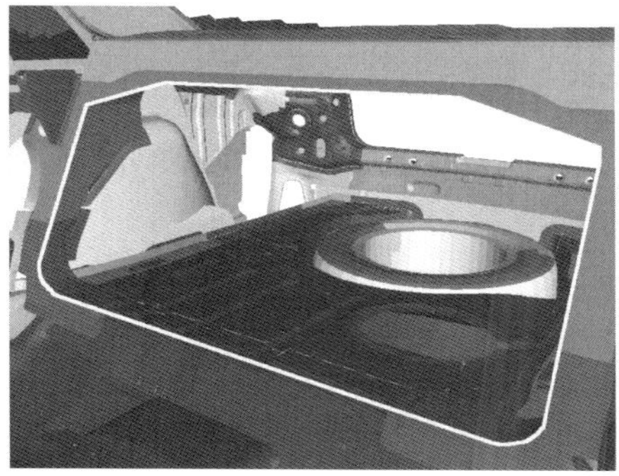

(b) Semantic hole (yellow curve) enclosed by several independent meshes (coloured).

Fig. 2. In contrast to topological holes, semantic holes may not be detected automatically if they do not provide a topological boundary – like this opening between a passenger compartment and a car trunk (yellow curve) – and need to be defined manually. The various colours define a separate mesh each.

independent meshes without any boundary edges at all, so that only the user can identify the presence of a hole by looking at the complete model. Semantic holes that span over several meshes but still consist of boundary edges can be detected if information about the connectivity of the meshes is provided. Algorithms using sophisticated heuristics may also be able to detect some of the semantic holes that do not contain boundary edges, however they are error-prone and it is unsure whether they produce satisfactory results when being applied to a model consisting of hundreds of independent meshes or even whether they will terminate at all. Thus, the user needs the possibility to manually define a loop of edges that can be used as the hole boundary, regardless whether these edges are boundary edges of the mesh or not.

Once all holes have been identified, they have to be closed by an appropriate patching mesh. This mesh either may be connected with the surrounding mesh or may be defined as a new and independent mesh. This procedure is described in Sec. 4.

The following requirements shall hold for the meshing algorithm patching the hole:

- Geometric features, such as constructive edges or design engineered curves, in the surrounding mesh should be preserved.
- If neighbourhood information of boundary edges is missing (e.g. at gaps) the algorithm still should produce acceptable results.
- The algorithm must provide interactive influence on the shape of the patching mesh, e.g. whether it shall smoothly blend into the surrounding mesh or not.

4 Definition of Semantic Holes

Topological holes and some of the semantic holes can be detected automatically, but, as described in the previous chapter, complex semantic holes require user interaction.We chose a compromise between a moderate effort for the user and decent speed of the definition process. The user has to select few nodes on the mesh, whereas the paths in between are calculated by a greedy algorithm, to run as straight and short as possible. Paths along feature lines or boundary edges are favoured by the algorithm. The most important functionality is the definition of semantic holes across several meshes, detailed below.

The path between two selected nodes that lie on the same mesh is chosen to run as straight and short as possible along element edges as mentioned above. In case the next selected target node v_t belongs to a different mesh an additional test is introduced and potential connecting node pairs are retrieved: Utilizing a bounding volume hierarchy, the minimum distance between the precedent selected path node v_s and the target mesh is calculated. If this distance is below a given threshold value, the nearest node on the target mesh is queried, again using the bounding volume hierarchy. This node v_{j_2} and the last node along the path on the current mesh v_{j_1} are then a potential connecting node pair between the two meshes (see Fig. 3).

To ensure an optimal connection, all valid connection candidates are stored and the one fitting best the users' expectations regarding the path run is chosen based on three criteria:

1. The angle ρ enclosed by the vector d_t (from the start node v_s to the target node v_t) and the vector d_j (from v_s to the first connecting node v_{j_1}) should be as small as possible, so that the resulting path runs as straight as possible towards the target node.
2. The path length $\|d_j\|$ from the start node to the first connecting node is preferred to be as short as possible.
3. The node pair distance Δ_j should be as short as possible.

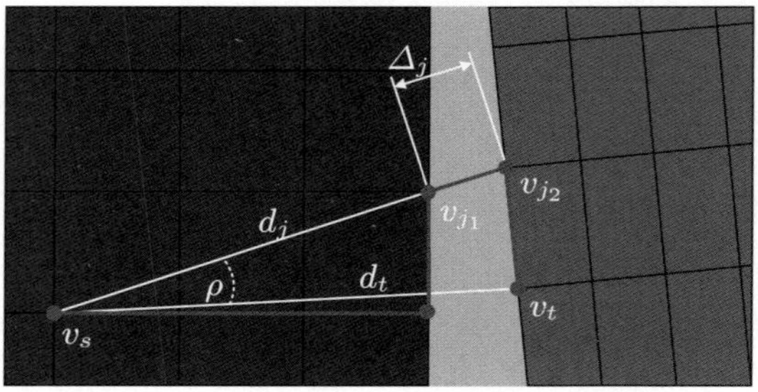

Fig. 3. Criteria for an ideal connection between separate meshes: 1) The angle ρ enclosed by the straight connection $d_t = (v_s, v_t)$ and the connection to the candidate node $d_j = (v_s, v_t)$. 2) The distance $\|d_j\|$ to the connecting node v_{j_1} and 3) The distance $\Delta_j = \|v_{j2} - v_{j1}\|$ between the candidate node pair.

The angle ρ is the primary criterion as it is also the most intuitive one. However, there still may be cases where a better angle does not result in a good path, e.g. if the chosen node lies on a completely different part of the mesh which has a slightly better angular deviation. Therefore, the second criterion is applied. The third one is least important criterion as it at best has aesthetic impact. Nevertheless, if the rating for several node pairs according to the criteria above is equal, Δ_j is used to choose the best pair.

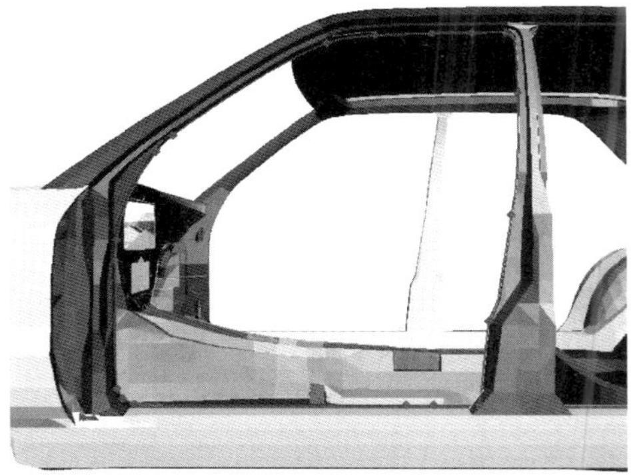

Fig. 4. Selection of a semantic hole: Red squares indicate user-selected nodes, the red line indicates the automatically generated path in between

Figure 4 gives an example of a user-defined selection of a semantic hole spanning over several independent meshes: The red squares mark the nodes defined by the user, the red line represents the automatically generated selection path in between and thus the border of the semantic hole.

5 Patching the Holes

The numerous small and simply shaped holes in the CAE models are usually plane bound and thus, satisfactory patching can be applied automatically, without any need of user interaction. Therefore, we adapted the approach [6], briefly described in Sec. 2, for filling these small holes with minimal surfaces.

Nevertheless, the more complicated shaped holes and especially semantic holes cannot be closed satisfactory with this algorithm as constructive features have to be conserved and perpetuated into the filling mesh. For these constellations a second approach using an Advancing Front Method was implemented and adapted to the needs of CAE meshes, as described in the following sections. In case the hole is too weirdly shaped for that method to be applied with moderate user interaction, the previously mentioned minimal surface algorithm is used to patch the hole – maybe imperfectly but sufficiently.

5.1 Filling Holes Using an Advancing Front Algorithm

For more complex shaped holes, especially those with feature lines to be perpetuated, another approach with supplementary user influence was needed in order to gain more satisfactory results.Therefore, we adapted and extended an existing Advancing Front Method (AFM) [11] in order to create a patching mesh to fill the holes. Advancing Front Methods mesh an area in an iterative manner by inserting new nodes and elements at the borders (the so called *front*) of the area.

This way the front advances into the inside of the area until it is closed. The basic concept of AFMs is pretty simple and advanced functionality and heuristics can easily be added without having to reimplement the whole algorithm. Thus these methods offer an easy possibility to be extended to suit a certain domain of problems, as in our case.

For simplicity, an *edge* in the following section always denotes an edge being part of the advancing front – if not stated otherwise. Without loss of generality, the presented algorithm creates a triangulation. It can be extended to create quadrilateral elements, desirable if the resulting mesh was used for FE simulation and not just as a volume delimiter as in the present case.

Basic Concept

The method as described in [11] consists of two basic steps, outlined below, denoted as *makeConvex* and *addVertices*:

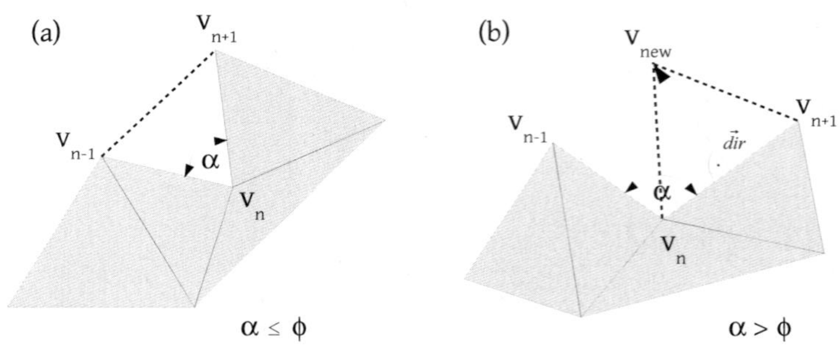

Fig. 5. The basic steps of the Advancing Front Method: Connecting adjacent edges (a) to make the front piecewise convex, and inserting new vertices to propagate the front (b)

- *makeConvex* searches for two adjacent edges that enclose an angle α smaller than a certain threshold value ϕ. In that case a new edge is introduced that completes those two edges to a triangle (see Fig. 5a) . Before committing the new triangle, a test is performed to assure that no other triangles are intersected, thus creating illegal geometry.
  This step is repeated until no adjacent edges enclose an angle smaller than ϕ.
- *addVertices* adds new nodes inside the hole and connects them with the current front in order to create new triangles and to make the front smaller (see Fig. 5b). For each edge the new node is placed along the perpendicular bisector of the edge. In order to fit the surrounding mesh the MLS Projection and a biquadratic polynomial is used to calculate the final position of the node on the surface.

These two steps are executed alternately until no more front edges exist and hence the hole has been closed.

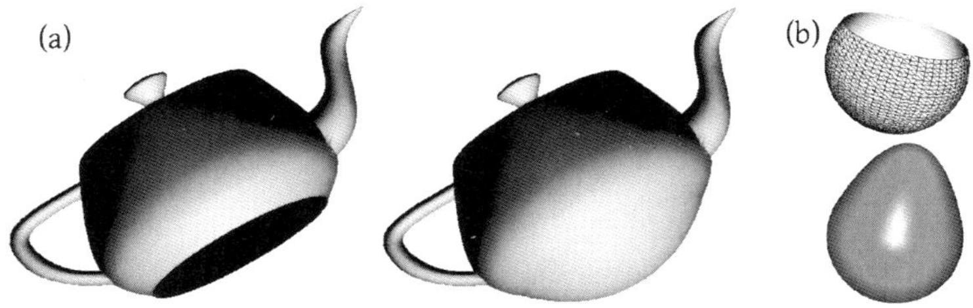

Fig. 6. Inappropriate meshes resulting from interpolation with the MLS projection method [11, 12]

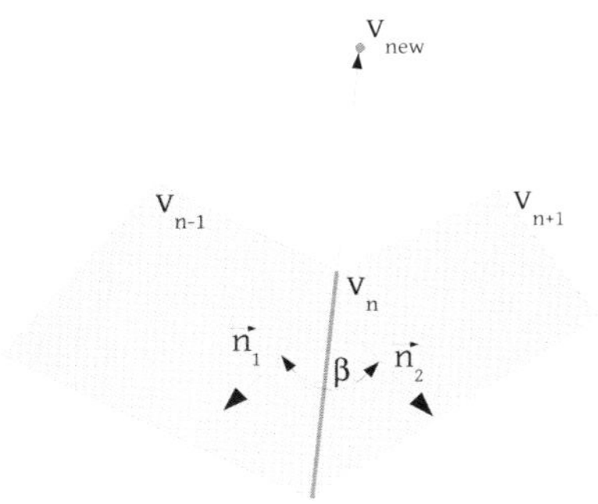

(a) Placement of a new node along a feature edge (red).

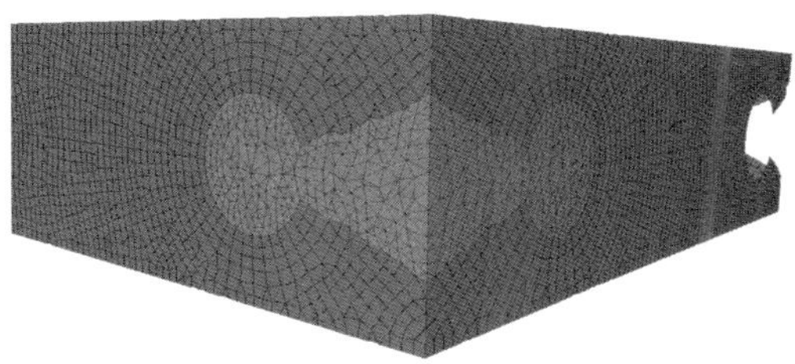

(b) Completely recovered feature at the missing edge of a box.

Fig. 7. Handling of feature lines during vertex insertion

Two problems arise from the method if applied as described: Enclosed feature lines are removed in the makeConvex-step and (nearly) perpendicularly aligned hole borders result in inappropriate meshes, as the MLS projection in the addVertices-step extends the patching mesh too far towards the surrounding mesh border elements (see Fig. 6).

In order to solve these problems, the following changes have been made to the original algorithm :

- During the makeConvex-step it is tested, whether the shared node of two adjacent edges is a feature node, to avoid the cutting of feature lines.
- Instead of using MLS projection, new nodes in the addVertices-step are at first simply placed in the plane of the adjacent element. The bending of the mesh is applied later and the angle is a user-defined parameter.
- New nodes at feature lines are placed along the feature line in order to perpetuate this important information.

The first extension simply introduces an additional test in which for two adjacent edges the angle β between the element normals of the adjacent elements is considered (see Fig. 7(a)). If this angle is larger than a threshold angle γ, a feature line has been detected and the two edges must not be closed by a triangle. The further extensions are more complex and will be described in the following sections.

5.2 Extensions to the Algorithm

Inserting Vertices with Bending Angle and Feature Line Handling

For edges that are not adjacent to a vertex on a feature line, a new node is placed along the perpendicular bisector of the edge. This vector is the result of the cross-product between the adjacent element normal and the edge itself and thus lies in the plane of the adjacent element. In order to prevent, noise in the mesh to is carried on or even increased, a user-defined support is applied. Therefore, also the bidirectional neighbouring elements are used to calculate the direction in which the new node is placed. These neighbours are weighted with weights w_i using a function similar to a Gaussian distribution:

$$w_i = \frac{1}{\sqrt{2\pi}}\, e^{\frac{1}{2}i}, \qquad i = 0..n-1$$

The support could be as large as the whole front, which however does annul the local approach of the algorithm and raises the computational complexity significantly. Due to the support, a slight smoothing is applied to the front which leads to a more robust algorithm.

In case an edge is adjacent to a feature node, this edge is treated special: Instead of placing a node relative to an edge, it is placed along the feature edge incident to the feature node (see Fig. 7(a)). The result of this method is exemplified in Fig. 7(b).

User-defined Bending Angle

After the new nodes have been placed, their positions need to be optimized in order to gain a satisfactory result. Without further treatment the algorithm would only work well for cases, where the hole borders are aligned in such a way, that their advancing fronts grow towards each other, e.g. for completely flat holes. Else, the problems shown in Fig. 6 can not be resolved. For more complex holes it is required that the patching mesh bends, either to blend well into the surrounding mesh or to meet the user expectations – which may also be, that the connection between the patching and the surrounding mesh shall only be C^0 continuous, e.g. when closing an open box.

The bending of the mesh is introduced by a rotation of the new nodes around the edge, to which a new node has been placed relative to (see Fig. 8). For feature nodes the axis is perpendicular to the feature direction, to align the rotation with the feature line. The new nodes do not simply bend by the given angle, but their positions are optimized relative to each other, so that the new front is as short as possible (resulting in a quasi-minimal area triangulation). The user-defined angle is an upper limit for the actual rotation of each node.

In order to optimize the position of each node, the neighbours are taken into account. Each node is rotated by at most the given angle to minimize the distance to its neighbours, i.e. the node is as close as possible to the average of the neighbours. Just as with the calculation of the growing direction described in section 5.2, a support is used. The node rotation is repeated until it converges towards a stable situation.

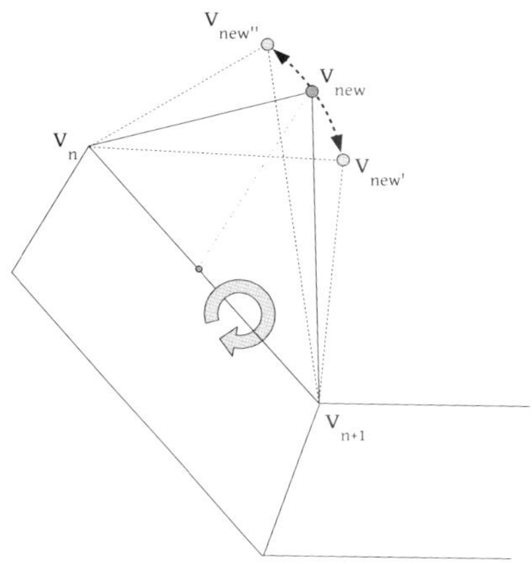

Fig. 8. Example of node rotation to optimize position. Depending on the neighbourhood, the new node v_{new} could e.g. rotate downwards ($v_{new'}$) or upwards ($v_{new''}$).

However, two problems occur using this approach: First, if many nodes are collinear, the local approach fails to optimize the node positions for the complete hole, as any movement of a node raises the distance to its neighbours (see Fig. 9). In fact, changes from other nodes will propagate, however only after a very large amount of iterations. The second problem is, that for complex holes with borders that are initially not oriented towards each other, front parts may miss each other as they grow, leading to either bad and unwanted results or even to unsolvable situations. Thus, for each node a point on the opposite part of the front is additionally used in the computation process. Therefore, a plane normal to the respective edge is placed through the node. The intersection point of this plane with the opposite part of the front is then considered as an additional neighbour (see Fig. 10). For concave bound holes special care has to be taken to avoid the use of intersection points lying behind a node. The opposite point is weighted by distance and the maximum distance where this point still has influence is the user-defined parameter $maxDistance$. The weight w then simply is calculated as

$$w = \frac{maxDistance - distance}{maxDistance}$$

with $maxDistance - distance$ being cropped into the range $[0, maxDistance]$.

Node and Triangle Validation

After the nodes have been rotated, they need to be inserted into the patching mesh along with the new triangles. First, nodes that are too close to each other become merged into a single node to avoid delicate situations and the appearance of cracks in the patching mesh. Then, the new triangles are tested to be valid, i.e. they must not intersect or encroach other triangles. It is also important, that new triangles are not too close to other ones without having nodes merged, as

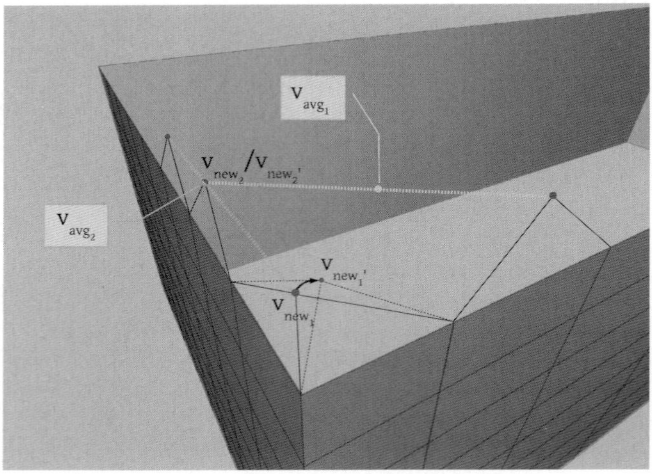

Fig. 9. Problems occurring at collinear nodes: The node v_{new_2} does not move as the average neighbour v_{avg_2} lies at the same position. Only the corner node v_{new_1} is able to move to a new position $v_{new'_1}$, close to the average neighbour v_{avg_1}.

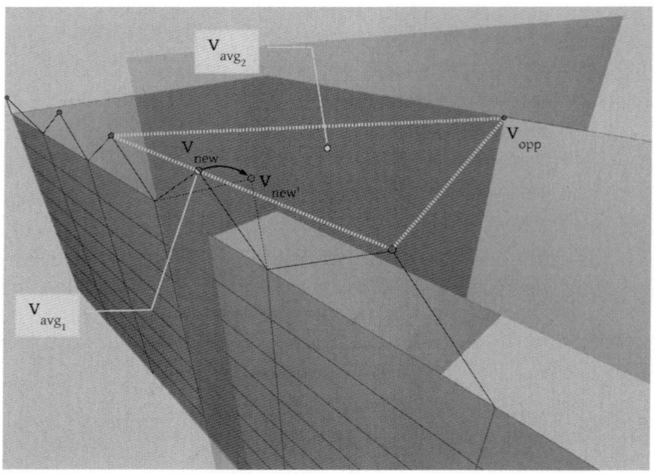

Fig. 10. Using the opposite point v_{opp} on the advancing front helps collinear nodes to bend correctly. Without that v_{new} would not be able to move, as the average v_{avg_1} lies at the same position. The new average v_{avg_2} lies further inside the hole and enables the node to move to position $v_{new'}$.

this also may lead to degenerated triangles or inappropriate situations in later steps of the algorithm.

To improve the intersection test, we adapted the concept of *fences*, as described by Schreiner et al. [9]. Fences are perpendicular extensions to an edge which are inserted for each front edge as auxiliary geometry. A test, whether a new triangle encroaches other boundary parts can now be implemented as an intersection test of the new triangle against the fences of nearby edges. The minimum distance of the node to the fences of nearby edges is computed to check whether a new node lies too close to other edges. If this distance is below a certain threshold value $\mathcal{T}$, the test fails and the new node is rejected.

The height of the fences is important for their efficient use. If they are too small, front parts will still encroach each other, if they are too large, the test might fail wrongly if fences from uninvolved parts of the front are intersected. Schreiner et al. propose an error metric which, however, can only be used during remeshing, where an original mesh is present the deviation can be calculated against. As in our applications there is no original data in the hole area, we use a factor of the average edge length of the hole boundary and a factor of the step size at which the front advances is used for the distance threshold $\mathcal{T}$.

After all tests have been passed, the new triangles are committed and the front is updated with the new boundary edges.

Initial Mesh Bending and Adjusting Edge Lengths

When closing holes with perpendicular oriented hole borders (e.g. an open box), the local approach would implicate a problem: Even if the user sets the bending angle to the maximum, the mesh will not bend down completely to close the hole evenly, due to the fact, that at some angle the nodes lie optimally to each

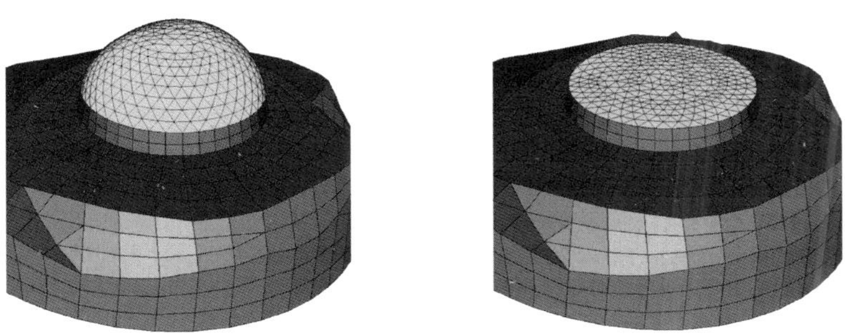

(a) Meshed without initial bending.

(b) Meshed with initial bending.

Fig. 11. Effect of user-interaction on bending: Meshing without (a) and with initial bending (b), causing the hole to be meshed flatly

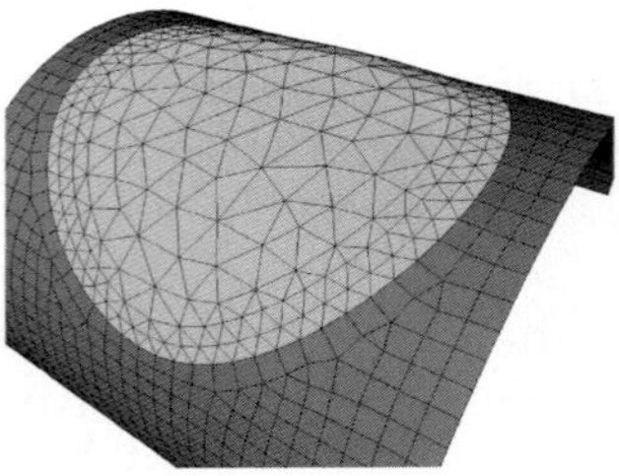

(a) Mesh with gradual adjustment of edge length.

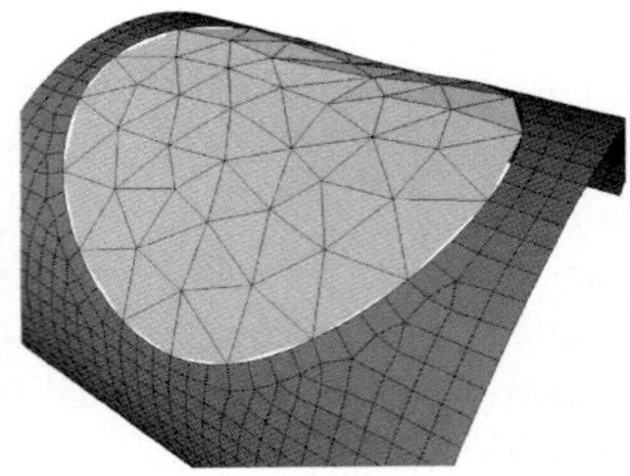

(b) Mesh with forced adjustment of edge length.

Fig. 12. Effect of user-interaction edge length adjustment: Gradual (c) and forced (d) edge lengths

other. However, the local approach is not sufficient for a real global optimization resulting in a new minimum border length (see Fig. 11(a)).

But even if it worked correctly, another problem remains: Often, the user wants the connection of the patching mesh with the original mesh to be only C^0 continuity but the patching mesh itself to have an overall smooth gradient. Using a large angle would not result in the expected surface, but in a quasi-minimum surface with possibly sharp bends as the new nodes might bend more or less freely during the optimization of their positions. Therefore, an option for an *initial maximum bend* is provided. In the first iteration step, it allows the mesh to bend freely and align in such a way, that the front parts point towards each other. After that, the algorithm is normally executed with the given parameters. The bending is gained by again using an opposite point of the front for each new node when optimizing the node positions. This point however is not weighted by distance but with a fixed, large weight to ensure that it is weighted stronger than all other neighbouring nodes. The result can be seen in Fig. 11(b).

Another important property that can be influenced by user interaction is the edge length of the new triangles. By default, the algorithm creates a triangulation with roughly the same tessellation as the surrounding mesh. This is satisfactory for most cases, however can be unnecessary. If the only requirement is a closed hole to be used as a volume boundary, there is no need for the mesh to have a fine tessellation. Especially for large holes, reducing the number of new triangles can additionally have a significant impact on the time needed to mesh a hole.

Therefore, two options are provided to adjust the edge lengths of the patching mesh. Either, the edge length is adjusted gradually, starting with the initial edge length of the hole boundary. This results in a visually smooth connection of the surrounding and the patching mesh (see Fig. 12(a)) and avoids gaps between the patching mesh and the enclosing one. The other option immediately forces the adjustment of the edge lengths by merging adjacent edges (or splitting single edges) until the desired length is gained (see Fig. 12(b)). This introduces a certain error which however is no problem for the volume growing process as long as the emerging gap is small enough, so that a volume element does not fit through.

6 Results

The hole definition interaction mechanism as well as the meshing algorithm described above have been implemented and evaluated with several CAE car models. In order to reduce the need of interaction as far as possible the user can apply different filters that classify holes for their minimum or maximum size, their flatness and their location. These filters allow to reduce the number of holes that need to be inspected by the user to less than 20% of the already filtered holes. Furthermore, convenient interaction mechanisms enable the user to review these holes. Since the meshing of even bigger holes is done in a few seconds the user can vary the control parameters if the created mesh did not meet the requirements. If any hole with a complex shape cannot be closed in one step because the generated mesh differs too much from the missing part this hole can be partitioned into multiple but less complex semantic holes which then can be closed (see meshed hole for missing trunk deck in Fig. 13).

It turned out that the overall process for hole detection, hole meshing and volume mesh generation now can be done in hours instead of days as before. For example, in a plain car body model without chassis consisting of 463 independent parts 3359 holes have been detected. Since we are able to specify filters for the minimum hole size (which equals the minimum size of the volume mesh hexahedra) and the hole position, the number of holes that need to be meshed could be reduced to 248. Our flatness criteria allowed to close 201 of them automatically. Only 47 holes had to be inspected interactively by the user before

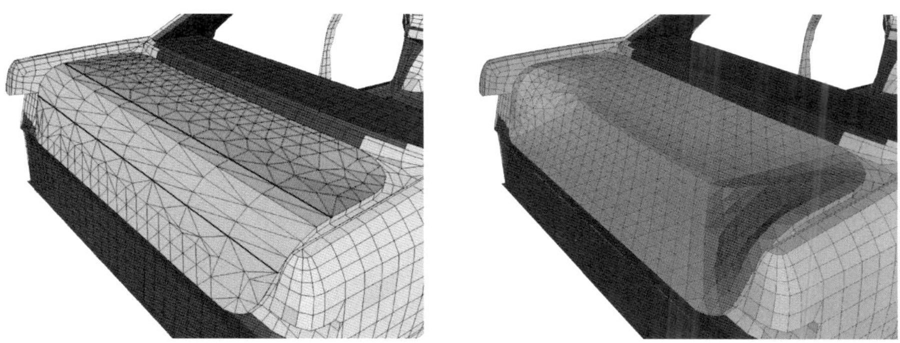

(a) Manual hole partition (thick lines) to optimally close a car trunk.

(b) Automatically closed car trunk.

Fig. 13. Car trunk meshed smoothly (a) by definition of three semantic holes that were meshed subsequently. The different shades of blue denote independent meshes of the CAE model. (b) shows the same hole but filled automatically.

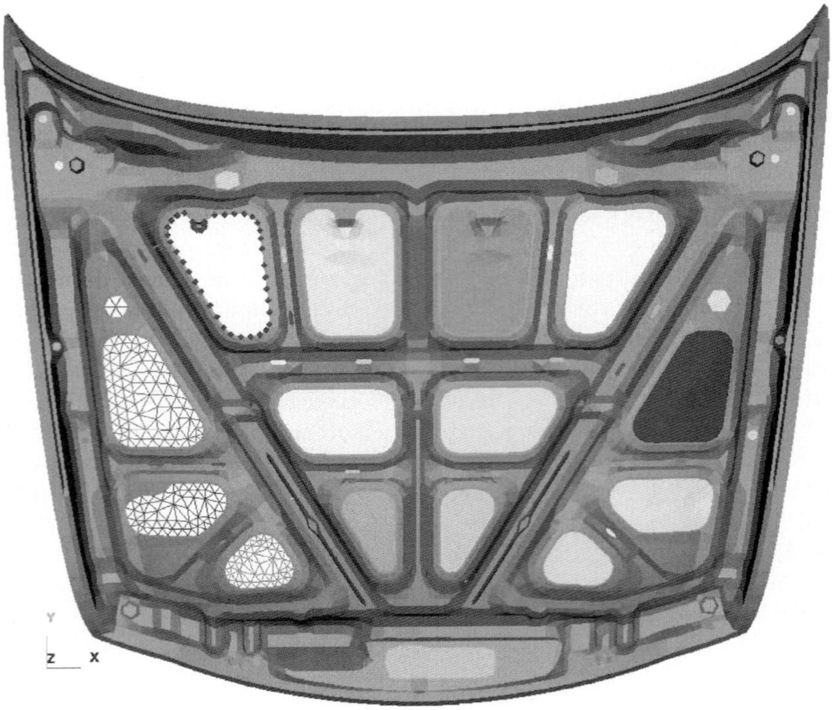

Fig. 14. The flatness classification of holes allows to automatically mesh the vast majority of all holes without any user interaction. In this case only 1 of 48 holes inside the engine hood required user interaction.

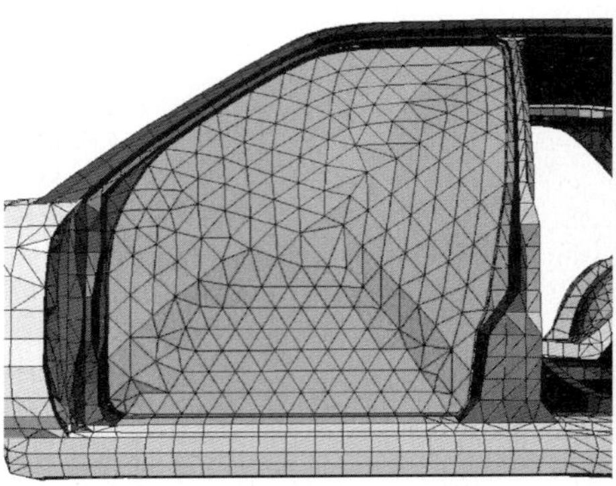

(a) Mesh generated for semantic hole manually defined in Fig. 4.

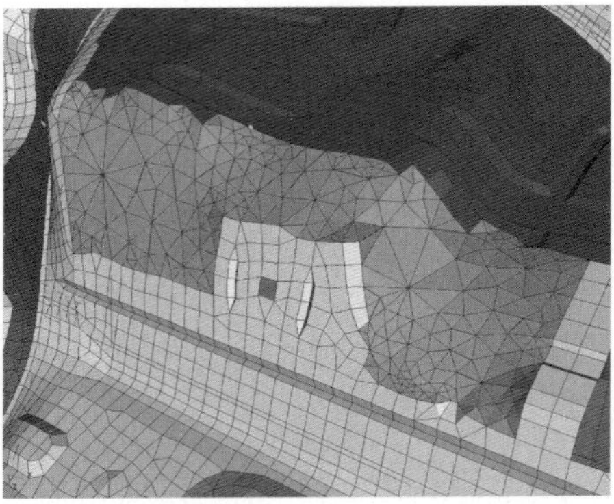

(b) Meshed hole between engine compartment and centre console.

Fig. 15. Examples of automatically filled semantic holes (green meshes) they could be closed, two of them needed to be partitioned as exemplified in Fig. 13. An example of a car part with both, automatically meshed holes and holes that require user interaction for ideal meshing is given in Fig. 14.

Another strong benefit of the presented method is the definition and filling of semantic holes. Beside the car trunk shown in Fig. 13 examples of filled semantic holes are given in Fig. 15. Figure 15(a) shows the automatically meshed hole selected previously in Fig. 4 which is enclosed by various independently meshed car parts. The generated geometry is depicted in green. The automatic meshing of a more complex shaped semantic hole is demonstrated in Fig. 15(b).

7 Conclusion

We presented a process to fill holes in FE models both, automatically and with user interaction. Topological holes in the meshes can be detected and closed automatically, still giving the user the possibility to influence the meshing method if desired. Semantic holes cannot be detected automatically as they span over several separate meshes and therefore are ignored by other approaches. We presented a method to define these holes interactively in a simple and time-saving way by the user. Once defined, these holes can be either filled automatically or more specifically according to the user's intention, whereas the algorithm still keeps in mind the special claims of meshes in CAE models.

References

1. Barequet, G., Kumar, S.: Repairing CAD models. In: VIS 1997: Proceedings of the 8th conference on Visualization, pp. 363–370 (1997)
2. Branch, J., Prieto, F., Boulanger, P.: Automatic hole-filling of triangular meshes using local radial basis function. In: 3DPVT 2006: Proceedings of the Third International Symposium on 3D Data Processing, Visualization, and Transmission, pp. 727–734 (2006)
3. Carr, J.C., Beatson, R.K., Cherrie, J.B., Mitchell, T.J., Fright, W.R., McCallum, B.C., Evans, T.R.: Reconstruction and representation of 3D objects with radial basis functions. In: SIGGRAPH 2001: Proceedings of the 28th annual conference on Computer graphics and interactive techniques, pp. 67–76 (2001)
4. Chen, C.-Y., Cheng, K.-Y.: A sharpness-dependent filter for recovering sharp features in repaired 3d mesh models. IEEE Transactions on Visualization and Computer Graphics 14(1), 200–212 (2008)
5. Davis, J., Marschner, S., Garr, M., Levoy, M.: Filling holes in complex surfaces using volumetric diffusion. In: First International Symposium on 3D Data Processing, Visualiztion and Transmission, vol. 11 (2002)
6. Liepa, P.: Filling holes in meshes. In: EG Symposium on Geometry Processing, pp. 200–205 (2003)
7. Pernot, J.-P., Moraru, G., Véron, P.: Filling holes in meshes using a mechanical model to simulate the curvature variation minimization. Computers & Graphics 30(6), 892–902 (2006)
8. Podolak, J., Rusinkiewicz, S.: Atomic volumes for mesh completion. In: EG Symposium on Geometry Processing, pp. 33–41 (2005)

9. Schreiner, J., Scheidegger, C.E., Fleishman, S., Silva, C.T.: Direct (Re)Meshing for Efficient Surface Processing. Computer Graphics Forum 25(3), 527–536 (2006)
10. Sharf, A., Alexa, M., Cohen-Or, D.: Context-based surface completion. In: Proceedings of the 2004 SIGGRAPH Conference, pp. 878–887 (2004)
11. Tekumalla, L.S., Cohen, E.: A hole-filling algorithm for triangular meshes. Technical report, School of Computing, University of Utah (2004)
12. Wang, J., Oliveira, M.: Filling holes on locally smooth surfaces reconstructed from point clouds. Image and Vision Computing 25(1), 103–113 (2007)
13. Wang, J., Oliveira, M.M.: A Hole-Filling Strategy for Reconstruction of Smooth Surfaces in Range Images. In: Proceedings of SIBGRAPI, pp. 11–18 (2003)
14. Yoo, D.-J.: Filling holes in large polygon models using an implicit surface scheme and the domain decomposition method. International Journal of Precision Engineering and Manufacturing 8(1), 3–10 (2007)

Generation of Spline Approximations to Tessellations

John F. Dannenhoffer III[1] and Robert Haimes[2]

[1] Syracuse University, Syracuse, NY, USA
jfdannen@syr.edu
[2] Massachusetts Institute of Technology, Cambridge, MA, USA
haimes@mit.edu

Summary. In geometrical modeling, one is often provided a description of a surface that is defined in terms of a triangulation, which is supported by a discrete number of nodes in space. These faceted surface representations are defined to be C-0 continuous, and therefore in general have slope and curvature discontinuities at the triangle sides, unless the tessellation is planar. Unfortunately, analytical and computational methods often require a surface description that has well-defined and smoothly-varying gradients and curvatures; in general spline surfaces possess such properties. Described herein is a process for generating a cubic spline surface that approximates, to within a user-specified tolerance, a given tessellated surface that may be non-convex or multiply-connected. The method combines a local least-squares technique for specifying knot properties as well as an adaptation technique of selecting the necessary knot spacings. This new technique is first described along a curve for illustrative purposes. It is then expanded to the case of the general surface. A reparameterization technique that is required for surfaces with non-smooth parameterizations is described next. Computed results for two configurations are then shown.

1 Introduction

1.1 Background

Creating spline fits of general data is a subject that has received much attention over the years. In fact, there are many books devoted to spline generation and applications[1, 2]. In most of the spline fitting literature, it is assumed that data is known at a specified number of knots, and the spline is constructed so as to pass through the knots and such that it has a specified level of smoothness (most frequently C-2 continuity).

There has been limited literature concerning the generation of splines that fit, in a least-squares sense, a set of data that are not located at the knots[3, 4, 5, 6]. In these cases, the knot locations and/or values can be modified to both minimize the distance between the resulting spline and the data, and to guarantee a specified level of smoothness. These techniques were developed primarily for cases with a larger number of data points than spline knots. Unfortunately, all these techniques are only presented with one independent parameter. Also, if the data are clumped (i.e., there are portions of the parameter's domain that contain no data points), the only way to apply these schemes is to move the knots to the regions where the data exist.

The extension of these techniques from a curve (one independent parameter) to a surface (two independent parameters) is not straightforward as there are a few other concerns that must be addressed. First, efficient spline evaluation techniques always operate on rectangular grids of knots. Therefore the ability to place knots is restricted; the only degrees of freedom that are available are the values of the dependent variables at the knots (as well as perhaps the total number of knots in each grid direction). Second, surfaces in general can be non-convex and can contain an arbitrary set of holes. Therefore along any constant parameter line there are likely to be gaps, which must be dealt with in an appropriate manner.

1.2 Problem Statement

The problem here is to create a spline surface that approximates, to within a user-specified tolerance, a surface that is defined in terms of a tessellation. In particular, assume that the surface to be fitted is defined in terms of N nodes, at which the space coordinates x_n, y_n, and z_n $(n = 1, \ldots, N)$ are known. Also known at each of the nodes are two parametric coordinates, u_n and v_n, which are defined prior to the execution of this algorithm. The nodes are connected into a tessellation that is defined in terms of T triangles, which are each defined in terms of its three nodes, $n1_t$, $n2_t$, and $n3_t$, $(t = 1, \ldots, T)$ as well as its three neighboring triangles $t1_t$, $t2_t$, and $t3_t$; triangle sides that are at the edge of the surface or which bound a hole within the surface have no neighbor. There is no assumption here that the surface is convex; additionally the surface can contain any number of arbitrarily-shaped and sized holes.

The objective then can be stated as: find the set of uniformly-spaced knots, $\hat{u}_{i,j}$ and $\hat{v}_{i,j}$ $(i = 1, \ldots, I$ and $j = 1, \ldots, J)$ and their corresponding space coordinates ($\hat{x}_{i,j}$, $\hat{y}_{i,j}$, and $\hat{z}_{i,j}$) that can be used to define a cubic spline surface. The cubic-spline surface should be such that the error associated with evaluating the spline at each of the surface's nodes (e_n) is within the user-specified tolerance e_{tol}. The definition of the node errors is given by

$$e_n = \left[(\tilde{x}_n - x_n)^2 + (\tilde{y}_n - y_n)^2 + (\tilde{z}_n - z_n)^2\right]^{1/2} \tag{1}$$

where $\tilde{x}_n$, $\tilde{y}_n$, and $\tilde{z}_n$ are the values that come from an evaluation of the spline at each of the nodes (u_n and v_n).

The problem has been stated in simple terms: given a set of triangles (with (x, y, z) and (u, v)), return a spline surface. The following are just some of the practical applications:

Topological Editing. The Boundary Representation (BRep) that is produced by modern CAD systems is both a function of the configuration and the modeling operations used by the CAD operator. Several modern tools have adopted the concept of a quilt, in which the faces in the original BRep are merged into quilts (or super-faces). One of the requirements of this merge operation is the ability to create an analytic representation of the merged faces; in fact, the generation of a spline surface for the merged faces is the driving motivation for the work described in this paper. It allows for the formation of a new BRep that stands on its own, where analytic representations of the geometry exist for each entity.

Converting a triangulation to a BRep. Sometimes the geometry of a configuration is only available via a legacy (triangular) representation. The ability to segment a triangulation, parameterize the segments, produce spline approximations, and then trim the resultant surfaces are all necessary techniques for converting a triangulation into the analytic representation that some analytical/computational tools require.

Retriangulation. The given triangulation may not be appropriate for 3D meshing or analysis. CAD systems can provide triangulations in conjunction with the BRep or they may be the result of a quilting operation. More importantly, if the source is a triangulation but the quality is not appropriate, then the tessellation will need to be reconstructed. Many quality surface triangulation schemes require knowledge of both the space coordinates (x, y, z) as well as parametric coordinates, (u, v). Additionally, many of these techniques require smoothly-varying surface slopes and curvatures (if the actual geometry is smooth). A spline surface that is fitted to the original surface naturally provides such a mechanism.

Multidisciplinary Coupling. One of the keys to multi-disciplinary analysis, such as fluid/structure interaction, is the transfer of analysis data on the surface between disciplines. Direct transfer of data from one analysis's grid nodes to the other's is not a well defined operation because the nodes are generally not co-located. If a smooth approximation to the analysis data can be constructed and tied to the analytic surface, then the transfer can be made simply. Therefore, given a surface parameterization and a scalar or vector field at the nodes, the goal is to provide an interpolant for the field data on the surface. This requires the same algorithms as described in this paper, but instead of (x, y, z), field values are used in the fitting process. The end result can be used to interpolate these quantities at a specific (u, v) on the surface and therefore onto the other discipline's mesh nodes.

Behavioral Morphing. The analysis of geometric shapes that change in response to the physical phenomena being modeled is a challenging problem. An obvious example is a structural member under stress, but the motivating example here is in turbomachinery, where the analysis is performed *hot* (i.e. at temperature and up to rotational speed) but the manufactured part (in the CAD system) is *cold*. Given a parameterization that defines the surface for an individual face in the BRep, it is possible to transform the given displacement field (however it is supplied) into an interpolating function, as described above, based on the surface's (u, v) space. It is then simple to create a new surface that is displaced from the original surface; $(x, y, z) = f(u, v)$ just becomes $(x, y, z) = f(u, v) + g(u, v)$ where g is the displacement interpolating function.

1.3 General Approach

The proposed solution to this problem is given by:

1. Pick a 2×2 array of knots $(\hat{u}_{i,j}, \hat{v}_{i,j})$ that surrounds the parametric coordinates in the original tessellated surface.
2. Find the space coordinates at the knots that minimizes, in a least-squares sense, the errors between each of the nodes and their spline counterparts.
3. If the maximum error if less than a user-specified tolerance, STOP.
4. Otherwise
   a) refine the number of knots in the i direction and then re-execute the least-squares algorithm (and its error)
   b) refine the number of knots in the j direction and then re-execute the least-squares algorithm (and its error)
   c) keep the knots (from a or b above) that have the smaller maximum error and repeat back to step 3.

1.4 Roadmap

Each of these steps is described below. This paper begins with a derivation of the proposed scheme along a curve (i.e., only one parametric coordinate, u), in order to both analytically and graphically describe the algorithm. Sample results of the basic algorithm are shown and the need for *smoothing* is highlighted; the smoothing formulation used here is then presented and more sample results are shown.

The section that follows re-derives the scheme for a surface with two parametric coordinates (u and v). The two-dimensional scheme can result in a very large sparse matrix; the technique used to solve the system is described. This section also concludes with some sample results.

One of the limitations of the proposed technique is that it requires that the provided (u, v) parameterization be relatively smooth. A section that shows what happens when you do not have smoothness, the method used to regularize the parameterization, and the resulting approximation follows.

Sample results for a two configurations follow. The paper concludes with a summary of the scheme and its benefits.

2 Spline Approximation for a Curve

In this section, the basic scheme is derived for the case of a curve in space that is defined in terms of one parametric coordinate, u. Although this result is sometimes useful by itself, it is presented here to simplify the algebra and so that a graphical interpretation of the results can be readily made.

2.1 Ferguson Splines

A Ferguson spline is a special form of a piece-wise cubic spline that was popularized by D. Ferguson[7]. In particular, consider a spline fit of one space coordinate x as a function of u. The characteristics of the Ferguson spline are:

- the spline is defined by a uniformly-spaced set of knots $\hat{u}_i$ $(i = 1, \ldots, I)$ so that $\hat{u}_i - \hat{u}_{i-1} =$ constant.
- as with all piece-wise cubic splines, the spline has C-2 continuity throughout;
- the particular flavor of Ferguson splines used here has natural end conditions (that is, the curvature vanishes at the end points $\hat{u}_1$ and $\hat{u}_I$. (In general, Ferguson splines for which the end slopes are prescribed can be generated, but are not considered here); and
- the spline creation process finds and stores the slopes $(\hat{X}_i)$ at each of the knots.

Given the above, the cubic in each interval i to $i+1$ can be written as

$$\tilde{x} = \tilde{A}\hat{x}_i + \tilde{B}\hat{x}_{i+1} + \tilde{C}\hat{X}_i + \tilde{D}\hat{X}_{i+1} \tag{2}$$

where the basis functions are given by

$$\tilde{A} = 1 - 3\tilde{s}^2 + 2\tilde{s}^3 \qquad \tilde{B} = 3\tilde{s}^2 - 2\tilde{s}^3$$
$$\tilde{C} = \tilde{s} - 2\tilde{s}^2 + \tilde{s}^3 \qquad \tilde{D} = -\tilde{s}^2 + \tilde{s}^3$$

and where $\tilde{s} = (\tilde{u} - \hat{u}_i)/(\hat{u}_{i+1} - \hat{u}_i)$ is the fractional distance of $\tilde{u}$ between $\hat{u}_i$ and $\hat{u}_{i+1}$.

To set up the spline, we want to make the curvatures match at the interior knots. Using the above, this can be written as

$$3\hat{x}_{i-1} - 3\hat{x}_{i+1} + \hat{X}_{i-1} + 4\hat{X}_i + \hat{X}_{i+1} = 0 \qquad (i = 2, \ldots, I-1) \tag{3}$$

The natural end conditions imply that the curvature vanishes at the endpoints. This can be written

$$3\hat{x}_1 - 3\hat{x}_2 + 2\hat{X}_1 + \hat{X}_2 = 0 \tag{4}$$
$$3\hat{x}_{I-1} - 3\hat{x}_I + \hat{X}_{I-1} + 2\hat{X}_I = 0 \tag{5}$$

The above I equations form a linear set for the $\hat{X}_i$ with a given set of $\hat{x}_i$. The resulting matrix equation is tridiagonal, it can be solved in $O(I)$ time.

2.2 Least-Squares Problem

In the above technique, it was assumed that the knot values $\hat{x}_i$ were known and we had to find the slopes $\hat{X}_i$ that yielded C-2 continuity. In this section, we do not *a priori* know the knot values and thus have to find the $\hat{x}_i$ as well as the $\hat{X}_i$.

Assume that we are given the desired number of knots, I. The first step is to define a vector of knot coordinates, $\hat{u}_i$, that are both evenly spaced and such all the u_n fall between $\hat{u}_1$ and $\hat{u}_I$. If we define a small buffer, b, to ensure that the spline always interpolates (and never extrapolates), then the above can be written

$$\hat{u}_1 = u_{\min} - b(u_{\max} - u_{\min})$$
$$\hat{u}_I = u_{\max} + b(u_{\max} - u_{\min})$$

where

$$u_{\min} = \min_{n=1}^{N} u_n \qquad \text{and} \qquad u_{\max} = \max_{n=1}^{N} u_n$$

For all cases shown here, $b = 0.10$.

Combining these, we finally get

$$\hat{u}_i = \left[\frac{(1+b)(I-i) - b(i-1)}{I-1}\right] u_{\min} + \left[\frac{(1+b)(i-1) - b(I-i)}{I-1}\right] u_{\max} \tag{6}$$

The sum-squared error can then be defined (using Eq. 1) as

$$E = \sum_{n=1}^{N} e_n^2 = \sum_{n=1}^{N} \left[(A_n \hat{x}_{in} + B_n \hat{x}_{in+1} + C_n \hat{X}_{in} + D_n \hat{X}_{in+1}) - x_n\right]^2 \tag{7}$$

where in in the knot immediately preceding u_n so that $\hat{u}_{in} \le u_n < \hat{u}_{in+1}$. Also, $A_n = 1 - 3s_n^2 + 2s_n^3$, where $s_n = (u_n - \hat{u}_{in})/(\hat{u}_{in+1} - \hat{u}_n)$.

To minimize E, we force the I partial derivatives, $\partial E/\partial \hat{x}_i$ to vanish

$$\frac{\partial E}{\partial \hat{x}_i} = \sum_{n=1}^{N} 2(A_n \hat{x}_i + B_n \hat{x}_{i+1} + C_n \hat{X}_i + D_n \hat{X}_{i+1}) F_n = 0 \tag{8}$$

where

$$F_n = \begin{cases} A_n & \text{if } \hat{u}_{i-1} \le u_n < \hat{u}_i \\ B_n & \text{if } \hat{u}_i \le u_n < \hat{u}_{i+1} \\ 0 & \text{otherwise} \end{cases}$$

The above I equations can be rearranged into the form

$$S_1 \hat{x}_{i-1} + S_2 \hat{x}_i + S_3 \hat{x}_{i+1} + S_4 \hat{X}_{i-1} + S_5 \hat{X}_i + S_6 \hat{X}_{i+1} = S_7 \tag{9}$$

with the shorthands

$$S_1 \equiv \sum_{n \in nm_i} A_n B_n \qquad S_2 \equiv \sum_{n \in nm_i} B_n^2 + \sum_{n \in np_i} A_n^2$$

$$S_3 \equiv \sum_{n \in np_i} A_n B_n \qquad S_4 \equiv \sum_{n \in nm_i} B_n C_n$$

$$S_5 \equiv \sum_{n \in nm_i} B_n D_n + \sum_{n \in np_i} A_n C_n \qquad S_6 \equiv \sum_{n \in np_i} A_n D_n$$

$$S_7 \equiv \sum_{n \in nm_i} B_n x_n + \sum_{n \in np_i} A_n x_n$$

where nm_i is the set of nodes for which $\hat{u}_{i-1} \leq u_n < \hat{u}_i$ and np_i is the corresponding set when $\hat{u}_i \leq u_n < \hat{u}_{i+1}$.

We can now form a large matrix equation from Eqs. 3 to 5 and Eq. 9, which can be written in the general form

$$\begin{bmatrix} M_1 & M_2 \\ M_5 & M_6 \end{bmatrix} \begin{bmatrix} \hat{x} \\ \hat{X} \end{bmatrix} = \begin{bmatrix} R_1 \\ 0 \end{bmatrix} \tag{10}$$

Here the sub-matrices M_1, M_2, and R_1 are taken directly from Eq. 9 and sub-matrices M_5 and M_6 are taken from Eqs. 3 to 5. The resulting matrix equation, which is quite sparse, can be solved in any convenient manner.

2.3 Numerical Solution

The above algorithm can be applied to a sample data set that was obtained by taking a constant-v-parameter cut through one of the surfaces of an example configuration. The original triangulation for the example surface is plotted in the (u, v) plane in Fig. 1, with the selected cut at $v = 2$ shown by the horizontal line. Fig. 2(a) shows the $x(u)$ data along this cut as a series of circles. One can

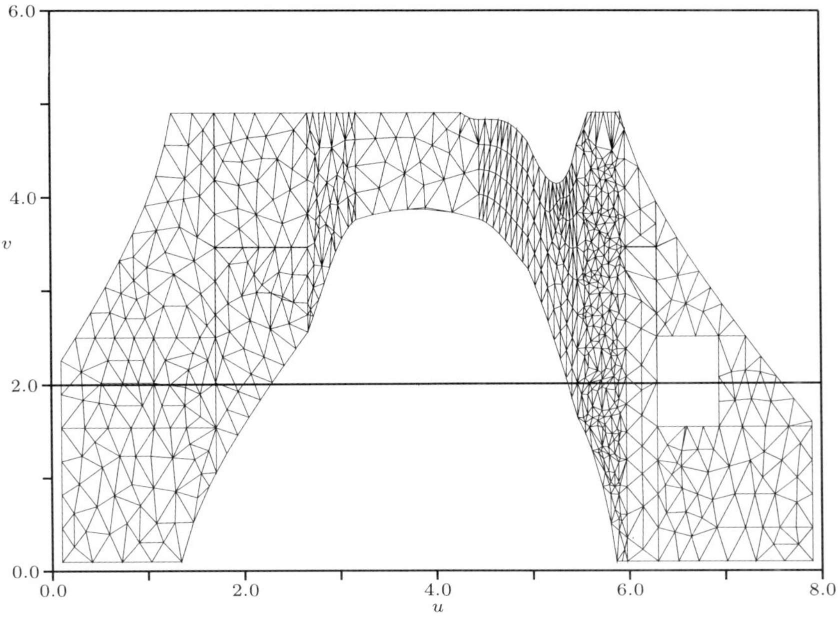

Fig. 1. Prescribed triangulation for a test configuration. The horizontal line shows the cut used in Figs. 2 and 3. This triangulation is the result of merging 13 faces in an original BRep.

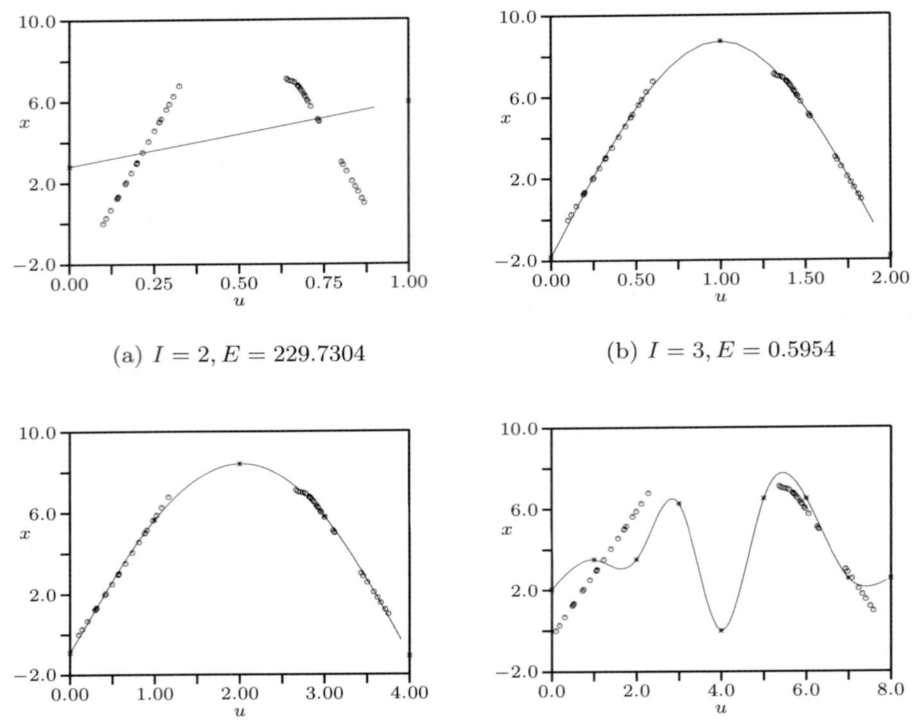

(a) $I = 2, E = 229.7304$

(b) $I = 3, E = 0.5954$

(c) $I = 5, E = 0.6366$

(d) $I = 9, E = 78.3642$

Fig. 2. Spline approximations along a curve without smoothing ($\mu = 0$)

note three regions in which data is defined; the gap in the middle comes from the shape of the outer boundary of the triangulation whereas the smaller gap near the right comes from the rectangular hole in the triangulation.

The first approximation, shown in Fig. 2(a), has only two knots, resulting in a spline that is simply a straight line. The caption of the figure states that when the number of knots, I is 2 the sum-squared error, E, is about 200; clearly this is not a good approximation. In part (b) of the figure, the number of knot-intervals was doubled (taking I from 2 to 3); here the fit is visually much better and the error has decreased to a little more than 1/2. When $I = 5$ the fit is about as good as before. Unfortunately when the number of knots is increased to $I = 9$ the spline becomes worse, with wild oscillations present. A careful look at the figure shows that the oscillation occurs in the region where there is insufficient data to support the spline. In other words, in the vicinity of the oscillations, the number of degrees of freedom in the spline knots is locally greater that the number of nodes for which an error is being minimized. Also, examination of the matrix equation 10 reveals that there are some rows that are not diagonally dominant (and in fact can become degenerate).

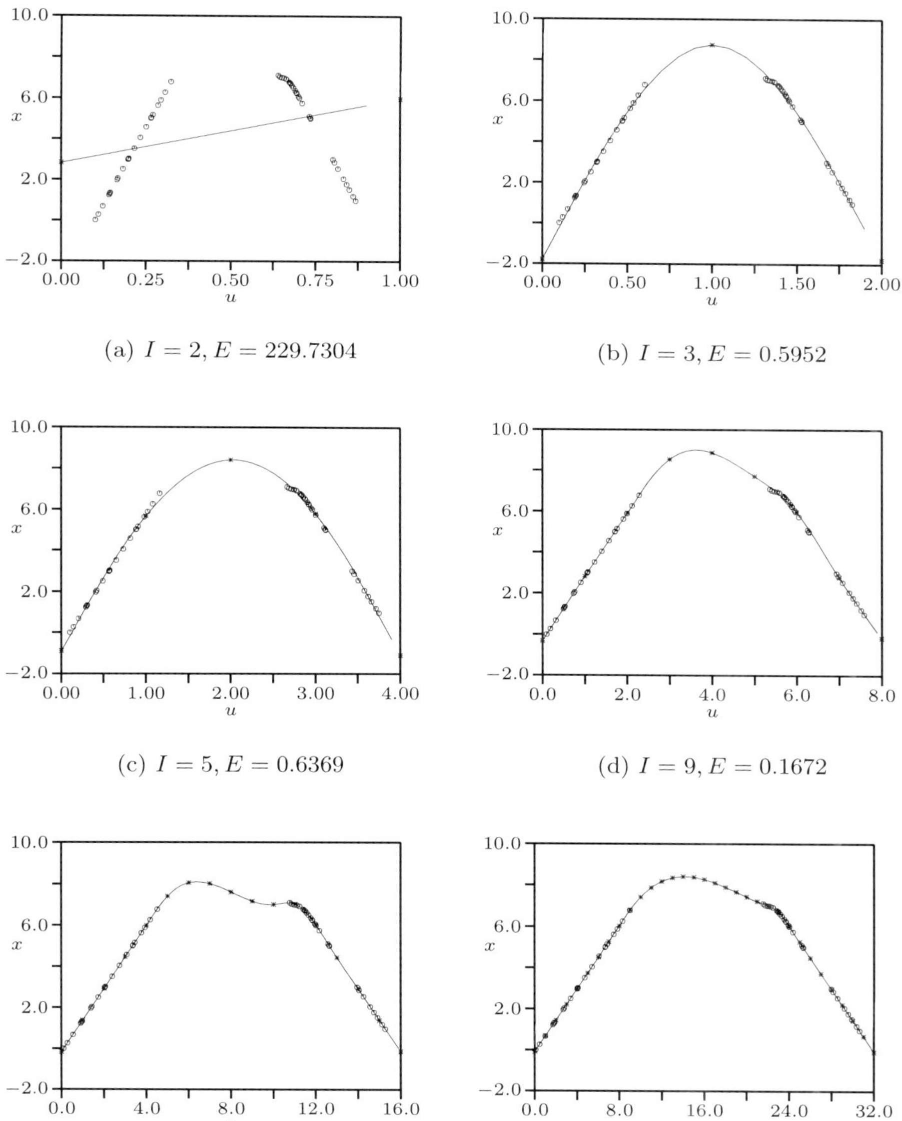

(a) $I = 2, E = 229.7304$

(b) $I = 3, E = 0.5952$

(c) $I = 5, E = 0.6369$

(d) $I = 9, E = 0.1672$

(e) $I = 17, E = 0.0013$

(f) $I = 33, E = 0.0003$

Fig. 3. Spline approximations along a curve with smoothing ($\mu = 0.001$)

2.4 Smoothing

To fix the above, there are two modifications to the basic least-squares equations that appear to be necessary. First for interior knots, it has been found useful to add a very small amount of Laplacian *smoothing* to the least-squares equations. In particular, adding

$$\mu\left(-\hat{x}_{i-1}+2\hat{x}_i-\hat{x}_{i+1}-\frac{1}{2}\hat{X}_{i-1}+\frac{1}{2}\hat{X}_{i+1}\right) \qquad (i=2,\ldots,I-1)$$

to the i-th Eq. 10 has been found to work very well, where μ is a very small parameter (here taken to be $\mu = 0.001$). Note that in regions where there exists enough data for least-square minimization, these smoothing terms are negligible. However in regions where there are no nodes (that is, both nm_i and np_i are empty sets), the smoothing restores the diagonal dominance in the matrix.

The other modification is associated with the knots at the ends of the spline. At those knots, the above smoothing cannot be used since there is not an outboard knot. For these cases, it has been found sufficient to transfer the *natural end condition* associated with the Ferguson splines to the first inboard knot. For example, if the first rows of M_1, M_2, and R_1 are all zero, we can change that equation to a zero-curvature condition at the inboard knot, or

$$3\hat{x}_1 - 3\hat{x}_2 + 2\hat{X}_1 + 4\hat{X}_2 = 0 \qquad (11)$$

A similar modification is added in the case where the last row of M_1, M_2, and R_1 are all zero, which occurs when there are no nodes between the last two knots.

Fig. 3 shows the results of applying this scheme, with smoothing, to the nodes that was shown previously in Fig. 2. Note that as more knots I are added, the shape of the spline better approximates the shape of the node data. Note in particular how smooth the spline is in the region that does not contains any nodes. Also shown in the figure is the sum-squared error, E, which clearly decreases as the number of knots increases.

3 Spline Approximation for a Surface

The extension of the above techniques from a curve to a surface is straightforward, but algebraically tedious. In this section, the most important aspects of the extension are discussed.

Surfaces are defined in terms of two parametric coordinates, u and v. Hence the knots at which the spline data is known form a uniformly-spaced two-dimensional array; the parametric coordinates at the knots are thus $\hat{u}_{i,j}$ and $\hat{v}_{i,j}$, where $i = 1,\ldots,I$ and $j = 1,\ldots,J$. The knot values of the dependent variables are $\hat{x}_{i,j}$, $\hat{y}_{i,j}$, and $\hat{z}_{i,j}$; for illustrative purposes, only $\hat{x}_{i,j}$ will be considered here.

3.1 Ferguson Splines

Creation of a Ferguson spline for a surface involves the determination of the slopes in both parametric directions, $\hat{X}^u_{i,j}$ and $\hat{X}^v_{i,j}$, as well as the cross-derivatives $\hat{X}^{uv}_{i,j}$. In practice when all the knot values $\hat{x}_{i,j}$ are known, the slopes

$\hat{X}^u_{i,j}$ are found by applying Eqs. 3 to 5 independently along each constant j line; then an analogous technique for finding $\hat{X}^v_{i,j}$ is used independently along each constant i line. Finally, the cross-derivatives $\hat{X}^{uv}_{i,j}$ can be found by applying Eqs. 3 to 5 along constant i lines to the $\hat{X}^v_{i,j}$ found previously (or alternatively by fitting $\hat{X}^u_{i,j}$ along constant j lines).

The evaluation of a Ferguson spline at $(\tilde{u}, \tilde{v})$ is performed using a two-dimensional form of Eq. 2, where there are now 16 basis functions that are polynomial functions of $\tilde{s}$ (the fractional distance of $\tilde{u}$ between $\hat{u}_i$ and $\hat{u}_{i+1}$) and $\tilde{t}$ (the fractional distance of $\tilde{v}$ between $\hat{v}_j$ and $\hat{v}_{j+1}$) .

3.2 Least-Squares Problem

The least-squares problem for finding the $\hat{x}_{i,j}$ that minimize the error E is also conceptually straightforward.

First, the parametric coordinates at each of the knots, $\hat{u}_{i,j}$ and $\hat{v}_{i,j}$, are found by independently scaling the u and v via equations similar to Eq. 6. Note that the sums used here are over all nodes. Then an equation that is analogous to Eq. 9 is formed, where the sums are over the set of nodes that are in the four quadrilateral cells that surround (i, j). Next, a large, sparse matrix equation

$$\begin{bmatrix} M_1 & M_2 & M_3 & M_4 \\ M_5 & M_6 & 0 & 0 \\ M_7 & 0 & M_8 & 0 \\ 0 & 0 & M_9 & M_{10} \end{bmatrix} \begin{bmatrix} \hat{x} \\ \hat{X}^u \\ \hat{X}^v \\ \hat{X}^{uv} \end{bmatrix} = \begin{bmatrix} R_1 \\ 0 \\ 0 \\ 0 \end{bmatrix} \tag{12}$$

is formed. Here, the sub-matrices M_1 to M_4 and R_1 are written directly from the analog of Eq. 9. The second set of sub-matrices (M_5 and M_6) come from the spline-fit of $\hat{x}$ in the i-direction (which yields $\hat{X}_u$). In a similar way, the third set of sub-matrices (M_7 and M_8) come from fitting $\hat{x}$ in the j-direction. Finally, M_9 and M_{10} come from splining $\hat{X}^v$ in the i-direction. Note that since the spline data $\hat{x}_{i,j}$ are not known prior to generating the least-square fit, the spline generation process requires that all three spline-generation sweeps be done simultaneously.

Smoothing is added as above, by adding the Laplacian operator

$$\mu(4\hat{x}_{i,j} - \hat{x}_{i-1.j} - \hat{x}_{i+1.j} - \hat{x}_{i,j-1} - \hat{x}_{i,j+1}) \\ + \frac{\mu}{2}(-\hat{X}^u_{i-1,j} + \hat{X}^u_{i+1,j} - \hat{X}^v_{i,j-1} + \hat{X}^v_{i,j+1})$$

to the matrix equation (as above). Finally, if any of the boundary knots ($i = 1$, or $i = I$, or $j = 1$, or $j = J$) have a degenerate equation in the matrix (because the are no nodes in the cells adjacent to the knot), the natural end condition is applied at the adjacent inboard knot (for example, at $(2, j)$).

The above matrix equation is rather large, but is very sparse. In fact each of the sub-matrices M is pentadiagonal. To see how large the matrix is, consider the generation of a 65×17 array of knots, which results in a square matrix that contains almost 20 million elements, fewer than one-half percent of which are non-zero. Clearly forming and solving the full matrix is unacceptable and thus sparse-matrix methods must be used. Here, the sparse-matrix data storage format and the bi-conjugate gradient solver described in [8] are used. The only delicate part of using this sparse-matrix technique is the direct creation of the matrix in sparse-matrix form.

4 Overall Algorithm

The above section describes the technique used to find the knot data, $\hat{x}_{i,j}$, for a spline that matches, in a least-squares sense, the data values prescribed at the nodes, x_n. Since in general we want to generate fits for all three space coordinates, (x, y, z), the above technique can be applied successively to x_n, y_n, and z_n. However, the objective is to find the knots that result in splines that are within a user-specified tolerance, e_{tol}, of the knot data. Above, a general technique is outlined; this section expands on that description.

As mentioned above, the method starts out with a 2×2 array of knots; the technique described above is then applied with $I = 2$ and $J = 2$ and the error at each node, e_n, is computed for each space coordinate. The maximum of all these errors

$$e_{\text{max}} = \max_{n=1}^{N} \left(e_n^x,\ e_n^y,\ e_n^z\right)$$

is compared with the tolerance, e_{tol} to determine if the knot set is sufficient. This coarse knot grid turns out to be sufficient for cases in which the tessellated surface is planar.

If e_{max} is larger than e_{tol}, then we do not have a sufficient number of knots. In this case, two refinements are made:

- First the number of knot intervals in the i-direction is doubled, or $I_{\text{new}} \leftarrow 2I - 1$. The least-squares algorithm is executed on a $I_{\text{new}} \times J$ grid and the maximum resulting error, called e_I, is computed.
- Next the number of knot intervals in the j-direction is doubled, or $J_{\text{new}} \leftarrow 2J - 1$. The least-squares algorithm is executed on a $I \times J_{\text{new}}$ grid and the maximum resulting error, called e_J, is computed.

The grid of knots with the smaller error becomes the new set of knots. For example, if $e_I < e_J$, then the new grid will be $I_{new} \times J$.

If the error associated with this new grid (just computed) is smaller than e_{tol}, the overall procedure terminates; otherwise the process is repeated until either a sufficient grid is found or until a maximum grid size is reached; in practice, the latter does not happen if the specified tolerance is reasonable. For all the cases shown here, e_{tol} is set to 10^{-4} of the largest dimension (in (x, y, z)) of the surface.

4.1 Improvements in the Parameterization

After exercising the above on over 2500 surfaces taken from both test and real-world configurations, it was found that the above algorithm sometimes results in a set of knots that exhibit large oscillations; other times, the algorithm produces good results within the trimming curves of the surface, but exhibits unwanted behavior beyond the surface's trimming curve, indicating that the spline would not be suitable for extrapolating a small distance beyond the edge of the surface. (This extrapolation might be needed for an algorithm such as surface-surface intersection.) Careful examination of these cases showed that this unwanted behavior was always associated with nearly-infinite slopes in the underlying parameterization.

Clearly the answer to this is to reparameterize so that these nearly-infinite slopes are eliminated. For a curve, this reparameterization is fairly easy to accomplish through stretching. Unfortunately, this reparameterization for a surface is not quite as straightforward.

Stated simply, the objective of the reparameterization is to make the spacings of the nodes in parametric (u, v) space to be proportional to the node spacing in terms of the space coordinates, (x, y, z). Here we have adopted the reparameterization proposed by Floater[9]. The first step is to find the mean value weights at each node due to its neighbors. For a triangle side between nodes $n1$ and $n2$, the mean-value weight is given by

$$w_{n1,n2} = \frac{\tan(\alpha/2) + \tan(\beta/2)}{d_{n1,n2}} \tag{13}$$

where α is the angle at node $n1$ in the triangle to the left of side $[n1, n2]$, β is the angle at node $n1$ in the triangle to the right of side $[n1, n2]$, and $d_{n1,n2}$ is the distance between the nodes. All of these angle and distance calculations are made using the space (x, y, z) coordinates.

Next two sparse matrix equations are formed. For each interior node in the surface, we can write

$$u_n \sum_m w_{m,n} + \sum_m (w_{m,n} u_m) = 0 \qquad (n = 1, \ldots, N) \tag{14}$$

$$v_n \sum_m w_{m,n} + \sum_m (w_{m,n} v_m) = 0 \qquad (n = 1, \ldots, N) \tag{15}$$

The parametric coordinates at boundary nodes are not allowed to move. These equations are solved using the sparse-matrix, bi-conjugate gradient technique mentioned above.

Fig. 4(a) show the tessellation that was provided; it was formed by projecting the nodes on a fuselage surface in the spanwise direction; that is $u_n = x_n$ and $v_n = y_n$. Note how the parameterization is squeezed near the top and bottom edges due to the fact that the surface is nearly parallel to the $x - z$ plane. A vertical cut was made through the parameterization at $u = 3$ and the corresponding spline fit of $z(v)$ is shown in part (b). Notice the oscillations that arise due to the nearly infinite slope, $\frac{dz}{dv}$, at the ends (which correspond to the top and bottom of the vertical cut).

The above reparameterization technique was then applied, resulting in the parameterization in part (c) of the figure. Since the nodes on the surface were originally placed such that the spacings between adjacent nodes was approximately

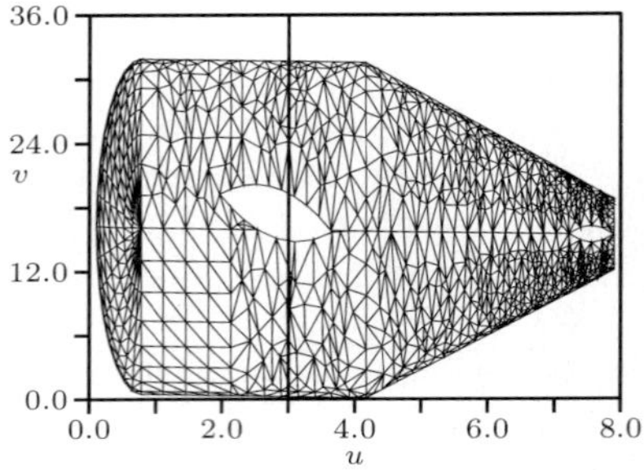

(a) original surface tessellation. The triangles near the bottom and top are highly stretched.

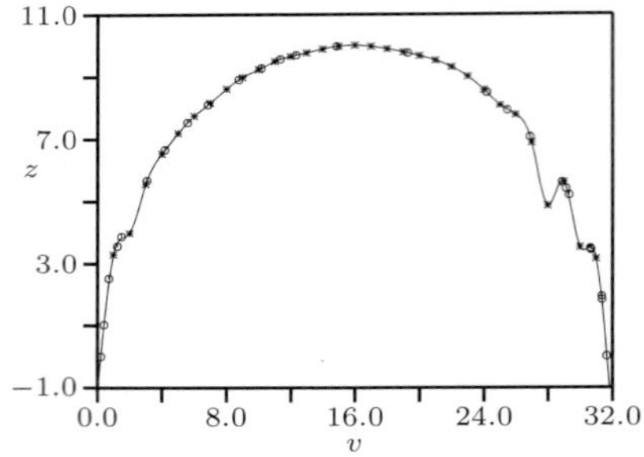

(b) spline approximation along $u = 3$ for the original surface tessellation.

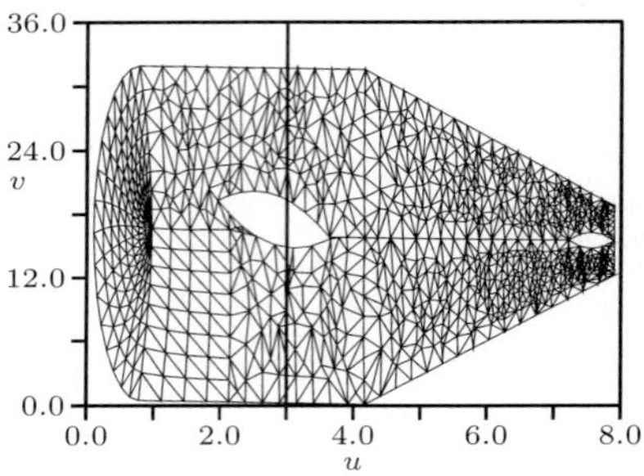

(c) reparameterized surface tessellation. The triangles near the bottom and top are more isotropic than before.

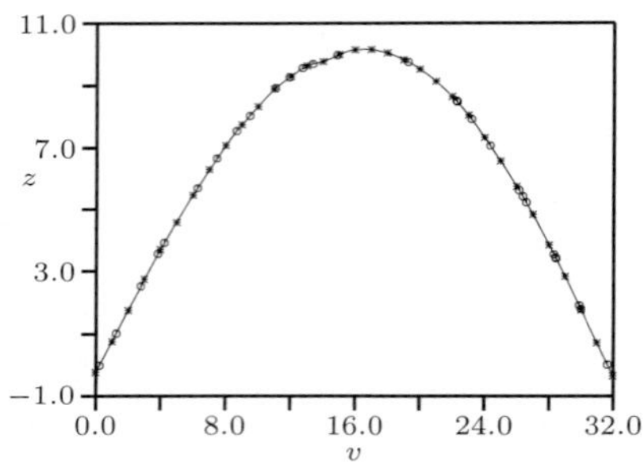

(d) improved spline approximation along $u = 3$ for the reparameterized tessellation

Fig. 4. Effect of reparameterizing a surface

constant, the reparameterization results in approximately-constant parametric spacings. The spline technique applied on this new parameterization is shown in Fig. 4(d). Here the oscillations have been eliminated and the error between the resulting spline and the nodes is negligible. Application of this reparameterization to the entire test suite shows that all the oscillations were eliminated.

5 Computational Examples

The spline generation scheme proposed here has been applied to over 100 configurations, comprised of over 2500 surfaces. Figs. 5 and 6 show two of these configurations. The BRep produced by the CAD systems, along with CAPRI's[10] tessellation is shown in part (a) of each figure. Shown in part (b) of each figure is the grid of knots that were automatically produced by this scheme. One can clearly

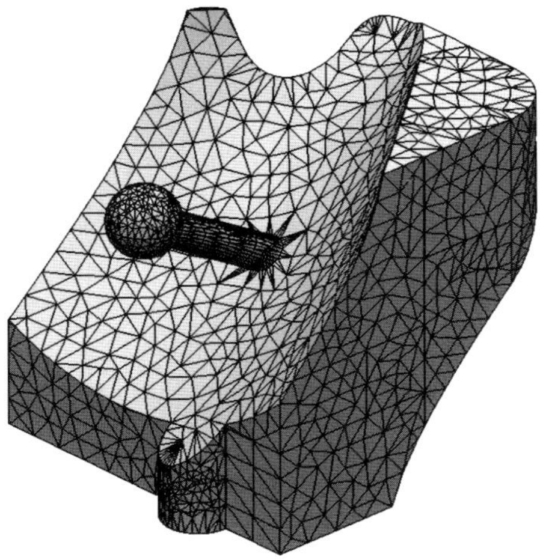

(a) BRep with given tessellation

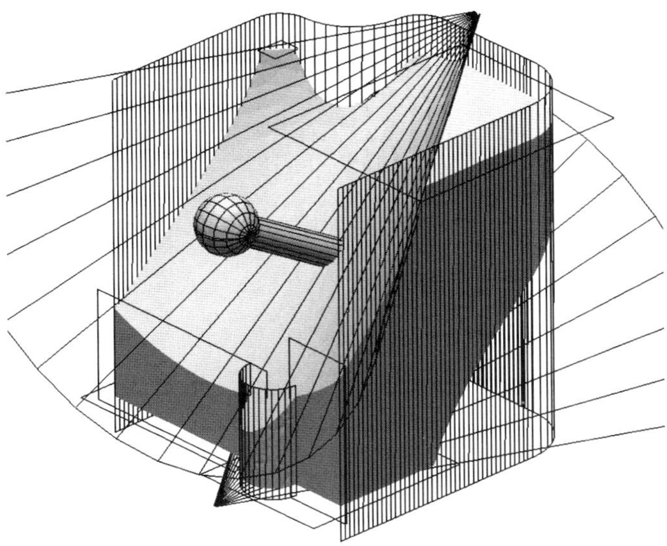

(b) BRep with grid of knots for each surface

Fig. 5. Test case 1, consisting of 17 surfaces (which were created by merging the original 82 faces in the Parasolid part). Note that the curved back surface is the surface used in Figs 1 to 3.

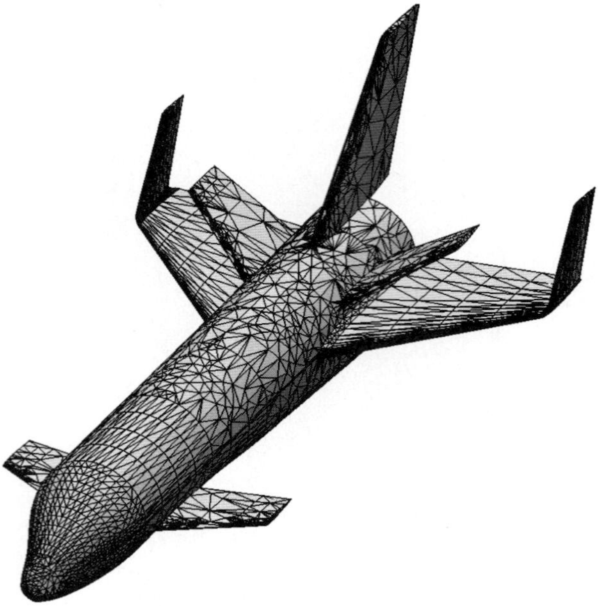

(a) BRep with given tessellation

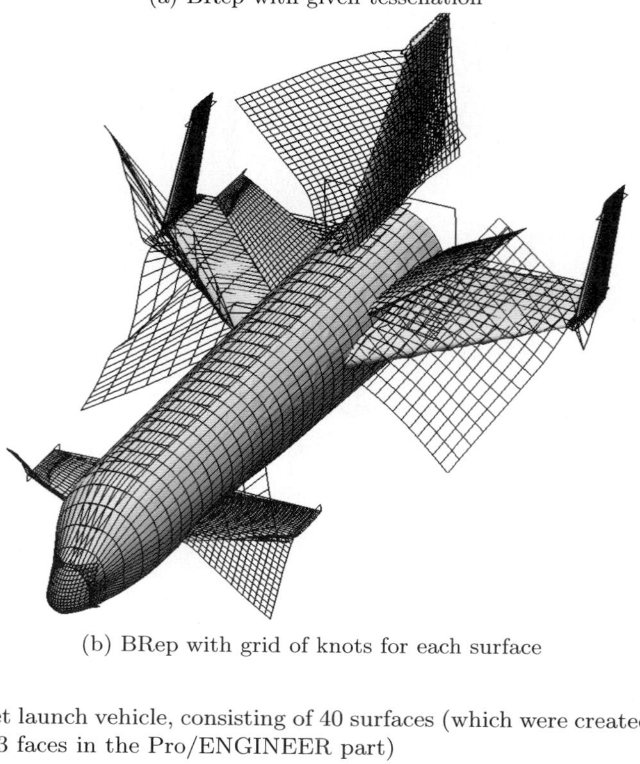

(b) BRep with grid of knots for each surface

Fig. 6. Rocket launch vehicle, consisting of 40 surfaces (which were created by merging the original 83 faces in the Pro/ENGINEER part) that most of the knots match the original configuration to within the specified tolerance, which was set to 10^{-4} times the configuration size. One can also see the extrapolation beyond the edges of the configuration that is necessary in order to have a rectangular grid of knots for each surface; this is especially apparent forward of the wing, canard, and empenage leading edges in Fig. 6. The total CPU time needed for the two cases was 4.3 and 49.9 seconds, respectively, on a MacBook-Pro with a 2.6 GHz Intel Core 2 Duo processor running under OSX 10.5.

6 Summary

A new process for generating a cubic spline surface that approximates, to within a user-specified tolerance, a given tessellated surface that may be non-convex or multiply-connected is described. The method combines a local least-squares technique for specifying knot properties as well as an adaptation technique of selecting the necessary knot spacings.

This new technique is fully described along curves, with all necessary equations. It is then extended to the case of the general surface; here the key equations are given together with an explanation as to how to derive the other equations by extending the curve method. Also, a reparameterization technique that is required for surfaces with poorly-defined parameterizations is described.

The new technique has been applied to over 100 configurations, composed of over 2500 surfaces. In all cases, the method was able to find spline approximations that were within the user-specified tolerance. Computed results for two configurations are shown.

The development of this technique was motivated by the need to create an approximation for surfaces that were created by merging other surfaces. However, this technique has wider applicability, such as to support retriangulation of surfaces, multi-disciplinary coupling, and behavior morphing.

Note that the development here used Ferguson splines as the approximating function, which are a subset of general NURBs. Therefore this technique provides a natural way for generating a NURB representation of a tessellated surface.

Acknowledgements

This work was funded by the NASA Fundamental Aeronautics Program under Cooperative Agreement NNX07AV29A. May-Fun Liou is the Technical Monitor.

References

1. Farin, G.: Curves and Surfaces for Computer Aided Geometric Design, 2nd edn. Academic Press, London (1990)
2. Piegl, L., Tiller, W.: The NURBS Book, 2nd edn. Springer, Heidelberg (1997)
3. Jupp, D.L.B.: Approximation to Data by Splines with Free Knots. SIAM Journal of Numerical Analysis 15(2), 328–343 (1978)

4. Eilers, P.H.C., Marx, B.D.: Splines, Knots, and Penalties (unpublished, 2004), `http://www.stat.lsu.edu/faculty/marx/splines_knots_penalties.pdf`
5. Schwetlick, H., Schutze, T.: Least Squares Approximation by Splines with Free Knots. BIT Numerical Mathematics 35(3), 361–384 (1995)
6. Ruppert, D.: Selecting the Number of Knots for Penalized Splines. Journal of Computational and Graphical Statistics 11(4), 735–757 (2002)
7. Ferguson, D.: Construction of curves and surfaces using numerical optimization techniques. Computer Aided Design 18(1), 15–21 (1986)
8. Press, W.H., Teukolsky, S.A., Vetterling, W.T., Flannery, B.P.: Numerical Recipes in Fortran, 2nd edn., pp. 71–82. Cambridge University Press, Cambridge (1992)
9. Floater, M.S.: Mean value coordinates. Computer Aided Geometric Design 20, 19–27 (2003)
10. Haimes, R., Aftosmis, M.J.: Watertight Anisotropic Surface Meshing Using Quadrilateral Patches. In: 13th International Meshing Roundtable (2004), `www.imr.sandia.gov/papers/imr13/haimes.pdf`

Conformal Assembly Meshing with Tolerant Imprinting

Brett W. Clark[1], Byron W. Hanks[2], and Corey D. Ernst[3]

[1] Sandia National Laboratories*, P.O. Box 5800, MS 0376
Albuquerque, New Mexico, 87185-0376
bwclark@sandia.gov
[2] Sandia National Laboratories, P.O. Box 5800, MS 0376
Albuquerque, New Mexico, 87185-0376
bwhanks@sandia.gov
[3] Elemental Technologies, Inc., 17 North Merchant Street, Suite 3
American Fork, Utah, 84003
corey@elemtech.com

Abstract. Solid assembly meshing has all of the same "dirty geometry" induced issues as single part meshing but also has the difficulty associated with generating a conformal mesh between solids where solid-solid interfaces are not obvious. Mesh generators usually don't have CAD assembly constraint information to identify interfacing solids and must therefore rely on geometric proximity to deduce these interactions. "Slop" in the positioning and alignment of parts in the assembly makes automatically discovering the interfaces and generating a conformal mesh at the interfaces very difficult. Most of the efforts in this area resort to some sort of discrete representation to deal with these issues losing the capability to do further solid modeling engine operations often necessary for all-hexahedral meshing. This paper presents a method for defining the non-manifold interfaces between volumes in an assembly required for generating a conformal mesh while maintaining the original solid modeling engine format.

1 Introduction

Conformal assembly meshing is a common requirement for finite element analysis of complex assemblies of solid parts. However, the solid model assemblies provided as input for mesh generation often contain inaccuracies or sloppiness resulting in a lack of clean or clear interfaces between adjacent volumes. Without clear definitions of the interfaces between adjacent volumes it is very difficult to generate a conformal assembly mesh.

In some cases small modifications can be made to individual volumes in the assembly to improve the mating interfaces between volumes but in general this will not work because of the global nature of modifying B-rep models composed of non-facet based surfaces. As a result, most efforts to solve this problem have

* Sandia is a multiprogram laboratory operated by Sandia Corporation, a Lockheed Martin Company, for the United States Department of Energys National Nuclear Security Administration under contract DE-AC04-94AL85000.

been made in some kind of a faceted representation whether it is a faceted representation of the input solid model or in the generated mesh itself (if the mesh is tetrahedral). In either case, very local modifications can easily be made to the facets for the desired effect. The drawback, however, of going to a faceted representation is the loss of the ability to do subsequent Boolean-type solid modeling operations on the volumes of the assembly. For tetrahedral meshes this may not be necessary, but for all-hexahedral meshing there is often decomposition of the volumes that is required prior to applying meshing algorithms. Therefore, it is desirable to be able to define the interfaces between adjacent volumes required for a conformal mesh and still be able to do solid modeling operations on the volumes for decomposition.

This work describes a "tolerant imprinting" capability and supporting tools that allow the definition of a clean interface between adjacent volumes in an assembly without having to go to a faceted representation. The result is the ability to stay within the solid modeling engine representation throughout allowing downstream solid modeling operations such as volume decomposition.

Section 2 will discuss related work in this area. Section 3 will describe conformal assembly meshing. Section 4 will discuss the difficulties in generating conformal assembly meshes. Section 5 will describe imprinting and its role in facilitating conformal assembly meshes. Section 6 will describe the tolerant imprinting algorithm. Section 7 describes some tools for determining the appropriate tolerance to use when tolerant imprinting. Section 8 will give some examples.

2 Related Work

White et al. [1] presented an algorithm that facets the bounding curves of the original surfaces at the interface of two volumes and generates an intersection graph between the two sets of facets using a user-specified tolerance. The resulting intersection graph defines the topology of the interface surface(s). "Virtual geometry" is used to represent the resulting interface surfaces and appropriate imprints on those surfaces. One advantage that White's approach has in common with the authors' approach is that by modifying the geometry to be meshed (as opposed to meshing the geometry and then modifying the mesh to make it conformal) the modifications are done only once and then multiple mesh instances can be generated for the modified geometry as opposed to doing the modifications each time a mesh is generated. A drawback of White's approach is the inability to do solid modeling engine operations on the geometry after the "virtual geometry" is applied.

Chouadria and Veron [2] showed an approach that works on polyhedral assemblies. The interfaces between adjacent polyhedral volumes were automatically discovered given a user-specified tolerance and then appropriate interfaces were generated by imprinting the interface faces onto one another. The adjacent polyhedral volumes were finally stitched up to the generated interface to create a non-manifold assembly. Although a related work, it does not help solve the problem of obtaining a non-manifold assembly while maintaining the original solid modeling format.

Various commercial codes advertise the ability to generate conformal solid assembly meshes [3, 4, 5]. However, the details of how this is accomplished are proprietary. Despite the apparent advancements in the area of commercial conformal tetrahedral assembly meshing there does not appear to be a solution for hexahedral meshing.

3 Conformal Meshes

In a conformal finite element mesh of an assembly, the solid-solid interface between two volumes is represented by a set of faces such that each face on the interface is a sub-element of the two adjacent solid elements corresponding respectively to the two volumes. Figure 1 A shows a conformal mesh and figure 1 B shows a non-conformal mesh of two volumes. Having a conformal mesh avoids the need for contact elements or additional constraint equations that must be solved with the analysis. This generally reduces the compute time when running an analysis. For some types of analysis, the lack of appropriate contact element algorithms may prevent achieving a solution without a contiguous mesh.

One approach to achieve a conformal assembly mesh is to define a non-manifold representation of the assembly model in which adjacencies between parts in the assembly are explicitly represented as shared or "merged" topology. The B-rep solid for each part contains a topological face representing the coincident boundary region of each part. To merge these two parts, a non-manifold representation of the combined parts is created and the two faces are replaced with a single shared face. Child entities of the two faces are also merged so that edges and vertices that bound the face are shared between the two parts.

During mesh generation, the shared face will be meshed before the volume meshes are created. Since the face is referenced by both volumes, the mesh on the shared face will be used in each volume mesh, resulting in a contiguous or conformal mesh. Figure 2 shows the shared mesh on one of the assembly volumes.

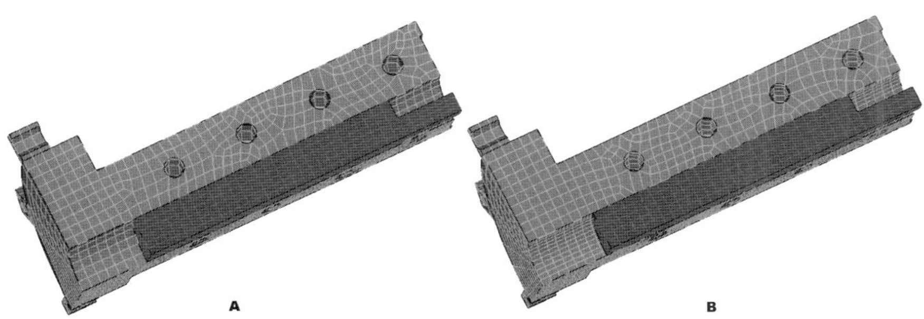

Fig. 1. Conformal and non-conformal assembly meshes

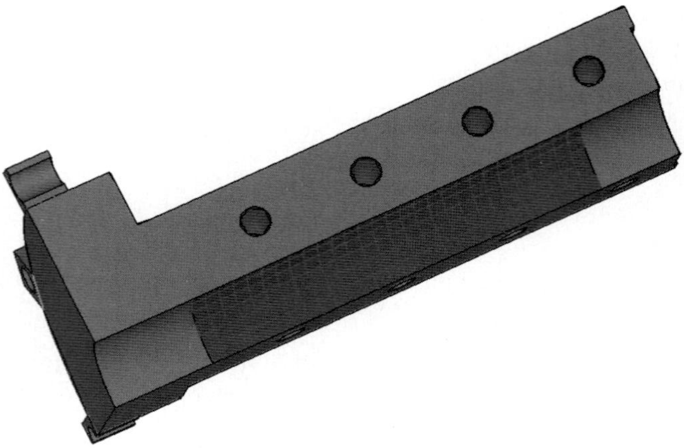

Fig. 2. Shared mesh between two volumes

It will be helpful here to discuss the requirements for merging to occur. At a minimum, the two topological subtrees to be merged must match exactly. Faces to be merged must have the same number of loops; corresponding loops must have the same number of edges; and corresponding edges must have the same number of vertices. Given such a topological match a merge can be forced, regardless of the geometric shape if desired. However, merging is usually only performed when the corresponding geometrical entities - the surface for each face, the curve for each edge, and the coordinates for each vertex - lie within a specified tolerance. This tolerance is referred to as the merge tolerance.

In general, the matching topology does not exist in an assembly where merging should occur between adjacent parts. To generate this topological match, a boolean imprint operation is used. The imprint operation is discussed further in Section 5.

The imprinting and merging approach described here is used in the CUBIT mesh generation software [6] to achieve conformal meshes. Other approaches to achieve a conformal mesh are mesh merging and mesh mirroring. Mesh merging can be manual, where the user is responsible to make the meshes match and merge the required nodes, or automatic, where an algorithm determines edge swaps and node insertions to make the meshes conformal. Mesh mirroring requires an imprint similar to mesh merging, but instead of a non-manifold data structure underlying the mesh, the mesh is mirrored to the corresponding geometry of an adjacent part such that the resulting mesh is conformal (see [1]).

4 Problem Definition

4.1 Assembly Models

Assembly models from a CAD system are typically represented with a B-rep solid model, or a reference to a solid model, for each part in the assembly. To position these parts within the assembly, a transformation matrix is kept for each part to maintain the transformation from the local coordinate system of the part to the global coordinate system of the assembly. The transformation matrix is often calculated from a set of mating relationships that define the position of a part with respect to other parts in the assembly. In calculating the transformation from the given mating relationships, each part is treated as a rigid body and the relationships are used to constrain the six degrees of freedom of the body.

In the general case, assembly constraint information is insufficient to capture the adjacencies within an assembly model. For example, if one face of part A should mate with two separate but coplanar faces of an adjacent part B, only one mating constraint between the face of part A and one of the coplanar faces of part B will be defined. The other mating constraint would be redundant and is generally not created by the user. Determining adjacencies therefore requires a calculation using geometric proximity to find all solid-solid interfaces in the assembly.

4.2 Problems with Solid-Solid Interfaces

Since assembly modelers focus on rigid-body positioning of parts and not on modeling the adjacencies between parts, it can be challenging to produce a model where the adjacencies are captured cleanly. Slight overlaps or gaps can easily occur between parts, or adjacent parts can be slightly misaligned. Gaps between parts may also be the result of excluding welds or adhesives from the assembly model. These situations can cause problems in the proximity calculation.

When an assembly model is translated from one format to another, different tolerances used by different modeling software can also cause problems. If an assembly is modeled in software that uses a large tolerance, corresponding geometry may lie within tolerance and be considered coincident. However, if this same model is translated into another modeler that uses very tight tolerances, geometry considered coincident by the first modeler may now lie outside of tolerance in the second modeler.

Model simplification for analysis purposes can also lead to problems in the solid-solid interfaces of the assembly. The assembly model may contain details which are geometrically accurate, but represent too much detail for the desired analysis. The analyst may simplify the model, which again can introduce inaccuracies or approximations in the solid-solid interfaces.

5 Imprinting

Merging has been described as a way in CUBIT to create a non-manifold representation of an assembly model for conformal meshing. The merging of two surfaces requires that the surfaces and their children entities (curves and vertices) match both topologically and geometrically. Imprinting is a common process for introducing topology (vertices and curves) from one entity onto another and is provided by most solid modeling engines. Figure 3 shows two adjacent volumes

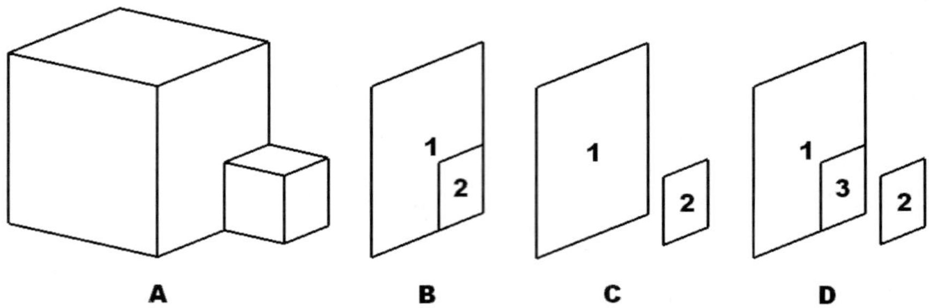

Fig. 3. Imprint example: A) two adjacent blocks, B) adjacent surfaces on blocks, C) exploded view of adjacent surfaces, D) exploded view of surfaces after imprint

and their overlapping surfaces. The surfaces lie right on top of each other but the boundaries of the surfaces don't match. By imprinting the surfaces (or volumes) onto one another curves are introduced into the larger surface creating a geometric and topological match between two of the surfaces as seen in Figure 3 D. After the imprint the two matching surfaces can be merged into one surface that is shared between the two volumes.

Imprinting is a solid modeling operation that uses a tolerance provided by the solid modeling engine. CUBIT's default solid modeling engine uses a tolerance of 1e-6 during imprinting. This tolerance is used to determine what should or should not be imprinted during an imprint operation. If the two volumes in Figure 3 had a gap between them that was larger than 1e-6 no imprinting would have taken place. Similarly, consider the two volumes in Figure 4 A that are misaligned by a very small amount. The adjacent surfaces between the volumes have a distance of zero between them so they are candidates for imprinting but depending on the size of the misalignment the results of the imprint operation will differ (see Figure 4).

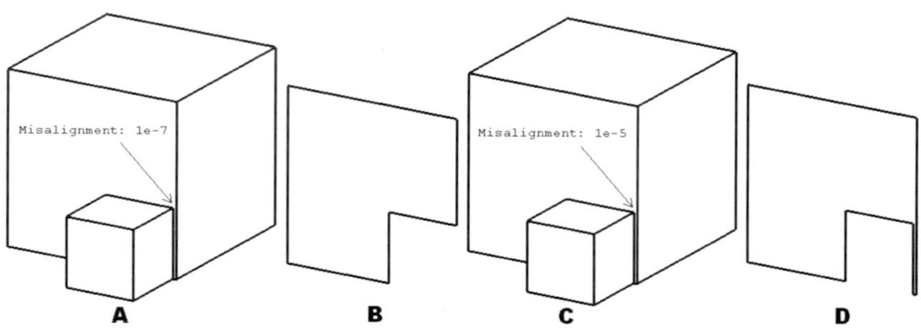

Fig. 4. Misalignment example: A) misalignment less than 1e-6, B) imprint result, C) misalignment greater than 1e-6, D) imprint result

The tolerance used when imprinting guarantees two things: 1) that entities will only be imprinted onto one another if they are within the imprint tolerance and 2) no features smaller than the imprint tolerance will be introduced into the model during the imprint operation. When considering an assembly in CUBIT (using the default solid modeling engine) this means that volumes must be within 1e-6 of one another to be imprinted and volume misalignments greater than 1e-6 will result in imprints and possibly sliver entities with size equal to the misalignment. Assembly constraints in CAD packages are often much looser than 1e-6 which can result in imprinting nightmares. For this reason "tolerant imprinting" has been developed.

6 Tolerant Imprinting

Tolerant imprinting is a combination of functionality provided by CUBIT and imprinting functionality provided by the solid modeling engine it uses. From the user's perspective tolerant imprinting is identical to regular imprinting in behavior except that it provides the user with a modifiable tolerance to be used during the imprinting process. This allows the user to set the imprint tolerance to be greater than the 1e-6 value provided by the solid modeling engine. As a result the user can set the imprint tolerance to match more closely the gap/misalignment sizes in his assembly and get the desired imprinting results.

The tolerance used in tolerant imprinting is the user-modifiable "merge tolerance" in CUBIT. Analogous to the way the solid modeling engine uses 1e-6 to determine what should or should not be imprinted CUBIT uses its merge tolerance value set by the user to decide what should or should not be imprinted. The tolerant imprinting algorithm has three main parts: 1) imprint vertices onto curves, 2) imprint curves onto surfaces, and 3) imprint vertices onto curves again.

6.1 Imprint Vertices onto Curves

The first step in the algorithm is to imprint curves with vertices. The algorithm will first search for all pairs of curves that overlap by at least merge tolerance.

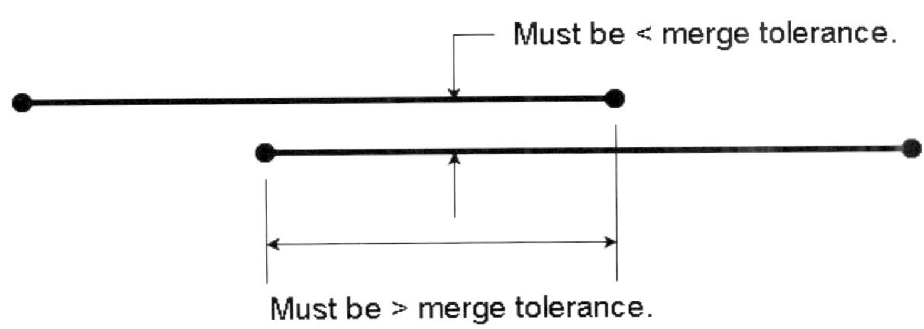

Fig. 5. Overlapping curve criteria

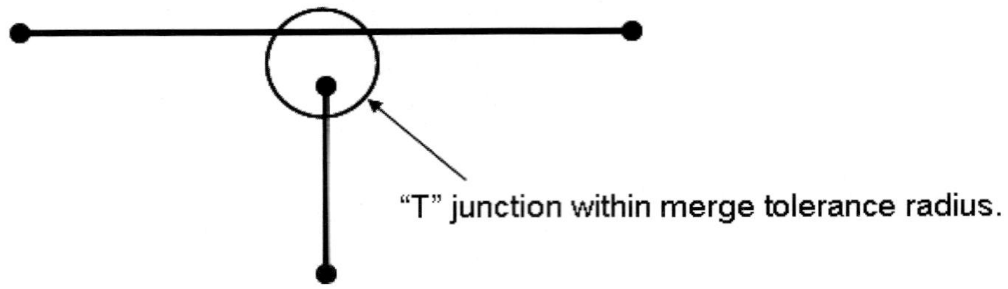

Fig. 6. T-junction criteria

Overlapping in this case means that the curves lay on top of one another spatially and have an overlapping region at least merge tolerance long (see Figure 5). Once these pairs are found each curve is imprinted by the vertices of the other curve where applicable. A common split curve operation provided by the solid modeling engine is used. "T" junctions (see Figure 6) between curves are also found and appropriate imprints are made there as well.

6.2 Imprint Curves onto Surfaces

The second part of the algorithm imprints curves onto surfaces. The algorithm will search for all pairs of surfaces that overlap within merge tolerance. Overlapping surfaces are defined as surface pairs that lay on top of one another spatially within merge tolerance and which have an overlapping region with area of at least merge tolerance squared. Once the overlapping surface pairs are found each curve from one surface is imprinted onto the other surface and vice versa. This part of the algorithm takes advantage of an api from the solid modeling engine that "tolerantly" embeds a curve into a surface. The advantage of using this api is that it will automatically lengthen/shorten the curve (within a specified tolerance) to match up with existing topology in the surface. In actuality it doesn't change the length of the curve but treats it as tolerant so that it can attach to existing topology in the surface. The default solid modeling engine in CUBIT provides this type of functionality but it would need to be implemented in CUBIT if the tolerant imprinting algorithm were ported to another solid modeling engine that didn't have this functionality.

It should be noted that when considering the curves to be imprinted onto a surface curves from one surface that are "overlapping" with curves in the other surface are excluded from the imprint operation as they have already been imprinted in the first step. One potential implementation would be to not do the first imprint overlapping curves step described above and just rely on the tolerant embed functionality to imprint the overlapping curves during the imprint overlapping surfaces step. However, the authors found that the tolerant embed functionality provided by the solid modeling engine was not robust in doing this and as a result the first step of imprinting overlapping curves is used.

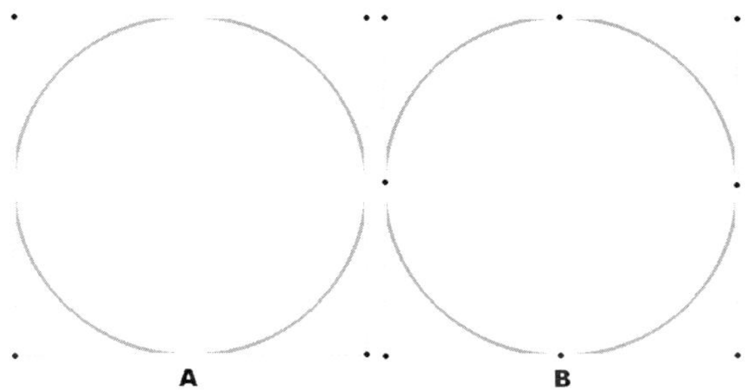

Fig. 7. Circular surface fitting tangentially inside a square surface

6.3 Imprint Vertices onto Curves Again

Finally, overlapping curves are once again imprinted in order to catch any remaining vertex-curve imprints that need to take place because of imprinting curves onto surfaces. One might expect that this final step is unnecessary. However, there are some geometry configurations that cause new vertices to be introduced during the "Imprint Curves Onto Surfaces" step that were not introduced during the first "Imprint Vertices Onto Curves" step and that subsequently need to be imprinted onto curves. One such configuration is a circular surface that fits tangentially into a square surface (see Figure 7 A). Vertices at the tangent locations are not introduced during the first "Imprint Vertices Onto Curves" step because the curves are not overlapping within tolerance. During the "Imprint Curves Onto Surfaces" step the curve of the circular surface is imprinted onto the square surface creating breaks in the curves of the square surface at the tangent locations. However, because the curves on the square surface are outside the circular surface they are not imprinted onto the circular surface so the circular curve does not get split. This requires the final "Imprint Vertices Onto Curves" step. The result is shown in Figure 7 B.

Just as regular imprinting will not introduce features smaller than 1e-6, tolerant imprinting will not introduce features smaller than merge tolerance.

7 Determining a Good Merge Tolerance

Determining an appropriate merge tolerance is the main key to effectively using tolerant imprinting. Too small a merge tolerance will result in missed imprints and too large a merge tolerance will result in unintended imprints. Unfortunately, automatically determining a merge tolerance for a given assembly is difficult without a user in the loop. Because of the variations in gap and misalignment sizes within a given assembly it is difficult to automatically determine what volumes should or should not be "touching." Therefore, in most cases the user needs to specify the merge tolerance after an examination of the assembly. Below are some important characteristics to consider when examining an assembly and

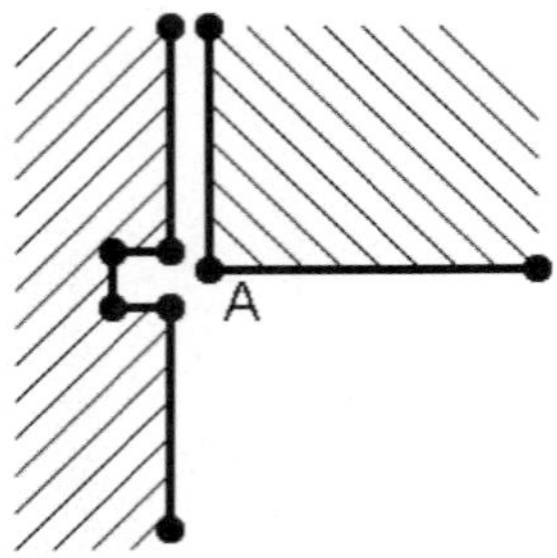

Fig. 8. Ambiguous merge example

a proposed process implemented in CUBIT to aid the user in determining a suitable merge tolerance.

7.1 Smallest Feature Size

The smallest feature size in the model helps to limit how large the merge tolerance can be. If the merge tolerance is larger than or on the order of the smallest feature size there can be ambiguities as to what should be merged between two volumes. For example consider the geometry configuration in Figure 8. The volume on the right contains vertex A. The volume on the left has a small feature near vertex A. Depending on the size of merge tolerance it will be unclear which vertices on the left volume could possibly be merged with vertex A. Merging will only allow one of the vertices from the volume on the left to merge with vertex A but there may be more than one candidate vertex based on the value of merge tolerance. To avoid these types of ambiguities merge tolerance should not be as large as the smallest feature the user wants resolved in the mesh. If gaps/misalignments in the model require a merge tolerance larger than the smallest feature the model really needs to be cleaned up with tighter assembly constraints or small features need to be removed.

Because of the potential ambiguities described above it is important to first know the size of the smallest feature you want to represent in the assembly. CUBIT provides tools for either finding the n smallest features in the model or for finding all of the features with a size less than some user-specified feature. When first becoming familiar with the model the user does not usually know a small size to specify so finding the n smallest features is very useful. This tool looks for proximities between different topological entities (vertex, curve, surface) within a given volume instead of just looking for "small curves" or "small surfaces." This was deemed a more general search. It is also important to note that this search is done on a per-volume basis; it is not a tool for finding proximities between two volumes. Figure 9 shows an example of the tool. In this example Vertex 42 and Curve 57 were found to be 0.254 units apart and are highlighted in orange in the graphics window. The user can select results from the search and zoom to

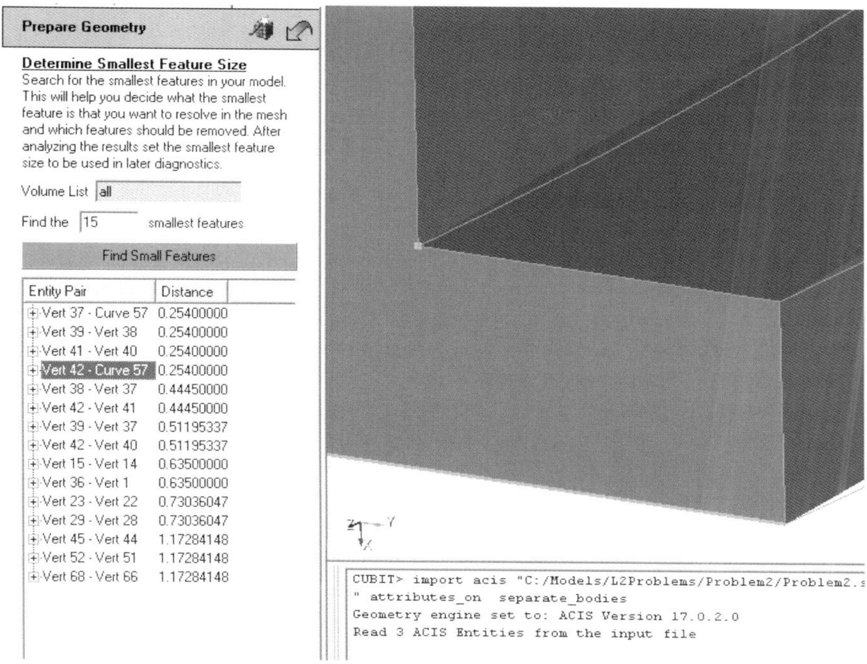

Fig. 9. Find smallest features tool

them quickly to determine which small features are intentional in the model and which ones need to be removed (sliver entities, etc.). Looking ahead to tolerant imprinting it is important to remove all features that are smaller than merge tolerance to eliminate any ambiguities when merging. Obviously, the user won't know the merge tolerance at this point but he can look at the small features that are in the model and remove any that are considered unnecessary for the analysis.

7.2 Merge Tolerance Estimator

Once the smallest feature size is established CUBIT provides a tool for automatically estimating the merge tolerance given the smallest feature size. This is useful in giving the user a starting point for the merge tolerance that can then be fine tuned. The estimator relies on there being a fairly consistent average gap/misalignment between volumes in the assembly. Obviously, this does not always hold true under which circumstances the estimation is not very good but in most cases there is a natural value at which most of the merging can take place in the model and after which not many new merges will take place. This can be

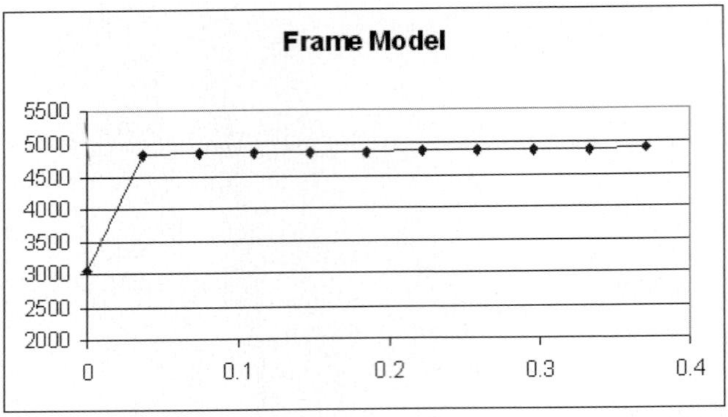

Fig. 10. Plot of number of merges vs. merge tolerance

seen in the plot in Figure 10 of values calculated from a real assembly containing about 900 volumes. The horizontal axis plots increasing merge tolerance values and the vertical axis plots the number of merges that would take place at that merge tolerance. Where the plot flattens out is usually a good guess of where the merge tolerance should be. The automatic estimator generates data similar to that in Figure 10 for the assembly and picks off a value where the plot starts to flatten out. To reiterate, this is only a rough estimation but usually gives a good starting point for the user.

7.3 Proximate Features

After an approximate merge tolerance has been decided upon either by the user or by the automatic estimator it is prudent to examine the assembly to see if the merge tolerance will be suitable to capture all of the desired merges. CUBIT provides a tool for doing this examination. The user can define a range around merge tolerance and search for entity-entity proximities in the assembly that fall within the range. For example, if the estimated merge tolerance was 0.005 the user could look at a range above merge tolerance, 0.005 to 0.05, to see if there are any entity pairs in that range that he would want to merge but that wouldn't because they are out of merge tolerance. Similarly, he could then look at the entity-entity proximities below merge tolerance, in the range 0.0005 to 0.005 for example, to see if there are some entities that would merge that really he wouldn't want to merge. In this way the user can fine tune what merge tolerance should be to capture the desired merges.

Figure 11 shows an example of the tool used to search a range for entity-entity proximities. Again, the tool allows the user to quickly zoom in on entity pairs to determine if merge tolerance should be adjusted.

Once the merge tolerance has been fine-tuned the user can run tolerant imprinting and then merging.

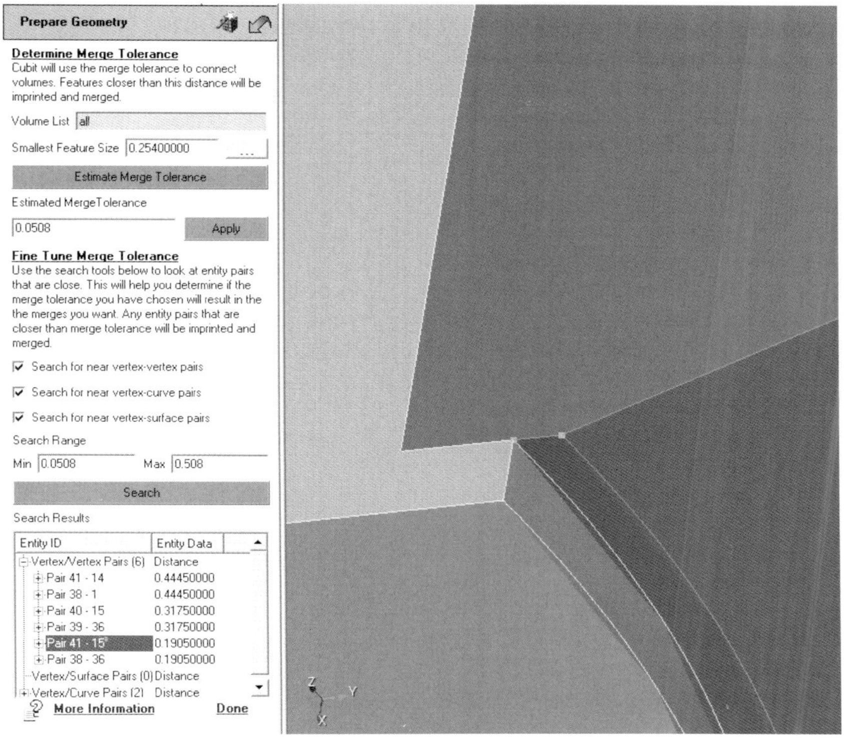

Fig. 11. Tool for determining merge tolerance

8 Example

The real benefit of the tolerant imprinting capability is the time saved in preparing a non-manifold assembly. Depending on the sloppiness and size of the assembly it can easily take days and even weeks to manually modify the assembly to have clean interfaces between adjacent volumes. Following is an example of the impact of having tolerant imprinting.

This example is of an AF&F (arming fusing and firing) assembly consisting of 240 volumes. This example is an Official Use Only model so pictures of the assembly are not included. The assembly is very compact in nature with a high number of volume-volume interfaces. This model is also considered quite "sloppy" in that many of the interfaces are out of tolerance when using the default imprinting tolerance of 1e-6. Two runs were made, first with regular imprinting and merging and then with tolerant imprinting and merging. Table 1 shows the results. The results show the time required during the imprinting step and the resulting number of overlapping surface pairs after imprinting and merging. An overlapping surface pair indicates two surfaces that did not merge correctly during the merge process. Overlapping surface pairs are usually resolved by the user

Table 1. AF&F data

	Time (min)	# Overlapping Surface Pairs
Regular Imprinting	3.5	169
Tolerant Imprinting	56	2

manually fixing the model so that the merge can take place. As can be seen in the results tolerant imprinting takes much longer to run. However, after running tolerant imprinting there are only 2 overlapping surface pairs to fix manually. Regular imprinting and merging resulted in 169 overlapping surface pairs that would each have to be analyzed and resolved manually. The time required to do this would easily outweigh the extra time required to run tolerant imprinting.

9 Conclusion

This paper has described the tolerant imprinting capability and other tools that can be used to generate a non-manifold representation of a volume assembly while maintaining the native solid modeling format of the assembly. This has the advantage of allowing the user to do subsequent solid modeling operations on the assembly (such as decomposition) to facilitate meshing. The real benefit of the tolerant imprinting capability is the time saved when the user does not have to modify the individual volumes in the assembly to define clean interfaces between adjacent volumes. An example of using tolerant imprinting on a complex assembly was given showing this potential time saving.

References

1. White, D.R.: An imprint and merge algorithm incoporating geometric tolerances for conformal meshing of misaligned assemblies. International Journal for Numerical Methods in Engineering 59, 1839–1860 (2004)
2. Chouadria, R., Veron, P.: Identifying and re-meshing contact interfaces in a polyhedral assembly for digital mock-up. Engineering with Computers 22, 47–58 (2006)
3. ALGOR, `http://www.algor.com`
4. Patran, `http://www.mscsoftware.com`
5. COSMOSWorks, `http://www.solidworks.com/pages/products/cosmos/cosmosworks.html`
6. CUBIT, `http://cubit.sandia.gov`

Automated Mixed Dimensional Modelling with the Medial Object

Trevor Robinson[1], Robin Fairey[2], Cecil Armstrong[1], Hengan Ou[1], and Geoffrey Butlin[2]

[1] School of Mechanical & Aerospace Engineering, The Queen's University of Belfast, Belfast, UK
[2] Transcendata Europe Ltd., Cambridge, UK

Abstract. This paper describes an automatic method for generating analysis models to be meshed with finite elements of more than one dimension, known as mixed dimensional models. Mixed dimensional models offer much reduced analysis times, while not compromising simulation accuracy to the same extent as fully dimensionally reduced models composed of 2D elements stiffened using beam elements, which are currently utilised in the aerospace industry. The techniques described make possible the automatic generation of mixed dimensional models directly from CAD, allowing for rapid iteration during early design.

1 Introduction

Fully detailed CAD models are often unsuitable for direct analysis. Analysis models can be produced from CAD using a simplification or transformation process such as defeaturing, in which small details are suppressed, or dimensional reduction. This paper will refer to all such transformations as 'idealisation'. The current idealisation method of choice in the aerospace industry is to create a stiffened shell model; where the mid-faces of the thin sheets in a component are shell meshed, and beam elements added to ensure the model has the appropriate stiffness. This reduces the original solid geometry to a collection of thin sheets, and is known as dimensional reduction since the model is now represented by entities of lower manifold dimension.

It has been demonstrated elsewhere [1] that it is beneficial to idealise thin walled structures for finite element analysis, especially during the early stages of the design process, since idealised models can be analysed more quickly because the number of degrees of freedom in the model is reduced. Most commercial finite element pre-processors constrain the continuum elements used to mesh thin regions to have lateral dimensions similar to the thickness of the region such that a large number of very small elements are required for large thin sheet regions. When a thin sheet region is approximated using its mid surface the thickness of the region does not constrain the size of shell elements used to mesh it. This means much larger elements can be used, which greatly reduces the number of elements, and therefore the number of degrees of freedom in the model. It has also been shown that shell elements are more accurate and have better numerical conditioning in thin sheet regions when compared to continuum elements.

Although stiffened shell models are extremely efficient to analyse, they do not have the ability to appropriately model the behaviour in areas with complex geometrical detail. Also, the preparation of these analysis models is extremely time consuming. According to industrial sources the idealisation process can take several months for some components. The quality of the idealisation is dependent on the skill of the analyst who creates it. It is their engineering judgment which is used to determine the section properties used for stiffeners, and how to treat local features e.g. fillet radii. Stiffened shell models are usually constructed from scratch, and as such are independent of the CAD model of the component. This hinders current ambitions for a fully integrated design process and prevents rapid iterative changes to a design.

This paper presents an alternative method of idealisation in which 3D models of the geometrically complex regions in the component are preserved, while the thin walled structures are identified and dimensionally reduced to their mid faces. The mid faces can then be shell meshed, while the complex regions are meshed using 3-dimensional elements.

Although these models are more computationally expensive than stiffened shell models, they have the ability to appropriately model the behaviour in the complex regions, while still offering considerable savings in analysis time over a fully 3D analysis model. Using the procedures in this paper, models of this kind can be created automatically. Coupling methods to ensure accurate results at the interface between the regions of differing dimension are also discussed.

2 Existing Research and Capabilities

[2] introduces an approach for creating mixed dimensional idealisations from solid models based on dividing a body into simple solids, and then selecting the solids with a thickness less than set value to dimensionally reduce to their mid-face. [3] presents a similar approach, but uses a different technique for creating the mid-faces. The procedures in both of these papers use the absolute thickness of a region to determine if it is thin, without taking account of the associated lateral dimensions. This can result in thin regions with small lateral dimensions being inappropriately dimensionally reduced, but regions with a thickness above the target value being retained as solid, even though their lateral dimensions are suitably large for the region to be dimensionally reduced. In this paper a relative measurement known as aspect ratio (AR) is used instead of absolute thickness limits. Aspect ratio is a measure of the size of a region's lateral dimensions relative to its thickness.

The requirement for integrated model idealisation has led to the inclusion of geometric idealisation and dimensional reduction tools in commercial finite element analysis pre-processors becoming more common. Many pre-processors offer "mid surface extraction" tools, allowing the user to extract the mid surface from a solid part, or from between two opposite faces. Some of the more advanced tools also have the ability to extract a surface from a group of user defined surfaces.

The idealisations offered by such tools can be inappropriate. Figure 1(a) shows a part and (b) the same part with the mid surface calculated using a commercial CAD package in grey. Clearly the calculated mid surface does not accurately represent the part. Indeed it is not obvious what form the mid surface should take in the vicinity of the change in thickness. Problems also exist for "T" shaped sections, where the faces from which the mid surface are to be extracted are ambiguous. When used on sections such as this many programs will produce an incorrect or incomplete idealisation. Another issue is that the idealisation tools cannot usually be used for regions within a part, and will reduce all of a component where perhaps only some regions are suitable for idealisation. In Figure 1 a full 3D analysis would be required at the change in section to model the complex local stresses that occur in such regions.

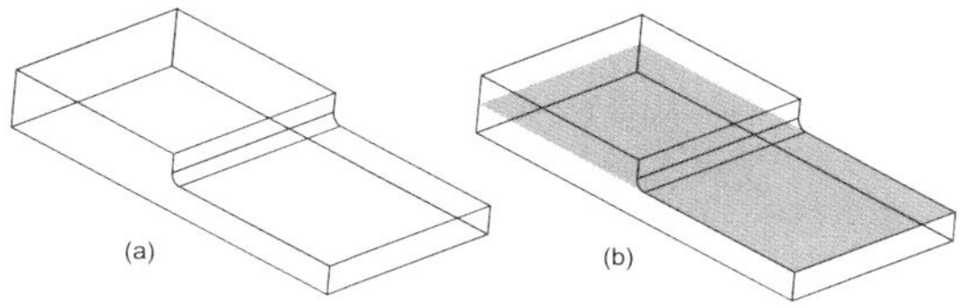

Fig. 1. Part body and part body with idealisation

The aim of the research presented in this paper is to automate the preparation of idealised 3D-2D mixed dimensional models from a 3D CAD model. This will allow the advantages of appropriate idealisation to be realised without the risk and expense of preparing the simplified models manually.

3 Mixed Dimensional Models

Figure 2 and Figure 3 respectively show a 2D aero engine casing profile and a 3D CAD model of an aero engine exhaust casing, along with their mixed dimensional idealisations.

For the 2D model the dimensionally reduced entities are shown as lines; the entities which are not dimensionally reduced are shown as faces. In Figure 3, the dimensionally reduced mid-faces are shown in light grey and the detail features are shown in dark grey.

Critical aspect ratio (CAR) controls how large the lateral dimensions of a region have to be relative to its thickness before it is considered suitable for dimensional reduction. Figure 4 shows the effect of varying critical aspect ratio on the dimensional reduction of the aero engine casing profile. As the critical aspect ratio is reduced the amount of the model which is dimensionally reduced increases. An increase in the amount of the model which is dimensionally reduced corresponds to a further reduction in analysis time; but possible increase in analysis error.

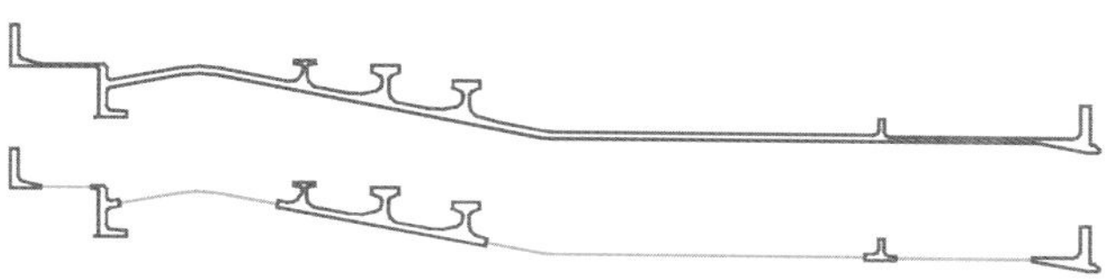

Fig. 2. Complex planar face and its mixed dimensional approximation

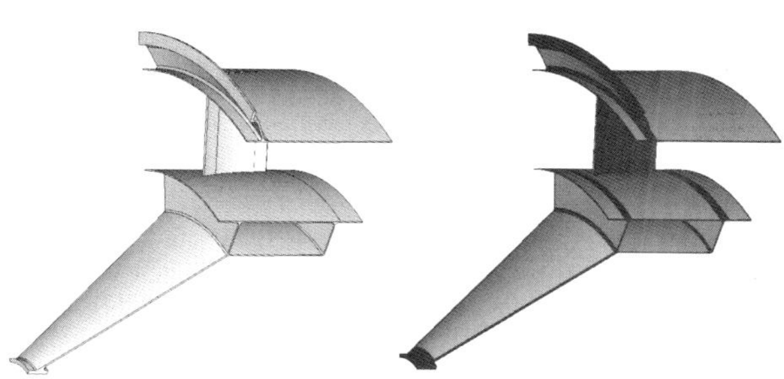

Fig. 3. Complex solid model and its mixed dimensional approximation

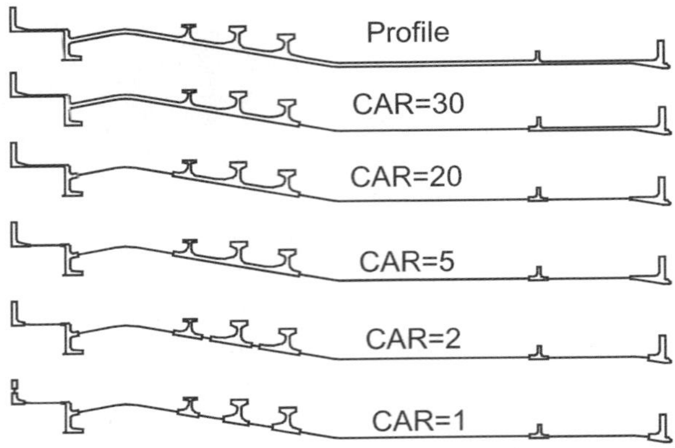

Fig. 4. Variation in critical aspect ratio

4 The Medial Axis Transform (MAT)

The medial axis transform (MAT) or medial object (MO), originally proposed by [4], is a property of geometric shape and is the underlying technology used in the procedures described in this paper to determine where the thin regions occur. The medial axis is a skeleton-like representation of geometric shape comprised of a set of geometric entities which are created by tracing out the centroid of the maximal inscribed disc as it rolls around the interior of a surface, Figure 5 (a); or of the maximal inscribed sphere as it rolls around the interior of a solid, Figure 5(b).

These geometric entities plus the associated radius function form the medial axis transform, which is a complete, unambiguous representation of the original shape.

The 2D MAT (Figure 5 (a)) is made up of medial edges and medial vertices. The edges are swept out by the centre of an inscribed disc that touches the face boundary in two places. Medial vertices occur where the disc touches in three places or more. Radius information is stored at each point on the medial axis, this is the radius function mentioned above. The 3D MAT (Figure 5 (b)) is a non-manifold collection of medial faces, edges and vertices. It is often referred to as the Medial Object or MO since it does not resemble an axis as the 2D case does. A medial face exists where the inscribed sphere is in contact with the boundary of the solid at two distinct points, as is the case for sphere A in Figure 5 (b) Since the inscribed sphere is defined by four quantities, the x, y and z coordinates of its centre plus the radius, the sphere centre has two degrees of freedom and therefore sweeps out a surface. Medial edges occur between medial faces where the inscribed sphere is in contact with the boundary at three distinct points, as in sphere B in Figure 5 (b). Medial vertices bound medial edges and occur where the medial sphere is in contact with the boundary at four distinct points, as is the case for sphere C in Figure 5 (b). Degenerate edges or vertices can occur where the inscribed sphere is simultaneously in contact with more than the minimum number of points on the boundary, for example where the inscribed sphere moves along the neutral axis of a square prism. The inscribed sphere radius reduces to zero at a convex edge of the model. A medial face where this occurs is sometimes known as a medial flap, Figure 5 (b). Medial edges and vertices can also be caused by curvature contact between the inscribed

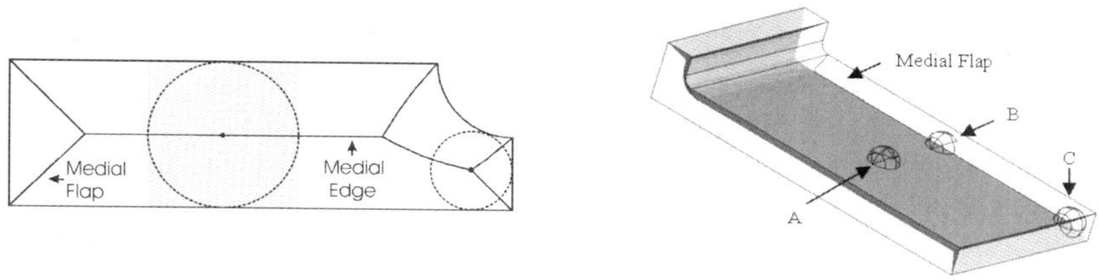

(a) (b)

Fig. 5. The medial axis transform; (a) The 2D medial axis with inscribed discs on a medial edge and at a medial vertex. The region represented by the disc on the medial edge is shown in grey; (b) The 3D medial axis with inscribed spheres on a medial face (A), on a medial edge (B) and at a medial vertex (C).

sphere and the boundary, for example where the sphere moves along the axis of a cylinder or terminates in a spherical surface, [5].

The set of object boundary entities the medial disc is in contact with as it traces each medial entity are known as its defining entities.

The TranscenData Europe Ltd implementation of the MAT [6], commercially available as the Medial Object Toolkit within the CADfix software, was used for the work outlined in this paper.

5 Dimensional Reduction of 3D Models

Thin sheet regions in solid bodies have large lateral dimensions relative to their thickness. The identification of thin sheets occurs in two steps. The first step eliminates all medial faces which are definitely not thin. The remaining medial faces, which may represent thin sheets of material suitable for dimensional reduction, are referred to as candidate medial faces. Candidate medial faces are defined by two object faces and are not medial flaps. For the thin walled example component shown in Figure 6 the candidate medial faces are shaded in Figure 7. The enlarged section of Figure 6 shows the complete medial axis, with additional medial flaps we wish to ignore, while Figure 7 shows only the seven candidate faces.

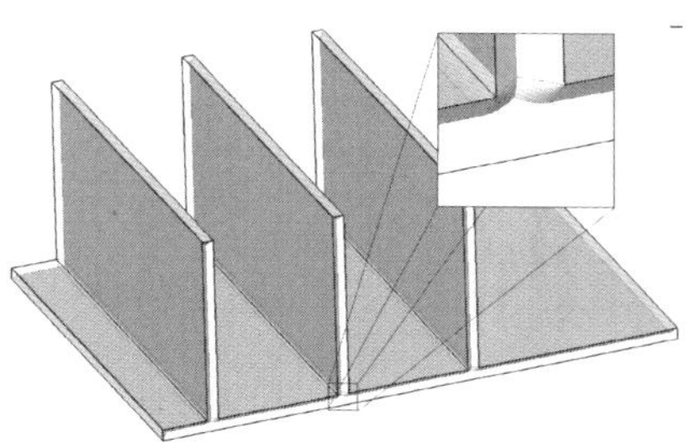

Fig. 6. A thin walled component, with 3D medial axis

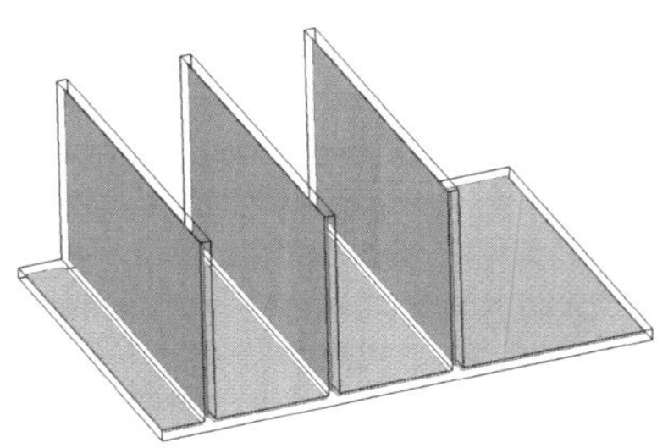

Fig. 7. Candidate thin faces

5.1 Insetting and Aspect Ratio

It is necessary to inset the proposed interface between the thin sheet and detail feature into the thin region to allow the complex stress contours which occur adjacent to detail features to be modelled using continuum elements. It is not possible to accurately model the complex behaviour in these regions using reduced dimensional elements. St Venant's principle [7] suggests that insetting the interface between the thin and complex regions into the thin region by a distance of the order of the region thickness is sufficient to capture the localised complex stresses in linear stress analysis problems.

Figure 8 (a) shows the bending stress for a section of a component which has traverse shear loading applied at one end. Figure 8 (a) shows the linear variation in stress contours through the sheet thickness at a mixed dimensional interface which has been inset by one thickness. In the model in Figure 8 (b) the interface has not been inset, and the resulting nonlinear variation in stress through the thickness at the 3D side is obvious. The stress contours are also not continuous between the top surface of the plate and the adjacent 3D surface. It is worth noting that even for the model where there is no inset the result is still accurate at a distance of one thickness from the mixed dimensional interface.

To effect insetting in 3D, an offset or shelling operation is carried out on the boundary of the thin regions. Shelling can be achieved by calculating the 2D MAT on a 3D medial face, and using the 2D radius function to create an offset profile. This offset profile can then be extruded through the thickness (i.e. projected onto the two defining faces of the medial face on which the offset profile lies) and used to slice the model into a cellular solid, as well as serving as the boundary for the mid-face representing the thin region.

Above, the bold black outline is a candidate 3D medial face that is possibly thin. The solid it represents is not shown. The solid lines in the interior are the 2D MAT of the 3D face. The dashed outline is the offset geometry, which will be extruded perpendicular to the 3D face and used to split the solid. Note that at convex corners of the face, the two offset edges found in that corner meet exactly on the 2D MAT, and that the 2D medial radius at these points is equal to the offset distance (OS). For concave corners, an arc can be inserted, since this is always equidistant from the concave vertex.

This process alone is not sufficient to determine if a medial face represents a slender region, since candidate faces may contain regions which differ in their suitability for dimensional reduction.

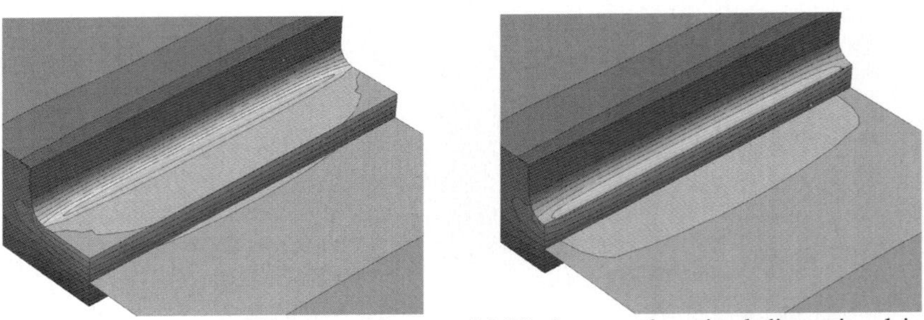

(a) Interface inset by one thickness. Linear stress distribution through thickness.

(b) No inset at the mixed dimensional interface. Non-linear stress distribution through thickness.

Fig. 8. Modelling the interfaces between thin sheets and details

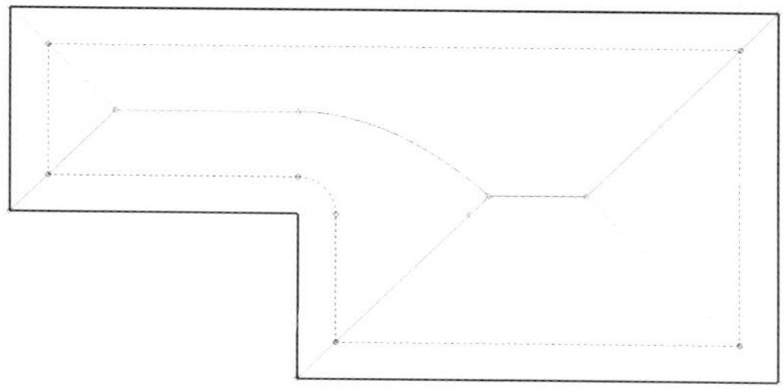

Fig. 9. Offset geometry to effect insetting in 3D

Figure 10 shows, from left to right, an example 3D solid, the single candidate medial face identified as possibly thin, and the resulting dimensional reduction (created using extruded offset geometry as just described). The narrow section does not have large lateral dimensions and should not be reduced; however the remaining section of the face is still suitable for reduction. The aspect ratio is not a property of the whole thin region, and must be calculated at a finer granularity.

The offset geometry is modified by using the 2D MAT to identify regions of suitable lateral dimension. Points are identified on the 2D MAT where the medial radius is equal to (CAR/2 + OS) x D_{3D}. Such a point is shown in Figure 11 (b) using a hollow black circle, and represents the region of the medial face which will have lateral dimensions equal to CAR x thickness after the insetting.

The touching points of these points are created on the defining edges. Vertices are created to represent the endpoints of a spline at the insetting of the touching points, depicted in Figure 11 (b) using filled black circles. A further spline control point (filled grey circle) is created at a distance of CAR/2 x D_{3D} from the radius threshold point (hollow black circle), positioned between the end points. Figure 11 shows this point on the medial edge, but this need not be the case.

Figure 11(a) shows the treatment of a concave corner. This is part of the offset algorithm itself as opposed to corner rounding. Figure 11 (b) shows an example of the process for a convex corner region. The spline is represented by a dashed black line. The inset touching points are used to trim the inset edges. The remainder of the inset edges and the splines are used to create edge loops, which are extruded into cutting faces for partitioning as well as being used to form mid-faces. Figure 11 (c) shows a section of an example medial face. The resulting offset geometry is shown in Figure 11 (d).

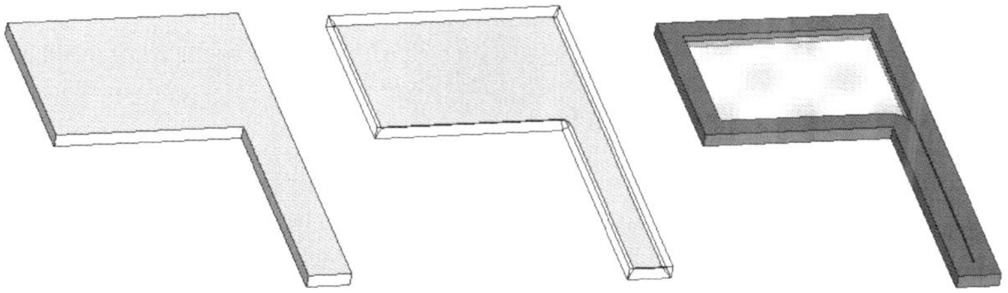

Fig. 10. Desirable/undesirable dimensional reduction in 3D

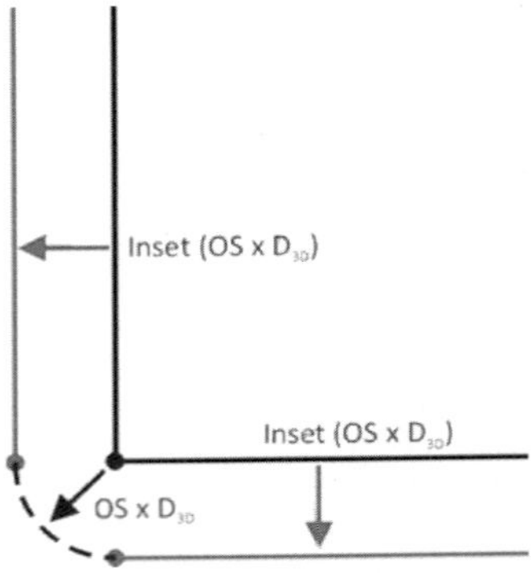

(a) Concave corner

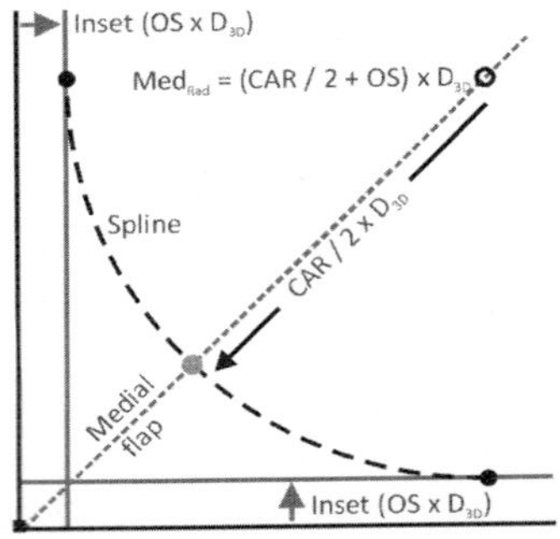

(b) Convex corner. Inset touching points as solid black circles. Hollow black circle where 2D medial radius = (CAR/2 + OS) D3D. Filled grey circle at vertex on medial flap. Spline in dashed black line.

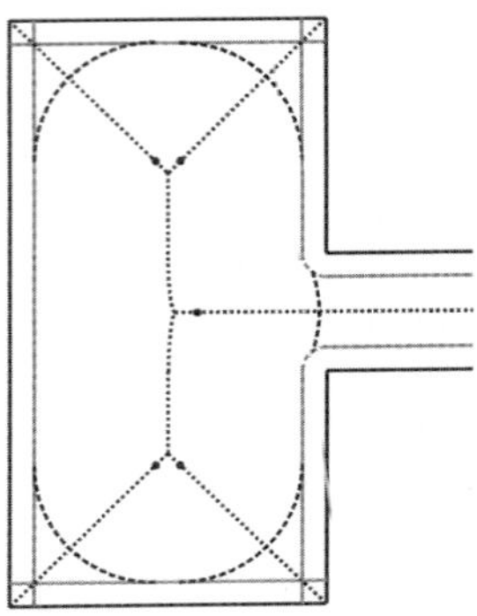

(c) Profile in black, MAT in black dotted lines, inset edges in grey, concave radius in dashed grey line, Splines in dashed black line, black dots where radius = (CAR/2 + OS) x D_{3D}.

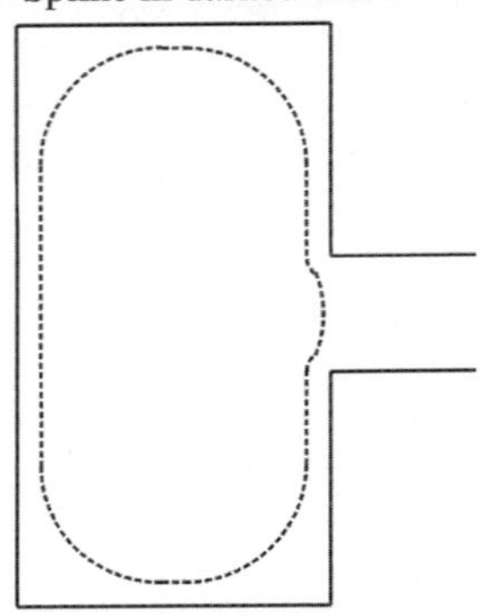

(d) Profile in black, and outline of new mid-face in dashed black lines.

Fig. 11. Creating an appropriate medial face

Variables used in Figure 11 are as follows:

- CAR – Critical aspect ratio (see Figure 4)
- D_{3D} – 3D Medial radius (thickness of the region to be reduced)
- OS – Offset factor (multiplier applied to D_{3D} to determine inset distance)

Although this process is referred to as corner rounding, it is not restricted to operation solely in the corners of the face. The defining entities of the medial edge on which the radius threshold point lies can be used to identify any two pieces of the offset geometry that can be bridged with a spline. In this way, narrow face regions that widen at both ends can be trimmed away as shown in Figure 12. In the top diagram is a dotted outline of the

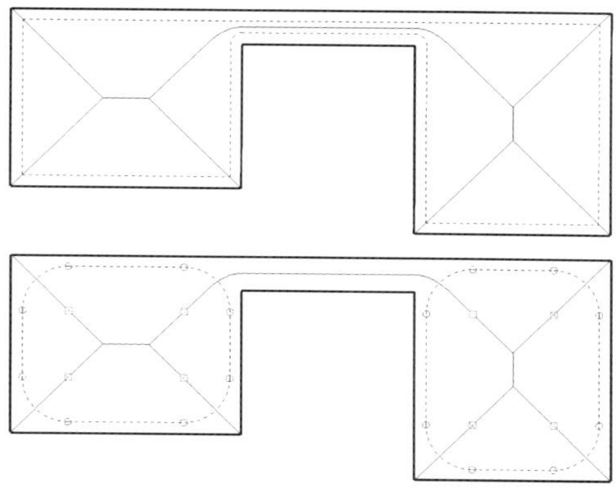

Fig. 12. Eliminating thin necks

offset geometry without corner rounding. In the bottom diagram is the result of corner rounding, showing the spline end points as circles, and the radius threshold points as boxes.

For an example on a 3D part, see Figure 14.

Rounding corners has the added benefit of easing the derivation of coupling equations at the 3D-2D interface, by removing sharp changes of direction in the cut faces (they all meet tangentially).

Using the rich information imparted by the 2D MAT, the desired insetting is achieved, while also calculating a useful aspect ratio measure for the 3D case, and creating tangential cut surfaces.

5.2 Creating the Cellular Model

In 3D a cellular model is created by extruding the offset geometry for each thin sheet mid face onto the defining faces of the region. The resulting cut faces are shown dashed in Figure 13 (c). The resulting cellular model (Figure 13 (d)) is a non-manifold solid, with the cut faces shared between neighbouring cells.

Figure 13 (c) and (d) and Figure 14 below also provide an illustration of the process described in Figure 11, in which the portion of the model above the central rib is considered to have too low an aspect ratio, and the opposing sides are bridged, creating two thin regions.

The mid-face representing the thin sheet region is also stored with the model as an attribute of the appropriate cell. Suppressing the thin sheet solid cells will result in a mixed dimensional model where the mid faces represent the thin sheet regions. Sup pressing the mid-faces and not the solid cells will result in the entities required for a detailed analysis model, but with the possibility of meshing the thin sheet solids using more efficient mesh structures than in the complex regions. Examples of alternative meshing techniques in the thin regions include using P-elements [8] or thick shell elements [9]. Anisotropic solid meshing is also an option (in which the elements have large lateral dimensions compared to thickness).

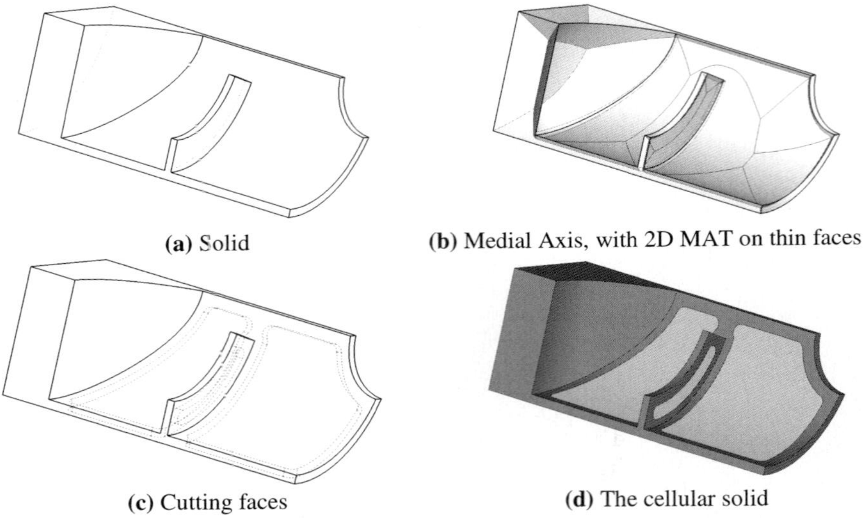

(a) Solid

(b) Medial Axis, with 2D MAT on thin faces

(c) Cutting faces

(d) The cellular solid

Fig. 13.

Fig. 14. Mixed dimensional model

6 Mixed Dimensional Coupling

To analyse mixed dimensional models coupling is required at the interfaces between elements of different dimension. It is the coupling equations that transfer the behaviour at one side of the mixed dimensional interface to the other. Most finite element pre-processors include mixed dimensional coupling tools, but many do not achieve accurate results. This section summarises previous successful research on this topic.

Many coupling techniques are inaccurate, which is demonstrated by the spurious stresses that occur at the mixed dimensional interfaces when they are utilised. These techniques are inaccurate because the majority of them are derived by considering the relative displacements of the nodes on either side of the interface.

The accurate coupling of regions of different dimension has been investigated by [10, 11, 12]. The approach adopted by these authors assumes the stress through the thickness adheres to the conventional strength of materials theories in dimensionally reduced regions i.e. that in-plane loading and/or plate bending causes stress which varies linearly through the thickness, whilst out-of-plane shear forces cause shear stress which varies parabolically through the thickness. It is the ability to determine the distribution of stress through the thickness of the plate that allows a more accurate coupling technique to be derived. St Venant's Principle [7] suggests the stress in a body will assume this distribution as the distance from a detail feature becomes large relative to thickness.

McCune introduced the procedures for coupling solids and shells (3D-2D) [10], Monaghan extended it to 3D-1D coupling [11] and Shim automated the procedure for 3D-2D models and for laminates of multiple materials [12]. The underlying principle of these techniques is to equate the work done on the boundary of the reduced dimensional side of the mixed dimensional interface to the work done by the stresses on the full dimensional side. Equation 1 shows the equation which equates the work done on both sides of a 1D/2D interface. The stresses on the 2D side of the interface are written in terms of forces and moments on the reduced side under the assumption that the mixed dimensional interface is within the slender region where the stresses are given by plate theory. P, Q and M are the axial force, shear force and bending moment respectively, as shown in Figure 15. u, υ and θ are the associated beam displacements, where as U and V are the continuum displacements.

$$Pu + Qv + M\theta = \int_A (\sigma_{xx} U + \sigma_{xy} V) dA = \int_A \left[\left(\frac{P}{A} + \frac{My}{I} \right) U + \frac{3}{2} \frac{Q}{A} \left(1 - \frac{4y^2}{d^2} \right) V \right] dA \quad (1)$$

This allows the forces and moments to be eliminated, resulting in multipoint constraint equations relating the displacements and rotations of the reduced dimensional side to the displacements on the 2D side (equation 2 below).

$$u = \frac{1}{A} \int_A U dA; \quad v = \frac{3}{2} \frac{1}{A} \int_A \left(1 - \frac{4y^2}{d^2} \right) V dA; \quad \theta = \frac{1}{I} \int_A y U dA \quad (2)$$

In practical terms this is achieved by writing multipoint constraint equations (MPCs) for each node on the reduced dimensional side of the interface, relating each degree of freedom for that node to the degrees of freedom of the nodes adjacent to it on the full dimension side of the interface.

The 3D-2D case is described in [12], and is not reproduced here since it is significantly more complex.

Figure 16 shows the spurious stresses occurring at the mixed dimensional interface when the example model from Figure 8 is coupled using the techniques in a commercial FEA package. The results shown in Figure 8 (a) are for a model coupled using the work described above.

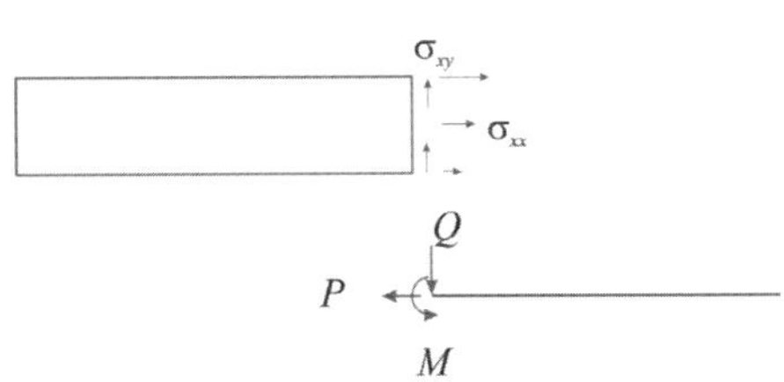

Fig. 15. A mixed dimensional interface

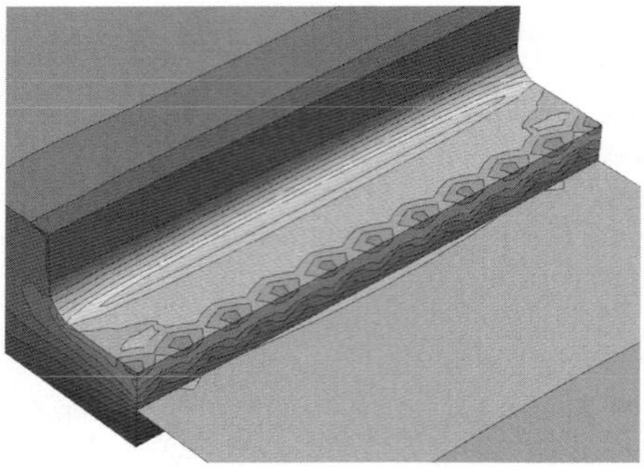

Fig. 16. Model coupled using element in commercial FEA package

7 Face Grouping

The procedures which have been described are only valid for simple topologies. Models which have more complex topology complicate the interpretation of the medial axis transform information. Here, the meaning of "complex" refers to the presence of additional edges and vertices between faces that join tangentially, for example score lines/points for load application, referred to hereafter as composite faces. When idealising models with composite faces, individual thin sheet regions may be represented by multiple medial entities, none of which are appropriate for dimensional reduction in their own right. A method is required for determining if a region can be reduced which is modelled using multiple medial entities.

To assess the suitability of a group of medial entities which represent the same region to dimensional reduction, the aspect ratio of the entire region is considered. For models with composite faces it is assumed that two medial entities define the same region if they are adjacent and have a similar thickness. For a group of adjacent medial entities, if *[(Max(D) – Min(D)) / Max(D)]* is small (<0.05) they are said represent the same region. *Min(D)* and *Max(D)* refer to the min/max medial radius values taken across all faces in the group. As before, entities which meet the criteria are labelled candidate medial entities.

Groups of medial entities which are determined to represent the same region are combined to create composite medial entities. Figure 17 (a) shows a profile in which the horizontal region is represented by five medial edges. There are two composite medial edges in

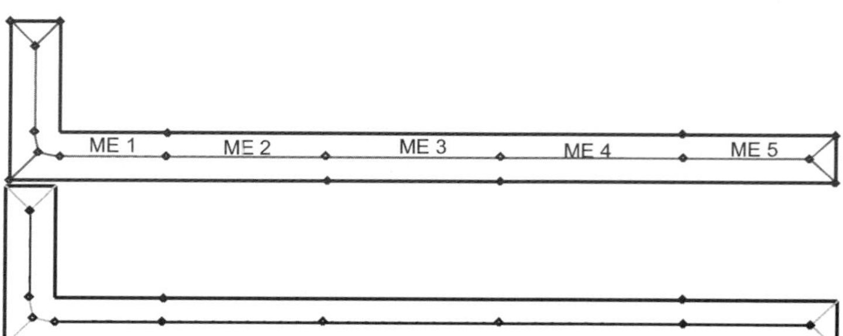

Fig. 17. (a) Dirty profile geometry and (b) candidate medial edges

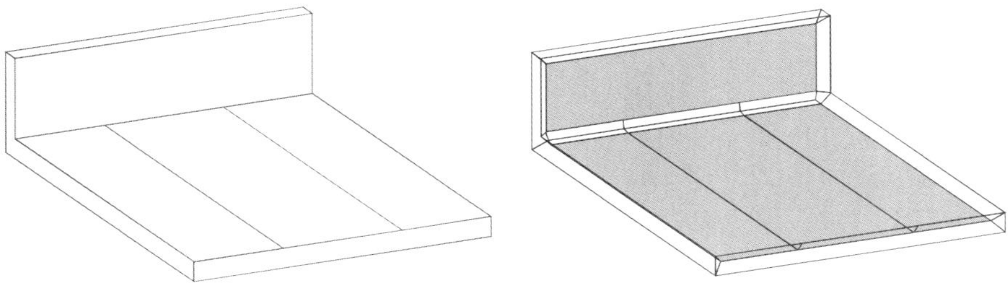

Fig. 18. (a) A solid model with composite faces and (b) candidate medial faces in grey

the model shown in black, one comprised of one vertical candidate edge, and one comprised of the five horizontal candidate edges (ME1 to ME5). The aspect ratio of a composite medial edge, calculated using the combined length of the edges and the diameter of the maximal inscribed disc used to trace any of the edges, is used to determine whether the region represented by the composite entity is suitable for dimensional reduction.

For the 3D model shown in Figure 18 there are four candidate medial faces; one in the vertical section, and three in the horizontal. This model has two candidate composite medial faces, one in the vertical member comprised of one candidate medial face, and one in the horizontal member comprised of three candidate medial faces. To dimensionally reduce the component the 2D MAT has to be computed for each composite medial face.

8 FE Models and Results

The results presented in this section are taken from [1], included here for completeness.

As part of the VIVACE Framework 6 European project [13] the mixed dimensional modelling procedures were used on a gas turbine engine component, the CAD model of which is shown in Figure 19 (a). A mixed dimensional model was compared to a densely meshed 3D model of the component used as the reference model, and a stiffened shell model, Figure 19 (b).

For the reference model an unstructured tetrahedral mesh was created; with curvature sensitive meshing used to provide smaller elements adjacent to complex features. The mixed dimensional geometry was created for a critical aspect ratio of 10; and the mixed dimensional interface was inset by the local thickness of the region. Two different mixed dimensional models were created using the same mixed dimensional geometry, but using different meshing techniques. For the first model, MD1, thin sheets of material, Figure 19 (d), were shell meshed and the remainder of the model, Figure 19 (c), was meshed using tetrahedral elements.

When the thin sheets of material were removed from casing, roughly half of the remaining solid was made up of long slender solids where one dimension was much longer than the other two. For the second mixed dimensional model, MD2, the long slender solids were isolated and meshed by extruding a triangular mesh along their length, creating wedge elements. The wedge elements are shown in middle grey in Figure 19 (e). Also shown in Figure 19 (e) is the shell mesh used for the thin sheets, in light grey, which were the same for MD1 and MD2. The unstructured tetrahedral

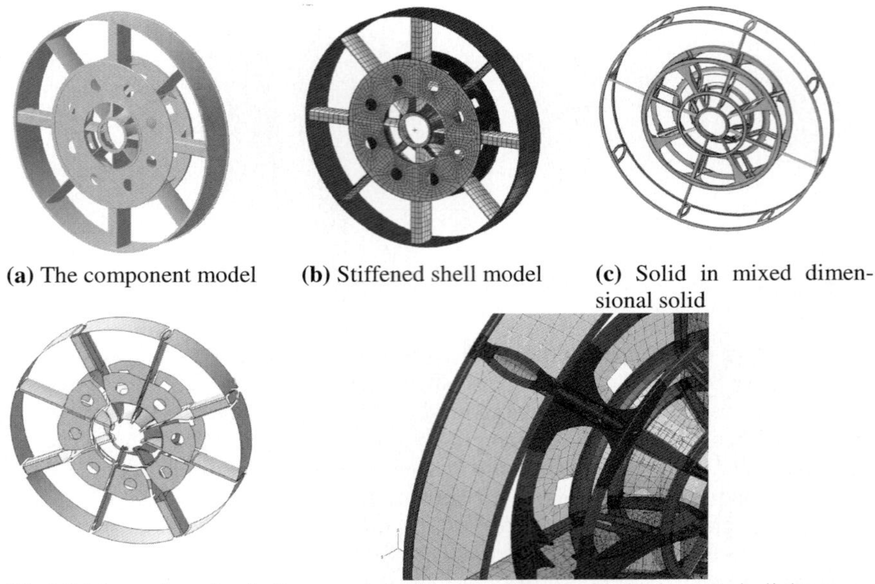

(a) The component model (b) Stiffened shell model (c) Solid in mixed dimensional solid

(d) Mid-faces in mixed dimensional model (e) Close up of mesh. Shell elements in lights grey, wedge elements in middle grey and tet elements in dark grey.

Fig. 19. Modelling the aero engine component (intermediate engine casing)

elements used for the remainder of the model are shown in dark grey. A modal analysis was carried out for all models. A static analysis was carried out for the 3D model and the mixed dimensional model MD2.

8.1 The Modal Analysis

A modal analysis is used to study the dynamic properties of a component when it is subjected to a vibration. A finite element analysis will return the frequency of vibration and the corresponding mode shape. A free-free modal analysis was carried out for each model, during which the first 40 modes of vibration were calculated, and the frequencies of the elastic modes compared. As frequency is dependent upon mass it is common practice to adjust the density applied to an idealisation so that its mass is correct, which for this investigation meant that it was the same as the 3D model. The mixed dimensional model had a mass difference of 0.2% compared to the solid model, and due to the small difference its density was not altered. The stiffened shell model had a mass difference exceeding 1.5%; therefore the density applied to the model was reduced from 2700 kg/m^3 to 2660.7 kg/m^3.

The computational efficiency is represented in this paper by the number of degrees of freedom (DoF) in the model. Unfortunately as each of the analysis models was created by different project partners, and analysed using different tools, a comparison of analysis times for the models is not meaningful. A summary of the results is shown in Table1.

At 9.4 million degrees of freedom the 3D model was extremely large, and at the limit of what was achievable using the available workstations. As well as being time consuming to analyse, a component model of this size cannot feasibly be used in the analysis of an assembly. The number of degrees of freedom in the stiffened shell model was three orders smaller than for the 3D model, and although an exact comparison of analysis times is not possible, the analysis time was reduced from in excess of 10 hours for the 3D model to less than 1 minute. Given this magnitude of the reduction in model size, the results for the stiffened shell model were surprisingly accurate; however this has to be offset by the considerable time taken to create the model, and the uncertainties in the approach.

The 60% reduction in degrees of freedom for mixed dimensional model MD1 was smaller than expected given that approximately half of the model was thin sheets and dimensionally reduced; however it is important to notice that there remain a lot of very small tet elements in the long thin regions which bound the thin sheets. The small tets were addressed in MD2 in which these long thin regions were sweep meshed, allowing the size of the model to be reduced to 8% of the size of the 3D model.

The mode shapes were the same for each of the models for all of the elastic modes of vibration. The mixed dimensional model MD2 had an average difference of 3.2% relative to the 3D model, which was less than half the 8.5% calculated for the stiffened shell model. The maximum difference in MD2 relative to the 3D model (8.2%) was less than the average difference for the stiffened shell model. The stiffened shell model had a maximum difference of 18%.

Clearly mixed dimensional models offer a significant advantage over stiffened shell models in terms of accuracy; whereas the stiffened shell models have considerably fewer DoF, and are therefore faster to analyse than mixed dimensional models. The reduction in analysis time, estimated to be less than one minute for the stiffened shell model compared to 30 minutes for the mixed dimensional model, has to be contrasted with the preparation time. The preparation of the stiffened shell model is believed to have taken in excess of two weeks, compared to the preparation of the mixed dimensional model which it is believed will take under half a day when the procedures have been properly implemented. Clearly when preparation time is considered mixed dimensional modelling offers significant advantages.

Table 1. Modal analysis data and results

	3D	MD1	MD2	Stiffened shell
Degrees of freedom (millions)	9.4	3.7	0.7	0.04
DoF (% of 3D)		40	8	0.5
Mean modal frequency (% diff. to 3D)		1.3	3.2	8.5
Max modal frequency (% diff. to 3D)		4.2	8.2	18

8.2 The Static Analysis

A static analysis is used to determine the equilibrium state of a part under a non-dynamic load. A static analysis was carried out for the 3D model and the mixed dimensional model MD2. A static analysis was not completed for MD1 as it was deemed to have too many degrees of freedom to be an acceptable approximation. The analysis consisted of clamping the face of one of the flanges, arrowed in Figure 20 (a), and applying a circumferential displacement of 0.1mm to the face arrowed in Figure 20 (b). The constraints, imposed using a cylindrical coordinate system, were u_r = 0mm, u_θ=0.1mm and u_z = 0mm. After the analysis the resultant torque on the face which had been displaced was measured. It is worth noting that both of these flanges are missing from the stiffened shell model, so if it was to be analysed in the same way the boundary conditions would also have to be approximated.

The results for the 3D model and the mixed dimensional model were compared. The resulting torque in the mixed dimensional model is 5.9% higher than in the 3D model. For both the static and modal analyses the mixed dimensional model had a higher stiffness than the 3D model.

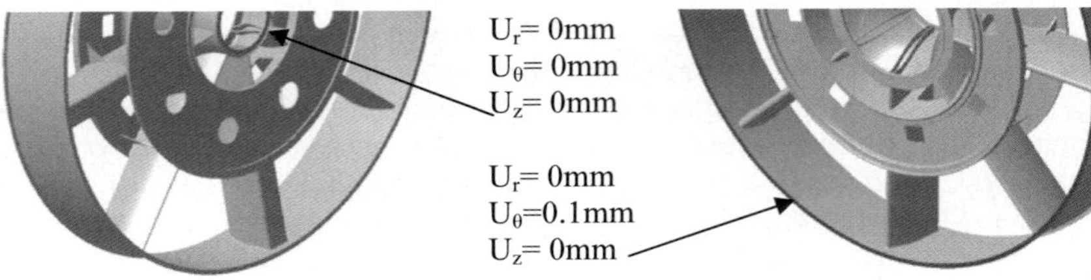

(a) clamped face **(b)** displaced face

Fig. 20.

9 Discussion

Stiffened shell models offer low analysis times, with reasonable global accuracy. This paper presented mixed dimensional modelling as an alternative, offering better accuracy than the stiffened shell model while retaining a significant reduction in analysis time over a full 3D model. The ability to create models of this type automatically enables rapid iteration during the design process.

The underlying technology for the procedures in this paper is the 3D medial axis transform which remains a topic of research in its own right. Although significant progress has been made in recent years, a robust implementation is not commercially available. The complex regions in the model, at the junctions where different solids meet, are the most difficult regions in which to compute the medial axis. The easiest are the medial faces which lie between two object faces. Fortunately it is the latter, easier to produce medial faces, which are required for the procedures outlined in this paper. Even though a robust implementation of the 3D MAT is not available at the time of writing, the information provided by the implementations that do currently exist is sufficient for the procedures in this paper.

The consequence of the medial axis failing to compute in a particular region is that the lack of medial information will stop the region from being dimensionally reduced, and the idealisation will not be as complete as it should be. This will have a detrimental effect on analysis time, but will not adversely affect the accuracy of the idealisation.

Even today's most powerful workstations have a limit to the number of degrees of freedom that can be successfully analysed, meaning it is not always possible to use a suitably dense mesh in the complex regions of the part to accurately model their behaviour. An alternative use offered by 3D-2D mixed dimensional modelling is that by reducing the number of degrees of freedom used to model the simple regions, more elements can be used to model the complex regions, providing more accurate results, while still remaining within the computational limits of the workstation.

Sometimes when the thin sheets of material are removed from a component the result is made up of long slender solids. This was demonstrated for the gas turbine engine component in Figure 3. A lot of elements can be required to mesh these regions due to the small element size imposed by the small lateral dimensions of the solid, and the constraint on element aspect ratio for unstructured tet meshing. A significant advantage can be achieved by appropriately meshing these regions using anisotropic elements, with greater dimensions along the length of the solid than on its cross section. This meshing structure can be achieved by stretching elements along the length of the solid, or by sweeping a mesh of the cross section along its length, with a large distance between where the layers of nodes are placed. This meshing structure requires the constraint on element aspect ratio to be removed. [14] discussed the identification of the "tubular" regions in a model, which are similar to the long slender solids.

An alternative approximation is to reduce these long thin regions to their midline and mesh them using 1D beam elements. If beam elements were used to model these long solids the global model would become a 3D-2D-1D mixed dimensional model. Appropriate mixed dimensional coupling would be required for this type of model. The coupling requirements would be greatly simplified if the 3D cells representing complex regions were replaced with 0D inertias, forming a 2D-1D-0D model, which would be more efficient still.

10 Conclusions

In this paper procedures have been described which can be used to create mixed dimensional idealisations from 3D CAD models suitable for finite element analysis.

Regions in a model are suitable for dimensional reduction when:

- In 3D the lateral dimensions are large relative to the thickness
- In mixed dimensional modelling:
- Medial axis transform information can be used to identify the regions suitable for dimensional reduction.
- Regions suitable for dimensional reduction can be idealised using the medial entity representing the region.
- Cellular models provide a suitable representation in which a detailed model and its structural idealisation can coexist.
- Mixed dimensional models can be created automatically, removing the need for a skilled analyst, and reducing model preparation time.
- Mixed dimensional modelling does not require a full 3D MAT. Any procedure which identifies the medial faces of thin sheet regions is acceptable.
- Finite element analysis is faster for a mixed dimensional model than for a solid model, and although stiffened shell models are faster still, their preparation time is much longer and they are subject to modelling errors.
- The increased analysis time required to analyse a mixed dimensional model compared to a stiffened shell model is acceptable, given the higher accuracy achieved and much lower preparation time.

Acknowledgements

The research reported in this paper was supported by the VIVACE Integrated Project (AIP3 CT-2003-502917) that is partly sponsored by the Sixth Framework Programme of the European Community (2002-2006) under priority 4 "Aeronautics and Space". The authors wish to express particular thanks to members of the Mixed Dimensional Modelling Task (led by Arnaud Quenardel) within the Whole Engine Modelling Workpackage (led by Graham Harlin). Alex Samson created the 3D reference model of Figures 19 and 20, and another model of which Figure 3 shows a section. Norbert Kill analysed mixed dimensional model MD1, whilst Edouard Dekytspotter created the stiffened shell model of Figure 19(b). This paper reflects only the author's views, and the Community is not liable for any use that may be made of the information contained therein.

References

1. Armstrong, C., Robinson, T., Ou, H.: QUB Recent Advances in CAD/CAE Technologies for Thin-Walled Structures Design and Analysis. In: Keynote address to Fifth International Conference on Thin-Walled Structures, Brisbane, Australia (2008)
2. Chong, C.S., Kumar, A.S., Lee, K.H.: Automatic solid decomposition and reduction for non-manifold geometric model generation. Computer Aided Design 36(13), 1357–1369 (2004)
3. Kim, S., Lee, K., Hong, T., Kim, M., Jung, M., Song, Y.: An integrated approach to realize multiresolution of B-rep model. In: Proceedings of the 2005 ACM symposium on Solid and physical modelling, Cambridge, Massachusetts, pp. 153–162 (2005)
4. Blum, H.: A transformation for extracting new descriptors of shape. Models for the Perception of Speech and Visual Form, 362–380 (1967)
5. Ramanathan, M., Gurumoorthy, B.: Constructing medial axis transform of planar domains with curved boundaries. Computer Aided Design 35(7), 619–632 (2003)
6. TranscenData, `http://www.transcendata.com/cadfix.htm`
7. Goodier, J.N.: On the problems of the beam and the plate in the theory of elasticity. Trans. Roy. Soc. Canada Sect. III, 65–88 (1938)
8. Yin, L., Luo, X., Shephard, M.S.: Identifying and Meshing Thin Sections of 3-D Curved Domains. In: Proceedings of the 14th International Meshing Roundtable, pp. 33–54 (2005)
9. Chapelle, D., Ferent, D., Bathe, K.J.: 3D-shell elements and their underlying mathematical model. Mathematical Models and Methods in Applied Sciences 14(1), 105–142 (2004)
10. McCune, R.W., Armstrong, C.G., Robinson, D.J.: Mixed-dimensional coupling in finite element models. Int. J. Numer. Methods Eng. 49(6), 725–750 (2000)
11. Monaghan, D.J., Lee, K.Y., Armstrong, C.G., Ou, H.: Mixed dimensional finite element analysis of frame models. In: Proceedings of the 10th International Offshore and Polar Engineering Conference, Seattle, WA, USA, May 28-June 2, 2000, vol. 4, pp. 263–269 (2000)
12. Shim, K.W., Monaghan, D.J., Armstrong, C.G.: Mixed Dimensional Coupling in Finite Element Stress Analysis. Engineering with Computers 18(3), 241–252 (2002)
13. VIVACE. Value Improvement through a Virtual Aeronautical Collaborative Enterprise (2007), `http://www.vivavceproject.com`
14. Goswami, S., Dey, T.K., Bajaj, C.: Identifying Flat and Tubular Regions of a Shape by Unstable Manifolds. In: ACM Symposium on Solid and Physical Modeling, Cardiff, Wales, ACM Solid and Physical Modeling 2006, pp. 27–37 (2006)

A Mesh-Geometry Based Approach for Mixed-Dimensional Analysis

Sylvain Bournival, Jean-Christophe Cuillière, and Vincent François

Université du Québec à Trois-Rivières
3351 Boulevard des Forges, CP 500 Trois Rivières G9A 5H7
Québec, Canada
bournisy@uqtr.ca, cuiller@uqtr.ca, francois@uqtr.ca

Summary. When conducting a finite element analysis, the total number of degrees of freedom can be dramatically decreased using finite elements such as beams and shells. Because of geometric complexities, entire models (or portions of models) must be meshed using volume elements in order to obtain accurate simulation results. If however some parts of these models fit the description of shells or beams, then a mixed-dimensional model containing shell, beam and volume elements side by side can be used. This approach can significantly reduce the time needed to mesh and solve the system. Unfortunately, problems arise when trying to connect elements of different dimensions in part due to the incompatible degrees of freedom and the independently created meshes. This paper presents a solution to these problems based on the generation of a compatible mesh composed solely of basic elements and without the requirement of constraint equations.

Keywords: mixed-dimensional analysis, model reduction, mesh, shell elements, beam elements.

1 Introduction

When conducting a finite element analysis (FEA) the engineer is often confronted with long calculation time or with a limit imposed on the number of elements. Dimensional reduction is one of the many techniques employed to reduce the computing time and/or the number of elements. Using plate, shell or beam elements can considerably decrease the total number of degrees of freedom (DOF) without loss of accuracy.

However, not all models are suitable for dimensional reduction. Indeed, portions of a model with a complex geometry must be simulated using volume elements. Some models, often structures, feature portions of their geometry which can be modeled by reduced-dimensional elements while other portions must be modeled using volume elements. Examples of these can be found in [1] and [2]. In these situations it is possible to use mixed-dimensional models which are models that contain more than one type of element. It is indeed possible to combine in a model shell and volume elements for example. The stiffness matrix has no restriction on the type of elements it can contain.

These mixed-dimensional analyses combine the advantages of each type of elements, a reduced computing time and decreased number of nodes for 1D and 2D element types and the ability to mesh complex geometries with volume elements.

Another advantage is the ability to easily modify the model when evaluating design scenarios. For example, in an optimization process, if a shell thickness is modified, only one parameter needs to be changed in a mixed dimensional analysis while, in a full 3D analysis, the CAD model must be changed.

Two major steps are needed to extract results from a mixed dimensional analysis. The first one is determining which component of an assembly, or which portion of a component can be modeled by using reduced dimensional elements. To make this task simpler and more automated, methods have been developed in order to determine automatically what can be reduced. The medial axis transform [3-5] and the work done by Chong in [2] are examples of the progress being made on this subject. Unfortunately, this task usually referred to as geometric idealization is far from being fully automated and a lot of work remains to be done on this subject. In the work presented here, idealized models are built by the designer in a CAD system.

Fig. 1 illustrates an example of a mixed dimensional model (or idealized model). A component is partly modeled by using 3D (volume) elements for complex geometry, beam elements for long and slender portions of the geometry and shell elements for portions of the geometry which can be accurately described by a face associated with a small wall thickness.

The next step in the mixed-dimensional analysis process is to take the idealized model, mesh it and solve the resulting model to obtain analysis results.

Unfortunately these analyses also face some problems, mainly at the interface between elements featuring different dimensions. In this paper, a zone where 1D (beams), 2D (shells) and 3D (volume) elements join is referred to as a *dimensional interface*. We will demonstrate here how these problems can be overcome solely by generating a mesh featuring specific arrangements of classical 1D, 2D and 3D finite elements. This paper focuses on these arrangements when coupling shell elements with volume elements.

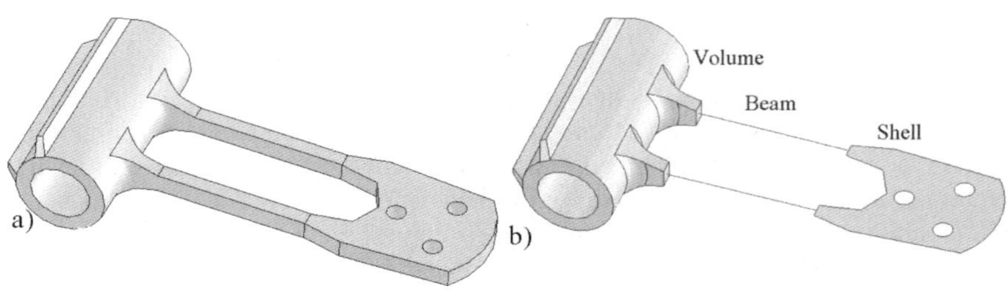

Fig. 1. Example of a mixed dimensional model a) Original model b) Idealized model

2 Problems with Mixed-Dimensional Analysis

When trying to join elements featuring different dimensions, two major problems arise. The first problem is meeting requirements of mesh continuity across a dimensional interface and the second is incompatible degrees of freedom (DOF) across a dimensional interface.

2.1 Mesh Continuity

A mixed dimensional model is created in a CAD system as a group of parts modeled with various dimensions. The mesh generator sees these parts as different entities and meshes them independently. This results in an incompatible mesh at dimensional interfaces. Even if the different parts are in contact with each other, if nodes are not coincident, these parts are not practically bounded together from a FEA perspective, which implies that the analysis will fail or that results will be inaccurate.

Problems also arise when only some nodes are coincident while others are not. Fig. 2 illustrates this specific situation. A W-beam is modeled using three 2D faces. These faces are then meshed independently with shell elements. Nodes of each face are not always coincident with other nodes at the interfaces. In the process, coincident nodes are merged whereas many nodes at the interfaces are left unmerged. Of course, such a model leads to inaccurate results at the interfaces. High stress concentrations appear near merged nodes since the model represents a situation where faces in contact are bounded together only at merged nodes locations. We will explain below how discontinuous binding between parts (with identical or different dimensions) can be avoided.

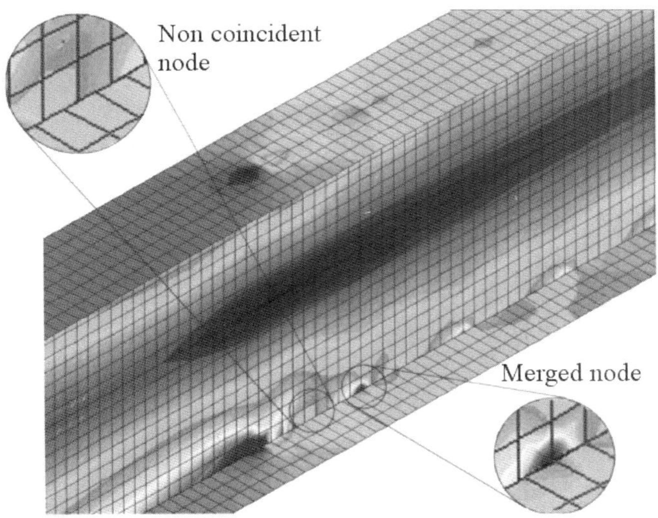

Fig. 2. Examples of a mesh continuity problem, coincident nodes are merged while others are not

2.2 Incompatible Degrees of Freedom

The second major problem faced when using mixed-dimensional analysis is incompatible degrees of freedom (DOF) at a dimensional interface. Typical 3D elements only feature three DOF (3 translations) per node while shell elements typically feature five or six DOF per node (three translations and two or three rotations).

Consequently, when coupling volume and shell elements, even if nodes of both parts in contact are merged, problems can still arise in some circumstances. Fig.3 illustrates a straight line contact between a shell and a 3D volume. The forces applied on the shell induce a bending moment in the shell. This bending moment is transmitted trough the shell thanks to the shell's rotational DOF. Nodes located at the straight line contact (common to shell and volume elements) bring about FEA modeling problems as volume elements do not "follow" rotations of shell element. Assembly of the stiffness matrix as is leads to a singular matrix or to inconsistent results.

Thus, coincident nodes directly merged this way are akin to ball-joint links. Since in the FE model, the shell and volume intersect as a series of coincident nodes, the incompatibility between DOF results in a situation where the shell and volume are joined by a hinge.

We will also explain below how this second type of inconsistency at a dimensional interface can be avoided.

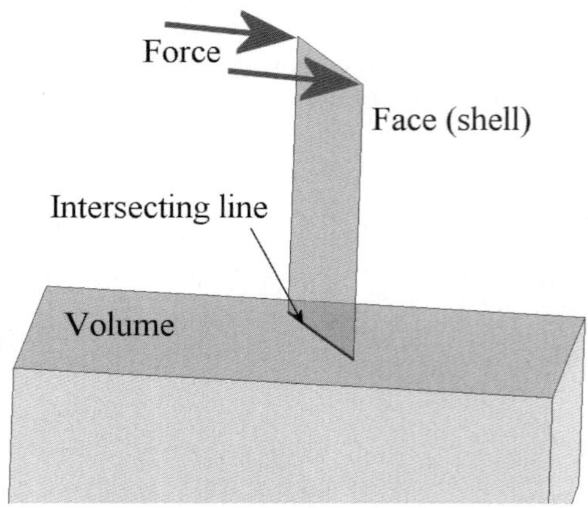

Fig. 3. A shell – volume intersection

3 Alternate solutions

Some techniques were designed to circumvent these problems. Shim [6] describes a way to couple shell and volume elements by adding constraints equations to the solver. These equations are based on equating the mechanical work for elements located on both sides of a dimensional interface and tend to overcome inconsistencies between DOF. Some FEA software feature functions used to couple shell and volume elements. These functions add constraints equations akin to those described above, but accuracy issues related to the use of this type of approaches have been outlined [7].

Another approach to overcome inconsistencies between DOF is to use volume elements with rotational DOF. Indeed, 3D elements have been designed, based on Allmans's triangle with rotational DOF [8]. This triangle, designed for 2D analysis, is a quadratic element with only three nodes, each with 3 DOF (two translations and one referred to as a pseudo-rotation). Since then, many tetrahedron element types have been designed, based on the same principles [9]. Sze [10] has referenced a large number of such elements. It should be noted however that these elements are designed to improve on the classical linear tetrahedral element and the additional rotational DOF is not designed with the intent of coupling elements of different dimensions. The use of this type of elements implies that the connection between beams or shells and solid elements is concentrated at one point or one line. Therefore, stress distributions in solid parts around these interfaces are completely independent of beam section shapes and shell thicknesses, which introduce significant inaccuracies.

4 Prior Work

Prior to the work presented here, a method was developed to couple beams and volume elements using only specific arrangements of typical 1D, and 3D FEA elements [11]. This method came from an adaptation and combination of two suggestions made by Craveur [12] and it results in solving both problems mentioned above (mesh continuity and incompatible DOF) through the two following steps: *face splitting* and *adding mini-beams*.

Face splitting aims at solving the mesh continuity problem with the objective of finding a solution that is independent from both the mesh generator and the solver. Therefore the best way to force the mesh generator to create nodes at correct locations is to use a technique which consists in splitting a face located at a dimensional interface, in several pieces. This operation generates one or more faces that fill the space of the original face. This is done to add at least one vertex at the intersecting point between the beam and the volume. A standard Delaunay or advancing front unstructured mesh generator will then automatically place a node at the newly created vertex. In order to improve accuracy, the shape of face splitting is the same as beam's section. This is illustrated in Fig. 4; a W-shaped beam intersects a volume at a specific location represented by a point. Two small internal faces (colored in the figure) are cut inside the volume's intersecting face. As it can be seen in Fig. 4, a vertex is located at the intersecting point with the beam, therefore insuring that the mesh generator will create a node at that location.

Adding mini-beams aims at solving the problem of incompatible DOF. Beam elements called mini-beams are placed between the main beam's node and nearby nodes on the volume's boundary. While there is still the ball-joint effect explained earlier, the fact that these supplementary beams are linked with the main beam solves the problem. Of course, the mini-beams should not all be collinear (located along a straight line) which introduces a free rotation around this line. Theoretically, a non-singular stiffness matrix can be obtained with only a few mini-beams and in the example illustrated in fig. 4, this would mean adding only four mini-beams on top of each triangular edge touching the intersection point. Nevertheless, as will be explained just below, the distribution of mini-beams has been studied carefully and optimized in order to obtain analysis results that are as close as possible to the real connection's mechanical behavior.

No simple equation could be found to adapt the mini-beams' elastic properties to the behavior of the beam's section at the dimensional interface. Therefore, in an attempt to improve accuracy, a great number of tests have been made to establish the best possible configuration with the layout of the mini-beams and their properties. However, no configuration was found to perfectly recreate the behavior in every circumstance.

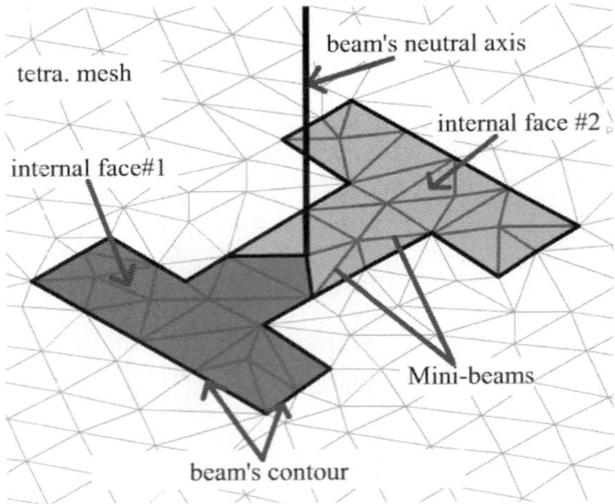

Fig. 4. A beam-volume intersection. Two faces are cut in the original face and mini-beams are overlaid on top of them.

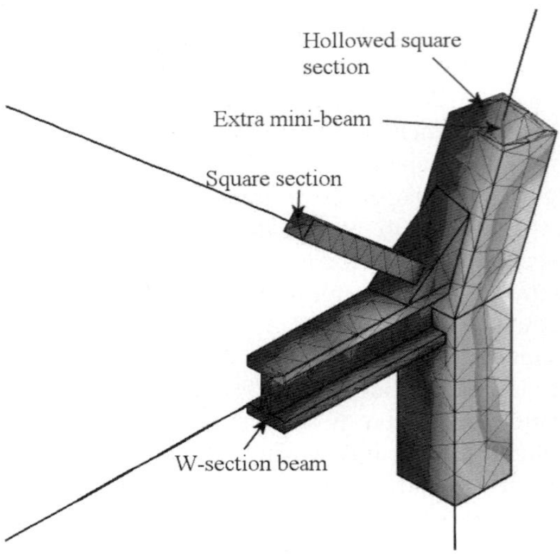

Fig. 5. Local stress results (Von Mises) for an idealized part with several beam-volume intersections

Nonetheless, an acceptable solution was found when using mini-beams with infinite stiffness. This hypothesis meets the Bernouilli-Euler beam element's formulation which states that a beam's section does not change under stress.

To make the entire beam's section rigid, mini-beams of infinite stiffness are laid on every edge of the mesh at the dimensional interface. In the example of fig. 4, this means that every triangle edge inside the two colored faces (obtained from the splitting face process) is overlaid with a mini-beam of infinite stiffness. This effectively translates in faces of infinite stiffness.

Fig. 5 illustrates a sample stress result on a model composed of four beams intersecting at odd angles. The geometry at the intersection of these four beams is modeled using 3D elements while most of the beams themselves are modeled using beam elements. Fig. 5 also illustrates how the specific case of hollow beams is processed. In fact, a specific scheme was developed in order to deal with beams of which neutral axis does not lie in the material of the beam, which is the case of most hollow beam as well as channel type beams. A supplementary mini-beam of infinite stiffness is used to link the node at the beam's extremity and a random node on the face representing the beam's section. This can be observed in fig. 5 at the intersection with the hollow square section. In this particular case no face splitting is required.

5 Extension to Shell-Volume Intersections

The method proposed in this paper aims at solving the same types of problems (mesh continuity and incompatibility of DOF) for shell-volume intersections. Thus, the work presented below is based on an extension of beam-volume intersections procedures to the case of shell-volume intersections. In the case of shell-volume intersections, dimensional interfaces are still processed following the two steps mentioned above (face splitting and mini-beams generation) but these steps are slightly modified.

5.1 Step One: Face Splitting

The aim of this operation is now to generate a common edge at the intersection between the shell and the volume. An example is illustrated in fig. 6 where a shell intersects a volume on one of its faces. Face splitting consists in cutting two smaller faces along the intersection between the shell and the volume (colored in the figure). Like in the case of beam-volume intersections, the shape of splitting face is consistent with the shell's section, which means the shell's thickness. In the CAD software, two edges are situated at the dimensional interface, one belonging to the volume (on both smaller faces) and one belonging to the shell. These edges and their associated vertices are merged to form only one edge. This results in a non-manifold model since this intersection edge features three adjacent faces.

Once launched, the mesh generator will start by generating nodes on all vertices. Then it will mesh all edges of the model, including the common edge at the dimensional interface, only once. Then, faces are meshed starting from the edges' mesh and therefore the nodes located at the intersection are necessarily coincident between the volume and the shell.

In some cases this procedure cannot be used directly in order to create a common edge between the shell and the volume. For example, this is the case when the intersection edge between the shell and the volume does not entirely lie on a single face bounding the volume. An example of this can be seen in fig. 7. A shell intersects a volume but the intersection edge lies at two different location and extends beyond the volume.

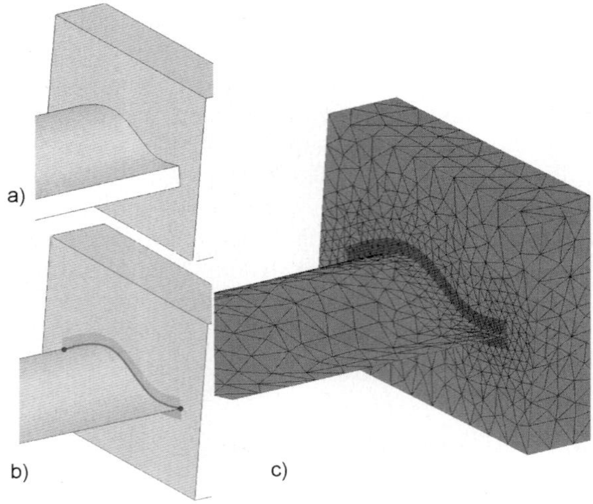

Fig. 6. The method applied to a shell-volume intersection. a) 3D Model b) Model after face splitting operation c) mesh with mini-beams (mini-beams and bounding triangles are represented using different colors also for clarity).

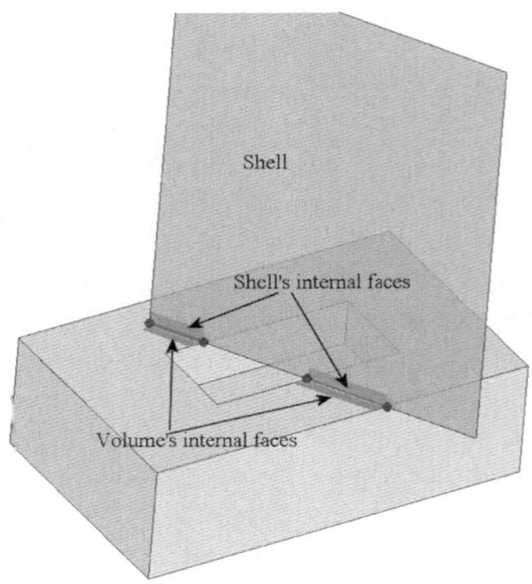

Fig. 7. In order to obtain a series of common edges, the shell's face has to be cut several times

In this case, the mesh continuity problem remains even if volume's intersection face is split correctly because the shell's edge at the intersection cannot be merged as is with an edge on the volume's boundary.

Since the Application Programming Interface (API) we are using does not feature any tool to split an edge of a shell, the solution in this case is to split the shell's face as well, as illustrated in fig. 7. Splitting the shell introduces two rectangular sub-faces, which leads to the division of the shell's intersection edge into four smaller edges. Two of these edges can now be merged with a coincident edge on the volume's boundary at the intersection. The height of these shell sub-faces is chosen with regard to the element size as specified in the size map[1] imposed in the dimensional interface. This is done in order to obtain well shaped triangular elements when meshing the shell, as the automatic mesh generation process is constrained by splitting faces.

5.2 Step Two: Adding Mini-beams

To prevent the hinge effect described earlier (a consequence of incompatible DOF) mini-beams are also added to both internal faces once the mesh is performed. While the FE solver would not fail in the case illustrated in fig. 6 as the edge is not a straight line, adding these mini-beams first models a rigid connection instead of a hinge connection at the intersection curve. Also, like in the case of beam-volume connections, it models the connection between the shell and the volume more accurately as the connection is made on a face (representing the shell thickness) instead of a curve, reducing by the way stress concentrations at the interface. The operation is rather straightforward as the list of all triangular segments belonging to the connecting faces (contour and inside) can be easily obtained. A mini-beam is then added between both nodes of each segment so that the entire connecting face is overlaid with mini-beams.

The principle of mini-beams of infinite stiffness used in the case of beam-volume intersections cannot be used here however as the shell's section at the interface must be able to bend. Fig. 8a illustrates this type of specific cases where three forces act to

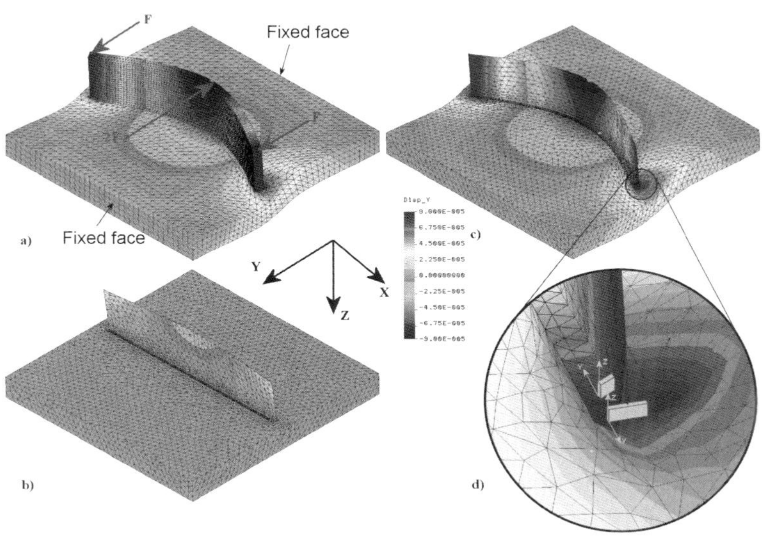

Fig. 8. Displacements results (Y direction) for a shell-volume intersection. a) 3D model b) Model with mini-beams of infinite stiffness c) Model with oriented mini-beams d) Illustration of two mini beam's orientation.

---

[1] In our work, the size map is a tensorial function describing the desired mesh size at each point in space.

bend the shell, but a section of infinite stiffness prevents the shell from deforming properly (Fig. 8b). In order to solve this problem, oriented mini-beams are used instead of mini-beams with infinite stiffness. Oriented mini-beams are configured with an infinite moment of inertia in one direction, but with no inertia in the other direction and a null section area. In the case of fig. 8, this means that the mini-beams have a moment of inertia Iyy infinitely large (see figure 8d). The stiffness in the other directions is supplied directly by the shell itself.

Fig. 9 illustrates Von Mises stress results obtained for a simple model where a portion of a shell is modeled using shell elements while the remaining portion uses volume elements. As it can be seen, the Von Mises stress field is quite continuous across

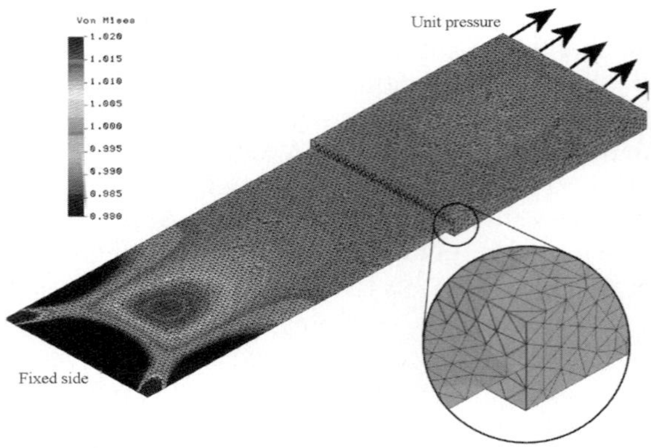

Fig. 9. Von Mises stress results on a shell (tension). Top face stresses are illustrated for shell elements.

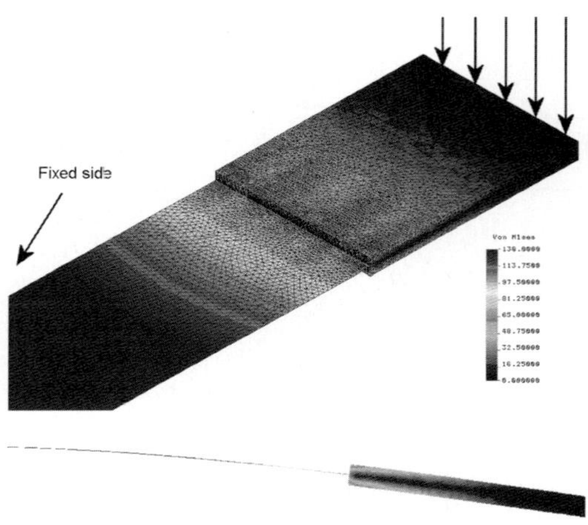

Fig. 10. Von Mises stress results on a shell (bending). Top face stresses are illustrated for shell elements. 3D view and side view pointing out the deformed shape (at the bottom).

the part except for very small discontinuities near the dimensional interface between volume and shell elements. In this particular case these discontinuities appear because of the inability to obtain mini-beams of real zero and infinite stiffness. Fig. 10 features the same model under a bending load.

6 Examples

The method described above has been successfully implemented through the design of a computer program. Practically, the whole process starts from an idealized model, designed in a commercial CAD system. Technical data such as wall thicknesses (for shells), section properties and orientation (for beams), boundary conditions and mechanical properties are also integrated into the model using an interactive interface, as well as an imposed size map for mesh generation. After this, the remainder of the process is fully automated. This automatic process starts with modifications to the idealized model (essentially face splitting) which are led directly in the CAD system using an Application Programming Interface (API). This API is also used to provide the geometric model with technical data mentioned above. This API is first used to process the identification of intersections between shells and volumes (for what concerns the scope of this paper), and then to apply face splitting algorithms. It is also used to flag internal faces to be overlaid by the mesh generator with mini-beams and to merge edges and vertices. Then, the addition of mini-beams is carried out by specific procedures integrated in the automatic mesh generation system developed by our research team [13-15].

It can also be mentioned here that the size map constraining mesh generation can be either specified by the analyst or by an automated process (referred to in our work as automatic mesh pre-optimization [16-18]). Also, in the context of the work presented here, the size map is adapted according to the shape of internal faces (resulting from face splitting).

Table 1. Comparison of the different models

	6 legged support	Coat hanger	Ventilation gate
		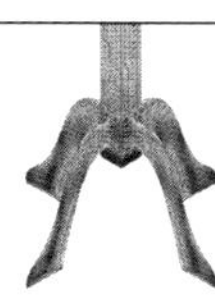	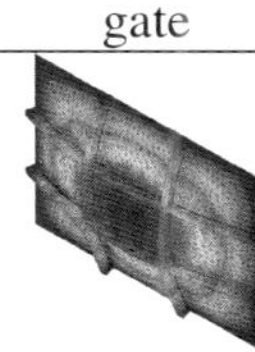
Mixed dimensional analysis			
Nb. elements	23 171	20 121	42 310
Nb. of DOF	20 295	17 900	47 646
Max displacement (m)	2.85 e-9	2.242 e-9	1.589 e-7
Full 3D model (quadratic tetrahedron)			
Nb. elements	88 048	26 608	34 373
Nb. of DOF	410 316	129 201	200 982
Max displacement (m)	2.995 e-9	2.248e-9	1.497 e-7

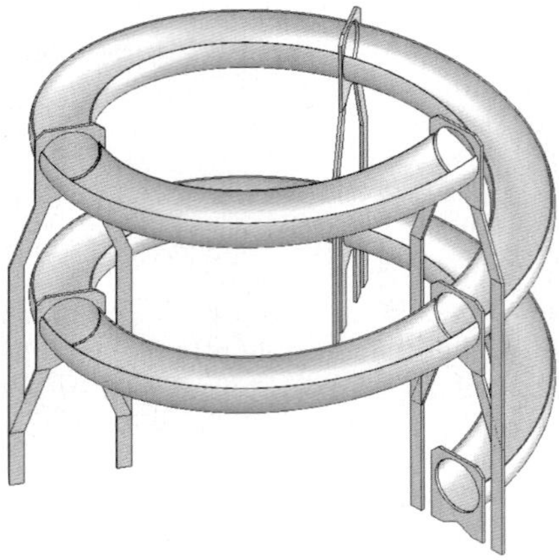

Fig. 11. Solid model of the waterslide

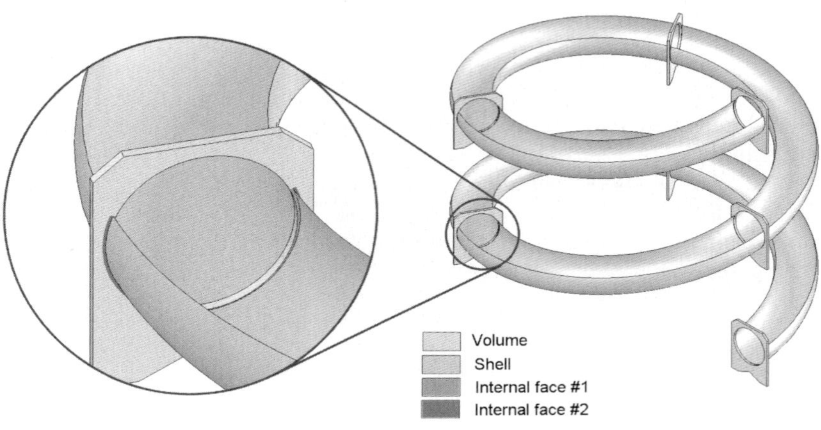

Fig. 12. The idealized model after face splitting

Table 1 summarizes a few simple examples modeled with the method described in this paper. The three different models are compared with results obtained from a full 3D model. The mixed-dimensional models use linear tetrahedrons, linear thin shells and oriented mini-beams while the 3D models use only quadratic tetrahedron. We were unable to compare these results with mixed-dimensional models using constraints equations as we do not have access to that code.

The use of oriented mini-beams is particularly noticeable in the case of the ventilation gate as the faces at the dimensional interface need to bend in order for the model to deform correctly. Note that the 3D models are not necessarily more accurate as tetrahedron elements are ill suited to model thin shells.

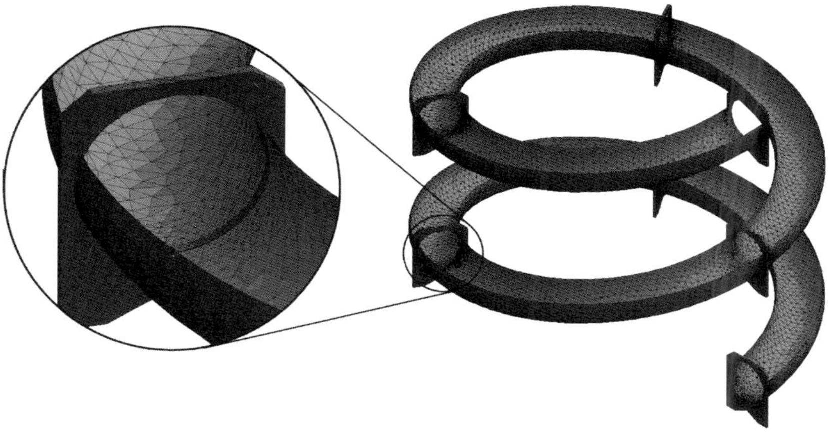

Fig. 13. Mixed-dimensional mesh of the waterslide

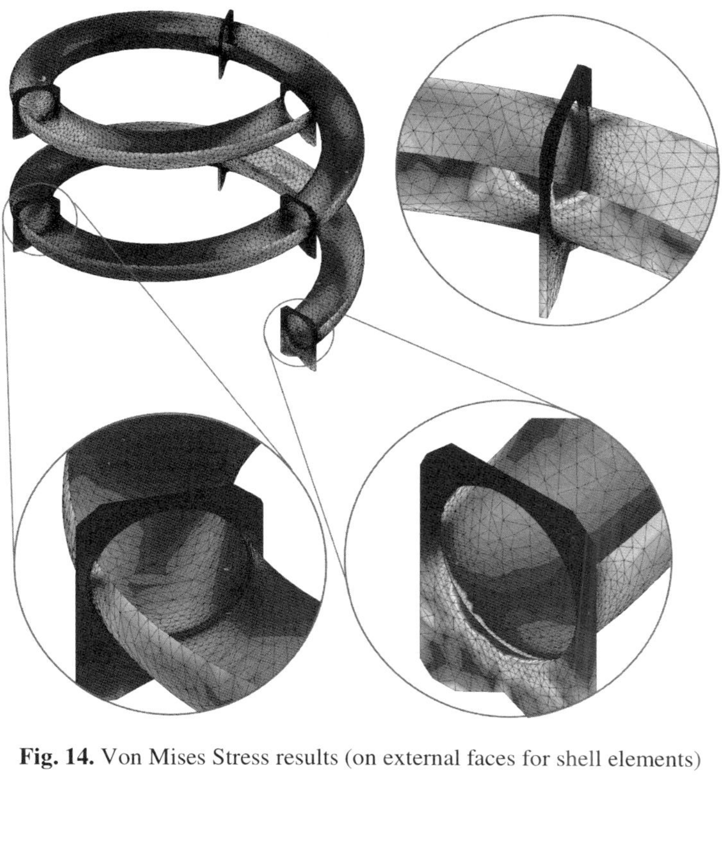

Fig. 14. Von Mises Stress results (on external faces for shell elements)

The next example illustrates the analysis of a waterslide structure in details. Fig. 12 illustrates the solid CAD model of the waterslide. This model is then idealized and provided with technical data (boundary conditions, shell thickness and material properties). For this example, support parts are modeled using 3D elements and the half-tube using shell elements. Fig. 13 illustrates this model after face splitting. As seen with the zoom, face splitting results are illustrated using different colors.

Fig 14 illustrates the resulting mixed-dimensional mesh. The 7 volumes and 6 shells are meshed independently. Oriented mini-beams are then added applying the proper orientation and the mesh is automatically graded around the dimensional interfaces.

The mesh, material properties, boundary conditions and shell properties are finally exported to a commercial FEA solver. The resulting Von Mises stress field can be observed in fig. 15. From this model, various design scenarios can now be easily investigated.

7 Conclusions

This paper presents a method intended to couple shell and volume elements for mixed-dimensional analysis only using specific arrangements of classical 2D and 3D finite elements. Since the technique is mesh-geometry based it presents the advantage of being independent from both the mesh generator and the solver. The mesh generator used in this work is based on advancing front techniques, but a Delaunay mesh generator could be used as well.

The software implementation of this work has proved that the concepts presented here are consistent and that the process can be fully automated from an idealized model added with relevant technical data (shell thicknesses and beam sections properties, boundary conditions and element size constraints).

In a design optimization context, this method generates reliable FEA results and in many occasions, the use of this method can provide more accurate results than using only tetrahedral elements. This is particularly the case when shell thickness is small in comparison with the size map. Generally, when using only tetrahedral or hexahedral meshing in the case of mechanical structures with beams and shells, resulting 3D meshes are either very poor (in terms of elements quality) by the way leading to poor FEA results, or excessively large (in terms of DOF), if compared to mixed-dimensional meshes.

Of course, the efficient use of mixed-dimensional analysis is also directly related to the ability to derive idealized models from 3D parametric and feature-based CAD models. Consequently, a significant research work is presently ongoing in order to provide efficient and robust tools for CAD models preparation for FEA and deriving idealized models is one of the subjects that are presently investigated.

Acknowledgments

This study was carried out as part of a project supported by research funding from Quebec Nature and Technology Research Fund and by the Natural Sciences and Engineering Research Council of Canada (NSERC).

References

1. Armstrong, C.G., Bridgett, S.J., Donaghy, R.J., Mc Cune, W., McKeag, R.M., Robinson, D.J.: Techniques for Interactive and Automatic Idealisation of CAD Models. In: Proceedings of the 6th International Conference on Numerical Grid Generation in Computational Field Simulations, July 1998, pp. 643–662 (1998)
2. Chong, C.S., Kumar, A.S., Lee, K.H.: Automatic solid decomposition and reduction for non-manifold geometric model generation. Computer Aided Design 36(13), 1357–1369 (2004)
3. Donaghy, R.J., McCune, W., Armstrong, C.G.: Dimensional Reduction of Analysis Models. In: Proceedings of the 5th International Meshing Roundtable, October 1996, pp. 307–320 (1996)
4. Monaghan, D.J., Doherty, I.W., McCourt, D., Armstrong, C.G.: Coupling 1D Beams to 3D Bodies. In: Proceedings of the 7th International Meshing Roundtable, October 1998, pp. 285–293 (1998)
5. Suresh, K.: Generalization of the Kantorovich Method of Dimensional Reduction. In: Proceedings of the 12th International Meshing Roundtable, September 2003, pp. 261–270 (2003)
6. Shim, K.W.: Mixed Dimensional Coupling in Finite Element Stress Analysis. In: Proceedings of the international meshing roundtable, October 2001, pp. 269–277 (2001)
7. McCune, W., Armstrong, C.G., Robinson, D.J.: Mixed-dimensional coupling in finite element models. International Journal for Numerical Methods in Engineering 49(6), 725–750 (2000)
8. Allman, D.J.: A compatible triangular element including vertex rotations for plane elasticity analysis. Computers and Structures 19(1), 1–8 (1984)
9. Pawlak, T.P., Yunus, S.M.: Solid elements with rotational degrees of freedom: Part II - tetrahedron elements. International Journal for Numerical Methods in Engineering 31(3), 593–610 (1991)
10. Sze, K.Y., Pan, Y.S.: Hybrid stress tetrahedral elements with Allman's rotational D.O.F.s. International Journal for Numerical Methods in Engineering 48(7), 1055–1070 (2000)
11. Bournival, S., Cuillière, J.C., Francois, V.: A Mesh based Method for Coupling 1D and 3D Finite Elements. In: Proceedings of the 2nd World Congress of Design and Modelling of Mechanical Systems (CMSM 2007) (March 2007)
12. Craveur, J.C., Marceau, D.: De la CAO au calcul, Paris, Dunod (2001)
13. Cuillière, J.C.: An adaptive method for the automatic triangulation of 3D parametric surfaces. Computer Aided Design 30(2), 139–149 (1998)
14. Béchet, E., Cuillière, J.C., Trochu, F.: Generation of a Finite Element Mesh from Stereolithography (STL) Files. Computer Aided design 34(1), 1–17 (2002)
15. Francois, V., Cuillière, J.C.: An a priori adaptive 3D advancing front mesh generator integrated to solid modeling. Recent Advances in Integrated Design and Manufacturing in Mechanical Engineering, 337–346 (2003)
16. Francois, V., Cuillière, J.C.: Automatic mesh pre-optimization based on the geometric discretization error. Advances in Engineering Software 31(10), 763–774 (2000)
17. Cuillière, J.C., Maranzana, R.: Automatic and a priori refinement of three-dimensional meshes based on feature recognition techniques. Advances in Engineering software 30(8), 563–573 (1999)
18. Boulet, A., Cuillière, J.C., Francois, V.: Automatic size map calculations for mesh pre-optimization in the context of CAD-FEA integration. In: Proceedings of the 2nd World Congress of Design and Modelling of Mechanical Systems (CMSM 2007) (March 2007)

Simple and Effective Variational Optimization of Surface and Volume Triangulations

Xiangmin Jiao[1,⋆], Duo Wang[1], and Hongyuan Zha[2]

[1] Department of Applied Mathematics, Stony Brook University, Stony Brook, NY 11794
[2] College of Computing, Georgia Institute of Technology, Atlanta, GA 30332
jiao@ams.sunysb.edu

Summary. Optimizing surface and volume triangulations is critical for advanced numerical simulations. We present a simple and effective variational approach for optimizing triangulated surface and volume meshes. Our method minimizes the differences between the actual elements and ideal reference elements by minimizing two energy functions based on conformal and isometric mappings. We derive simple, closed-form formulas for the values, gradients, and Hessians of these energy functions, which reveal important connections of our method with some well-known concepts and methods in mesh generation and surface parameterization. We then introduce a simple and efficient iterative algorithm for minimizing the energy functions, including a novel asynchronous step-size control scheme. We demonstrate the effectiveness of our method experimentally and compare it against Laplacian smoothing and other mesh optimization techniques.

Keywords: Mesh optimization; isometric mapping; conformal mapping; inverse mean ratio; variational methods.

1 Introduction

Optimization of mesh quality is a fundamental problem for numerical simulations in science and engineering. It has become increasingly important in recent years, because modern computational applications often involve dynamic interfaces or moving boundaries. These applications require on-line optimization of the meshes to maintain the validity and quality within the simulation codes as the physical domains deform. Therefore, the mesh-optimization algorithms must be *effective* and *robust* without user intervention, and at the same time be *efficient* and *easy to implement*. These requirements pose decidedly nontrivial new challenges. For problems with very large deformation or motion, one may have to change the mesh connectivity by adding or removing points or flipping edges for best robustness. However, it is often desirable to maintain the mesh connectivity as far as possible to minimize numerical errors, such as in the Arbitrary Lagrangian Eulerian (ALE) methods [15]. In this paper, we focus on mesh optimization with fixed connectivity, i.e. the *mesh smoothing* problem.

⋆ Corresponding author.

1.1 Related Work

Mesh smoothing has a vast amount of literature. The most popular methods for optimizing two-dimensional meshes are probably Laplacian smoothing and its variations [6, 13], which move each point to a weighted average of its neighbors. Their popularity is largely due to their simplicity and efficiency, though they are ineffective at concave regions. A simple but more effective approach is the angle-based smoothing in [27], which provides no theoretical guarantee either but often works well in practice. The most effective methods are optimization based, such as those in [21, 24]. However, these methods are much more expensive and difficult to implement than other simpler alternatives. An ideal method should be simple, efficient, and effective, which we strive to achieve in this paper.

Most methods for optimizing triangular meshes are designed for meshes in $\mathbb{R}^2$. They can be applied to surface meshes by first parameterizing the surface locally or globally and then optimize the flattened mesh (see e.g. [12, 19]). To preserve geometry, these methods typically require a smooth or discrete CAD model and associated point-location procedure to project the points, which increase implementation complexity and computational cost. We propose a simpler approach in this paper.

Mesh smoothing of volume or even higher-dimensional meshes is far more complex than smoothing 2-D meshes. Many methods fail to generalize. For example, Laplacian smoothing often produces inverted elements in 3-D, angle-based method [27] no longer applies, and condition-number based optimization [8, 21] becomes increasingly cumbersome. One of more successful methods is the minimization of the inverted mean ratio [10, 23]. A side product of this paper is to derive the inverted mean ratio from a differential geometry point of view, shows its intimate connection with the aspect ratio of tetrahedra [25], and in turn provide a justification and a simpler and more efficient algorithm for minimizing it.

In mesh optimization, there have been some notable endeavors to *unify* different methods and concepts, such as the study in [1] from an algorithmic point of view, the algebraic framework in [19, 20], and the finite-element-based methods in [14]. One of the objectives of this paper is also such a unification, but in the aspect of unifying the optimization methods for different dimensions. A notable benefit of such a unification is to enable simultaneous optimization of surface and volume meshes, which we will demonstrate to be critical for effective mesh optimization.

1.2 Contributions and Overview

In this paper, we strive to meet the needs of mesh smoothing within numerical simulations. We make contributions in both theoretical and practical aspects. Theoretically, our first contribution is to establish a unified framework of variational mesh optimization in arbitrary dimensions (including surfaces and manifolds) based on the theory of isometric mappings in differential geometry. The core of this framework is two energy functions that capture angle and volume preservation, respectively. We show the connection of the angle-preservation energy with the well-known harmonic smoothing in mesh generation (e.g., [17]) and conformal parameterization in computer graphics (e.g., [7]).

Our second theoretical contribution is a set of simple, closed-form formulas for the gradient and Hessian of the energy functions for surface and volume meshes. These formulas provide us direct proofs of the convexity of our energy functions and enable efficient algorithms and simple implementations for solving the minimization problems. In addition, they reveal the equivalence of angle-preserving energy with the inverse mean ratio in Euclidean space and establish an important connection of these energies with the aspect ratios of tetrahedra [25]. These results substantially enhance the fundamental understandings of variational mesh optimization.

In the practical aspect, our main contribution is a new algorithm for mesh optimization. The most notable features of our method include 1) its ability to smooth surface and volume meshes simultaneously for improved effectiveness, and 2) a simple asynchronous step-size control scheme. Other virtues of our algorithm include ease of implementation and parallelization (comparable with Laplacian smoothing), effectiveness (at least as comparable as inverse mean ratio), and high efficiency (significant improvement in computational cost and memory requirements compared to other optimization-based methods). These properties make our algorithm especially well suited for direct implementation within numerical simulation codes.

The remainder of this paper is organized as follows. Section 2 defines the notation and reviews some fundamental concepts for variational mesh optimization. Section 3 describes our formulation for optimizing surface meshes. Section 4 generalizes our method to volume meshes. Section 5 contains some experimental results of our method and comparisons with other methods for optimizing static and dynamic meshes. Section 6 concludes the paper with discussions of future research directions.

2 Unified Variational Framework for Mesh Optimization

In mesh optimization, there are typically desired shapes and/or sizes of the elements in the resulting mesh. Therefore, we envision each element has a corresponding "ideal" reference element of the desired shape and size. We focus on isotropic (instead of anisotropic) meshes. In this setting, the desired shape is typically regular (i.e., equilateral), and the desired size may be determined by a density function over the domain. The elements of an "optimal" mesh should be as "close" to their corresponding ideal elements as possible. To achieve this, we must define the measures for the "closeness" and then optimize them over the mesh. We seek inspirations from the continuous mappings between surfaces (or manifolds in general), which have been well studied in differential geometry.

2.1 Harmonic, Conformal, and Isometric Mappings

Mappings between surfaces (or manifolds in general) are central in differential and Riemannian geometry (see e.g. [4, 5, 22]). Before delving into the details, we first review the intuitive meanings of three types of mappings to motivate our method.

The most well-known type of mapping is probably *harmonic mappings*. Such mappings are "smooth," as they satisfy some elliptic partial differential equation (PDE), namely the Laplace equation. Harmonic mappings are popular especially for optimizing structured meshes for their simplicity and efficiency [17], but unfortunately less so for unstructured meshes.

Another important type is *conformal mappings*, which preserve angles locally within a infinitesimal neighborhood of every point on the surface. Conformal mapping is a special type of harmonic mapping. The two types are equivalent under some conditions (such as for simply connected surfaces without boundaries and holes).

The former two types are more widely used. However, the most classical type is in fact *isometric mapping*, which is the foundation of intrinsic differential geometry. Such mappings preserve the first fundamental forms, and in turn preserve the length measures along all directions. In other words, an isometric mapping preserves angles and areas/volumes locally at all points, so it is a special type of conformal mapping and also of harmonic mapping.

We develop our method based on conformal and isometric mappings, as they are well suited for controlling the shapes and/or sizes of isotropic meshes. As we will explain, isometric mappings provide a powerful and flexible framework that can supersede conformal mappings in a variational framework.

2.2 Computational Aspects of Conformal and Isometric Mappings

We now give a more detailed and rigorous treatment of the mappings. Consider two m-dimensional manifolds Γ_1 and Γ_2 embedded in $\mathbb{R}^{n_1}$ and $\mathbb{R}^{n_2}$, respectively, where $n_1 \geq m$ and $n_2 \geq m$. Let $\boldsymbol{u}(\boldsymbol{\xi}) : \Omega \subseteq \mathbb{R}^m \to \mathbb{R}^{n_1}$ and $\boldsymbol{x}(\boldsymbol{\xi}) : \Omega \subseteq \mathbb{R}^m \to \mathbb{R}^{n_2}$ denote the corresponding coordinate functions of two parametric patches on Γ_1 and Γ_2, where $\boldsymbol{\xi} = [\xi_1, \ldots, \xi_m]^T$, $\boldsymbol{u} = [u_1, \ldots, u_{n_1}]^T$ and $\boldsymbol{x} = [x_1, \ldots, x_{n_2}]^T$. Let $\boldsymbol{J}_1(\boldsymbol{\xi})$ and $\boldsymbol{J}_2(\boldsymbol{\xi})$ denote the Jacobian matrices of $\boldsymbol{u}(\boldsymbol{\xi})$ and $\boldsymbol{x}(\boldsymbol{\xi})$, respectively. For brevity, we will omit the argument $\boldsymbol{\xi}$ in the following discussions. The *first fundamental tensors* of Γ_1 and Γ_2 are $\boldsymbol{G}_1 = \boldsymbol{J}_1^T \boldsymbol{J}_1$ and $\boldsymbol{G}_2 = \boldsymbol{J}_2^T \boldsymbol{J}_2$, respectively. Let $\boldsymbol{G}$ denote the matrix $\boldsymbol{G}_2 \boldsymbol{G}_1^{-1}$, which is in general nonsymmetric. Let $\boldsymbol{F}$ denote the mapping from $\boldsymbol{u}(\boldsymbol{\xi})$ to $\boldsymbol{x}(\boldsymbol{\xi})$ over the local patches. The mapping $\boldsymbol{F}$ is said to be *isometric* if $\boldsymbol{G}_2 = \boldsymbol{G}_1$ or equivalently $\boldsymbol{G} = \boldsymbol{I}$, i.e., if the *first fundamental tensor* is preserved by the mapping. Similarly, the mapping $\boldsymbol{F}$ is *conformal* if $\boldsymbol{G}_2 = c\boldsymbol{G}_1$ or equivalently $\boldsymbol{G} = c\boldsymbol{I}$, where $c > 0$ and in general varies from point to point. These definitions of the mappings are classical but inconvenient for numerical computations. It turns out to be revealing to consider the eigenvalues of $\boldsymbol{G}$.

Lemma 1. *Let $\boldsymbol{G}_1$ and $\boldsymbol{G}_2$ denote the first fundamental tensors of local patches on Γ_1 and Γ_2 with parameterization $\boldsymbol{u}(\boldsymbol{\xi}) : \Omega \subseteq \mathbb{R}^m \to \mathbb{R}^{n_1}$ and $\boldsymbol{x}(\boldsymbol{\xi}) : \Omega \subseteq \mathbb{R}^m \to \mathbb{R}^{n_2}$, respectively. Let $\boldsymbol{F}$ denote the mapping from $\mathbf{u}(\boldsymbol{\xi})$ onto $\boldsymbol{x}(\boldsymbol{\xi})$. (a) The eigenvalues λ_i of $\boldsymbol{G} = \boldsymbol{G}_2 \boldsymbol{G}_1^{-1}$ are all positive. (b) $\boldsymbol{F}$ is conformal iff $\lambda_1 = \cdots = \lambda_m$. (c) $\boldsymbol{F}$ is isometric iff $\lambda_i = 1$ for $i = 1, \ldots, m$.*

Due to length limitations we omit all the proofs in this paper. The presence of eigenvalues in Lemma 1 may appear to be daunting computationally in high dimensions. Note that the sum of the eigenvalues of a matrix is equal to its trace (i.e., $\sum_{i=1}^m \lambda_i = \text{tr}(\boldsymbol{G}_2 \boldsymbol{G}_1^{-1})$), and the product of the eigenvalues is equal to its determinant (i.e.,

$\prod_{i=1}^{m} \lambda_i = |\boldsymbol{G}_2 \boldsymbol{G}_1^{-1}|$). We use $|\boldsymbol{A}|$ to denote the determinant of a matrix $\boldsymbol{A}$ in this paper. We then obtain the following lemma.

Lemma 2. *The mapping $\boldsymbol{F}$ is conformal if and only if*

$$\frac{tr(\boldsymbol{G}_2 \boldsymbol{G}_1^{-1})}{\sqrt[m]{|\boldsymbol{G}_2|/|\boldsymbol{G}_1|}} = m. \tag{1}$$

$\boldsymbol{F}$ is isometric if and only if $\boldsymbol{F}$ is conformal and

$$|\boldsymbol{G}_2| = |\boldsymbol{G}_1|. \tag{2}$$

Based on Lemma 2, we can verify whether a mapping is conformal or isometric from $\mathrm{tr}(\boldsymbol{G})$, $|\boldsymbol{G}_2|$, and $|\boldsymbol{G}_1|$, which are easy to compute as we will explain in later sections. Eqs. (1) and (2) are essentially measures for angle- and area-preservation, respectively. This separation of concerns will prove to be very convenient.

2.3 Variational Formulation for Discrete Isometric Mapping

Our preceding discussion focused on continuous surfaces. In meshing or more generally in most numerical computations, the domain is discretized (i.e., triangulated or tesselated), so the surfaces are discrete. Fortunately, variational calculus provides us exactly what we need for generalizing the formulations from continuous to discrete surfaces. The basic idea is to define the mappings over each element, where the formulas for the continuous surfaces apply. We refer to the mapping for each element as the *elemental mapping*, which maps from an *ideal element* to the *actual element* through a *parametric element*. As in the continuous case, we use $\boldsymbol{u}$, $\boldsymbol{x}$, and $\boldsymbol{\xi}$ to denote the coordinates of these elements, respectively. Fig. 1 illustrates the elemental mappings between triangles in 2-D and tetrahedra in 3-D.

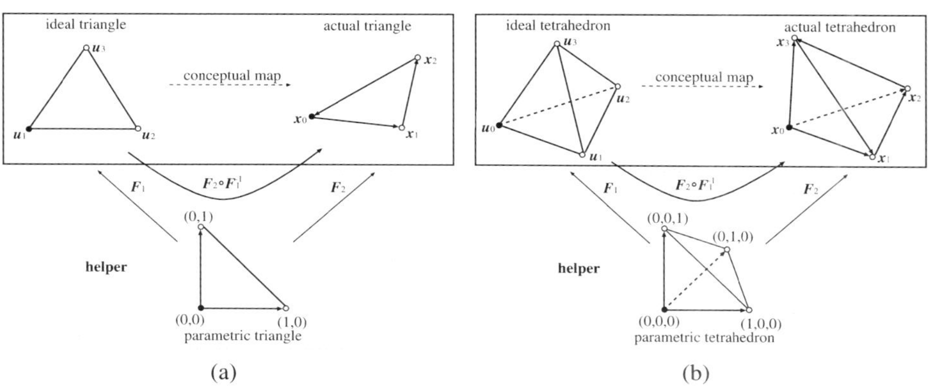

Fig. 1. Schematic of elemental mapping for (a) triangles and (b) tetrahedra

Ideally, we hope all of the elemental mappings are isometric or conformal. Unfortunately, these requirements in general are over-determined and in turn would not yield a solution. We solve the problem in an approximate sense by minimizing a potential energy function (or objective function). This is in spirit similar to solving an over-determined linear system using linear least squares approximations, except that our equations are more complex and nonlinear.

A fundamental question in a variational method is the definition of the energies. For best effectiveness, we require an energy to have the following properties:

Optimality: It must be positive and be minimized for an "ideal" mesh;
Barrier: It should be infinite for degenerate elements with zero area/volume;
Convexity: It should be free of local minimum;
Simplicity: It should be simple and efficient to evaluate and differentiate.

Based on Lemma 2, for each element τ we define two energies E_θ and E_l for angle preservation and area/volume preservation:

$$E_\theta(\tau) = \frac{\mathrm{tr}(\boldsymbol{G}_2\boldsymbol{G}_1^{-1})}{\sqrt[m]{|\boldsymbol{G}_2|/|\boldsymbol{G}_1|}} \tag{3}$$

$$E_l(\tau) = (|\boldsymbol{G}_2|/|\boldsymbol{G}_1|)^p + (|\boldsymbol{G}_1|/|\boldsymbol{G}_2|)^p\,. \tag{4}$$

In E_l, $p > 0$, and the larger p is, the steeper the gradient of E_l becomes. In particular, we choose $p = 0.5$, as we explain later. Per Lemma 2, E_θ is minimized for conformal mappings, and E_θ and E_l are minimized simultaneously for isometric mappings. If the volume of an element is zero, then $|\boldsymbol{G}_2| = 0$ and in turn $E_\theta = E_l = \infty$. Therefore, the first two requirements are satisfied. The convexity and simplicity of these functions are far more involved, which we address in later sections.

We remark that in Euclidean spaces, the *inverse mean ratio* [10, 23] is

$$\frac{\|\boldsymbol{J}_2\boldsymbol{J}_1^{-1}\|_F^2}{m\,(|\boldsymbol{J}_2|/|\boldsymbol{J}_1|)^{2/m}}, \tag{5}$$

which is equal to E_θ/m, because $\mathrm{tr}(\boldsymbol{G}_2\boldsymbol{G}_1^{-1}) = \mathrm{tr}(\boldsymbol{J}_1^{-T}\boldsymbol{J}_2^T\boldsymbol{J}_2\boldsymbol{J}_1^{-1}) = \|\boldsymbol{J}_2\boldsymbol{J}_1^{-1}\|_F^2$ and $|\boldsymbol{G}_i| = |\boldsymbol{J}_i|^2$. However, the energy (3) is more general as it is directly applicable to surfaces and manifolds. Given the energy of elemental mappings, we define the total energy over a mesh M as

$$E(\Gamma,\mu) = \sum_{\tau\in M} \left(\mu E_\theta(\tau) + (1-\mu)E_l(\tau)\right), \tag{6}$$

where $\mu \in [0, 1)$ controls the significance of angle versus volume conservation.

Overall, our method computes the energy and their first and second derivatives with respect to the vertex positions, and then moves each vertex to reduce the energy subject to constraints (such as boundary conditions, validity of the mesh, and preservation of sharp features). We repeat this process for a number of iterations or until the mesh quality no longer improves. This framework applies to meshes in any dimensions. It produces approximations to special types of harmonic maps, so it is justified to refer to our method as *variational mesh optimization* or *mesh smoothing*. However, we have not yet to derive efficient formulas for evaluating and differentiating the energy functions and define simple and effective numerical solutions for solving the resulting nonlinear optimization. We address these issues for surface and volume meshes in the next two sections, respectively.

3 Optimizing Triangular Surface Meshes

We now present our formulation and algorithm for optimizing a triangulated surface mesh embedded in $\mathbb{R}^3$. Obviously, this algorithm would also be applicable to 2-D meshes, but we derive it directly for curved surfaces for generality. As illustrated in Fig. 1(a), we construct a mapping from the ideal to the actual triangle through a parametric triangle. In this specific setting, $\boldsymbol{x}$ has three components, namely $\{x, y, z\}$, while $\boldsymbol{u}$ and $\boldsymbol{\xi}$ has two components, namely $\{u, v\}$ and $\{\xi, \eta\}$, respectively.

3.1 Energy for Angle Preservation

We first derive the formulas for E_θ over a triangle $\tau = \boldsymbol{x}_1\boldsymbol{x}_2\boldsymbol{x}_2$. For efficient minimization of the energy over the mesh, we need to evaluate its value and compute its gradient and Hessian with respect to the vertex positions of the triangle. Fortunately, these can be computed efficiently from the angles (or the first fundamental tensor) of the ideal triangle as well as the edge lengths and area of the actual triangle. As illustrated in Fig. 2, let θ_i denote the angle at the ith vertex of the ideal triangle. Let $a = \sqrt{|\boldsymbol{G}_1|}$ and $A = \sqrt{|\boldsymbol{G}_2|}$, i.e., twice of the areas of the ideal and actual triangle, respectively. Let $\boldsymbol{l}_i$ denote the opposite edge of the ith vertex of the actual triangle, $\widehat{\boldsymbol{n}}$ is its unit normal in $\mathbb{R}^3$, and $\boldsymbol{l}_i^\perp = \widehat{\boldsymbol{n}} \times \boldsymbol{l}_i$, the $90°$ counter-clockwise rotation of $\boldsymbol{l}_i$. We define the shorthands

$$i+ = \begin{cases} i+1 & i<3 \\ 1 & i=3 \end{cases} \text{ and } i- = \begin{cases} i-1 & i>1 \\ 3 & i=1 \end{cases}. \tag{7}$$

Given a function $\boldsymbol{f} = [f_1|f_2|f_3]^T$, let $\nabla_{\boldsymbol{x}_1}\boldsymbol{f} = [\nabla_{\boldsymbol{x}_1}f_1|\nabla_{\boldsymbol{x}_1}f_2|\nabla_{\boldsymbol{x}_1}f_3]$. Let $\|\boldsymbol{x}\|$ denote the 2-norm of a vector $\boldsymbol{x}$. We obtain the following theorem.

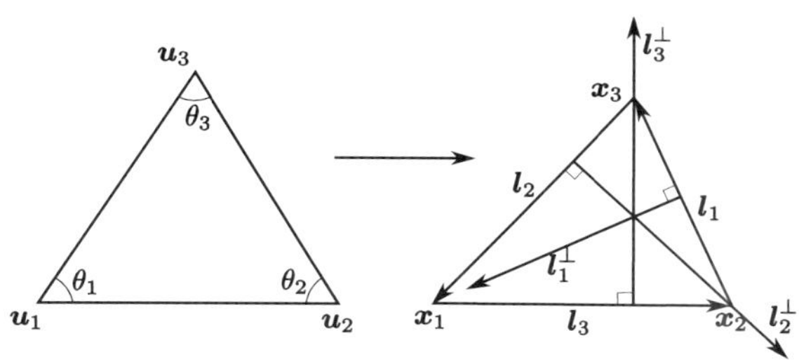

Fig. 2. Naming convention for triangles

Theorem 1. *The energy $E_\theta(\tau)$ in (3) for a triangle $\tau = \boldsymbol{x}_1\boldsymbol{x}_2\boldsymbol{x}_3$ in $\mathbb{R}^3$ is*

$$E_\theta = \frac{1}{A}\sum_{i=1}^{3} \omega_i \|\boldsymbol{l}_i\|^2, \tag{8}$$

where $\omega_i = \cot\theta_i$. Its gradient $\nabla_{\boldsymbol{x}_i} E_\theta$ and Hessian $\nabla^2_{\boldsymbol{x}_i} E_\theta$ with respect to the xyz coordinates of vertex $\boldsymbol{x}_i$ for $i = 1, 2, 3$ are

$$\nabla_{\boldsymbol{x}_i} E_\theta = \frac{2}{A}(\omega_{i+}\boldsymbol{l}_{i+} - \omega_{i-}\boldsymbol{l}_{i-}) - \frac{E_\theta}{A}\boldsymbol{l}_i^{\perp}, \tag{9}$$

$$\nabla^2_{\boldsymbol{x}_i} E_\theta = \frac{2}{A}(\omega_{i+} + \omega_{i-})\boldsymbol{I} - \frac{1}{A}\boldsymbol{B}_i - \frac{E_\theta\|\boldsymbol{l}_i\|^2}{A^2}\hat{\boldsymbol{n}}\hat{\boldsymbol{n}}^T, \tag{10}$$

where $\boldsymbol{I}$ is the 3×3 identity matrix and $\boldsymbol{B}_i = (\nabla_{\boldsymbol{x}_i} E_\theta)\,\boldsymbol{l}_i^{\perp T} + \boldsymbol{l}_i^{\perp}(\nabla_{\boldsymbol{x}_i} E_\theta)^T$.

We note that in $\mathbb{R}^2$, the energy E_θ has re-occurred many times in the literature: It is essentially equivalent to the functional for harmonic mapping in [17], the "MIPS energy" for conformal parameterization in [16], and the condition-number-based metric in [12]. The notation using cotangent is due to Pinkall and Polthier [26], but we generalized it to triangles in $\mathbb{R}^3$. Our formulas for the gradient and Hessian in (9) and (10) appear to be new. In (13), the $\hat{\boldsymbol{n}}\hat{\boldsymbol{n}}^T$ term has no effect if the surface is flat and is negligible if a vertex moves only nearly tangentially. In surface mesh optimization, all vertices move nearly tangentially, so we can omit this term. Under this assumption, for any reference triangle with acute angles, $-\boldsymbol{B}_i$ is positive definite, so is $\nabla^2_{\boldsymbol{x}_i} E_\theta$, and E_θ is therefore convex. A different analysis in $\mathbb{R}^2$ can be found [23].

For isotropic meshing, the ideal triangle is regular, and $\omega_i = \cot 60° = 1/\sqrt{3}$ for $i = 1, 2, 3$. It is more convenient to use $\tilde{E}_\theta = \sqrt{3}E_\theta$ instead of E_θ for the energy. In this case, the energy is simply the ratio between the sum of the squared edge lengths versus twice of the area, and we then have the following simplified formulas:

$$\tilde{E}_\theta = \frac{1}{A}\sum_{i=1}^{3}\|\boldsymbol{l}_i\|^2, \tag{11}$$

$$\nabla_{\boldsymbol{x}_i}\tilde{E}_\theta = \frac{1}{A}(2\boldsymbol{l}_{i+} - 2\boldsymbol{l}_{i-} - \tilde{E}_\theta\boldsymbol{l}_i^{\perp}), \tag{12}$$

$$\nabla^2_{\boldsymbol{x}_i}\tilde{E}_\theta \approx \frac{1}{A}\left(4\boldsymbol{I} - \tilde{\boldsymbol{B}}_i\right), \tag{13}$$

where $\tilde{\boldsymbol{B}}_i = \left(\nabla_{\boldsymbol{x}_i}\tilde{E}_\theta\right)\boldsymbol{l}_i^{\perp T} + \boldsymbol{l}_i^{\perp}\left(\nabla_{\boldsymbol{x}_i}\tilde{E}_\theta\right)^T$.

3.2 Energy for Area Preservation

For triangles, our definition of E_l in (4) reduces to

$$E_l = \frac{A^{2p}}{a^{2p}} + \frac{a^{2p}}{A^{2p}}, \tag{14}$$

where $a = \sqrt{|\boldsymbol{G}_1|}$, $A = \sqrt{|\boldsymbol{G}_2|}$, and $p > 0$. This energy controls the triangle areas. Assume a is independent of $\boldsymbol{x}_i$. It can be shown that the gradient and Hessian of E_l with respect to $\boldsymbol{x}_i$ are

$$\nabla_{\boldsymbol{x}_i} E_l = \frac{2p}{A}\left(\frac{A^{2p}}{a^{2p}} - \frac{a^{2p}}{A^{2p}}\right)\boldsymbol{l}_i^{\perp} = \frac{2p\left(A^{4p} - a^{4p}\right)}{a^{2p}A^{1+2p}}\boldsymbol{l}_i^{\perp}, \tag{15}$$

$$\begin{aligned}\nabla^2_{\boldsymbol{x}_i} E_l &\approx \left(\frac{8p^2A^{4p}}{a^{2p}A^{2+2p}} - \frac{(1+2p)2p\left(A^{4p} - a^{4p}\right)}{a^{2p}A^{2+2p}}\right)\boldsymbol{l}_i^{\perp}\boldsymbol{l}_i^{\perp T} \\ &= \frac{(4p^2 - 2p)A^{4p} + (4p^2 + 2p)a^{4p}}{a^{2p}A^{2+2p}}\boldsymbol{l}_i^{\perp}\boldsymbol{l}_i^{\perp T},\end{aligned} \tag{16}$$

where the residual in $\nabla^2_{\boldsymbol{x}_i} E_l$ is a negligible $\hat{\boldsymbol{n}}\hat{\boldsymbol{n}}^T$ term. $\nabla^2_{\boldsymbol{x}_i} E_l$ is positive semi-definite if $p \geq 0.5$. For $p = 0.5$, the formulas simplify to

$$\nabla_{\boldsymbol{x}_i} E_l = \frac{\left(A^2 - a^2\right)}{aA^2}\boldsymbol{l}_i^{\perp} \text{ and } \nabla^2_{\boldsymbol{x}_i} E_l \approx \frac{2a}{A^3}\boldsymbol{l}_i^{\perp}\boldsymbol{l}_i^{\perp T}. \tag{17}$$

Therefore, we choose $p = 0.5$ in practice. This choice of E_l coincides with the energy used for area distortion for surface parameterization in [3].

3.3 Energy Minimization

The total energy is the sum of the elemental energies over all the triangles. The gradient and Hessian of the total energy with respect to a vertex $\boldsymbol{x}_v$ is equal to the summation of those in the incident triangles of the vertex, i.e.,

$$\nabla_{\boldsymbol{x}_v} E = \sum_{\{\tau | v \in \tau\}} \left(\mu \nabla_{\boldsymbol{x}_v} \tilde{E}_\theta(\tau) + (1-\mu)\nabla_{\boldsymbol{x}_v} E_l(\tau)\right), \tag{18}$$

$$\nabla^2_{\boldsymbol{x}_v} E = \sum_{\{\tau | v \in \tau\}} \left(\mu \nabla^2_{\boldsymbol{x}_v} \tilde{E}_\theta(\tau) + (1-\mu)\nabla^2_{\boldsymbol{x}_v} E_l(\tau)\right). \tag{19}$$

Because $\nabla^2_{\boldsymbol{x}_v} \tilde{E}_\theta(\tau)$ is positive definite and $\nabla^2_{\boldsymbol{x}_i} E_l$ is positive semi-definite, $\nabla^2_{\boldsymbol{x}_v} E$ is positive definite if $0 \leq \mu < 1$ and all the triangles have positive areas. In other words, E is convex locally at each vertex. In practice, we choose $\mu = 0$ for conformal optimization and $\mu = 0.5$ for isometric optimization. After obtaining the gradient and Hessian, one could apply one step of Newton's method to determine a displacement $\boldsymbol{d}_v$ for the vertex, i.e., by solving $\left(\nabla^2_{\boldsymbol{x}_v} E\right)\boldsymbol{d}_v = -\nabla_{\boldsymbol{x}_v} E$. To preserve the geometry, we must constrain the displacement to be nearly tangential. We compute the tangent vectors using an eigenvalue analysis as described in [18]. For a vertex on a smooth surface, let $\hat{\boldsymbol{t}}_1$ and $\hat{\boldsymbol{t}}_2$ denote two orthonormal tangent vectors at v, and let $\boldsymbol{T} = [\hat{\boldsymbol{t}}_1 \mid \hat{\boldsymbol{t}}_2]$. Instead of solving for $\boldsymbol{d}_v$ in $\mathbb{R}^3$, we reformulate the problem to solve for $\boldsymbol{d}_v = \boldsymbol{T}\boldsymbol{u}$, where $\boldsymbol{u} \in \mathbb{R}^2$. The gradient and Hessian of E with respect to $\boldsymbol{u}$ are then $\nabla_{\boldsymbol{u}} E = \boldsymbol{T}^T \nabla_{\boldsymbol{x}_v} E$ and $\nabla^2_{\boldsymbol{u}} E = \boldsymbol{T}^T\left(\nabla^2_{\boldsymbol{x}_v} E\right)\boldsymbol{T}$, respectively. Therefore, Newton's equation for minimizing E with respect to $\boldsymbol{u}$ becomes

$$\left(\boldsymbol{T}^T\left(\nabla^2_{\boldsymbol{x}_v} E\right)\boldsymbol{T}\right)\boldsymbol{u} = -\boldsymbol{T}^T\nabla_{\boldsymbol{x}_v} E,$$

and therefore

$$\boldsymbol{d}_v = -\boldsymbol{T}\left(\boldsymbol{T}^T\left(\nabla^2_{\boldsymbol{x}_v}E\right)\boldsymbol{T}\right)^{-1}\boldsymbol{T}^T\nabla_{\boldsymbol{x}_v}E. \tag{20}$$

Note that $\nabla^2_{\boldsymbol{u}}E$ is symmetric positive definite for any reasonable approximations of $\boldsymbol{T}$. For a point on a ridge curve, we replace $\boldsymbol{T}$ in (20) by the unit tangent vector to the ridge curve. If a point is at a corner, we fix the vertex and set $\boldsymbol{d}_v = \boldsymbol{0}$. This procedure preserves sharp features and preserve smooth surface geometry to second-order accuracy. If higher-order accuracy is required, the displacement $\boldsymbol{d}_v$ can be adjusted by projecting onto a high-order reconstruction of the surface as in [11].

3.4 Overall Algorithm

For the overall algorithm, it is most efficient and convenient to compute the gradient and Hessian of E_θ and E_l over all the triangles, accumulate them to vertices, and then compute the displacements for all the vertices concurrently. This is similar to a typical finite-element code. When applying the displacements, we determine a relaxation factor α_v for each vertex and then add $\alpha_v\boldsymbol{d}_v$ to the vertex, where the relaxation factor is chosen to ensure mesh validity, as we will explain shortly. Algorithm 1 summarizes the core of this variational algorithm, which is essentially a block-Jacobi solver for one step of Newton's method. Our overall algorithm repeatedly invokes this procedure for a desired number of iterations or until the mesh no longer improves.

Algorithm 1. One step of variational smoothing of surface triangulation.

Initialize vector grad and matrix H to zero for each vertex;
for each triangle τ **do**
  for each vertex $\boldsymbol{x}_i$ of τ **do**
    $v \leftarrow$vertex id of $\boldsymbol{x}_i$
    $\text{grad}[v] \leftarrow \text{grad}[v] + \mu\nabla_{\boldsymbol{x}_i}\tilde{E}_\theta(\tau) + (1-\mu)\nabla_{\boldsymbol{x}_i}E_l(\tau)$;
    $\text{H}[v] \leftarrow \text{H}[v] + \mu\nabla^2_{\boldsymbol{x}_i}\tilde{E}_\theta(\tau) + (1-\mu)\nabla^2_{\boldsymbol{x}_i}E_l(\tau)$;
  end for
end for
for each vertex v **do**
  let $\boldsymbol{T}$ be composed of tangent vector to surface at v;
  $\boldsymbol{x}_v \leftarrow \boldsymbol{x}_v - \alpha_v\boldsymbol{T}\left(\boldsymbol{T}^T\left(\text{H}[v]\right)\boldsymbol{T}\right)^{-1}\boldsymbol{T}^T\text{grad}[v]$;
end for

In Algorithm 1, we compute the displacements for all the vertices concurrently for better efficiency and ease of implementation. However, concurrent vertex motions may lead to mesh folding if checking is not performed. To address this problem, we introduce a safeguard via an asynchronous step-size control: For each triangle $\boldsymbol{x}_1\boldsymbol{x}_2\boldsymbol{x}_3$, we solve for the maximum $\alpha \leq 1$ such that the triangle $\boldsymbol{x}_1^{(\alpha)}\boldsymbol{x}_2^{(\alpha)}\boldsymbol{x}_3^{(\alpha)}$ does not fold, where $\boldsymbol{x}_i^{(\alpha)} = \boldsymbol{x}_i + \alpha\boldsymbol{d}_i$. We scale the displacement $\boldsymbol{d}_v$ at v by a factor α_v equal to the minimum α among the incident faces of v. After rescaling the displacement of all the vertices, we recompute α and repeat the rescaling process until $\alpha_v = 1$ for all vertices. Finally, we move all the vertices by the scaled displacement.

Note that this rescaling scheme can be used with any smoothing algorithms to prevent mesh folding. It is particularly effective for our method because the "barrier" property of the energy functions prevents the formation of poor-shaped elements. In practice, the rescaling is rarely invoked except during the first few iterations when optimizing an extremely poor-shaped input mesh.

4 Optimizing Tetrahedral Meshes

We now extend our algorithm to optimize tetrahedral meshes. We reuse the same framework, including the definitions of the energies and the control flow in Algorithm 1. In the elemental mapping, as illustrated in Fig. 1(b), all the coordinates, namely, $\boldsymbol{x}$, $\boldsymbol{u}$, and $\boldsymbol{\xi}$, now have three components, denoted by $\{x, y, z\}$, $\{u, v, w\}$, and $\{\xi, \eta, \zeta\}$, respectively. Our main focus here will be on the efficient evaluation and differentiation of the energies.

4.1 Energy for Angle Preservation

For tetrahedral elements, the angle-preserving energy of the elemental map for a tetrahedron $\tau = \boldsymbol{x}_0\boldsymbol{x}_1\boldsymbol{x}_2\boldsymbol{x}_3$ in $\mathbb{R}^3$ is

$$E_\theta(\tau) = \frac{\text{tr}(\boldsymbol{G})}{|\boldsymbol{G}_2/\boldsymbol{G}_1|^{1/3}}. \tag{21}$$

As illustrated in Fig. 3, we label the edges as $\boldsymbol{l}_i$ and $\boldsymbol{r}_i$ for $i = 1, 2, 3$, where $\boldsymbol{l}_i = \boldsymbol{x}_i - \boldsymbol{x}_0$ and $\boldsymbol{l}_i = \boldsymbol{x}_{i-} - \boldsymbol{x}_{i+}$, where $i+$ and $i-$ are defined in (7). Let $\boldsymbol{n}_0 = \boldsymbol{r}_2 \times \boldsymbol{r}_1$, the normal to the opposite triangle of $\boldsymbol{x}_0$ pointing toward $\boldsymbol{x}_0$ with magnitude equal to twice of the triangle area. Let

$$\boldsymbol{G}_1 = \begin{bmatrix} g_{11} & g_{12} & g_{13} \\ g_{21} & g_{22} & g_{23} \\ g_{31} & g_{32} & g_{33} \end{bmatrix} \text{ and } \boldsymbol{G}_2 = \begin{bmatrix} h_{11} & h_{12} & h_{13} \\ h_{21} & h_{22} & h_{23} \\ h_{31} & h_{32} & h_{33} \end{bmatrix},$$

where $g_{ij} = g_{ij}$ and $h_{ij} = h_{ji}$, and let $\boldsymbol{g}_j$ and $\boldsymbol{h}_j$ denote the jth columns of $\boldsymbol{G}_1$ and $\boldsymbol{G}_2$, respectively. Let $\nu = \sqrt{|\boldsymbol{G}_1|}$ and $V = \sqrt{|\boldsymbol{G}_2|} | = -\boldsymbol{n}_0^T \boldsymbol{l}_i$, i.e., six times of the

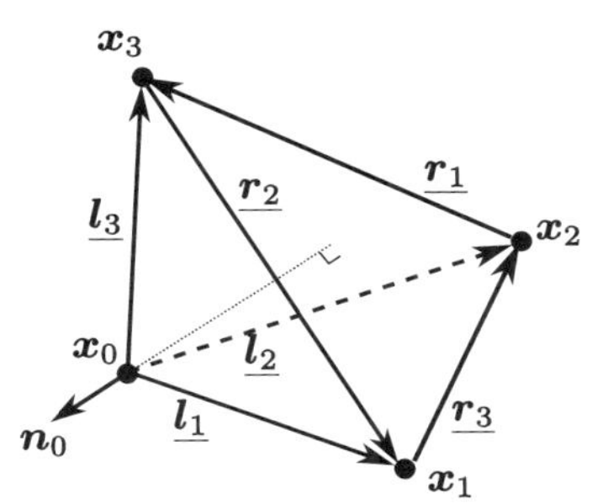

Fig. 3. Naming convention for tetrahedron. Underlined symbols indicate edges.

volumes of the ideal and actual tetrahedra, respectively. Let $\text{sum}(\boldsymbol{a}) = a_1 + a_2 + a_3$ for any $\boldsymbol{a} \in \mathbb{R}^3$. We obtain the following theorem.

Theorem 2. *The energy $E_\theta(\tau)$ for a tetrahedron $\tau = \boldsymbol{x}_0\boldsymbol{x}_1\boldsymbol{x}_2\boldsymbol{x}_3$ in $\mathbb{R}^3$ is*

$$E_\theta = \frac{1}{V^{2/3}} \sum_{i=1}^{3} \left(\alpha_i \|\boldsymbol{l}_i\|^2 + \beta_i \|\boldsymbol{r}_i\|^2 \right), \tag{22}$$

where $\alpha_i = \text{sum}(\boldsymbol{g}_{i+} \times \boldsymbol{g}_{i-})/\nu^{4/3}$ and $\beta_i = (g_{ii}g_{i-,i+} - g_{i,i-}g_{i,i+})/\nu^{4/3}$. Its gradient $\nabla_{\boldsymbol{x}_0} E_\theta$ and Hessian $\nabla^2_{\boldsymbol{x}_0} E_\theta$ with respect to $\boldsymbol{x}_0$ are

$$\nabla_{\boldsymbol{x}_0} E_\theta = -\frac{2}{V^{2/3}} \left(\sum_{i=1}^{3} \alpha_i \boldsymbol{l}_i \right) - \frac{2}{3V} E_\theta \boldsymbol{n}_0 \tag{23}$$

$$\nabla^2_{\boldsymbol{x}_0} E_\theta = \left(\frac{2}{V^{2/3}} \sum_{i=1}^{3} \alpha_i \right) \boldsymbol{I} - \frac{2}{3V} \boldsymbol{B} - \frac{2}{9V^2} E_\theta \boldsymbol{n}_0 \boldsymbol{n}_0^T, \tag{24}$$

where $\boldsymbol{I}$ denotes the 3×3 identity matrix and $\boldsymbol{B} = (\nabla_{\boldsymbol{x}_0} E_\theta)\, \boldsymbol{n}_0^T + \boldsymbol{n}_0 \,(\nabla_{\boldsymbol{x}_0} E_\theta)^T$.

The theorem gives only the derivatives with respect to $\boldsymbol{x}_0$, but it is easy to adapt the formulas for the other vertices by symmetry. To the authors' knowledge, these closed-form formulas are all new. For isometric mesh smoothing where the ideal tetrahedra are regular, $\alpha_i = \beta_i = 2^{-4/3}$, and it is more convenient to omit these constants and use the alternative function

$$\tilde{E}_\theta = \frac{1}{V^{2/3}} \sum_{i=1}^{3} \left(\|\boldsymbol{l}_i\|^2 + \|\boldsymbol{r}_i\|^2 \right), \tag{25}$$

i.e., the ratio between the sum of squared edge lengths versus $V^{2/3}$. The gradient and Hessian of $\tilde{E}_\theta$ with respect to $\boldsymbol{x}_0$ are then

$$\nabla_{\boldsymbol{x}_0} \tilde{E}_\theta = -\frac{2}{V^{2/3}} \sum_{i=1}^{3} \boldsymbol{l}_i - \frac{2}{3V} \tilde{E}_\theta \boldsymbol{n}_0, \tag{26}$$

$$\nabla^2_{\boldsymbol{x}_0} \tilde{E}_\theta = \frac{6}{V^{2/3}} \boldsymbol{I} - \frac{2}{3V} \tilde{\boldsymbol{B}} - \frac{2}{9V^2} \tilde{E}_\theta \boldsymbol{n}_0 \boldsymbol{n}_0^T, \tag{27}$$

where $\tilde{\boldsymbol{B}} = \left(\nabla_{\boldsymbol{x}_0} \tilde{E}_\theta\right) \boldsymbol{n}_0^T + \boldsymbol{n}_0 \left(\nabla_{\boldsymbol{x}_0} \tilde{E}_\theta\right)^T$. By plugging (26) into (27), it is easy to show that $\nabla^2_{\boldsymbol{x}_0} \tilde{E}_\theta$ is positive definite as long as $\tilde{E}_\theta > 0$.

Note that the aspect ratio of a tetrahedron [25] is defined as

$$\rho = \frac{c}{V} \left(\sum_{i=1}^{3} \left(\|\boldsymbol{l}_i\|^2 + \|\boldsymbol{r}_i\|^2 \right) \right)^{3/2},$$

where $c \approx 0.048113$. When a regular tetrahedron is used as the reference element, $\tilde{E}_\theta$ and the inverse mean ratio (5) are both a constant factor of $\rho^{2/3}$. This connection provides another justification for minimizing $\tilde{E}_\theta$ for isotropic meshes.

4.2 Energy for Volume Preservation

For tetrahedra, the energy for volume preservation is

$$E_l = \frac{V^{2p}}{\nu^{2p}} + \frac{\nu^{2p}}{V^{2p}} \tag{28}$$

for $p > 0$. For simplicity, we assume ν is independent of $\boldsymbol{x}_i$. The gradient and Hessian of E_l with respect to $\boldsymbol{x}_0$ are

$$\nabla_{\boldsymbol{x}_0} E_l = \frac{2p}{V}\left(\frac{V^{2p}}{\nu^{2p}} - \frac{\nu^{2p}}{V^{2p}}\right)\boldsymbol{n}_0 = \frac{2p\left(V^{4p} - \nu^{4p}\right)}{\nu^{2p}V^{1+2p}}\boldsymbol{n}_0, \tag{29}$$

$$\nabla^2_{\boldsymbol{x}_0} E_l = \frac{(4p^2 - 2p)V^{4p} + (2p + 4p^2)\nu^{4p}}{\nu^{2p}V^{2+2p}}\boldsymbol{n}_0\boldsymbol{n}_0^T. \tag{30}$$

As for surfaces, $\nabla^2_{\boldsymbol{x}_0} E_l$ is positive semi-definite if $p \geq 0.5$ and all the volumes are positive. We choose $p = 0.5$ in practice, which lead to simplified formulas

$$\nabla_{\boldsymbol{x}_0} E_l = \frac{\left(V^2 - \nu^2\right)}{\nu V^2}\boldsymbol{n}_0 \text{ and } \nabla^2_{\boldsymbol{x}_0} E_l = \frac{2\nu}{V^3}\boldsymbol{n}_0\boldsymbol{n}_0^T. \tag{31}$$

Let E be the total energy defined in (6) for tetrahedral meshes. $\nabla^2_{\boldsymbol{x}_i} E$ is symmetric positive definite for any tetrahedral mesh with positive volume, so E is convex locally at each vertex. To smooth a tetrahedral mesh, it is now a simple matter of plugging in the above formulas for E_θ and E_l into Algorithm 1, except that the triangles would be replaced by the tetrahedra. Regarding to the matrix $\boldsymbol{T}$ in the algorithm, we construct it from the tangent vectors for the vertices on the boundary but replace $\boldsymbol{T}$ by the identity matrix for interior vertices.

5 Experimental Study

In this section, we present some experimental results using our methods for static and dynamic meshes. We report results only for tetrahedral meshes. Because our volume mesh optimization contains integrated surface mesh optimization, these results are also representative for surface meshes. To assess the methods, we measure mesh qualities using the maximum and minimum dihedral angles of tetrahedra. These measures are intuitive and can capture most bad elements, including slivers.

5.1 Optimizing Static Meshes

For static meshes, we used two meshes of a unit cube generated using GAMBIT of Fluent/ANSYS Inc., and also meshes of a hand and a dragon from [2], as shown in Fig. 4. The latter generated the meshes by enhancing some initial meshes through mesh modification and Laplacian smoothing. We optimized the meshes using conformal optimization with fixed boundary, conformal optimization with surface optimization, and

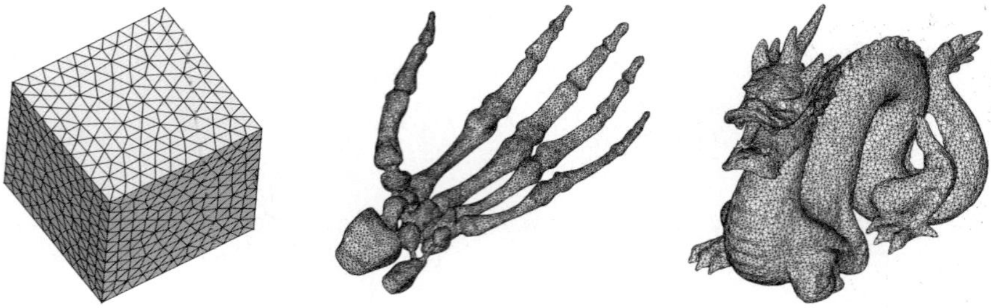

Fig. 4. Sample test tetrahedral meshes

Table 1. Optimization of tetrahedral meshes

Case	#vertices	#tets	Min and max dihedral angles in degrees			
			Original	Fixed-boundary	Conformal	Isometric
Cube-1K	1,603	7,417	18.9 / 149.3	22.1 / 141.7	22.1 / 141.7	22.4 / 14.1.3
Cube-10K	10,246	53,068	14.4 / 157.0	19.6 / 152.7	19.6 / 152.7	19.9 / 151.9
Hand-10K	3,144	10,000	7.9 / 164.4	7.9 / 163.5	13.7 / 160.9	13.9 / 160.9
Hand-100K	16,649	99,995	3.6 / 174.6	3.6 / 174.6	6.1 / 171.0	6.4 / 170.8
Dragon-5k	1490	4,999	8.0 / 161.2	8.6 / 160.3	12.1 / 160.3	12.1 / 161.1
Dragon-100k	26,436	100,000	3.4 / 174.7	3.4 / 174.7	4.8 / 171.7	4.8 / 171.7

isometric optimization with surface optimization. In isometric optimization, for each tetrahedron we used the average volume of its neighbors as the desired volume.

Table 1 shows the quality measures of the initial and optimized meshes. We observe that the optimization with fixed boundary delivered virtually no improvement for the hand and dragon meshes. This is likely because the worst mesh qualities are constrained by the vertices on the boundary. On the other hand, simultaneous surface and volume mesh optimization improved the dihedral angles in all cases by about 3 to 7.7 degrees. Furthermore, we observe that isometric optimization produced roughly the same results as conformal optimization in terms of the dihedral angles while it provides better control of element sizes and improved the volume ratios by about 50% in our tests.

5.2 Optimizing Moving Meshes

One of the motivating applications of our methods is mesh motion. To demonstrate the effectiveness for such applications, we compressed the smaller mesh of the unit cube progressively up to 40% of its dimension along the x direction. We adapted the time steps to prevent mesh folding and performed up to 10 iterations of mesh smoothing per time step. Before smoothing, the worst minimum dihedral angle was $0.43°$ and the maximum was $179.3°$ among all the time steps, which were improved to $18.2°$ and $154.4°$ respectively after smoothing. Fig. 5(a) shows the cross-section view of the mesh after 40% compression along with the original unit cube. Fig. 5(b) and (c) show the profiles of the minimum and maximum dihedral angles of the original mesh and

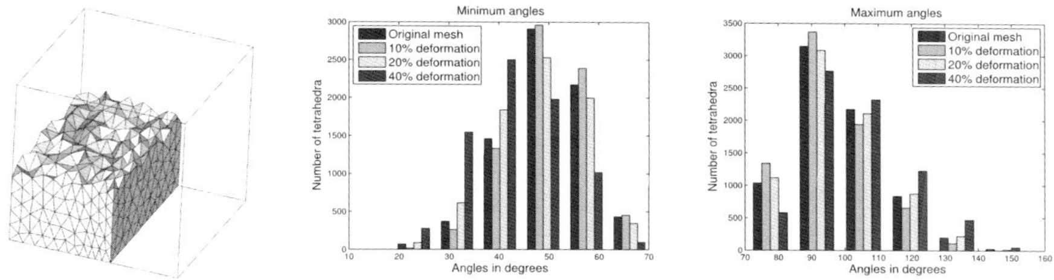

(a) (b) (c)

Fig. 5. Cross section of cube after 40% compression (a) and profiles of dihedral angles (b-c)

after 10%, 20%, and 40% compressions. Our method delivered high quality meshes throughout the simulation. Furthermore, our optimized mesh after 10% compression eevn has a higher quality than the original undeformed mesh.

As another example, we deform a sphere of radius 0.15 centered at $(0.5, 0.75, 0.5)$ in a velocity field given by

$$u(x, y, z) = \sin^2(\pi x)(\sin(2\pi z) - \sin(2\pi y)), \tag{32}$$

$$v(x, y, z) = \sin^2(\pi y)(\sin(2\pi x) - \sin(2\pi z)), \tag{33}$$

$$w(x, y, z) = \sin^2(\pi z)(\sin(2\pi y) - \sin(2\pi x)). \tag{34}$$

We move the boundary vertices by integrating the velocity using the fourth-order Runge-Kutta scheme and move the interior vertices by mesh smoothing. Fig. 6(a) shows the deformed shape of the sphere at $t = 0.3$. As in the previous example, we constrained the time steps so that the tetrahedra do not become inverted. We compare our variational smoothing with Laplacian smoothing. Before smoothing, the worst minimum dihedral angle was as small as $0.004°$ and the maximum was as large as $179.987°$ among all the time steps, which were improved to $1.74°$ and $175.66°$ by Laplacian smoothing, respectively. In contrast, our variational smoothing improved the angles to $5.32°$ and $171.08°$, respectively. Fig. 6(b) and (c) show the histograms of the dihedral angles at times $t = 0.15$ and 0.3 along with the original mesh. As we highlighted in the figure,

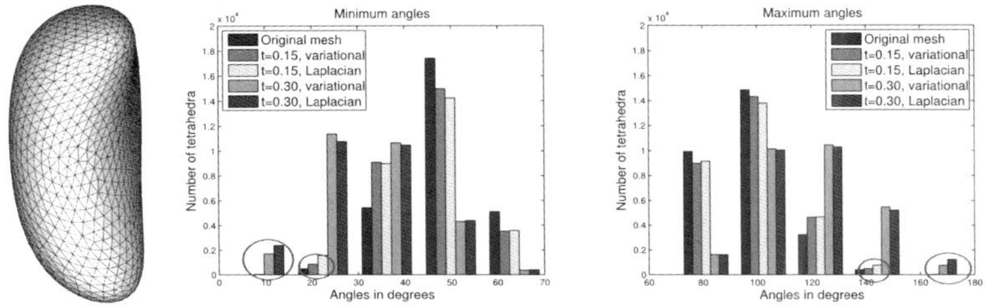

(a) (b) (c)

Fig. 6. Sphere after moderate deformation (a) and profiles of dihedral angles (b-c)

Laplacian smoothing produced far more poor-shaped elements than variational smoothing, although both methods produced similar distributions for well-shaped elements. In this test, our variational smoothing was about twice slower than Laplacian smoothing. However, it allowed time steps twice as large as those allowed by Laplacian smoothing, so it can result in a net saving in the total computational time in a dynamic simulation, where the physics solvers dominate the overall computational cost.

5.3 Comparison of Computational Cost

For tetrahedral meshes, computational cost of mesh smoothing is often a major concern. The most computational intensive steps are the evaluation and differentiation of the energies. We use the methods in [10], including the block coordinate decent method and inexact Newton method, as the basis of a comparison. The block coordinate decent method requires 480 floating-point operations (flops) to compute the gradient and Hessian of the elemental energy with respect to each vertex, which amounts to $480 \times 4 = 1920$ flops per element. The inexact Newton's method requires 20% extra operations. In contrast, our formulas require 312 flops per element to compute the gradient and Hessian of the elemental energy with respect to all its vertices (including 159 multiplications, 142 additions, and 1 power operation), which are more than six times faster than their counterparts in [10]. In terms of memory requirement, our method requires nine floating-point numbers per vertex, which is nearly an order of magnitude less than what is required by the inexact Newton's method. Besides these reductions in cost, our algorithm is easy to implement, without requiring any sophisticated packages. Therefore, we consider our algorithm as simple, efficient, and effective, and it is promising for direct integration into simulation codes.

6 Conclusions and Discussions

In this paper, we presented a new variational method for optimizing surface and volume meshes. We defined the energy functions based on conformal and isometric mappings between actual elements with ideal reference elements and derived closed-form formulas for their gradient and Hessian. In addition, we developed a simple and efficient algorithm for optimizing the surface and volume meshes by minimizing the energies. We compared our method with some existing methods and demonstrated its effectiveness for optimizing static and moving meshes.

Our method is not without limitations. First, it requires the input mesh to be valid (i.e., with no folding). This is not a severe constraint for numerical simulations where mesh validity is required by other numerical computations. However, it may limit the applicability of our method for other purposes. This limitation can be alleviated by applying a mesh untangling procedure, such as in [9]. For future research directions, we plan to extend our method to optimize quadrilateral meshes and hybrid volume meshes and extend it to anisotropic mesh optimization.

Acknowledgement. This work was supported by National Science Foundation under award number DMS-0809285 and also by a subcontract from the Center for Simulation of Advanced Rockets of the University of Illinois at Urbana-Champaign funded by the U.S. Department of Energy through the University of California under subcontract B523819. We thank Dr. Eric Shaffer of UIUC for pointers to some test meshes.

References

[1] Amenta, N., Bern, M., Eppstein, D.: Optimal point placement for mesh smoothing. J. Algorithms 30, 302–322 (1999)

[2] Cutler, B., Dorsey, J., McMillan, L.: Simplification and improvement of tetrahedral models for simulation. In: Eurographics / ACM SIGGRAPH Symposium on Geometry Processing, pp. 93–102 (2004), http://people.csail.mit.edu/bmcutler/PROJECTS/SGP04/index.html

[3] Degener, P., Meseth, J., Klein, R.: An adaptable surface parameterization method. In: Proc. 12th Int. Meshing Roundtable, pp. 227–237 (2003)

[4] do Carmo, M.P.: Differential Geometry of Curves and Surfaces. Prentice-Hall, Englewood Cliffs (1976)

[5] do Carmo, M.P.: Riemannian Geometry. Birkhäuser, Boston (1992)

[6] Field, D.A.: Laplacian smoothing and Delaunay triangulation. Comm. Appl. Numer. Meth. 4, 709–712 (1988)

[7] Floater, M.S., Hormann, K.: Surface parameterization: a tutorial and survey. In: Advances in Multiresolution for Geometric Modelling, pp. 157–186. Springer, Heidelberg (2005)

[8] Freitag, L.A., Knupp, P.M.: Tetrahedral mesh improvement via optimization of the element condition number. Int. J. Numer. Meth. Engr. 53, 1377–1391 (2002)

[9] Freitag, L.A., Plassmann, P.: Local optimization-based simplicial mesh untangling and improvement. Int. J. Numer. Meth. Engr. 49, 109–125 (2000)

[10] Freitag Diachin, L., Knupp, P.M., Munson, T., Shontz, S.: A comparison of two optimization methods for mesh quality improvement. Engrg. Comput. 22, 61–74 (2006)

[11] Frey, P., Borouchaki, H.: Geometric surface mesh optimization. Comput. Visual. Sci, 113–121 (1998)

[12] Garimella, R.V., Shashkov, M.J., Knupp, P.M.: Triangular and quadrilateral surface mesh quality optimization using local parametrization. Comput. Meth. Appl. Mech. Engrg. 193, 913–928 (2004)

[13] Hansbo, P.: Generalized Laplacian smoothing of unstructured grids. Comm. Numer. Meth. Engrg. 11, 455–464 (1995)

[14] Hansen, G.A., Douglass, R.W., Zardecki, A.: Mesh Enhancement. Imperial College Press (2005)

[15] Hirt, C.W., Amsden, A.A., Cook, J.L.: An arbitrary Lagrangian-Eulerian computing method for all flow speeds. J. Comput. Phys. 14, 227–253 (1974); Reprinted in 135, 203–216 (1997)

[16] Hormann, K., Greiner, G.: MIPS: An efficient global parametrization method. In: Laurent, P.-J., Sablonniere, P., Schumaker, L.L. (eds.) Curve and Surface Design: Saint-Malo 1999, pp. 153–162 (2000)

[17] Ivanenko, S.A.: Harmonic mappings. In: Thompson, J.F., Soni, B.K., Weatherill, N.P. (eds.) Handbook of Grid Generation. CRC, Boca Raton (1999)

[18] Jiao, X.: Face offsetting: A unified approach for explicit moving interfaces. J. Comput. Phys. 220, 612–625 (2007)

[19] Knupp, P.M.: Achieving finite element mesh quality via optimization of the Jacobian matrix norm and associated quantities. Part i: a framework for surface mesh optimization. Int. J. Numer. Meth. Engr. 48, 401–420 (2000)

[20] Knupp, P.M.: Algebraic mesh quality metrics. SIAM J. Sci. Comput. 23, 193 (2001)

[21] Knupp, P.M., Robidoux, N.: A framework for variational grid generation: conditioning the Jacobian matrix with matrix norms. SIAM J. Sci. Comput. 21, 20–29 (2000)
[22] Kreyszig, E.: Differential Geometry. Dover (1991)
[23] Munson, T.: Mesh shape-quality optimization using the inverse mean-ratio metric. Math. Program. Ser. A 110, 561–590 (2007)
[24] Parthasarathy, V.N., Kodiyalam, S.: A constrained optimization approach to finite element mesh smoothing. Finite Elem. Anal. Des. 9, 309–320 (1991)
[25] Parthasarathy, V.N., Graichen, C.M., Af, A.F.H.: A comparison of tetrahedron quality measures. Finite Elem. Anal. Des. 15, 255–261 (1993)
[26] Pinkall, U., Polthier, K.: Computing discrete minimal surfaces and their conjugates. Exper. Math. 2, 15–36 (1993)
[27] Zhou, T., Shimada, K.: An angle-based approach to two-dimensional mesh smoothing. In: Proc. 9th Int. Meshing Roundtable, pp. 373–384 (2000)

Simultaneous Aligning and Smoothing of Surface Triangulations

J.M. Escobar, R. Montenegro, E. Rodríguez, and G. Montero

Institute for Intelligent Systems and Numerical Applications in Engineering,
University of Las Palmas de Gran Canaria, Campus Universitario de Tafira,
35017 Las Palmas de Gran Canaria, Spain
{jmescobar,rmontenegro,erodriguez,gmontero}@siani.es
http://www.dca.iusiani.ulpgc.es/proyecto0507

Summary. In this work we develop a procedure to deform a given surface triangulation to obtain its alignment with interior curves. At present, we consider that these curves are defined by the orthogonal projection from plane cubic splines to the initial surface triangulation. For example, the curves can represent interfaces between different materials or boundary conditions, internal boundaries or feature lines. Another possibility of this procedure is the adaption of a reference mesh to changing curves in the course of an evolutionary process (for example, aligning of mesh nodes and edges to moving shocks in compressible flows). Specifically, we propose a new method that moves the nodes of the mesh, maintaining its topology, in order to achieve two objectives simultaneously: the piecewise approximation of the curves by edges of the surface triangulation and the optimization of the resulting mesh. We will designate this procedure as *projecting/smoothing* method and it is based on the smoothing technique that we have introduced for surface triangulations in previous works. The mesh quality improvement is obtained by an iterative process where each *free node* is moved to a new position that minimizes a certain objective function. The minimization process is done on a surface projection plane attending to the surface piece-wise approximation and to an algebraic quality measure (*mean ratio*) of the set of triangles that are connected to the *free node*. So, the 3-D local projecting/smoothing problem is reduced to a 2-D optimization problem. Several applications of this method are presented.

Keywords: Mesh alignment, Moving meshes, Mesh adaptation, Surface mesh smoothing, Node movement, R-adaptivity.

1 Introduction

The numerical simulation of physical problems requires the internal boundaries and discontinuities to be properly represented. Usually, the largest errors are introduced in a neighborhood of such discontinuities. These errors are often greatly reduced if the mesh is aligned with the discontinuities. That is why it is desirable to have a procedure capable of deforming a given triangulation to get its alignment with a curve. Although there are numerous works dealing with surface mesh optimization, see for example [9, 10], only a few of them address the problem of the exact mesh alignment with interior curves. In fact, the only paper that we have found in the bibliography tackling this question in similar terms, but for quadrilateral grids is [12]. The authors consider the problem of aligning a planar grid with multiple embedded curves defined by basic segments as straight lines or arcs of circle. A different approach to the problem can be found in [3], where the curve is approximated by a polygonal line included in the surface triangulation, but in this case the segments are not edges of the mesh. The paper [19] presents a variant of Ruppert's algorithm for producing a 2-D Delaunay triangulation of a domain containing arbitrary curved inputs. Nevertheless, this algorithm does not allow a dynamical adaption of the mesh without remeshing.

The procedure that we describe in this paper aligns a given surface triangulation with an arbitrary curve and it is based on the surface mesh smoothing technique proposed in [5]. An analytical representation of the curve is not usually available. Instead, it is approximately known by a sequence of interpolating data points. We have chosen a parametric cubic spline as interpolating curve due it is C^2 continuous and it has other interesting properties that will be used later. Obviously, the grade of approximation of the curve depends on the element sizes, therefore, a good strategy is to combine the projecting/smoothing technique with a local mesh refinement [11]. Our procedure is specially indicated for evolutionary problems where the boundaries change their shape or position with time; for example, the ones related to fluid-structure interactions involving large displacement (see, for example [21]), or crack modeling. The projecting/smoothing technique could be also applied to domain decomposition, definition of material interfaces, free boundary problems, etc.

The organization of the paper is as follows. In section 2 a rough description of the proposed method is presented. In section 3 we propose an objective function, and the corresponding modification, able to untangle and smooth plane triangulations simultaneously. The projecting/smoothing method is initially analyzed for plane triangulations in section 4 and, afterward, it is extended to triangulations defined on curved surfaces in section 5. Section 6 is devoted to applications. The paper concludes with a brief discussion of the work and its possible extensions.

2 Statement of the Projecting/Smoothing Method

Let C be a curve, and suppose that it is embedded in a surface triangulation T_Σ (see Figure 1). The basic idea of the projecting/smoothing method lies in relocating the nodes of T_Σ closest to C in positions just sited in the curve. This operation, which we will refer to as node projection onto the curve, goes on until getting an approximate representation (interpolation) of C by linked edges of T_Σ. A node of T_Σ is considered *projectable* if we can displace it from its initial position to any point of C in such a way the local mesh does not get *tangled.* This projection implies an enforced alteration of the original positions of the nodes and, in general, has a negative effect on the quality of the triangles close

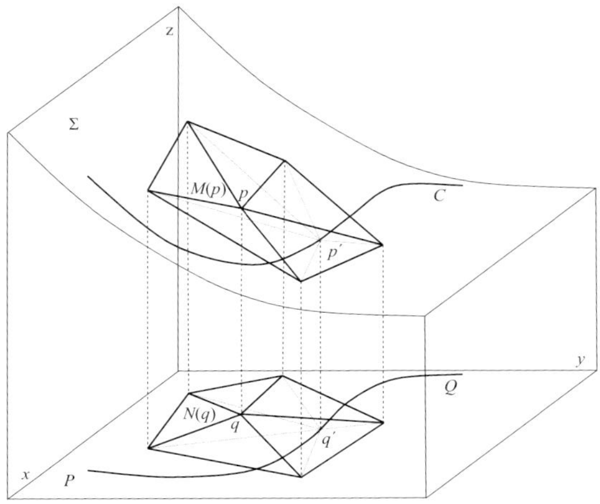

Fig. 1. The relocation of node $p \in \Sigma$ is performed in the plane P by projecting q on Q and, consequently, p on C

to C. To avoid this drawback, the remaining nodes are also displaced to new positions following the smoothing procedure proposed in [5].

For 2-D (or 3-D) meshes, the quality improvement can be obtained by an iterative process where each node of the mesh is moved to a new position that minimizes an objective function derived from certain *algebraic quality measure* of the local mesh [15, 8]. The objective function presents a barrier in the boundary of the *feasible region* associated to the free node. In this context the 2-D (or 3-D) feasible region is the set of points where the free node could be placed to get a *valid* local mesh, that is, without *inverted elements*. The barrier has an important role because it avoids the optimization algorithm to create a tangled mesh when it starts with a valid one. We show in [5] a procedure for smoothing surface triangulations taking into account these aspects. The basic idea lies in transforming the original problem on Σ into a two-dimensional one on a plane P. To do this, the local mesh $M(p)$, belonging to T_Σ, is orthogonally projected onto a plane P performing a local mesh $N(q)$, where p is the free node on Σ and q is its orthogonal projection onto P. The plane P is suitably chosen in terms of $M(p)$ in order to get a valid mesh on P (see Figure 1). Thus, the optimization of $M(p)$ is got by the appropriate optimization of $N(q)$. It involves the construction of *ideal* triangles in $N(q)$ that become near equilateral in $M(p)$.

When Σ is a curved surface, each triangle of $M(p)$ is placed on a different plane. Therefore, it is not possible to define the feasible region associated to the free node p. Nevertheless, the feasible region associated to node q is perfectly defined in plane P and it is denoted as $\mathcal{H}_q$. Furthermore, its associated objective function has a barrier at the boundary of $\mathcal{H}_q$ (see [6]). This is a crucial reason for working on P instead of on Σ.

In the present work the curve C is defined as the image of a curve Q sited on a plane P. Specifically, if we define a plane curve by the parametrization $Q(u) = (x(u), y(u))$ and we consider that $f(x, y)$ is the z coordinate of the underlaying surface (the true surface, if it is known, or the piece-wise linear interpolation, if it is not), then the curve C is given by $C(u) = (x(u), y(u), f(x(u), y(u)))$ (see Figure 1). We note that this type of parametrization can be straightforward introduced in the new *meccano* method which has been recently developed by the authors [4]. We remark also that, although the surface mesh smoothing process can be carried out in different planes chosen in terms of $M(p)$ [5], the particular way in which C is defined in the present paper demands a unique plane. A general parametric curve $C(u) = (x(u), y(u), z(u))$ will be considered in future works.

Since the problem of getting a piecewise approximation of C by edges of T_Σ is translated to the plane P, the task to determine if a node q can be projected onto Q and, that being the case, which is its optimal position, is undertaken by an objective function derived from algebraic quality measures of the local mesh $N(q)$. This objective function incorporates the modifications proposed in [6] in order to deal with tangled meshes. Obviously, a control of the allowed distance between $M(p)$ and $M(p')$ is done in the analysis of the movement of node q.

3 Smoothing and Untangling of Plane Triangulations

Firstly, we will focus our attention on finding an objective function to smooth a valid plane triangulation. As it is shown in [8], [13], and [14] we can derive optimization functions from algebraic quality measures of the elements belonging to a local mesh. Let us consider a triangular mesh T_P defined in $\mathbb{R}^2$ and let t be a triangle in the physical space whose vertices are given by $\mathbf{x}_k = (x_k, y_k)^T \in \mathbb{R}^2$, $k = 0, 1, 2$. To start with, we introduce an algebraic quality measure for t. Let t_R be the reference triangle with vertices $\mathbf{u}_0 = (0,0)^T$, $\mathbf{u}_1 = (1,0)^T$, and $\mathbf{u}_2 = (0,1)^T$. If we choose $\mathbf{x}_0$ as the translation vector, the affine map that takes t_R to t is $\mathbf{x} = A\mathbf{u} + \mathbf{x}_0$, where A is the Jacobian matrix of the affine map referenced to node $\mathbf{x}_0$, given by $A = (\mathbf{x}_1 - \mathbf{x}_0, \mathbf{x}_2 - \mathbf{x}_0)$. We will denote this type of affine maps as $t_R \xrightarrow{A} t$. Let now t_I be an *ideal* triangle whose vertices are $\mathbf{w}_k \in \mathbb{R}^2$, $(k = 0, 1, 2)$ and let $W_I = (\mathbf{w}_1 - \mathbf{w}_0, \mathbf{w}_2 - \mathbf{w}_0)$ be the Jacobian matrix, referenced to node $\mathbf{w}_0$, of the affine map $t_R \xrightarrow{W_I} t_I$; then, we define $S = AW_I^{-1}$ as the weighted Jacobian matrix of the affine map $t_I \xrightarrow{S} t$. In the particular case that t_I was the equilateral triangle t_E, the Jacobian matrix $W_I = W_E$ will be defined by $\mathbf{w}_0 = (0,0)^T$, $\mathbf{w}_1 = (1,0)^T$ and $\mathbf{w}_2 = (1/2, \sqrt{3}/2)^T$.

We can use matrix norms, determinant or trace of S to construct algebraic quality measures of t. For example, the Frobenius norm of S, defined by $|S| = \sqrt{tr\,(S^T S)}$, is specially indicated because it is easily computable. Thus, it is shown in [15] that $q_\eta = \frac{2\sigma}{|S|^2}$ is an algebraic quality measure of t, where $\sigma = \det(S)$. We use this quality measure to construct an objective function. Let $\mathbf{x} = (x, y)^T$ be the position vector of the free node q, and let S_m be the weighted

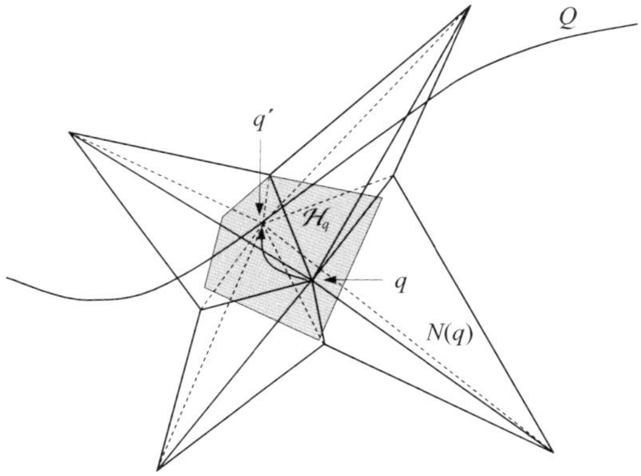

Fig. 2. The curve Q intersects the feasible region $\mathcal{H}_q$ (in gray) and, therefore, the node q is projectable, being q' its optimal position on the curve

Jacobian matrix of the m-th triangle of a valid local mesh $N(q)$ composed of M triangles, see Figure 2. The objective function associated to m-th triangle is $\frac{|S_m|^2}{2\sigma_m}$, and the corresponding objective function for the local mesh is

$$|K_\eta|_n(\mathbf{x}) = \left[\sum_{m=1}^{M}\left(\frac{|S_m|^2}{2\sigma_m}\right)^n(\mathbf{x})\right]^{\frac{1}{n}} \tag{1}$$

being n an integer number, typically $n = 1$ or $n = 2$.

The feasible region for the local mesh is defined as the interior of the polygonal set $\mathcal{H}_q = \bigcap_{m=1}^{M} H_m$, where H_m are the half-planes defined by $\sigma_m(\mathbf{x}) \geq 0$. We say that a triangle is *inverted* if $\sigma < 0$. The objective function (1) presents a barrier in the boundary of the feasible region. This barrier avoids the optimization method to create a tangled mesh when it starts with a valid one, but, on the other hand, it prevents the algorithm to untangle it when there are inverted elements. Therefore, this objective function is only appropriate to improve the quality of a valid mesh, not to untangle it. To construct an objective function applicable to deal with tangled meshes we propose to modify it following the criteria developed in [6]. This modification lies in substituting σ in (1) by the positive and increasing function

$$h(\sigma) = \frac{1}{2}(\sigma + \sqrt{\sigma^2 + 4\delta^2}) \tag{2}$$

where the parameter $\delta = h(0)$ is an appropriate small value.

In this way, the barrier associated with the singularities of $|K_\eta|_n(\mathbf{x})$ will be eliminated and the modified objective function will be smooth all over $\mathbb{R}^2$

$$\left|K'_{\eta}\right|_{n}(\mathbf{x}) = \left[\sum_{m=1}^{M} \left(\frac{|S_m|^2}{2h(\sigma_m)}\right)^n (\mathbf{x})\right]^{\frac{1}{n}} \tag{3}$$

This new objective function strongly penalizes the negative values of σ, so that the minimization process of (3) leads to the construction of a local mesh $N(q')$ without inverted triangles, provided it is possible. Note that the minimum of original and modified functions are nearly identical when $\mathcal{H}_q \neq \emptyset$ and δ tends to zero. With this approach, we can use any standard and efficient unconstrained optimization method to find the minimum of the modified objective function, see for example [2].

4 Alignment of Plane Triangulations

Node movement provides the mesh with the ability to align with an arbitrary curve. Suppose that Q is a curve defined on a 2-D triangulation T_P, our objective is to move some nodes of T_P, projecting them onto Q, to get an interpolation of Q by linked edges of T_P. To achieve this objective we have to decide which nodes of T_P can be projected onto Q without inverting any triangle of its local mesh. More accurately, we say that the free node q is projectable onto Q if there are points of this curve belonging to the feasible region $\mathcal{H}_q$ (see Figure 2).

In general, if q is projectable, its possible placement on Q is not unique. The projecting/smoothing method must determine if q can be projected onto Q and, if so, which is its optimal position. The last question can be answered by using the objective function (3) subject to the constrained $\mathbf{x} \in Q$. Thus, the problem of finding the optimal position to project the free node onto the curve is

$$\text{minimize} \left|K'_{\eta}\right|_{n}(\mathbf{x}), \text{ subject to } \mathbf{x} \in Q \tag{4}$$

If $\bar{\mathbf{x}}$ is the position vector of the minimizing point q' of (4) and $\sigma(\bar{\mathbf{x}}) > 0$ for all triangle of $N(q')$, we conclude that q is projectable onto Q and $\bar{\mathbf{x}}$ is its optimal position. Otherwise, we say that node q is not projectable.

4.1 Curve Definition

The previous criterion allows us to determine whether q is projectable onto Q or not, but it involves a high computational cost because it needs to solve the constrained minimization problem (4). Nevertheless, it is clear that most nodes of T_P are not projectable because they are very far from any point of the curve. Therefore, it is convenient to have an efficient method to select those nodes, close to some segment of Q, expected to be projectable.

In many situations of practical interest we do not have an analytical representation of Q, but Q is approximately known by a sequence of interpolating data points. Among the options to define an interpolating curve, we have chosen a parametric cubic spline as it has many desired properties: it is a C^2 continuous function, it has a very simple local form, small oscillations, etc. Moreover, each segment of the spline is a degree 3 Bézier curve that lies within the convex hull of its four defining control points (see, for example [1]). We will use this property in order to know if a given node is close to some segment of Q.

Let $\{P_0, P_1, \ldots, P_m\} \subset \mathbb{R}^2$ be a set of interpolating points belonging to plane P. The parametric cubic spline

$$Q(u) = (x(u), y(u)), \text{ where } u \in [u_0, u_m] \tag{5}$$

is an interpolating curve that satisfies $Q(u_i) = P_i$ for $i = 0, \ldots, m$ and two additional constraints in order to be fully defined. Usually, these constraints are imposed at the ends of the curve. For example, it is well known that the conditions $Q''(u_0) = 0$ and $Q''(u_m) = 0$ define a spline known as *natural*.

Every segment of the spline delimited by two consecutive interpolating points is a degree 3 polynomial. Suppose that $Q_i(t) = \mathbf{a}^i + \mathbf{b}^i t + \mathbf{c}^i t^2 + \mathbf{d}^i t^3$, with $\mathbf{a}^i$, $\mathbf{b}^i$, $\mathbf{c}^i$ and $\mathbf{d}^i$ in $\mathbb{R}^2$, is the polynomial associated to the segment Q_i $(i = 0, 1, \ldots, m-1)$ that runs from P_i to P_{i+1}, being $t \in [0, 1]$ the local parameter, see Figure 3. This one is related with the parameter of the entire curve by $t = (u - u_i) / (u_{i+1} - u_i)$.

4.2 Node Projection onto the Curve

The Q_i segment also is a degree 3 Bézier curve, given by $Q_i(t) = \sum_{j=0}^{3} \mathbf{u}_j^i B_j^i(t)$ with $t \in [0, 1]$, where $B_j^i(t)$ are the Berstein polynomials and $\mathbf{u}_j^i \in \mathbb{R}^2$ are the control points. The relation between the polynomial coefficients and the control points are given by

$$\begin{pmatrix} \mathbf{u}_0^i \\ \mathbf{u}_1^i \\ \mathbf{u}_2^i \\ \mathbf{u}_3^i \end{pmatrix} = \frac{1}{3} \begin{pmatrix} 3 & 0 & 0 & 0 \\ 3 & 1 & 0 & 0 \\ 3 & 2 & 1 & 0 \\ 3 & 3 & 3 & 3 \end{pmatrix} \begin{pmatrix} \mathbf{a}^i \\ \mathbf{b}^i \\ \mathbf{c}^i \\ \mathbf{d}^i \end{pmatrix} \tag{6}$$

As we have already said, an interesting property of the Bézier curves establishes that the Q_i segment lies within the convex hull of its control points. If CH denotes the convex hull of a set of points, we have $Q_i \subseteq \mathrm{CH}\left(\mathbf{u}_0^i, \ldots, \mathbf{u}_3^i\right)$. Note that a necessary (but not sufficient) condition for the node q to be projectable onto Q is that its feasible region $\mathcal{H}_q$ intersects the convex hull of some segment of the curve. In other words, it must exist a segment Q_i such that $\mathcal{H}_q \cap \mathrm{CH}\left(\mathbf{u}_0^i, \ldots, \mathbf{u}_3^i\right) \neq \emptyset$. This property allows us to know beforehand which nodes are not projectable, because they yield an empty intersection for all segments of the curve. Nevertheless, calculating the set $\mathcal{H}_q$ and, moreover, its intersection with a convex set, is not a trivial problem, so it is more advisable to deal with a simplified version.

Let R_q and R_{Q_i} be the minimal rectangles, with sides parallel to the axes, enclosing the sets $N(q)$ and $\mathrm{CH}\left(\mathbf{u}_0^i, \ldots, \mathbf{u}_3^i\right)$, respectively. Then, due to $\mathcal{H}_q \subset R_q$, it is clear that q is projectable onto Q_i only if $R_q \cap R_{Q_i} \neq \emptyset$ (see Figure 3). The

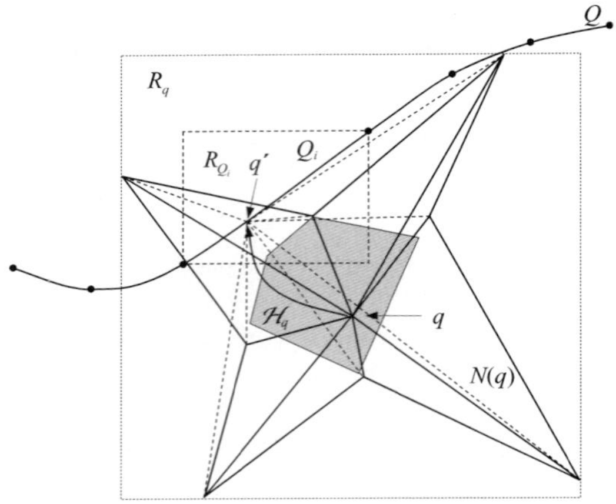

Fig. 3. The figure shows the situation in which $R_q \cap R_{Q_i} \neq \emptyset$, but node q is not projectable because the optimal position for the free node, q', is outside the feasible region

computation of this intersection allows us to take a quick decision about if a node is candidate to be projected onto the curve.

The algorithm to determine if q is projectable onto Q and, if it is so, which is its optimal position, can be summarized as follows. For each segment of the curve analyze $R_q \cap R_{Q_i}$ and, if this set is not empty, solve the minimization problem

$$\text{minimize} \left|K'_\eta\right|_n (Q_i(t)), \text{ for } t \in [0, 1] \tag{7}$$

Let $\bar{t}$ be the global minimum of (7) and $\bar{\mathbf{x}}_i = Q_i(\bar{t})$ the corresponding position of the free node q on the segment Q_i. We say that $\bar{\mathbf{x}}_i$ is an *admissible* optimal position for the free node if $\sigma_m(\bar{\mathbf{x}}_i) > 0$ for $m = 1, \ldots, M$. In order to determine the optimal position of the free node on Q, we take $\bar{\mathbf{x}}_{opt}$ as the best admissible position for all segments Q_i. Obviously, if no admissible position exists, the conclusion of previous algorithm is that node q is not projectable onto Q.

Take into account that an admissible projection of a free node on Q can give rise to a local mesh with very poor quality. Although this effect is partly palliated after smoothing the remainder nodes, following the procedure described in section 3, it is appropriate to tighten the condition $\sigma_m(\bar{\mathbf{x}}_i) > 0$ enforcing $\sigma_m(\bar{\mathbf{x}}_i) > \varepsilon$, with $\varepsilon > 0$ a prescribed tolerance. Nevertheless, this more restrictive condition makes it difficult for the nodes to be projected onto the curve and it could produce situations where some sections of the curve are not interpolated by edges of T_P. This drawback will be studied in the next subsection but, for that purpose, it needs further clarification.

Up to now, we have accepted that parameter t pertains to the closed interval $[0, 1]$ and, in consequence, the problem (7) admits a global minimum. But, with this consideration, the ends of the consecutive segments are shared and, therefore, a projected point can belong to two segments at the same time. In order to avoid this ambiguity, we will assume that each segment $Q_i(t)$ is defined for $t \in [0,1)$, except the last one, that it is for $t \in [0,1]$ if the curve is open. In this way, each point of the curve belongs to a unique segment.

4.3 Discontinuities of the Mesh Alignment with the Curve

It can happen that, after repositioning all the nodes of the mesh, the piecewise approximation of Q by edges of T_P is not continuous. We can detect this discontinuity if we take into account that the projected nodes are arranged in the curve. Thus, a section of the interpolated curve among two consecutive projected nodes is discontinuous if they are not connected by an edge of T_P.

As the parameter $t \in [0,1)$ induces an order relation in each segment of the curve and, in turn, each segment is ordered by its subindex, we can say that the node $p \in Q_i$ precedes $p' \in Q_j$ if $i < j$ or, in case of $i = j$, if the corresponding parameters satisfies $t_p < t_{p'}$. A possibility to correct a detected discontinuity in the piecewise approximation of Q is to relax the condition $\sigma_m(\bar{\mathbf{x}}_i) > \varepsilon$, by decreasing the value of ε. However, there are situations in which, even taking ε equal to zero, there are discontinuities impossible to avoid without removing some of the projected nodes. The Figure 4(a) shows a scheme of this problem. It can be seen that it is impossible to project the node q (neither r) without

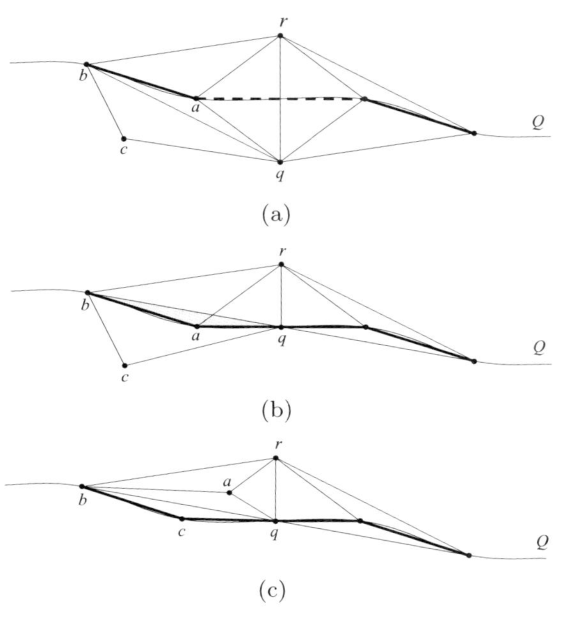

Fig. 4. The dashed line is non-recoverable without tangling the mesh (a). The free node q is enforced to be projected (b). The tangled triangle abq is untangled and the node c is also projected (c).

tangling the mesh. We propose a solution to this conflict by enforcing the free node q to be projected, even if a tangled mesh is created. The Figure 4(b) shows how the movement of q produces the tangled triangle abq.

Afterward, the position of q is fixed for subsequent iterations of the projecting/smoothing algorithm, but the surrounding nodes are free to move in search of their optimal positions that untangle the mesh and complete the interpolation of the curve (see 4(c)). So, the algorithm extracts nodes from the curve if their current positions are not admissible (see the new position of node a in figure 4(c)).

Sometimes the curve represented by splines has sharp features that we want to preserve in the piecewise interpolation. To reach this objective we select, from the interpolating points, a set of *prescribed* points sited in strategic locations. Once the projecting/smoothing process has finished, the algorithm searches among the nodes projected on Q, which one is the optimal candidate, say q, to be relocated in the position of each prescribed node. If $\mathbf{x}_{pres}$ is the position of certain prescribed point, the node q is chosen, among the nodes projected on Q and close to $\mathbf{x}_{pres}$, as the one that maximizes the quality of $N(q)$ when q is enforced to take the position $\mathbf{x}_{pres}$. Obviously, if $N(q)$ is not valid after the relocation of q, a new iteration of the projecting/smoothing procedure must be done.

5 Extension to Curved Surfaces

We are interested in extending the projecting/smoothing method to curved surfaces. As we pointed in section 2, the original problem on Σ is transformed into another one on the plane P. The more significant difference with respect to the former method lies in searching *ideal* triangles in $N(q)$ that become equilateral in $M(p)$. Due to restriction of the extension of this paper, we summarize in this section our main result relative to this aspect. A detailed analysis can be carried out in the surface mesh smoothing procedure that is presented in [5]. Its connection with the problem of surface mesh aligning with curves could be easily implemented.

Consider that triangle $t \in N(q)$ (located in plane P) is the orthogonal projection of triangle $\tau \in M(p)$ (located in plane π), see Figure 5. Let t_R and τ_R be the references triangles defined in planes P and π, respectively. Suppose that we chose as ideal triangle in π the equilateral one ($\tau_I = \tau_E$). Our goal is to find the *ideal* triangle $t_I \subset P$, moving q on P, such that t_I is mapped into an equilateral one, $\tau_E \subset \pi$. For this purpose, the following similarity transformation between the matrices S and S_I was proved in [5]

$$S = S_E^{-1} S_I S_E \tag{8}$$

where S is defined on the plane π as the 2×2 weighted Jacobian matrix of the affine map that transforms the equilateral triangle into the physical one, that is, $\tau_E \xrightarrow{S} \tau$, S_E is defined on plane P as the *equilateral-weighted* Jacobian matrix

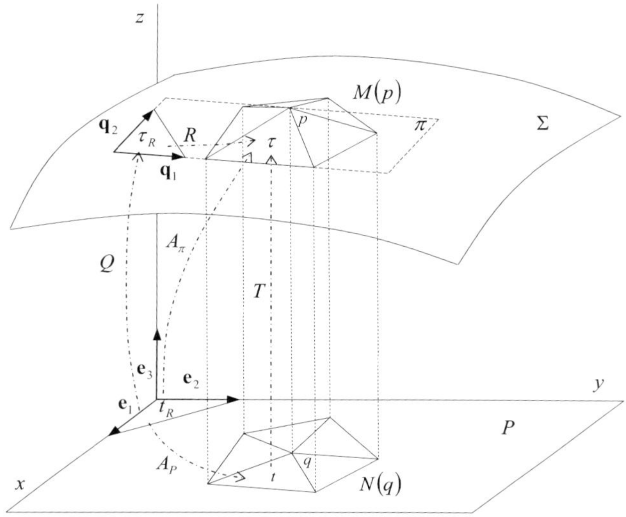

Fig. 5. Local surface mesh, $M(p)$, and its projection, $N(q)$, on plane P

of the affine map $t_E \overset{S_E}{\to} t$ and S_I is defined on plane P as the *ideal-weighted* Jacobian matrix of the affine map $t_I \overset{S_I}{\to} t$.

Matrices S may be used, as it is defined in (8) for any triangle $\tau \in M(p)$, to construct the objective function associated to $M(p)$ and, then, solve the optimization problem. Nevertheless, this procedure has important disadvantages. First, the optimization of $M(p)$, working on the true surface, would require the imposition of the constraint $p \in \Sigma$. It would complicate the resolution of the problem because, in many cases, Σ is not defined by a smooth function. Moreover, when the local mesh $M(p)$ is on a curved surface, each triangle is sited on a different plane and the objective function, constructed from S, lacks barriers. It is impossible to define a feasible region in the same way as it was done at section 3. Indeed, all the positions of the free node, except those that make $\det(S) = 0$ for any triangle, produce correct triangulations of $M(p)$. However, there are another unacceptable positions of the free node.

To overcome these difficulties we proposed in [5] to carry out the optimization of $M(p)$ in an indirect way, working on $N(q)$. With this approach the movement of the free node will be restricted to $\mathcal{H}_q \subset P$, which avoids unacceptable surface triangulations to be formed. Then, the original smoothing problem is transformed in a two-dimensional approach on P. The algorithm to determine the optimal projection of a free node $p \in \Sigma$ onto the curve C is reduced to the one presented in section 4 for reaching the optimal projection of q onto Q. If $\bar{\mathbf{x}}_{opt}$ is the optimal projection of the free node q onto Q, then the corresponding position on the surface is given by $\bar{\mathbf{y}}_{opt} = (\bar{\mathbf{x}}_{opt}, f(\bar{\mathbf{x}}_{opt}))^T$, where $f(\mathbf{x})$ is the z coordinate of the underlaying surface. If this one is not known analytically, we take the initial triangulation as reference surface. The algorithm follows the usual smoothing procedure when the free node is not projectable on C. The discontinuities of the mesh alignment with the curve are solved by using the same idea of section 4.3.

In order to prevent a loss of the details of the original geometry when we are smoothing the mesh, our algorithm evaluates the difference of heights ($[\Delta z]$) between the centroid of the triangles of $M(p)$ and the reference surface, every time a new position of the free node is calculated. If this distance exceeds a threshold, $\Delta(p)$, the movement of the node is aborted and the previous position is stored. This threshold $\Delta(p)$ is established attending to the size of the elements of $M(p)$. That is, the algorithm evaluates the average distance between the free node and the nodes connected to it, and takes $\Delta(p)$ as percentage of this distance. Other possibility is to fix $\Delta(p)$ as a constant for all local meshes. In the particular case in which we have an explicit representation of the surface by a function $f(x, y)$, $\Delta(p)$ can be established as a percentage of the maximum difference of heights between the original surface and the initial mesh.

6 Applications

In this section we present an application that shows the satisfactory behavior of the projecting/smoothing technique. In particular, we have applied the projecting/smoothing technique to the Igea surface triangulation (see Figure 6) obtained from *http://www.cyberware.com/*. The mesh contains 67170 triangles and 33587 nodes.

Our goal is to obtain a new triangulation (maintaining the initial mesh topology) after applying the projecting/smoothing procedure to reach the alignment of the new mesh with the contours of a mask and of a star drawn on the face of Igea. These curves are defined by the spline piece-wise interpolation of a few points that are placed on a front view of Igea. In Figure 7 we show the polylines that connect the fixed points for the definition of the mask and the star. Moreover, we construct several reference windows to define surface patches and to evaluate the coordinate transformation from the image parametric space to the physical one. In order to keep the sharp angles of the star drawing, we have prescribed 10 points as the extremes of 10 splines. We propose the following strategy to get our objective.

Initially, we apply the smoothing technique [5] to the whole triangulation. In this case, the projection plane is chosen in terms of the local mesh to be optimized. The resulting mesh, after 4 iterations of our optimization procedure, is shown in Figure 8. The value of the average mesh quality (measured with the algebraic quality metric based on the condition number proposed in [8]) increases from 0.794 to 0.913. A more significant data is that average quality of the worst 100 triangles increases from 0.379 to 0.575. We have fixed $\Delta(p)$ as 10% of average distance between the free node and the nodes connected to it. More details about this application can be seen in [17].

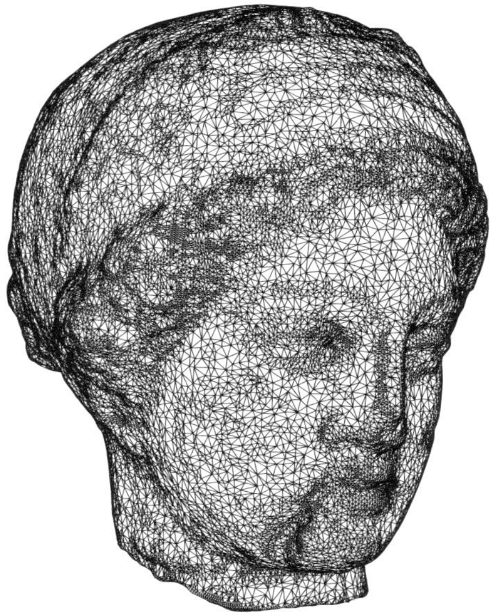

Fig. 6. Original mesh of Igea obtained from *http://www.cyberware.com/*

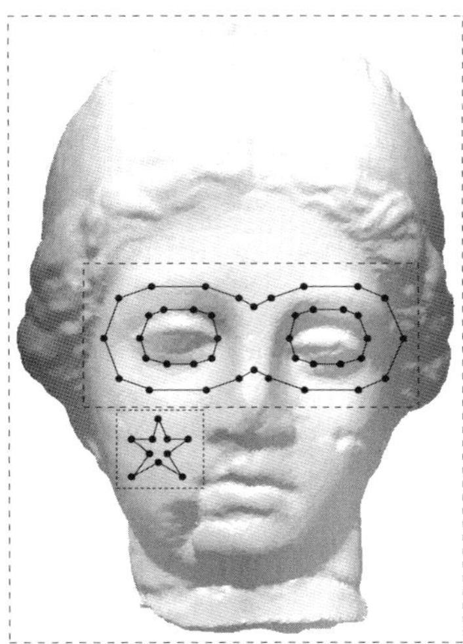

Fig. 7. Point input data for the definition of the curves, approximation of the splines as polylines and reference windows

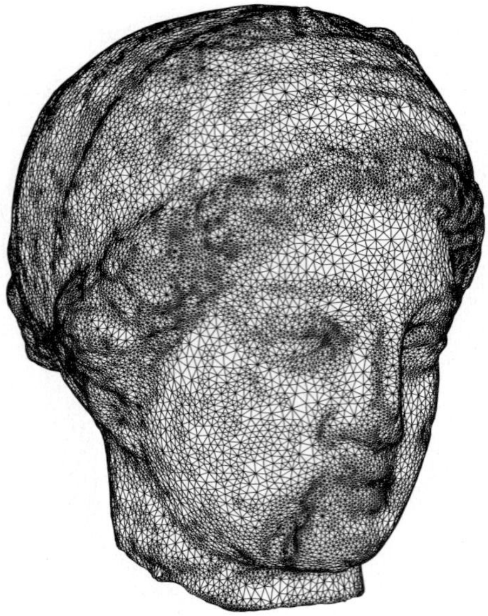

Fig. 8. Optimized mesh of Igea after 4 iterations of our smoothing procedure

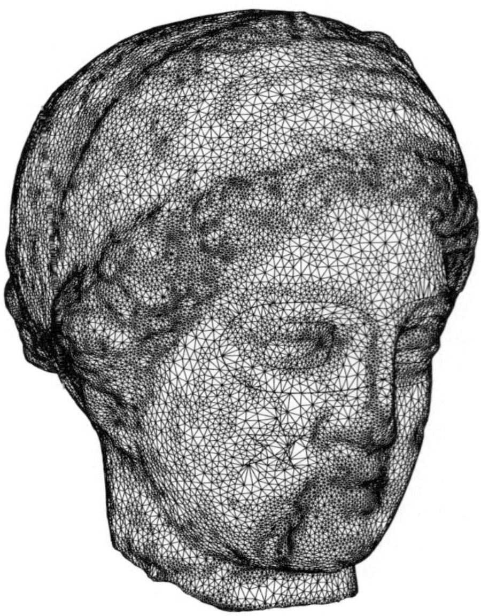

Fig. 9. Aligned and optimized mesh of Igea after 12 iterations of the local projecting/smoothing procedure

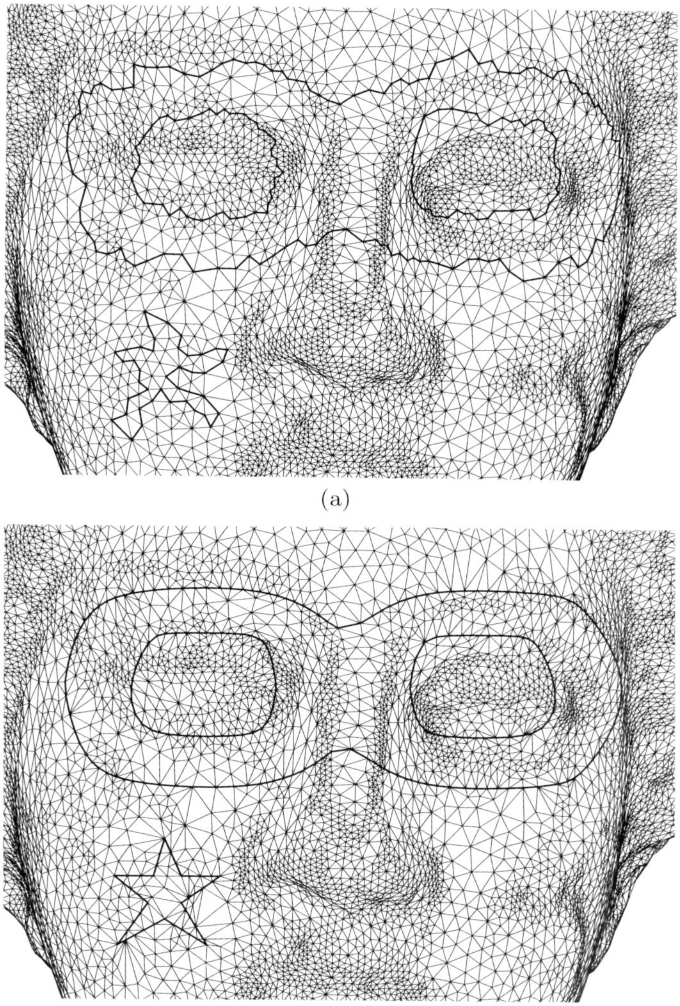

(a)

(b)

Fig. 10. Detail of the initial mesh of Igea with marked edges before projection (a). The same edges are remarked after the algorithm has projected them onto the contours of the mask and the star (b).

In order to reduce the computational cost of the alignment step, the projecting/smoothing process is carried out on the surface patches associated to the mask and to the star. For this purpose, we select the set of triangles whose centroids are included in the reference windows of the mask and of the star. Then, we apply the projecting/smoothing procedure to these two sets separately. We note that, in the former process, the boundary of each patch triangulation is fixed. So, we obtain an appropriate connection of the modified local meshes with the rest of the surface triangulation. These ideas could be used for a parallel implementation of the simultaneous aligning and smoothing local technique.

In Figure 9 it is presented a general view of the surface triangulation of Igea after applying 12 iterations of the aligning and smoothing procedure. The approximation of the contours by edges of the resulting triangulation is marked. After the application of our algorithm the values of minimum and average qualities become 0.100 (same value than in previous meshes) and 0.911, respectively. The average quality of the worst 100 triangles is 0.519. Therefore, the mesh qualities are similar before and after the application of the projecting/smoothing technique.

Two details of the initial and final meshes (Figures 8 and 9) are shown in Figures 10(a) and (b), respectively. We represent the same marked edges before and after the local projecting/smoothing process. We note that it is very difficult to determine *a priori* which are the best edges for a suitable approximation of the contours of the mask and the star. However, the algorithm finds appropriate nodes (and consequently edges) automatically.

7 Concluding Remarks and Future Research

In this paper we have introduced the projecting/smoothing technique which is able to align a surface triangulation with arbitrary curves without producing, in general, a significant decrease in the minimum quality of the mesh. Indeed, the average quality is increased in many cases as the remainder part of the mesh undergoes a smoothing process.

In present work the curves have been defined by splines whose interpolating points are fixed on a plane. Applications of this technique can be done in a straightforward manner, for example, in environmental modeling [16, 18] for aligning topographic surface meshes to significant contours, as coastlines, river banks, etc. In addition, this particular curve definition can be applied on different patches of a more general surface.

Our method for aligning and smoothing of surface triangulations could be generalized by using a global parametric space (in similar terms as it is proposed in [7, 20]) which makes the projection on a plane unnecessary. So, general parametric curves embedded on the surface will be considered. Another more ambitious generalization lies in extending the present method to align a tetrahedral mesh with interior surfaces. This is an open problem. It is clear that the mesh alignment problem is not always possible to solve. Generally, the existence of an admissible solution can not be assured. Moreover, several admissible solutions may exit. Obviously, the existence of solution of the problem depends on size, quality and topology of the initial mesh and regularity of the embedded curves or surfaces.

Acknowledgments

This work has been supported by the *Secretaría de Estado de Universidades e Investigación* of the *Ministerio de Educación y Ciencia* of the Spanish Government and FEDER, grant contracts: CGL2007-65680-C03-01 and CGL2008-06003-C03-01.

References

1. Bartels, H.R., Beatty, J.C., Barsky, B.: An introduction to Splines for use in computer graphics & geometric modeling. Morgan Kaufmann, Los Altos (1987)
2. Bazaraa, M.S., Sherali, H.D., Shetty, C.M.: Nonlinear programing: theory and algorithms. John Wiley and Sons Inc., New York (1993)
3. Bonneau, G.P., Hahmann, S.: Smooth polylines on polygon meshes. In: Brunnett, G., Hamann, B., Mueller, H. (eds.) Geometric modeling for scientific visualization, pp. 69–84. Springer, Berlin (2003)
4. Cascón, J.M., Montenegro, R., Escobar, J.M., Rodríguez, E., Montero, G.: A new *meccano* technique for adaptive 3-D triangulations. In: Proc 16th Int Meshing Roundtable, pp. 103–120. Springer, Berlin (2007)
5. Escobar, J.M., Montero, G., Montenegro, R., Rodríguez, E.: An algebraic method for smoothing surface triangulations on a local parametric space. Int. J. Num. Meth. Eng. 66, 740–760 (2006)
6. Escobar, J.M., Rodríguez, E., Montenegro, R., Montero, G., González-Yuste, J.M.: Simultaneous untangling and smoothing of tetrahedral meshes. Comput. Meth. Appl. Mech. Eng. 192, 2775–2787 (2003)
7. Floater, M.S.: Parametrization and smooth approximation of surface triangulations. Comput. Aid. Geom. Design 14, 231–250 (1997)
8. Freitag, L.A., Knupp, P.M.: Tetrahedral mesh improvement via optimization of the element condition number. Int. J. Num. Meth. Eng. 53, 1377–1391 (2002)
9. Frey, P.J., Borouchaki, H.: Geometric surface mesh optimization. Comput. Visual Sci. 1, 113–121 (1998)
10. Garimella, R.V., Shaskov, M.J., Knupp, P.M.: Triangular and quadrilateral surface mesh quality optimization using local parametrization. Comput. Meth. Appl. Mech. Eng. 193, 913–928 (2004)
11. González-Yuste, J.M., Montenegro, R., Escobar, J.M., Montero, G., Rodríguez, E.: Local refinement of 3-D triangulations using object-oriented methods. Adv. Eng. Soft. 35, 693–702 (2004)
12. Hyman, J.M., Li, S., Knupp, P.M., Shashkov, M.: An algorithm to align a quadrilateral grid with internal boundaries. J. Comput. Phys. 163, 133–149 (2000)
13. Knupp, P.M.: Achieving finite element mesh quality via optimization of the Jacobian matrix norm and associated quantities. Part I - A framework for surface mesh optimization. Int. J. Num. Meth. Eng. 48, 401–420 (2000)
14. Knupp, P.M.: Achieving finite element mesh quality via optimization of the Jacobian matrix norm and associated quantities. Part II - A framework for volume mesh optimization and the condition number of the Jacobian matrix. Int. J. Num. Meth. Eng. 48, 1165–1185 (2000)
15. Knupp, P.M.: Algebraic mesh quality metrics. SIAM J. Sci. Comput. 23, 193–218 (2001)

16. Montenegro, R., Montero, G., Escobar, J.M., Rodríguez, E., González-Yuste, J.M.: Tetrahedral mesh generation for environmental problems over complex terrains. In: Sloot, P.M.A., Tan, C.J.K., Dongarra, J., Hoekstra, A.G. (eds.) ICCS-ComputSci 2002. LNCS, vol. 2329, pp. 335–344. Springer, Heidelberg (2002)
17. Montenegro, R., Escobar, J.M., Montero, G., Rodríguez, E.: Quality improvement of surface triangulations. In: Proc 14th Int Meshing Roundtable, pp. 469–484. Springer, Berlin (2005)
18. Montero, G., Rodríguez, E., Montenegro, R., Escobar, J.M., González-Yuste, J.M.: Genetic algorithms for an improved parameter estimation with local refinement of tetrahedral meshes in a wind model. Adv. Eng. Soft. 36, 3–10 (2005)
19. Pav, S.E., Walkington, N.J.: Delaunay refinement by corner looping. In: Proc 14th Int Meshing Roundtable, pp. 165–181. Springer, Berlin (2005)
20. Sheffer, A., De Sturler, E.: Smoothing an overlay grid to minimize linear distortion in texture mapping. ACM Trans. Graph 21, 874–890 (2002)
21. Stein, K., Tezduyar, T.E., Benney, R.: Automatic mesh update with the solid-extension mesh moving technique. Comput. Meth. Appl. Mech. Eng. 193, 2019–2032 (2004)

Accurate Interpolation of Terrain Surfaces from Over-Sampled Grid Data

Narcís Coll[1], Marité Guerrieri[1], Maria-Cecilia Rivara[2], and J. Antoni Sellarès[1]

[1] Institut d'Informàtica i Aplicacions - Universitat de Girona - Spain
{coll,mariteg,sellares}@ima.udg.edu*

[2] Department of Computer Science - Universidad de Chile - Chile
mcrivara@dcc.uchile.cl

Summary. For grid terrain data, we propose a Lepp-surface triangulation method that starts with a coarse initial triangulation of the input data, and incrementally adds data points that reduce the worst edge approximation error in the mesh. The method generalizes a previous Lepp-centroid method in two dimensions as follows: for the edge E, having highest error in the mesh, one or two points close to (one or two) terminal edges associated to E, are inserted in the mesh. The edge error is computed by adding the triangle approximation errors of the two triangles that share E, while each triangle error in L_2-norm is computed by using a curvature tensor (good approximation of the surface) at a representative point associated to both triangles. The method produces triangular approximations that capture well the relevant features of the terrain surface by naturally producing well-shaped triangles.

1 Introduction

Techniques for the efficient and timely management of huge terrain grid data produced by satellites have been intensively developed in the last ten years. Problems related with storage requirements, real-time rendering, data transmission, and visualization arise when dealing with these huge data sets. Data representation methods that minimize the data size by modeling well the important features of the terrain are highly desirable.

For visualization purposes, a fast and efficient large scale rendering with minimal memory and hardware requirements is always desirable, since the grid data set is usually too large to be displayed as a whole. Two main approaches have been mainly used to efficiently deal with the rendering task: the use of different levels of detail (LODs) depending on the view focus, and the reduction of the data size without loosing visual representation. View dependent LOD methods display the surfaces with adaptive resolution depending on the distance to the view focus [7, 17]. Data reduction methods basically consider two alternative strategies for the treatment of grid data, top-down approach [2, 4, 10, 14, 15, 23] and bottom-up approach [6, 9, 11]. The top-down algorithms, start with a few

* Authors partially supported by the Spanish Ministerio de Educación y Ciencia under grant TIN2007-67982-C02-02.

triangles and incrementally add vertices until obtaining a triangulation that accomplishes requirements previously established. This kind of process is known as refinement process. The second strategy starts with a mesh of the whole data and deletes vertices or edges and sometimes moves vertices to a new better location until it achieves the desired requirements. These methods have been called decimation or simplification. The vertices added or moved in both strategies can belong to the initial data set or can be extrapolated from them. Different criteria to finish the process have been used: number of elements, faces or vertices, error estimation between the initial and final model, element quality, etc.

For numerical simulations, additional constraints need also to be taken into account. The triangles of the fitting triangulated surface must be reasonably good-shaped (with bounded smallest angle) which guarantees the convergence of the numerical method [24]. A triangulation of a terrain surface can be built either by using only the xy-coordinates of the input points, or the 3D information of the input points. In the latter case it is called a data-dependent triangulation [5]. When the fitting surface is created from a triangulation in the xy-plane, usually the Delaunay triangulation is used since this maximizes the minimum angle. However, having well-shaped triangles in the xy-plane does not guarantee well-shaped triangles on the the fitting surface. Moreover, it is also known that when a data-dependent triangulation is used to fit a triangulated surface by minimizing the L_2 norm of the difference between the original and the approximated surfaces, long and skinny triangles can be obtained throughout the fitting process [18].

Figure 1 shows two possible triangulations of a half cylinder. The triangulation of Figure 1(a) minimizes the L_2 error, while the triangulation in Figure 1(b) was obtained with our method that considers the L_2 norm error and produces well-shaped triangles. Even when the first triangulation approximates the surface with less mean error (2.9e-05) than the second one (0.1), the second triangulation is more desirable for numerous applications requiring of the use of numerical methods.

In this paper we center on the problem of fitting a data-dependent triangulation with a small number of triangles to a set of grid data points so that the triangles are well-shaped and the L_2 error remains low. Note that the decimation

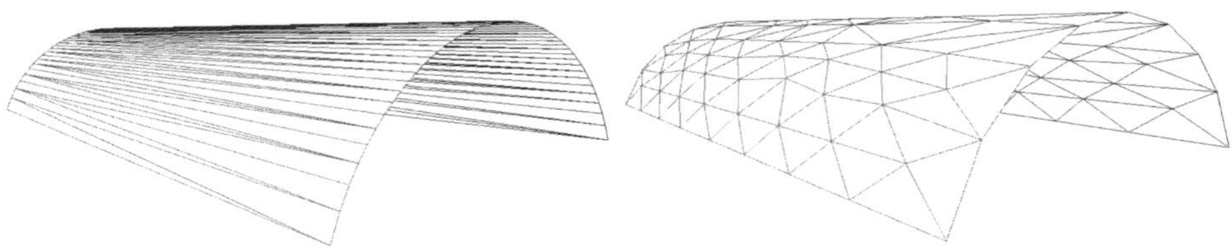

(a) (b)

Fig. 1. Comparison of two possible triangulations of a half cylinder surface with the same quantity of final vertices. In (a) the triangulation minimizes the L_2 error and in (b) the triangulation tries to maximize the quality of the triangles.

strategy centers on the elimination of biggest possible number of input data. On the contrary, our goal is to obtain an accurate approximation by progressively adding the smallest possible number of input data points.

1.1 Our Contribution

We consider a finite set V of grid data from a terrain surface F. The output is an interpolating data-dependent triangulation (piecewise linear surface defined by a set of triangles) $\overline{F}$ whose vertices belong to a subset $\overline{V} \subset V$. Out method selects relevant points of V in such a way that: 1) $\overline{F}$ fits well the original surface F; 2) the triangles of $\overline{F}$, when possible, are well-shaped; 3) the number of points in $\overline{V}$ (triangles of $\overline{F}$) remains low.

The proposed method estimates the edge approximation error as the sum of the L_2 error of the two triangles that share the edge by using a curvature tensor that locally approximates the surface at each point of the model. The algorithm gives more importance to the L_2 metric but when it is possible the aspect-ratio of the elements is taken also into account, achieving in this way an approximation of the terrain with good quality elements. This is the main difference of our algorithm with previous works in this area [2, 6, 10, 11].

The remaining part of this paper is organized as follows. Basic concepts necessary to understand the proposed method like linear interpolation, surface curvature and the Lepp Delaunay method are introduced in Section 2. Section 3 details the theoretical tools needed to approximate a terrain from a grid: edge error, point insertion, two lepps strategy and surface-midpoint or surface-centroid selection. In Section 4 we describe the algorithm that integrates the different tools. In section 5 we discuss empirical testing as well as the advantages of our method. The document finishes with same conclusions and future work issues.

2 Preliminaries

Regular grids are widely used techniques for modeling terrain surfaces because of their simple structure and the availability of satellite technology to get the data. A disadvantage of this structure is that this uses more information than necessary to represent the irregularity of the terrain. Irregular triangulations obtained by sampling of irregularly spaced relevant points over the terrain surface are more adequate to adaptively represent the terrains. In this way density of the sample points can be adapted to the local variation in elevation: high density is only needed in regions where the elevation varies a lot.

In the following sections we introduce the theoretical concepts we need to develop our approximation methods.

2.1 Linear Interpolation

Given a function $F(x, y)$, $(x, y) \in \mathbb{R}^2$, and a set V of n arbitrary points $(x_i, y_i) \in \mathbb{R}^2$, the Linear Interpolation consist of constructing a piecewise linear interpolation function $\overline{F}(x, y)$ such that

$$\overline{F}(x_i, y_i) = F(x_i, y_i) \, , i = 1, \ldots, n \, .$$

The common method is to construct a triangulation T of V and to define $\overline{F}_T(x, y)$ by

$$\overline{F}_T(x, y) = \sum_{T_k} \overline{F}_{T_k}(x, y) \, ,$$

where T_k is a triangle of the triangulation T and $\overline{F}_{T_k}(x, y)$ is the function with value 0 outside T_k and inside T_k is equal to the unique linear polynomial function defined by the values of F at the three vertices of T_k. Since for fixed V the quality of the interpolation depends on the specific triangulation T, T has to be chosen in an optimal manner.

The L_p norm of a function $f(x, y)$ is defined by

$$\|f\|_{L_p} = \left(\int_{\mathbb{R}^2} |f(x, y)|^p \, dxdy \right)^{\frac{1}{p}} \, .$$

Thus, the triangulation T minimizing

$$\|F - \overline{F}_T\|_{L_p}$$

is the best triangulation to obtain a good linear interpolation of a function $F(x, y)$.

An expression for the L_2 error for a quadratic function F with hessian $H(F)$ on a single triangle T_k with vertices v_1, v_2 and v_3 was computed by Nadler [16]:

$$E_2(F, \overline{F}_{T_k}) = \int_{T_k} |F - \overline{F}_{T_k}|^2 \, dxdy = \frac{|T_k|}{180} \left((d_1 + d_2 + d_3)^2 + d_1^2 + d_2^2 + d_3^2 \right) \, ,$$

where $|T_k|$ is the area of T_k and

$$d_i = \frac{1}{2} (v_{i+1} - v_i)^T H(F) (v_{i+1} - v_i) \, .$$

From this expression Nadler proved that, when the number n of points goes to infinity, the orientation of the triangles for best L_2 interpolation is given by the eigenvectors of the hessian matrix, while their relative size in each principal direction is given by the reciprocal square root of the absolute value of the corresponding eigenvalue. Later Rippa [18] indicated how to construct the triangulation T for best L_p interpolation of a quadratic function F with $\det(H(F)) > 0$ as follows. Let λ_1, λ_2 be the eigenvalues of the hessian $H(F)$, and e_1, e_2 the corresponding eigenvectors. Consider also the transformation $(x, y) = \sqrt{\frac{\lambda_2}{\lambda_1}} x' e_1 + y' e_2$. The optimal triangulation is obtained from the Delaunay triangulation of the transformed set of input points V. Given an edge $e = v_1 v_3$ adjacent to triangles $\triangle v_1 v_2 v_3$, $\triangle v_3 v_4 v_1$, the Delaunay criterion replaces e by the edge $e' = v_2 v_4$ if and only if

$$\angle v_1 v_2 v_3 + \angle v_3 v_4 v_1 > \angle v_2 v_3 v_4 + \angle v_4 v_1 v_2 \, .$$

This condition of the Delaunay criterion applied to the (x', y') coordinates is equivalent to the same condition applied to the (x, y) coordinates by computing the angles using the metric with eigenvalues $|\lambda_1|$, $|\lambda_2|$ and corresponding eigenvectors e_1, e_2.

2.2 Surface Curvature

Let $S \subset \mathbb{R}^3$ be a smooth surface. Given a point $p \in S$, let t be a unit vector in the tangent plane at p and let n be the unit normal vector at p. The normal curvature $k_n(t)$ is the curvature of the planar curve that results from intersecting S with the plane through p spanned by n and t. The minimal normal curvature k_1 and the maximal normal curvature k_2 are called principal curvatures. The associated tangent vectors t_1 and t_2 are called principal directions and are always perpendicular to each other (if $k_1 = k_2$, it is sufficient to pick two arbitrary orthogonal tangent vectors). The normal curvature function $k_n(t)$ is a quadratic form and satisfies

$$k_n(t) = \cos^2(\theta)k_1 + \sin^2(\theta)k_2 \,,$$

where θ is the angle between t and t_1.

The *curvature tensor* T is the symmetric 3×3 matrix with the eigenvalues k_1, k_2, 0 and the corresponding eigenvectors t_1, t_2, n. Then, the normal curvature can be computed by

$$k_n(t) = t^T T t \,.$$

When surfaces are approximated by using polygonal meshes, the goal is to compute approximations of the previous concepts directly from the mesh data. Many methods have been proposed to this purpose. The general idea of the techniques is to compute discrete differential properties as spatial averages over a local neighborhood of a point p on the mesh. The size of the local neighborhood critically affects the stability and accuracy of the discrete operators. A complete survey of all the proposed methods and a robust statistical estimation of curvature on sampled surfaces can be found in [12]. In our approach the curvature tensor is approximated by a quadratic patch fitted with the least-squares technique.

2.3 Lepp Delaunay Methods for Quality Triangulation

The following concepts were introduced and used in references [19, 21, 22]. An edge E is called a terminal edge in a triangulation if E is the longest edge of every triangle that shares E, while the triangles that share E are called terminal triangles. Note that in 2-dimensions either E is shared by two terminal triangles t_1, t_2 if E is an interior edge, or E is shared by a single terminal triangle t_1 if E is a boundary (constrained) edge. See Figure 2 where edge AB is an interior terminal edge shared by two terminal triangles t_2, t_3, while edge CD is a boundary terminal edge with associated terminal triangle t_3.

For any triangle, the longest edge propagating path of t_0, called *Lepp(t_0)*, is the ordered sequence $\{t_j\}_0^{N+1}$, where t_j is the neighbor triangle on a longest edge of t_{j-1}, and longest-edge $(t_j) >$ longest-edge (t_{j-1}), for *j=1,... N*. Edge E = longest-edge(t_{N+1}) = longest-edge(t_N) is a terminal edge and this condition determines N. Consequently either E is shared by a couple of terminal triangles (t_N, t_{N+1}) if E is an interior edge, or E is shared by a unique terminal triangle t_N with boundary (constrained) longest edge. See Figure 2.

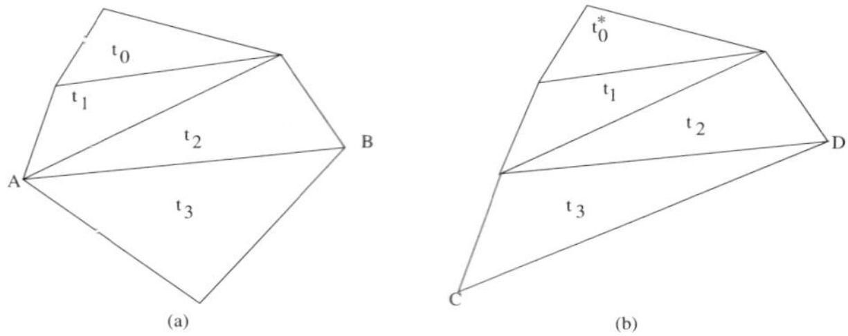

Fig. 2. (a) AB is an interior terminal edge shared by terminal triangles (t_2, t_3) associated to Lepp(t_0) $=\{t_0, t_1, t_2, t_3\}$; (b) CD is a boundary terminal edge with unique terminal triangle t_3 associated to Lepp(t_0^*) $= \{t_0^*, t_1, t_2, t_3\}$

Lepp-midpoint and Lepp-centroid algorithms work essentially in the same way. For any general (planar straight line graph) input data, and a quality threshold angle θ, the algorithm constructs constrained Delaunay triangulations that have all angles at least θ as follows: for every bad triangle t with smallest angle less than θ, a Lepp-search is used to find an associated convex terminal quadrilateral formed by the union of two terminal triangles which share a local longest edge (terminal edge) in the mesh. Then, associated to this terminal quadrilateral, a point p is selected for point insertion, and Delaunay inserted in the mesh. The process is repeated until the triangle t is destroyed in the mesh. In the case of Lepp-midpoint algorithm, p is selected as the terminal edge midpoint, while for the Lepp-centroid algorithm the centroid of this terminal quadrilateral is computed. In [20] it has been proved that the centroid version produces triangulations both with average smallest angles greater than those obtained with the midpoint version, and with bigger smallest edges without suffering of a rare looping case associated to the midpoint method.

3 Terrain Approximation from Grid Data Points

Consider a set V of size n of grid data of a terrain surface F. Given a fixed size data $\overline{n} \ll n$, we want to construct a triangulation $\overline{F}$ over a subset $\overline{V} \subset V$ of size $\overline{n}$ such that $\overline{F}$ fits well to F. We base our approach on an iterative refinement methods. These methods take as input a triangulated mesh $\overline{F}^0$ of a very small subset of V and compute a sequence of meshes, $\overline{F}^i$, $i = 1, 2, \ldots$, by successive insertions of new vertices. The new vertex of a mesh $\overline{F}^i$ is a related point of the worst element, edge or face, of the mesh $\overline{F}^{i-1}$. In the following sections we present the concepts required by our refinement method: the error associated to each edge based on the L_2 norm, the point insertion mechanism used to improve the fitting of the approximated surface to the original surface or to obtain triangles with good aspect-ratio, and the two lepps strategy in order to determine the data points, surface-midpoints or surface-centroids, to be selected and inserted into the triangulation $\overline{F}$.

3.1 Edge Error

Given a point p on F, we can approximate the neighborhood of F around p by a surface patch of the form

$$\mathrm{x}(u,v) = \begin{pmatrix} u \\ v \\ \frac{1}{2}(k_1 u^2 + k_2 v^2) \end{pmatrix},$$

using the frame coordinates with origin at p and axes determined by the principal directions t_1, t_2 and the surface normal n at p. Then, locally, the surface F is a quadratic function $F(u,v) = \frac{1}{2}(k_1 u^2 + k_2 v^2)$ with hessian as follows:

$$\begin{pmatrix} k_1 & 0 \\ 0 & k_2 \end{pmatrix}.$$

In the canonical $\mathbb{R}^3$ frame coordinates the quadratic form defined by this hessian coincides with the curvature tensor.

Given an input vertex v, let T_v be the curvature tensor at v. Then, each vertex v defines a local interpolation error for a triangle $t = \triangle v_1 v_2 v_3$

$$E_2(t,v) = \frac{|t|}{180}\left((d_1 + d_2 + d_3)^2 + d_1^2 + d_2^2 + d_3^2\right),$$

where $|t|$ is the area of the projection of t onto the tangent plane at v and

$$d_i = \frac{1}{2}(v_{i+1} - v_i)^T T_v (v_{i+1} - v_i).$$

For a given edge e adjacent to triangles t_e and t'_e we consider the grid data point c_e whose projection onto the xy-plane is the closest data point to the centroid of the quadrilateral determined by union of the projections onto the xy-plane of t_e and t'_e among the projections of input points in V. The interpolation error of the edge e is defined by

$$E_2(e) = E_2(t_e, c_e) + E_2(t'_e, c_e).$$

The curvature tensor T_{c_e} needed to compute $E_2(e)$, as discussed in Section 2.2, is approximated by a quadratic patch fitted with a least-squares technique. The size of the neighborhood of c_e is taken proportionally to the length of the shortest edge of triangles t_e and t'_e.

3.2 Point Insertion

When a new point p is selected for insertion in the triangulation, a simple valid mesh is obtained by joining p with vertices of the triangle whose xy-projection contains the xy-projection of p. Then, a sequence of edge flips are used either for error improvement or for triangle quality improvement as follows:

(i) if the error of one of the two configurations is far bigger than the error associated to the other configuration, the configuration of less error is selected.
(ii) if both configurations have approximately the same error, the configuration of better angles on the terrain surface is selected.

Notice that:

- triangles with small angles can be obtained even in regions with low curvature by only applying this strategy,
- vertices need to be well fitted by the surface patch defined by the local curvature tensor in order to properly apply the error criterion.

Consequently, adjustment and planarity measures are needed for an accurate point insertion. Following sections are dedicated to define these measures and to specify how the edge flipping operator works.

Adjustment and planarity measures

For each triangle t of area $A(t)$ adjacent to an edge e let

- $\overline{A}(t)$ be the area of the triangle determined by the projection of vertices of t onto the surface patch defined by the local curvature tensor at c_e,
- $\overline{\overline{A}}(t)$ be the area of the projection of t onto the tangent plane at c_e.

Note that $A(t) \cong \overline{A}(t)$ when t is well fitted by the local surface patch, $\overline{\overline{A}}(t) \leq A(t)$ and $A(t) \cong \overline{\overline{A}}(t)$ when t is within a region with low curvature. We define the adjustment measure of t by

$$Ad(t, e) = \frac{|A(t) - \overline{A}(t)|}{A(t)},$$

and the planarity measure by

$$Pl(t, e) = \frac{A(t) - \overline{\overline{A}}(t)}{A(t)}.$$

Consequently, the adjustment measure of e is defined by

$$Ad(e) = \max\{Ad(t_e, e), Ad(t'_e, e)\},$$

and the planarity measure by

$$Pl(e) = \max\{Pl(t_e, e), Pl(t'_e, e)\}.$$

Edge flipping operator

We say that an edge $e = v_1v_3$ is *2D-flippable* if the xy-projections of its two adjacent triangles $t_e = \triangle v_1v_2v_3$, $t'_e = \triangle v_3v_4v_1$ form a convex quadrilateral.

Now we specify when a 2D-flippable edge e is replaced by the edge $e' = v_2v_4$. First we compute the adjustment measures of e and e'. If at least one of these measures is considered unacceptable, i.e.:

$$\max\{Ad(e), Ad(e')\} > \epsilon_1 \,,$$

edge e is replaced by edge e' only when

$$Ad(e) > Ad(e') \,.$$

Otherwise, we compute the planarity measures and the edge errors of e and e'. When the edges are within a low curved region, i.e.:

$$\max\{Pl(e), Pl(e')\} < \epsilon_2 \,,$$

or when the edge errors are similar, i.e.:

$$\frac{|E_2(e) - E_2(e')|}{\max\{E_2(e), E_2(e')\}} < \epsilon_3 \,,$$

edge e is replaced by edge e' only when e' improves the triangle quality, i.e.:

$$\angle v_1v_2v_3 + \angle v_3v_4v_1 > \angle v_2v_3v_4 + \angle v_4v_1v_2 \,.$$

On the contrary, e is replaced by edge e' if e' improves the edge error, i.e. when

$$E_2(e) > E_2(e') \,.$$

By varying the thresholds ϵ_1, ϵ_2 and ϵ_3 we can prioritize the adjustment, planarity and L_2 error measures. For all the results in this paper, we use $\epsilon_1 = 1.0$, $\epsilon_2 = 0.2$ and $\epsilon_3 = 0.2$

We say that a non boundary edge e is called *3D-flippable* when the angle criterion can be applied to it.

3.3 Two Lepps Strategy

Our approach is based on destroying edges with higher error by inserting one or two points close to (one or two) associated terminal edges. In this way we create well-shaped triangles, like the Lepp Delaunay methods do. The three possible cases for each processing bad edge e with adjacent triangles t_e and t'_e are as follows:

1. Edge e is terminal (see Figure 3(a)).
2. Edge e is the longest edge for t_e but not for t'_e (o viceversa). In this case we use the lepp of triangle t'_e for finding a terminal edge e' (see Figure 3(b)).
3. Edge e is not the longest edge for both t_e and t'_e. In this case we use the lepps of triangles t_e and t'_e for finding two terminal edge e'_1 and e'_2. Figure 3(c) shows the two lepps used for finding e'_1 and e'_2 (in order to make clear how the method works, only interior triangles and insertion of midpoints are considered).

For the three cases, we insert a point inside the terminal quadrilateral defined by the associated terminal edges e, e', e'_1, e'_2.

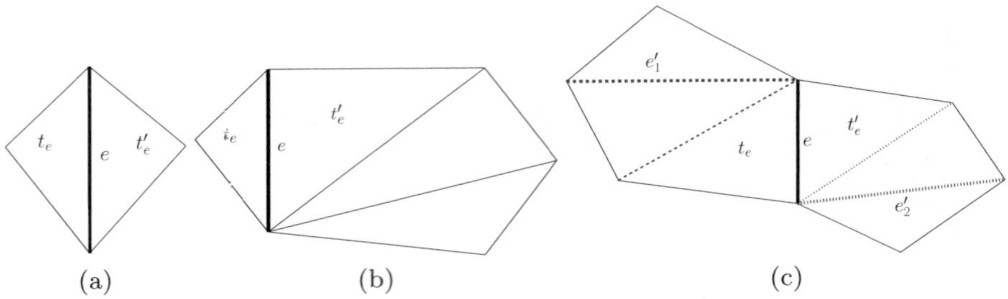

Fig. 3. (a) Special case where e is terminal. (b) Special case where e is the longest edge of t_e but not of t'_e (or viceversa). (c) The two lepp for edge e.

3.4 Surface-Midpoint or Surface-Centroid Selection

In this section we determine the point to be inserted for each reached terminal edge e. We denote by π_e the vertical plane containing e and by F_e the intersection between the surface F and π_e.

The *surface-midpoint* sm_e is the point of F_e equidistant from the endpoints v_i and v_j of e. The point sm_e can be computed as the intersection between F and the line contained in π_e, orthogonal to e and passing through the point $(v_i + v_j)/2$. Since in the common case this computation can not be done exactly, the point sm_e can be approximated by a bipartition algorithm.

Given four points p_1, p_2, p_3, p_4 in a plane, the centroid of the quadrilateral $p_1p_2p_3p_4$ is defined by $(p_1 + p_2 + p_3 + p_4)/4$ or equivalently by $((p_1 + p_3)/2 + (p_2 + p_4)/2)/2$. We define the *surface-centroid* sc_e of an edge based on the last expression. Let e' be the edge determined by two vertices of the triangles adjacent to e not contained in e. The point sc_e is defined as the surface midpoint of the edge determined by sm_e and $sm_{e'}$.

Our method inserts the surface-midpoint when e is a non 3D-flippable edge, and the surface-centroid when e is a 3D-flippable edge. After the point insertion the link edges of the point are tested to be flipped with the edge-flip operator in a similar way as in Lawson [13] algorithm.

4 Lepp-Surface Refinement Algorithm

In what follows we explain the details of our algorithm summarized in Algorithm 1. The process starts with a triangulation $\overline{F}$ of a subset of representative points of the real terrain. Then, new points of the real terrain are progressively added and triangulated using our algorithm until a predetermined number of vertices is inserted. The set of representative points is composed by a reduced subset of boundary points and some singular interior points which represent peaks, valleys, passes, etc. The algorithm iterates over the set of edges E sorted in decreasing order of approximation error. For each processing edge e_f the two lepps strategy is applied, and one or two terminal edges are obtained. If a terminal edge e is

Algorithm 1. LEPP-SURFACE REFINEMENT ALGORITHM

Input = Grid data V, number of desired final points $\overline{n}$
Output = Irregular triangulation $\overline{F}$
Select a small set of points of V and construct a surface triangulation $\overline{F}$
Find E = the set of edges of $\overline{F}$ in descendent order by error
while (Number of vertex of $\overline{F}$ is less than $\overline{n}$) **do**
  Consider processing e_f the worst edge of E
  Let $t_i, i = 1, 2$ the two adjacent triangles of e_f
  if longest edge of t_1 and t_2 is equal to e_f **then**
    v = Select-Point(t_1, t_2, e_f)
    Insertion of v in $\overline{F}$
  else
    for $i = 1, 2$ **do**
      if longest edge of t_i is not equal to e_f **then**
        Use lepp(t_i) for finding terminal triangles t_a, t_b and terminal edge e_i
        v = Select-Point(t_a, t_b, e_i)
        Insertion of v in $\overline{F}$
      end if
    end for
  end if
  Update E
end while

Algorithm 2. SELECT POINT (t_1, t_2, e)

if (second longest edge of t_1 is not constrained and second longest edge of t_2 is not boundary) or e is boundary **then**
  if e is 3D-flippable **then**
    Select p equal to surface centroid sc_e
  else
    Select p equal to surface midpoint sm_e
  end if
else
  for j = 1,2 **do**
    if t_j is not null and has constrained second longest edge e^* **then**
      Select p equal to surface midpoint sm_{e^*}
    end if
  end for
end if
Select v equal to the closest vertex to p in V
RETURN v

3D-flippable, we select the point p as the surface centroid sc_e, otherwise we select p as the surface midpoint sm_e (see Algorithm 2). Next the closest input vertex v to p in V is determined and inserted to $\overline{F}$. The determination of v has a low computational cost because of the subjacent grid data input. Since each vertex insertion changes the triangulation, the list of edges E needs to be updated.

5 Results

In this section we present empirical testing performed to evaluate the practical behavior of our method. The software was implemented in C++ language and using Qt libraries, to real terrain data (Cervino, Kilimanjaro, Como lake, Maggiore Lake, and Dolomiti). All our test problems were run with threshold parameters $\epsilon_1 = 1.0$, $\epsilon_2 = 0.2$ and $\epsilon_3 = 0.2$(see Section 3.2).

We compare our method with the simplification method of QSlim [8] which is a widely known decimation algorithm that starting with the original surface, iteratively collapses edges from the model by minimizing the surface error of the approximation by using a quadric error metric at each vertex.

Table 1. Number of triangles and vertices in the original model and in the triangulation obtained with the Lepp-surface and the QSlim methods

Model	Initial model	Lepp-surface triangulation		QSlim triangulation	
	Vertices	Triangles	Vertices	Triangles	Vertices
Cervino	9801	949	505	948	508
Kilimanjaro	9801	940	505	939	502

Table 2. L_2 norm errors computed by using Mesh1.3 software. Minim, maxim, and mean errors are presented for triangulations of the same size obtained with Lepp-surface method and QSlim method.

Model	Lepp-surface method			QSlim method		
	Min. Error	Max. Error	Mean Error	Min. Error	Max. Error	Mean Error
Cervino	0	1.93	0.2	0	1.40	0.13
Kilimanjaro	0	0.74	0.1	0	0.84	0.07

Table 3. Distribution of the minimum angles obtained with Lepp-surface method and QSlim method

Model	Lepp-surface method						QSlim method					
	Angles				Mean angle	1st Quartile	Angles				Mean angle	1st Quartile
	$< 15°$	$< 20°$	$< 25°$	$< 30°$			$< 15°$	$< 20°$	$< 25°$	$< 30°$		
Cervino	6	22	60	127	40.31	34.55	45	128	245	402	32.90	24.68
Kilimanjaro	0	0	9	31	42.50	38.45	50	122	241	431	32.23	24.74

Table 4. Distribution of the minimum angles and error approximation obtained applying a smoothing post-processing

Model	Laplacian smoothing				HC-Laplacian smoothing			
	Max. Error	Mean Error	Mean angle	1st quartile	Max. Error	Mean Error	Mean angle	1st quartile
Cervino	2.56	0.43	40.05	33.5	1.71	0.35	37.56	30.97
Kilimanjaro	1.10	0.21	39.98	33.77	0.85	0.17	37.81	30.77

Table 5. Distribution of the minimum angles and error approximation obtained with Lepp-surface method for different models

Model	Vertices original model	Lepp-surface method									
		Triangles	Vertices	Angles				Mean angle	1st Quartile	Max. error	Mean error
				$< 15^\circ$	$< 20^\circ$	$< 25^\circ$	$< 30^\circ$				
Como Lake	810000	15985	8105	38	290	915	2078	40.40	34.74	0.57	0.001
Dolomiti	810000	15974	8105	31	277	842	1993	40.39	34.67	0.9	0.002
Maggiore lake	810000	15957	8105	44	259	765	1750	40.76	35.70	0.64	0.0009

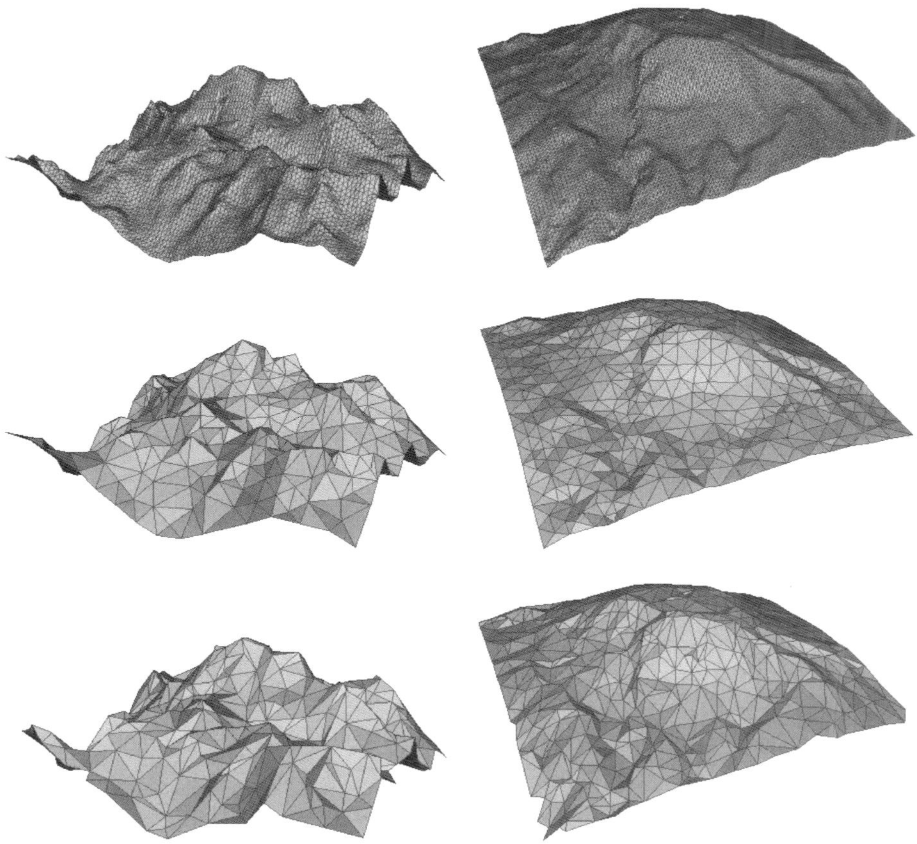

(a) (b)

Fig. 4. Cervino mountain in column (a) and Kilimanjaro mountain in column (b). By rows from top to down, original model, refinement and simplification.

For both triangular approximations, we have used the freely available Mesh1.3 [1] package to estimate the L_2 approximation error between the the triangular approximation computed and the grid data. The L_2 error is expressed as a percentage over the length of the bounding-box diagonal of the model.

The half cylinders presented in the introduction in Figure 1 were obtained using QSlim (Figure 1(a)) and our approach (Figure 1(b)). As can be appreciated in this example our approach obtains triangles with better quality meanwhile the L_2 error is nearly the same as in QSlim. Both examples contain 100 vertices over a sample of 4950 points.

We have used real terrain test data corresponding to Cervino and Kilimanjaro mountains both to compare the Lepp-surface and the QSlim methods, and for illustrating the the ability of our method for modeling complex terrain features. The comparison performed between the Lepp-surface and the QSLIM methods, is illustrated in Tables 1, 2, 3 and Figures 4, 5 for the Cervino and Kilimanjaro mountains. Table 1 summarizes the number of elements of the original and of the computed models, while Table 2 summarizes the errors associated to the results of Table 1. In addition, Table 3 and Figure 5 show the minimun angle distribution for the resulting triangulations obtained with the Lepp surface and QSlim methods.

In order to test the behavior of our method in presence of sharp features we have multiplied the original highs of the terrain two times, except Kilimajaro model.

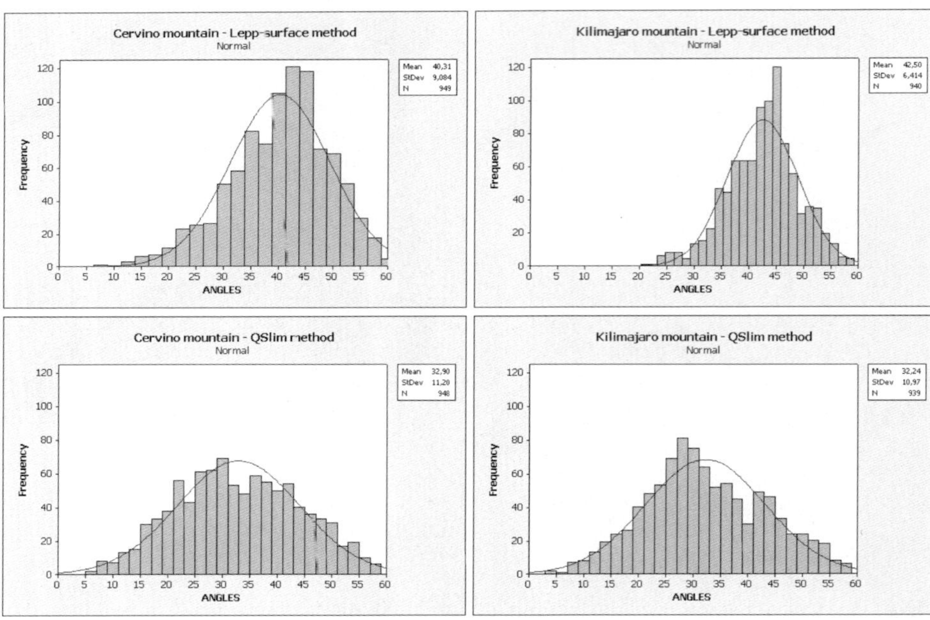

(a) (b)

Fig. 5. Minimum angle distribution for the Lepp-surface (first row) and QSlim (second row) triangulations for Cervino mountain in column (a) and Kilimanjaro mountain in column (b)

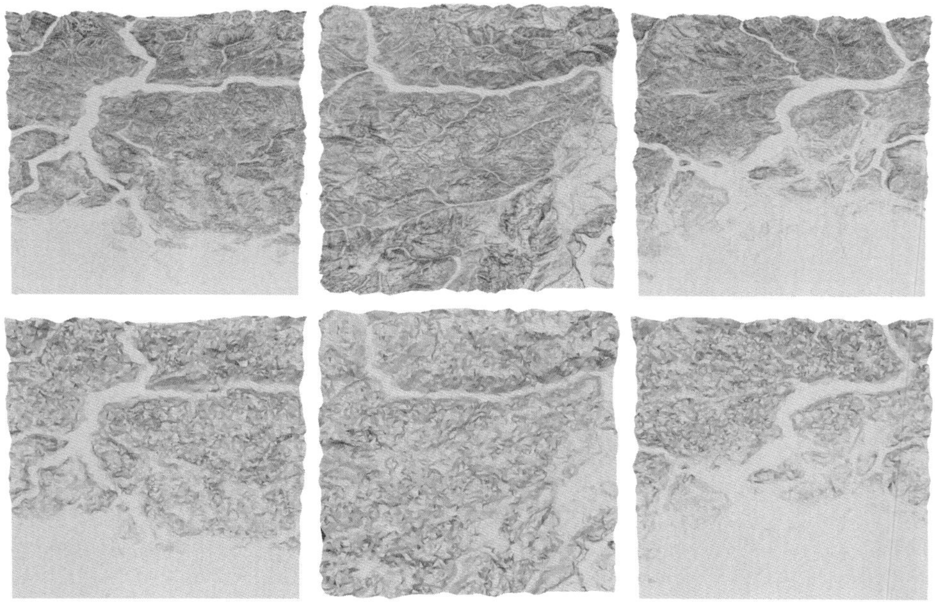

(a) (b) (c)

Fig. 6. Como lake in column (a), Dolomiti in column (b) and Maggiore lake in column (c). By rows from top to down, original model and refinement (1%).

These results show that the refinement process achieves the main goals proposed: 1) the approximation of a terrain surface considering only original grid data; 2) with triangles of good aspect ratio. The error obtained with the refinement process is comparable with that obtained with a simplification process, as it is shown in Table 2 and Figure 4. The statistical study of the minimum angle allows us to take several favorable conclusions to our method. As it can be appreciated in the histograms of Figure 5 the distribution of the minimum angles in refinement process is better than when a simplification process is applied. Table 3 shows that the mean minimum angle, the quantity of triangles with angles less than 15°, 20°, 25° and 30°, and the first quartile obtained with the refinement are better than with the simplification.

The quality of the triangles can be optimized in post-processing being the Laplacian smoothing the commonly used technique. This technique lacks motivation because the input grid data is not respected and may not even remain a valid triangulation. The freely available software MeshLab [3] offers the Laplacian and the HC-Laplacian [25] smoothing techniques. We have applied these techniques to the simplified surfaces of Cervino and Kilimanjaro examples (see

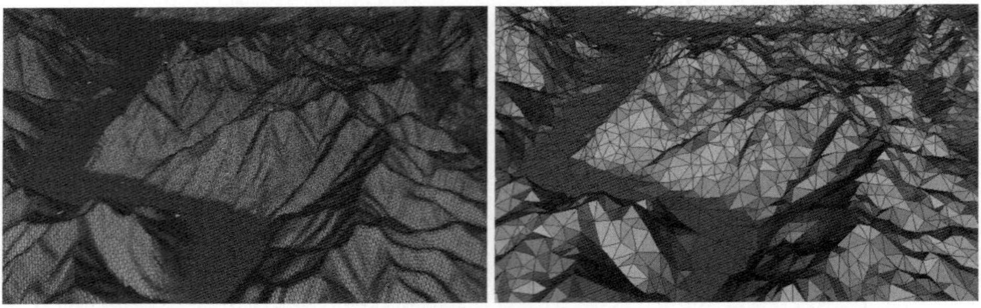

(a) (b)

Fig. 7. Detail of the Como lake. In (a) the original model and in (b) the refined model (5%).

Table 4). In error approximation as well as in angle quality our method gives better results than the considered post-processing techniques.

Table 5 presents results obtained applying Lepp-surface method to other models: Como lake, Maggiore Lake and Dolomiti. Here, error approximation and statistical data of angles are shown for each model. This table is complemented by Figure 6 in which views of original models and their refinement can be compared. As can be appreciated in this figures in spite of the lower quantity of points used in the refinement (1% of the original set) the main features of the terrains can be distinguished. The algorithm takes about 9 seconds in generating this models in an Intel Pentium 3.00 GHz with 1.00 GB of RAM.

Finally, to illustrate that local details are preserved Figure 7 shows a detail both of the original data and of the Lepp-surface approximation for the Como Lake with 5% of the original data set points with a total running time of about 60 seconds.

6 Conclusions and Future Work

We have presented a refinement method that interpolates grid input data based on estimating the L_2 error by using the curvature tensor and a two lepp strategy in order to create a well-shape triangulation. The L_2 metric is prioritized but when it is possible the aspect-ratio of the triangles is taken also into account, achieving in this way an irregular triangulation of the terrain with good quality elements that fits well the original input data.

The method has been tested on several models giving very good results in the approximation of terrain surfaces with sharp features. The main goals pursued in this research, that is, good error approximation and good aspect-ratio of the triangles were achieved.

The basic ideas of the method presented in this paper can be easily adapted to work with irregular terrain input data, and generalized to deal with input data obtained by using 3D scanners.

References

1. Aspert, N., Santa-Cruz, D., Ebrahimi, T.: Mesh: Measuring errors between surfaces using the hausdorff distance. In: Proc. of the IEEE International Conference in Multimedia and Expo., pp. 705–708 (2002)
2. Bertram, M., Barnes, J.C., Hamann, B., Joy, K.I., Pottmann, H., Wushour, D.: Piecewise optimal triangulation for the approximation of scattered data in the plane. Computer Aided Geometric Design 17(8), 767–787 (2000)
3. Cignoni, P., Corsini, M., Ranzuglia, G.: Meshlab: an open-source 3d mesh processing system. ERCIM News (73), 45–46 (2008)
4. DeFloriani, L., Magillo, P., Puppo, E.: Variant: A system for terrain modeling at variable resolution. GeoInformatica 4(3), 287–315 (2000)
5. Dyn, N., Levin, D., Rippa, S.: Data dependent triangulations for piecewise linear interpolation. IMA Journal of Numerical Analysis 10, 137–154 (1990)
6. Frey, P.J., Borouchaki, H.: Geometric surface mesh optimization. Computing and Visualization in Science 1(3), 121–133 (1998)
7. Garland, M.: Multiresolution modeling: Survey & future opportunities. Eurographics State of The Art Report (1999)
8. Garland, M.: Quadric-based polygonal surface simplification, Ph.D. thesis, Carnegie Mellon University (1999)
9. Garland, M., Heckbert, P.: Surface simplification using quadric error metrics. In: SIGGRAPH 1997: Proceedings of the 24th annual conference on Computer graphics and interactive techniques, pp. 209–216 (1997)
10. Garland, M., Heckbert, P.-S.: Fast polygonal approximation of terrains and height fields, Tech. Report CMU-CS-95-181 (1995)
11. Hoppe, H., DeRose, T., Duchamp, T., McDonald, J., Stuetzle, W.: Mesh optimization. Computer Graphics 27, 19–26 (1993)
12. Kalogerakis, E., Simari, P., Nowrouzezahrai, D., Singh, K.: Robust statistical estimation of curvature on discretized surfaces. In: SGP 2007: Proceedings of the fifth Eurographics symposium on Geometry processing, Eurographics Association, pp. 13–22 (2007)
13. Lawson, C.-L.: Software for $\mathcal{C}^1$ surface interpolation. In: Rice, J.R. (ed.) Mathematical Software III, pp. 161–194 (1977)
14. Lindstrom, P., Pascucci, V.: Visualization of large terrains made easy. In: Proceedings of IEEE Visualization 2001, pp. 363–370, 574 (2001)
15. Lindstrom, P., Pascucci, V.: Terrain simplification simplified: a general framework for view-dependent out-of-core visualization. IEEE Transactions on Visualization and Computer Graphics 8(3), 239–254 (2002)
16. Nadler, E.: Piecewise linear best L_2 approximation on triangulations. In: Chui, C.K., et al. (eds.) Approximation Theory V, pp. 499–502 (1986)
17. Pajarola, R., Gobbetti, E.: Survey on semi-regular multiresolution models for interactive terrain rendering. The Visual Computer 23(8), 585–605 (2007)
18. Rippa, S.: Long and thin triangles can be good for linear interpolation. SIAM Journal on Numerical Analysis 29(1), 257–270 (1992)
19. Rivara, M.-C.: New longest-edge algorithms for the refinement and/or improvement of unstructured triangulation. International Journal for Numerical Methods in Engineering 40, 3313–3324 (1997)
20. Rivara, M.-C., Calderon, C.: Lepp terminal centroid method for quality triangulation: A study on a new algorithm. In: Chen, F., Jüttler, B. (eds.) GMP 2008. LNCS, vol. 4975, pp. 215–230. Springer, Heidelberg (2008)

21. Rivara, M.-C., Hitschfeld, N., Simpson, R.-B.: Terminal edges delaunay (small angle based) algorithm for the quality triangulation problem. Computer-Aided Design 33, 263–277 (2001)
22. Rivara, M.-C., Palma, M.: New lepp algorithms for quality polygon and volume triangulation: Implementation issues and practical behavior. 220, 543–562 (1997)
23. Röttger, S., Heidrich, W., Slussallek, P., Seidel, H.-P.: Real-time generation of continuous levels of detail for height fields. In: Proceedings of 1998 International Conference in Central Europe on Computer Graphics and Viaualization, pp. 315–322 (1998)
24. Shewchuk, J.-R.: What is a good linear element? interpolation, conditioning, and quality measures. In: Proceedings of the 11th International Meshing Roundtable (2002)
25. Vollmer, J., Mencl, R., Môller, H.: Improved laplacian smoothing of noisy surface meshes. Computer Graphics Forum (Eurographics 1999) 18(3), 131–138 (1999)

GPU-Assisted Surface Reconstruction on Locally-Uniform Samples

Yong Joo Kil and Nina Amenta

University of California, Davis
kil@cs.ucdavis.edu,
amenta@cs.ucdavis.edu

Summary. In point-based graphics, surfaces are represented by point clouds without explicit connectivity. If the distribution of the points can be carefully controlled, surface reconstruction becomes a much easier problem. We present a simple, completely local surface reconstruction algorithm for input point distributions that are locally uniform. The locality of the computation lets us handle large point sets using parallel and out-of-core methods. The algorithm can be implemented robustly with floating-point arithmetic. We demonstrate the simplicity, efficiency, and numerical stability of our algorithm with an out-of-core and parallel implementation using graphics hardware.

1 Introduction

The idea of point-based graphics is that a point sample can be the primary representation of a surface, simplifying computation and saving space by doing without explicit connectivity information. In some situations, for instance a geometric modeling system or the simulation of a moving front, the distribution of the sample points is entirely under the control of the application, which upsamples, downsamples and smooths the distribution as necessary. What kind of a distribution should a point-based application seek to maintain? One criterion, among others, is that the distribution should make later geometry processing operations, such as surface reconstruction, easier. A surface reconstruction algorithm which will be applied to a carefully maintained distribution can be simpler, faster, and easier to implement and to parallelize, and it should scale better.

In this paper we explore how simple and local a surface reconstruction algorithm can be when applied to well-distributed points. Specifically, we consider distributions which are locally uniform, and we give a very easy surface reconstruction algorithm which works well in practice on such distributions. We emphasize that this scenario is very different from the problem of surface reconstruction from acquired data. Rather than designing the algorithm for robustness to noise and sampling irregularity, we can design for simplicity, speed, and locality of computation.

Outline of algorithm. We construct a octree for the sample points and download it to the GPU. In parallel, we find the k-nearest neighbors of each sample point. Then at each sample point we construct an umbrella of possible surface triangles. Edges that are consistent between adjacent umbrellas are defined to be *consensus edges*. On arbitrary distributions it is possible that the consensus edges will form a very sparse graph, possibly not even connected. But on distributions that are locally uniform, we observe that the consensus edges define a polygonalization of the surface, in which all faces have a small number of edges. We triangulate any non-triangular faces in this graph, using a deterministic rule, completing the triangulation. Our implementation can triangulate a sample consisting of 6 million points on a surface in $\mathbb{R}^3$ in about 35 seconds (not including the CPU octree computation). It is written in CUDA, and uses native GPU floating point arithmetic.

Related work. Surface reconstruction is a huge field and we focus here on only the work most related to ours. The idea of reconciling local umbrellas at each sample point is the central concept in the algorithm of Gopi et al. [1], in which the umbrellas are computed on an advancing front. While the high-level idea of Gopi's algorithm is similar to ours, it is essentially sequential. A recent GPU-assisted algorithm based on Gopi's method is due to appear [2]. This algorithm apparently computes local umbrellas in parallel on the GPU, with both the earlier step of finding nearest-neighbors and a post-processing reconciliation of the umbrellas done sequentially. Also in this same vein, Adamy et al. [3] select umbrellas at each point (using a criterion similar to ours) and reconcile them using simple topological processing followed by a global linear programming step. Our algorithm differs from all of this previous work in that we reconcile the umbrellas locally and in parallel; we do not have to resort to a sequential global process.

Computing and maintaining an umbrella (or "star") at every vertex is also central to Shewchuk's "star splaying" algorithm [4] for maintaining a Delaunay triangulations under perturbations.

Dey et al. [5] and Funke et al. [6] considered surface reconstruction given a locally uniform distribution, and give an almost linear time algorithm. While candidate triangles are found independently at each sample, a final reconciliation stage requires a sequential sweep through the surface. Finally, Dumitriu et al. [7] give a sequential surface reconstruction algorithm which takes a very dense, possibly poorly sampled point cloud as input, and constructs a triangulation of a sparse locally uniform subsample. Their algorithm is interesting in that the only geometric computation required is the construction of a neighborhood graph on the original dense sample. Like us, they find a graph of "good" edges, within which the minimal cycles form a polygonization of the surface; they report cycles of length at most five in practice. In an unpublished manuscript [8], they give a (large) constant upper bound on the size of the cycles produced by their algorithm, which unfortunately cannot be applied to ours. In contrast, our algorithm requires only the sparse locally uniform input, not the dense input point set as well.

A very recent GPU surface reconstruction algorithm by Zhou et al. [9] implements the Poisson surface reconstruction algorithm of Kazhdan et al. [10].

The paper emphasizes that a major contribution is the introduction of an octree which is not only searched on the GPU, like ours, but also constructed in parallel on the GPU. They report an excellent running time of 0.2 seconds for an input of about 500K points. Poisson reconstruction requires normal information and is a global algorithm; the output size is limited by the resolution of the grid achievable on the GPU. Local algorithms like ours can handle larger inputs by breaking them up into smaller parts.

Locally uniform point samples. We define locally uniformity for a point distribution on a surface using a parameter α.

Definition 1. *Let P be a sample of points from a surface in $\mathbb{R}^3$. For a sample point $v \in P$, let δ_v be the minimum distance from v to any other sample. Let ϵ_v be the maximum distance from v to any* surface *point in the Voronoi cell of v. We say that P is* locally uniform *with parameter α if, for all $v \in P$, $\epsilon_v \leq \alpha\delta_v$.*

This measure of the distribution of a point set is similar to the well-known circumradius to shortest edge criterion for a triangulation. Computing ϵ_v is not possible if we are not given the surface itself. But it is easy to estimate ϵ_v once we have an output triangulation; we use such estimates to illustrate the range of α within which our algorithm is successful in our experiments.

Local uniformity of a point sample does not guarantee that it is dense enough to allow for a correct reconstruction of the surface. Besides being locally uniform, our algorithm also requires the input point distribution to be everywhere sufficiently dense; as noted by Amenta et al. [11], the minimum required density varies over the surface linearly with distance to the medial axis. Any distribution which everywhere exceeds this minimum density and is locally uniform is an appropriate input to our surface reconstruction algorithm.

Producing and maintaining locally uniform samples. It is indeed possible to produce locally uniform point samples on surfaces, and to maintain them as the surface changes. A perfect triangular grid of points in the plane achieves the best possible $\alpha = 1/\sqrt{3} \approx 0.577$ everywhere, but we cannot expect to do so well on an arbitrary surface. A simple greedy sampling strategy [12], however, does achieve $\alpha \leq 1$ on an arbitrary surface: choose a small enough δ, and then while there is some point of the surface farther than δ from any sample, place a new sample at any such point. This procedure can also be adapted to handle varying sampling density. The distribution is improved if the new point is chosen uniformly at random (the Poisson disk sampling method; see Dunbar et al. [13]). Another strategy is to insert the point furthest from any previously-placed sample (eg. Boissonnat and Oudot [14]).

The greedy algorithm creates locally uniform point sets. Among heuristics which can be used to maintain local uniformity after a perturbation (e.g. during a modeling operation or simulation step), Lloyd's algorithm works well in practice. It moves every sample to the center of mass of its Voronoi cell, iteratively, until convergence. Surazhsky et al. [15] uses local parameterizations and Llyod's relaxation to produce locally uniform sampling on arbitrary genus surfaces. Recent work on optimal Delaunay triangulations [16] uses a similar idea to produce even nicer-looking distributions on the plane. On implicit surfaces, a classic paper by Witkin and Heckber [17] uses repulsion and attraction forces to produce and maintain locally uniform point distributions; the distribution in Figure 9 was produced using our implementation of their algorithm. More recently, Meyer et al. [18] gave an improved force-based algorithm along similar lines.

2 Geometric Intuition and Groundwork

Let's begin by considering the uniform unit triangular grid in a plane embedded in R^3, the vertices of which form a distribution P with the minimum $\alpha = 0.577$. An *empty circumsphere* of a triangle has the vertices of the triangle on its boundary and has no other point of P in its interior. The minimal empty circumspheres of the triangles of the grid have radius α, while any other triangle with vertices on the grid has no empty circumsphere of non-infinite radius. This motivates our umbrella selection criterion: we will choose an umbrella in which each of the triangles has a small empty circumsphere.

In more realistic situations, the surface is curved and the distribution is not so nice. We are however always guaranteed that every sample point in P has at least one umbrella consisting of triangles with small circumspheres. The *restricted Delaunay triangulation* [19] is the set of triangles with empty circumspheres centered at points on the surface; these centers are the points at which the edges of the 3D Voronoi diagram of P pierce the surface.

Observation 1. *A triangle t of restricted Delaunay triangulation has circumspheres of radius at most ϵ_v, for any vertex v of t.*

Since there is an umbrella of triangles at each vertex, each of which has a small circumsphere, we are always able to select such an umbrella. Not all triangles with small circumspheres will be restricted Delaunay triangles, however, even on distributions that are everywhere dense and locally uniform. The difficulty has to do with *slivers*: very flat tetrahedra, the vertices of which are nearly co-circular. It is possible (although not necessary) for all of the triangular faces of a sliver tetrahedron in the 3D Delaunay tetrahedarlization of P to have small circumspheres. Also, even when P is locally uniform, five, six or more points can be nearly co-circular and define many triangles with small circumspheres; the number of points that can be nearly co-circular increases with α. If v and several of its neighbors are nearly co-circular, they might choose umbrellas that are inconsistent with each other. On the other hand, any subset of these triangles chosen for the umbrella at v will include the two edges connecting v to its neighbors u, w along the "circle". Similarly, these edges will also appear in any subsets of the triangles chosen for the umbrellas at u and w. This intuition leads us to the definition of consensus edges.

Definition 2. *We define the Delaunay edge uv to be a* consensus edge *if it appears in every umbrella which contains the two vertices u, v. Note that this includes the umbrellas of points other than u and v.*

We conclude with some observations about the triangles with small circumspheres (of radius at most $\alpha\delta_v$). These observations will be useful in selecting the triangles, and they also clarify the connection between α and the quality of the output triangles.

Since every triangle in P is enclosed in a circumsphere of radius at most $\epsilon_v \leq \alpha\delta_v$, we know that each triangle has a circumcircle of radius at most ϵ_v. Since every edge must have length at least δ_v, the smallest possible circumcircle (and hence the smallest possible circumsphere), is the one surrounding an eqilateral triangle with side length δ_v.

Observation 2. *The smallest triangle circumcircle has radius* $1/\sqrt{3}\delta_v \approx 0.577\delta_v$.

Consider the figure below.

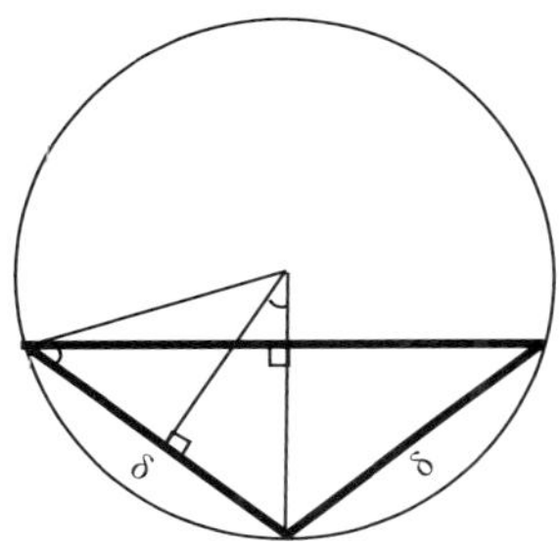

The triangle with the largest possible angle has to look like the bold triangle; the radius of the circumcircle is as large as possible, that is, ϵ_v. To get as large an angle as possible at v, we spread the two adjacent edges as far apart as possible, until they are as short as possible, that is, δ_v. The two marked angles are the same, and must be $\arcsin((\delta_v/2)/\epsilon_v) = \arcsin(1/2\alpha)$. So the opposite angle in each of the right triangles is $\arccos(1/2\alpha)$, and the large angle at the bottom is $2\arccos(1/2\alpha)$. Also, notice that to make as small an angle as possible, we also need to use as large a circumcircle as possible, and the smallest angle will be opposite the smallest edge of the triangle. So any angle of a triangle with circumradius ϵ_v, opposite an edge of length δ_v, has the smallest possible angle. Hence:

Observation 3. *The largest angle of any triangle is* $2\arccos((1/2\alpha)$ *and the smallest angle of any triangle is* $\arcsin(1/2\alpha)$. *For instance, with $\alpha = .9$, we have largest angle of* 112.5 *and a smallest of* 33.75 *degrees.*

3 Algorithm

We now describe the parallel GPU part of our algorithm. Each step in the following description is run in parallel over each vertex $v \in P$.

Algorithm 1. Parallel surface reconstruction for each $v \in P$

1: Retrieve the k nearest neighbors N_v of v.
2: Compute Delaunay faces F_v connected to v using N_v
3: Compute umbrella U_v by choosing a subset from F_v.
4: Determine if out-going edges of U_v are consensus edges C_v by searching through all umbrellas in N_v.
5: Find cycles for every edge in C_v and output polygons.

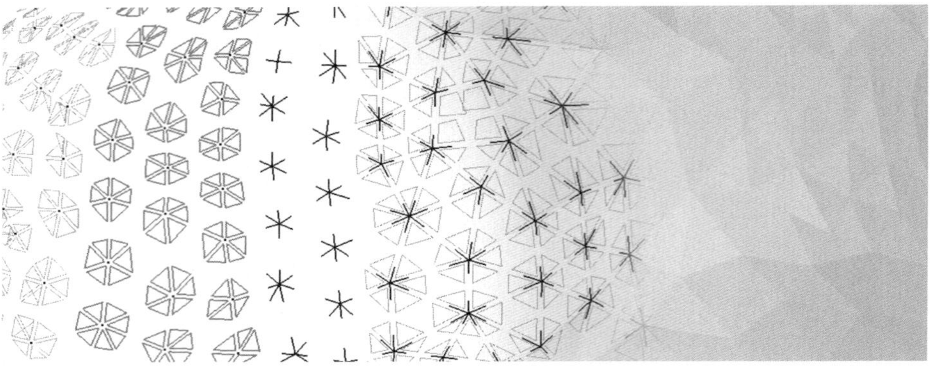

Fig. 1. A visualization of the steps of our algorithm. We draw the simplicies by scaling them towards their respective vertices. Starting from the left, we compute Delaunay and nearly-Delaunay triangles (light green), umbrellas (dark green), consensus edges (dark blue), and cycles (orange). We then render the resulting triangles.

Let us discuss these steps in detail. The first step retrieves the k nearest neighbors using a search structure; we use an octree which is pre-computed on the CPU. The second step computes Delaunay faces connected to v by considering samples in N_v, which is sufficient as long as the k nearest neighbors contain all points within $2\epsilon_v$ distance from v. The choice of k depends on the uniformity of the point-set.

We defer discussion of the second parallel step, which produces a set of triangles at each vertex v, until Section 4. From this set, we extract an umbrella of triangles, each with small circumball radius in the third step, using the following algorithm. An *outgoing* edge of v is an edge connected to v. Triangles in F_v are *adjacent* if they share an outgoing edge. Outgoing edges of an umbrella U_v are *manifold edges*, meaning that edge is connected to exactly two triangles. To form an umbrella from F_v and to ensure that the resulting umbrella if formed by triangles with small circumball radii, we look for triangles with large circumball radii that can be removed. A triangle is a *flake* if one of its outgoing edges does not meet another triangle at a large dihedral angle. A flake is never a restricted Delaunay triangle and can always be removed immediately. Second, a triangle is

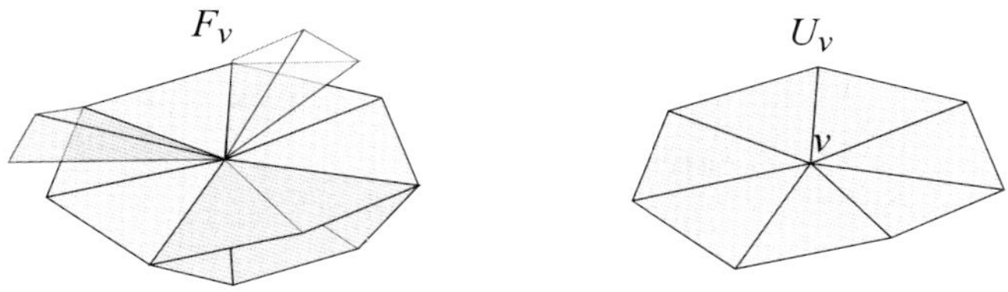

Fig. 2. The left figure is a F_v with flake triangles highlighted in yellow and pocket triangles highlighted in blue. The right figure is a resulting umbrella U_v. Note that removable of a pocket triangle can change pocket triangles into flake triangles.

Algorithm 2. UMBRELLA(v, F_v)

$U_v \leftarrow F_v$
while U_v is not an umbrella **do**
  Remove all flakes in U_v.
  Remove the pocket triangle with the largest circumball radius in U_v.
return U_v.

a *pocket* triangle if it is not a flake and if the dihedral angle between it and any of its adjacent triangles are small; adjacent triangles with manifold edges are also pockets. Figure 2 shows an example of flakes and pockets and Algorithm 2 describes the umbrella filtering.

We may remove a restricted Delaunay triangle which is a pocket triangle, but if so the remaining umbrella will consist only of triangles with even smaller minimal circumspheres.

The fourth step determines if the outgoing edges of U_v are consensus edges. For a vertex v, let us define L_v as the *link* of U_v, which contains all edges in U_v not touching v. Let O_v be outgoing edges of U_v. In order for an edge $vw \in O_v$ to be consensus, we check two conditions. First, we check if $wv \in O_w$. Second, we check for all $x \in N_v$ such that if vertices $\{v, w\} \in U_x$, then $vw \in L_x$. If either of these conditions fail, then vw is not a consensus edge.

We can easily extend the consensus definition to handle boundaries. A point on a boundary cannot have an umbrella, so the UMBRELLA algorithm will produce the empty set (see Figure 3). In order to produce triangles that partially cover a boundary sample, we simply ignore consensus tests that involve empty U_w or U_x. This works quite well when the boundary is nicely sampled, as in Figure 3.

Finally, in the last step, we extract the polygonal faces by detecting cycles in the graph of the consensus edges. This is complicated by the fact that we do not have vertex normals or a consistent orientation. Following a consensus edge from v to w, we orient w consistently with v as follows. We consider the triangle t_w that is on the same side as t_v of the plane that is perpendicular to t_v and passing through vw (t_w may or may not be t_v). If t_w has a consensus edge wz opposite

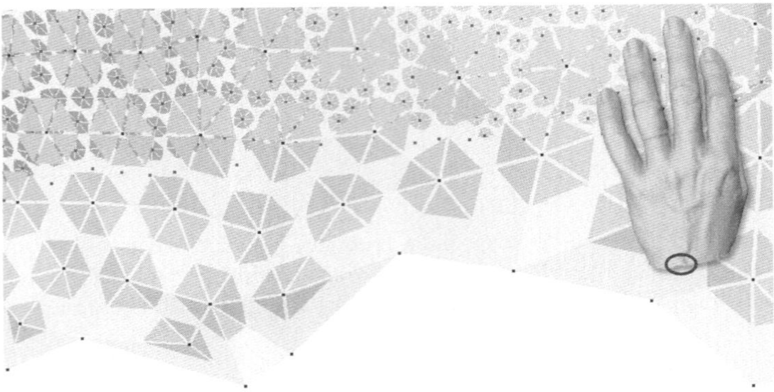

Fig. 3. A closeup of the boundary of the hand model. Our algorithm handles boundaries gracefully.

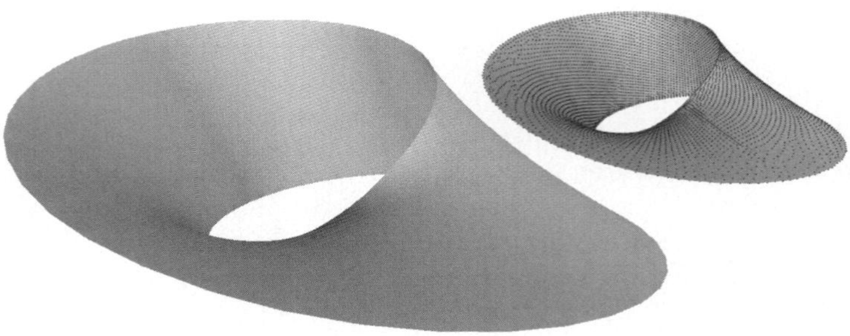

Fig. 4. Our algorithm works on non-orientable surfaces. Here, the input point samples are shown at the top right.

of vw, we choose wz as the next consensus edge in the cycle. If t_w does not have a consensus edge, we traverse the adjacent triangles in the umbrella until we do find a consensus edge; if we get back to vw before finding another consensus edge we quit and do not output a cycle. For a consensus edge touching a boundary sample b, we do something similar to define a face: we start a path from the edge containing b and terminate the search once the path reaches another boundary sample (which might be b).

Once a cycle is found, we can return the cycle as a polygon or triangulate it if necessary. To make sure all vertices compute the same triangulations, we deterministically break each cycle into a triangle fan with the center located at the vertex with the smallest memory address. By not relying on any orientation information, we can extract non-orientable surfaces, such as a Mobius strip as shown in Figure 4.

4 Nearly-Delaunay Computation with Floating-Point Arithmetic

In this section, we discuss the computation of the set of Delaunay and nearly-Delaunay triangles from which the umbrellas are selected. This computation is done with the GPU's floating-point arithmetic, which is generally single-precision and does not conform to the IEEE floating-point standard. Our approach is to be tolerant of numerical error.

The umbrella selection algorithm requires at each vertex v a set of triangles, each labeled with an estimate of the radius of its smallest empty circumsphere. In the interests of maximizing the number of consensus edges, we would like this set to be as small as possible. In this section we consider the geometric computation we use to find that set. We first select a list of possible candidate triangles, and then we check each candidate. We can eliminate a candidate if we determine that it is definitely not Delaunay, or if we determine confidently that its smallest empty circumsphere is much larger than the distance from v to its nearest neighbor.

We begin with the candidate set of all well-shaped triangles T_v in N_v connected to v. To estimate the radius of the smallest circumsphere for a triangle touching $t \in T_v$ we consider the Voronoi edge dual to t. We parameterize the dual edge as $e(\lambda) = c + \lambda n$, where c is the circumcircle center of t, n is the unit normal perpendicular to the plane of t and $\lambda \in \mathbb{R}$. Let B_λ be the circumsphere centered at $e(\lambda)$, touching the vertices of t. For each point $q \in \{N_v \backslash t\}$, we find the interval Λ_q that contains values of λ for which B_λ is empty of q. This idea is elaborated in Algorithm 3.

The fundamental operation in the algorithm is computing the interval Λ_q, which is equivalent to finding the center of the sphere touching the three vertices of triangle t and the fourth point q. This circumcenter operation is notoriously numerically unstable: if the four points are nearly co-circular, then round-off error in fixed-precision floating-point computation can lead to wildly incorrect results. We use our assumption that the input P is locally uniform in several

---

Algorithm 3. DELAUNAY_FACES(v, N_v)

$F_v \leftarrow \{\}$
Let T_v be the set of all possible triangles from N_v, touching v, and without very large or very small angles.
for all $t \in T_v$ **do**
  $\Lambda_t \leftarrow (-\infty, +\infty)$
  for all $q \in \{N_v \backslash t\}$ **do**
    Compute the valid interval Λ_q
    $\Lambda_t \leftarrow \Lambda_t \cap \Lambda_q$
  if Λ_t is not empty **then**
    Find the smallest $|\lambda|$ in Λ_t to determine the smallest circumball radius r_t.
    Add $\{t, r_t\}$ in F_v.
return F_v.

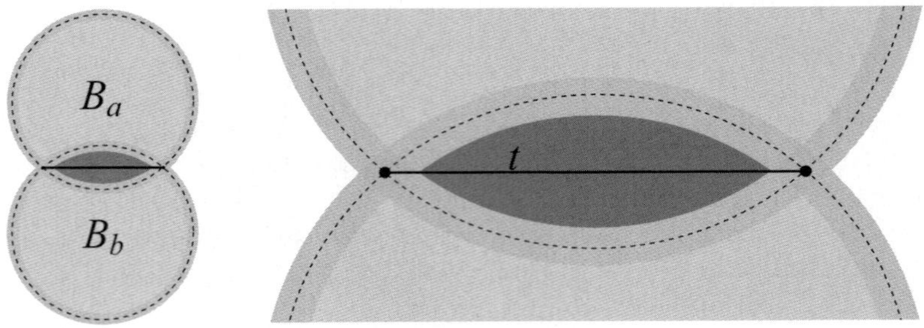

Fig. 5. This figures shows the regions considered in our robust floating-point computation. The triangle t and its points are shown in black, the two balls B_a and B_b are shown in dotted lines, and the shells are shown in blue. The right figure is a closeup of the area near the triangle. Points in the orange region are considered to be co-circular with the vertices of t. Points in the blue region are considered to be co-spherical with either B_a or B_b. Points in the red region are considered to have circumball radius larger than ϵ. Points in the gray region are numerically safe to use Equation 1.

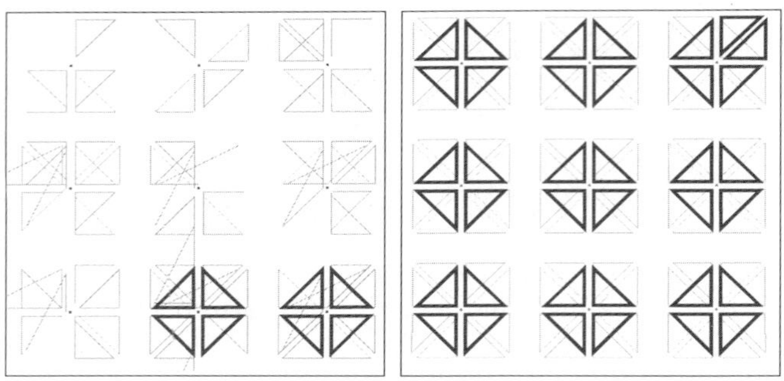

Fig. 6. The computed Delaunay faces for each vertex are highlighted in green and the darker green triangles are the resulting umbrellas. The left figure shows triangles computed using without using the careful numerical approach described in Section 4, and the right figure shows the triangles using the approach.

ways to deal with the problems caused by numerical instability. Our goal is to avoid eliminating restricted Delaunay triangles, while filtering out triangles that cannot possibly be restricted Delaunay triangles, thus producing better umbrellas and more consensus edges.

First, we eliminate all triangles that have any angle that is too large or too small, which is justified by Observation 3. In our experiments, we simply ignore triangles with angles smaller than one degree.

Second, we observe that because triangle t is well-shaped, the computation of its normal n and the center of its circumcircle c are, while not exact, at least numerically stable. Instability in these computations occur when the three triangle vertices are nearly co-linear, and these situations have been eliminated.

So we compute c and n once for each triangle and use them in the computation of each Λ_q. In computing n and c, we improve the stability by taking v as the origin, and then in the computation of Λ_q we use c as the origin. Also, we ensure that the vertices of a triangle are always used in the same (memory address) order when computing n and c, so that n and c will be identical when t is processed at each of its vertices.

To solve for the endpoint in the interval Λ_q, we solve for λ such that $|e(\lambda)-p| = |e(\lambda) - q|$, where p is a vertex of t. If c is at the origin, then

$$\lambda = \frac{(p-q)^T(p+q)}{2(p-q)^T n} \tag{1}$$

This computation will *not* be stable if q and the vertices of t are nearly co-circular or nearly co-planar. We do some special processing to identify these cases. We take a conservative approach and set Λ_q to $(-\infty, \infty)$ if the four points are nearly co-circular, passing the triangle (as far as q is concerned) as nearly-Delaunay. At the same time, we eliminate triangles with large minimal empty circumspheres.

First, let us consider an idea for avoiding unnecessary evaluations of λ. Recall that Observation 2 tells us that the radius of the circumcircle of each well-shaped triangle is no smaller than $\sqrt{3}\delta_v$, while Observation 1 tells us that any restricted Delaunay triangle has a smallest circumsphere of radius at most $\epsilon_v = \alpha\delta_v$. We consider two balls, B_a and B_b, passing through the vertices of t with the center of B_a above and the center of B_b below the plane of t (see Figure 5), each of radius $\sqrt{3}\alpha$ multiplied by the circumcircle radius of t. Any sample q_{out} outside $B_a \cup B_b$ cannot improve our lower bound on the size of the smallest empty circumsphere, so immediately conclude that t is acceptable with respect to q_{out} and continue on to the next q. Any sample q_{in} inside $B_a \cap B_b$ forces the minimum empty circumsphere to have radius radius larger than ϵ, so we immediately reject t for F_v.

Now we consider the unstable case. Notice that in Equation 1, if the vector $(p-q)$ is very small or is almost perpendicular to n, then the denominator will be close to zero. The dot product between these two vectors may produce a catastrophic cancelation that results in a large relative error. To avoid doing the computation in the unstable case, we consider the *shells* S_a and S_b of B_a and B_b, respectively. A shell of a ball B is formed by the points between the two balls sharing the same center as B with radii $r(1-\mu)$ and $r(1+\mu)$, where r is the radius of B and $0 < \mu \ll 1$. If $q \in S_a \cap S_b$, then we assume q is co-circular with the vertices of t and we treat Λ_q as unbounded. If $q \in (S_a \cup S_b) - (S_a \cap S_b)$, we handle it as described above for B_a and B_b. Only if $q \notin (S_a \cup S_b)$ and $q \in (B_a \cup B_b) - (B_a \cap B_b)$ do we compute λ using the formula above, and in that case the computation is stable. To make sure the resulting Λ_t has an valid interval when its interval is very small, we also scale each interval Λ_q by $(1+\mu)$. In our experiments, we set μ as low as 10^{-5} without noticing any problems with the 32-bit floating point arithmetic on our graphics hardware.

Figure 6 shows an example of a difficult input, containing many truly co-circular quadruples of points, with Λ_q computed using a naive floating-point

implementation and also with our more careful conservative strategy. Notice that we retain triangles which are possibly not Delaunay. The later umbrella computation and the choice of consensus edges tolerate this behavior.

5 Parallel and Out-of-Core Implementation

We wrote our parallel algorithm using NVIDIA's CUDA GPU Computing environment [20], using OpenGL for rendering. Each of the steps in Algorithm 1 runs in parallel as a CUDA kernel. In order to communicate between kernel stages, we create workspace buffers to hold intermediate data, such as the pointers to the k neighbors of each vertex in P. If these workspace buffers collectively take up too much memory, we partition P into smaller groups, which are then computed serially. The groups are bounded by chains of input samples, which are included in both groups. To make sure the samples in these boundary groups produce consistent triangles, the neighboring samples on either side should also be included in both groups. In our experiments so far, however, we treat the group boundaries as described in Section 3, and we have not noticed holes in the resulting outputs.

For the Delaunay computation of F_v, each vertex v iterates over all possible candidate triangles formed by v and two other samples in N_v. To speed this up, we create a thread for each of v's neighbors. We allow these threads to communicate by caching vertex positions into the on-chip memory (shared memory) that is shared amongst v's neighbors.

Reducing the number of nearest-neighbors k of course speeds up the computation of the nearly-Delaunay triangles. But small values of k can lead to trouble when we have varying sampling density, as in Figure 7. In this situation it might happen that some of v's neighbors find triangles including v which are not found by the computation at v itself. To remedy this, we share near-Delaunay faces between neighbors after they are computed. We do this by searching through each

Fig. 7. The left figure shows the Delaunay faces each v finds by considering just its k nearest neighbors. Notice that with the quickly varying sampling density, some triangles at v are missed. The right figure shows the results after each v searches for faces connected to v from its k^+ neighbors.

v's k^+-nearest neighbors. In our experiments, we chose $k^+ = 2k$. This allows us to reduce the size of k.

6 Results

We used an NVIDIA GeForce 8800 GTX GPU with 768MB connected via PCI Express 16x to an Intel Core2 Quad 2.4GHz CPU. Our applications are built on top of Windows XP and the CUDA Toolkit 1.1.

To see how running time varies as function of the input size, we produced inputs of varying size using subdivision. The timing results are shown in Figure 8. To focus on our algorithm, we do not include rendering time nor the time required to compute the octree on the CPU, which takes less than a minute for 6 million points. In most cases, rendering adds an additional 20% to the overall time. The first three stages of Algorithm 1 take the majority of time: getting the neighbors takes about 20%, making nearly-Delaunay triangles takes 40%, and selecting umbrellas takes 20%. We optimized our code for the Delaunay computation by caching vertex positions into the shared memory space, which reduced the timing by a factor of four. We have not applied similar optimizations to the other stages yet, which we expect would yield similar speedups.

A method to generate our sampling requirement is shown in Figure 9. We begin with a uniform random distribution, and use our implementation of the Witkin-Heckbert technique [17] to improve the distribution. Any vertex that does not have an umbrella or has less than three consensus edges is considered to have failed the reconstruction. The percent of vertices for which the reconstruction fails appears in the figure. The α value is estimated at each v by finding δ_v, the distance to the closest other sample, and then estimating ϵ_v by the largest circumcircle radius in v's umbrella. Histograms of these estimated alpha values can be seen in Figures 12-11.

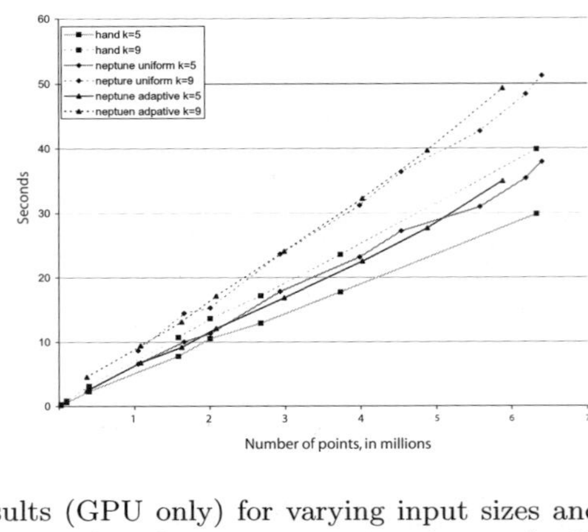

Fig. 8. Timing results (GPU only) for varying input sizes and for different choices of k. As expected, we see that that reconstruction time scales linearly with the input size.

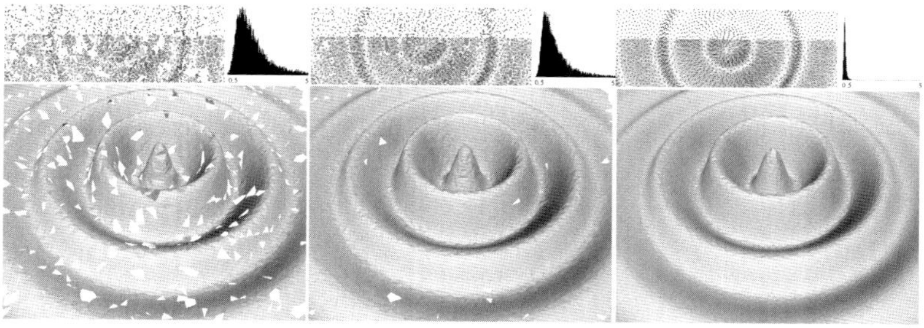

Fig. 9. The left figure shows the initial uniform-random distribution on a sinusoidal surface. Using $k = 9$ the reconstruction works on 92% of the vertices. The middle figure shows the distribution after one iteration of the Witkin and Heckber particle smoothing algorithm with 96% correct reconstruction. The right figure shows the perfect reconstruction after the particles converge to a stable configuration, which takes about ten iterations.

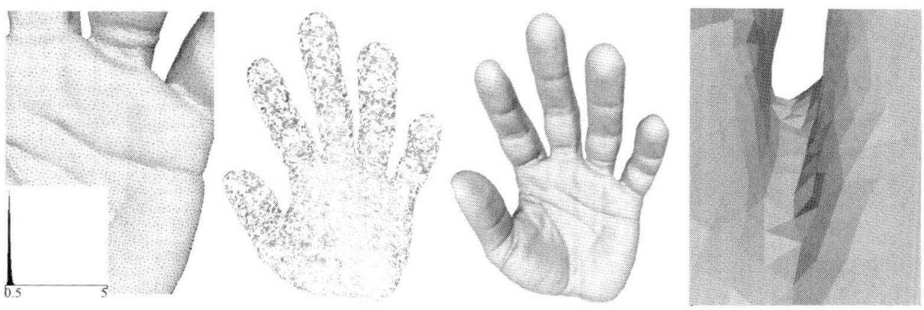

Fig. 10. We consider the Oliver hand model. We see from the left figure that the hand has very uniform sampling. The two hands in the middle show the reconstruction using using $k = 3$ with 10% correct (left) and $k = 7$ with 100% correct (right). Notice from the most right figure that the regions between the fingers with high curvature are reconstructed properly.

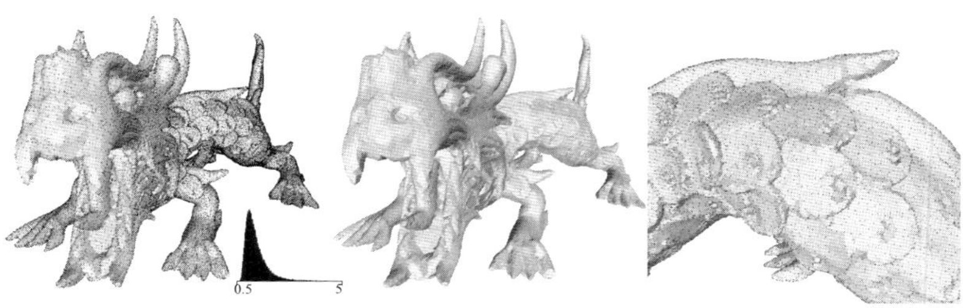

Fig. 11. To consider a very non-uniform sampling, we decimated the dragon model from 3.6 million points to 100k points. With $k = 16$ the reconstruction works on 95% of the vertices. Notice from the right figure that the scales are reconstructed reasonably well. The spike at the top of this figure is sampled too sparsely to have a meaningful reconstruction.

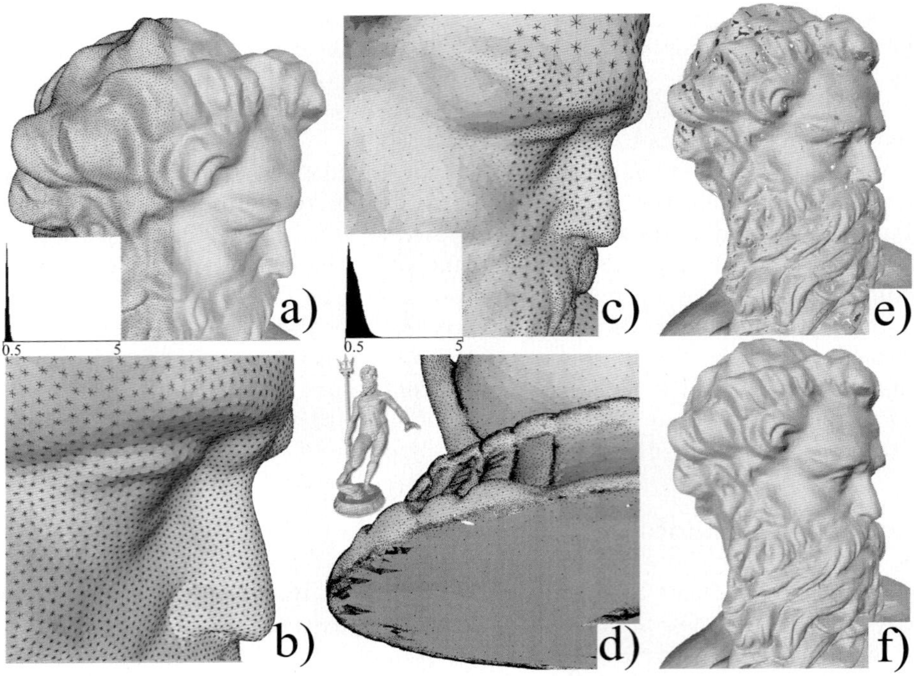

Fig. 12. These figures show the uniform and non-uniform Neptune models. Histogram of their estimated alpha values are also shown. (a) The uniformly sampled Neptune model. (b) A close up of the uniform model with consensus edges. Using $k = 9$, the reconstruction fails on 13 of the 1.6 million vertices. (c) The non-uniform Neptune model. (d) A close up of the base. Notice that regions near extremely varying sampling density are not reconstructed properly. (e) Reconstruction of the non-uniform model using $k = 9$ with 97% correct (f) and $k = 15$ with 98% correct.

We considered the overall reconstruction successful if a very large percentage of the vertices produce at least three consensus edges. The number of nearest neighbors required to get a successful reconstruction varies with α. The uniformly sampled Neptune only requires $k = 9$, while the non-uniformly sampled Neptune model requires $k = 15$. For the very uniform hand model, $k = 7$ produces perfect reconstruction. We also tried our algorithm on a very non-uniform sampling with the dragon model shown in Figure 11. The dragon model works on 95% of the vertices with $k = 15$.

7 Future Work

This work is one example of a geometry processing problem which is easier when applied to locally uniform data. We emphasize that our algorithm is not meant to reconstruct noisy data from laser scanners, but instead meant to represent

surfaces purely from point clouds that can be carefully controlled. If some vertices do not meet the sampling condition, one can explicitly maintain their umbrellas. However, we hope that research in this direction will inspire more work on producing and especially maintaining nice sampling distributions, and showing that these distributions have useful properties.

We are currently working on a proof that an exact version of this algorithm would always produce cycles with small constant maximum length on inputs that are locally uniform and sufficiently dense. On the practical side, we plan to incorporate this surface reconstruction algorithm into an application that maintains a locally uniform distribution on a moving surface.

Our algorithm fails to be completely parallel since we begin by computing an octree on the CPU. While this is not the bottleneck, developing a good octree implementation on the GPU is an important issue, recently addressed in the algorithm of Zhou et al. [9]. We plan on incorporating similar data structure in order to make our algorithm run completely in the GPU.

Acknowledgments

We would like to thank Shubho Sengupta and Brian Budge for their suggestions with CUDA. The Oliver hand and Nepture models are provided in courtesy of INRIA in the AIM@SHAPE Shape Repository. The Asian Dragon Model is provided in courtesy of XYZ RGB Inc. This work was supported by NSF grants CCF–0331736 and CCF–0635250.

References

1. Gopi, M., Krishnan, S.: A fast and efficient projection based approach for surface reconstruction. International Journal of High Performance Computer Graphics (2002)
2. Buchart, C., Borro, D., Amundarain, A.: Gpu local triangulation: an interpolating surface reconstruction algorithm. Computer Graphics Forum 27(3) (2008)
3. Adamy, U., Giesen, J., John, M.: Surface reconstruction using umbrella filters. Comput. Geom. Theory Appl. 21(1), 63–86 (2002)
4. Shewchuk, R.: Star splaying: an algorithm for repairing delaunay triangulations and convex hulls. In: SCG 2005: Proceedings of the twenty-first annual symposium on Computational geometry, pp. 237–246. ACM Press, New York (2005)
5. Dey, T., Funke, S., Ramos, E.: Surface reconstruction in almost linear time under locally uniform sampling (2001)
6. Funke, S., Ramos, E.: Smooth-surface reconstruction in near-linear time. In: Proc. 15th Annu. ACM-SIAM Sympos. Discrete Algorithms (2002)
7. Dumitriu, D., Funke, S., Kutz, M., Milosavljevic, N.: How much geometry it takes to reconstruct a 2-manifold in r3. In: Workshop on Algorithm Engineering and Experiments (2008)
8. Dumitriu, D., Funke, S., Kutz, M., Milosavljevic, N.: On the locality of reconstructing a 2-manifold in r3 (unpublished manuscript, 2008), `http://www.mpi-inf.mpg.de/~funke`

9. Zhou, K., Gong, M., Huang, X., Guo, B.: Highly parallel surface reconstruction. Technical report (April 2008)
10. Kazhdan, M., Bolitho, M., Hoppe, H.: Poisson surface reconstruction. In: SGP 2006: Proceedings of the fourth Eurographics symposium on Geometry processing, Aire-la-Ville, Switzerland, pp. 61–70. Eurographics Association (2006)
11. Amenta, N., Bern, M.: Surface reconstruction by voronoi filtering. Discrete and Computational Geometry 22, 481–504 (1999)
12. Gonzalez, T.F.: Clustering to minimize the maximum intercluster distance. Theoretical Computer Science 38(2–3), 293–306 (1985)
13. Dunbar, D., Humphreys, G.: A spatial data structure for fast poisson-disk sample generation. In: SIGGRAPH 2006: ACM SIGGRAPH 2006 Papers, pp. 503–508. ACM Press, New York (2006)
14. Boissonnat, J.-D., Oudot, S.: Provably good sampling and meshing of surfaces. Graph. Models 67(5), 405–451 (2005)
15. Surazhsky, V., Alliez, P., Gotsman, C.: Isotropic remeshing of surfaces: a local parameterization approach. In: Proceedings of 12th International Meshing Roundtable (2003)
16. Chen, L.: Mesh smoothing schemes based on optimal delaunay triangulations. In: In Proceedings of 13th International Meshing Roundtable, pp. 109–120 (2004)
17. Witkin, A.P., Heckbert, P.S.: Using particles to sample and control implicit surfaces. Computer Graphics 28(Annual Conference Series), 269–277 (1994)
18. Meyer, M., Georgel, P., Whitaker, R.T.: Robust particle systems for curvature dependent sampling of implicit surfaces. In: Proceedings of the International Conference on Shape Modeling and Applications (SMI), June 2005, pp. 124–133 (2005)
19. Edelsbrunner, H., Shah, N.R.: Triangulating topological spaces. In: SCG 1994: Proceedings of the tenth annual symposium on Computational geometry, pp. 285–292. ACM, New York (1994)
20. NVIDIA Corporation. NVIDIA CUDA compute unified device architecture programming guide (January 2007), http://developer.nvidia.com/cuda

Application of Smooth Mixed Meshes Based on Metric Identity in Aerospace Analysis and Design

Jochen Wild

DLR, Institute for Aerodynamics and Flow Technology, Lilienthalplatz 7,
D-38108 Braunschweig, Germany

Abstract. An applicable method for generating mixed meshes around simple and medium complex configurations is presented. Core part of the mesh generation strategy is a smooth meshing of the boundary layer region by structured hexahedrons and prisms applying smoothing based on solving face-weighted Laplacian equations. A preconditioning of the point set for triangulation by shear and edge-weighted average affine invariant transformation is proposed to stabilize end accelerate constrained Delaunay volume triangulation. An examples of the mesh generation strategy is shown for a generic transport aircraft in high-lift configuration.

1 Introduction

Numerical design optimization is an emerging design tool for real world applications in aerodynamics. The state of the art of 3D Navier Stokes solvers already show a good predictive capability even for more complex 3D aircraft configurations as shown for example in [1][2][3].

Optimization based on high-level computational fluid dynamics (CFD) tools using Reynolds-averaged Navier-Stokes equations (RANS) in the meantime is widely used for 2D airfoil design. Today the step towards 3D wing optimization is to be done. Although already practical for simple configurations like wing-body, the effort of calculating more complex configurations is still too high for application within an optimization environment. The main limiting factor hereby is the number of grid points necessary for accurate flow simulation using hybrid unstructured grids consisting of tetrahedral and prismatic elements. E.g. for the configuration used in the Drag Prediction Workshop [1] the medium hybrid unstructured grid used giving promising results contained approximately 12 million grid points. This resulted in a computational effort of 100 hours CPU-time on a NEC SX5 high performance computer for a given flow condition. Even the grid generation for this case took about 1 day turn-around time. This computational effort is clearly unacceptable in terms of an optimization process, where multiple configurations have to be evaluated. As a rough estimate, when assuming 200 evaluations for a converged optimization, this will lead linearly to 200 days grid generation and 800 CPU-days flow calculation. Even assuming a parallel processing of the flow on 4 processors will result in more than 1 year turn-around time for the optimization.

Regarding applications in high-lift optimization the situation worsens. Today a hybrid unstructured mesh for a transport aircraft configuration contains about 20 Million grid points. A mesh is usually generated over night using an advancing front method for high quality meshes. Even with the new computer technology providing massively parallel LINUX cluster systems, 1 to 5 days are needed for a fully converged flow simulation. Using such meshes would require 40 to 1000 days of computing, depending on the optimization algorithm.

The big shortcoming of hybrid unstructured grids is the low anisotropy of surface triangles resulting in a large number of grid points, which is agglomerated through the number of prismatic layers for the boundary layer resolution. This low anisotropy leads to an unnecessary high resolution in spanwise direction, especially for high aspect ratio wings. Recovering the experience with the application of structured grids, it is known, that the aspect ratios of the surface quadrilaterals can be much higher, additionally resulting in well aligned body conforming meshes. The shortcoming of an overall use of structured meshes is the increasing complexity of the targeted configurations, where structured meshing reaches its limitations, mainly due to grid topology issues.

In the past the author *et al.* [4] reported a new strategy to generate mixed meshes containing smooth structured hexahedral and prismatic elements for the resolution of boundary layers by a parabolic marching routine based on the face-weighted Laplace equation. In the following the method is briefly described. New extensions are presented concerning the interface to the field triangulation and a preconditioning method for the Delaunay field triangulation. These extensions have been crucial in order to accelerate and stabilize the process. Finally the method is applied to a high-lift wing-body configuration including the use within an optimization environment.

2 Mixed Mesh Generation Methods

As long as the variations of the field variables to be resolved by the numerical solution to a set of differential equations are of the same order of magnitude in all directions the use of hybrid unstructured meshes is appropriate and furthermore avoids the manual labor of subdividing the field into blocks usually needed for structured meshes. On the other hand an adequate resolution of very thin layers with large normal gradients requires a layered structure of the mesh. Such meshes may be generated by marching away from a fixed mesh face, say the body surface in case of a boundary layer, or away from a floating mesh face in case of a wake. When the mesh on the initial face is unstructured, the resulting three-dimensional mesh extruded from it consists of prisms of triangular cross-sections and is called hybrid. A structured mesh on the initial face, of course, leads to a fully structured three-dimensional mesh.

Structured meshes are easily clustered differently in all grid line directions. Since they need far less points for directional clustering than hybrid unstructured meshes, they add efficiency to the numerical solution procedure of the differential equations. On the other hand their generation suffers from the amount of work to be invested to achieve the blocking for complex geometries. To ease the situation we combine both kinds of surface meshes starting the three-dimensional marching generation of mixed meshes simultaneously from neighboring patches with structured and unstructured surface meshes.

For the generation of these mixed meshes we integrated surface and volume triangulation methods into the DLR structured mesh generator MegaCads [5]. We used available methods, in fact the TRIANGLE code of Shewchuk [6] and the DELAUNDO code of Müller et al. [7] for surface mesh generation in parametric space. For Volume triangulation we chose the NETGEN code of Schöberl [8].

2.1 Structured Hexahedral Boundary Layer Meshes

Like in most structured mesh generation methods the 3D structured mesh blocks are initially generated by an algebraic interpolation method. Depending on the block boundaries we use either tri-linear or transfinite interpolation [9]. Since our objective is to generate smooth meshes we additionally apply smoothing with the use of elliptic equations. In the work of Niederdrenk [10] beginning with the metric identity in its differential conservative form the well known Poisson equations for structured meshes were derived including a precise analytical definition of all 9 control functions. The metric identity in its differential form simply states that the sum over the face normal vectors in the grid conforming $+\xi$ and $-\xi$ direction together with the sum over the remaining face normal vectors in η- and ζ-directions must vanish. A decomposition using the face normal vectors leads to a system of differential equations that can be separated into the face weighted Laplace equation in computational space and the so-called control functions

$$\begin{aligned}
&\underline{S}_1^2\left\{\underline{r}_{\xi\xi}+\frac{V}{\underline{S}_1^2}\left[\left(\frac{\underline{S}_1^2}{V}\right)_\xi+\left(\frac{\underline{S}_1\underline{S}_2}{V}\right)_\eta+\left(\frac{\underline{S}_1\underline{S}_3}{V}\right)_\zeta\right]\underline{r}_\xi\right\}+\\
&\underline{S}_2^2\left\{\underline{r}_{\eta\eta}+\frac{V}{\underline{S}_2^2}\left[\left(\frac{\underline{S}_2\underline{S}_1}{V}\right)_\xi+\left(\frac{\underline{S}_2^2}{V}\right)_\eta+\left(\frac{\underline{S}_2\underline{S}_3}{V}\right)_\zeta\right]\underline{r}_\eta\right\}+\\
&\underline{S}_3^2\left\{\underline{r}_{\zeta\zeta}+\frac{V}{\underline{S}_3^2}\left[\left(\frac{\underline{S}_3\underline{S}_1}{V}\right)_\xi+\left(\frac{\underline{S}_3\underline{S}_2}{V}\right)_\mu+\left(\frac{\underline{S}_3^2}{V}\right)_\zeta\right]\underline{r}_\zeta\right\}+\\
&\qquad 2\underline{S}_1\underline{S}_2\underline{r}_{\xi\eta}+2\underline{S}_2\underline{S}_3\underline{r}_{\eta\zeta}+2\underline{S}_3\underline{S}_1\underline{r}_{\zeta\xi}=\mathbf{0}.
\end{aligned} \tag{1}$$

with the $\underline{S}_i$ being the face normal vectors

$$\underline{S}_1=\underline{r}_\eta\times\underline{r}_\zeta\,,\quad \underline{S}_2=\underline{r}_\zeta\times\underline{r}_\xi\,,\quad \underline{S}_1=\underline{r}_\xi\times\underline{r}_\eta \tag{2}$$

Prescribing the control functions, the metric identity becomes a mesh generating equation. This is used either for elliptic smoothing or for an advancing layer method through these differential equations parabolized with respect to the marching direction and subsequently discretized [11]. By this the generation of the initial algebraic mesh is not necessary.

2.2 Smooth Prismatic Boundary Layer Meshes

For unstructured meshes there exists no globally underlying computational space. Therefore, starting from the same basic principle we begin with the metric identity in its discrete form to directly derive from it the set of algebraic equations to be solved numerically. The derivation of the equations has been shown in a previous paper of the author *et al.* [4]. Proceeding as in the structured case we decomposed the face normal vectors into the line vector directions and rewrote the metric identity in terms of line vectors with coefficients consisting of scalar products of face vectors. Compared with the expanded differential form of the identity for structured meshes, which split into the Laplacian terms and the control functions, the corresponding decomposition of the discrete form of the identity yields nothing similar, still leaving the problem to isolate the Laplacian smoothing part from the pure identity. The problem area is, of course, confined to the unstructured mesh in the body conforming surfaces. Decomposing the discrete identity for planar surfaces results in the definition of suited averages, such that the discrete Laplace equation is represented. To verify this we compared them to the discretized form of the Laplace equation for structured quadrangles and hexagonal cells [12]. So our averaging procedure is not proven but only believed to hold for any collection of triangles forming an *N*-cornered cell.

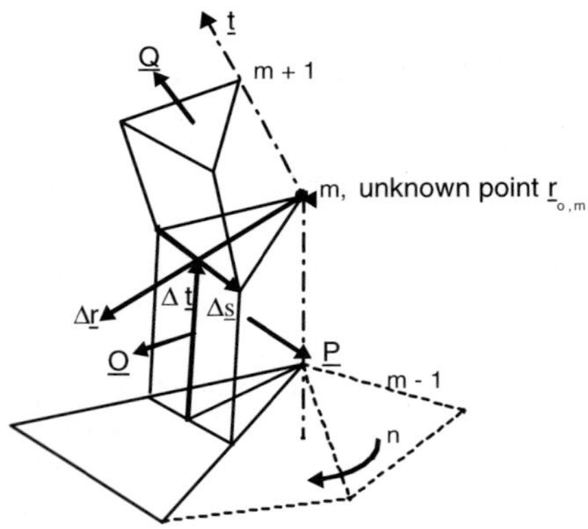

Fig. 1. Discrete triangular prism

Finally we solve equation (3) for the unknown central point $\underline{r}_0$ on level *m*:

$$\underline{r}_{o,m}\sum\left(\bar{\underline{O}}_n\bar{\underline{O}}_n+\frac{1}{2}\bar{\underline{Q}}\bar{\underline{Q}}_m\right)=$$

$$\sum\Big[\bar{\underline{O}}_{n,m}\bar{\underline{O}}_{n,m}\underline{rm}_{n,m}+\bar{\underline{O}}_{n,m}\bar{\underline{P}}_{n,m}\Delta\underline{s}_{n,m}+\frac{1}{4}\bar{\underline{Q}}_m\bar{\underline{Q}}_m\left(\underline{t}_{m-1/2}+\underline{t}_{m+1/2}\right)\tag{3}$$

$$+\frac{1}{4}\bar{\underline{O}}_{n,m}\bar{\underline{Q}}_{n,m}\left(\Delta\underline{r}_{m+1/2}-\Delta\underline{r}_{m-1/2}\right)+\frac{1}{4}\bar{\underline{Q}}_{n,m}\bar{\underline{P}}_{n,m}\left(\Delta\underline{s}_{m+1/2}-\Delta\underline{s}_{m-1/2}\right)\Big],$$

where $\Delta\underline{r}_{n,m}$ is defined as the midpoint between two adjacent corner points

$$\Delta\underline{r}_{n,m}=\underline{rm}_{n,m}-\underline{r}_{0,m}\quad\textbf{with}\quad\underline{rm}_{n,m}=1/2\left(\underline{rc}_{n,m}+\underline{rc}_{n+1,m}\right).\tag{4}$$

and the face and edge vectors according to **Fig. 1**.

2.3 Intermediate Layer to Connect to Volume Triangulation

In order to be able to triangulate the volume the outer hull of the hexahedrons of the structured mesh must be converted to a triangulation. First attempts to generate pyramidal elements on top of the hexahedrons led to intersections in concave regions that were only removed by using very flat pyramidal cells.

The way chosen now is a division of the outermost hexahedrons into pyramids and tetrahedrons using the center point. By this method the interface surface grid formed by the outer layer is as smooth as the hexahedral layer itself. **Fig. 2** shows the

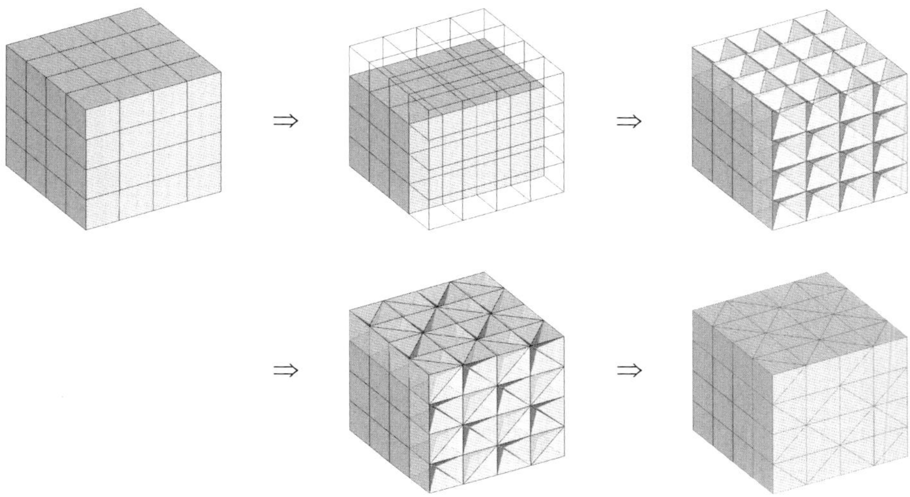

Fig. 2. Division of outermost structured hexahedrons into pyramids and tetrahedrons to achieve a smooth interface surface for tetrahedral mesh generation

sequence of dividing the outer layer first into pyramids, afterwards into tetrahedrons to form a smooth outer hull for the volume triangulation.

2.4 Preconditioned Delaunay Meshing of Tetrahedrons

For the meshing of the outer field with tetrahedrons we use the NETGEN code [8], which is a constrained Delaunay method with an advancing front for boundary recovery. One major difficulty of using structured hexahedral cells in unstructured meshes is the interface to the outer tetrahedral cells, especially if Delaunay triangulation is applied. The Delaunay criterion drives the mesh towards isotropic tetrahedrons, and by this the height of the triangle interfacing to the hexahedrons is much larger than the height of the hexahedrons. This normally results in a jump of volume size and is therefore unfavorable for the flow simulation. Another difficulty arises due to the fact that although the interface grids at the outer hexahedral layer looks smooth and well aligned, it does not fulfill the Delaunay criterion. This usually degrades the performance of the triangulation algorithm.

In order to be able to generate a suited tetrahedral mesh we use a geometric preconditioning of the surface mesh for the Delaunay algorithm. We call it preconditioning since it is a global transformation of the initial boundary triangulation in order to make it more suitable for Delaunay meshing. The volume triangulation is applied on the modified point set without major changes.

A first transformation is a shear transformation, which is important for the application to swept high aspect ratio wings of transport aircraft.

$$\boldsymbol{P}_1 = \boldsymbol{S}^{-1}\boldsymbol{P} \tag{5}$$

where $\boldsymbol{P}$ is the set of boundary points and $\boldsymbol{S}$ is the shear transformation for the wing sweep. Assuming the spanwise coordinate to be y the shear transformation is given by

$$\boldsymbol{S} = \begin{bmatrix} 1 & \sin\varphi & 0 \\ 0 & 1 & 0 \\ 0 & 0 & 1 \end{bmatrix} \tag{6}$$

where φ is the wing sweep angle. This transformation shears the surface elements and results in an intermediate mesh that is more Delaunay than the original one.

The second item to deal with is the high aspect ratio of the surface triangles. At this time anisotropic Delaunay meshing is not available. We therefore apply an edge-weighted mean affine invariant transformation. For each surface triangle an affine invariant transformation

$$\tilde{\boldsymbol{p}}_i = \boldsymbol{A}^{-1}\boldsymbol{p}_i \tag{7}$$

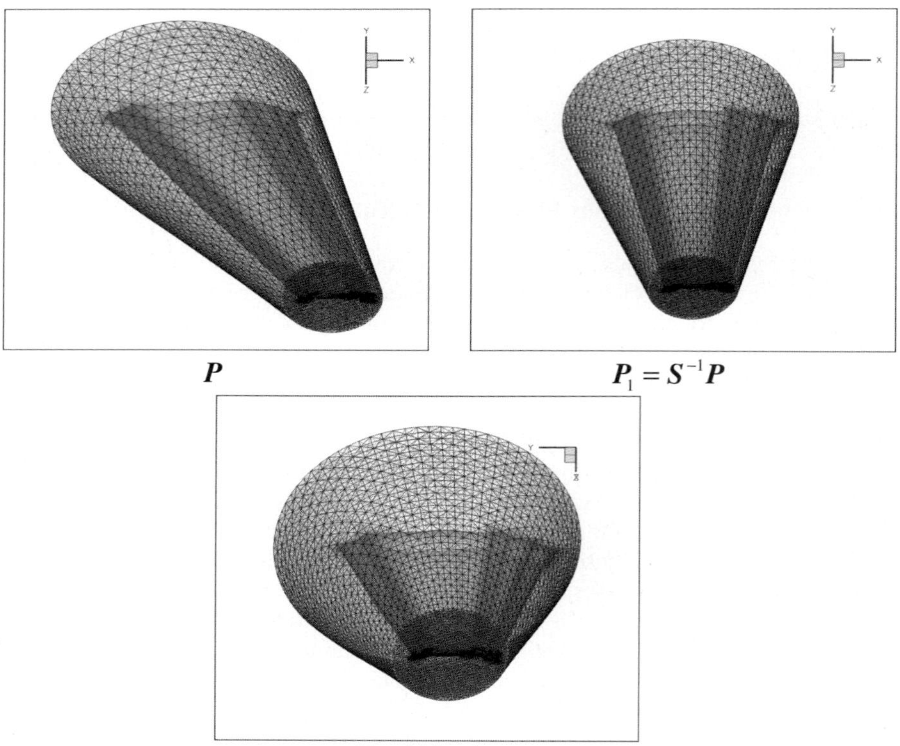

$\boldsymbol{P}$

$\boldsymbol{P}_1 = \boldsymbol{S}^{-1}\boldsymbol{P}$

$\overline{\boldsymbol{P}} = \overline{\boldsymbol{A}}^{-1}\boldsymbol{S}^{-1}\boldsymbol{P}$

Fig. 3. Meshing box around outboard high-lift wing: top left) original geometry; top right) shear transformation applied; bottom) shear and edge-weighted affine invariant transformation applied is defined by a matrix A reflecting the anisotropy of the surface element

$$A_{ij} = \frac{1}{n}\sum_{k=1}^{n}\left(x_{k,i} - x_{c,i}\right)\left(x_{k,j} - x_{c,j}\right) \tag{8}$$

where n is the number of points of the surface element and $\boldsymbol{x}_c$ is the center point. In fact the affine invariant is the quadratic form of an ellipsoid passing through the points of the surface element. By this the matrix reflects the size and stretching of the element. We now calculate a mean edge-weighted average of these matrices of all surface elements.

$$\overline{\boldsymbol{A}} = \sum_{i=1}^{m} \frac{n}{\sum_{j=1}^{n}\left(x_j - x_c\right)^2} \boldsymbol{A}_i \tag{9}$$

The advantage of the edge weighting is that the large regular triangles at the outer domain boundary don't mask the high anisotropy of the small surface elements at the interface. The final point array used for the Delaunay triangulation is now

$$\overline{\boldsymbol{P}} = \overline{\boldsymbol{A}}^{-1}\boldsymbol{S}^{-1}\boldsymbol{P} \tag{10}$$

In **Fig. 3** the meshing box around the outboard portion of a transport aircraft high-lift wing is shown for all steps of the preconditioning. The points-of-view are identical excluding a necessary swapping of the axes due to the affine invariant transformation where the sorting of the axes is according to the eigenvalues.

3 Application for a Generic Transport Aircraft in High-Lift Configuration

For demonstration of the mixed mesh approach using the described parabolic marching algorithms the meshing of a generic transport aircraft wing-body configuration with deployed high-lift devices is shown. This type of configuration is often too complicated for structured meshing. Especially the geometric details at the device ends and their cut-out of the main wing are very hard to be meshed with the block structured approach. On the other hand purely hybrid unstructured meshes lead to point numbers of 20 million grid points or more. For design applications these grids are therefore not suitable.

With the mesh generation strategy described in this work we are able to generate a mesh for a generic transport aircraft in high-lift configuration that contains only about 2 Mio. grid points. The wakes of all three wing elements are resolved by aligned hexahedral elements up to one half flap chord behind the flap trailing edge. Due to the parametric grid generation process, the mesh generation is applicable to geometric changes within an optimization context. The grid generation process takes about 50 minutes from scratch. Compared to the time needed to generate a classic hybrid unstructured mesh of at least 6-8 hours, there remains a significant acceleration. This time frame is acceptable for the application within an optimization framework.

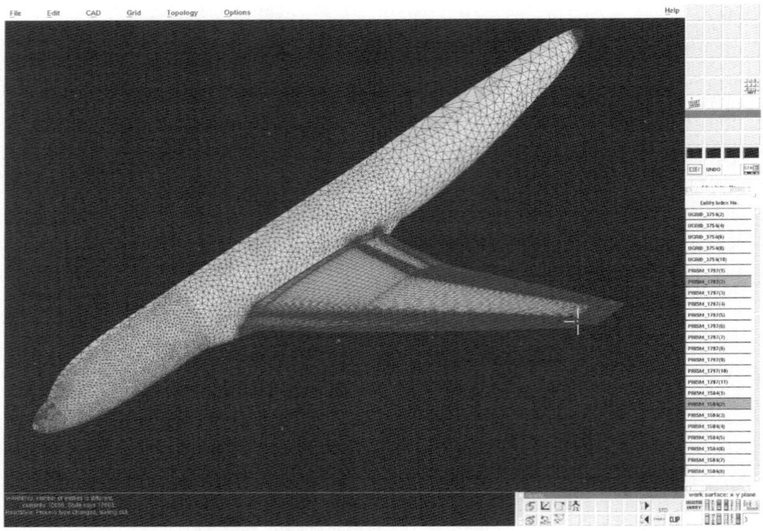

Fig. 4. View on boundary layer grid of generic transport aircraft in high-lift configuration, wing meshed with structured hexahedrons, wing tip and fuselage meshed with prisms

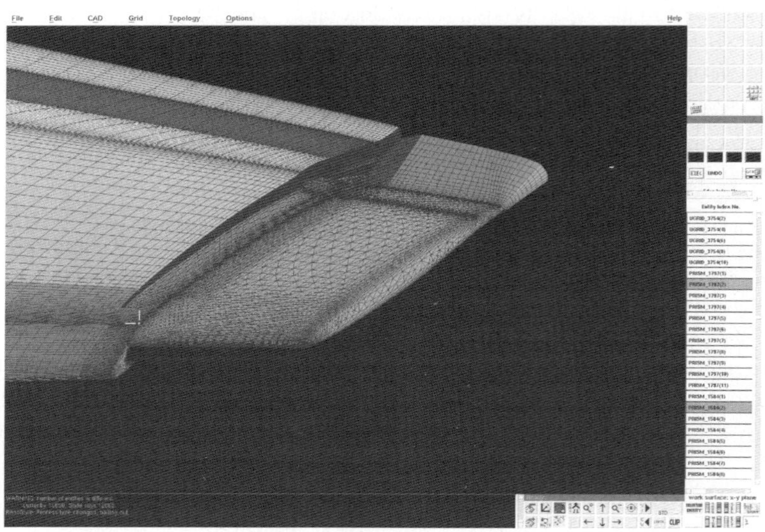

Fig. 5. Zoom on wing boundary layer grid of the generic transport aircraft in high-lift configuration with structured cells are covering boundary layers and wakes of all three wing elements and the wing tip resolved by smooth prismatic mesh.

Fig. 4 shows an overview of the boundary layer grid. Nearly the complete wing is meshed with structured type meshes. Prismatic elements are applied around the fuselage, on the wing tip and at the device ends. Additionally the wakes of all 3 wing

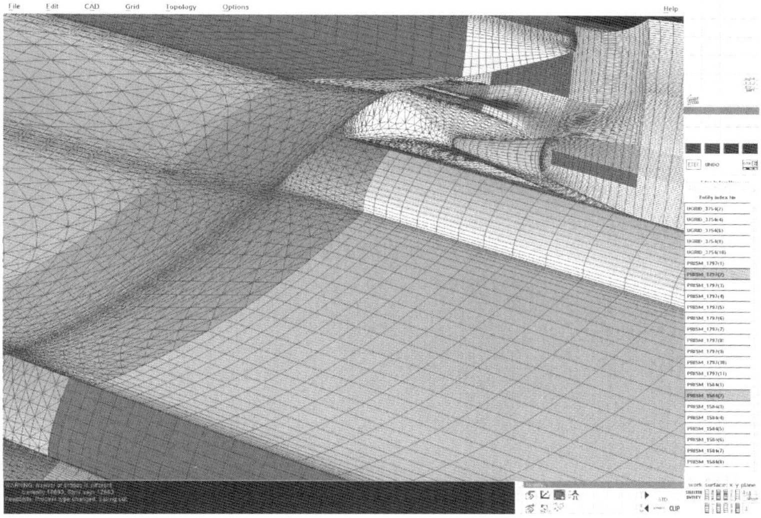

Fig. 6. Zoom on junction of main wing and outboard flap end of the generic transport aircraft showing smooth transition from hexahedral to prismatic elements and the structured wake grids

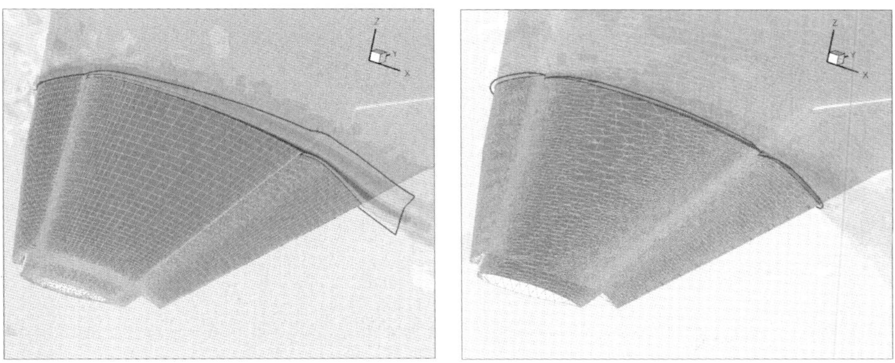

Fig. 7. Comparison of flow field around high-lift configuration at AoA 10°: left) mixed mesh; right) hybrid mesh. Contours are shown for Mach number in field and pressure coefficient on surface mesh. Solid line indicates boundary of area resolved with structured elements.

elements are captured by structured wake grids (**Fig. 5**). This is very relevant for high-lift flows, since the mixing of the shear layers has a big influence on the aerodynamic performance. **Fig. 6** shows a zoom on the device ends of the slat and the flap together with the wing cut-out in this area. A smooth transition from the structured hexahedral to the prismatic elements is observed.

For this case a comparison to a hybrid unstructured mesh containing approximately the same number of points on the surface is done. Due to the missing resolution of the wakes the overall point number of 1.5 Mio. is lower than for the mixed mesh. **Fig. 7** shows a comparison of a cut through the flow field. The contours in the field are the local Mach number while the coloring of the surface mesh is made by the pressure coefficient. Especially the wakes of the wing elements are much better resolved with the mixed mesh.

For this configuration an aerodynamic optimization the of the positioning of slats and flaps has been performed applying the presented meshing algorithms. Within this optimization meshes were generated for 151 different configurations, while only in 13 cases the grid generation process failed. The overall optimization including mesh generation and flow simulation on a 96 CPU parallel PC cluster took about 15 days turn-around time.

4 Conclusion

An applicable method for generating mixed meshes around simple and medium complex configurations has been shown. Core part of the mesh generation strategy is a smooth meshing of the boundary layer region by structured hexahedrons and prisms applying smoothing based on solving face-weighted Laplacian equations. Special emphasis was put on the interface layer and the generation of tetrahedral elements in the outer field. A preconditioning of the point set for triangulation by shear and edge-weighted average affine invariant transformation has been proposed to stabilize end accelerate constrained Delaunay volume triangulation.

An example of the mesh generation strategy is shown for a generic transport aircraft in high-lift configuration. A superior grid accuracy compared to standard hybrid unstructured hybrid meshes with an order of magnitude less points was shown. The improved resolution of wake areas due to structured elements has been demonstrated. Finally the mesh generation strategy was successfully applied within an optimization environment.

Acknowledgements

The author dedicates this paper to Peter Niederdrenk, who has developed and implemented the theory of the face weighted Laplacian equations based on metric identity. He also is the father of the parabolic marching procedures the author uses throughout the mesh generation strategy. The author wishes Peter Niederdrenk all the best for his retirement.

References

[1] Levy, D., Zickuhr, T., Vassberg, J., Agrawal, S., Wahls, R., Pirzadeh, S., Hemsch, M.: Data summary from the first AIAA computational fluid dynamics drag prediction workshop. Journal of Aircraft 40(5), 875–882 (2003)

[2] Brodersen, O., Rakowitz, M., Amant, S., Larrieu, P., Destarac, D., Sutcliffe, M.: Airbus, ONERA and DLR results from the 2nd AIAA drag prediction workshop. AIAA Paper 2004-0391 (2004)

[3] Rudnik, R., Heinrich, R., Eisfeld, B., Schwarz, T.: DLR contributions to code validation activities within the European high lift project EUROLIFT. In: Breitsamer, C., et al. (eds.) New Results in Numerical and Experimental Fluid Mechanics IV. Notes on Numerical Fluid Mechanics, vol. 87, pp. 42–49. Springer, Heidelberg (2004)
[4] Wild, J., Niederdrenk, P., Gerhold, T.: Marching Generation of Smooth Structured and Hybrid Meshes Based on Metric Identity. In: Hanks, B.W. (ed.) Proceedings of the 14th Int. Meshing Roundtable, San Diego, CA, pp. 109–128. Springer, Berlin (2005)
[5] Brodersen, O., Hepperle, M., Ronzheimer, A., Rossow, C.-C., Schöning, B.: The Parametric Grid Generation System MegaCads. In: Proc. 5th Intern. Conf. On Numerical Grid Generation in Comp. Field Simulation, National Science Foundation (NSF), pp. 353–362 (1996)
[6] Shewchuk, J.R.: Triangle: Engineering a 2D Quality Mesh Generator and Delaunay Triangulator. In: First Workshop on Applied Computational Geometry. ACM, New York (1996)
[7] Müller, J.D., Roe, P.L., Deconinck, H.: A frontal approach for internal node generation in Delaunay triangulations. International Journal for Numerical Methods in Fluids 17(3), 241–255 (1993)
[8] Schöberl, J.: NETGEN - An advancing front 2D/3D mesh generator based on abstract rules. Computing and Visualization in Science 1, 41–52 (1997)
[9] Eriksson, L.E.: Generation of boundary-conforming grids around wing-body configurations using transfinite interpolation. AIAA Journal 20(10), 1313–1320 (1982)
[10] Niederdrenk, P.: On the Control of Elliptic Grid Generation. In: Cross, M., et al. (eds.) Proc. 6th Intern. Conf. On Numerical Grid Generation in Comp. Field Simulations, London, July 1998, p. 257. NSF Eng. Center, Mississippi (1998)
[11] Niederdrenk, P., Brodersen, O.: Controlled Parabolic Marching Grid Generation. In: Soni, B., et al. (eds.) Proc. 7th Intern. Conf. On Numerical Grid Generation in Comp. Field Simulations, Whistler, British Columbia, Canada, September 2000, p. 29. NSF Eng. Center, Mississippi, (2000)
[12] Winslow, A.: Numerical Solution of the Quasi-Linear Poisson Equations in a Non-Uniform Triangle Mesh. J. Comp. Phys. 2, 149 (1967)

High Quality Realtime Tessellation of Trimmed NURBS Surfaces for Interactive Examination of Surface Quality on Car Bodies

Gerd Sußner, Quirin Meyer, and Günther Greiner

Computer Graphics Group, University of Erlangen-Nuremberg, Germany
{sussner,siqumeye,greiner}@cs.fau.de

Summary. Virtual interactive examination of the quality of car body surfaces is an important issue in the development process of a car. The method is based on simulating reflection lines using striped environment maps and strong specular highlights. For this purpose high quality meshes are created from the NURBS surfaces. However, the meshes have a fixed resolution, hence a closer examination requires a finer tessellation. In this paper we present a method which allows a view-dependent, arbitrary fine tessellation of trimmed NURBS surfaces at interactive frame rates.

Keywords: realtime tessellation, view-dependent tessellation, quality control.

1 Introduction

An important issue in the development process of a car is the examination of car bodies by reflection lines and specular highlights. This methods allows detecting curvature discontinuities within the surfaces. Furthermore, reflection lines also reveal the character of the car body, no matter whether it is for the overall shape or details of certain parts. For the sake of a short development cycle, designers and construction engineers meet frequently to examine the car shape and tune the character of the car.

In order to avoid expensive prototypes this quality process is done virtually as soon as possible in the development cycle. For this purpose, high quality meshes are generated offline from CAD-models and are examined with standard graphic tools [5]. However, since the meshes have a fixed number of triangles, the examination process is limited to a certain distance due to mesh resolution. A closer view would require a finer tessellation. Unfortunately it is not a good choice to provide meshes with a very fine resolution because the parts of a whole car body exceed the graphics hardware limit of displaying triangles.

There are a large number of algorithms for generating triangular meshes from NURBS patches ([1], [2], [11], [15], [16], [17], [18]). They are either optimized for speed, for triangle count, for best approximation or for surface examination. But none of them is capable of tessellating several CAD parts in reasonable mesh quality in realtime due to the expensive trimming of NURBS surfaces.

Guthe and Balázs [7] present a method which uses the graphics hardware to evaluate NURBS surfaces. The trimming is done by a binary trimming texture which is created by also evaluating the NURBS trimming curves on the GPU. The approximation error is choosen to be below half a pixel of the screen for each single patch of a model. In [8] the method is extended to be appearance preserving in order to visualize reflection lines. However these methods are based on cubic surfaces only. Surfaces of higher degrees are reduced to cubic degree. If the approximation error of the reduction is too coarse, the surfaces are subdivided until the overall error is below half a pixel. In [12] the authors extend the methods above to evaluate the original, i.e. non-reduced, surfaces directly within the fragment program by an GPGPU-approach with several render passes.

The visualization of reflection lines as well as of specular highlights is heavily based on the surface normal and is very sensitive to even small displacements of the normal's direction. Therefore it is very important for examining surfaces that a vertex and its normal are computed *exactly* from the original surface and not from an approximation as it is done in [7] and [8]. All methods above share the same disadvantage: Since the surfaces are evaluated at the GPU, the special case of degenerated surface corners cannot be handled properly. An example is given in Fig. 1. This is a common construction method when several beveled edges meet at a corner.

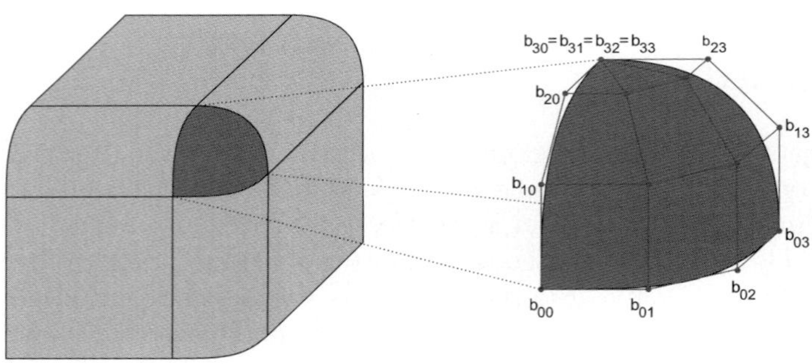

Fig. 1. When three beveled edges meet at a corner, the usual construction is a degenerated NURBS surface where one row or column of the control points are collapsing to a single control point. At this point one of the partial derivatives vanishes and hence the surface normal cannot be computed. In [?] a solution for that special case is proposed.

Contributions

In this paper we present a hybrid method for view-dependent realtime tessellation. The NURBS surfaces and trimming curves are tessellated separately by the CPU and stored in a dynamically extendable hierarchy. This ensures an appropriate evaluation of a surface and its normals. The trimming of the surfaces is done by the GPU using trimming textures while rendering the hierarchy. We extend the trimming texture approach of [7] by using 28 bits of an RGBA-texture. Furthermore, we present an efficient texture atlas for each bit-plane of the RGBA-texture to reduce the number of fragment-program changes as much as possible.

2 Preliminaries

In order to detect surface errors and design issues, the car industry used to have cube like rooms with many parallel lights rows on the wall and on the ceiling to detect surface errors and design issues. The reflected light rows on a shiny car surface are called *reflection lines*. Since this method requires real hardware, it is necessary to cut out the current CAD model from a block of solid material. This is a very expensive and time consuming process. Thus virtual examination at a workstation is a very attractive alternative. It is even possible to have several sessions a day, i.e. examining several different designs and comparison between them. Detected errors may be corrected and re-examined instantly.

The use of reflection lines is described in [14],[9], or [5] and the principle is sketched in Fig. 2. In this paper we use sphere environment textures with a pixel-based texture lookup in order to avoid visualization artecfacts due to texture interpolation (in contrast to the per-vertex-based lookup in [5]).

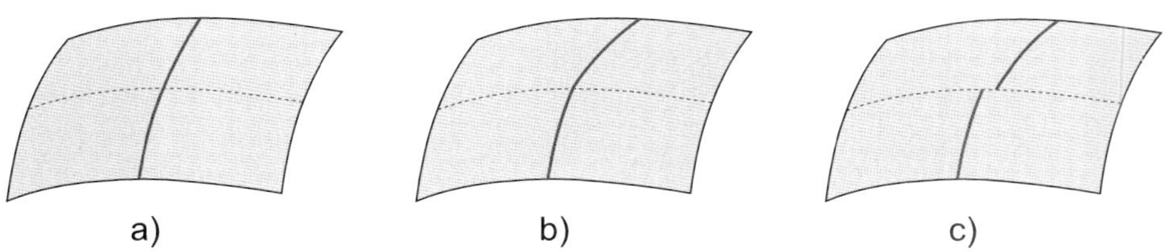

Fig. 2. The reflected line on a shiny G^n-continuous surface is exactly G^{n-1}-continuous. In a) the surfaces have at least a G^2-continuous joint and thus the line is G^1 as well. In b) the surfaces have the same tangent plane but not the same curvature at the joint (G^1), hence the reflected line changes direction immediatly. In c) the surfaces only join at a common boundary (G^0) which breaks the reflected line.

2.1 Estimates on NURBS Curves and Surfaces

Since the hierarchy we build later in this paper is extended dynamically, a fast tessellation of the untrimmed NURBS surfaces and trimming curves is needed. In [4] the authors present some estimates on how many uniform samples are necessary to approximate a curve or surface within a given tolerance.

The number of samples needed for trimming curve $C(t)$ is:

$$T = \left\lceil (t_1 - t_0)\sqrt{\frac{\sup_{t\in[t_0,t_1]} \|C''(t)\|}{8\delta}} \right\rceil + 1. \tag{1}$$

where δ is the tolerance given in domain space of the surface $S(u,v)$ the trimming curve belongs to. The number of samples in both directions needed for a surface $S(u,v)$ are:

$$N = \left\lceil (u_1 - u_0)(v_1 - v_0)\sqrt{\frac{\frac{1}{8}\left(S_{uu} + S_{uv}\right)}{2\varepsilon}} \right\rceil + 1 \tag{2}$$

$$M = \left\lceil (u_1 - u_0)(v_1 - v_0)\sqrt{\frac{\frac{1}{8}\left(S_{vv} + S_{uv}\right)}{2\varepsilon}} \right\rceil + 1 \tag{3}$$

with

$$S_{uu} = \sup_{(u,v)\in[u_0,u_1]\times[v_0,v_1]} \left\| \frac{\partial^2 S(u,v)}{\partial u^2} \right\| \tag{4}$$

$$S_{uv} = \sup_{(u,v)\in[u_0,u_1]\times[v_0,v_1]} \left\| \frac{\partial^2 S(u,v)}{\partial u \partial v} \right\| \tag{5}$$

$$S_{vv} = \sup_{(u,v)\in[u_0,u_1]\times[v_0,v_1]} \left\| \frac{\partial^2 S(u,v)}{\partial v^2} \right\|, \tag{6}$$

where ε is the given tolerance in world space.

The maximum deviation of $S(u,v)$ is bound to the absolute value of the partial derivatives $\|\frac{\partial S(u,v)}{\partial u}\|$ and $\|\frac{\partial S(u,v)}{\partial v}\|$. To specify a tolerance ε for a curve in world space the maximum deviation is scaled with the occupied range of the curve in the domain of $S(u,v)$:

$$\delta_u = \left\lceil \frac{\varepsilon}{(u_1 - u_0)\sup_{(u,v)\in[u_0,u_1]} \left\| \frac{\partial S(u,v)}{\partial u} \right\|} \right\rceil \tag{7}$$

$$\delta_v = \left\lceil \frac{\varepsilon}{(v_1 - v_0)\sup_{(u,v)\in[u_0,u_1]} \left\| \frac{\partial S(u,v)}{\partial v} \right\|} \right\rceil \tag{8}$$

$$\delta = \min(\delta_u, \delta_v). \tag{9}$$

3 Related Work

Guthe [7] introduced a GPU-based realtime tessellation scheme. It is based on cubic Bézier surfaces which are defined on $[0,1]^2$. Hence NURBS data are converted to single Bézier patches by knot insertion. In a preprocessing step, a number of predefined grids with uv-coordinates in the range $[0,1]^2$ are loaded on the graphics card as display lists. The trimming curves are preprocessed in a similar manner. For each frame the following steps have to be done for each single patch.

3.1 Adapting Patches to the Camera

For each patch the distance to the camera determines the approximation tolerance ε and the number of needed samples is determined according to equations 2 and 3. If a single patch is of higher degree then it is geometrically continuously reduced to cubic degree. Since degree reduction causes geometric distortion, the patch may be subdivided into several sub-patches until the combined approximation error of degree reduction and sampling is below the approximation tolerance for each sub-patch (Fig. 3).

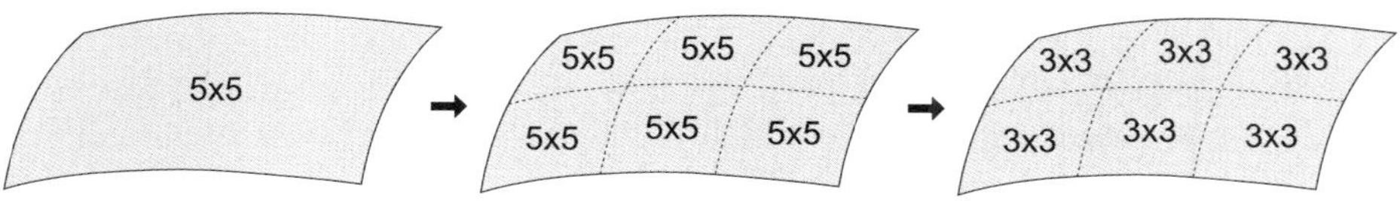

Fig. 3. If the combined approximation error due to degree reduction and sampling exceeds given tolerance, the single patch is subdivided in advance. The resulting sub-patches are then processed like single patches.

Finally, the size of the trimming texture has to be determined. The boundary for the maximum deviation from Section 2.1 is scaled to fit into one texel unit in the domain of $S(u, v)$, i.e the texture size in u- and v-direction is:

$$T_u = \left\lceil \frac{u_1 - u_0}{\varepsilon} \sup_{(u,v)\in[u_0,u_1]} \left\| \frac{\partial S(u,v)}{\partial u} \right\| \right\rceil \tag{10}$$

$$T_v = \left\lceil \frac{v_1 - v_0}{\varepsilon} \sup_{(u,v)\in[v_0,v_1]} \left\| \frac{\partial S(u,v)}{\partial v} \right\| \right\rceil . \tag{11}$$

Since the requested texture size may exceed the maximum texture size of the graphics hardware, further subdivision of the Bézier patch may be necessary.

Degree reduction and subdivision of the Bézier patches have to be done in each frame, and may be cached for repetitive viewing of the same region. However the cache is invalidated if a face needs to be further subdivided in the case of a closer view.

3.2 Creating the Trimming Texture

The main idea is to render closed trimming polygons as triangle fans in XOR-mode (Fig. 4). For each curve of a trimming loop the number of samples is determined according to Equation 9. Then a vertex shader is loaded which evaluates a cubic Bézier curve using one ordinate of the vertex position.

Since a single triangle fan must be used for the whole closed polygon, for each polygon vertex the four control points of its Bézier curve must be provided as vertex attributes. This dramatically increases the amount of transferred data to the GPU.

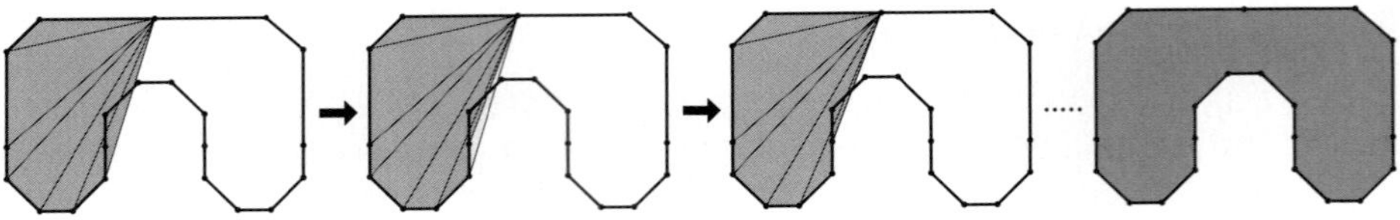

Fig. 4. A trimming polygon rendered as triangle fan in XOR-mode. In the second picture the last drawn triangle is partially overwritten by the current triangle (yellow). The XOR-mode causes the pixels to be unset.

3.3 Applying the Trimming Texture

After the trimming texture has been created, the vertex program is changed to evaluate cubic Bézier patches. The control points of the patch are set as uniform variables of the vertex program. Then the number of samples for a patch is determined and the display list of an appropriate predefined grid is rendered.

Within the fragment program a zero texel value causes a *fragment-kill* operation, i.e. the current fragment is not processed anymore and hence no pixel in the framebuffer is overwritten.

3.4 Thick Border Lines

Since this method does not use any topology information, neighbored surfaces are tessellated independently. This results in so-called *T-vertices* which cause rendering artefacts. At those boundaries, some pixel are not set by either surface and the background shines through within the whole surface.

In order to avoid this the authors propose drawing the boundary polygons with a line thickness of two and finally draw the trimmed nodes with a polygonoffset (Fig. 5).

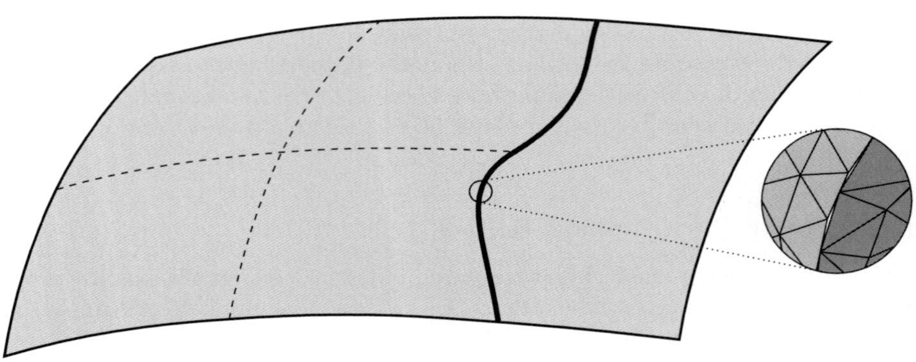

Fig. 5. Due to rounding errors while rendering boundary triangles the background may shine through the gaps between patches. Drawing the boundary as thick lines avoids this. The triangles are rendered with polygonoffset the border lines appear behind the triangles which just fills the gaps.

4 Hybrid Tessellation

For surface interrogation it is important, that a displayed vertex and its normal are exactly evaluated. Due to the approximative degree reduction the above method is not useful for examining surfaces.

In this paper we build a hierarchy where each node has its pretessellated untrimmed mesh and the corresponding trimming polygons. While rendering the active nodes a trimming texture is created similar to Section 3.2 and is applied to the 3D mesh. Since changing vertex/fragment programs causes a pipeline stall, we present two efficient methods for processing several patches with one RGBA-texture.

4.1 Domain Hierarchy

In contrast to [7] the NURBS data remains as is, i.e. no conversion to Bézier representation is done. Furthermore subdivision is done by subdividing the domain which is far more memory efficient than subdividing the surface geometrically. E.g. a bi-quintic Bézier patch has 36 control points in $\mathbb{R}^3$. Subdividing the geometry at $u = 0.5$ results in 72 control points. The domain is defined by two coordinates in $\mathbb{R}^2$, no matter what degree the surface has. Thus subdivision of the domain always adds only two more coordinates of $\mathbb{R}^2$ (Fig. 6).

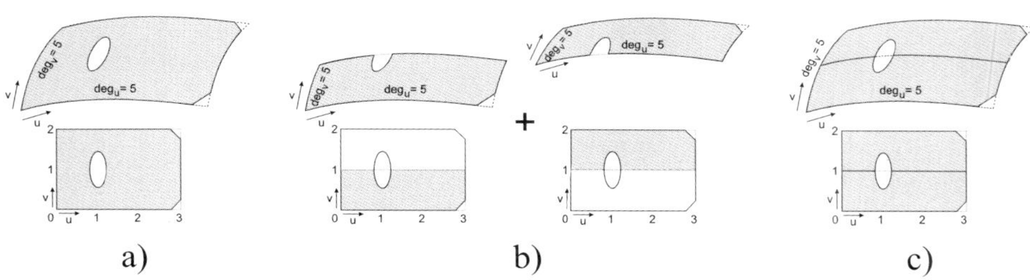

Fig. 6. Regular subdivision of a surface (a) produces two new separate surfaces within the hierarchy. Each of them has exactly as much control points as the original surface (b). By subdividing the domain only the boundaries of the new sub-domains has to be stored within the hierarchy while the surface remains as it is (c).

Motivation

The hierarchy is organized so that each level belongs to a certain tolerance for approximating surfaces. In a preprocessing step, the hierarchy is built up to a fixed user-specified level as a cache. The idea is if the whole data is viewed at far distance, only a coarse level is needed. Thus moving the camera around the car always use cached nodes. A close examination of a part of the car would cull many surfaces out of the viewing frustum which will not processed. So the CPU has enough time to extend the hierarchy dynamically. In practice a tolerance of 0.5 units for the fixed levels is reasonable.

Building the Fixed Hierarchy

Building the hierarchy starts with one node for each surface at level 0. Each node is recursively tested whether the trimming texture is suitable for that tolerance. If not, the domain is split in u- or in v-direction at its midpoint (Fig. 7). Which direction is finally split is determined by testing the remaining half domains. The split which produces more positive tested half domains is used. The resulting two nodes replace the input node in the same level. For the next level all nodes of the parent level are copied and processed as described in the previous paragraph.

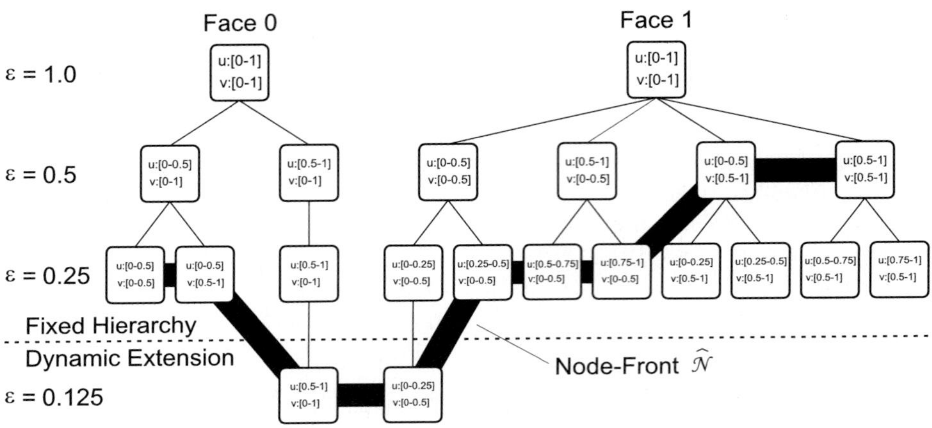

Fig. 7. The hierarchy is organized in a fixed number of levels of tolerances and may be dynamically extended. If the computed trimming texture size exceeds the size of the real trimming texture the domains are split accordingly and inserted into the hierarchy at the same level. A *Node-Front* $\widehat{N}$ marks the valid nodes needed to render the object at a projected approximation error of less than half a pixel.

Tessellating the Fixed Hierarchy

After all fixed levels have been built, each node in the hierarchy is tessellated. For fast tessellation the formulas of Section 2.1 are used. The mesh vertices and normals of the untrimmed surface are stored as points in $\mathbb{R}^3$. The corresponding uv-texture coordinates are mapped to texture size ($[u_0, u_1] \times [v_0, v_1] \rightarrow [0, \text{maxTexSize} - 1]^2$) and stored as 16-bit unsigned integers. According to the Nvidia Vertex-Buffer-Objects (VBO) whitepaper, using 16-bit integers save bus bandwidth between CPU and GPU.

Finally, the trimming loops are tessellated as well and mapped as explained above. The trimming polygons are then clipped against the trimming texture boundaries after a method from [6]. This algorithm creates separate disjoint closed sub-polygons while the classic clipping method [19] is not able to properly separate the resulting sub-polygons of a concave trimming polygon (Fig. 8).

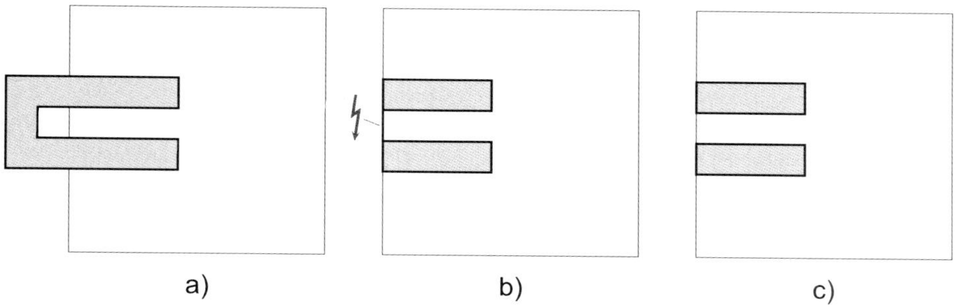

Fig. 8. Clipping the concave trimming polygon in a) against the rectangle with the classic clipping method [19] produces overlapping line segments in b). The algorithm of Glassner [6] is able to create separate trimming sub-polygons (c).

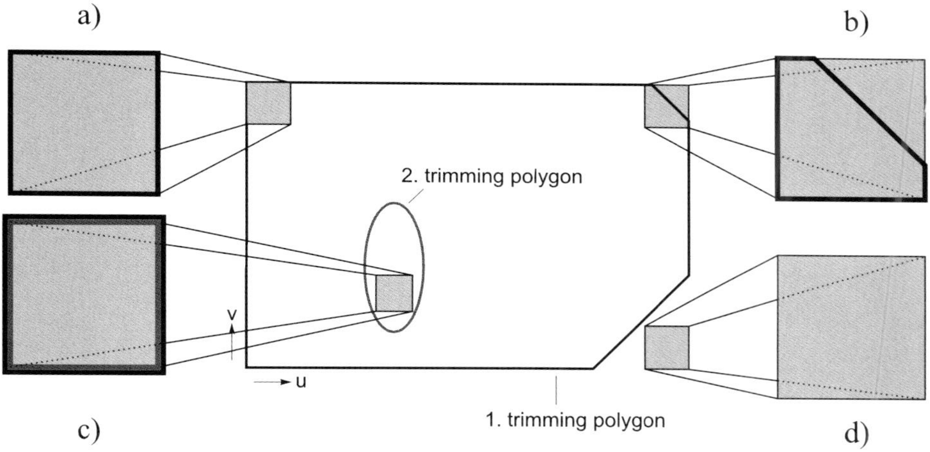

Fig. 9. Clipping of trimming polygons against the area of the trimming texture allows classification of nodes. a) The node is completely visible - no trimming texture is needed. b) The surface mesh need to be trimmed. c,d) The node is located completely in a trimmed hole or outside of the first trimming polygon - it is not rendered at all.

The resulting inner polygons are then tested for lying exactly on the boundary of the trimming texture. If there is exactly one polygon on the boundary and no other polygon intersects the trimming texture, then this part of the surface is marked as completely visible and thus the node can be drawn without trimming texture. In the case of two polygons lying exactly on the boundary, this part of the surface lies completely within a trimmed hole. The node is marked as invisible and should not be rendered at all. This is also the case if none of the polygons intersect the trimming texture (Fig. 9). In any other cases the polygons are stored as 16-bit unsigned short integers as well. For drawing thick border lines, the intersecting segments are lifted into 3D and both points and normals in $\mathbb{R}^3$ are stored.

4.2 Adaptive Refinement

Through the hierarchy runs a so-called *Node-Front* $\widehat{N}$ similar to the *Vertex-Front* in [10]. This data structure assumes the the camera view has not changed dramatically with respect to the previous frame. So the *Node-Front* can be adapted very quickly from frame to frame by a simple test of each node in the front.

Adapting *Node-Front*

With this hierarchy the test is pretty simple. If the node is outside of the viewing frustum, the front is pushed upwards. If it is visible the distance d of the current node $N_i \in \widehat{F}$ to the camera position is used determine the projected tolerance ε of N_i. If ε_i is smaller or equal than $\varepsilon_{\text{level}}$ of the current level the front must be pushed down one level and the test starts again. Otherwise the parent node and the sister nodes have to be tested for pushing the front upwards.

Extending Hierarchy

If the front has reached the leaves of the tree and more refinements are needed, the hierarchy is dynamically extended by the same procedure used while building the fixed hierarchy. When the *Node-Front* is pushed upwards again and the dynamically created nodes are added to the *Obsolete-Node*-list. The nodes in this list are removed from the hierarchy when they have not been visited by a user-specified time. This lazy clean-up of the hierarchy allows a re-examination of a specific part within a short time avoiding continuously creating the same nodes.

4.3 RGBA-Trimming-Textures

Each node of the hierarchy represents a specific detail from a single trimmed surface. The trimming is done using binary trimming textures (see Section 3.2). Since texel values are fixed-point values in the range $[0, 1]$ the method uses a complete channel of eight bit to test a fragment within the fragment shader, i.e. seven bits of a channel are wasted.

Since creating the trimming textures always means changing shader programs and thus causing pipeline stalls, a more efficient use of trimming textures is required. In order to use all bits of a channel, the trimming polygons of a bunch of nodes within $\widehat{N}$ are rendered into one RGBA-texture. Each polygon is drawn in XOR-mode with a single 32-bit RGBA-color value with just one bit set. A parameter given to the fragment shader states which bit of the texel has to be used for the fragment-kill-test.

Since comparing values of a texture in the fragment shader treats texels as fixed-point float values, it is not possible to use logical bit-operations for the fragment-kill-test. We present a method which is capable of testing seven bits of a channel by a very small set of fixed-point operations, i.e. 28 bits of a RGBA-texture can be used for trimming at almost the same speed. Using the remaining four sign bits would demand a more complex fragment shader.

The test of a single bit in a channel c exploits the bit pattern of 8-bit fixed-point values shown in Tab. 1. To test the channels k'th bit of the current texel value T, the dot product D_{Tt} of T and the shader parameter t, which is a RGBA-color consisting of just bit k of channel c, are computed.

Table 1. Table of color values used for setting a single bit of a color channel in the trimming texture. The last column contains the test values used for testing that bit in the fragment shader.

Bit k	Color 2^k	Fixpoint Representation	Test Value 2^{6-k}
6	01000000_2	$\frac{1}{2^1} = \frac{1}{2} = 0.5$	1
5	00100000_2	$\frac{1}{2^2} = \frac{1}{4} = 0.25$	2
4	00010000_2	$\frac{1}{2^3} = \frac{1}{8} = 0.125$	4
3	00001000_2	$\frac{1}{2^4} = \frac{1}{16} = 0.0625$	8
2	00000100_2	$\frac{1}{2^5} = \frac{1}{32} = 0.03125$	16
1	00000010_2	$\frac{1}{2^6} = \frac{1}{64} = 0.015625$	32
0	00000001_2	$\frac{1}{2^7} = \frac{1}{128} = 0.0078125$	64

If all higher bits of c and bit k are **not** set then $c < \frac{1}{2^{7-k}}$ and hence the dot product with the test vector component 2^{6-k} is below 0.5:

$$c \cdot 2^{6-k} < \frac{1}{2^{7-k}} \cdot 2^{6-k} = \frac{1}{2} \tag{12}$$

If additionally bit k is set, then $\frac{1}{2^{7-k}} \leq c < \frac{1}{2^{7-(k+1)}}$ and the dot product is above or equal to 0.5 but below 1.0:

$$\frac{1}{2} = \frac{1}{2^{7-k}} \cdot 2^{6-k} \leq c \cdot 2^{6-k} < \frac{1}{2^{7-(k+1)}} \cdot 2^{6-k} = 1 \tag{13}$$

Furthermore, if any bit above k is set then $c \geq \frac{1}{2^{7-(k+1)}}$ and the resulting dot product is greater or equal than 1.0:

$$1 = \frac{1}{2^{7-(k+1)}} \cdot 2^{6-k} \leq c \cdot 2^{6-k} \tag{14}$$

The last equation indicates that the bits above k only affect the integer part of the dot product. Hence testing bit k is done by comparing only the fractional part. The fragment is discarded if the fractional part is below 0.5. A simple fragment program which implements this algorithm is shown in Fig. 10.

4.4 Efficient Trimming Texture Atlas

With RGBA-Trimming-Textures 28 nodes of $\widehat{N}$ can be processed at once. However each node occupies a whole bit-plane of the trimming texture, no matter

```
uniform sampler2D trimTexture
uniform vec4 testValue

void main() {
  // Get texel according to uv-coordinates
  vec4 tex = texture2D(trimTexture, vec2(gl_TexCoord[0]));
  // Compute dot product of texel and testValue
  // Kill fragment if fractional part of the dot product is < 0.5
  if (fract(dot(tex, testValue)) < 0.5)
    discard;
  // Set fragment's color
  gl_FragColor = gl_Color;
}
```

Fig. 10. A simple fragment shader for using trimming textures

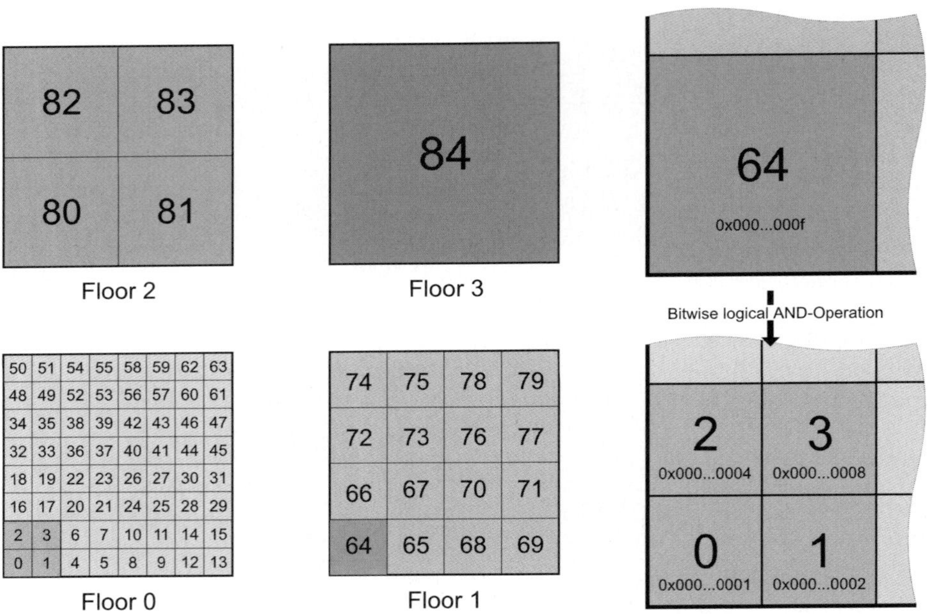

Fig. 11. One bit-plane of the texture atlas consists of 64 sub-textures located in *Floor 0*. Four unused sub-textures of a floor may be clustered to form a bigger sub-texture of the next floor. The test for unoccupied sub-textures is done by masking the corresponding bits with an bitwise logical AND-Operation.

what size is needed by the trimming polygon. Texture atlases may be used to increase the number of nodes per trimming texture. They are usually created offline and thus the time for organizing textures of different size within the atlas is insignificant. However, a fast administration of free/occupied texture slots is inevitable for this application since the texture atlas has to be filled several times per frame in the case of large models.

The texture atlas presented here consists of 64 quadratic equally sized sub-textures per bit-plane, i.e. up to 2048 nodes may processed at once for drawing the trimming texture. In order to incorporate sub-textures of different sizes each bit-plane is organized in four floors. The above 64 slots are located in floor 0. Four of them can be clustered to form a sub-texture of doubled size and are located in floor 1 (Fig. 11) until floor 3 is reached.

It is important that a slot in the upper floors may be only used if all of the corresponding slots in floor 0 is available. This is ensured by using a 64-bit integer value - one bit for each slot in floor 0. Requesting a slot of an upper floor simply consists of a bitwise logical AND-operation of 4, 16 or 64 bits. After a successful request the corresponding bits of the integer are set by an bitwise logical OR-operation.

4.5 Rendering

After the *Node-Front* $\widehat{N}$ has been adapted, it is traversed node by node. Those nodes marked as invisible are immediately skipped. Nodes whose projected bounding box occupies less than a half pixel are skipped as well. Nodes marked as completely visible are drawn in a separate traversal at the end. For each trimmed node a free slot within the texture atlas is requested, which determines the position within the texture and the drawing color of the trimming polygon. If there is no slot available, the traversal stops and the trimming texture is created by drawing the nodes' trimming polygons as triangle fans in XOR-mode using streaming *Vertex Buffer Objects* (VBOs) and very simple and fast shaders. The trimming texture is used as render target.

After creation of the trimming texture, the render target is set back to the framebuffer and the regular shaders are set. The fragment shader *must* include the trimming texture test as shown in Fig. 10. Then the surface mesh of the trimmed nodes are drawn as streaming VBOs as well and the corresponding test-color for trimming is set according to Tab. 1. In order to use Thick-Border-Lines from 3.4 the triangles are drawn with polygon offset. After this, the texture atlas is reset and the traversal continues.

Finally $\widehat{N}$ is traversed again. Now the surface meshes of all nodes marked as completely visible are drawn together with the 3D trimming polygons of the trimmed nodes as Thick-Border-Lines.

5 Results

The method was implemented and tested on a regular PC with a Intel Core2 Duo processor 2.6 GHz and a Nvidia Quadro FX3500 graphics card. The program uses only one of the two CPUs.

Table 2 shows the average frames per second and some statistics for the node hierarchy. Since we are using streaming data instead of display lists, the

Table 2. This table shows some statistics about three single models and an assembly of a car body at a total view rotating around their centers. The first two columns show the number of trimmed surfaces each model consists of and the average frame rate. The other columns show average values per frame of the number of drawn nodes, the number of drawn triangles in 2D plus 3D, the number of line segments for thick borders and finally the number of draw texels used while filling the trimming texture. The size of the render window is 1550 × 888 pixels.

	# NURBS	avg. fps	# nodes	# tris (2D+3D)	# 3D lines	# Mio. Texels
Hood	251	73	320	10564+3841	17480	61
Frame	1730	21	1237	36164+8209	56138	132
Mirror	3970	6	3900	108893+26400	170143	247
Car Body	19781	6	7500	198191+74333	348925	230

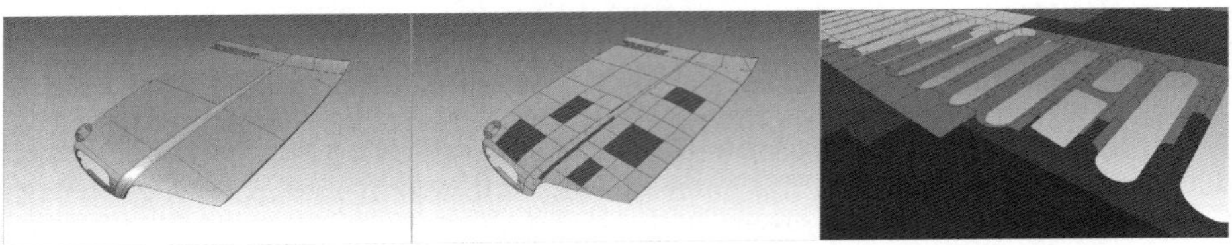

Fig. 12. The hood of a car is rendered so that the approximation error is less than half a pixel on screen. On the left side the boundaries of the single face are shown as well. In the middle the nodes are drawn color-coded. The hue-value corresponds to the hierarchy-level of the node. Dark shaded sub-patches indicate that this node is completely visible, i.e. no trimming textures is used. On the right side a close-up is shown with nodes closer to the viewer using finer approximation (violett) than nodes at the horizon (turquois). Data is courtesy of BMW AG.

frame rates may be higher when rendering single parts even with high resolution meshes. However, our method allows to render a complete car body at an arbitrary resolution at considerable frame rates.

Basically the frame rate of this method is limited by two factors.

- Number of hierarchy nodes to render: For models consisting of thousands of small faces, e.g. mechanical parts modeled by Constructed Solid Modeling, the overhead for trimming each single face is tremendous. The reason is the large number of batches sent to the graphics card (trimming polygon, triangular mesh and Thick-Border-Lines) for each trimmed node in the hierarchy.
- GPU fill rate while creating the trimming texture: The texture size is usually 4096^2 or 2048^2 texels. A node which occupies almost the full size of a bit-plane will set lots of texels. If many of such nodes are rendered, the fillrate-limit of the GPU is reached pretty fast and slows down rendering.

Since the hierarchy is extended dynamically a fast evaluation of NURBS surfaces is inevitable. In our implementation we are able to evaluate more than 5 Mio. arbitrary points and normals of a bi-cubic NURBS surface per second. For

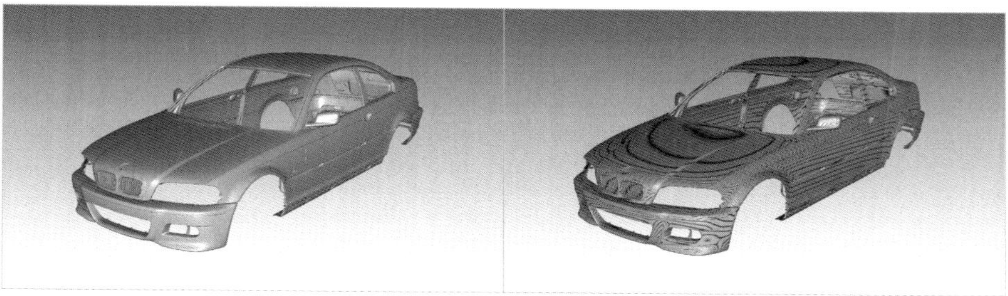

Fig. 13. Here the application of reflection lines on a whole car body is shown. Data is courtesy of BMW AG.

bi-quintic surfaces this value reduces to about 1.8 Mio samples. A fast implementation of the polygon clipping algorithm is needed as well. Using a deep fixed hierarchy revealed that extending the hierarchy is neclectable compared to the other limiting factors mentioned above.

This method also allows a mixture between static and dynamic rendering. Examination of a whole car can be done by rendering only those parts dynamically which are of special interest. All others are rendered statically. This gives an overall impression of the whole car while important features can be viewed in detail.

6 Future Work

This paper demonstrates a powerful method for view-dependent realtime tessellation of large NURBS models. Yet there is still room for improvements. Avoiding the use of Thick-Border-Lines would reduce the number of batches for GPU. This can be achieved by a combination of two methods

- Dilatation of the trimming polygons of one texel while filling the trimming texture causes a small overlap in 3D of two neighbored trimmed regions closing the small gaps.
- Dilatation does not work on trimming polygon segments lying on the domain boundary. Fractional tessellation [13] of the untrimmed surface ensures a tight tessellation between neighbored regions at the domain boundaries.

Another improvement concerns the quality of the reflection line visualization. If the difference of approximation errors between hierarchy levels is too large, popping artefacts may occur. This is especially noticable when the viewed nodes change from a coarse mesh, e.g. 4mm, to the next level, e.g. 2mm. Due to the different approxmation the displayed normals may change abruptly. Since reflection lines are very sensitive to normals this causes a popping artefact on screen. This can be solved by either choosing smaller appromation differences at coarse levels which increases the depth of the hierarchy or by introducing geo-morphing between two levels.

References

1. Abi-Ezzi, S.S., Subramaniam, S.: Fast Dynamic Tessellation of Trimmed NURBS Surfaces. In: Proceedings of Eurographics 1994, pp. 107–126 (1994)
2. Chew, L.: Guaranteed-Quality Mesh Generation for Curved Surfaces. In: Proceedings of Nineth Annual ACM Symposium on Computational Geometry, pp. 274–280 (1993)
3. Farin, G.E.: Curves and Surfaces for Computer Aided Geometric Design, 4th edn. Academic Press, London (1997)
4. Filip, D., Magedson, R., Markot, R.: Surface algorithms using bounds on derivatives. In: Computer Aided Geometric Design, vol. 3, pp. 295–311. Elsevier Science Publishers B.V., Amsterdam (1987)
5. Sußner, G., Greiner, G., Augustiniack, S.: Interactive examination of surface quality on car bodies. In: Computer-Aided Design, pp. 425–436 (2004)
6. Glassner, A.: Clipping a concave polygon. In: Paeth, A.W. (ed.) Graphics Gems, pp. 50–54. Academic Press, Inc., Orlando (1995)
7. Guthe, M., Balázs, A., Klein, R.: GPU-based trimming and tessellation of NURBS and T-Spline surfaces. In: ACM Transactions on Graphics, August 2005, vol. 24, pp. 1016–1023 (2005)
8. Guthe, M., Balázs, A., Klein, R.: GPU-based Appearance Preserving Trimmed NURBS Rendering. In: Skala, V. (ed.) Journal of WSCG, vol. 14, UNION Agency-Science Press (February 2006)
9. Hahmann, S.: Surface interrogation: Visualization techniques for surface analysis. In: Bajaj, C. (ed.) Data Visualization Techniques, pp. 44–74. John Wiley & Sons Ltd., Chichester (1999)
10. Hoppe, H.: View-dependent refinement of progressive meshes. In: SIGGRAPH 1997: Proceedings of the 24th annual conference on Computer graphics and interactive techniques, pp. 189–198. ACM Press/Addison-Wesley Publishing Co., New York (1997)
11. Klein, R.: Linear approximation of trimmed surfaces. In: Proceedings of the 6th IMA Conference on the Mathematics of Surfaces, pp. 201–212. Clarendon Press, New York (1996)
12. Krishnamurthy, A., Khardekar, R., McMains, S.: Direct evaluation of nurbs curves and surfaces on the gpu. In: SPM 2007: Proceedings of the 2007 ACM symposium on Solid and physical modeling, pp. 329–334. ACM, New York (2007)
13. Moreton, H.: Watertight tessellation using forward differencing. In: HWWS 2001: Proceedings of the ACM SIGGRAPH/EUROGRAPHICS workshop on Graphics hardware, pp. 25–32. ACM, New York (2001)
14. Pottmann, H.: Visualizing curvature discontinuities of free-form surfaces. In: Proceedings of Eurographics 1989, pp. 529–536 (1989)
15. Schumaker, L.L.: Computing optimal triangulations using simulated annealing. In: Comput. Aided Geom. Des., vol. 10, pp. 329–345. Elsevier Science Publishers B. V., Amsterdam (1993)
16. Schumaker, L.L.: Triangulations in cagd. IEEE Comput. Graph. Appl. 13, 47–52 (1993)
17. Sheng, X., Hirsch, B.E.: Triangulation of trimmed surfaces in parametric space. In: Computer-Aided Design, vol. 24, pp. 437–444 (1992)
18. Sußner, G., Greiner, G., Grosso, R.: Generating high quality meshes for interactive examination of surface quality on car bodies. In: IMR, pp. 99–109 (2002)
19. Sutherland, I.E., Hodgman, G.W.: Reentrant polygon clipping vol. 17, pp. 32–42 (1974)

Online Triangulation of Laser-Scan Data

Klaus Denker, Burkhard Lehner, and Georg Umlauf

Department of Computer Science, University of Kaiserslautern, Germany
{denker,lehner,umlauf}@cs.uni-kl.de

Summary. Hand-held laser scanners are used massively in industry for reverse engineering and quality measurements. In this process, it is difficult for the human operator to scan the target object completely and uniformly. Therefore, an interactive triangulation of the scanned points can assist the operator in this task.

Our method computes a triangulation of the point stream generated by the laser scanner online, i.e., the data points are added to the triangulation as they are received from the scanner. Multiple scanned areas and areas with a higher point density result in a finer mesh and a higher accuracy. On the other hand, the vertex density adapts to the estimated surface curvature. To assist the human operator the resulting triangulation is rendered with a visualization of its faithfulness. Additionally, our triangulation method allows for a level-of-detail representation to reduce the mesh complexity for fast rendering on low-cost graphics hardware.

1 Introduction

In industry, the scanning of surfaces of 3d objects is used for measurement and analysis of manufactured objects and for reverse engineering. Most scanning devices use a laser to sample points on the surface. Some scanners move the object while others move the laser device. While some scanning devices measure the sample points in a regular pattern, hand-held laser scanners have a movable scanning device that is moved along the surface by a human operator. These scanning devices generate a vast amount of data in very short time with very high precision. To process and triangulate this data the used method must preserve the precision while reducing the data to an adequate level, see Figures 1 and 2. Thus, it is necessary to allow for heterogeneous triangulations and point densities, especially when areas are scanned multiple times. This is particularly important for hand-held devices, where the operator most likely will scan the object from different directions and with different speeds. This generates disconnected triangulation fragments with highly different point densities.

Since the local point density can become arbitrarily high by multiple scans, the feature size that can be reconstructed is theoretically arbitrary small. Of course, the measuring accuracy/error sets a lower bound for the feature size.

For these hand-held laser scanners, the operator has to cover the complete surface of the object. Because the scanning may take a long time, it is difficult

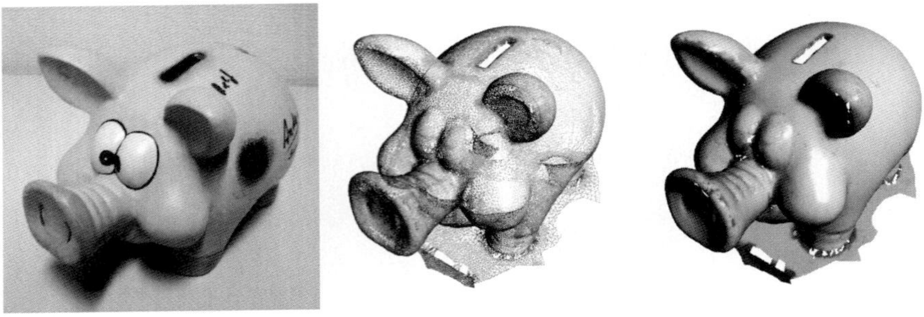

Fig. 1. A piggy bank. Original object (left), wire-frame (center), and smooth shaded triangulation with uncertainty visualization (right).

(a) Original object. (b) Raw points. (c) Vertices and normals.

Fig. 2. Different levels of data reduction shown with a little bronze bird

for the operator to keep track of the already scanned area. Furthermore, there is no feedback to tell the operator to re-scan a region to increase the point density to improve the quality of the reconstructed surface. Therefore, a full-automatic real time triangulation and visualization of the scanned points is crucial to assist the operator to improve the scans in less time. Thus, the crucial constraints for our triangulation task are:

A. Handling of large point sets,
B. handling of heterogeneous point densities of incoherently scanned regions,
C. handling of high precision point data with predefined measurement errors,
D. handling of point streams of arbitrary order, i.e., online triangulation,
E. triangulating full-automatically,
F. triangulating in real-time, and
G. assisting the human operator during the scanning process.

In the rest of the paper we first discuss related work in Section 2 and describe the principle of our method in Section 3. Subsequently we discuss various aspects of our method in detail, i.e., in Sections 4–6 the neighborhood and normal calculations and local triangulation, in Section 7 our octree-based data structure, and in Section 8 local improvements of the approximation quality. In Section 9 we present a method to assist the human operator during the scanning process and in Section 10 the integrated level-of-detail representation. Finally we show results of our method in Section 11 and close with an outlook on our future research plans in Section 12.

2 Related Work

To contrast our approach to other methods for surface reconstruction from unorganized point clouds, we briefly describe alternative methods and discuss their pros and cons with respect to the constraints A.–G. given above.

One of the first methods in this field was proposed in [14]. Here, for every point a surface normal is estimated from its k nearest neighbors and its orientation is propagated from the orientation of one particular normal to all other normals using a global minimal spanning tree of the points. This allows to estimate tangent planes defining an estimated signed distance function to the surface. Its zero-set is used to compute a triangulation of the surface using marching cubes. This method can deal with surfaces with boundaries and holes, and no additional information (such as surface normals) is necessary. On the other hand, it is not capable of dealing with incremental insertion of data points, since the orientation propagation is global, and the density of the points on the surface is pre-defined, otherwise spurious holes are introduced. The minimum feature size that can be reconstructed is fixed a priori by the edge length of the marching cubes algorithm and increasing the density of points does not reveal more details. Therefore, at least constraints B. and D. are not satisfied.

Alpha shapes were defined in [9,10]. For a given real number $\alpha > 0$, the alpha shape S_α of a point set P is the set of all k-simplices $T \subset P$ $(k < d)$ with vertices lying on a sphere with radius α that does not contain any other point of P. The alpha shape can efficiently be determined from the Delaunay triangulation where α controls how many "details" of the point cloud are "cut" out of the convex hull of P. If α is too large, details remain hidden under larger faces, if it is too small, the object may be cut into disconnected pieces. Therefore, the choice of α is crucial for the optimal reconstruction of a surface from a point cloud. If the variation of the point density is too high, there may not even exist a suitable α value, violating constraint B.

For the so-called *weighted alpha shapes* of [1] every point of P gets an associated weight. This permits using different values of α for different regions of P. The weights have to be tuned to the point density very accurately to achieve a good reconstruction of the underlying surface, making it difficult to achieve B. Another extension was proposed in [20]. The sphere with radius α is deformed anisotropically into an ellipsoid, achieving a more accurate separation of surfaces close to each other. But their approach relies on user input, violating constraint E.

All methods based on alpha shapes can be used to reconstruct surfaces with borders and holes, an no additional information per vertex other than its position is necessary. But, the correct choice of α or the point weights, respectively, is crucial for the quality of the final triangulation, which is in contrast to constraint C. Furthermore, the computed triangulation is not guaranteed to be a

2-manifold with border. It can contain edges with more than two adjacent faces, or isolated edges and vertices. Therefore, a postprocessing clean-up step is necessary, violating also constraint F.

The so-called *power crust* of [4] is based on an approximation of the medial axis transform of the point set. It is computed from the Voronoi diagram and the poles of the input points. From this, it calculates in an inverse transformation the original surface using the power diagram of the poles and taking the simplices dividing the interior and exterior cells of the power diagram from each other as triangulation for the surface. The power crust approach produces connected surfaces possibly with intentional holes. So, it is not suitable for online triangulations (constraint D.) where the triangulation may consist of disconnected fragments (constraint B.).

The *eigencrust* method proposed in [16] is specialized to produce high quality surface reconstructions on noisy point clouds. It labels all tetrahedra in the 3d Delaunay triangulation of the sample points as either being inside or outside the surface based on a global optimization. The triangulation of the reconstructed surface is the set of faces that are adjacent to one inside and one outside tetrahedron. Because of the global optimization step, the results are of high quality even with the presence of noise and outliers. The final surface is always a 2-manifold without border. Therefore, the eigencrust method cannot be used with constraint D.

The geometric convection approach for surface reconstruction described in [6] starts with a Delaunay triangulation of the point set, and shrinks the boundary surface by removing tetrahedra containing a boundary triangle that does not fulfill the *oriented Gabriel property*, i.e., the half-sphere centered at the triangle's circumcenter and oriented to the inside contains point of the point set. This procedure is repeated until all boundary triangles fulfill the oriented Gabriel property. In some cases, cavities are not opened by this algorithm, so another property has to be defined to remove the involved tetrahedra. The approach requires the Delaunay triangulation of the complete point set, therefore it contradicts constraint D.

An extension of [6] for streams of point sets is proposed in [3]. The point set is divided into slices, and only a limited number of slices is kept in memory. A slice that cannot have impact on the current slice can be removed from memory, storing the triangles found for that slice. For the division into slices, all points have to be known in advance, because they have to be ordered according to one of the spatial coordinates. Therefore, this method is not suitable for constraint D.

A method suitable not only for surface reconstruction but also for re-meshing of an existing mesh is presented in [18]. Using an advancing front approach, triangles are constructed that fulfill two user-defined constraints: the maximum edge length with respect to the curvature, and ratio bounds of adjacent edges. For surface reconstruction, a projection operator $\mathcal{P}$ and a guidance field g have to be defined, contradicting constraints B., D., and F.

A common problem of the methods [10,1,20,4,16,6,3] is their computational complexity, which is too high for real-time applications (constraint F.). Another disadvantage of these methods is the fact that the sample points or the same number of points are used to create the surface mesh. Thus, the complexity of the meshes increases rapidly while scanning, and the measurement errors are not corrected. Furthermore, if a region is scanned multiple times, the additional vertices decrease the area of the mesh faces, but the noise remains constant, leading to a bumpier surface after every scan pass. These aspects are in contrast to constraints A. and C.

In [5] an interactive online triangulation method is proposed. The sampled points are processed in a pipeline. In the first stage the number of points is reduced by dropping every point that is closer than a specified radius from an already existing point. In the second stage, the normal at the point is estimated using the points in a local neighborhood. After another reduction stage with a larger radius, the points with a stable normal are inserted into the surface mesh which is re-triangulated locally with a shortest edge criterion to decide which edges to keep. This approach works well and is fast, but has some major drawbacks:

- A large fraction of the input is ignored and not used to reduce the noise of the input data, violating C.
- The size of the smallest features that can be modeled is fix, violating C.
- The size of the mesh triangles is not adapted to the density of sample points or the curvature of the surface, violating B. and E.

The method proposed in this paper uses some of the ideas of the approach of [5], but satisfies all constraints A.-G.

3 Online Triangulation

Our method is based on a laser scanner like the FARO Laser ScanArm [11] as described in Section 11. Scanners of this type generate a stream $D = (d_1, d_2, \dots)$ of *data points* d_i. Each data point is a pair $d_i = (p_i, h_i) \in \mathbb{R}^3 \times \mathbb{R}^3$ of a *raw point* p_i, that is measured by the scanner on the scanned object, and the *scan position* h_i of the laser scanner at the moment of scanning p_i.

A laser scanner of this kind scans an object line by line measuring a certain number of data points per scan line. The scanner we used scans up to 30 lines per second measuring up to 640 data points per scan line, see e.g. Figure 2(b). These scan lines are arranged in scan passes that the human operator triggers by pressing a button. The pauses between two scan passes are usually used by the operator to reposition the scanner for a different scan direction.

In order to triangulate this huge data stream online, the data points need to be reduced. For this the data points are classified by their distance and added to so-called *neighborhood balls* b_j that represent a subset of data points within a certain radius and similar scan positions. The radius depends on the point density and estimated curvature. Subsequently only the averages of data points of the neighborhood balls are used as vertex positions in the triangulation T approximating $\{p_1, p_2, \dots\}$. So, the overall process is described schematically as follows (for the used data structure see Section 7)

ONLINE-TRIANGULATION($d_1, d_2, \dots$)

Input: Data point stream $D = (d_1, d_2, \dots)$;
Output: Triangulation T approximating $\{p_1, p_2, \dots\}$.
while (D not terminated) **do {**
  ADD-TO-NEIGHBORHOOD-BALLS(d_i); \\ see Section 4
  Update normals of affected neighborhood balls; \\ see Section 5
  Update local approximation; \\ see Section 8
  Triangulate area of affected neighborhood balls; \\ see Section 6
  Render triangulation with uncertainty visualization; \\ see Section 9
}

4 Neighborhood Balls

A bounding cube of edge length R enclosing the maximal scanning range is

$$O = [x_{\min}, x_{\min} + R] \times [y_{\min}, y_{\min} + R] \times [z_{\min}, z_{\min} + R],$$

i.e., $p_i \in O$ for all i. Furthermore, a *ball* with center $c \in \mathbb{R}^3$ and radius $r \in \mathbb{R}, r \geq 0$, is for the Euclidian norm $\|\cdot\|$ defined as the set

$$\beta(c, r) = \{x \in \mathbb{R}^3 : \|x - c\| \leq r\}.$$

As in [5] we use *neighborhood balls* $b_j = (c_j, r_j, D_j)$ to represent a set of n_j data points $D_j = \{d_{j,1}, \dots, d_{j,n_j}\} \subset D$ contained in the ball $\beta_j = \beta(c_j, r_j)$. Every neighborhood ball corresponds to a local estimate N_j for the oriented surface normal, which might be undefined (cf. Section 5), and a vertex v_j of T. Thus, neighborhood balls can intersect and serve three purposes:

- Collecting n_j data points to reduce the number of visualized data points.
- Estimating a local oriented surface normal.
- Averaging its data points gives the position of a vertex of the triangulation.

The minimal ball radius $R_{\min} = 0.75\,\text{mm}$ prevents ball sizes below scanner accuracy. A neighborhood ball may contain up to $n_{\text{split}} = 40$ data points. Later this value is depending on curvature, see Section 8.2. The set of all neighborhood balls b_j is denoted by B. It is initialized with the first data point $B = \{(p_1, R, \{d_1\})\}$. Then new neighborhood balls are generated by adding one data point after the other with ADD-TO-NEIGHBORHOOD-BALLS(d_i), using the following steps:

1. To add data point $d_i = (p_i, h_i)$, first all k neighborhood balls $b_j = (c_j, r_j, D_j)$ are determined that contain $p_i \in \beta_j$ with normals N_j aligned to the scanning direction, i.e.

$$N_j^T \cdot (h_i - p_i) \geq 0 \qquad \text{(if } N_j \text{ is defined).} \qquad (1)$$

   a) If $k = 1$, d_i is added to D_j.
   b) If $k > 1$, d_i is added to D_j of the neighborhood ball b_j with largest radius and smallest distance $\|c_j - p_i\|$.

c) If $k = 0$, a new neighborhood ball $b = (p, r, \{d\})$ is generated, with radius $r = 2^{-\mu}R$ where μ is the smallest integer such that $\beta(p, r)$ does not contain the center of any other b_j. The new ball b is added to B.

2. If in cases a) and b) n_j equals n_{split} after d_i is added and $r > R_{\min}$, the neighborhood ball b_j is removed from B and all data points in D_j are added using Step 1. If in this process a data point $d_l \in D_j$ is not contained in any other neighborhood ball of $B \setminus \{b_j\}$, a new ball $b = (p_l, r_j/2, \{d_l\})$ is generated and added to B.

Remark 1. This definition of neighborhood balls has two advantages over the method of [5]: First all data points are collected in neighborhood balls and not only the first one to support C. (see Sections 5 and 8.1), and second the radii of the balls can be adapted to the density of the data points and the estimated curvature of the surface to support B. and E. (see Section 8.2).

For later triangulation we use the *average* of a neighborhood ball b_j

$$\bar{b}_j = \frac{1}{n_j} \sum_{l=1}^{n_j} p_{j,l},$$

as position of vertex v_j representing b_j in the triangulation, see Figure 2(c).

5 Normal Estimation

For every neighborhood ball b_j an estimated surface normal N_j is calculated. First all n_l data points d_l contained in $\beta(\bar{b}_j, 2r_j) \ni p_l$ are determined. This provides a more stable normal estimation than using only the data points in D_j. These data points are used for a principal component analysis as in [5, 15]. Computing the eigenvalues $0 \leq e_1 \leq e_2 \leq e_3$ of the 3×3 covariance matrix

$$C = \sum_l (p_l - \bar{b}_j)(p_l - \bar{b}_j)^T$$

using [19], yields for N_j the direction of the eigenvector v_{e_1} of C corresponding to the smallest eigenvalue e_1

$$N_j = v_{e_1} / \|v_{e_1}\|.$$

To get a stable normal estimate it is necessary that $e_2 \geq 2e_1$. Otherwise b_j does not have a normal estimate and all subsequent computations requiring a N_j are rejected. Thus, highly curved regions with too low point density are either not triangulated or marked as regions that need a further scan pass, see Section 9. This ensures a locally rather planar point distribution. To get the orientation of N_j, the average scan direction of all determined d_l

$$\bar{s}_j = \frac{1}{n_l} \sum_l (h_l - p_l).$$

is used. The normal orientation is correct if $N_j^T \cdot \overline{s}_j \geq 0$. Otherwise the orientation is inverted. Finally, all data points of D_j that do not satisfy (1) are removed from D_j and re-inserted using ADD-TO-NEIGHBORHOOD-BALLS. Figure 2(c) shows the estimated normal of each neighborhood ball.

6 Local Triangulation

Because every neighborhood ball corresponds to one vertex in the triangulation, the latter is updated in five steps if a neighborhood ball is added or removed from B:

1. Collect potential neighbor vertices in a set V_j. (see Section 6.1)
2. Project V_j onto the estimated tangent plane. (see Section 6.2)
3. Adapt the border of V_j. (see Section 6.3)
4. Determine the triangulation T_j of V_j. (see Section 6.4)
5. Insert T_j into the triangulation T. (see Section 6.5)

6.1 Collecting Potential Neighbor Vertices

Every neighborhood ball b_j corresponds to a vertex v_j in T with position $\overline{b}_j$. Thus, if b_j is added or removed from B, the corresponding vertex v_j is added or removed from T. In both cases the local neighborhood of v_j needs to be re-triangulated. To determine the geometric neighbors of v_j we define a ball

$$\eta_j(r) = \{x \in \mathbb{R}^3 : \|(x - \overline{b}_j) + ((x - \overline{b}_j)^T N_j)(f_\eta - 1)N_j\| \leq r\}.$$

of radius r around $\overline{b}_j$ flattened along the normal N_j by f_η to provide a better separation of close parallel surfaces sheets. Then, the geometric neighbors are all b_l with $\overline{b}_l \in \eta_j(5r_j)$ for $f_\eta = 3$ and $N_l \cdot N_j \geq 0.5$. This yields a set $V_j = \{v_{j_1}, \ldots, v_{j_m}\} \subset V \cup \{v_j\}$ of vertices that will be re-meshed.

6.2 Projection onto the Estimated Tangent Plane

Because the triangulation of the area around b_j is computed in the plane perpendicular to N_j all vertices $v_{j_i} \in V_j$ are projected along N_j onto this plane, i.e. $\overline{b}_{j_i}$ is projected to t_{j_i}. Then, because of the one-to-one correspondence of v_{j_i} to t_{j_i}, triangulating the t_{j_i} is equivalent to triangulating the v_{j_i}. Therefore, we will speak of a triangulation of V_j although the triangulation is computed in the local estimated tangent plane.

6.3 Adapting the Border of the Local Triangulation

Triangulating V_j yields a triangulation T_j of the local neighborhood of v_j. Because most vertices in T_j are also in T, the edges in T_j should match edges in T. Thus, the *border* ∂T_j of T_j has to match edges in T. The border

$\partial T_j = (v_{x_1}, \ldots, v_{x_{n_x}})$ is represented as a counter-clockwise oriented, ordered sequence of n_x *border vertices* $v_{x_l} \in V_j$, $l = 1, \ldots, n_x - 1$ with $v_1 = v_{x_{n_x}}$, i.e., the index of border vertices is understood modulo n_x. Every pair $(v_{x_l}, v_{x_{l+1}})$ is a so-called *border edge* and the border ∂T_j is initialized as the convex hull of V_j using "Jarvis' March" [7]. Subsequently V_j and ∂T_j are modified until the border edges match edges in T as good as possible.

If a border edge $e_b = (v_{x_k}, v_{x_{k+1}})$ does not match any edge in T, determine the edge $e_s = (v_{x_k}, v_s)$ or $e_{\overline{s}} = (v_s, v_{x_{k+1}}) \in V_j \times V_j$ in T inside ∂T_j with smallest angle φ to e_b. Then v_s is inserted to ∂T_j between v_{x_k} and $v_{x_{k+1}}$, if e_s respectively $e_{\overline{s}}$ is either an inner edge or $\varphi < 20°$ and if this does not cause proper intersections of the interior of two edges of ∂T_j or any loops containing more than two border edges, see Figure 3. This approach can lead to a border of the form $(\ldots, v_{x_l}, v_{x_{l+1}}, v_{x_{l+2}}, \ldots)$ with $v_{x_l} = v_{x_{l+2}}$, see Figure 3, which is handled by removing v_{x_l} and $v_{x_{l+1}}$ from ∂T_j and $v_{x_{l+1}}$ from V_j.

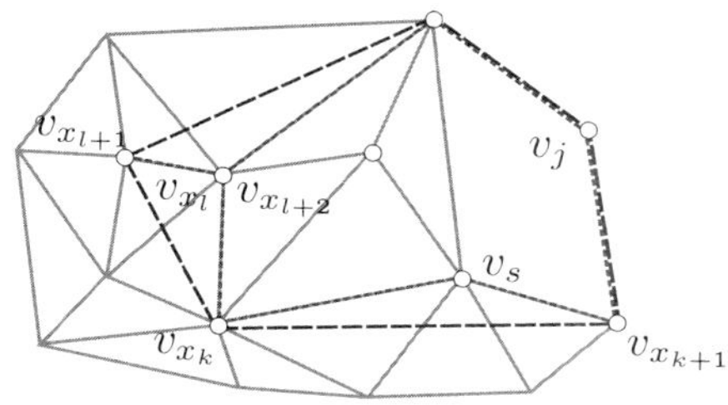

Fig. 3. The border ∂T_j before (blue) and after (red) the modification

Repeating these operations until there are no more edges that can be removed results in a border that fits the existing triangulation T better. This process terminates because the border shrinks monotonically in each step.

6.4 Triangulation of the Border

To triangulate V_j first ∂T_j is split into monotone sub-polygons which are triangulated individually, see e.g. [7]. This determines a triangulation T_j' of ∂T_j. Second all vertices of $v_l \in V_j \setminus \partial T_j$ are added to T_j' successively by splitting the triangle of T_j' that contains t_l at t_l into three new triangles. This is repeated for every vertex of $V_j \setminus \partial T_j$ yielding a triangulation T_j''. Finally, the Delaunay criterion [13] is applied repeatedly constrained by ∂T_j to improve the triangle quality generating a triangulation T_j of V_j.

6.5 Insertion of the Local Triangulation

Before T_j can be inserted into T, the triangles of T in conflict with T_j must be removed. We first remove all triangles of T incident to a vertex of $V_j \setminus \partial T_j$.

Remark 2. Note that this also removes triangles from vertices in $V_j \setminus \partial T_j$ to vertices in $V \setminus V_j$ outside of $\eta_j(5r_j)$. Thus, the global topology of the surface is corrected due to the increased local point density.

At this stage T contains no triangles connected to vertices of $V_j \setminus \partial T_j$. All triangles remaining in T conflicting with T_j involve only vertices on ∂T_j. For such a vertex

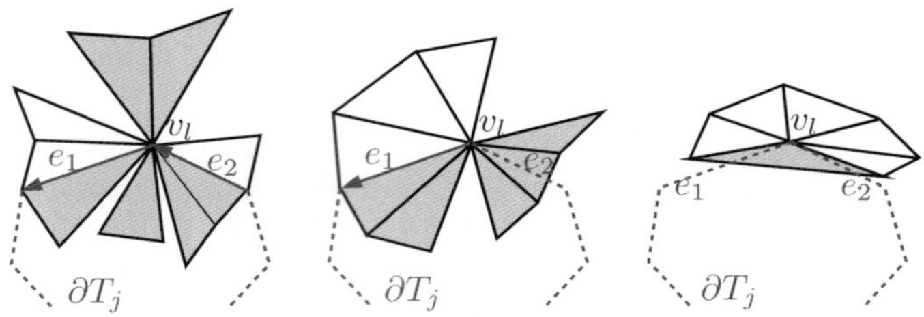

Fig. 4. Topologically preparing the border ∂T_j (red) for the new triangulation T_j by deleting the blue faces. All triangles shown in the figures are in T, T_j is not shown.

$v_l \in \partial T_j$ with border edges $e_1 = (v_{l-1}, v_l)$ and $e_2 = (v_l, v_{l+1})$, all faces potentially pointing to the inside of the border are deleted. Thus, there are three different cases that are solved topologically:

1. If both edges e_1 and e_2 belong to T, all faces of the one-ring of v_l are removed if they are left of e_1 or e_2 or if they are not connected to the right faces of e_1 or e_2. This removes triangulated areas of T that are also covered by T_j and reduces the complexity of the one-ring, see Figure 4 (left).
2. If only e_1 belongs to T, the face of the one-ring of v_l left of e_1 and all faces not connected to the right face of e_1 are deleted, see Figure 4 (middle).
3. If both e_1 and e_2 do not belong to T and v_l has a closed one-ring, the face pointing the most inside the triangulated area is removed, which is e.g. the triangle of the one-ring of v_l intersected by the bisector of e_1 and e_2 in the local estimated tangent plane, see Figure 4 (right).

Finally, if there are two vertices v_{l_1} and v_{l_2} from ∂T_j that are connected by an edge e that does not belong to T_j and lies inside of the polygon spanned by ∂T_j, the corresponding triangles are removed. To test if such an edge is inside the polygon spanned by ∂T_j the angle β, the inner angle of the polygon at v_{l_1}, must be larger than the angle α between the incoming border edge at v_{l_1} and e, see Figure 5.

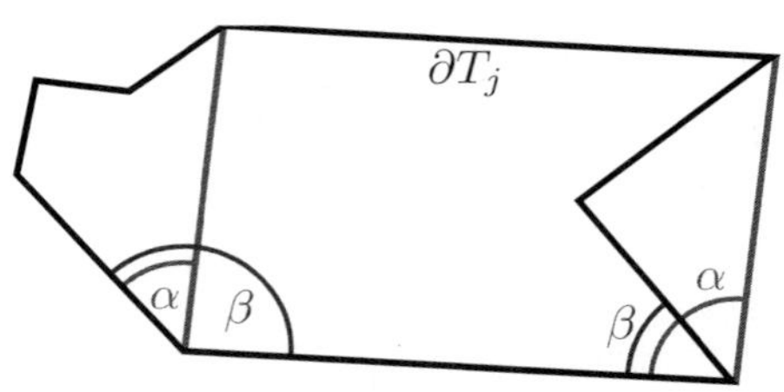

Fig. 5. Border ∂T_j (black) with edges (red) not connected by faces to the border

Remark 3. This also changes the global topology of the surface.

7 The Octree

To support the geometric neighborhood searches in Sections 5 and 6.1 efficiently we use an octree data structure to manage the neighborhood balls. The root node of the octree represents the cube O. Every node in the tree represents a sub-cube

$$o = [x, x+\delta] \times [y, y+\delta] \times [z, z+\delta]$$

of O and has no or exactly eight child-nodes holding the eight sub-cubes $o_1, \ldots, o_8$ with side length $\delta/2$, see Figure 6. To accelerate searches in the local neighborhood, every node stores additional links to the 26 face-, edge- and corner-neighbor nodes $o_{n_1}, \ldots, o_{n_{26}}$ on the same level, as in [5].

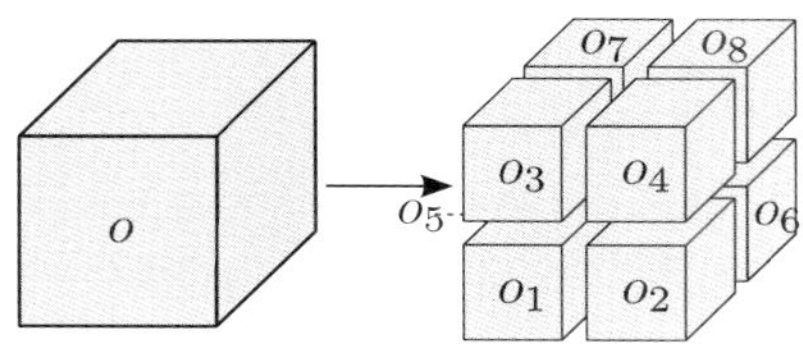

Fig. 6. The cube o of an octree node and its child-node's sub-cubes $o_1, \ldots, o_8$

Remark 4. The main advantage of using an octree instead of a grid as in [5] is the use of different levels of detail corresponding to the levels of the octree to create a finer triangulation in regions with higher point density.

Every neighborhood ball b_i belongs to one cube in the octree which contains its center point c_i. The radius r_i equals the edge length of the cube. Every cube in the octree has a list of the neighborhood balls it contains.

Every data point $d_j = (p_j, h_j)$ is inserted into the octree by searching for the neighborhood ball b_i containing the raw point $p_j \in \beta_i$ and, if b_i is found, inserting it to that ball as in Section 4. To search for b_i the octree is descended from the root node O traversing on any level of the octree the cube o containing p_j. For the traversal all 26 neighbor cubes are tested to find the ball with c_i closest to p_j. If no b_i is found, the search descends one level in the tree and repeats the neighbor traversal. This is repeated until a ball is found or the leaf nodes are searched unsuccessfully. In the latter case a new neighborhood ball b containing d_j is created in leaf node cube containing p_j, see Section 4.

To make b as large as possible the highest level in the octree with sufficient space is determined, such that the ball β_i does not contain the center of any other ball of B. These centers can only be in the siblings of the 26 indirect neighbor cubes, which share at least one corner with the actual cube o. The set of these cubes is denoted by $S(o)$. In $S(o)$ the center c_k that is closest to p_j is determined with $\Delta := \|c_k - p_j\|$.

1. If Δ is smaller than the radius of b, i.e. the edge length ℓ of o, the cube o is split into sub-cubes until the radius of b is small than Δ. Thus, b is added to a sub-cube o' that is $\lceil \log_2(\ell/\Delta) \rceil$ levels below the node of o in the octree.
2. If Δ is larger than the radius of b, it can be added to the cube o or one of its ancestors. Thus, the ancestors o' of o are tested if their siblings of $S(o')$ contain a center too close to p_j. Finally, b is added to the highest ancestor above o for which this test is negative.

8 Improving the Approximation

For every neighborhood ball b_i a least square fit f_i for its raw points is computed to smooth the triangulation and to reduce noise in the raw points. The fit f_i is a cubic approximation of the raw points of b_i parametrized over the estimated tangent plane of b_i as in [2] computed by a singular value decomposition. It serves two purposes:

- correction of vertex positions and
- curvature dependent ball sizes.

8.1 Correction of Mesh Points

The position of a vertex v corresponding to a neighborhood ball b_i is approximated by the arithmetic mean of all its raw points $\overline{b}_i$. On curved surfaces this results in a displaced position. In local coordinates of the estimated tangent plane $\overline{b}_i$ has coordinates $(0,0,0)^T$. Thus, the point $\widetilde{b}_i$ with local coordinates $(0,0,f_i(0,0))^T$ is a better approximation of the raw points. In order to guarantee that the vertices of T are within scanner precision the raw point p_j closest to $\widetilde{b}_i$ is determined. If ε is the scanner precision and $\widetilde{b}_i$ is not contained in $\beta(p_j,\varepsilon)$, $\widetilde{b}_i$ is projected onto $\beta(p_j,\varepsilon)$ in direction $p_j-\widetilde{b}_i$. This new point is the position of the corresponding vertex v_i.

Remark 5. Because the laser scanner has different precisions in different directions, i.e. along the scan line, between scan lines, and in laser beam direction, $\widetilde{b}_i$ is projected onto an ellipsoid around p_j .

8.2 Curvature Dependent Ball Size

The fits f_i are also used to estimate the curvature of T in the vertex v_i. Then the size of the neighborhood balls is controlled by the curvature measure $C_i := (|\kappa_1|+|\kappa_2|)/2$ at v_i, where κ_1 and κ_2 are the principle curvatures of f_i at $\widetilde{b}_i$ before the projection onto $\beta(p_j,\varepsilon)$. A small value of C_i indicates that the region is rather flat. So, for each neighborhood ball with a valid normal the curvature C_i is computed and the neighborhood ball is split if

$$\arctan(4r_jC_j)\cdot 2n_j/\pi \geq n_{\text{split}}.$$

This results in larger triangles in flat regions.

9 Uncertainty Visualization

Regions with a high uncertainty in the triangulation should be highlighted, to enable the operator of the laser scanner to increase the point density by multiple scan passes. To measure the uncertainty, the stability of the normal estimation is used. It can be calculated for each neighborhood ball b_i by the two smallest eigenvalues e_1 and e_2 of the principal component analysis in Section 5. The uncertainty u_i is defined as

$$u_i = \arctan\left(\left(e_2/e_1 - 2\right)/20\right) \cdot 2/\pi.$$

It is restricted to $[0, 1]$ and visualized by coloring the vertices using a transition from red ($u_i = 0$), yellow ($u_i = 0.25$), green ($u_i = 0.5$) to white ($u_i \geq 0.75$).

10 Level of Detail

For very complex or large objects it may be necessary to use a further reduced mesh for fast rendering on low cost graphics hardware. The octree contains different levels, each containing neighborhood balls b_j of a certain size. These levels can be used for a reduced level of detail on the mesh.

To make use of the levels in the octree, the data points d_i are also added to so-called *LOD balls* b_k^{LOD} on the levels above the enclosing neighborhood ball. These LOD balls b_k^{LOD} are modified neighborhood balls containing all data points d_i of the neighborhood balls b_i on the levels below. Each data point d_i belongs to one neighborhood ball $b_j = (c_j, 2^{-l}R, D_j)$ on the level l and one LOD ball $b_k^{\text{LOD}} = (c_k, 2^{-\lambda}R, D_k)$ for all $0 \leq \lambda < l$ in each level above.

The *global level of detail* l^{LOD} is the depth of the deepest level in the octree used to determine T. If l^{LOD} changes, the complete mesh has to be *re-triangulated.* First all faces of the mesh are deleted. Then the neighborhood balls or LOD balls necessary for re-triangulation are selected according to their depth in the octree. An ordinary neighborhood ball is used if its depth is less or equal to l^{LOD}. If the depth of a neighborhood ball is larger than l^{LOD} its corresponding LOD ball on level l^{LOD} is used. With these balls the triangulation is computed as described above. Figure 7 shows a schematic illustration of neighborhood balls (white) and LOD balls (gray) used for re-triangulation on different levels of detail.

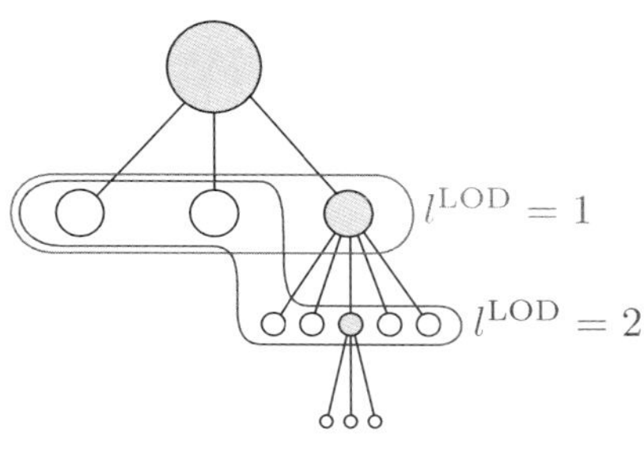

Fig. 7. Schematic illustration of the level of detail re-triangulation in the octree

11 Results

We used a hand-held laser scanner "Laser ScanArm" from Faro [11] (Figure 8). That is a measurement arm with seven joints and an assembled laser scanner "Laser Line Probe". The scanner driver provides 3d point data relative to the foot of the measurement arm. Lines of up to 640 points can be scanned up to 30 times per second. For each of these lines the position and the viewing direction of the laser scanner is tracked.

In Figure 1 a scanned piggy bank is shown, while Figure 11 shows the result of scanning the bronze bust "Bildnis Theodor Heuss" by Gerhard Marcks. The uncertainty visualization in the wire-frame and smooth shaded representations reveal the regions that can be improved by additional scan passes.

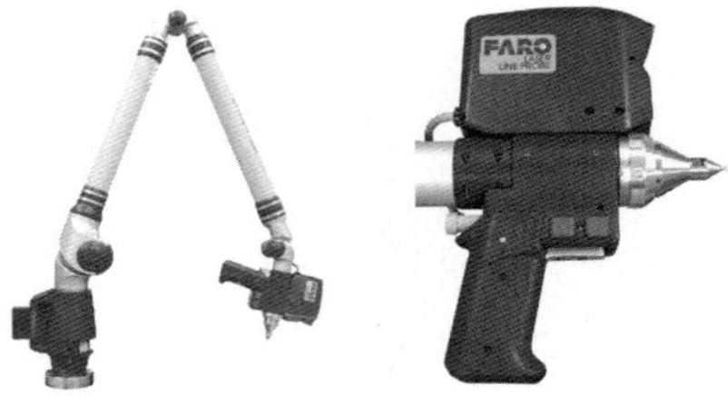

Fig. 8. The measuring arm "FaroArm" with laser scanner "Laser Line Probe" [11]

The reduction of the input points is demonstrated in Figure 2,

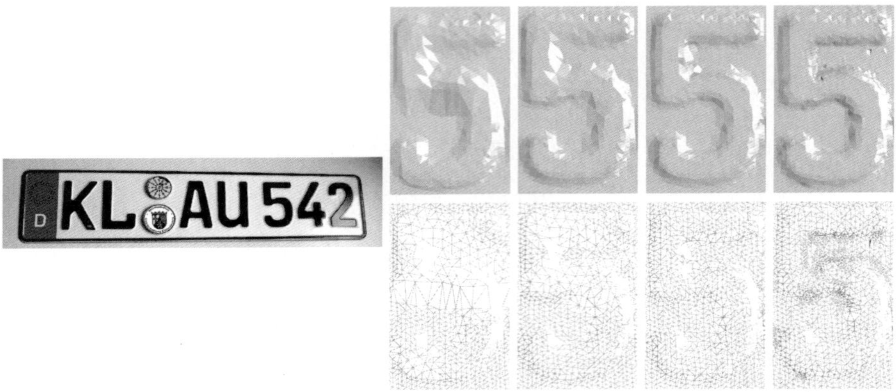

Fig. 9. Flat shading (top) and wire-frame (bottom) of triangulation of a license plate after 250, 2000, 5000, 9000 scan lines with uncertainty visualization

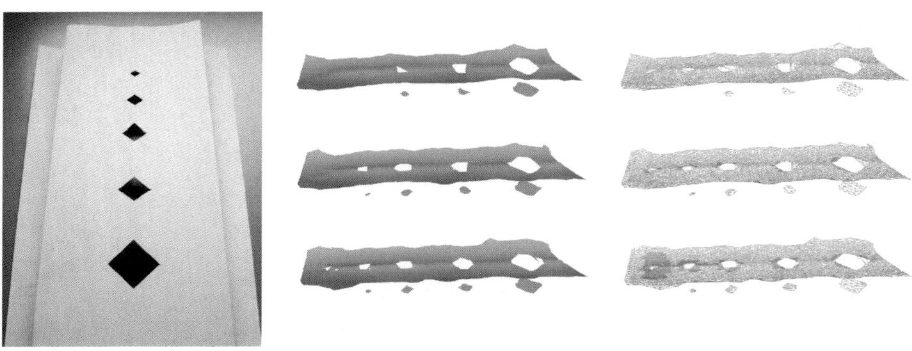

Fig. 10. Smooth shading and wire-frame representation of triangulation of a sheet of paper with holes after 585, 1895, and 3149 scan lines

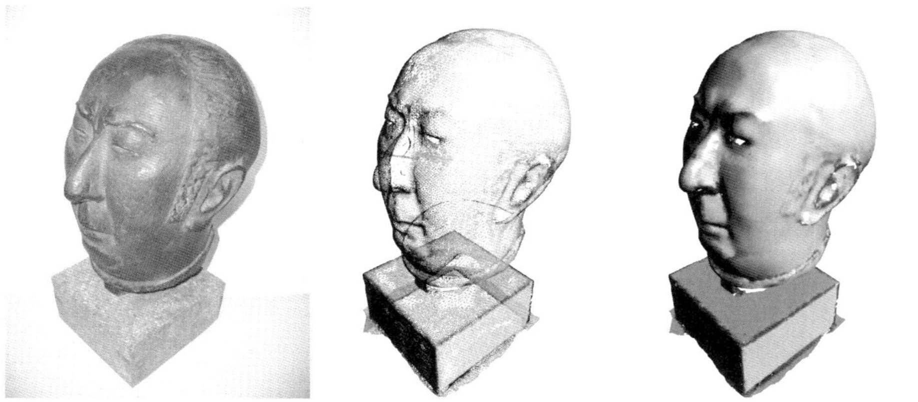

Fig. 11. Gerhard Marcks - Bildnis Theodor Heuss, [17]. Original object (left), wire-frame (center), smooth shaded triangulation with uncertainty visualization (right)

Fig. 12. Flat shaded mug with different levels of detail (top row) and wire-frame (bottom row)

where Figure 2(b) shows the data generated by the scanner, and Figure 2(c) shows the averages $\bar{b}_j$ of the neighborhood balls, together with the estimated normal for every ball.

Figure 9 shows how the triangulation is adaptively refined for several scan passes of the same region. The edge of the protruded digit becomes more precise after each scan pass. Furthermore, the wire-frame representation (bottom row) shows that the triangles near the protruded edge are smaller, because the neighborhood balls are dissolved earlier in this region of higher curvature.

Another example of increasing accuracy for multiple scan passes is shown in Figure 10. The reconstructed surface of a piece of paper with holes of different size is shown after the first, the second and the third scan pass. While after the first scan pass two small holes are still closed by the triangulation procedure, after the third scan pass all holes are correctly detected.

Four levels of detail of a scan of a mug are shown in Figure 12. The number of triangles is 20,758 for the finest level of detail (left), 17,678 for the next coarser level, and 11,277 and 3,584 triangles for the two subsequent levels. Choosing a coarser level of detail reduces the complexity of the triangulation, but can result in more reconstruction errors, especially at sharp edges.

All examples are computed on an Intel Core2 Quad Q6600, 2.4 GHz computer with 4 GB RAM. To demonstrate the efficiency of the proposed method we list in Table 1 the sizes of the models on the fines level in terms of data points d_i and vertices v_i and the times spend for the different computations: number of data points processed per second, number of vertices processed per second, and the overall times for the computation of the triangulation and the scanning process. The times for the scanning process do not include the pauses between scan passes. It is apparent from Table 1 that our method works in real time even for complex objects. A video of a live scanning process is available at [8].

Table 1. Table of number of data points d_i and vertices v_i in the final triangulation, times for processing data points and vertices, overall time for the computation of the triangulation and the scanning process for all models presented in this paper.

	Fig.	Data points	Vertices	Data points per second	Vertices per sec.	CPU time [sec]	Scan time [sec]
Piggy bank	1	1280387	30130	4198	98.9	304.6	318
Bronze bird	2	181731	6247	4432	152.0	41.1	143
License plate	9	1866413	29703	4392	69.9	424.8	504
Sheet of paper	10	478900	2538	7483	39.9	63.6	105
Theodor Heuss	11	2451014	35412	4300	62.1	570.0	623
Mug	12	358232	8866	5778	143.7	61.7	105

12 Conclusion and Outlook

Our experiments show that the proposed method is suitable to assist the operator of a hand-held laser scanner to produce fast and complete high quality triangulations satisfying all constraints A.–G. of the Introduction. The online visualization helps to reduce the time for the scanning process significantly. The final surface mesh is a correct triangulation that can be used without further postprocessing for measurement, surface analysis or reverse engineering.

The robustness of our method is derived from the fact that the human operator can increase the point density by additional scan passes in regions that are not yet reconstructed topologically correct, or where important features are still missing.

Nevertheless, there are some requirements for the proposed method to work satisfactorily. Only those parts of the object that are covered by scan lines can be reconstructed. Because of the interactive rendering the human operator can easily detect uncovered regions, fill the remaining holes and scan a topologically correct reconstruction of the target object. Regions of the object that cannot be scanned, because of occlusions or limitations of the scan arm, cannot be reconstructed and remain as holes in the triangulation.

Some extensions to improve the usability are worth further investigation. The detection of sharp features and surface borders could improve the final mesh. The method of [12] can be used to detect sharp features within the scan and move the vertices onto these features.

Experiments show that the orientation of the scanning device during the scan has an impact on the quality of the input data. Scanning the same region with different orientations improves the stability of the computation significantly. Additionally to the uncertainty visualization the optimal orientation for the next scan of an uncertain region could be visualized to aid the operator to achieve better and faster results.

Acknowledgments

This work was supported by DFG IRTG 1131 "Visualization of large and unstructured data sets." We also thank FARO Europe for lending out their "Laser ScanArm" and "Pfalzgalerie Kaiserslautern" and "Vereinigung Pfälzer Kunstfreunde (VKP)" for their generous support with the sculpture [17].

References

1. Akkiraju, N., Edelsbrunner, H., Facello, M., Fu, P., Mücke, E.P., Varela, C.: Alpha shapes: Definition and software. In: 1st Int. Computational Geometry Software Workshop, pp. 63–66 (1995)
2. Alexa, M., Behr, J., Cohen-Or, D., Fleishman, S., Levin, D., Silva, C.T.: Computing and rendering point set surfaces. IEEE Trans. Vis. Comput. Graphics 9(1), 3–15 (2003)
3. Allègre, R., Chaine, R., Akkouche, S.: A streaming algorithm for surface reconstruction. In: Symp. Geom. Proc., pp. 79–88 (2007)
4. Amenta, N., Choi, S., Kolluri, R.K.: The power crust. In: 6th ACM Symp. on Solid Modeling and Applications, pp. 249–266 (2001)
5. Bodenmüller, T., Hirzinger, G.: Online surface reconstruction from unorganized 3d-points for the DLR hand-guided scanner system. In: 2nd Symp. on 3D Data Processing, Visualization and Transmission, pp. 285–292 (2004)
6. Chaine, R.: A geometric convection approach of 3-d reconstruction. In: Symp. Geom. Proc., pp. 218–229 (2003)
7. de Berg, M., van Kreveld, M., Overmars, M., Schwarzkopf, O.: Computational Geometry. Springer, Heidelberg (2000)
8. Denker, K., Lehner, B., Umlauf, G.: Live scanning video (2008), `http://www-umlauf.informatik.uni-kl.de/~lehner/scanning.avi`
9. Edelsbrunner, H., Kirkpatrick, D., Seidel, R.: On the shape of a set of points in the plane. IEEE Trans. Inf. Theory 29(4), 551–559 (1983)
10. Edelsbrunner, H., Mücke, E.P.: Three-dimensional alpha shapes. ACM Trans. Graph. 13(1), 43–72 (1994)
11. Faro Europe GmbH & Co. KG (2008), `http://www.faro.com`
12. Fleishman, S., Cohen-Or, D., Silva, C.T.: Robust moving least-squares fitting with sharp features. ACM Trans. Graph. 24(3), 544–552 (2005)

13. Hjelle, Ø., Dæhlen, M.: Triangulations and Applications (Mathematics and Visualization). Springer, Heidelberg (2006)
14. Hoppe, H., DeRose, T., Duchamp, T., McDonald, J., Stuetzle, W.: Surface reconstruction from unorganized points. Comput. Graph. 26(2), 71–78 (1992)
15. Jolliffe, I.T.: Principal Component Analysis. Springer, Heidelberg (2002)
16. Kolluri, R., Richard Shewchuk, J., O'Brien, J.F.: Spectral surface reconstruction from noisy point clouds. In: Symp. Geom. Proc., pp. 11–21 (2004)
17. Marcks, G.: Bildnis Theodor Heuss. In: Pfalzgalerie Kaiserslautern, permanent loan from Vereinigung Pfälzer Kunstfreunde (VKP), Bronze sculpture, 31.5 x 21 x 23.5 cm (1952)
18. Schneider, J., Scheidegger, C.E., Fleishman, S., Silva, C.T.: Direct (re)meshing for efficient surface processing. Comp. Graph. Forum 25(3), 527–536 (2006)
19. Smith, O.K.: Eigenvalues of a symmetric 3×3 matrix. Comm. ACM 4, 168 (1961)
20. Teichmann, M., Capps, M.: Surface reconstruction with anisotropic density-scaled alpha shapes. In: IEEE Conf. Vis., pp. 67–72 (1998)

Quality Meshing of a Forest of Branching Structures

Chandrajit Bajaj[1] and Andrew Gillette[2]

[1] Department of Computer Sciences & Institute for Computational Engineering and Sciences, University of Texas at Austin, USA
bajaj@cs.utexas.edu

[2] Department of Mathematics & Institute for Computational Engineering and Sciences, University of Texas at Austin, USA
agillette@math.utexas.edu

Abstract. Neurons are cellular compartments possessing branching morphologies, with information processing functionality, and the ability to communicate with each other via synaptic junctions (e.g. neurons come within less than a nano-meter of each other in a specialized way). A collection of neurons in each part of the brain form a dense forest of such branching structures, with myriad inter-twined branches, inter-neuron synaptic connections, and a packing density that leaves only 5% - 10% volume fraction of exterior-cellular space. Small-scale variations in branching morphology of neurons and inter-neuron spacing can exert dramatically different electrical effects that are overlooked by models that treat dendrites as cylindrical compartments in one dimension with lumped parameters. In this paper, we address the problems of generating topologically accurate and spatially realistic boundary element meshes of a forest of neuronal membranes for analyzing their collective electrodynamic properties through simulation. We provide a robust multi-surface reconstruction and quality meshing solution for the forest of densely packed multiple branched structures starting from a stack of segmented 2D serial sections from electron microscopy imaging. The entire 3D domain is about 8 cubic microns, with inter-neuron spacing down to sub-nanometers, adding additional complexity to the robust reconstruction and meshing problem.

1 Introduction

Biological modeling problems have long been an inspiration for geometry processing and meshing research. A relatively recent technique known as serial section transmission electron microscopy (ssTEM) presents a new set of challenges to the computational geometry community. The technique employs a CCD camera to capture high resolution images of parallel slices of a collection of cells, usually neurons. The resolution on a single image slice is 5-10 nanometers, small enough to identify features on the boundaries of the cells, as well as organelles within the cells. [18, 36] Trained neuro-anatomy aware users can approximate the cell boundaries by visual inspection and label their polygonal contours consistently through the stack so that slice images of a single cell (dendrite, glial, axon, etc.) can be traced through the entire volume, as shown in Figure 1. This suggests the following computational problem: develop a method to create a watertight mesh

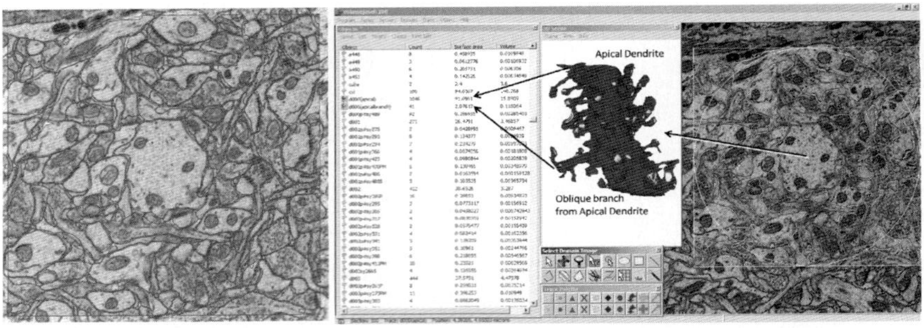

Fig. 1. Neuronal modeling from imaging uses serial section transmission electron microscopy (ssTEM), producing a stack of image slices. [18, 36] The imaging data is courtesy of Dr. Kristen Harris from the Section of Neurobiology at UT Austin. On the left, a single slice is shown with contours representing cell boundaries shown in red. The contours are tagged based on the neuronal process they belong to - purple for dendrites, green for axons, yellow for glial - as shown on the right. The tagged contours are used for accurate surface reconstruction, such as the purple dendrite model shown here, so that surface area, volume, and other properties of the cells can be calculated.

of a multi-component, compact 2-manifold passing through the labeled boundaries on each slice with biologically-accurate topology and spatially-accurate geometry. A sample of meshes created by our lab solving this problem are shown in Figure 2. The meshes will be used to research quantitative morphology [21] and to calibrate time-dependent electrophysiological simulation of voltage potentials [33]. We discuss this further in Section 5.

The unique challenge in processing and modeling neuronal cell data is the multi-component nature of the problem. Reconstruction of a single component surface from 2D slices has been heavily researched in recent decades. A seminal paper in this vein is the work of Fuchs, Kedem, and Uselton [22] which focused on triangulating a stack of polygonal contours using a toroidal graph to guide construction. Christiansen and Sederberg [15] helped to characterize the branching problem and potentially ambiguous situations that may arise. Meyers, Skinner and Sloan [32] identified the subproblems of correspondence, tiling, branching and surface-fitting and gave a resolution based on a minimum spanning tree. Barequet and Sharir [7] developed an algorithm focused on medical applications such as Computerized Tomography (CT) and Magnetic Resonance Imaging (MRI); this was expanded upon by Bajaj, Coyle and Lin [4] who introduced the use of the medial axis to aid in topologically accurate surface meshing. Similar approaches used the Voronoi diagram [34] or a discrete distance function [28] to guide reconstruction. Barequet et al. [6] have also used the straight skeleton to aid in the case of nested contours. Related to these surface meshing approaches are techniques for volume meshing [9, 24, 14], non-linear surface fitting [11, 27, 37, 8], and recently non-parallel plane methods [10, 31].

Unfortunately, resolving the multi-component problem by separating it into independent single-component problems is not sufficient to guarantee an appropriate solution for neuronal models. Image data reveals that neurons have widely varying cross-sectional areas and boundary shapes and are packed very densely (an example is shown in Figure 3), with as little as 5-10% of the image representing extra-cellular space [26, 35]. Even if the contours approximating the boundaries on each slice are non-overlapping, an under-constrained surface reconstruction might create intersections in three dimensions.

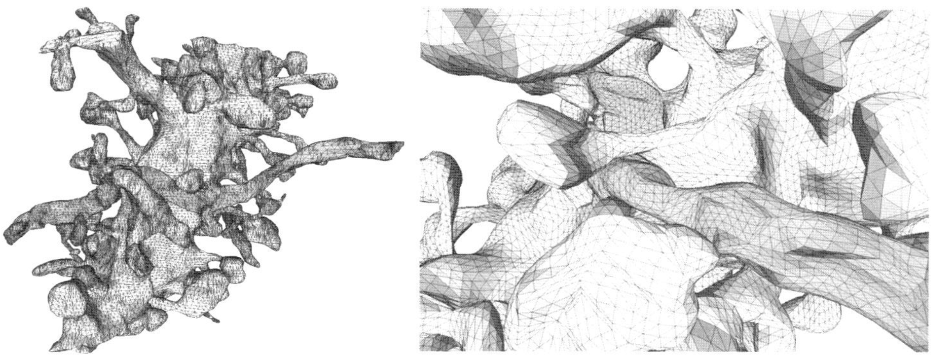

(a) (b)

Fig. 2. The approach provided in this paper allows for quality and accurate meshing of intertwined and branching structures, such as the neuronal meshes shown here. **(a)** A meshed dendrite is shown in grey with an adjacent axon passing nearby, shown in green. **(b)** A zoomed in portion of the meshes shows the triangles of the mesh are of good quality, making the mesh fit for electro-physiological simulations.

In this paper, we present a solution to the general problem of simultaneously reconstructing a densely packed forest of intertwined branching surfaces from point samples lying on a finite set of parallel slices through the volume. Our approach uses the medial axis on each slice to help construct a 3D fencing between contours, thereby preventing the intersection of distinct surfaces. In Section 2, we formalize the problem and provide additional explanation of background concepts. In Section 3, we describe in detail an algorithm to resolve the problem and prove its correctness. In Section 4, we show some initial mesh results. Finally, in Section 5 we conclude and describe the various uses these types of meshes will have in our future work.

2 Problem Statement and Background

In Section 2.1, we formally state the multi-component reconstruction problem we aim to solve and fix notation for use in the description of our solution. In Section 2.2 we describe the three main problems of the single component problem - correspondence, tiling, and branching - which are relevant to the multi-component problem.

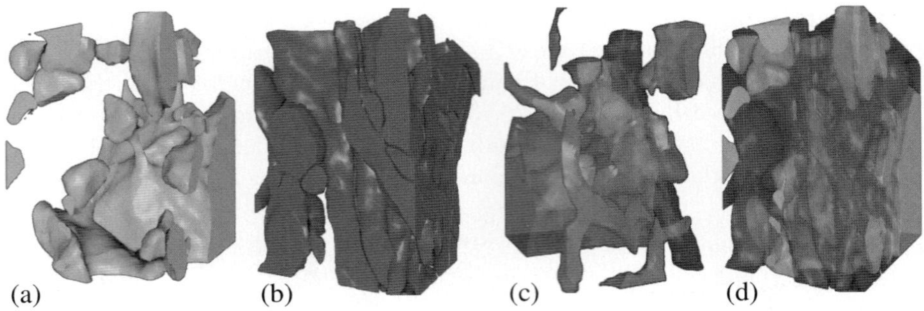

Fig. 3. A visualization of a portion of neuronal data indicating the difficulty in the forest of branching structures problem. The reconstruction visualization was prepared by Justin Kinney, Thomas Bartol, and Terrence Sejnowski of the Salk Institute using our tiling program [4]. **(a,b)** Surfaces representing dendrites and axons, shown in yellow and green, respectively. **(c)** Surfaces representing glial cells are shown in purple in this view, along with some of the axonal and dendritic surfaces for reference. **(d)** All three types of surfaces share this densely packed volume. Since each surface is reconstructed independently of the others, we have to resolve any possible intersections that result when the individual models are assembled in the same volume. Here, the green surface has been made partially transparent to show how the surfaces wind and branch around each other. The glial cells are present but not visible in this view.

In Section 2.3 we explain the basic theory behind the medial axis which is used prominently in our solution.

2.1 Formal Problem Statement

With the goal in mind of designing an algorithm for neuronal modeling, we now describe the more general surface reconstruction problem we aim to solve.

Suppose we are given a positive integer M and the input ($\{Z_i\}$, $\{g_i\}$, $\{c_i^j\}$, G) as follows:

- **Input 1:** A set of M horizontal planes $Z_1, \ldots Z_M$ where Z_i is the plane $z = z_i \in \mathbb{R}$ with $z_1 < \cdots < z_M$.
- **Input 2:** A set of curves (contours) in each Z_i plane described implicitly by $g_i(x, y) = 0$.
- **Input 3:** A list $\{c_i^j\}$ of the contours of the set $g_i(x, y) = 0$ where c_i^j denotes the jth contour on the plane Z_i.
- **Input 4:** A directed graph G with vertices $\bigcup_{i,j} c_i^j$ and edges only pointing to an element of list index incremented by one. That is, every edge of G can be written as $(c_i^{j_1}, c_{i+1}^{j_2})$ for some indices i, j_1, and j_2.

Our goal is to construct a smooth function $g : \mathbb{R}^3 \to \mathbb{R}$ such that the following properties hold:

- **Property 1:** The surface $g(x, y, z) = 0$ is a compact 2-manifold.
- **Property 2:** The function g restricts to g_i on Z_i. That is, for all i, $g(x, y, z_i) = g_i(x, y)$ for all $(x, y) \in \mathbb{R}^2$.

- **Property 3:** The surface $g(x, y, z) = 0$ has local connectivity corresponding to the graph G. That is, if $(c_i^{j_1}, c_{i+1}^{j_2})$ is an edge in G, then $c_i^{j_1}$ and $c_{i+1}^{j_2}$ are homologous (meaning one smoothly deforms into the other) on the surface

$$\{(x, y, z) : g(x, y, z) = 0, \quad z \in [z_i, z_{i+1}]\}$$

We note that the set $\{g^{-1}(0)\}$ is the desired surface passing through the contours with the correct connectivity. A general analytical solution to this problem is difficult and unnecessary for implementation, hence we make the following simplifying assumptions based on the application context.

Definition 1. *(adapted from [1]) A homeomorphism h of $\mathbb{R}^3$ is* ***isotopic to the identity*** *if there is a homotopy $H : \mathbb{R}^3 \times [0, 1] \to \mathbb{R}^3$ such that for each $t \in [0, 1]$, $h_t := H(\cdot, t) : \mathbb{R}^3 \to \mathbb{R}^3$ is a homeomorphism, h_0 is the identity and $h_1 = h$. A mesh $M \subset \mathbb{R}^3$ of a surface $S \subset \mathbb{R}^3$ is called an* ***isotopic mesh*** *if and only if there exists a homeomorphism h of $\mathbb{R}^3$ carrying M to S with h isotopic to the identity.*

- **Assumption 1:** The contours are described by simple polygons whose vertices lie on the contour. If the geometry of a particular polygon needs to be refined, the appropriate g_i can be referenced to increase the sampling density.
- **Assumption 2:** If M is a compact piecewise linear 2-manifold such that it restricts to the contours on each slice and has local connectivity corresponding to the graph G, then M is an isotopic mesh of the surface $g(x, y, z) = 0$.
- **Assumption 3:** The distance between consecutive z_i values is small enough that the slice sampling of each component in the surface $g(x, y, z) = 0$ meets the requirements of the single-component meshing algorithm.

Each assumption is based on practical considerations from neuronal data. To satisfy Assumption 1, we fit each polygonal contour data as tagged by biologists with a regular algebraic spline curve called an A-spline. By using the error-bounded spline method described in [5], we can construct smooth approximations of the contours while preventing overlaps between adjacent contours. The splines are defined locally based on a scaffolding mesh, allowing us to efficiently increase the sampling of a contour in a particular region if needed. Assumption 2 is made to distinguish the surface meshing problem from the smooth surface construction problem. In this paper, we are only interested in how to create an isotopic mesh of the smooth surface approximation and leave the surface fitting to future work. Assumption 3 is stated so that we can tackle the multi-component problem without inheriting existing difficulties from the single component problem. As is discussed in [4], it is desirable to produce a mesh such that any vertical line between two adjacent slices passes through the mesh at most once. We clarify this criterion in the next subsection.

2.2 Correspondence, Tiling, and Branching

The single component surface reconstruction problem faces three major subproblems: the correspondence problem, the tiling problem and the branching problem. Our solution to the multi-component problem requires an effective single component solver, hence we will review the three problem types here. The single component solver we use is denoted SINGLESURFRECON and resolves the problems in full generality under the assumptions previously stated. The method is summarized below, but is given in full detail in [4]. To describe the problems, we consider two planes Z_1 and Z_2 with $z_1 < z_2$ and let $\{c_1^j\}$, $\{c_2^k\}$ denote the lists of contours in the respective planes. Figure 4 illustrates some of the difficulties in meshing such intricate data.

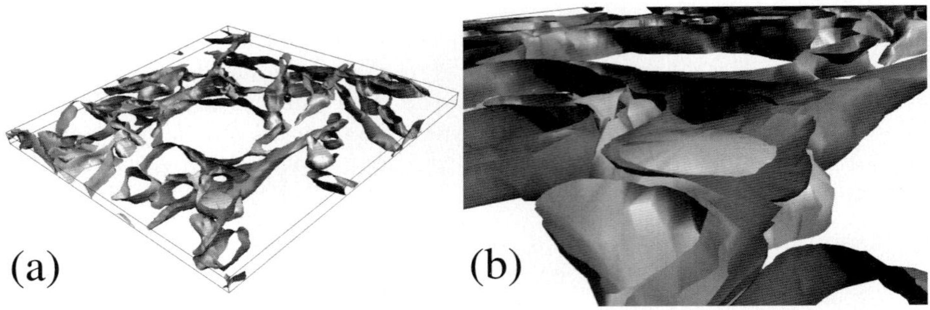

Fig. 4. (a) The region between two consecutive slices is shown, with portions of many dendrites (grey) and axons (green) present. The mesh has been filled in and smoothed slightly for the purpose of visualization. The contours on the top plane must be tiled to the contours on the bottom, subject to the correspondence constraints of the data. **(b)** Zooming in on a portion of the previous image reveals that branching has occurred in an axon and dendrite in close geometric proximity, resulting in this unwanted surface intersection. We will present a method for detecting and removing such intersections in Section 3.

The *correspondence problem* is to decide how the contours of $\{c_1^j\}$ will match up to the contours of $\{c_2^k\}$. This requires either *a priori* knowledge of the contours' connectivity or a geometric criterion for declaring correspondence. In the context of our problem domain, the correspondence problem for SINGLESURFRECON is resolved by consulting the graph G as it prescribes the connectivity between contours.

The *tiling problem* is to decide, given contours $c_1 \in \{c_1^j\}$ and $c_2 \in \{c_2^k\}$, how c_1 and c_2 will be joined in the interslice region. In the context of meshing, c_1 and c_2 are given as polygons and the problem is to decide how to add edges between them so that a suitable mesh of the ribbon surface between the contours is produced. A line segment connecting c_1 to c_2 is called a *slice chord* and a triangle formed by two slice chords an edge of a contour is called a *tiling triangle*. Even in very simple cases, there are many choices available for how to choose slice chords and tiling triangles. One resolution to this problem is to define a quality measure on possible tilings and seek a tiling with the optimal quality [22, 15]. A summary of different quality metrics used in early methods is given in [32]. Alternative approaches, such as the one we use, project contours from Z_2 to Z_1

and use planar geometry properties to lift a watertight surface mesh from the projection [4, 34, 28, 6].

The choice of tiling method is highly relevant to the types of guarantees one can provide on the output meshes. We have selected the method of [4] because it only outputs surfaces that meet the following three criteria.

- **Criterion 1:** The reconstructed surface forms a piecewise closed surface of polyhedra.
- **Criterion 2:** Any vertical (meaning parallel to the z axis) line segment between two adjacent slices intersecting the reconstructed surface does so at exactly one point or along exactly one line segment.
- **Criterion 3:** Resampling of the reconstructed surface on any slice reproduces the original contours.

Criterion 1 ensures that self-intersecting surfaces and other incorrect meshes are not formed. Criterion 3 ensures that all the contours are interpolated and no new ones are created. Criterion 2 is especially important because it ensures that the reconstructed surface is functional from its nearest planes. That is, barring the case of vertical surface patches, the projection of any triangle in the mesh to either of its two nearest Z_i planes is a one-to-one mapping. This prevents unlikely topologies from being formed and aids in the proof of the correctness of the multi-component method as discussed in Section 3.7. We note that Criterion 2 may not be satisfiable if the distance between consecutive z_i values is too large relative to the surface feature size, however, by Assumption 3, this is not the case. In practice, exceptions to Assumption 3 are rare and therefore Criterion 2 is not particularly restrictive on the input type.

We will now summarize the approach that the SINGLESURFRECON algorithm uses to resolve the tiling problem. We will appeal to intuitive notions of the *left side* and *right side* of a vertex on a contour relative to its clockwise (CW) or counterclockwise (CCW) orientation; these definitions are made explicit in [4]. Let $\{c_2^{k'}\}$ denote the projection of $\{c_2^k\}$ vertically onto Z_1. We compute the points of intersection between $\{c_1^j\}$ and $\{c_2^{k'}\}$; without loss of generality we assume that these points of intersection are vertices on contours in both sets. Such vertices that are common to both sets are called an *overlapping* vertices; all other vertices are called *non-overlapping*. If a tiling of $\{c_1^j\}$ and $\{c_2^k\}$ satisfies the three Criteria, the following two theorems must hold.

Theorem 1. *([4], Theorem 2) Let v be a vertex on a contour in $c_1 \in \{c_1^j\}$ and T a slice chord from v. (i) If v is a non-overlapping vertex, the projection of T onto Z_1 is contained entirely on one side (right or left) of c_1. The side is determined by the orientation of the nearest enclosing contour from the set $\{c_2^k\}$ when v is projected to Z_2. (ii) If v is an overlapping vertex, v also belongs to some $c_2' \in \{c_2^{k'}\}$ by hypothesis. In this case, the projection of T onto Z_1 does not intersect the region to the left of both c_1 and c_2' at v, nor the region to the right of both c_1 and c_2' at v.*

Theorem 2. *([4], Theorem 4) Let T be a slice chord and $c_1 \in \{c_1^j\}$. Then the projection of T onto Z_1 cannot intersect both the inside and outside of c_1*

The tiling algorithm proceeds as follows. For each vertex $v \in \{c_1^j\}$, make a list of all the slice chords that could be formed to a vertex of $\{c_2^k\}$ (based on the resolution of the correspondence problem). Select the shortest length chord from this list which satisfies the results of Theorems 1 and 2. If no chord from the list satisfies both Theorems, tag the vertex as "untiled." The boundaries of untiled regions are later collected and meshed separately while resolving the branching problem.

The *branching problem* arises when the result of the correspondence problem yields a matching of more than one contour in $\{c_1^j\}$ to any number of contours in $\{c_2^k\}$ (or vice versa). Solutions to the problem include adding edges to the contours on one of the planes or adding vertices in between the planes to create an appropriate mesh. Since the former approach violates Criterion 3, we employ the latter. The branching problem occurs only in regions previously tagged as "untiled" by SINGLESURFRECON, so we collect the boundaries of these regions and project them to a plane half way between the two planes in consideration. We triangulate these untiled regions based on their Edge Voronoi Diagram using the algorithm of Lee [29]. Any new vertices added are given a z value of $.5(z_1+z_2)$. A summary of alternative approaches to the branching problem and further details are given in [4].

The output of SINGLESURFRECON is a mesh with the desired Properties and satisfying the given Criteria. In practice, SINGLESURFRECON can be implemented in a numerically stable manner and will provide an output for many types of real data.

2.3 Medial Axis

The main tool employed by our algorithm to ensure accurate surface reconstruction is the medial axis of the region exterior to the contours. We give a thorough explanation of the medial axis in Appendix A. The medial axis of a set O is often approximated based on a point sampling P of the boundary ∂O and the Voronoi and Delaunay diagrams Vor P and Del P. A definition of these diagrams and

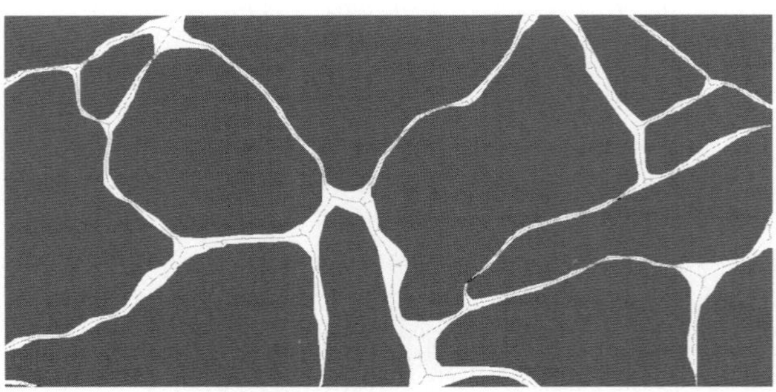

Fig. 5. A collection of contours representing cell boundaries and their interiors are shown in red along with the approximate medial axis of intracellular space $\mathfrak{M}_P$ shown in blue

how to compute them can be found in any standard computational geometry textbook, e.g. [19]. The medial axis is approximated by the graph of the vertices of Vor P along with those edges of Vor P not intersecting the contours. We denote this graph $\mathfrak{M}_P$; an example is shown in Figure 5. For 2D cellular contour data, $\mathfrak{M}_P$ will have one component for each cell interior plus a component for the extracellular region, assuming P is a sufficiently dense sampling of ∂O. We discuss sampling density and our robust medial axis computation method further in Section 3.2.

3 Algorithm

In this section we describe an algorithm to solve the multi-component problem under the Assumptions given in Section 2.1. In the next six subsections, we will explain the input to the algorithm and its five major steps. We conclude this section with a proof of the correctness of the approach.

3.1 Algorithm Input

As stated in Section 2.1, our input is the following objects: $(\{Z_i\}, \{g_i\}, \{c_i^j\}, G)$. By our stated Assumptions, we may suppose that the contours $\{c_i^j\}$ are simple polygons with sufficiently dense sampling of the level sets $g_i(x,y) = 0$. Define $P_i = \{v : \exists j \text{ such that } v \text{ is a vertex on } c_i^j\}$. We pass these polygonal contours and the associated point sets P_i to our algorithm instead of the functions g_i.

For ease of description and to maintain consistency with our neuronal data, we explain the algorithm for the case where G has just two components, with one colored yellow and one colored green. Each contour can thus be uniquely assigned as belonging to the set of yellow contours YC or green contours GC, i.e.

$$\{c_i^j\} = YC \coprod GC$$

The notation we will use for calling the algorithm with this input is

$$\textsc{MultiSurfRecon}\left(\{Z_i\}, \left\{c_i^j\right\}, \{P_i\}, G\right)$$

3.2 Step 1: Construct and Partition 2D Medial Axes

Overview

Compute the approximated medial axis $\mathfrak{M}^{P_i}$ for each $1 \le i \le M$ and discard the portions interior to the contours. Each remaining edge e separates two unique contours, call them c_{e1} and c_{e2}. Partition these edges into three sets:

$$E_{YY} = \{e : c_{e1}, c_{e2} \in YC\}, \quad E_{GG} = \{e : c_{e1}, c_{e2} \in GC\},$$

$$E_{GY} = \{e : c_{e1} \in YC, c_{e2} \in GC \text{ or } c_{e1} \in GC, c_{e2} \in YC\}$$

Details

It is crucial that the medial axes on each slice be computed robustly and with topological accuracy for the remainder of the algorithm to work. Thus, we turn to a criterion established by Edelsbrunner and Shah known as the *closed ball property* [20] which we state here for polygonal contours in a plane. Let P be the vertices of all the contours and Σ the union of all the contours on a particular plane. A Voronoi object V from Vor P of dimension k satisfies the closed ball property if and only if $V \cap \Sigma = \emptyset$ or $V \cap \Sigma$ is homeomorphic to a closed ball of dimension $k - 1$. If all $V \in$ Vor P satisfy the closed ball property and Σ intersects each V transversally, then Σ is a subgraph of Del P [20]. This implies that $\mathfrak{M}_P$ is a subgraph of Vor P. Cheng, Dey, Ramos, and Ray used this criterion to create Delaunay-conforming meshes without approximating local feature size [13]. We provided an efficient algorithm exploiting their approach for use with image data [25] and we adopt that algorithm for computation of $\mathfrak{M}_P$. Figure 5 shows a zoomed in picture of a slice of extracted contours with $\mathfrak{M}_P$ displayed.

By construction, $\mathfrak{M}_P$ is a multi-component piecewise linear graph. We discard all components of $\mathfrak{M}_P$ which are interior to the contours, leaving a single medial axis approximation of extracellular space. From this point on, we will take $\mathfrak{M}_P$ to mean this single connected graph. Each edge of $\mathfrak{M}_P$ is a Voronoi edge and thus its dual Delaunay edge connects two points of P. We assign each edge to E_{GG}, E_{YY}, or E_{GY} based on whether its associated points of P are both on green contours, yellow contours, or one of each, respectively.

3.3 Step 2: Construct Medial Surface

Overview

Dilate the edges in E_{GY} so that they form contours and color these contours red. Run SingleSurfRecon on the red contours, producing a partial, approximate medial surface $\mathfrak{M}S$.

Details

On each plane, we consider only those edges of $\mathfrak{M}_P$ which belong to E_{GY}. We dilate E_{GY} by setting a distance parameter ϵ and defining the contours to be the locus of points at distance ϵ from E_{GY}. For small enough values of ϵ, this produces a collection of contours to which we assign the color red. We denote the dilated contours RC. Once we have done this for all the slices, we call SingleSurfRecon on the whole set RC to produce a partial medial surface approximation denoted $\mathfrak{M}S$. In this computation, we also compute the intersection of $\mathfrak{M}S$ with each intermediate plane $z = .5(z_i + z_{i+1})$.

3.4 Step 3: Create Single Component Surfaces

Run SingleSurfRecon on YC and GC separately, producing meshes of the yellow and green contours. In this run, we also keep a list of those i values for which vertices were added on the intermediate plane $z = .5(z_i + z_{i+1})$. This occurs whenever there are dissimilar contours or branching in a portion of the tiling.

3.5 Step 4: Remove Overlaps on Intermediate Planes

Overview

For those i values on the list created in Step 3, compute the yellow, green, and red contours that occur on plane $z = .5(z_i + z_{i+1})$, which we denote $Z_{i,i+1}$. Do a plane sweep to detect overlaps between contours and remove them so that the mesh now avoids overlaps on each intermediate plane. Output the modified meshes of YC and GC.

Details

From Steps 2 and 3, we have calculated red, yellow, and green contours on each intermediate plane $Z_{i,i+1}$. Geometrical problems may arise on those $Z_{i,i+1}$ planes which have yellow and/or green vertices. For those planes, we do a plane sweep to detect overlap regions. We distinguish between two types of overlaps: simple transversal overlaps and exotic overlaps. A simple transversal overlap is one in which the border of the region is exactly two colors and each of these colors forms a connected component of the border. We resolve simple transversal overlaps by first computing the medial axis associated to the interior of the overlap. This is then used as a dividing line to guide vertex movement; the overlap is resolved by moving each vertex on the boundary across this axis. Note that this method does not add or remove vertices, does not change the mesh connectivity, and does not change the z value of any vertex.

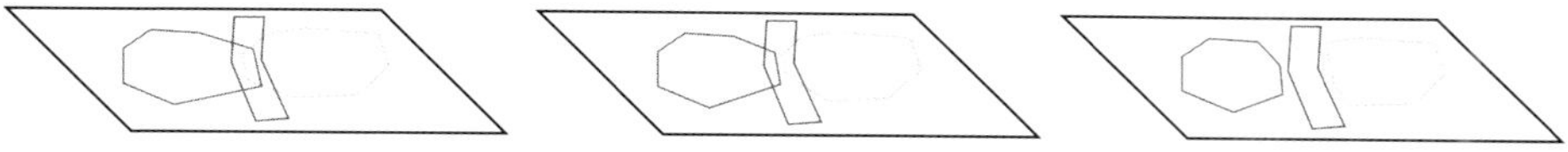

(a) (b) (c)

Fig. 6. **(a)** A few overlaps have occurred on an intermediate plane $Z_{i,i+1}$ between green contours, yellow contours, and red medial surface contours. We attempt to resolve them by searching for simple transversal overlaps such as the green/yellow overlap in this case. **(b)** After removing the green/yellow overlap, we are left with only red/yellow and red/green overlaps. **(c)** Removing the final overlaps, all the contours are separated.

An exotic overlap in a region Q can often be resolved by finding simple transversal overlaps within Q and resolving them in the manner described above. Such a case is shown in Figure 6. If this is not possible, we resolve the exotic overlap by changing the z values of vertices in the overlap region to either $.25(z_i + z_{i+1})$ or $.75(z_i + z_{i+1})$. An example of such a case and how it is resolved is shown in Figure 7.

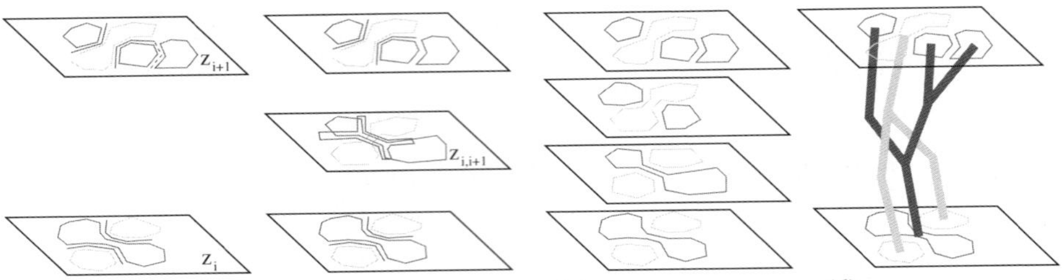

(a) (b) (c) (d)

Fig. 7. A toy example of how our algorithm works in the case of exotic overlaps (see Section 3.5). **(a)** Green and yellow contours are detected in adjacent planes Z_i, Z_{i+1} and the medial axis $\mathfrak{M}_P$ in each plane is computed, shown in red. Those portions of the medial axis which lie between same color contours are discarded, such as the dashed red portion in the upper plane. **(b)** An approximate medial surface $\mathfrak{M}S$ is computed by dilating the $\mathfrak{M}_P$ into contours and calling SINGLESURFRECON. The intersection of $\mathfrak{M}S$ and the separately computed green and yellow surfaces with the intermediate plane $Z_{i,i+1}$ is shown. The overlap of the contours cannot be resolved by removing simple transversal overlaps. **(c)** To remove the overlap, we move the green vertices on $Z_{i,i+1}$ to a lower z value and the yellow vertices to a higher z value, without changing connectivity of the mesh. We show the intersections of the yellow and green surfaces on the two intermediate planes where the vertices have been moved. **(d)** The skeletons of the yellow and green meshes are shown to illustrate that the overall topology is now correct and no geometrical overlaps occur.

3.6 Step 5: Quality Improvement

We decimate and improve the quality of our output meshes in the following manner. For decimation, we first do edge contractions based on the Delaunay diagram, using the QSlim software by Michael Garland [23]. We then do normal-based triangle decimation based on the method presented in [3]. Finally, we improve the shape of triangles in the mesh using a geometric flow technique [38] which is a library in Level Set Boundary Interior and Exterior Mesher (LBIE) [16], part of the Volume Rover (VolRover) [17] software developed by our lab.

3.7 Correctness of the Algorithm

The goal of MULTISURFRECON is to output an isotopic mesh M of the multi-component surface S representing neuronal membranes such that the Hausdorff distance between M and S is small. Since SINGLESURFRECON produces such meshes for individual components, the main question is whether the unioning and separating processes employed by MULTISURFRECON truly remove all intersections between the component surfaces in three dimensions. We will show that removing mesh intersections on certain horizontal planes suffices to preclude any 3D intersections in the complete mesh. To state this more precisely, we introduce the following definitions and lemmas.

Definition 2

- *An **original contour** is any contour c_i^j from Input 3 (described in Section 2.1). By Assumption 1, each c_i^j is a simple polygon.*
- *A **branching point** is a vertex added by* MultiSurfRecon *whose z-value is not one of the z_i values from Input 1.*
- *An **intermediate plane** is a horizontal plane which, after running* MultiSurfRecon*, contains at least one branching point.*
- *The **branching set** of a mesh component C on an intermediate plane P is the collection of branching points on P which belong to C, along with any edges of C between these points.*
- *An **intermediate contour** is a contour formed by intersecting M with an intermediate plane P. Note that an intermediate contour may contain some of the branching set of the component in which the contour lies.*
- *Let c_1, c_2 be contours in the same plane and $I(c_1), I(c_2)$ their respective interiors. Then c_1 and c_2 are said to **overlap** if and only if $(I(c_1) \cup c_1) \cap c_2 \neq \emptyset$ and $(I(c_2) \cup c_2) \cap c_1 \neq \emptyset$. Note that if one contour is completely contained in the other, they do not overlap.*

Lemma 1. *Suppose that for each i, there are no pairwise overlaps between the original contours on plane Z_i. Then the output of* SingleSurfRecon *run on any subset of the original contours is not self-intersecting.*

Lemma 1 is a consequence of the three Criteria laid out in Section 2.1 and the tiling method; if the original contours do not overlap, SingleSurfRecon cannot create self-intersections.

Next, observe that for the output of MultiSurfRecon with components YC and GC, there are three types of overlap that may occur on an intermediate plane P:

- An intermediate contour of YC intersects an intermediate contour of GC.
- An intermediate contour of YC intersects the branching set of GC (or vice versa).
- The branching sets of YC and GC intersect.

This characterization of types of overlaps extends naturally to outputs with more than two components. The following lemma shows that detecting and removing these types of overlaps is equivalent to removing self-intersections of the entire mesh. The lemma makes use of Assumption 2 from Section 2.1, without which, problems could arise from contours on consecutive planes being nested inside each other in contradictory ways. Since this should not happen with real data, we feel we are justified with the Assumption as stated.

Lemma 2. *If there are no overlaps of any of the three types on any intermediate plane, then the output of* MultiSurfRecon *has no self-intersections in the entire volume. Conversely, if the output has no self-intersections in the volume, it does not have any overlap on any intermediate plane.*

Proof. For the forward direction, consider the stack of all the original and intermediate planes in the output. By hypothesis, there are no overlaps and hence no surface intersections on any of these planes. Further, between any pair of consecutive planes there are no branching problems, as otherwise the planes would not be consecutive in the stack. Therefore, the surface between consecutive planes is a linear interpolation of the contours by Criterion 2 from Section 2.2. By Assumption 2 from Section 2.1, the true surface is a 2-manifold in this region meaning the linear interpolation is not self-intersecting. Therefore, the entire output is not self-intersecting. For the converse, if the output is not self-intersecting, the components do not intersect each other and therefore do not overlap on any intermediate plane. □

By Lemma 2, the removal of simple transversal and exotic overlaps done in Step 4 of the algorithm suffices to separate the components of the forest of branched surfaces in the entire volume. In future work, we plan to examine whether knowing the type of overlap on an intermediate plane can be used to simplify the procedure of Step 4.

4 Implementation and Results

We show examples of our final, quality improved meshes in Figures 2 and 8. The entire computational pipeline - imaging → contours → tiling → quality improvement - is handled by the Volume Rover (VolRover) [17] software developed by our lab, including a library for the Level Set Boundary Interior and Exterior Mesher (LBIE) [16].

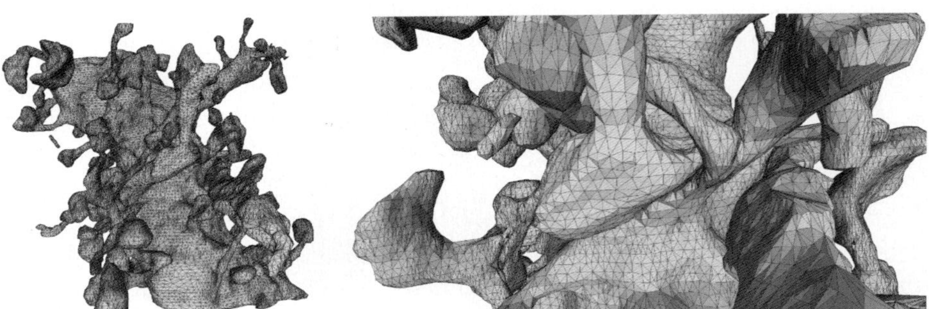

(a) (b)

Fig. 8. (a) A quality-improved mesh of three components - a dendrite shown in grey and two axons. **(b)** A zoomed in portion of the meshes.

5 Conclusion and Future Work

In this paper, we have presented a solution for creating isotopic meshes resolving the forest of branching structures problem and have demonstrated the feasibility of our approach. This is only the first step, however, in the simulation-based morphological studies we plan to explore. With topologically accurate and error-bounded meshes of neuronal forests, we can begin to quantify properties of these forests in a variety of manners, e.g. computing the packing density of the cells in the volume and measuring the size and dimension of various spines along the dendritic structures. These quantitative measurements can be used for comparison among different brain samples; studies have already found that certain neurological disorders correlate with atypical dendritic spine formations and densities [21]. Additionally, the meshes we create can be used to simulate voltage potentials traveling along a dendrite or axon. The models will be calibrated with real electrophysiological measurements and then used to study morphological dependence on neuronal potentials.

Acknowledgements. We would like to thank our collaborators at UT Austin, Drs. Kristen Harris, Daniel Johnston, and Clifton Rumsey from the Section of Neurobiology and the Center for Learning and Memory for their high resolution data and the numerous discussions on the dependence of neuronal morphology and plasticity on synaptic activity where this meshing effort would have utility. Thanks also to Drs. Justin Kinney, Thomas Bartol, and Terrence Sejnowski of the Salk Institute for their help and encouragement in producing quality meshing results. Finally, thanks to Jose Rivera for his assistance in generating many of the images in this paper. This research was supported in part by NSF grants 0636643, IIS-0325550, CNS-0540033 and NIH contracts P20-RR020647, R01-EB00487, R01-GM074258, R01-GM07308.

References

1. Armstrong, M.A.: Basic Topolgoy. Springer, New York (1983)
2. Attali, D., Boissonnat, J.-D., Edelsbrunner, H.: Stability and computation of medial axes: a state of the art report. In: Möller, B.H.T., Russell, B. (eds.) Mathematical Foundations of Scientific Visualization, Computer Graphics, and Massive Data Exploration. Mathematics and Visualization. Springer, Heidelberg (2007)
3. Bajaj, C., Xu, G., Holt, R., Netravali, A.: Hierarchical multiresolution reconstruction of shell surfaces. Comput. Aided Geom. Des. 19(2), 89–112 (2002)
4. Bajaj, C.L., Coyle, E.J., Lin, K.-N.: Arbitrary topology shape reconstruction from planar cross sections. Graph. Models Image Process. 58(6), 524–543 (1996)
5. Bajaj, C.L., Xu, G.: Regular algebraic curve segments (iii): applications in interactive design and data fitting. Comput. Aided Geom. Des. 18(3), 149–173 (2001)
6. Barequet, G., Goodrich, M.T., Levi-Steiner, A., Steiner, D.: Straight-skeleton based contour interpolation. In: SODA 2003: Proceedings of the fourteenth annual ACM-SIAM symposium on Discrete algorithms, Philadelphia, PA, USA, pp. 119–127. Society for Industrial and Applied Mathematics (2003)
7. Barequet, G., Sharir, M.: Piecewise-linear interpolation between polygonal slices. In: SCG 1994: Proceedings of the tenth annual symposium on Computational geometry, pp. 93–102. ACM, New York (1994)
8. Barequet, G., Vaxman, A.: Nonlinear interpolation between slices. In: SPM 2007: Proceedings of the 2007 ACM symposium on Solid and physical modeling, pp. 97–107. ACM, New York (2007)

9. Boissonnat, J.-D.: Shape reconstruction from planar cross sections. Comput. Vision Graph. Image Process. 44(1), 1–29 (1988)
10. Boissonnat, J.-D., Memari, P.: Shape reconstruction from unorganized cross-sections. In: SGP 2007: Proc. of the fifth Eurographics symposium on Geometry processing, pp. 89–98 (2007)
11. Bresler, Y., Fessler, J.A., Macovski, A.: A bayesian approach to reconstruction from incomplete projections of a multiple object 3d domain. IEEE Trans. Pattern Anal. Mach. Intell. 11(8), 840–858 (1989)
12. Chazal, F., Lieutier, A.: Stability and homotopy of a subset of the medial axis. In: Proc. 9th ACM Sympos. Solid Modeling and Applications, pp. 243–248 (2004)
13. Cheng, S.-W., Dey, T., Ramos, E., Ray, T.: Sampling and meshing a surface with guaranteed topology and geometry. In: SCG 2004: Proc. of the 20th Annual Symposium on Computational Geometry, pp. 280–289 (2004)
14. Cheng, S.-W., Dey, T.K.: Improved constructions of delaunay based contour surfaces. In: SMA 1999: Proceedings of the fifth ACM symposium on Solid modeling and applications, pp. 322–323. ACM, New York (1999)
15. Christiansen, H.N., Sederberg, T.W.: Conversion of complex contour line definitions into polygonal element mosaics. SIGGRAPH Comput. Graph. 12(3), 187–192 (1978)
16. CVC. LBIE: Level Set Boundary Interior and Exterior Mesher, `http://cvcweb.ices.utexas.edu/ccv/projects/project.php?proID=10`
17. CVC. Volume Rover, `http://ccvweb.csres.utexas.edu/ccv/projects/project.php?proID=9`
18. Dailey, M.E., Smith, S.J.: The Dynamics of Dendritic Structure in Developing Hippocampal Slices. J. Neurosci. 16(9), 2983–2994 (1996)
19. de Berg, M., van Kreveld, M., Overmars, M., Schwarzkopf, O.: Computational Geometry: Algorithms and Applications. Springer, Berlin (1997)
20. Edelsbrunner, H., Shah, N.: Triangulating topological spaces. Intl. Journal of Comput. Geom. and Appl. 7, 365–378 (1997)
21. Fiala, J., Spacek, J., Harris, K.M.: Dendritic spine pathology: Cause or consequence of neurological disorders? Brain Research Reviews 39, 29–54 (2002)
22. Fuchs, H., Kedem, Z.M., Uselton, S.P.: Optimal surface reconstruction from planar contours. Commun. ACM 20(10), 693–702 (1977)
23. Garland, M.: QSlim, `http://graphics.cs.uiuc.edu/~garland/software/qslim.html`
24. Geiger, B.: Three-dimensional modeling of human organs and its application to diagnosis and surgical planning. Technical report, INRIA (1993)
25. Goswami, S., Gillette, A., Bajaj, C.: Efficient delaunay mesh generation from sampled scalar functions. In: Proceedings of the 16th International Meshing Roundtable, pp. 495–511. Springer, Heidelberg (2007)
26. Harris, K., Perry, E., Bourne, J., Feinberg, M., Ostroff, L., Hurlburt, J.: Uniform serial sectioning for transmission electron microscopy. J Neurosci. 26(47), 12101–12103 (2006)
27. Herman, G.T., Zheng, J., Bucholtz, C.A.: Shape-based interpolation. IEEE Comput. Graph. Appl. 12(3), 69–79 (1992)
28. Klein, R., Schilling, A., Strasser, W.: Reconstruction and simplification of surfaces from contours. Graph. Models 62(6), 429–443 (2000)
29. Lee, D.T.: Medial axis transformation of a planar shape. IEEE Trans. Pattern Anal. Mach. Intell. PAMI-4(4), 363–369 (1982)
30. Lieutier, A.: Any open bounded subset of has the same homotopy type as its medial axis. Computer-Aided Design 36, 1029–1046 (2004)

31. Liu, L., Bajaj, C., Deasy, J.O., Low, D.A., Ju, T.: Surface reconstruction from non-parallel curve networks. Computer Graphics Forum 27(2), 155–163 (2008)
32. Meyers, D., Skinner, S., Sloan, K.: Surfaces from contours. ACM Trans. Graph. 11(3), 228–258 (1992)
33. Narayanan, R., Johnston, D.: Long-term potentiation in rat hippocampal neurons is accompanied by spatially widespread changes in intrinsic oscillatory dynamics and excitability. Neuron 56, 1061–1075 (2007)
34. Oliva, J.-M., Perrin, M., Coquillart, S.: 3d reconstruction of complex polyhedral shapes from contours using a simplified generalized voronoi diagram. Computer Graphics Forum 15(3), 397–408 (1996)
35. Park, M., Salgado, J.M., Ostroff, L., Helton, T.D., Robinson, C.G., Harris, K.M., Ehlers, M.D.: Plasticity-induced growth of dendritic spines by exocytic trafficking from recycling endosomes. Neuron 52(5), 817–830 (2006)
36. Sorra, K.E., Harris, K.M.: Overview on the structure, composition, function, development, and plasticity of hippocampal dendritic spines. Hippocampus 10(5), 501–511 (2000)
37. Turk, G., O'Brien, J.F.: Shape transformation using variational implicit functions. In: SIGGRAPH 2005: ACM SIGGRAPH 2005 Courses, p. 13. ACM, New York (2005)
38. Zhang, Y., Bajaj, C., Xu, G.: Surface smoothing and quality improvement of quadrilateral/hexahedral meshes with geometric flow. Communications in Numerical Methods in Engineering 24 (in press, 2008)

A Medial Axis Definition

There is some disagreement in the literature about the definition of the medial axis as its intuitive notion is not the most general or accurate formulation. We will give the most general definition of the medial axis as used in [30, 12]. Let O be an open subset of $\mathbb{R}^n$. The *medial axis* $\mathfrak{M}$ is defined to be the set of points $x \in O$ for which there are at least two closest points to x on the complement O^c. For the 2D images of cells we consider, O will denote the union of all inter- and intracellular regions. This makes O^c exactly the same as ∂O, the boundary of O, which is the collection of closed curves meant to represent the cellular boundaries. Note that authors frequently refer to the "medial axis of ∂O" to mean $\mathfrak{M}$ (or sometimes just a subset of $\mathfrak{M}$) so our definition encompasses commonly held notions. The *skeleton* S is defined to be the locus of centers of maximal inscribed balls in $\mathbb{R}^n \backslash \partial O$ where a *maximal ball* is an open ball in $\mathbb{R}^n \backslash \partial O$ which is not contained in any other open ball in $\mathbb{R}^n \backslash \partial O$. In two dimensions, $\mathfrak{M}$ and S are often nearly identical and so we will treat them as such; we refer readers to [2] for a careful comparison of the two concepts.

A New Set of Hexahedral Meshes Local Transformations

Frédéric Hecht[1], Raphaël Kuate[2], and Timothy Tautges[3]

[1] Laboratoire Jacques-Louis Lions, Université Pierre et Marie Curie
hecht@ann.jussieu.fr
[2] Laboratoire Jacques-Louis Lions, Université Pierre et Marie Curie
kuate@ann.jussieu.fr
[3] Sandia National Laboratories, Computational Mechanics and Visualization Department
tautges@mcs.anl.gov

Abstract. The modification of hexahedral meshes is difficult to perform since their structure does not allow easy local refinement or un-refinement such that the modification does not go through the boundary. In this paper we prove that the set of hex flipping transformations of Bern et. al. [1] is the only possible local modification on a geometrical hex mesh with less than 5 edges per vertex. We propose a new basis of local transformations that can generate an infinite number of transformations on hex meshes with less than 6 edges per vertex. Those results are a continuation of a previous work [9], on topological modification of hexahedral meshes. We prove that one necessary condition for filling the enclosed volume of a surface quad mesh with compatible hexes is that the number of vertices of that quad mesh with 3 edges should be no less than 8. For quad meshes, we show the equivalence between modifying locally the number of quads on a mesh and the number of its internal vertices.

1 Introduction

The using of hexahedral or quadrilateral meshes has its advantages. For example, the smaller number of elements obtained for a given number of vertices than tetrahedral or triangle meshes. But one difficulty within hexahedral meshes is to transform it locally without touching a large region of the mesh, due to its structure as set of layers. In [1] Bern et. al. propose the set of hex flipping transformations as local transformations of hex meshes i.e., the modifications does not change the boundary quads of the local domain where they are applied.

In the second section, we prove that modifying locally quad meshes is equivalent to modifying the number of its internal vertices. In the third section, we propose a necessary condition for meshing a volume with hexes from its boundary quad mesh. We study the problem of local transformations on hex meshes with less than 5 edges per vertex. That brings us to the next section where we look at the hex flipping transformations of Bern et. al. as the unique set of local transformations on that type of hex meshes. In the fifth section, we extend those local transformations to hex meshes with less than 6 edges per vertex and propose a new set of geometrical hex local transformations.

1.1 Definitions and Notations

A *polyhedron* is a set of closed planar polygonal faces in $\mathbb{R}^3$ that intersect only at shared vertices and edges, and such that each edge is shared by exactly two faces and each vertex is shared by exactly one cycle of faces.[1] A polyhedron divides $\mathbb{R}^3$ into a bounded *interior* and an unbounded *exterior*, each of which may contain more than one connected component.

A *quad surface mesh* is a polyhedron whose faces are all quads. A *hex mesh* is a quad surface mesh, along with interior quads that subdivide the polyhedron's interior into hexes. The intersection of any pair of hexes is either empty, a single vertex, a single edge, or a single quad. Quad surface meshes and hex meshes are both examples of *cubical complexes* [2, 8, 10], the analogue of simplicial complexes in which each k-dimensional face is a k-dimensional cuboid.

In the following, if not mentioned, we'll consider only *non singular* simply connected meshes (without holes): simply connected meshes where each element (quad or hex) has at least one common facet (edge for quad mesh or quad for hex mesh) with the rest of the mesh.

We call *elementary hex mesh* a conformal hex mesh without holes where every vertex has at the most 4 edges.
We call *interior* of a hex mesh the rest of the hex mesh obtained without its boundary vertices (boundary quads).
On a mesh, let us denote:

P the number of Hexes (polyhedrons), F the number of faces (quads),
$\bar{F}$ the number of boundary faces, $\mathring{F}$ the number of internal faces,
E the number of edges, $\bar{E}$ the number of boundary edges,
$\mathring{E}$ the number of internal edges, V the number of vertices,
$\mathring{V}$ the number of internal vertices, $\bar{V}$ the number of boundary vertices,
V_j^k the number of vertices with k boundary edges and j internal edges.

We'll use Euler's and Cauchy's formulas [7, 3]:

$$F - E + V - P =, 1 \tag{1}$$

for a complex of polyhedrons, and

$$F - E + V = 2, \tag{2}$$

for a single polyhedron.

2 Quad Mesh and Local Modifications

In hexahedral and quadrilateral mesh generation, the existence of a mesh need some conditions such as an even number of facets on its boundary [5]; we give a generalization of this property to any mesh whose elements have an even number of boundary facets even if they are of different type.

[1] This definition of polyhedron allows multiple connected components (two cubes can be one polyhedron!), as well as voids and tunnels model objects made from two different materials.

Theorem 1. *Any conformal mesh whose elements (of the same type or not) have each an even number of facets, has an even number of facets on its boundary.*

Proof. Let us consider a mesh with n elements.This is true for $n = 1$.
We suppose that this is the case for any mesh with n elements. For a mesh with $n + 1$ elements, we have to add one element to some previous n-element mesh. So the added element intersects the boundary of the previous mesh at m facets, $1 \leq m \leq k - 1$, if k is the number of facets of the added element. If b is the number of facets of the previous mesh, the boundary of the new mesh has $b - m + (k - m) = b + k - 2m$. Which is even.

It is known that quad meshes give a smaller number of elements than triangle meshes for a given set of vertices. Knowing the number of boundary and internal vertices, we have exactly the number of quads and edges in a quad mesh.

Proposition 1. *In a 2D simply-connected mesh with h holes and k vertices per element $(k \geq 3)$ the number F of elements, $\bar{V}$ of boundary vertices, $\mathring{V}$ of internal vertices and E of edges satisfies the relations:*

$$F = \frac{\bar{V} + 2(\mathring{V} + h - 1)}{k - 2}.$$

$$E = \frac{(k-1)\bar{V} + k(\mathring{V} + h - 1)}{k - 2}.$$

Proof. In a $2D$ mesh each internal edge $\mathring{E}$ belongs to two elements and each boundary edge $\bar{E}$ belongs to one element. So

$$2\mathring{E} + \bar{E} = kF.$$

As the boundary is a union of closed lines (only one if there are no holes), $\bar{E} = \bar{V}$. Thanks to the Euler formula [3],

$$F + V - E = 2 - h.$$

A quad mesh is solution of the following equations:

$$\begin{cases} V - E + F = 2 - h \\ 2\mathring{E} + \bar{V} = kF \end{cases} \tag{3}$$

whose solutions give the formulas.

For quad meshes, to locally transform a mesh (i.e., a sub-domain of the mesh) when keeping the boundary of that sub-domain unchanged, amounts to moving only its internal vertices and re-meshing the sub-domain.

Theorem 2. *The variation of the number of quads equal the variation of the number of internal vertices between two quad meshes ($mesh_a$ and $mesh_b$), if and only if the two meshes have the same number of boundary vertices:*

$$\mathring{V}_a - \mathring{V}_b = F_a - F_b \iff \bar{V}_a = \bar{V}_b.$$

In other words, adding or removing some elements in a quad mesh without changing its boundary is equivalent to adding or removing exactly the same number of its internal vertices.

Proof

• Let us suppose that $\bar{V}_a = \bar{V}_b$. As we have said, on a quad mesh each internal edge is shared by exactly two quads so we can write:

$$4F = 2\mathring{E} + \bar{E}. \tag{4}$$

Thanks to the first equation of (3) the difference between two meshes that have the same boundary satisfies

$$(F_a - F_b) + (V_a - V_b) - (E_a - E_b) = 0,$$

so

$$(F_a - F_b) + (\mathring{V}_a - \mathring{V}_b) - (\mathring{E}_a - \mathring{E}_b) = 0.$$

Thanks to equation (4),

$$\mathring{E}_a - \mathring{E}_b = 2(F_a - F_b),$$

so

$$(\mathring{V}_a - \mathring{V}_b) - (F_a - F_b) = 0.$$

• Let us suppose that $\mathring{V}_a - \mathring{V}_b = F_a - F_b$. Then

$$(F_a - F_b) + (V_a - V_b) - (E_a - E_b) = 0$$

gives

$$2(\mathring{V}_a - \mathring{V}_b) + (\bar{V}_a - \bar{V}_b) - (E_a - E_b) = 0.$$

Equation (4) gives $4(\mathring{V}_a - \mathring{V}_b) = 2(\mathring{E}_a - \mathring{E}_b) + (\bar{E}_a - \bar{E}_b)$, so

$$2(\bar{E}_a - \bar{E}_b) - 2(\bar{V}_a - \bar{V}_b) = \bar{E}_a - \bar{E}_b.$$

But on each quad mesh, $\bar{E} = \bar{V}$ so

$$\bar{E}_a = \bar{E}_b.$$

3 Hexahedral Meshing and Local Modifications

As a complex of polyhedrons, a hex mesh satisfies the equation (1). Its boundary is a polyhedron so it satisfies (2). On its boundary, each edge belongs exactly to two quads and the number of hex times the number of faces per hex equals the number of boundary faces plus two times the number of internal faces. Then any hes mesh is subject to the following system:

$$\begin{cases} V - E + F - P = 1 \\ \bar{V} - \bar{E} + \bar{F} = 2 \\ \bar{E} = 2\bar{F} \\ \bar{F} + 2\mathring{F} = 6P. \end{cases} \tag{5}$$

Proposition 2. *A simply-connected hex mesh satisfies the following equations:*

$$\bar{V} = \bar{F} + 2,$$

$$\sum_{j\geq 0,\ k\geq 3} (4-k)\, V_j^k = 8,$$

$$P = \frac{1}{2}\left(\mathring{E} - \mathring{V} - 1\right) + \frac{\bar{F}}{4}.$$

Proof

- The second equation of system (5) combined with the third gives

$$\bar{V} = \bar{F} + 2.$$

- The number of boundary edges satisfies $\sum_{j\geq 0,\ k\geq 3} kV_j^k = 2\bar{E} = 4\bar{F}$. As $\bar{F} = \bar{V} - 2$, we have

$$\sum_{j\geq 0,\ k\geq 3} (4-k)\, V_j^k = 8.$$

- The first minus the second equation of system (5) gives

$$\mathring{V} - \mathring{E} + \mathring{F} - P = -1,$$

so with $\mathring{F} = 3P - \frac{\bar{F}}{2}$, we obtain

$$P = \frac{1}{2}\left(\mathring{E} - \mathring{V} - 1\right) + \frac{\bar{F}}{4}.$$

Mitchell [11] gives some necessary conditions for meshing a domain with compatible hexes from an existing boundary quad mesh. This proposition gives another one.

Theorem 3. *The enclosed volume of a surface quad meshed with an even number of quads can be mesh with compatible hexes only if the number of vertices with 3 edges of that surface mesh is greater than or equal to 8.*

Proof. The second equation of the previous proposition gives

$$\sum_{j\geq 0} V_j^3 = 8 + \sum_{j\geq 0,\ k\geq 5} (k-4)\, V_j^k.$$

Since $k - 4 > 0\ \forall k \geq 5$, we have $\sum_{j\geq 0} V_j^3 \geq 8$.

3.1 Elementary Hex Meshes Local Modifications

Considering a hex mesh, we want to modify it locally such that in some region we modify a sub-mesh without changing the boundary. We also wish to investigate the changes of parity (i.e., a transformation between two configurations, one with an even number of hexes and the other with an odd number). As this is not an easy problem, we consider first a simple case.

We want to deal with the simplest (in terms of connectivity between vertices) sub-domain (sub-mesh) with more than one hex of any mesh. A mesh with only 3 edges per vertex is reduced to only one hex, so the simplest mesh should have some vertices with more than 3 edges. That is why we've define the *elementary hex mesh* and it implies some consequences.

Lemma 1. *The interior of an elementary hex mesh is a mesh (internal mesh) with at most one hex, one quad, one edge or one vertex.*

Proof. We suppose that the interior of an elementary hex mesh is a union of non connected (single) edges, vertices or faces. Then there is at least one boundary vertex sharing one edge with at least one single part of that interior, and one other edge with another boundary vertex or another single part of the interior of that elementary hex mesh. So that boundary vertex will have five edges which is not permitted. So, the interior of an elementary hex mesh has no single part; it is an *internal mesh* (1 dimension mesh, quad mesh or hex mesh).

If the interior of an elementary hex mesh is a mesh with more than one hex, there will be at least one boundary vertex of that internal mesh with four edges (a vertex shared by two hexes) plus the edge connecting that vertex to the boundary of that elementary hex mesh. This will give five edges, which is not admitted. If it has no hex but more than one quad, each face of those quads should belong to at least two different hexes (one per side). So each vertex of the intersection of two of those quads will have at least five edges. This is not admitted. If there is more than two aligned vertices, three for example, there will be at least two internal faces around the middle vertex which should be coplanar. So there is no way to have hexes above and below those faces by adding only one edge to that middle vertex.

As each internal vertex has 4 edges, we have

$$4(\mathring{V} + V_0^4 + V_1^3) + 3V_0^3 = 2E.$$

Each internal vertex belongs to exactly 4 hexes and to 6 faces, each boundary vertex with no internal edge belongs to two hexes and to five faces, each boundary vertex with one internal edge belongs to 3 hexes and to six faces and each boundary vertex with only 3 edges belongs to only one hex and to 3 faces. So we have

$$4\mathring{V} + V_0^3 + 2V_0^4 + 3V_1^3 = 8P,$$

$$6\mathring{V} + 3V_0^3 + 5V_0^4 + 6V_1^3 = 4F.$$

Euler's and Cauchy's formulas (1),(2) give:

$$F - E + V - P = 1, \ \bar{F} - \bar{E} + \bar{V} = 2.$$

Then

$$\mathring{F} - \mathring{E} + \mathring{V} - P = -1.$$

We can say that an elementary hex mesh is a solution of the following system

$$\begin{cases} 2\mathring{F} + \bar{F} = 6P \\ \mathring{F} - \mathring{E} + \mathring{V} - P = -1 \\ 4(\mathring{V} + V_0^4 + V_1^3) + 3V_0^3 = 2E \\ 4\mathring{V} + V_0^3 + 2V_0^4 + 3V_1^3 = 8P \\ 6\mathring{V} + 3V_0^3 + 5V_0^4 + 6V_1^3 = 4F \\ V_0^3 + V_0^4 + V_1^3 = 2 + \bar{F}. \end{cases} \tag{6}$$

It solution satisfies

$$\begin{cases} \mathring{F} = 3P - \frac{\bar{F}}{2} \\ \mathring{V} = 2P + 1 - \frac{\bar{F} + V_1^3}{2} \\ \mathring{E} = 4P + 2 - \frac{2\bar{F} + V_1^3}{2} \\ V_0^4 = \bar{F} - 6 \\ V_0^3 = 8 - V_1^3, \end{cases} \tag{7}$$

or

$$\begin{cases} \mathring{F} = \frac{6(\mathring{V} - 1) + \bar{F} + 3V_1^3}{4} \\ P = \frac{2(\mathring{V} - 1) + \bar{F} + V_1^3}{4} \\ \mathring{E} = 2\mathring{V} + \frac{V_1^3}{2} \\ V_0^4 = \bar{F} - 6 \\ V_0^3 = 8 - V_1^3. \end{cases} \tag{8}$$

Theorem 4

1. *No elementary hex mesh with more than 7 hexes can be transformed into another elementary hex mesh without changing its boundary quads.*
2. *The parity of an elementary hex mesh can't be changed without modifying its boundary quads.*

Proof. Let us suppose that we have an elementary hex mesh_a that we want to transform into another elementary hex mesh_b with the same boundary quads.

The number of transformations of any elementary hex mesh is the sum of the number of different ways to connect the boundary of that mesh to its internal mesh, each way times the number of possibles transformations of the interior of the internal mesh associated. This because if we modify the position of any boundary edge it will modify boundary quads. Thanks to the previous lemma, the interior of any internal mesh of an elementary hex mesh is empty (no vertex).

So the number of transformations is only the number of different possibilities to connect the boundary of the elementary hex mesh to each of its possible internal mesh. Let us examine how to modify the connection between the boundary and the interior of an elementary hex mesh.

Of course we have that $V_0^4 = \bar{F} - 6$ as part of the solution of the system (6), any transformation will keep the number of V_0^4 constant. The only possible modifications that can change the connection between the boundary and the interior, is to interchange V_1^3 and V_0^3. This means that any transformation will modify any boundary vertex of the first configuration (mesh_a) that has 3 edges to a boundary vertex of the second configuration (mesh_b) that has 3 boundary edges and one internal edge. If there is a boundary vertex that has 3 boundary edges not modified while the others are modified, the hexes that it belong to will not be transformed. So all the vertices that have 3 edges on the boundary will be modified by any transformation. Then there is at the most one possible transformation of the mesh_a into the mesh_b with the same boundary quads, which satisfies

$$\begin{cases} {V_0^3}_a = {V_1^3}_b \\ {V_1^3}_a = {V_0^3}_b \end{cases} \tag{9}$$

or

$$\begin{cases} {V_0^3}_a + {V_0^3}_b = 8 \\ {V_1^3}_a + {V_1^3}_b = 8 \end{cases}$$

Thanks to the previous lemma any internal mesh has at most one hex, so $\mathring{V} \leq 8$.

Remark 1

•${V_1^3}_a$ is always even because the number of edges going from the internal mesh (previous lemma) to the boundary is always even and when a boundary vertex of the mesh a that has one internal edge is not connected to an internal vertex, it is connected to the same type of boundary vertex.

• Thanks to the solutions (8) of the system (6), we have

$$P_a - P_b = \frac{\mathring{V}_a - \mathring{V}_b + {V_1^3}_a - 4}{2}.$$

This means that $\mathring{V}_a - \mathring{V}_b$ have the same parity because ${V_1^3}_a$ is always even. Since $\bar{F}$ is always even (by theorem 1), let us look at the possible values of $\mathring{V}$ in the solutions (7).

- If $\mathring{V}_a = 8$,
  then ${V_1^3}_a = 8$ (each internal vertex will have one boundary edge). ${V_0^3}_a = 0, \mathring{V}_b = 0, P_a = 7$; one internal hex + six around its faces; since those 7 hexes have already 4 edges per vertex there is no way to add another one. In this case, we have $P_a - P_b = 6$, so $P_a + P_b = 8$.
- If $\mathring{V}_a = 4$,
  those internal vertices form one face around which there are 6 hexes: one above, one bellow and four around the four edges. Since each internal vertex

has four edges we will have ${V_1^3}_a = 8$ and $V_3^1 = 0$. We have ${V_1^3}_b = 0$ so $\mathring{V}_b = 0$. As there is no way to add another hex to the first six hexes, $P_a = 6$; we have $P_a - P_b = 4$ so $P_a + P_b = 8$.

- If $\mathring{V}_a = 2$,
each internal vertex has four edges. We will have ${V_1^3}_a = 6$. Since $\mathring{V}_a$ and $\mathring{V}_b$ have the same parity, $\mathring{V}_b = 0$ or 2; the cases $\mathring{V} = 4$ and $\mathring{V} = 8$ have been examined above. If $\mathring{V}_b = 2$, then ${V_1^3}_b = 6$ which is absurd because of system (9) and the fact that ${V_1^3}_a = 6$. So $\mathring{V}_b = 0$ gives $P_a - P_b = 2$. For the same reasons as in the previous cases $P_a = 5$ (around 2 internal vertices there cannot be more or less than five hexes in an elementary hex mesh) then $P_a + P_b = 8$.
- If $\mathring{V}_a =, 1$
then $\mathring{V}_b = 1$ (parity of both internal vertices); ${V_1^3}_a = {V_1^3}_b = 4$, $P_a = 4$, $P_a - P_b = 0$. With one internal edge, there is no way to have more or less than four hexes in an elementary hex mesh; so $P_a + P_b = 8$.
- If $\mathring{V}_a = 0$,
We will examine only the case $\mathring{V}_b = 0$ because the other cases are contained in the previous ones. Let us examine the number of ${V_1^3}_a$:
  - If ${V_1^3}_a = 0$ we have $P_a - P_b = -2, {V_1^3}_b = 8$ and $\mathring{E}_b = 4$ (see (8)). So $P_b = 5$ (around four internal edges there cannot be more or less than five hexes in an elementary hex mesh), so $P_a + P_b = 8$.
  - If ${V_1^3}_a = 2$,
Then $P_a - P_b = -1$, ${V_1^3}_b = 6, \mathring{E}_a = 1$ and $\mathring{E}_b = 3$ (see (8)) so $P_a = 3$. With one internal edge, we have at least 3 internal faces and 3 hexes. Each hex has four V_0^4 (two with each of the other hexes) and the two V_1^3, then we can't have four V_0^3 on the same hex. This means that there is no way to add another hex (you can extend an elementary hex mesh by adding a hex to its boundary if and only if there is at least four V_0^3 on the same boundary's face). As $P_a - P_b = -1$ we have $P_b = 4$ which is absurd because $\mathring{E}_b = 3$ with $\mathring{V}_b = 0$.
  - If ${V_1^3}_a = 4$,
then $P_a - P_b = 0, \mathring{E}_a = 2$. There is no way to have $P_a \neq 4$ because of two internal edges. So once more $P_a + P_b = 8$.
  - The cases ${V_1^3}_a = 6$ or 8 correspond to ${V_1^3}_b = 2$ or 0. They were already examined for ${V_1^3}_a$.

Finally, in an elementary hex mesh transformation into another elementary hex mesh with the boundary unchanged,

$$P_a + P_b = 8.$$

4 Flipping Quad and Hex Meshes

For Hexahedral and quadrilateral meshes, Bern and Eppstein's operations described in [1] as *flipping* for Hexahedral meshes and which are the extension in 3D of some of the quad refinement operations of Canann, Muthukrishnan and Phillips [4] (figure 2), can be use to modify locally hexes inside a mesh (figure 1). One property suggested in [1] for 2D flipping is that together with the parity-changing operation, it should form a complete set. Here we show that those flipping operations for quad or hex meshes are not linked.

Theorem 5. *[9]*
Each set of flipping transformations [1] represented in figures 2 and 1 is topologically free; i.e., you can't go from one configuration to the other in a transformation using a combination of other transformations in a topological or geometrical mesh.

Proof. The dual theory of mesh transformations and its sheet diagrams projections (see figure 3), [12], [13],[1] are more useful in this context, but it is not easy to use them in this case for relationships between different flipping operations. Our way is more algebraic.

Considering the number of new internal edges without a boundary vertex n_e, internal vertices n_v, internal faces without any boundary edge n_f, hexes n_h in 3D and quads n_q in 2D created or removed within each transformation, we model them as follows:

$$
\begin{array}{lcl}
(0,0) \longleftrightarrow (0,0) & \Longrightarrow & \pm \ (0_h, 0_f, 0_e, 0_v) \\
(1,1) \longleftrightarrow (1,1) & \Longrightarrow & \pm \ (0_h, 0_f, 0_e, 0_v) \\
(1,0) \longleftrightarrow (0,1) & \Longrightarrow & \pm \ (2_h, 0_f, 1_e, 2_v) \\
(2,0) \longleftrightarrow (0,2) & \Longrightarrow & \pm \ (4_h, 1_f, 4_e, 4_v) \\
(3,0) \longleftrightarrow (0,3) & \Longrightarrow & \pm \ (6_h, 6_f, 12_e, 8_v) \\
(2,1) \longleftrightarrow (1,2) & \Longrightarrow & \pm \ (2_h, 0_f, 0_e, 0_v) \quad \text{in 3D,}
\end{array}
$$

$$
\begin{array}{lcl}
(0,0) \longleftrightarrow (0,0) & \Longrightarrow & \pm \ (0_q, 0_e, 0_v) \\
(1,1) \longleftrightarrow (1,1) & \Longrightarrow & \pm \ (0_q, 0_e, 0_v) \\
(1,0) \longleftrightarrow (0,1) & \Longrightarrow & \pm \ (2_q, 1_e, 2_v) \\
(2,0) \longleftrightarrow (0,2) & \Longrightarrow & \pm \ (4_q, 4_e, 4_v) \quad \text{in 2D.}
\end{array}
$$

So if each set of flipping operations is not free, we can write one of the right right-hand side of the previous relationships as a linear combination of some others with coefficients in $\mathbb{Z}$. Let us look at these operations except $(0,0) \longleftrightarrow (0,0)$ and $(1,1) \longleftrightarrow (1,1)$ because those change nothing numerically; we have to solve the following equations in $\mathbb{Z}$ for each of the unknowns fixed at 1 or -1 (for example $x = 1$ means that you are going from the $(1,0)$ to the $(0,1)$ configuration and $x = -1$ is for the $(0,1)$ to the $(1,0)$ configuration). We obtain

$$
\left\{
\begin{array}{c}
x + 2y + 3z + t = 0 \\
y + 6z = 0 \\
x + 4y + 12z = 0 \\
x + 2y + 4z = 0
\end{array}
\right.
\tag{10}
$$

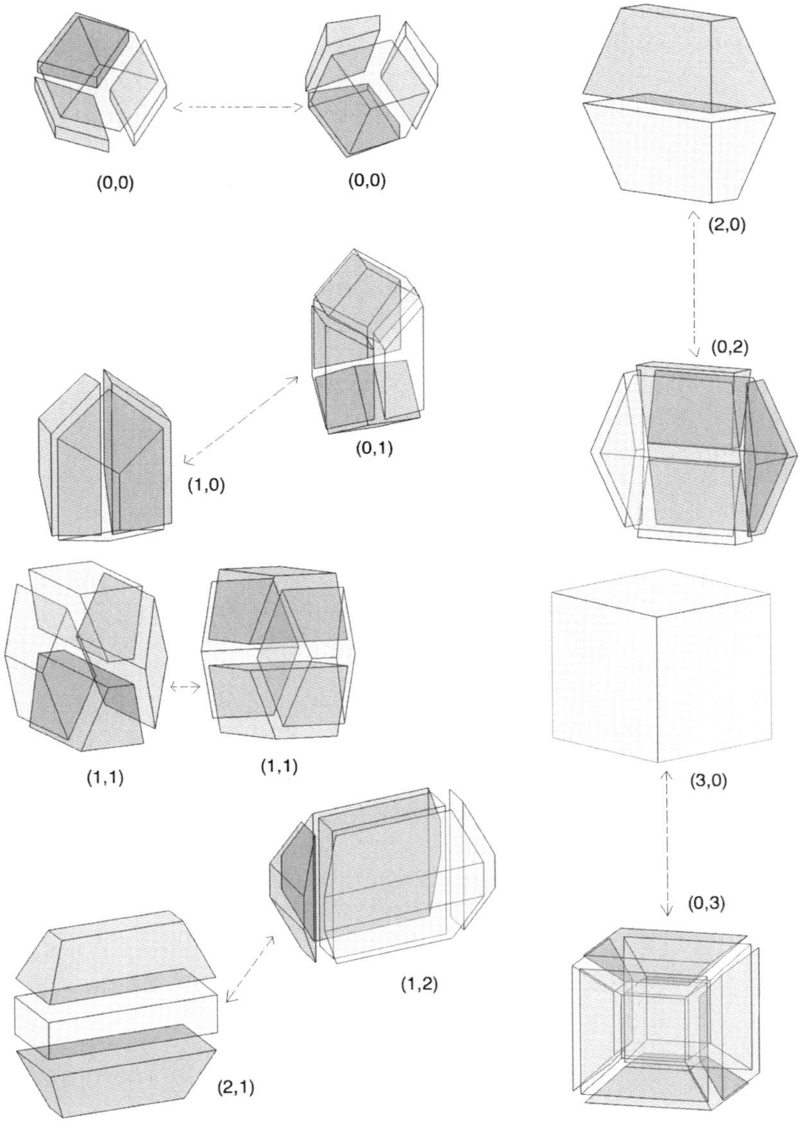

Fig. 1. The Flips for hex meshes

for the hexes flipping, and

$$\begin{cases} x + 2y = 0 \\ x + 4y = 0 \end{cases} \tag{11}$$

for the quads flipping.

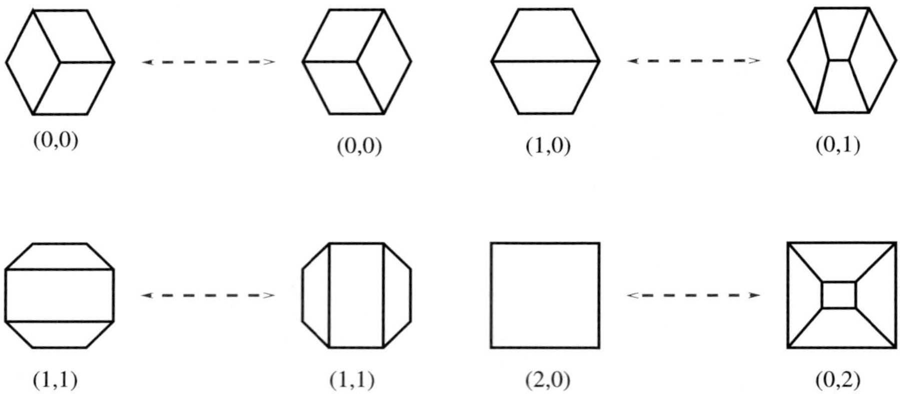

Fig. 2. The Flips for quad meshes

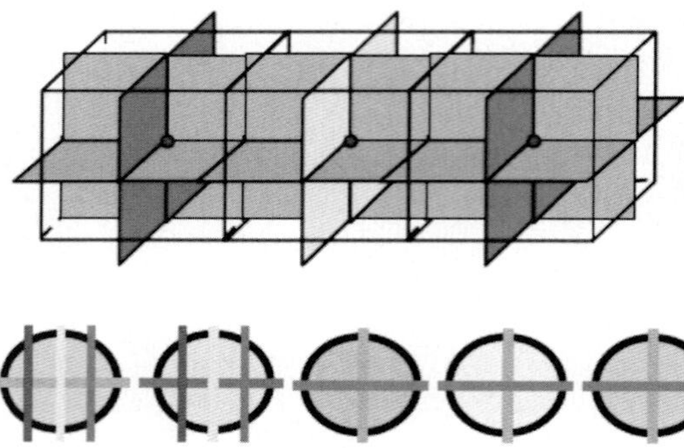

Fig. 3. Hex mesh with three elements (top); dual surfaces (sheets) and dual vertices shown. Two dimensional projection into "sheet diagrams" shown below.

For these two systems, $x = y = z = t = 0$ is the only solution. So there is no linear combination with those transformations. Then each set of operations without $(0,0) \longleftrightarrow (0,0)$ and $(1,1) \longleftrightarrow (1,1)$ is free. If we add the two other operation, it will not change the equations, nor their solutions. The possible solutions of those systems could have been considered as result of one of the two previous operations when combining all the others, so we can conclude that each set of flipping transformations is topologically free.

Theorem 6. *(existence of elementary mesh transformations)*
The set of hexahedral flipping operations [1] figure 1 form the set of transformations that keep unchanged the boundary of any mesh in the subset of elementary hex meshes.

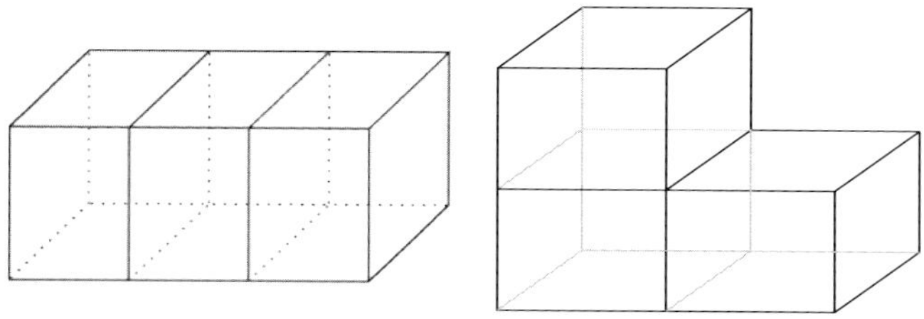

Fig. 4. some 3 hexes configurations

Proof. As the Theorem 4 says that any elementary mesh transformation needs at most 7 hexes and the sum of the hexes of the two configuration of the transformation is always 8, this is the case with all the operations of the set of hex flipping. Let us examine what happens for each number P of hexes between 1 and 7:

- If $P = 1$ we have only one possibility, $\bar{F} = 6$ and it gives the $(3,0) \longleftrightarrow (0,3)$ flipping transformation.
- If $P = 2$ we have only one possibility, $\bar{F} = 10$ and it gives the $(2,0) \longleftrightarrow (0,2)$ flipping transformation.
- If $P = 3$ we have two possibilities:
  1. $\bar{F} = 12$ and it gives the $(1,0) \longleftrightarrow (0,1)$ flipping transformation.
  2. $\bar{F} = 14$ for the left configuration of the previous figure, there is no possible transformation without having five edge on a boundary vertex. But numerically, the solution of system (6) gives
  $$P_b = 5, \quad \mathring{F}_b = 8, \ \mathring{V}_b = 0, \ \mathring{E}_b = 4$$
  which corresponds to the $(2,1) \longleftrightarrow (1,2)$ flipping transformation.
- If $P = 4$, we have $\bar{F} \leq 18$:
  1. If $\bar{F} = 18$, the four hexes should be aligned, the solution of system (6) tell us that the second configuration has a negative number of internal edges, so there is no possible transformation.
  2. If $\bar{F} = 16$, we will have $\mathring{V} = 1 - \frac{V_1^3}{2}$. It means that $\mathring{V} = 0$ or 2 any way $\mathring{V}$ will be negative in one of the two configurations, so there is no solution.
  3. If $\bar{F} = 14$, we obtain the $(1,1) \longleftrightarrow (1,1)$.
  4. If $\bar{F} = 12$, we obtain the $(0,0) \longleftrightarrow (0,0)$.
  5. If $\bar{F} \leq 10$, we will have $\mathring{F} \geq 7$. But a mesh with 4 hexes could not have more than 6 internal faces, so there is no solution.
- If $P = 5$ we have $\bar{F} \leq 22$:
  1. if $22 \leq \bar{F} \leq 16$, there is one configuration where $\mathring{V}$ is negative.
  2. For the case 14 and 12, it corresponds to the solutions obtained with $P = 3$.
  3. As a mesh with 5 hexes couldn't have less than 10 boundary faces, if $\bar{F} = 10$ the second configuration will have 3 hexes with 10 boundary faces which is not realizable.

- If $P = 6$, we have $\bar{F} \leq 26$:
  1. If $26 \leq \bar{F} \leq 12$ there is one configuration where $\mathring{V}$ is negative.
  2. For the case $\bar{F} = 10$, it corresponds to the solution obtained with $P = 2$.
  3. If $\bar{F} < 10$ the second configuration will have 2 hexes with less than 10 boundary faces which is not realizable.
- If $P = 7$, we have $\bar{F} \leq 30$:
  1. If $30 \leq \bar{F} \leq 8$, there is one configuration where $\mathring{V}$ is negative.
  2. For the case 6, it corresponds to the solution obtained with $P = 1$.

So apart from the set of hex flips, there is no other transformation of an elementary hex mesh into another elementary hex mesh that keeps the boundary unchanged.

5 Some New Hex Mesh Local Modifications

We have previously observed that there is no other local hex transformation with an elementary hex mesh. So we are trying here to see what happens on a general hex mesh. Let us call $M_{e/v}$-hex mesh a hex mesh where the maximal number of edges per vertex is M.

We tried to build local hex transformations for a $5_{e/v}$-hex mesh and obtained four basic transformations: See figure 5, where (x, y) denotes x the number of boundary faces and y the number of hexes. Those mesh files in medit [6] format, are available at `http://www.ann.jussieu.fr/~kuate/meshes.html`

These basic transformations can be extended by adding k layers of suitable hexes, $k \in \mathbb{N}$, between the group of hexes on each side of those transformations (figure 6).

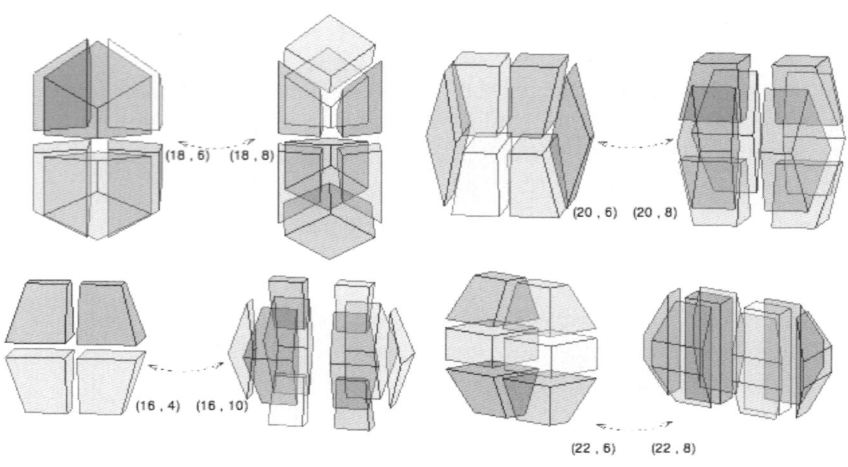

Fig. 5. Some basic $5_{e/v}$-hex meshes local transformations

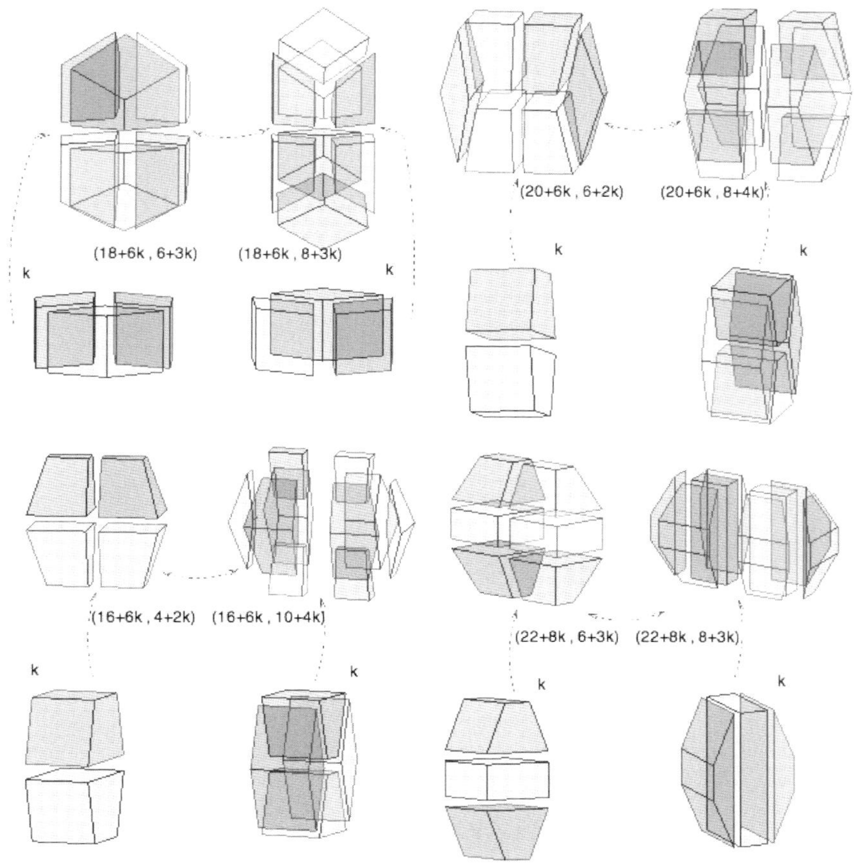

Fig. 6. Some $5_{e/v}$-hex meshes local transformations, $k \in \mathbb{N}$

Theorem 7. *For each fixed value of k, $k \in \mathbb{N}$, the set of four transformations obtained by adding k layers of suitable hexes on each configuration of the four basic $5_{e/v}$-hex meshes local transformations (figure 6), is topologically free.*

Proof. The same reasoning done in Theorem 5 gives the result.

6 Conclusion

We have described some properties of local transformations on quad meshes and essentially of the hex flipping transformations of Bern et. al. [1], from a geometrical and combinatorial point of view. This view allow us to propose a new set of basic local transformations for hexahedral meshes. We find that the sum of the number of hexes on both sides of each existing hexahedral local transformation equals $2 + 6k$, $k \in \mathbb{N}$, $k = 1$ corresponds to the set of Bern et. al. and $k \geq 2$ to ours. Then all those geometrical transformations cannot change the parity of the mesh. It would be interesting to prove that the previous relationship is the only one that is possible to have on any geometrical hex mesh with less than 6 edges per vertex. Therefore the problem of parity changing on a geometrical hex mesh in more complicated situations (some vertices should have more than 5 edges) may be considered.

We thank the reviewers for their comments which helped to improve the quality of this paper. We thank Paulo Amorim who helped us. We also thank the CEMRACS 2007 and its organizing committee P. Frey et. al. who where at the origin of this collaboration between Laboratoire Jacques-Louis Lions and Sandia National Laboratories, which produced this work.

References

1. Bern, M., Eppstein, D., Erickson, J.: Flipping cubical meshes. ACM Computer Science Archive (2002), `http://arXiv.org/abs/cs/0108020`
2. Blind, G., Blind, R.: The almost simple cubical polytopes. Discrete Math. 184, 25–48 (1998)
3. Boissonnat, J.-D., Yvinec, M.: Géométrie Algorithmique. Ediscience International (1995)
4. Canann, S.A., Muthukrishann, S.N., Phillips, R.K.: Topological refinement procedures for quadrilateral finite element meshes. Engineering with Computers 12, 168–177 (1998)
5. Eppstein, D.: Linear complexity hexahedral mesh generation. Department of Information and Computer Science University of california, Irvine, CA 92717 (1996)
6. Frey, P.: Medit, scientific visualization. Université Pierre et Marie, Laboratoire Jacques-Louis Lions, `http://www.ann.jussieu.fr/~frey/logiciels/medit.html`
7. Gergonne, J.: Géométrie de situation sur théorème d'euler relatif aux polyèdres. Annales de Mathématiques Pures et Appliquées, 333–338, tome 19 (1828-1829)
8. Joswig, M., Ziegler, G.M.: Neighborly cubical polytopes. Discrete & Computational Geometry 24(2–3), 325–344 (2000); arXiv:math.CO/9812033
9. Jurkova, K., Kuate, R., Ledoux, F., Tautges, T.J., Zorgati, H.: Local topological modification of hexahedral meshes using dual-based operations. CEMRACS 2007 (preprint, 2007)
10. Matveev, S., Polyak, M.: Cubic complexes and finite type invariants. arXiv:math.GT/0204085 (2002)
11. Mitchell, S.A.: A characterization of the quadrilateral meshes of a surface which admits a compatible hexaedral mesh of the enclosed volume. In: Puech, C., Reischuk, R. (eds.) STACS 1996. LNCS, vol. 1046. Springer, Heidelberg (1996), `http://www.cs.sandia.gov/~samitch/STACS-slides.pdf`
12. Müller-Hannemann, M.: Hexahedral mesh generation by successive dual cycle elimination. Techniche Universität Berlin (1998)
13. Tautges, T., Knoop, S.E.: Topology modification of hexahedral meshes using atomic dual-based operations. Sandia National Laboratories, Albuquerque, NM & University of Wisconsin-Madison, WI, USA (2002)

Mesh Matching – Creating Conforming Interfaces between Hexahedral Meshes

Matthew L. Staten[1,2], Jason F. Shepherd[1], and Kenji Shimada[2]

[1] Sandia National Laboratories*, P.O. Box 5800, Albuquerque, NM 87185
{mlstate,jfsheph}@sandia.gov
[2] Carnegie Mellon University, Pittsburgh, PA
{mstaten,shimada}@cmu.edu

Summary. This paper presents a new method for handling non-conforming hexahedral-to-hexahedral interfaces. One or both of the adjacent hexahedral meshes are locally modified to create a one-to-one mapping between between the mesh nodes and quadrilaterals at the interface allowing a conforming mesh to be created. In the finite element method, non-conforming interfaces are currently handled using constraint conditions such as gap-elements, tied contacts, or multi-point constraints. By creating a conforming mesh, the need for constraint conditions is eliminated resulting in a smoother, more precise numerical solution. The method presented in this paper uses hexahedral dual operations, including pillowing, sheet extraction, dicing and column collapse operations, to affect the local mesh modifications. In addition, an extension to pillowing, called sheet inflation, is introduced to handle the insertion of self-intersecting and self-touching sheets. The quality of the resultant conforming hexahedral mesh is high and the increase in number of elements is moderate.

1 Introduction

The finite element method is an indispensable part of the design through analysis process. Mesh generation is often a key bottleneck preventing broader use of the finite element method. The method utilized to handle interface conditions between assembly components can have a dramatic impact on the quality of the solution. Commonly, two spatially adjacent geometric volumes must behave as a single component, and under ideal conditions, a conforming mesh will be created between components. A conforming mesh ensures a smooth and accurate interpolation of the solution to the governing equations over the interface, and also improves solution efficiency by minimizing the number of equations that must be solved. Although conforming meshes are preferable, non-conforming meshes are regularly encountered for a variety of reasons including:

1. Different engineers created the mesh on the different components.
2. The meshing algorithm used on the model did not honor boundary meshing constraints [1, 2].

* Sandia is a multiprogram laboratory operated by Sandia Corporation, a Lockheed Martin Company, for the United States Department of Energy under Contract DE-AC04-94AL85000.

3. Difficulties in generating the mesh required a different mesh topology on the interfacing surfaces for the two components. For example, utilizing hexahedral sweeping [3, 4], one interface surface may be required to be a linking-surface requiring a mapped mesh, while the other interface surface may be a source surface allowing a paved mesh.
4. The desired density of elements is different in the two components.

The current state of the art is to artificially constrain the non-conforming meshes with multi-point constraints, tied contacts, or gap elements [5, 6, 7, 8] to maintain solution continuity across the interface. However, these methods typically result in solution quality degradation, disjoint solution fields, and/or adverse effects on solution convergence. Thus, these non-conforming interface conditions should only be used in non-critical regions of the model. A conforming interface is preferred, whenever possible.

In this paper, we present a new algorithm, entitled 'mesh matching', which converts non-conforming hexahedral-to-hexahedral interfaces into conforming interfaces. This new method locally modifies the topology of the hexahedral elements in one or both of the adjacent hexahedral meshes to create a one-to-one pairing of nodes and quadrilaterals on the interface surfaces so that the meshes can be merged into a conforming mesh across the interface. As with any mesh modification procedure, the quality of the modified elements may be reduced from the initial mesh quality; however, assuming the element quality remains above prescribed element quality thresholds, the benefits of having a conforming mesh may compensate for the reduction in element quality.

This paper is organized as follows: Section 2 reviews existing hexahedral mesh topology and modification theory used during mesh matching. In Section 3 a new mesh topology operator is defined. In Section 4, the mesh matching algorithm is presented. In Section 5, two examples of mesh matching are provided. Finally, in Section 6, we provide some concluding remarks along with some areas of current and future efforts.

2 Previous Research – The Hexahedral Mesh Dual

The dual of a hexahedral mesh [9, 10] is an alternate representation of the mesh composed of sheets and columns of hexahedral elements in the interior of the mesh, and chords and vertices on the boundary of the mesh.

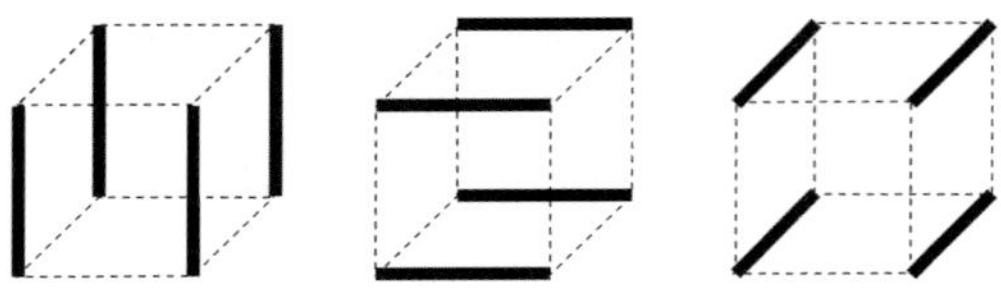

Fig. 1. A hexahedral element has 12 edges organized as 3 sets of 4 parallel edges

Dual Sheets and Columns

Figure 1 illustrates how the 12 edges on a hexahedral element can be divided into three sets of four edges. The four edges in each set are topologically parallel to each other (i.e. do not share any nodes, but have one or more common adjacent hexahedra). Given one edge, the other three topologically parallel edges in each adjacent hexahedra can be identified. From each of these edges a similar set of topologically parallel edges can be recursively gathered from each of the adjacent hexahedra extending through the mesh. Thus, a dual sheet, S_i, can be defined as a set of topologically parallel edges. Alternatively, S_i can also be defined as the set of hexahedral elements traversed to build this set of edges. Figure 2b shows a single dual sheet uniquely defined by traversing starting from edge A in Figure 2a. A dual sheet is self-intersecting if any hexahedron in the sheet has more than one of its three edge sets in the definition of the sheet (Figure 2c). A dual sheet is self-touching if two or more edges defining the sheet use the same mesh node (Figure 9d and Figure 10c).

A hexahedral element contains six quadrilateral faces, grouped into three pairs of topologically opposite quadrilaterals. From a single quadrilateral, a column of hexahedra is defined by traversing adjacent hexahedra through their topologically opposite quadrilaterals. Thus a dual column, C_i, is defined as the set of topologically opposite quadrilaterals of adjacent hexahedra. Alternatively, C_i is defined as the set of hexahedral elements traversed to locate this set of quadrilaterals. Figure 3b illustrates the dual column defined by quadrilateral face A specified in Figure 3a. An important link between sheets and columns is that a column defines the intersection of two sheets (Figure 3c). A column is self-intersecting if any hexahedron in the column has more than one of its quadrilateral pairs in the definition of the column (Figure 3d).

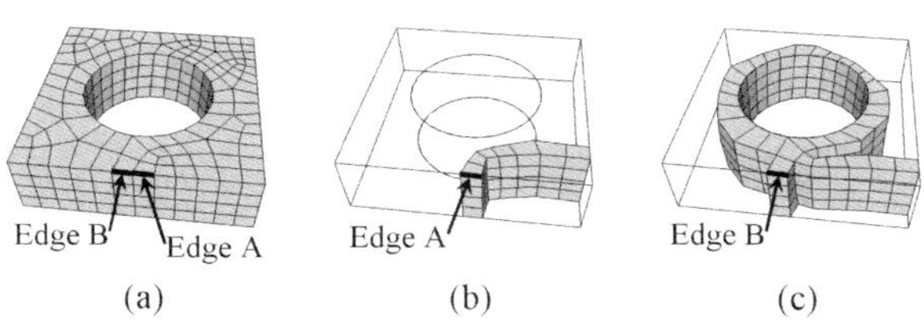

Fig. 2. Hexahedral Dual Sheets: (a) A simple mesh with two edges, A and B, identified. (b) The dual sheet uniquely identified by edge A. (c) A self-intersecting sheet identified by edge B.

Dual on the Boundary of a Hexahedral Mesh

The boundary of a mesh is the set of quadrilaterals which have exactly one adjacent hexahedron. These quadrilaterals can be grouped based on their associated geometric surface. Quadrilateral meshes have a dual representation of dual chords and vertices. The four edges on a quadrilateral are grouped into two

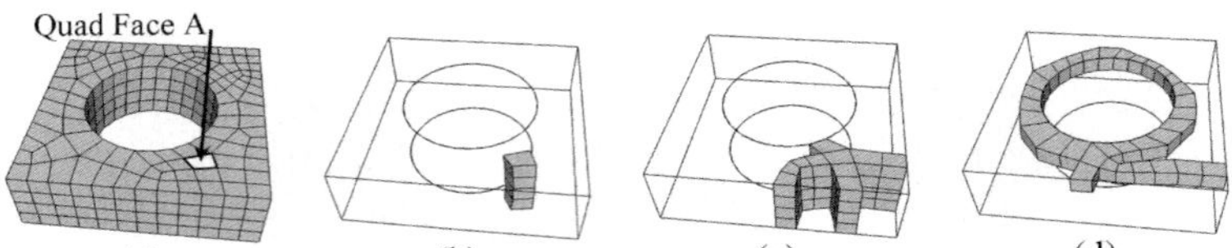

Fig. 3. Hexahedral Dual Columns: (a) Hex mesh with highlighted quadrilateral A. (b) The column of hexahedra defined by quadrilateral A. (c) The sheets which intersect to form the column in (b). (d) A self-intersecting column.

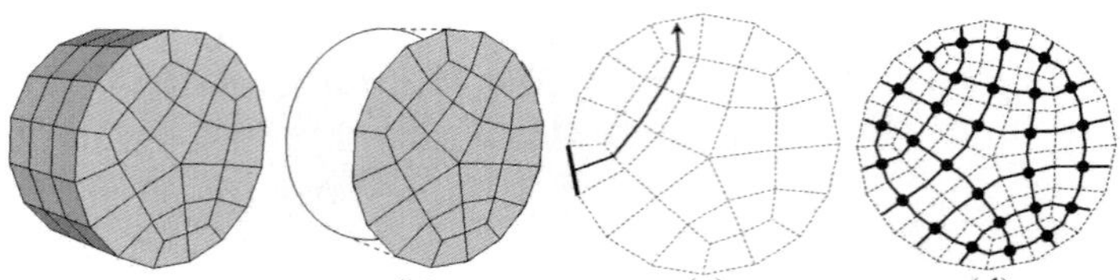

Fig. 4. Boundary quad meshes and their dual: (a) A hexahedral mesh. (b) One boundary surface mesh. (c) A single dual chord. (d) The complete surface dual.

pairs of topologically opposite edges. A dual chord is uniquely defined starting from a single edge and traversing adjacent quadrilaterals through opposite edges (Figure 4). This process is repeated until every edge in the quadrilateral mesh has been associated with a dual chord. Thus, a dual chord, c_i, can be defined as a set of the topologically opposite edges on a quadrilateral mesh. Alternatively, c_i can also be defined as the quadrilaterals that were traversed to build this set of edges. Finally, a dual chord c_i can also be defined as the collection of line segments connecting the centroids (dual vertices, v_i) of this set of quadrilaterals. A dual chord is self-intersecting if any quadrilateral in the chord has all four of its edges in the definition of the chord. Associated with each dual chord, c_i, is the dual sheet, S_i, defined by traversing topologically parallel edges from any edges in c_i. Likewise, associated with each dual vertex, v_i, is a dual column, C_i, defined by traversing topologically opposite quadrilaterals from the quadrilateral associated with v_i.

2.1 Dual Topological Operators

The matching procedure described in Section 4 performs a series of topological operations on hexahedral dual sheets and columns. Sheet extraction [11] removes a dual sheet by collapsing all edges that define it, reducing it to a continuous set of quadrilateral faces (Figure 5). Any sheet topology, including self-intersecting and self-touching sheets, can be extracted. Sheet extraction is not always possible due to geometric nodal associativity. That is, when collapsing the edges that define a sheet, the two nodes on each edge are merged. If two edge nodes have

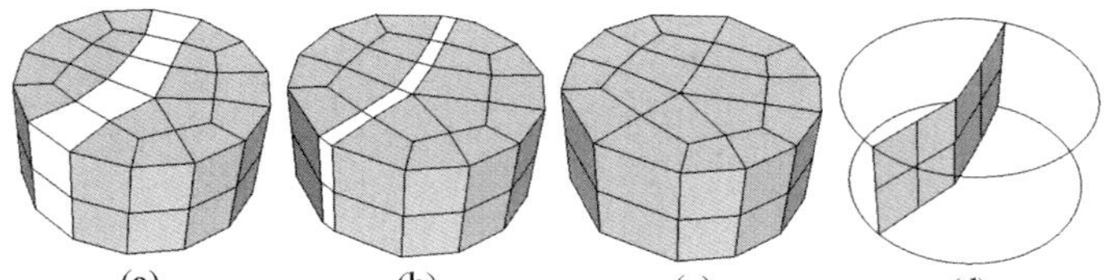

Fig. 5. Hexahedral sheet extraction: (a) A hexahedral mesh with one dual sheet. (b) The edges are collapsed to extract the sheet. (c) The sheet is extracted. (d) The sheet becomes a continuous set of quadrilaterals.

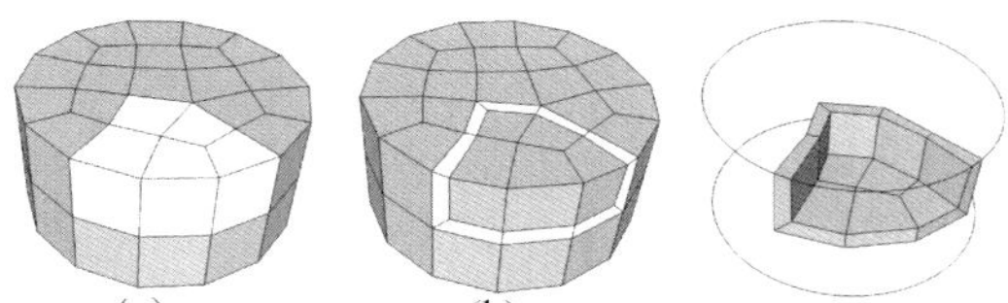

Fig. 6. Pillowing: (a) A hexahedral mesh with a shrink set of five hex elements identified. (b) A pillow is inserted. (c) The pillow sheet.

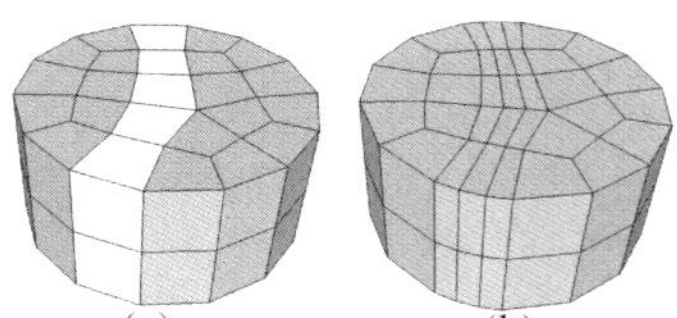

Fig. 7. Dicing: (a) A hexahedral mesh with one dual sheet indicated. (b) The sheet is diced into three sheets.

conflicting geometric associativities, the edge cannot be collapsed. However, the sheet can be extracted if preceded by a sheet insertion to add sufficient mesh topology. In addition, low node valency in the region of the sheet extraction can sometimes lead to doublets [12], resulting in ill-shaped elements with zero or negative scaled Jacobians [13].

Pillowing [10, 12, 14] is a method of inserting new dual sheets into a mesh. Pillowing is performed by identifying a set of hexahedral elements as the *shrink set* (Figure 6a). The hexahedra in the shrink set are separated from the remainder of the mesh by a 'shrink' distance, allowing the placement of a new sheet between the shrink set and the other hexahedra in the mesh (Figures 6b and 6c). The new sheet is always non-self-intersecting and non-self-touching. In contrast to sheet extraction, pillowing is always possible given a well-defined shrink set. Since its introduction by Mitchell [12], it has been applied in many mesh modification procedures [10, 15, 16, 17, 18].

Dicing [19] is another method of inserting dual sheets into a mesh. Dicing is performed by splitting the edges that define an existing dual sheet. Dicing can insert multiple sheets at once by splitting each edge multiple times (Figure 7). The new sheets inserted with dicing are duplicates of the input sheet; if the input sheet is self-intersecting, the new sheets will also be self-intersecting. The disadvantage of dicing is that it can only copy existing sheets (i.e., it cannot create new sheets that did not already exist in the mesh). In addition, dicing cannot create self-touching sheets.

The column collapse operation is also an important dual operation (Figure 8). A column is collapsed by merging one pair of opposite nodes of each quadrilateral defining the column. As described in Section 2, a dual column defines the intersection of two dual sheets. This intersection is removed by collapsing the column. In addition, the paths of the two sheets is altered. Collapsing self-intersecting columns creates doublets [12] and should be avoided.

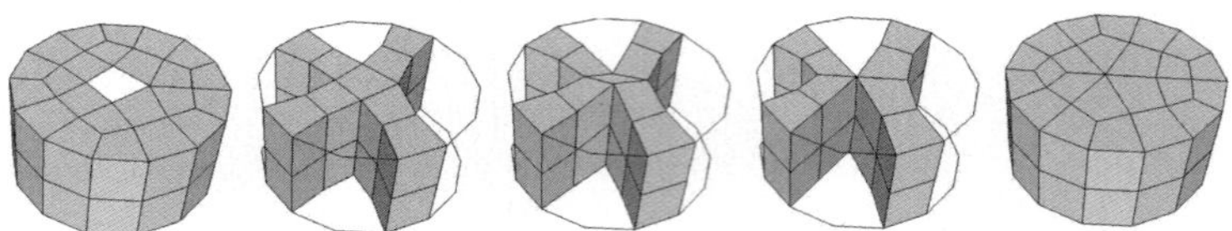

(a) (b) (c) (d) (e)

Fig. 8. Column collapse: (a) A hexahedral mesh with one dual column indicated. (b) The sheets that intersect to define the column drawn separate from the mesh. (c) Opposite nodes are merged to collapse the column. (d) The sheets defining the column no longer intersect. (e) The entire mesh after the column collapse.

3 Sheet Inflation – Generalized Sheet Insertion

The mesh matching algorithm presented in Section 4 requires the ability to insert any kind of sheet including both self-intersecting and self-touching sheets. Previous research allows the insertion of sheets through pillowing and dicing. However, neither pillowing nor dicing can insert self-touching sheets, and pillowing is unable to insert self-intersecting sheets. Dicing can insert self-intersecting sheets, but only if an existing self-intersecting sheet exists in the correct location of the mesh. Hence, a sheet insertion operator which inserts both self-intersecting and self-touching sheets is required.

Sheet inflation can be thought of as the reverse of sheet extraction. In sheet extraction, a dual sheet is reduced to a continuous set of quadrilaterals. This process can be reversed by *inflating* the quadrilaterals to re-introduce the extracted sheet. Knupp et al. [14] introduced a similar operator with the inflate hex ring which inflates a set of quadrilaterals into a new dual column. For sheet inflation, the boundary of the quadrilateral set must lie on the boundary of the hex mesh. Self-intersecting and self-touching sheets are inserted by inflating quadrilateral sets with non-manifold edges. A set of non-manifold edges with four adjacent quadrilaterals (i.e. 4NMEsets) can be inflated as either a self-intersecting (Figure 9c)

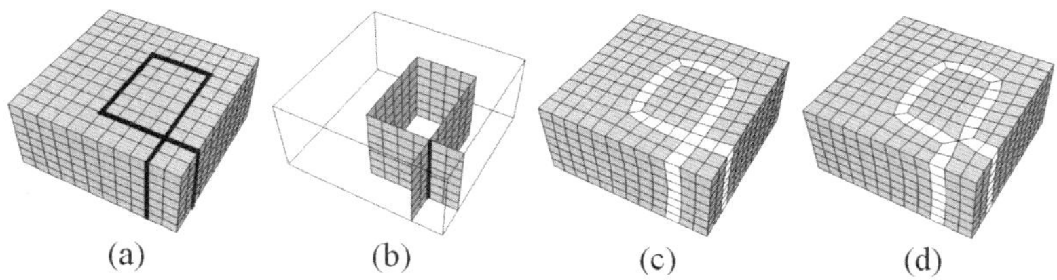

Fig. 9. Sheet inflation example #1: (a) A simple hex mesh with boundary edges indicated. (b) A non-manifold continuous set of quadrilaterals bounded by the indicated boundary edges in (a). The highlighted edges form a 4NMEset. (c) A self-intersecting sheet inflation option. (d) A self-touching sheet inflation option.

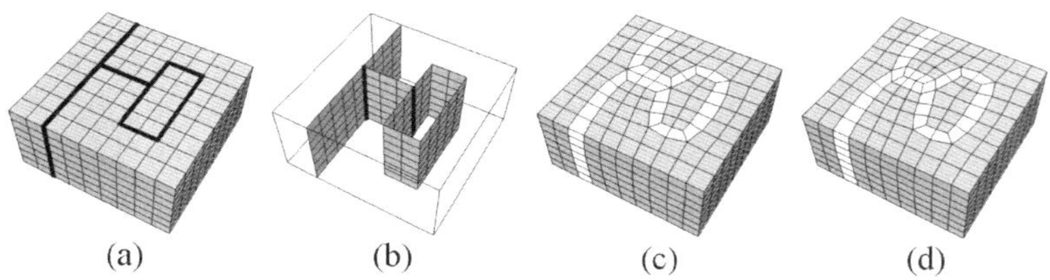

Fig. 10. Sheet inflation example #2: (a) A simple hex mesh with boundary edges indicated. (b) A non-manifold continuous set of quadrilaterals bounded by the indicated boundary edges in (a). The highlighted edges form two 3NMEsets. (c) A self-touching sheet inflation option. (d) A self-touching and self-intersecting sheet inflation option.

or a self-touching sheet (9d). A set of non-manifold edges with three adjacent adjacent quadrilaterals (i.e. 3NMEsets) can be inflated as either self-touching (Figure 10c), or self-touching and self-intersecting (Figure 10d). Thus each non-manifold edge set can be inflated in two different ways. The input to sheet inflation requires each non-manifold edge set have a flag indicating which option should be performed. 3NMEsets must appear in the quadrilateral set in pairs, or be paired with a boundary.

For manifold sets of quadrilaterals, sheet inflation is the same as pillowing the hexahedra on one side of the quadrilaterals set. Thus, sheet inflation can be implemented in a manner similar to pillowing, with the following three differences caused by the non-manifold edge sets:

1. Multiple shrink sets are required, partitioned from each other by the non-manifold edge sets.
2. Nodes along non-manifold edge sets must be duplicated either twice for self-touching sheets, or three times for self-intersecting sheets.
3. Quadrilaterals which lie between two 3NMEsets must be duplicated twice, and the resulting gap is filled with two hexahedra instead of one.

4 Hexahedral Mesh Matching

4.1 Mesh Matching Input Requirements

The input requirements of the mesh matching algorithm are:

1. Two geometric surfaces, A and B, that are:
   a) Topologically identical (The number of boundary curves, loops, and vertices defining the two surfaces must be the same),
   b) Geometrically similar (each boundary curve/vertex on Surface A must have a corresponding boundary curve/vertex on Surface B that is within a tolerance, β), and
   c) Both adjacent to hexahedral mesh elements.
2. An integer value for a *depth* parameter indicating how many layers into the adjacent hexahedral meshes the modifications can propagate.
3. A flag indicating which (or both) of the surfaces can have their mesh topology modified. The simplest case is that both meshes can be modified, but all changes can be done on one side of the interface if necessary.

If input requirement 1a or 1b are not met by the initial hexahedral meshes, the Graft Tool [18] can be used to imprint the boundaries of the interface surfaces onto each other.

4.2 Mesh Matching Procedure

Figure 11a illustrates a two-volume model positioned such that Surface A on Volume A overlaps exactly with Surface B on Volume B meeting the input requirements in Section 4.1. However, as seen in Figures 11b and 11c, the quadrilateral meshes on Surfaces A and B do not match. In this case, the non-conforming mesh was created because the topology of Volume B requires Surface B to be a linking surface for sweeping, while the topology of Volume A requires Surface A to be a source surface for sweeping. The resulting mesh is non-conforming as shown in Figure 11d and 11e. The objective is to modify the mesh topology of one or both of the adjacent hexahedral meshes such that the quadrilateral meshes on Surfaces A and B match, node-for-node and quad-for-quad. The node pairs can then be merged resulting in a conforming mesh across the interface.

We make the following assertion:

Assertion 1: *Given two geometric surfaces, A and B, which meet the input requirements stated in Section 4.1, the topology of the two adjacent hexahedral meshes can always be modified to create identical quadrilateral mesh topology on both Surfaces A and B, which can then be merged to create a conforming hexahedral mesh across the interface.*

Rationale: Two quadrilateral meshes will be topologically identical *iff* the duals of the two quadrilateral meshes are identical. Let Ω_{cA} and Ω_{cB} be the sets of chords, c_i, in the quadrilateral meshes on Surfaces A and B respectively. Initially $\Omega_{cA} \neq \Omega_{cB}$. However, through sheet insertion and extraction, dual chords can

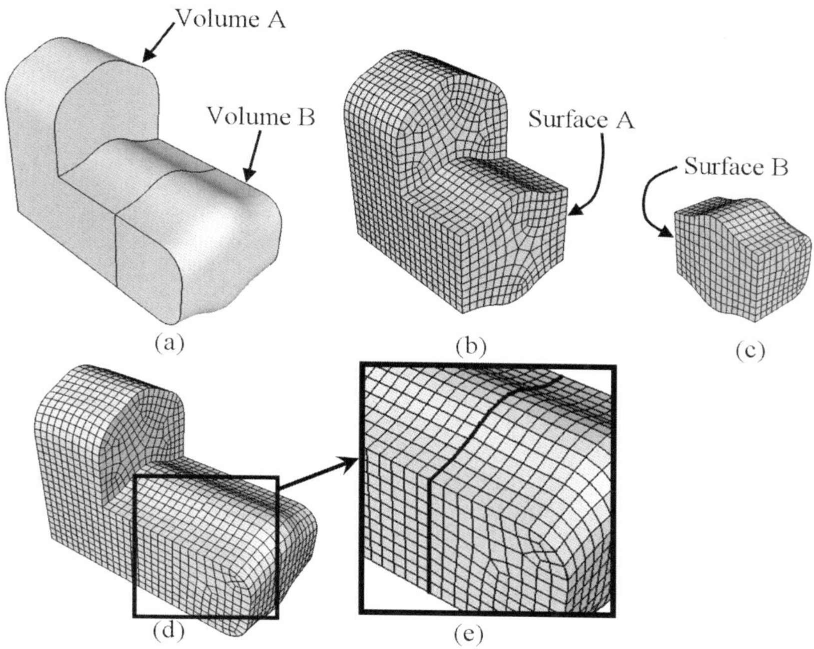

Fig. 11. Non-Conforming mesh example: (a) Geometric model composed to two volumes, A and B. (b) The mesh on Volume A. (c) The mesh on Volume B. (d) The complete mesh showing the discontinuity in the mesh. (e) Zoom-in of mesh discontinuity.

be inserted and extracted from boundary quadrilateral meshes. Thus, one or both of Ω_{cA} and Ω_{cB} can be modified such that they do match. The algorithm of mesh matching is then:

1. For each $c_i \in \Omega_{cA}$:
   a) Search for a $c_j \in \Omega_{cB}$ such that $c_j = c_i$ within a tolerance, δ. If an equal c_j is found, insert c_i and c_j into $\Omega_{c-pairs}$, else, insert c_i into $\Omega_{cA-unmatched}$.
2. For each $c_j \in \Omega_{cB}$:
   a) If $c_j \notin \Omega_{c-pairs}$ insert c_j into $\Omega_{cB-unmatched}$.
3. For each $c_i \in \Omega_{cA-unmatched}$:
   a) Use the following rules to determine if c_i should be inserted into Ω_{cB} or extracted from Ω_{cA}:
      i. No c_i should be extracted if doing so violates geometric associativity (see Section 2.1).
      ii. No c_i should be extracted if doing so creates doublet topology.
      iii. Sheet insertions should be done if mesh density has already been decreased bordering upon acceptability thresholds.
      iv. Sheet extractions should be done if mesh density has already been increased bordering upon acceptability thresholds.

v. If insertion is done:
   A. Project c_i onto Surface B, and find a continuous set of mesh edges, E_B, on Surface B which approximate this projection.
   B. Use dicing if $\exists\ c_j \in \Omega_{cB}$ such that $c_j = E_B$ with a tolerance, γ, where $\gamma >> \delta$.
   C. Use pillowing if a well-connected shrink set can be defined behind the quadrilaterals on one side of E_B.
   D. Otherwise, use sheet inflation on a continuous quadrilateral set which contains E_B on its boundary.

vi. If extraction is done, consider doing a column collapse (see example 1 in Section 5.1) in order to avoid global changes.

b) If c_i is to be extracted, remove c_i from Ω_{cA} and $\Omega_{cA-unmatched}$

c) If c_i is to be inserted, remove c_i from $\Omega_{cA-unmatched}$, and insert it along with the newly inserted chord into $\Omega_{c-pairs}$.

4. Repeat Step 3 for each $c_j \in \Omega_{cB-unmatched}$
5. Smooth all nodes local to the interface surface modifications to improve element quality [21].

The 'Chord Equals' Operator

In order to perform Step 1a and Step 3avB, $c_i = c_j$ must be defined:

Definition 1. *$c_i = c_j$, within a tolerance, δ, iff the maximum distance between c_i and all v_k in c_j is less than δ AND the maximum distance between c_j and all v_k in c_i is also less than δ.*

$c_i = c_j$ does not require c_i and c_j to have the same number of dual vertices. Rather, $c_i = c_j$ if the two chords are spatially close to each other. For example, in Figure 12 the two indicated chords have 16 and 15 chord vertices. However, when the two surfaces are overlaid, the maximum distance between the two chords is less than δ. Thus, these two chords are equal.

In short, mesh matching implements Shepherd's [20] 2nd assertion which states: *There exists a transformation that converts one set of fundamental sheets*

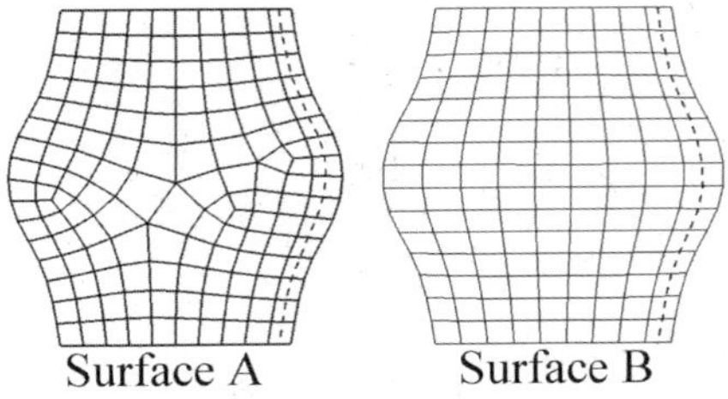

Fig. 12. The mesh topology of example Surfaces A and B. Although the indicated chords have different number of dual vertices, they are still spatially within δ when overlaid.

into an alternative set of fundamental sheets. This assertion can be generalized as follows:

Assertion 2: *There exists a transformation that converts one hexahedral mesh into any other hexahedral mesh on a given geometry.*

Rationale: Any sheet can be extracted from a mesh. If geometric associativity would be violated by sheet extraction, the sheet can be extracted after a sheet insertion to adequately add the appropriate mesh topology. In addition, any sheet can be inserted, including self-touching and self-intersecting sheets. Thus any sheet not matching the goal topology can be extracted, and any missing sheet can be inserted. In the case of mesh matching, the hexahedral mesh to convert to is a mesh which matches across the interface.

5 Examples

5.1 Simple Example

We now illustrate mesh matching on the simple example from Figure 11. Figure 13a identifies a chord, c_i, in Surface A, which has no pair in Surface B. In Figure 13b, a string of edges, E_B, on Surface B is identified which roughly matches the projection of c_i. E_B partitions the surface quadrilaterals into two sets, of which, one is chosen (normally the smaller set). A pillow shrink set is then defined as the hexahedral elements behind the chosen quadrilateral set. The input *depth* parameter is used to determine how far into the volume to propagate to build the shrink set. Figures 13c and 13d show the mesh after the pillow is inserted using depth=2, followed by appropriate smoothing [21]. The resulting new chord in Surface B is then paired with the identified unpaired chord in Figure 13a. Figure 14 repeats this process for another unpaired chord in Surface A.

Figure 15 illustrates the use of dicing to introduce topology required for mesh matching. In Figure 15a, three topologically parallel chords are indicated. One of these three chords is paired with the chord indicated in Figure 15b, which is diced, followed by smoothing, introducing the required topology to match all three chords. After these three operations, the topology on the left side of Surface B is beginning to match the topology on the left side of Surface A. Additional sheets are inserted and extracted until the mesh topology on these two surfaces matches.

Sheet Extraction for Mesh Matching

Although all required topology can be introduced with sheet insertion, doing so will have the potentially undesirable side-effect of increasing the density of the mesh local to the interface surfaces. Sheet extraction is useful in reducing or eliminating the increase in mesh density. For example, Figure 16a shows the mesh topology on Surface A with one unpaired chord indicated. Figure 16b shows the mesh topology of Surface B; clearly the chord indicated in Figure 16a has

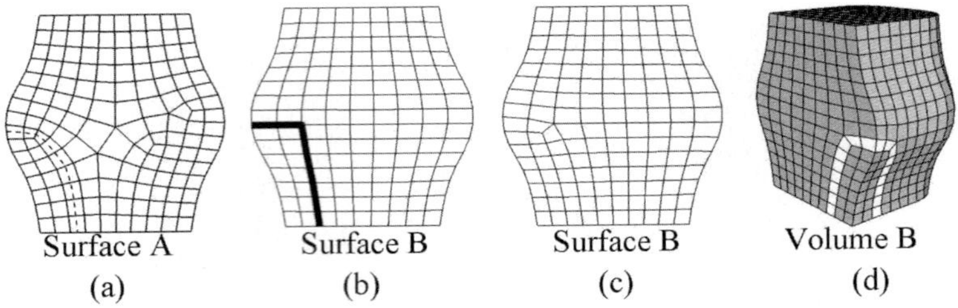

Fig. 13. Pillowing during mesh matching: (a) The mesh topology on Surface A with one non-paired chord, c_i, indicated. (b) The mesh topology on Surface B with a string of mesh edges, E_B, indicating where pillowing will be performed. (c) Surface B after pillowing is performed. (d) Volume B after pillowing is performed.

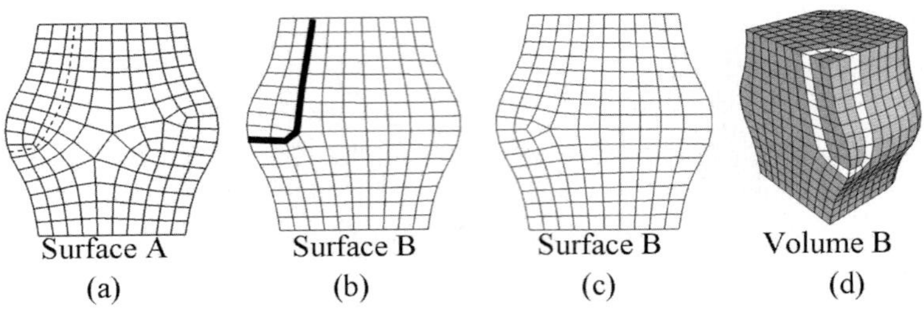

Fig. 14. Pillowing during mesh matching: (a) The mesh topology on Surface A with one non-paired chord, c_i, indicated. (b) The mesh topology on Surface B with a string of mesh edges, E_B, indicating where the 2nd pillow will be inserted. (c) Surface B after pillowing is performed. (d) Volume B after pillowing is performed.

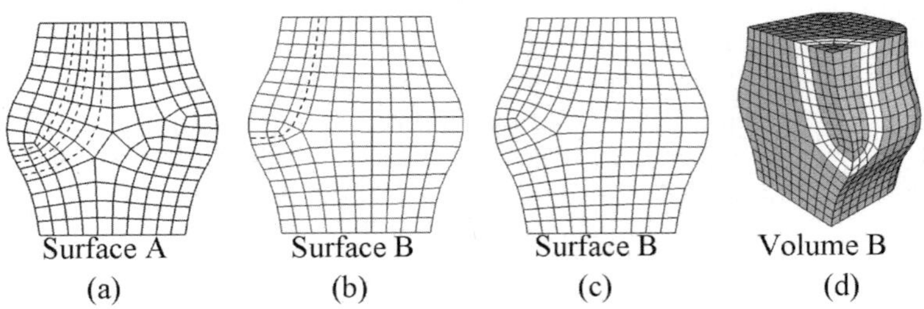

Fig. 15. Dicing during mesh matching: (a) The mesh topology on Surface A with three chords indicated. (b) The mesh topology on Surface B with a single chord to be diced indicated. (c) Surface B after dicing is performed. (d) Volume B after dicing is performed.

no match in Surface B. Figure 16c shows the mesh in Volume A. The indicated sheet is extracted from the mesh as shown in Figure 16d which removes the unpaired chord in 16a from Surface A as shown in Figure 16e.

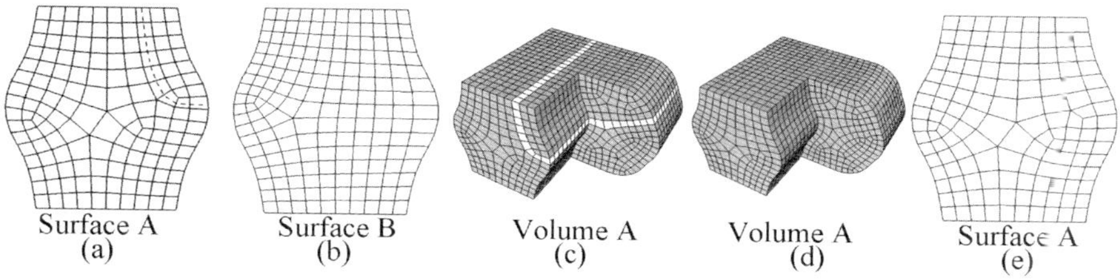

Fig. 16. Sheet extraction during mesh matching: (a) The mesh topology on Surface A with one unpaired chords indicated. (b) The mesh topology on Surface B. (c) Volume A with the sheet extending from the chord indicated in (a) highlighted. (d) Volume A after indicated sheet is extracted. (e) Surface A the sheet is extracted.

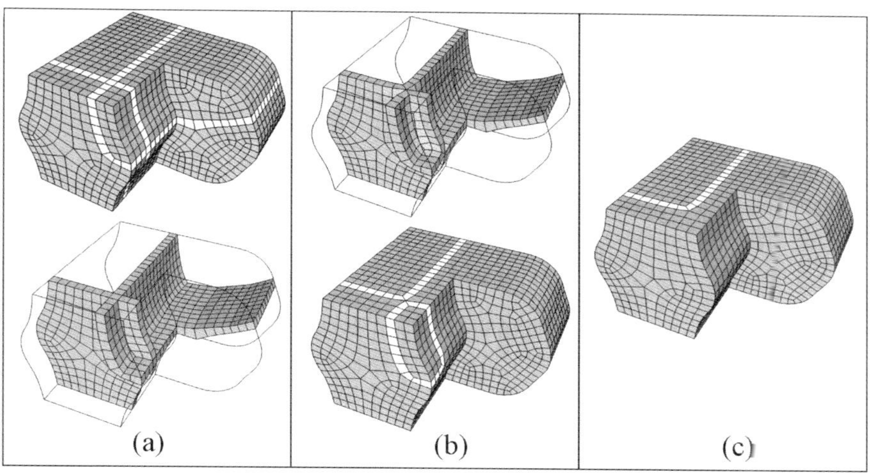

Fig. 17. Localized sheet extraction during mesh matching: (a) The mesh in Volume A with two sheets identified, the one to extract, and one that remains local to the interface surface. (b) The column where the two indicated sheets intersect has been collapsed, redirecting the sheet to extract to remain local to the interface region. (c) After sheet extraction; all changes are local to interface surfaces.

One potentially undesirable side-effect of sheet extraction is that the entire sheet must be extracted in order to maintain a conforming all-hexahedral mesh. Figure 16c clearly shows that the sheet to be extracted extends far away from the interface surfaces, resulting in a global change. However, the changes can be kept local to the region around Surface A if we first perform a column collapse operation (see Section 2.1). For example, in Figure 17a one additional sheet, which remains local to the interface surfaces, is identified. If such a local sheet does not exist in the mesh, one can be inserted by pillowing a few layers of hexahedra away from the interface surfaces. As described in Section 2, the two sheets indicated in Figure 17a intersect in a column of hexahedra. By collapsing

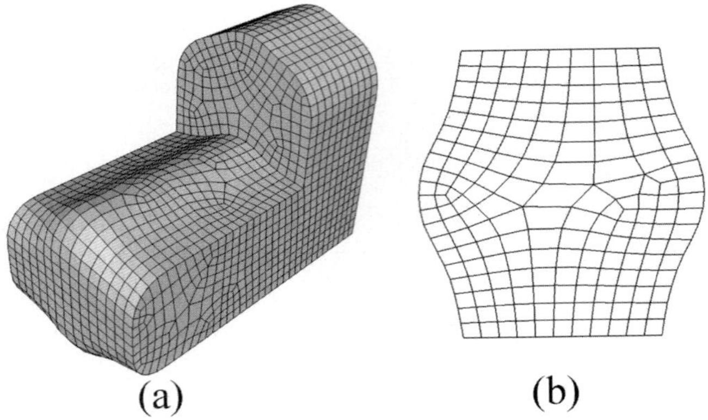

Fig. 18. Final mesh after mesh matching: (a) Final conforming mesh on example model after all topology on interface surfaces was matched. (b) The final mesh topology on the interface surface.

Table 1. Element Quality Results for Simple Example Model

	Number of Elements	Minimum Scaled Jacobian
Before Matching	5,925	0.7334
After Matching	6,024	0.5335

this column, we redirect the sheet to extract in such a way that it now remains local to the interface surfaces as illustrated in Figure 17b. The extraction sheet can then be extracted as illustrated in Figure 17c keeping all changes local to the interface surfaces.

The pillow, sheet inflation, sheet extraction, dicing and column collapse operations can be applied repeatedly until the topology on the interface surfaces matches allowing the mesh to be merged into a single conforming mesh. Figure 18a shows the final mesh with the final interface quadrilateral mesh shown in Figure 18b. Table 1 shows the element counts and element quality before and after mesh matching. As with any hexahedral mesh modification, mesh matching introduces irregular nodes into the mesh topology which will tend to decrease element quality. In this case, the resulting mesh has a minimum scaled Jacobian [13] of 0.5335, which is still well suited for analysis.

In the case that one side of the interface cannot be changed, the dicing, pillowing, and sheet extraction operations are restricted to be performed only where changes are allowed. For example if Surface A cannot be modified, all unpaired chords in Surface B are removed through sheet extraction in Volume B. Likewise, any unpaired chords in Surface A are inserted into Surface B through sheet insertion in Volume B. Thus, Surface and Volume A remain unchanged. If extraction of a sheet is required that would result in invalid geometric associativity,

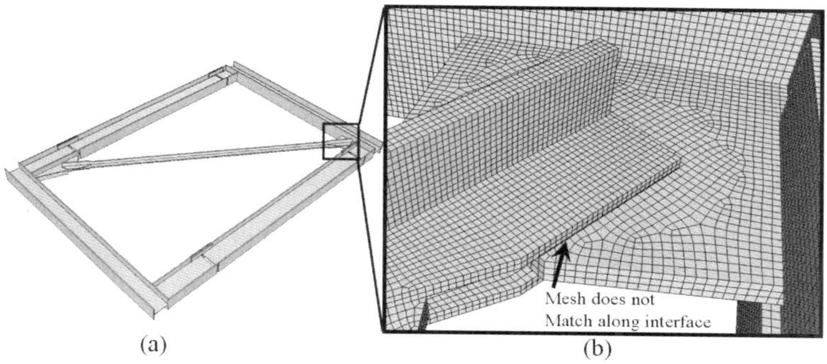

Fig. 19. Industrial example: (a) An I-beam structure for a civil engineering application. (b) Close look at interface between diagonal stiffener and corner I-Beams. The meshes do not matches at the interface. This model provided courtesy of Tyler Josephson and Professor Paul Richards from the Civil and Environmental Engineering Department at Brigham Young University.

this sheet can be redirected before the extraction using a column collapse thus allowing it to be extracted.

5.2 Industry Example

Figure 19a shows an I-beam structure used in a civil engineering application. The critical component to be analyzed is the diagonal stiffener. In fact, as part of the analysis, several different designs of the stiffener as well as adaptive studies using different size elements will be used. The mesh on the rectangular I-beams has approximately one million hexahedral elements, and required significant effort to generate. Ideally each time a new stiffener is introduced, the existing mesh on the rectangular I-beam structure can be re-used rather than requiring it to be re-meshed. The concepts of mesh matching presented in this paper apply to this application since every time a new mesh is generated for the stiffener, the mesh matching algorithm can be run on the connection with the rectangular I-beam structure to create a conforming mesh.

Figure 19b shows a close up of the corner of the structure where the stiffener connects to the corner plate. The stiffener is meshed with hexahedral elements that are slightly smaller than that of the corner plate. As a result, we have a

Table 2. Element Quality Results for I-beam Model before and after mesh matching

	Number of Element in Corner Plate	Minimum Scaled Jacobian
Before Mesh Matching	4,611	0.6624
After Mesh Matching - course Stiffener	7,221	0.5872
After Mesh Matching - fine Stiffener	17,883	0.4924

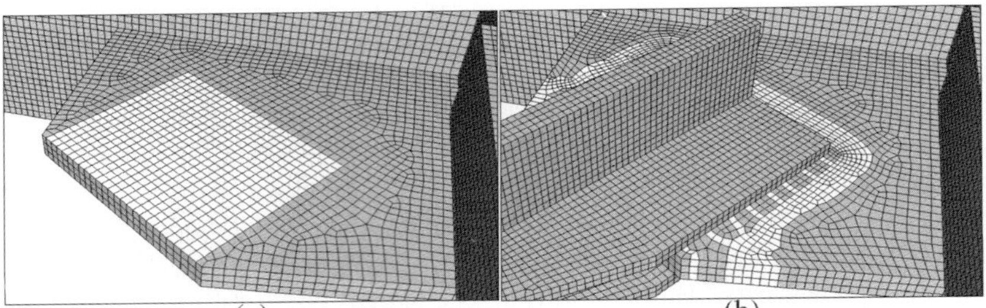

(a) (b)

Fig. 20. Industrial example: (a) The mesh on the interface surface on the corner plate. (b) The mesh after mesh matching. The mesh topology was matched and merged into a single all-hexahedral mesh. The highlighted elements indicate the sheets that were inserted during mesh matching.

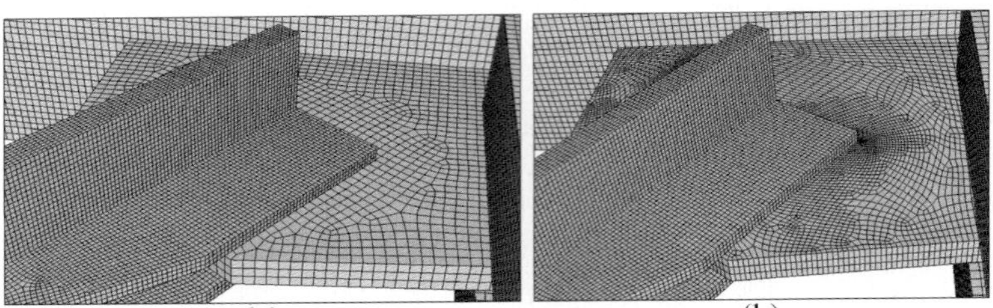

(a) (b)

Fig. 21. Industrial Example: (a) The same I-beam model before mesh matching, this time stiffener is meshed at a much higher density. (b) After mesh matching.

non-conforming mesh. Figure 20a shows the interface surface on the corner plate. It is meshed with 21×24 mapped quadrilateral mesh. The interface surface on the stiffener is meshed with a 29×30 mapped quadrilateral mesh. The mesh matching algorithm will need to increase the density of elements in the corner plate so that it also has a 29×30 mapped quadrilateral mesh so that the mesh on the interface can be merged. Figure 20b illustrates the mesh after successfully creating a conforming mesh using mesh matching.

Figure 21a shows the same I-beam model, however, the stiffener has been meshed at a much higher density of elements. The element size difference between the stiffener and the corner plate is 2.3 to 1. Rather than remeshing the rectangular I-beam structure, mesh matching is used to enforce a conforming mesh across the interface. Figure 21b shows the mesh after mesh matching has successfully matched the mesh topology on the interface creating a conforming all-hexahedral mesh through the interface. Table 2 shows the element count and element qualities before and after mesh matching. As with the first example, although element quality is reduced some by mesh matching, the resulting element qualities are still suitable for analysis.

6 Conclusion and Future Work

A new computational method, called mesh matching, for converting non-conforming hexahedral-to-hexahedral mesh interfaces into conforming interfaces utilizing localized hexahedral topology modification methods has been presented. Mesh matching eliminates the need for artificial constraint conditions, such as tied contacts, gap elements, and multi-point constraints. Mesh topology is modified using the dual operators of column collapse, dicing, pillowing, and sheet extraction, along with a new operator, sheet inflation.

Many meshing algorithms require entire assemblies to be meshed at once in order to have conforming meshes between components [1, 2]. Mesh matching relaxes this requirement by creating conforming meshes between assembly components after each component is meshed individually. Further, Tautges [22] asserts that hex meshing would be greatly simplified if global coupling between assemblies could be reduced or eliminated. Mesh matching reduces global coupling by enforcing a conforming mesh after the initial meshes have been created. Thus, mesh matching has the potential to greatly simplify the generation of conforming assembly hexahedral meshes.

Research on mesh matching continues with focus in the following areas. First, mesh matching requires repeated dicing, pillowing, sheet extraction and sheet inflation. The example in Figure 21 required more than 50 pillowing operations. Manually specifying 50 shrink sets is a tedious task that is best automated. Second, mesh matching requires the interface be limited to two topologically identical and spatially similar surfaces. In practice, interface conditions often include multiple adjacent surfaces on each component. Compared to a serial approach, simultaneous matching of all interface surfaces will likely result in a more optimal mesh. Third, mesh matching theory depends upon *Assertion 1* in Section 4.2. Future research to prove this assertion will guarantee the usefulness of mesh matching. The unproven elements of *Assertion 1* involve guaranteeing that sheet inflation is always possible for any possible hexahedral topology.

References

1. Zhang, Y., Bajaj, C.L., Sohn, B.S.: 3D Finite Element Meshing from Imaging Data. The special issue of Computer Methods in Applied Mechanics and Engineering (CMAME) on Unstructured Mesh Generation 194(48-49), 5083–5106 (2005)
2. Staten, M.L., Kerr, R.A., Owen, S.J., Blacker, T.D.: Unconstrained Paving and Plastering: Progress Update. In: Proceedings 15th International Meshing Roundtable, pp. 469–486 (2006)
3. Blacker, T.D.: The Cooper Tool. In: Proc. 5th International Meshing Roundtable, pp. 13–29 (1996)
4. Staten, M.L., Canann, S.A., Owen, S.J.: BMSweep: Locating Interior Nodes During Sweeping. Engineering with Computers 15, 212–218 (1999)
5. Dohrmann, C.R., Key, S.W.: A Transition Element for Uniform Strain Hexahedral and Tetrahedral Finite Elements. International Journal for Numerical Methods in Engineering 44, 1933–1950 (1999)

6. Trujillo, D.M., Pappoff, C.G.: A General Thermal Contact Resistance Finite Element. Finite Elements in Analysis and Design 38(3), 263–276 (2002)
7. Dohrmann, C.R., Key, S.W., Heinstein, M.W.: Methods for Connecting Dissimilar Three-Dimensional Finite Element Meshes. International Journal for Numerical Methods in Engineering 47, 1057–1080 (2000)
8. Puso, M.A., Laursen, T.A.: Mesh Tying on Curved Interfaces in 3D. Engineering Computations 20(3), 305–319 (2003)
9. Murdoch, P., Benzley, S.E.: The Spatial Twist Continuum. In: Proc. 4th International Meshing Roundtable, pp. 243–251 (1995)
10. Merkley, K., Ernst, C., Shepherd, J.F., Borden, M.J.: Methods and Applications of Generalized Sheet Insertion for Hexahedral Meshing. In: Proc. 16th International Meshing Roundtable, pp. 233–250 (2007)
11. Borden, M.J., Benzley, S.E., Shepherd, J.F.: Hexahedral Sheet Extraction. In: Proc. 11th International Meshing Roundtable, pp. 147–152 (2002)
12. Mitchell, S.A., Tautges, T.J.: Pillowing Doublets: Refining a Mesh to Ensure That Faces Share At Most One Edges. In: Proc. 4th International Meshing Roundtable, pp. 231–240 (1995)
13. Knupp, P.M.: Algebraic Mesh Quality Measures. SIAM Journal on Scientific Computing 23, 193–218 (2001)
14. Knupp, P.M., Mitchell, S.A.: Integration of Mesh Optimization with 3D All-Hex Mesh Generation, SAND99-2852, Sandia National Laboratories (1999)
15. Borden, M., Benzley, S.E., Mitchell, S.A., White, D.R., Meyers, R.: The Cleave and Fill Tool: An all-Hexahedral Refinement Algorithm for Swept Meshes. In: Proc. 9th International Meshing Roundtable, pp. 69–76 (2000)
16. Harris, N.J., Benzley, S.E., Owen, S.J.: Conformal Refinement of All-Hexahedral Element Meshes Based on Multiple Twist Plane Insertion. In: Proc. 13th International Meshing Roundtable, pp. 157–167 (2004)
17. Tchon, K.-F., Dompierre, J., Camarero, R.: Conformal Refinement of All-Quadrilateral and All-Hexahedral Meshes According to an Anisotropic Metric. In: Proc. 11th International Meshing Roundtable, pp. 231–242 (2002)
18. Jankovich, S.R., Benzley, S.E., Shepherd, J.F., Mitchell, S.: The Graft Tool: an All-Hexahedral Transition Algorithm for Creating a Multi-Directional Swept Volume Mesh. In: Proc. 8th International Meshing Roundtable, pp. 387–392 (1999)
19. Melander, D.J., Taugtes, T.J., Benzley, S.E.: Generation of Multi-Million Element Meshes for Solid Model-Based Geometries: the Dicer Algorithm. AMD - Trends in Unstructured Mesh Generation 220, 131–135 (1997)
20. Shepherd, J.: Topologic and Geometric Constraint-Based Hexahedral Mesh Generation. Dissertation - University of Utah (2007)
21. Knupp, P.M.: A Method for Hexahedral Mesh Shape Optimization. International Journal for Numerical Methods in Engineering 58, 319–332 (2003)
22. Tautges, T.J.: The Generation of Hexahedral Meshes for Assembly Geometry: Survey and Progress. International Journal for Numerical Methods in Engineering 50, 2617–2642 (2001)

Conformal Meshing of Multiple Domains in Contact

Franjo Juretić

AVL-AST d.o.o., Avenija Dubrovnik 10, 10020 Zagreb, Croatia
franjo.juretic@avl.com

Summary. The paper presents an inside-out method for conformal meshing of multiple domains in contact applicable to meshes consisting of arbitrary polyhedra. The method is automatic and does not require high-quality surface triangulation. It is an extension of AVL's automatic polyhedral mesh generator FAME with an additional step during which the cells are distributed to their domains and the internal interfaces are captured. Several industrial examples are provided to demonstrate the potential of the method.

1 Introduction

Problems occurring in engineering practice often include two or more materials interacting with each other. Some examples of such problems are: heat exchange between coolant in the cooling jacket of an engine and the engine block, flow and heat transfer in a heat exchanger, multiphase flows, fluid-structure interaction, balance of masses in crankshafts, etc. The geometries of such devices are very complex and their simulation often requires high amount of manual work, which is often tedious and time consuming. The main reason for these problems lies in the mesh generation process, which is usually the bottleneck of the simulation [22].

Automatic mesh-generation methods can be classified into two main groups, namely, boundary-to-interior and inside-out methods. Boundary-to-interior methods start the mesh-generation process at the boundary and then proceed to the generation of the volume mesh. These methods fall into two categories: Advancing front methods [9, 12] and Delaunay methods [5, 6]. On the contrary, inside-out methods start the meshing process by generating the interior mesh template first which is then connected with the boundary. These methods [21, 1, 11, 19, 15, 16, 14] have received a great deal attention during the last decade, and have enjoyed a great deal of success in the industrial environment due to the fact they do not require high quality surface triangulations and complex algorithms to analyse the topology of the geometry under consideration. The method implemented in FAME is based on the template, generated from a background octree, mainly consisting of hexahedral cells, while the transition regions between the fine and coarse mesh are filled with prisms and pyramids. Treatment of outer boundaries follows the approaches of Wang *et. al.* [21] and Tchon *et. al.* [19] because they are not very sensitive to the quality of the surface mesh.

Methods for capturing internal surfaces have mainly been used in the context of adaptive re-meshing. Some examples where the interface is tracked during the calculation can be found for mould casting [13], two-phase flows on Voronoi meshes [2], rising of air bubbles in the water [20] and many more. However, the literature on mesh generation methods for multiple domains in contact are quite limited. Sullivan *et. al.* [18] have presented a method generating tetrahedral meshes in multiple complex domains. Surface facets are determined based on nodal distances from the surface, which is not reliable in case of bad quality surface meshes. Kim *et. al.* have discussed their approach in [10] aimed for modelling of composite materials and cloths. The position of the internal surfaces is prescribed by using analytical functions for the threads in the composite matrix, and the internal surfaces are captured by using element splitting and node relocations. Togashi *et. al.* [8] developed a method for multiple bodies in contact using overset meshes. Shepherd [17] has implemented a similar method for capturing internal surfaces to the one presented in this paper. The method finds the template cells intersected by the internal surface and distributes all of them to one of the materials at inner boundaries. The approach presented in this paper distributes the intersected cells depending on the number of neighbours an intersected cell shares with each material, and distributes the cell to the material with most neighbours. This results in the lowest number of cells with more than one face at the inner boundary and the lowest number of deformed cells after the mapping stage.

This paper presents an inside-out method, applicable to arbitrary polyhedral meshes, which generates a mesh for multiple domains in contact in a single mesh generation process. The paper is organised as follows. Section 2 gives an overview of the mesh generation process. The procedure generating internal boundaries is discussed in section 2.5. The examples of problems under consideration are shown in section 3. Conclusions and suggestions for future work are given in section 4.

2 Methodology

2.1 Input Data

The method requires the following input:

1. A surface triangulation and a mesh of edges user wants to capture. All the facets in the surface triangulation belonging to each material have to be collected in a selection and given a name. By doing this, it is possible to identify facets at internal surfaces because they must appear in exactly two such selections. This information is used later in the meshing process.
2. The second part are the settings enabling the user to set cell sizes in different parts of domains, number of boundary layers in each domain or at specified parts of the boundary, etc. The settings are stored in a xml file and can be generated using a GUI application.

2.2 Octree Creation

The meshing process starts by generating an octree based on the settings given in the xml file. The octree is needed to generate the mesh template and to perform inside/outside calculations applicable to bad quality surface triangulations. The octree is refined level-wise until no further refinement is requested. In order to achieve the desired cell size for every domain it is necessary to refine the octree accordingly. The leaves belonging a particular domain, which shall be refined, are selected using the following procedure:

1. Mark all leaves with a box type. Available types are: OUTSIDE, DATA (leaves containing some surface elements), INSIDE and ENCLOSED (domains having no connection with the outer boundary). The marking process starts by marking all OUTSIDE leaves first, followed by INSIDE leaves and then the ENCLOSED ones.
2. Once the box types are known, it is possible to find leaves belonging to the domain of interest. Leaves in partially-enclosed domains, containing a portion of some outer boundary, are of INSIDE type and are marked using the frontal-marking algorithm starting from an initial INSIDE leaf near a DATA leaf containing a portion of domain under consideration. Leaves in enclosed domains are also marked using the frontal-marking algorithm starting from an initial ENCLOSED leaf near a DATA leaf containing a portion of the enclosed domain.

2.3 Mesh Template

Once the octree is created it is used for generating the initial mesh template. The template mesh consists of standard cell types, namely, hexahedra, prisms and pyramids. Please note that it encompasses all domains of interest and the existence of internal surfaces is also ignored at this stage. An example of the surface mesh with two domains and the resulting template is shown in figures 1(a) and 1(b), respectively.

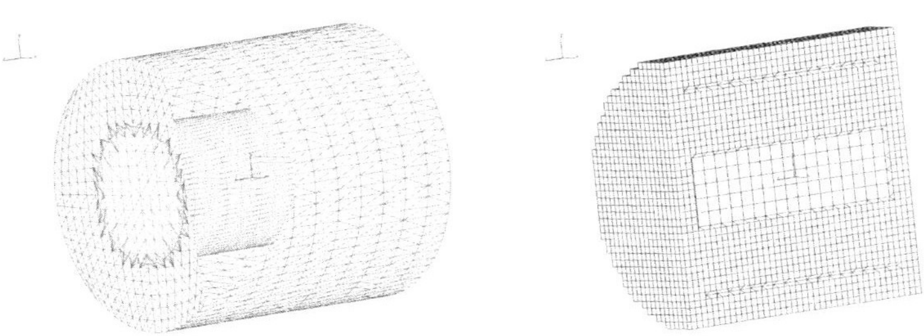

(a) Surface mesh (b) Mesh template

Fig. 1. An example with internal surfaces

2.4 Fitting the Mesh to Outer Boundaries

The current step is aimed at creating a mesh fitting the external boundaries; from the existing mesh template. This is performed by pre-smoothing the outer surface and mapping each outer vertex to its nearest points at the outer boundaries of the surface triangulation as described in [21, 15]. Surface facets belonging to internal boundaries are not considered for proximity searches. In addition, capturing edges and corners sometimes requires some topological changes of the quadrilateral boundary faces, which are decomposed into triangles and their cell owners are decomposed into tetrahedra and pyramids [7]. The mesh at this stage may become tangled as a consequence of the boundary-fitting procedure. In order to speed-up the untangling process and improve the near-boundary topology, an O-topology boundary layer is generated. The process of fitting the mesh to the outer boundary finishes with the smoothing/untangling procedure which is a combination of Laplacian smoothing [4] and metric-based untangling [3]. An example of boundary-fitting mesh for the example shown in figures 1(a) and 1(b) is given in figure 2.

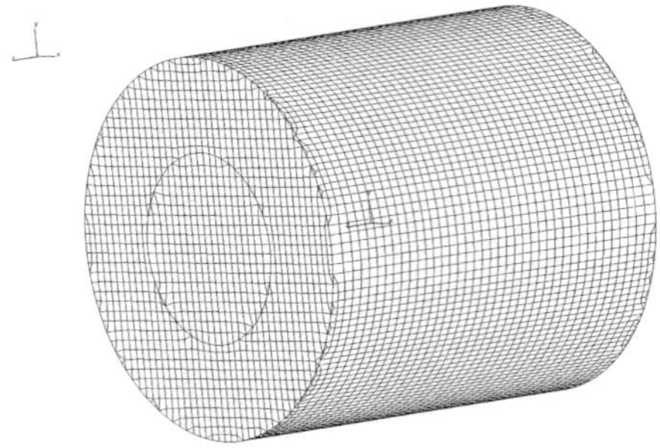

Fig. 2. A mesh fitted to the outer boundary

2.5 Capturing of Internal Surfaces

The process for capturing internal surfaces starts once the mesh is fitted to the outer boundary and has the following workflow:

1. Find and mark the cells intersected by the internal-surface facets. The quality of the surface is not critical here as long as the cell size is larger than the possible gaps in the surface. The marked cells are not assigned to any domain here. The figures 3(a) and 3(b) show the intersected cells coloured in white.
2. Start marking cells belonging to partially-enclosed domains using the frontal-marking algorithm. The process is performed on domain by domain basis. For each domain it starts from a cell at the outer boundary and stops when the front reaches cells intersected by the internal surface, see figure 3(a). Cells in fully-enclosed domains are not marked yet. Those cells are also marked using the frontal-marking algorithm starting from a cell which has

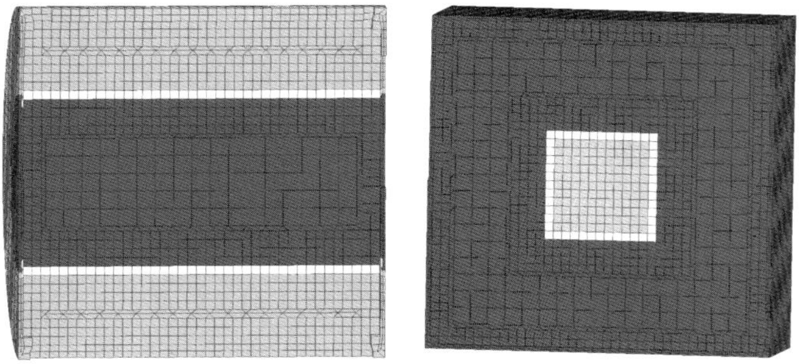

(a) Partially-enclosed materials (b) Enclosed material

Fig. 3. Meshes after the second step of the algorithm for capturing internal surfaces

a neighbour cell intersected by a facet belonging the corresponding enclosed domain. And example with a partially and a fully-enclosed domain is given in figure 3(b).

3. Distribute the cells intersected by the internal surfaces to their domains. It is performed on a cell-by-cell basis unlike the method of Shepherd [17] which assigns all those cells to a single domain. The main criterion for distributing an intersected cell is the number of neighbours already assigned to the neighbouring domains. The cell is assigned to the domain which contains most of its neighbours. Cells having equal number of neighbours in other materials are not assigned in that sweep. The process is iterative in nature and is repeated as long as some cells matching this criterion are found. The remaining undetermined cells, if there are any, are assigned to the domain with the smallest number of cells. Please note that this procedure does not require any floating point calculations and is therefore applicable to bad-quality surface triangulations. Figures 4(a) and 4(b) show an example of the mesh before and after the cells are distributed to their domain. It is clearly visible that the mesh would get tangled in case if all intersected cells were distributed to the same domain. However, the procedure presented here has generated the interface with the smallest possible number of faces.

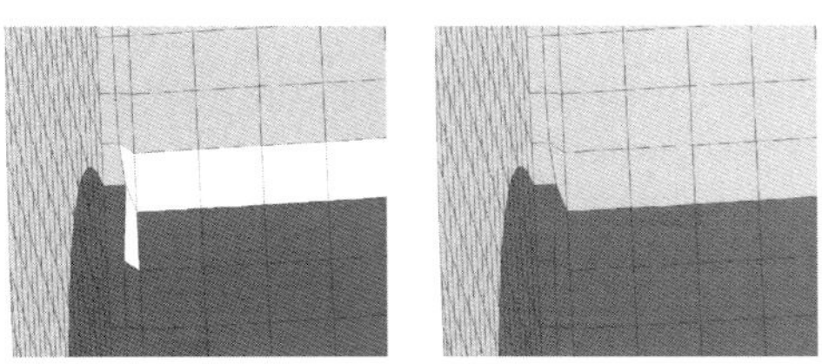

(a) Before distribution (b) After distribution

Fig. 4. Meshes before and after distribution of intersected cells

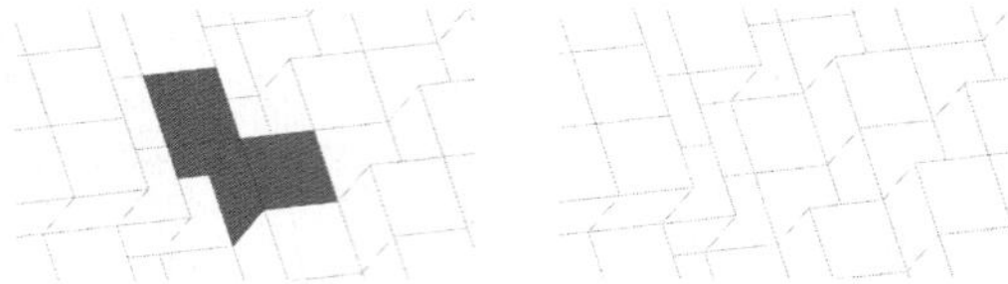

(a) Invalid connection (b) After correction

Fig. 5. Internal surface of the volume mesh

4. Check if the internal surface of the volume mesh is singly-connected. If any such problems occur, see figure 5(a), cells are locally-transferred from the domain with higher number of cells to the other one, until the surface becomes singly-connected, see figure 5(b). This check has proven sufficient in the tests carried so far.
5. Pre-smooth the internal surface of the volume mesh and project the points onto the nearest location on the surface mesh. Internal corners and edges are also captured at this stage. This may also require some topological changes of quadrilateral faces; such faces are decomposed into two triangles and the cells sharing that face are decomposed into tetrahedra and pyramids. Please note that this procedure for treating surface faces is the same as the procedure for outer boundaries [7]. However, in case a quad face is decomposed into triangles; cells at both sides of the internal surface must be decomposed. With the currently used mesh templates, this approach is robust for internal surfaces where two domains meet, but does not guarantee that the edges and corners where three or more domains meet will be preserved in the final mesh. Figure 6 shows an example where the domains, which should only share a common edge, share common faces, too. The reason for this is the insufficient number of degrees of freedom in the mesh template, which shall be increased in order to capture such edges accurately. Even though,

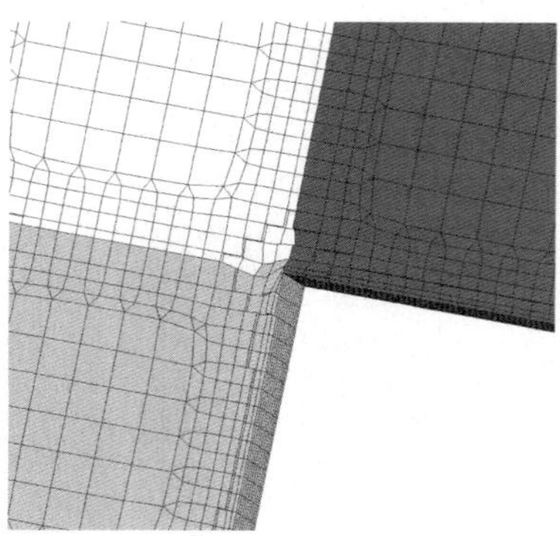

Fig. 6. An example with a poorly captured edge

the approach is still considered acceptable for most engineering applications where the modelling and the discretisation errors are still larger than the error introduced by the inaccurate inner-boundary description.

Once the domains are created and the internal surfaces captured, it is still necessary to generate boundary layers for each domain, in order to create a sufficient number of degrees of freedom needed by the smoothing/untangling procedure to produce a valid mesh. The layers are generated separately for each domain. The process is started by activating the cells in the domain of interest and deactivating the other ones. Once the domain cells are active, it is possible to find boundary faces of the active domain which will serve as the extrusion front. Create a new co-located vertex for every vertex in the front and move the original vertices inside the active domain in the direction of their normals. This is followed by the generation of the layer cells, one per boundary face, from the face vertices and the newly created vertices. Inactive cells near the internal surface are renumbered to use the newly created vertices instead of the original ones. An example of boundary layers of O-topology is shown in figure 7(a), and it is not preferred for inlet and outlet boundaries. However, it is possible to change the mesh topology by simply removing the first layer of cells at the selected boundaries, see figure 7(b). This topology performs better in terms of higher accuracy and the number of computational cells is slightly decreased.

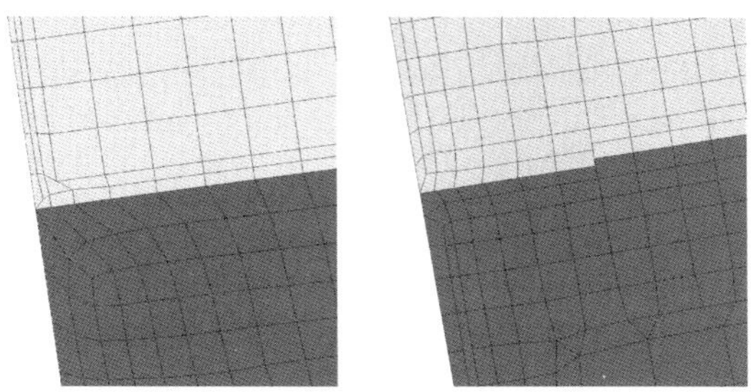

(a) O-topology (b) Improved topology

Fig. 7. Mesh with improved topology at the outer inflow/outflow boundary

After the layers are created, the mesh is checked if any negative volume cells exist. The smoothing/untangling procedure, mentioned before, is not allowed to move nodes at the internal surface during the initial iterations. If the procedure is not capable of untangling all problematic regions, it may be allowed to move constrained nodes as long as the problems persist. The experience shows that the constrained nodes are moved mainly in the case of coarse settings which are anyway prone to high modelling and discretisation errors.

Finally, a figure showing the final mesh with two domains in contact is given in figure 8.

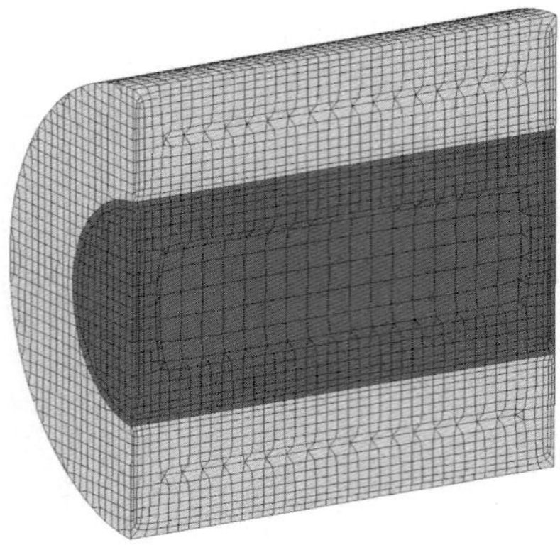

Fig. 8. A view through the final mesh with two domains

3 Examples

This section presents the behaviour of the method on two examples of industrial interest. The first example presents the mesh of a crankshaft in order to analyse the stiffness and displacements of each part. The second example is a mesh of an engine needed to analyse the exchange of heat between the coolant and the metal parts.

The surface mesh for the crankshaft, consisting of three parts, is shown in the figure 9(a) and the section through the resulting volume mesh is given in figure

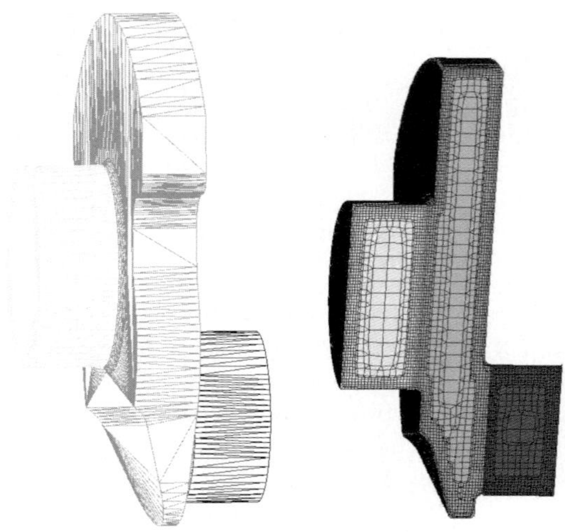

(a) Surface mesh (b) Volume mesh

Fig. 9. A mesh for the crankshaft example

Table 1. Mesh quality measured in minimal scaled jacobians for the crankshaft case

	Number of cells	Minimum	Percentage of cells below 0.1	Average
All cells	336923	0.0015	0.01	0.915
Cell type				
Hexahedra	198892	0.1204	0.00	0.949
Prisms	24290	0.0015	0.09	0.646
Pyramids	13375	0.1318	0.00	0.650
Tetrahedra	366	0.0544	11.5	0.424

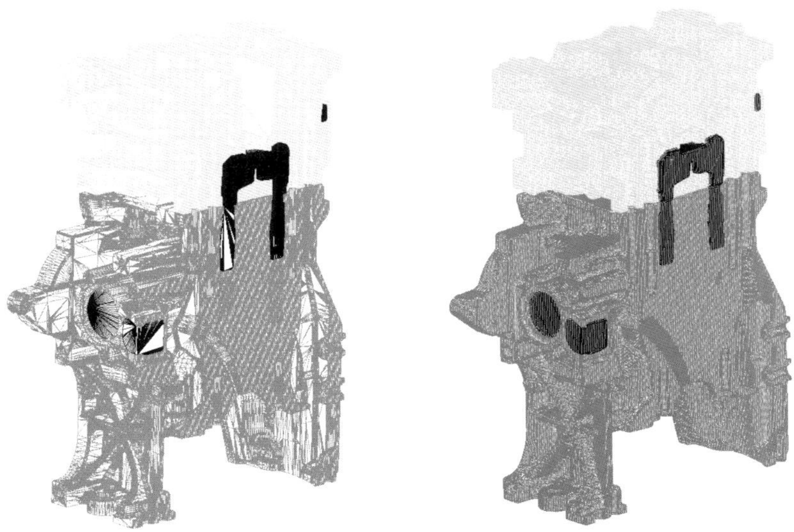

(a) Surface mesh (b) Volume mesh

Fig. 10. A mesh for the engine example

9(b). The resulting volume mesh consists of 336923 cells, out of which 47392 belong the shaft connected to the conrod (coloured in dark grey), 224606 are used to mesh the counterweight, and 64925 are in the shaft domain coloured in light grey. The meshing process lasted 2 minutes on a HP nx9420 notebook with 3 Gb of RAM and a Centrino Duo processor with the clock speed of 2.16 GHz and the peak virtual memory usage was 250 Mb. The quality of the mesh is measured in minimal scaled jacobians, defined in [23], and is given in table 1. The table shows that the average cell quality is high. The analysis by cell type indicates that high percentage of tetrahedral cells is of low quality. The tetrahedra are only generated as a consequence of the edge-capturing procedure, and are therefore located near edges where the smoother is limited by the geometric constraints.

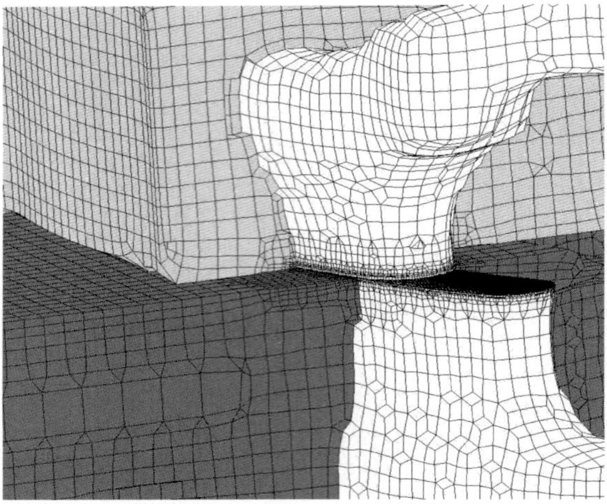

Fig. 11. A cut through the mesh for the engine example

Table 2. Mesh quality measured in minimal scaled jacobians for the engine case

	Number of cells	Minimum	Percentage of cells below 0.1	Average
All cells	3884472	8.91e-6	0.5	0.847
Cell type				
Hexahedra	3572221	2.04e-5	0.4	0.871
Prisms	200547	4.54e-4	0.5	0.581
Pyramids	103416	8.91e-6	2.2	0.573
Tetrahedra	8288	1.22e-3	23.5	0.283

The surface of the engine example is shown in figure 10(a) where three domains can be identified, namely, the cylinder head, the block and the cooling jacket. The surface is of bad quality with many high aspect ratio triangles and some holes due to missing parts. The volume mesh, shown in figures 10(b) and 11, consists of 3884472 cells, 574469 belonging to the cooling jacket, 1321737 to the cylinder head and 1988266 to the block. The figure 11 shows the cooling jacket, coloured in white, and a cut through the head (light grey) and block (dark grey), and the conformal interfaces between the domains. Quality measures of the mesh are given in table 2 and are again measured in minimal scaled jacobians. The table shows that the average cell quality is high. However, a considerable percentage of tetrahedra are of low quality due to very complex geometry and coarse settings which can barely resolve some feature edges of the geometry.

4 Conclusions

The paper presents a new mesh-generation method for multiple domains in contact. It is designed for industrial problems, and the resulting meshes are not dependent on the quality of the surface triangulation. The meshes are of acceptable quality for AVL's CFD solver FIRE. The procedure for generating internal surface is applicable for any mesh type, and could also be applied standalone tool for generating internal interfaces in a mesh generated by some other meshing tool.

The method can be further improved by developing a method for automatic generation of surface selections belonging to each domain. This will be tackled using the "octree" approach for inside-outside calculations due to its applicability to bad-quality surface triangulations. Treatment of internal edges and corners where three or more domains meet shall also undergo some development in order to be able to capture them more accurately.

References

1. Aftosmis, M.J., Berger, M.J., Melton, J.E.: Robust and Efficient Cartesian Mesh Generation for Component-Based Geometry. In: 35th AIAA Aerospace Sciences Meeting, no. AIAA-97-0196 (1997)
2. Dukowitz, J.K., Cline, M.C., Addesso, F.L.: A General Topology Godunov Method. J. Comput. Phys. 82, 29–63 (1989)
3. Escobar, J.M., Rodriguez, E., Montenegro, R., Montero, G., Gonzalez-Yuste, J.M.: Simulataneous untangling and smoothing of tetrahedral meshes. Comput. Methods Appl. Mech. Engrg. 192, 2775–2787 (2003)
4. Field, D.: A Laplacian Smoothing and Delaunay Triangulation. Comm. Appl. Numer. Meth. 4, 709–712 (1988)
5. George, P.L., Hecht, F., Saltel, E.: Automatic parallel generator with specified boundary. Comput. Methods Appl. Mech. Engrg. 92, 269–288 (1991)
6. George, P.L., Hermeline, F.: Delaunay's mesh of a convex polyhedron in dimension d. Application to arbitrary polyhedra. Int. J. Numer. Meth. Engng. 33, 975–995 (1992)
7. Masayuki, H., Nishigaki, I., Kataoka, I., Hiro, Y.: Automatic Hexahedral Mesh Generation with Feature Line Extraction. In: Proceedings, 15th International Meshing Roundatable, pp. 453–468. Springer, Heidelberg (2006)
8. Togashi, F., Ito, Y., Nakahashi, K., Obayashi, S.: Extensions of Overset Unstructured Grids to Multiple Bodies in Contact. Journal of Aircraft 43, 52–57 (2006)
9. Jin, H., Tanner, R.I.: Generation of unstructured tetrahedral meshes by Advancing Front Technique. Int. J. Numer. Meth. Engng. 36, 1805–1823 (1993)
10. Kim, H.J., Swan, C.C.: Algorithms for automated meshing and unit cell analysis of periodic composites with hierarhical tri-quadric tetrahedral elements. Int. J. Numer. Meth. Engng. 58, 1683–1711 (2003)
11. Kraft, P.: Automatic Remeshing with Hexahedral Elements: Problems, Solutions and Applications. In: Proceedings, 8th International Meshing Roundtable, pp. 357–367 (1999)
12. Löhner, R., Camberos, J., Merriam, M.: Parallel unstructured grid generation. Comput. Methods Appl. Mech. Engrg. 95, 343–357 (1992)

13. Maliska, C.R., Flávio, J., Vasconcellos, V.: An unstructured finite volume procedure for simulating flows with moving fronts. Comput. Methods Appl. Mech. Engrg. 182, 401–420 (2000)
14. Oaks, W., Paoletti, S.: Polyhedral mesh generation. In: Proceedings, 9th International Meshing Roundtable, Sandia National Laboratories, pp. 57–67 (2000)
15. Putz, N., Rebhandl, H., Sampl, P.: FAME Flexible Automatic Meshing Environment. In: Proceedings, 4th World Conference in Applied Fluid Dynamics (WUA-CFD 1998) (1998)
16. Schneiders, R., Bünten, R.: Automatic generation of hexahedral finite element meshes. Computer Aided Geometric Design 12, 693–707 (1995)
17. Shepherd, J.F.: Topologic and Geometric Constraint-Based Hexahedral Mesh Generation. PhD Thesis, University of Utah (2007)
18. Sullivan, J.M., Charron, G., Paulsen, K.D.: A three-dimensional mesh generator for arbitrary multiple material domains. Finite Elements in Analysis and Design 25, 219–241 (1997)
19. Tchon, K.F., Hirsch, C., Schneiders, R.: Octree-based Hexahedral Mesh Generation for Viscous Flow Simulations. In: 13th AIAA Computation Fluid Dynamics Conference, no. AIAA-97-1980 (1997)
20. Tuković, Ž., Jasak, H.: Simulation of Free-rising Bubble with Soluble Surfactant using Moving Mesh Finite Volume/Area Method. In: Proceedings of 6th International Conference on CFD in Oil & Gas, Metallurgical and Process Industries, no. CFD08-072 (2008)
21. Wang, Z.J., Srinivasan, K.: An adaptive Cartesian grid generation method for 'Dirty' geometry. Int. J. Numer. Meth. Fluids 39, 703–717 (2002)
22. White, D.R., Saigal, S., Owen, S.: Meshing Complexity of Single Part CAD Models. In: Proceedings, 12th International Meshing Roundtable, Sandia National Laboratories, pp. 121–134 (2003)
23. Yamakawa, S., Shimada, K.: Hex-Dominant Mesh Generation with Directionality Control via Packing Rectangular Solid Cells. In: Proceedings, Geometric Modeling and Processing, pp. 107–118 (2002)

Subdivision Templates for Converting a Non-conformal Hex-Dominant Mesh to a Conformal Hex-Dominant Mesh without Pyramid Elements

Soji Yamakawa, Iacopo Gentilini, and Kenji Shimada

Department of Mechanical Engineering
Carnegie Mellon University

Abstract. This paper presents a computational method for converting a non-conformal hex-dominant mesh to a conformal hex-dominant mesh without help of pyramid elements. During the conversion, the proposed method subdivides a non-conformal element by applying a subdivision template and conformal elements by a conventional subdivision scheme. Although many finite element solvers accept mixed elements, some of them require a mesh to be conformal without a pyramid element. None of the published automated methods could create a conformal hex-dominant mesh without help of pyramid elements, and therefore the applicability of the hex-dominant mesh has been significantly limited. The proposed method takes a non-conformal hex-dominant mesh as an input and converts it to a conformal hex-dominant mesh that consists only of hex, tet, and prism elements. No pyramid element will be introduced. The conversion thus considerably increases the applicability of the hex-dominant mesh in many finite element solvers.

1 Introduction

This paper presents a computational method for converting a non-conformal hex-dominant mesh to a conformal hex-dominant mesh without introducing a pyramid element. The input hex-dominant mesh can include hex, tet, and prism elements (also known as wedge elements), and some quadrilateral faces of the input mesh can be non-conformal; i.e., a triangular face can directly be connected to a quadrilateral face. The proposed method subdivides non-conformal hex-elements by conformal conversion templates and other elements by a conventional subdivision scheme. The conversion makes the hex-dominant mesh fully conformal without introducing a pyramid element, which is often rejected by a finite element solver. The output mesh thus can be used in a finite element solver even when it accepts neither a non-conformal face nor a pyramid element.

The accuracy of the finite element simulation depends on three factors: (1) mesh resolution, (2) mesh element quality, and (3) type of the mesh. A high-resolution mesh with good-quality elements yields an accurate solution. In addition, the mesh type also has a considerable impact on the accuracy of the finite element simulation. The type of a mesh depends on the type of elements included in the mesh. A volume mesh typically consists of four types of elements, tet, hex, prism, and/or pyramid elements as shown in Figure 1. Each type of element performs differently in the finite element simulation due to the difference of the shape functions, which interpolate physical quantity within the element.

A physical quantity within an element is interpolated as a linear combination of the shape functions, and higher-order shape functions can approximate a complex finite element solution better than lower-order shape functions. A shape function of an 8-node hexahedral element includes tri-linear, bi-linear, and linear terms, whereas 6-node prism and 5-node pyramid elements consist of bi-linear and linear terms, and a 4-node tetrahedral element consists only of linear terms. Therefore, a hexahedral element yields the most accurate solution, and an all-hexahedral mesh is preferred for a finite element analysis when available. Although a tetrahedral mesh can be created easily by an automatic mesh generation scheme [1-3], it requires more elements than an all-hexahedral mesh to obtain an equally accurate finite-element solution.

An all-hexahedral mesh, however, turned out to be difficult to create automatically for a complicated shape. Despite numerous attempts, none of the known methods can create an all-hexahedral mesh of adequate quality for an arbitrary geometric domain [4-18].

A hybrid mesh is an alternative to an all-hexahedral mesh and consists of hexahedral, tetrahedral, prism, and pyramid elements. The most common form of a hybrid mesh is a hex-dominant mesh [19-22], in which most of the volume is filled with hexahedral elements, and the rest with prism and tetrahedral elements, and pyramid elements are inserted between a quadrilateral face of a hexahedral or prism element and triangular faces of tetrahedral or prism elements as shown in Figure 2. However, some finite-element solvers do not accept a pyramid element. During the finite element calculation, an element is transformed into a master element, in which partial differential equations are numerically integrated [23]. Such a transformation utilizes Jacobian determinant, which is defined at every corner of an element where three edges converge. However, four edges converge at one of the nodes of the pyramid, where Jacobian determinant becomes ambiguous. Due to this ambiguity, some finite element solvers do not accept a pyramid element, and it becomes necessary to connect two triangular faces to a quadrilateral face. Such a connection is called a non-conformal connection. Some solvers accept neither pyramid element nor non-conformal transition, and applicability of a hex-dominant mesh is significantly limited.

This paper proposes a set of mesh conversion templates that are applied to hex elements with non-conformal faces. The templates are based on HEXHOOP templates [13] and subdivide a non-conformal face into eight triangles (triangular pattern) and a conformal face into four quadrilaterals (rectangular pattern) as shown in Figure 3. Other elements are subdivided by a conventional subdivision scheme so that the elements become conformal after the subdivision. The conversion templates are constructed by a family of modular sub-templates that can be assembled to form conformal conversion templates for hex elements. The template design uses two types of modular sub-templates, called a *core* and a *cap*. A core module defines the subdivision patterns of two opposite faces of a hex element. Four caps define the subdivision patterns of the rest of the faces. The advantage of this method is that two subdivision patterns, rectangular and triangular patterns, can be mixed and matched freely on the exterior surfaces of a hex element.

The conversion makes a fully-conformal hex-dominant mesh that consists of tet, hex, and prism elements, and does not include pyramid elements. The mesh can be

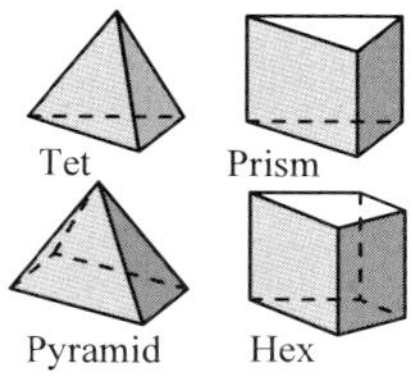

Fig. 1. Four types of volume element

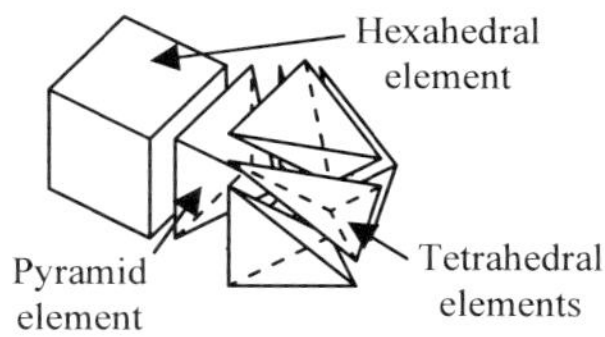

Fig. 2. Inserting a Pyramid Element between Hex and Tet Elements

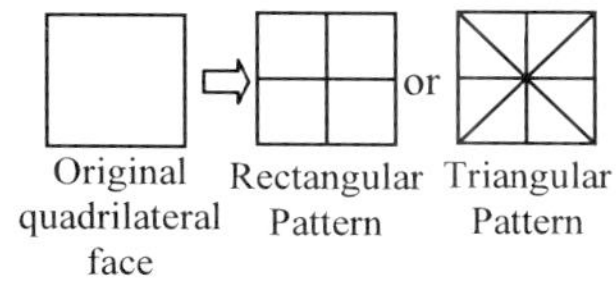

Fig. 3. Two subdivision patterns of a quadrilateral face

used by a finite element solver that rejects both non-conformal faces and pyramid elements. Hence, the conversion significantly increases the applicability of the hex-dominant mesh.

In the rest of the paper, Section 2 explains the construction of the subdivision templates, and Section 3 explains the non-template subdivision. Section 4 discusses the requirement of the input non-conformal hex-dominant mesh. Section 5 describes a post-process called cap-suppression. Section 6 shows some conversion examples, and Section 7 demonstrates the performance of the conformal hex-dominant mesh in finite element simulation. Section 8 discusses potential improvements, and Section 9 concludes the paper.

2 Construction of the Conversion Templates

The goal of the proposed method is to develop a system of conformal conversion templates for hex elements. The difficulty is that two face subdivision patterns, rectangular pattern and triangular pattern, need to be freely mixed on the exterior faces. Since a hex element consists of six faces, there are $2^6=64$ of the combinations of the subdivision patterns per element. Due to symmetry, the total number of combinations is reduced to ten as shown in Figure 4. Among the ten cases, there are three cases for which a known simple solution exists. H0 template can be achieved by a basic mid-point subdivision. H3 template can be constructed by sweeping a triangular pattern

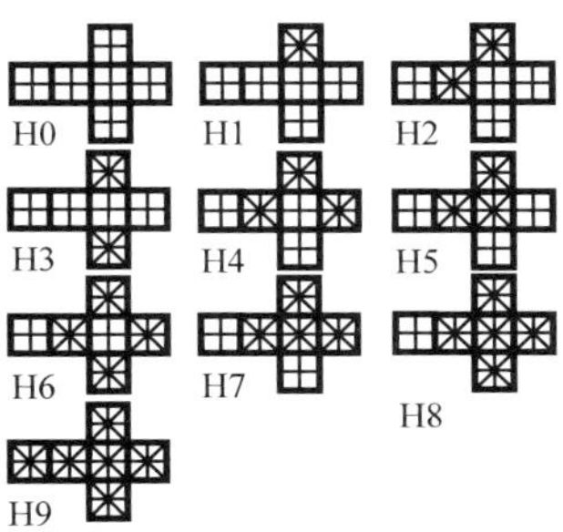

Fig. 4. Ten required conversion templates

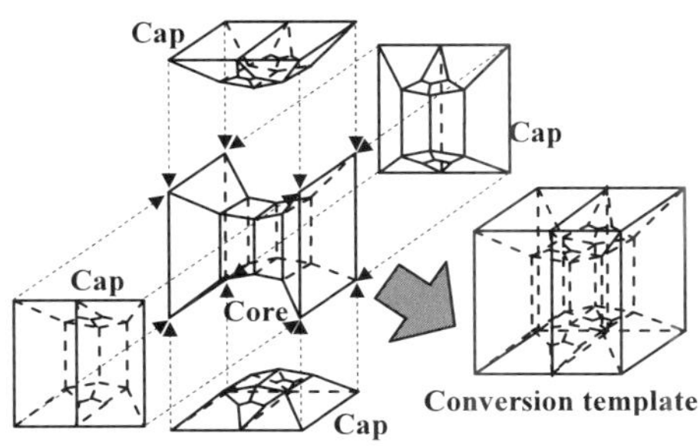

Fig. 5. Modular approach to a all-hex template

for two layers. H9 template can be constructed by adding a node in the middle of the template and making tet elements by connecting the mid node and exterior triangles. All other seven cases are non-trivial.

To construct all non-trivial conversion templates, the proposed method constructs a conversion template by assembling sub-templates, each of which has either rectangular or triangular pattern on its external faces. The new template design uses two types of modular sub-templates, called *core*, and *cap*. A core module defines subdivision patterns of two opposite faces, and four caps define subdivision patterns of the other four faces as shown in Figure 5. Two subdivision patterns can be freely mixed and combined on the exterior faces of a hex element.

2.1 Construction of a Cap Module

Exterior faces of a cap module are grouped into four groups: T-face, which is exposed to the exterior after assembled together with other modules, B-face, which is connected to a slot of a core module, and F-face, which is connected to an adjacent cap module as shown in Figure 6.

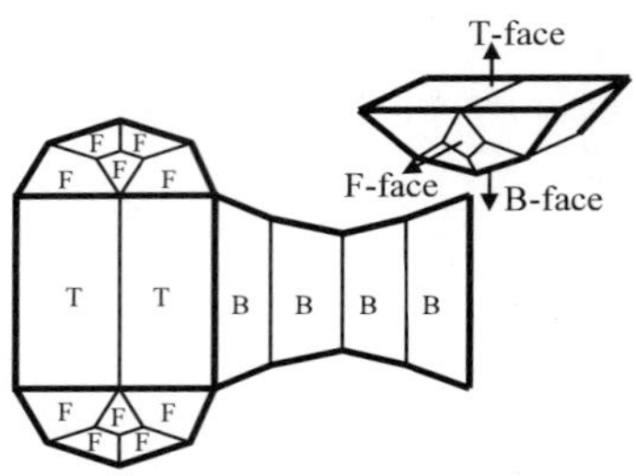

Fig. 6. Faces of a Cap

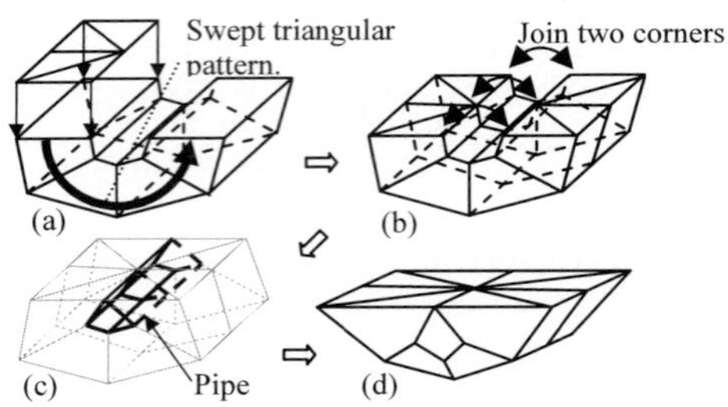

Fig. 7. Construction of a triangular cap

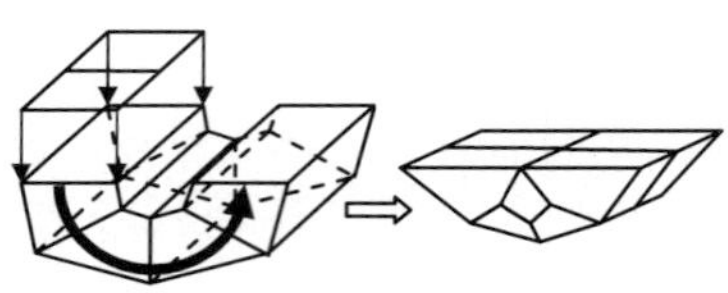

Fig. 8. Construction of a rectangular cap

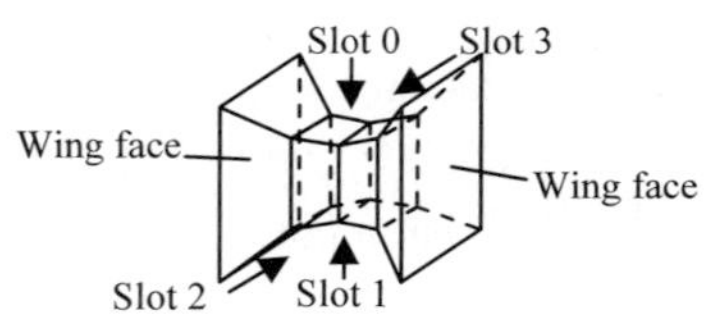

Fig. 9. Two wing faces and four slots of a core module

A cap that exposes a triangular pattern on the T-face can be constructed as follows. First, sixteen prism elements are created by sweeping four triangles as shown in Figure 7 (a). Next, the two corners of the two ends are joined together as shown in Figure 7 (b). Joining two corners yields a volumetric region surrounded by inside faces of the mesh as shown in Figure 7 (c). The region is called a *pipe* and is filled by two hex elements. The mesh is then deformed so that neither gaps nor overlaps remain when the cap is assembled with a core and other caps as shown in Figure 7 (d). This particular cap module consists of 16 prism elements and 2 hex elements. It is called a triangular cap and subdivides a corresponding quadrilateral face into eight triangles. Similarly, a rectangular cap, which exposes the rectangular pattern to the corresponding quadrilateral face, is constructed by applying a rectangular pattern to the T-face as shown in Figure 8.

2.2 Construction of a Standard Core

A core module has two *wing faces* and four *slots* as shown in Figure 9. Two wing faces are exposed to the exterior of the conversion template, and define subdivision patterns of the two opposite faces of the template. A slot is connected to a B-face of a cap module.

There are two types of standard cores, in which the subdivision patterns of the two wing faces are identical, called triangular core and rectangular core as shown in Figure 10. A standard core is created by sweeping a wing-face subdivision pattern four times and then deforming the middle section so that a cap module fits in a slot. A rectangular core shown in Figure 10 (a) consists of 16 hex elements, and a triangular core shown in Figure 10 (b) consists of 32 prism elements.

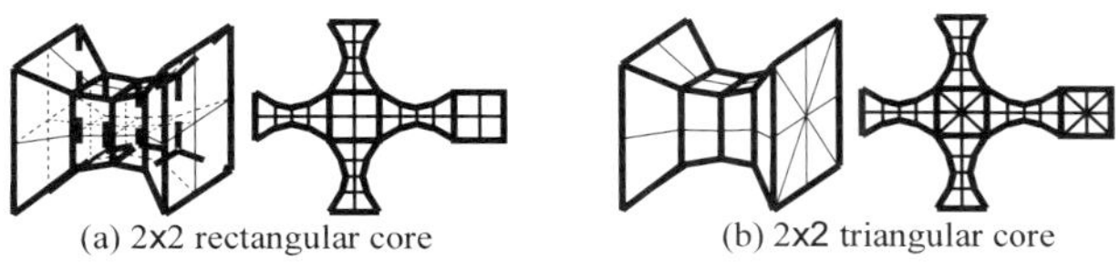

Fig. 10. Rectangular and Triangular cores

2.3 Assembling a Conversion Template

A core module and four cap modules can be assembled to form a conversion template for a hex element as shown in Figure 5. A B-face of a cap module is connected to a slot of a core module, and an F-face of a cap module is connected to an F-face of a neighboring cap module. After assembling, only T-faces of the four cap modules and wing faces of the core module are exposed to the exterior. This assembly is possible because the subdivision pattern of the B-face of a cap module matches the subdivision pattern of a slot of a core module, and the subdivision pattern of an F-face of a cap module always matches the F-face of the neighboring cap module regardless of the combination of four cap modules.

By combining appropriate core and caps, all templates except H5 shown in Figure 4 can be assembled. In H5 template, however, none of the opposite faces has an identical subdivision pattern, and it cannot be assembled with a standard core. H5 template requires a special type of core called a double-hoop core as explained in the next section.

2.4 Constructing a Double-Hoop Core

A standard core constrains at least one pair of opposite faces to have identical subdivision pattern. Since none of the opposite faces in H5 template shown in Figure 4 has identical subdivision pattern, standard core cannot be used for assembling H5 template. H5 template requires a core that has a triangular pattern on one of the wing faces and a rectangular pattern on the other.

A double-hoop core is created as follows. First, a H1 template is assembled from a rectangular core, a triangular cap, and three rectangular caps. The H1 template is oriented so that a triangular pattern is on the top face. The H1 template can be split into top and bottom halves as shown in Figure 11 (a). The cross-section between the top and bottom halves lies on a plane and exposes interior faces of cap and core modules as shown in Figure 11 (b). Two layers of hex elements are then created by sweeping the cross-sectional pattern and inserted between top and bottom halves as shown in Figure 11 (c). The result is a conformal mesh consisting of hex and prism elements. The top face of the mesh has a triangular pattern, and the bottom face a rectangular pattern. Four side faces are also rectangular pattern, however, each face has four rectangles vertically and two horizontally as shown in Figure 11 (d). The mesh is then re-oriented so that the triangular pattern face to the left, and then the middle section of the mesh is shrunk so that four caps fit to the mesh as shown in Figure 11 (e). Interior nodes should be re-located by an appropriate mesh-smoothing scheme.

Since this double-hoop core exposes a triangular pattern on one of the two wing faces, and a rectangular pattern on the other, the H5 template can be assembled by a double-hoop core, two rectangular caps, and two triangular caps. This core is called a double-hoop core because two 'hoops' of pipe volumes exist in the H5 template as shown in Figure 12.

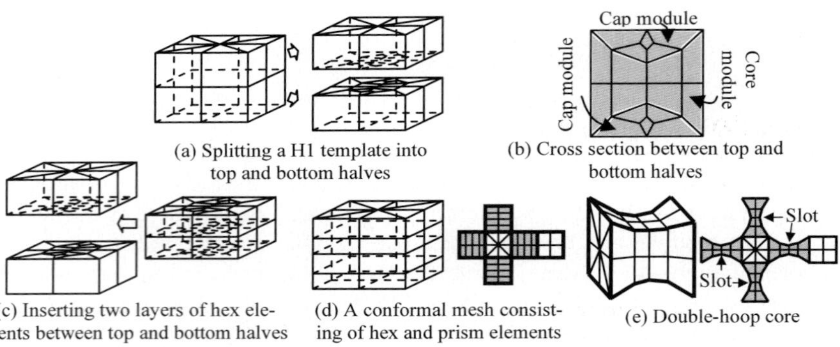

Fig. 11. Inserting two layers of hex elements in the middle of the H1 template

3 Subdividing Other Elements

3.1 Subdividing Prism Elements

A prism element is subdivided into six hex elements as shown in Figure 13 (c), or eight prism elements as shown in Figure 13 (b) or Figure 13 (c). The subdivision pattern is chosen based on the constraint on the triangular faces. A triangular face of a prism element is constrained if it is connected to a non-conformal face of the neighboring hex element or a triangular face of the neighboring tet element. Since the subdivision patterns on both triangular faces of a prism element needs to be identical, when one triangular face of a prism element is constrained, the opposite triangular face is also constrained. If multiple prism elements are chained by triangular faces, the constraint on one end of the chain propagate through the chain of the prism element to the other end, and all triangular faces of the prism chain are constrained.

If the two ends of a prism chain are exposed to the exterior of the mesh, each of the prism elements in the chain is subdivided into six hex elements as shown in Figure 13 (a). If one of the ends of a prism chain is connected to a tet element, and the other end is also connected to a tet element or exposed to exterior of the mesh, each of the prism elements in the chain is subdivided into eight prism elements as shown in Figure 13 (b).

If one of the ends of a prism chain is connected to a non-conformal face of a hex element, each of the prism elements in the chain is subdivided into eight prism elements as shown in Figure 13 (c). In this case, the subdivision pattern of the triangle connected to the non-conformal face must match half of the subdivision pattern of the non-conformal face as shown in Figure 14. However, due to the asymmetry of the triangle subdivision pattern, if both ends of a prism chain are connected to non-conformal faces, constraints from the two ends may conflict to each other as shown in Figure 15. Such a conflict can be resolved by subdividing one of the prism elements in the chain into two tet elements and three prism elements by adding two nodes inside the original prism element. However, this resolution often yields a low-quality element. The hex-dominant mesh generator thus needs to avoid such conflict when it creates an input hex-dominant mesh. Discussion on the requirement of the input hex-dominant mesh is also found in Section 4.

3.2 Subdividing Tet Elements

A tet element is subdivided by (1) adding a node at the center of the element, (2) subdividing triangular faces so that it satisfies the constraints imposed by the neighboring prism element or a non-conformal face, and then (3) connecting each triangle to the node created at the center of the original tet element.

A triangular face of a tet element can be constrained by the neighboring prism element, or a non-conformal face connected to the triangular face. All non-constrained faces of tet elements are subdivided into the pattern shown in Figure 16 (a), and constrained faces into the pattern shown in Figure 16 (b) so that the faces conform to the neighboring elements after the subdivision.

The nodes added at the center of the original tet element will be deleted by mesh smoothing [24-27], local transformation [28, 29], and edge collapse [30] as many as possible after the conversion.

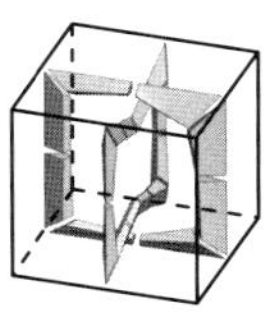

Fig. 12. Double hoop in the H5 template

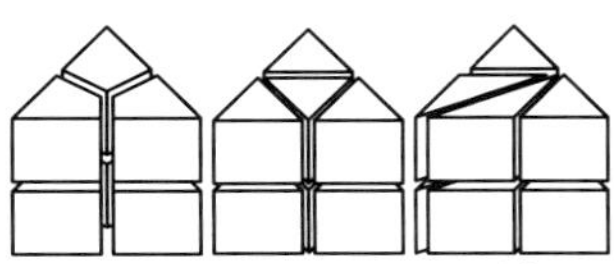

Fig. 13. Three possible subdivision patterns of a prism element

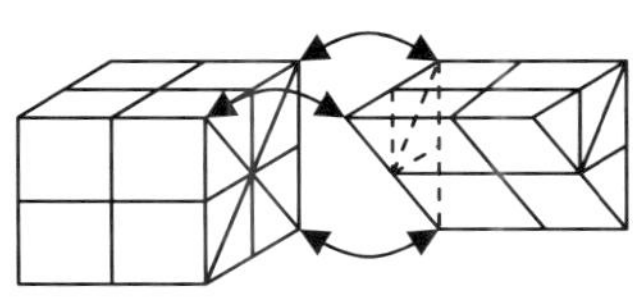

Fig. 14. Subdivision pattern of a triangular face connected to a non-conformal face

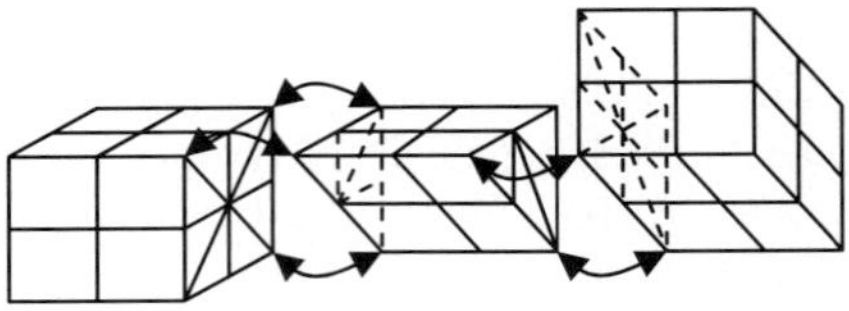

Fig. 15. Conflict of triangle subdivision patterns from the two ends of a prism chain

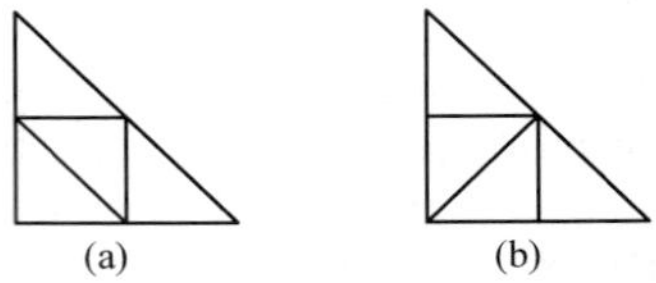

Fig. 16. Two subdivision patterns of a face of a tetrahedral element

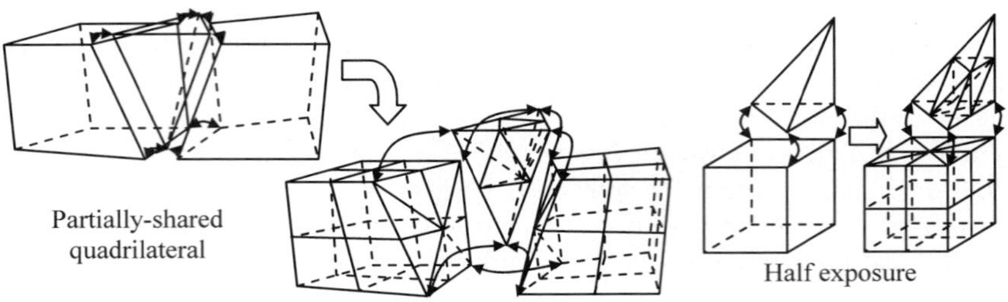

Fig. 17. Permissible non-conformal connections

4 Requirement of the Input Hex-Dominant Mesh

The input to the proposed method is a non-conformal hex-dominant mesh, which can include a triangular face directly connected to a quadrilateral face of a hex element. The conversion templates are flexible and can convert even more complex connection such as a partially-shared quadrilateral and half-exposed quadrilateral to a conformal connection as shown in Figure 17. Such flexibility helps a hex-dominant mesh generator to create more hex elements in the input mesh.

Nonetheless, the input hex-dominant mesh needs to satisfy certain conditions discussed in this section. A hex-dominant mesh that satisfies those conditions can easily be created by a conventional algorithm [19, 21, 31] with a slight modification.

4.1 Hex Element

A quadrilateral face of a hex element can be connected to a quadrilateral face of a neighboring hex or prism element, or one or two triangular faces of neighboring tet and/or prism elements.

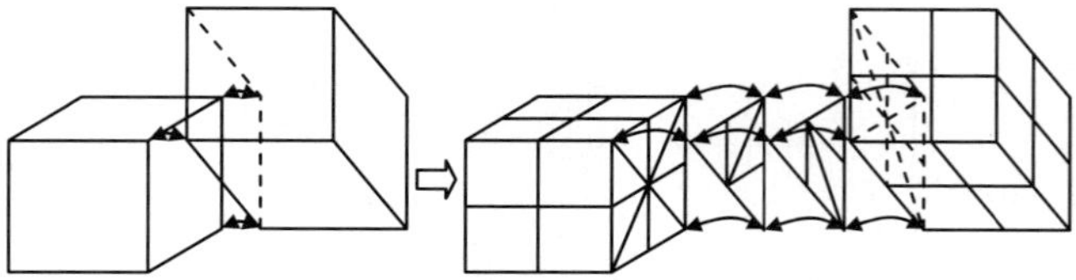

Fig. 18. Impermissible connection: Two quadrilateral faces sharing three nodes are not sharing a diagonal

However, when a quadrilateral face is sharing three nodes with a quadrilateral face of a neighboring hex element, two quadrilateral faces must share the diagonal. The connection shown in Figure 18 is thus impermissible.

4.2 Prism Element

All quadrilateral faces of prism elements need to be conformal. I.e., a quadrilateral face of a prism element needs to be completely shared by the neighboring prism or hex element, or completely exposed to the exterior. No triangular face can be connected to a quadrilateral face of a prism element.

A HEXHOOP template for a prism element can also be assembled as described in the original HEXHOOP paper [32], and such a template can subdivide a non-conformal prism element into conformal elements. However, the experiments performed in this research showed that applying a prism template often yields very low-quality elements. The possible reason why a prism template yields a low-quality element is because prism element are created in the difficult region where hex elements cannot be placed and the quality of the original prism element is often not high enough. The original element needs to be of reasonably high quality to have reasonably good elements after the conversion. Therefore, the current implementation does not utilize prism templates.

As already discussed in Section 1, the subdivision pattern of one of the two triangular faces of a prism element must match the pattern of the other triangular face. The hex-dominant mesh generator must avoid a prism element with two conflicting subdivision patterns on the two triangular faces.

If the input non-conformal hex-dominant mesh satisfies the above-mentioned conditions, the proposed conversion templates create a topologically-valid conformal hex-dominant mesh.

5 Cap Suppression

Some elements from cap modules may be deleted after the conversion. If two neighboring cap modules are sharing T-face as shown in Figure 19 (a), elements included in both cap modules can be eliminated by joining corresponding nodes as shown in Figure 19 (b). The subdivision pattern between the two conversion templates is shown in Figure 19 (c). Since the quality of the elements in cap modules tends to be relatively lower than for other elements, the overall mesh quality can be improved by eliminating cap modules.

(a) Two cap modules facing each other (b) Joining corresponding nodes (c) After suppressing caps

Fig. 19. Suppressing two cap modules facing each other

6 Conversion Examples

Figure 20 and Figure 21 show conformal hex-dominant meshes created by the proposed method. The input non-conformal hex-dominant meshes are created by first converting a tet mesh into a prism-tet hybrid mesh by the method described in [33], and then merging pairs of neighboring prism elements into hex elements. Table 1 shows the mesh statistics, including Scaled Jacobian, which is defined at each node as a triple scalar product of the edge vectors of three converging edges of an element divided by the product of the lengths of the three vectors.

Table 1. Statistics of the sample meshes

	# node	# element			Volume Ratio (%)			Minimum Scaled Jacobian		Max Tet Radius Ratio	Max Quad Warpage
		Hex	Prism	Tet	Hex	Prism	Tet	Hex	Prism		
Leg-bone	10,001	3,288	3,872	16,352	36	23	41	0.20	0.17	59.32	46 deg
AAA	13,462	7,468	6,728	2,772	68	30	2	0.20	0.20	59.86	54 deg

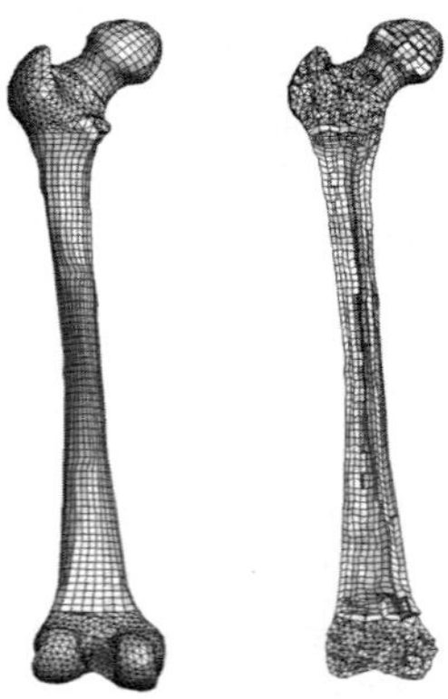

Fig. 20. Leg-bone model

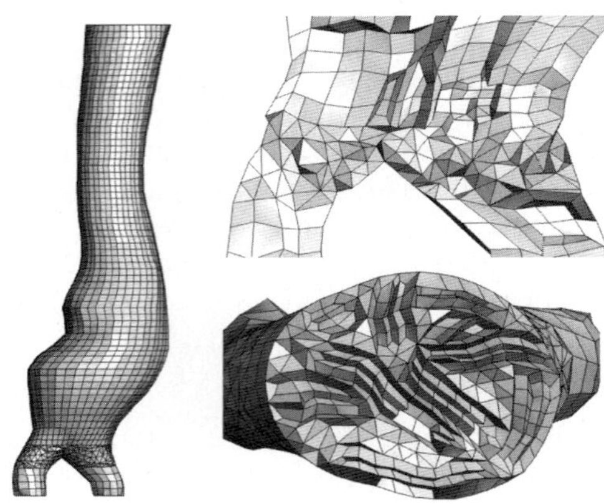

Fig. 21. Abdominal aortic aneurysm model

7 Experimental Finite Element Analysis

This section presents results from a series of structural analyses of a connecting rod of an automotive engine. Four sets of analyses, each with (1) linear hex-dominant, (2) linear tet, (3) quadratic hex-dominant, and (4) quadratic tet meshes have been performed with different node density to compare the characteristics of different types of meshes. Statistics of the meshes used in the analyses are shown in Table 2.

The analyses are calculated by Abaqus/CAE Version 6.7 [34] on a PC with a Dual Core AMD Opteron Processor at 2.0GHz and with 3GB RAM. A load of 18,000N is applied to the internal surface of the crankshaft bearing housing in form of uniform pressure, and one half of the surface of the wrist pin seat is constrained against displacements in the XY-plane, and symmetry boundary conditions are also applied. Figure 22 shows a von Mises stress plot calculated with one of the hex-dominant meshes.

Figure 23 and Figure 24 plot in-body maximum displacement and von Mises stress vs. the number of nodes calculated by using linear and quadratic elements, respectively. In both plots, the solutions from tet meshes show a smoother convergence, and the solutions from hex-dominant mesh show an oscillatory behavior. Nonetheless, the solutions from tet meshes and hex-dominant meshes both converge to a similar values as the number of nodes increases, and virtually no difference was observed between tet mesh and hex-dominant meshes if the number of node reaches 25,000 (linear) and 110,000 (quadratic).

Figure 25 depicts von Mises stresses along a path starting in the middle of the rod and moving toward the surface at the fillet location, calculated from quadratic tet and hybrid meshes. The path plot also indicates that the results from a tet mesh and a hybrid meshes converge to the same solution as the mesh resolution increases.

The results imply that hybrid meshes exhibit unsmooth convergence pattern when the mesh resolution is relatively low. However, the solution becomes as accurate as tet meshes when node density becomes high.

When the number of nodes is equal, a hex-dominant mesh is expected to take shorter computational time to obtain a solution than a tet mesh. The dominant component of the finite element computation comes from the matrix solver. If the number of nodes is equal, the size of the stiffness matrix is equal. However, as can be seen from Table 2, average number of elements per node is much smaller in a hybrid mesh than a tet mesh, and the bandwidth of the stiffness matrix from a hex-dominant mesh is thus smaller. Hence, the matrix solver should take shorter time to solve a stiffness matrix from a hex-dominant mesh.

The expectation was confirmed for quadratic meshes. As can be seen from Figure 26, a quadratic hex-dominant mesh required shorter CPU time than a quadratic tet mesh. The difference becomes significant as the number of nodes increases. Results from the linear

Table 2. Statistics of the meshes used in the analyses

Hex dominant mesh						Tet mesh				
# node		# elem	# elem. / # node		Hex vol. ratio (%)	# node quad.		# elem	# elem. / # node	
lin.	quad.		lin.	quad.		lin.	quad.		lin.	quad.
4,196	17,850	5,571	1.33	0.31	57	4,182	28,260	17,588	4.21	0.62
5,208	22,151	6,894	1.32	0.31	57	5,522	37,785	23,844	4.32	0.63
6,075	25,599	7,778	1.28	0.30	59	7,355	50,998	32,689	4.44	0.64
8,692	37,909	12,517	1.44	0.33	51	10,156	71,456	46,556	4.58	0.65
9,977	42,406	13,166	1.32	0.31	64	12,927	91,542	60,106	4.65	0.66
12,849	54,304	16,591	1.29	0.31	67	14,616	104,280	69,014	4.72	0.66
15,835	66,332	19,804	1.25	0.30	75	16,783	120,133	79,835	4.76	0.66
16,173	69,292	21,900	1.35	0.32	62	19,371	139,655	93,528	4.83	0.67
16,456	70,053	21,824	1.33	0.31	66	22,504	162,980	109,694	4.87	0.67
23,751	101,915	32,367	1.36	0.32	68	24,075	174,708	117,851	4.90	0.67
24,456	104,610	32,930	1.35	0.31	70	26,249	190,928	129,169	4.92	0.68
29,991	125,860	37,610	1.25	0.30	75	28,979	211,373	143,414	4.95	0.68
30,914	130,977	40,222	1.30	0.31	73	30,126	220,152	149,666	4.97	0.68
31,952	135,154	41,306	1.29	0.31	75	31,171	227,889	155,031	4.97	0.68

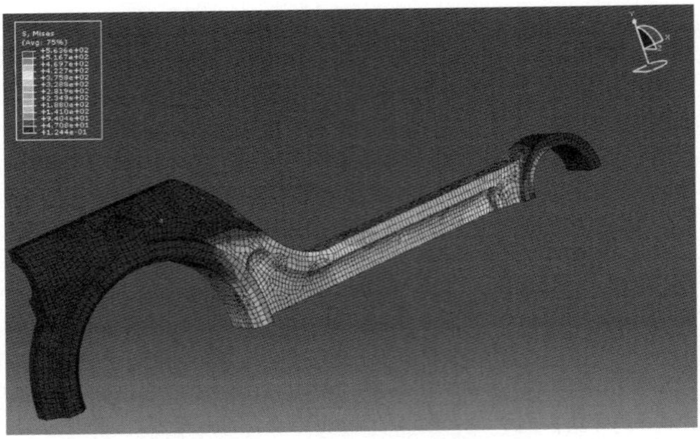

Fig. 22. Connecting rod

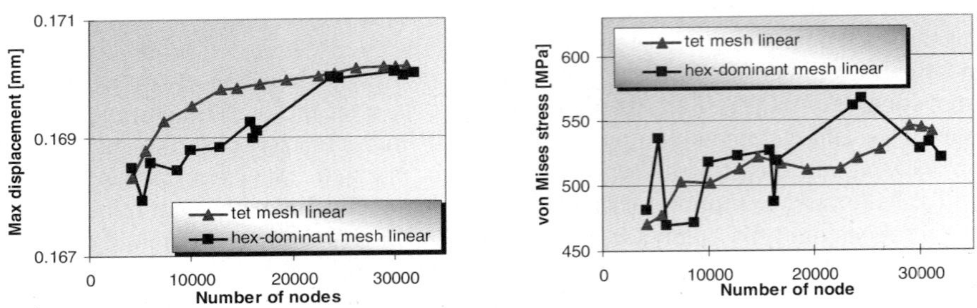

Fig. 23. Results with linear elements

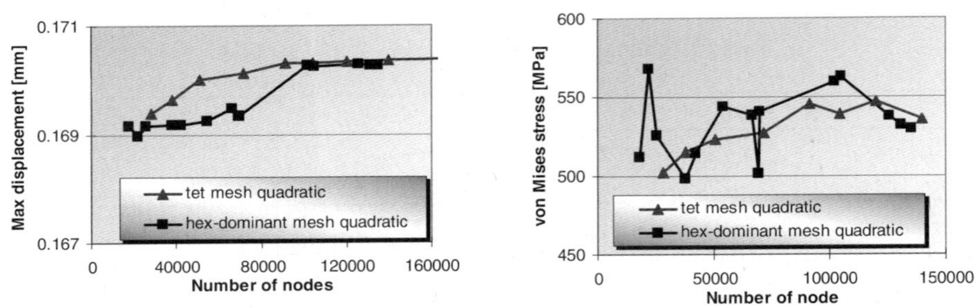

Fig. 24. Results with quadratic elements

meshes, however, contradicted the expectation; a linear hex-dominant mesh took more CPU time than a linear tet mesh. It could be due to the characteristic of the solver used in the experiment. Further study is required to find the cause of the contradiction.

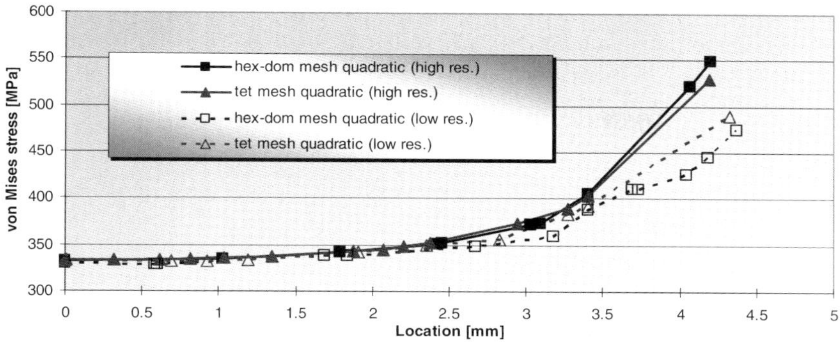

Fig. 25. Path plot of the von Mises stress through the rod

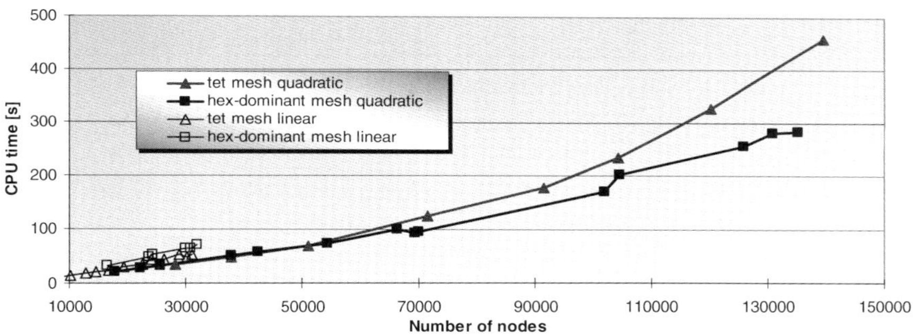

Fig. 26. CPU time plot

The experimental results imply that a quadratic hex-dominant mesh is advantageous especially when the mesh resolution is high compared to a tet mesh. Since the difference between a linear element and a quadratic element becomes significant in a non-linear analysis, a quadratic hex-dominant mesh should be particularly effective for a large-scale non-linear analysis. Further research is needed to test this speculation.

8 Potential Improvements

The quality of the output conformal hex-dominant mesh can be further improved by taking as input a higher quality non-conformal hex-dominant mesh. If a subdivision template is applied to a low-quality element, the output mesh will include an even lower-quality element. Such a low-quality element should be avoided by eliminating low-quality elements included in the input non-conformal mesh.

The quality of the output mesh can also be improved by avoiding certain configuration of the input mesh can be avoided. The quality of the elements included in non-trivial templates (H1, H2, H5, H6, H7, and H8 templates) is relatively low due to the complex structure of the templates - the element quality of the non-trivial templates is shown in Table 3. Values in parentheses indicate the mesh quality after applying conventional smoothing schemes. In particular, H5 template has the lowest quality element among the non-trivial templates. The quality of the output mesh therefore can be improved by developing and applying a new non-conformal hex-dominant meshing scheme that reduces the usage of the H5 template.

The quality of the output mesh can also be improved by applying mesh-smoothing schemes. Note, however, that not all conventional mesh-smoothing schemes perform well on a hybrid mesh.

Table 3. The quality of the non-trivial templates

	H1	H2	H5	H6	H7	H8
Minimum Scaled Jacobian	0.34 (0.28)	0.34 (0.30)	0.20 (0.20)	0.34 (0.21)	0.34 (0.20)	0.34 (0.20)
Worst Aspect Ratio	2.9 (3.0)	2.9 (3.1)	5.3 (4.6)	2.9 (3.0)	2.9 (3.1)	2.9 (3.2)
Largest Quadrilateral Warpage	28.2deg (1.1deg)	28.2deg (1.9deg)	62.3deg (18.1deg)	28.2deg (1.4deg)	28.2deg (1.3deg)	28.2deg (0.8deg)

9 Conclusions

This paper has presented a method for converting a non-conformal hex-dominant mesh into a conformal hex-dominant mesh by applying a set of subdivision templates. The method always creates a conformal hex-dominant mesh if the input non-conformal hex-dominant mesh satisfies a certain condition, where a non-conformal hex-dominant mesh can be easily created by an existing non-conformal hex-dominant meshing scheme. Experimental results indicate the method creates a good-quality conformal hex-dominant mesh. A series of finite element simulations have been performed and the results imply that a conformal hex-dominant mesh can be effective for a large-scale non-linear analysis.

References

[1] George, P.L.: Tet Meshing: Construction, Optimization and Adaptation. In: Proceedings of 8th International Meshing Roundtable, pp. 133–141 (1999)

[2] Yamakawa, S., Shimada, K.: Anisotropic Tetrahedral Meshing via Bubble Packing and Advancing Front. International Journal for Numerical Methods in Engineering 57, 1923–1942 (2003)

[3] Shewchuk, J.: Constrained Delaunay Tetrahedralizations and Provably Good Boundary Recovery. In: Proceedings of 11th International Meshing Roundtable, pp. 193–204 (2002)

[4] Blacker, T.D., Meyers, R.J.: Seams and Wedges in Plastering: A 3-D Hexahedral Mesh Generation Algorithm. Engineering with Computers 2, 83–93 (1993)

[5] Schneiders, R.: A Grid-based Algorithm for the Generation of Hexahedral Element Meshes. Engineering with Computers 12, 168–177 (1996)

[6] Tautges, T.J., Blacker, T., Mitchell, S.A.: The Whisker Weaving Algorithm: A Connectivity-Based Method for Constructing All-Hexahedral Finite Element Meshes. International Journal for Numerical Methods in Engineering 39, 3327–3349 (1996)

[7] Ledoux, F., Weill, J.-C.: An Extension of the Reliable Whisker Weaving Algorithm. In: Proceedings of 16th International Meshing Roundtable, pp. 215–232 (2007)

[8] Staten, M.L., Owen, S.J., Blacker, T.D.: Unconstrained Paving & Plastering: A New Idea for All Hexahedral Mesh Generation. In: Proceedings of 14th International Meshing Roundtable, pp. 399–416 (2005)

[9] Taghavi, R.: Automatic Block Decomposition Using Fuzzy Logic Analysis. In: Proceedings of 9th International Mething Roundtable, pp. 187–192 (2000)

[10] Kwak, D.-Y., Im, Y.-T.: Remeshing for Metal Forming Simulations-Part II: Three-Dimensional Hexahedral Mesh Generation. International Journal for Numerical Methods in Engineering 53, 2501–2528 (2002)

[11] Hariya, M., Nishigaki, I., Kataoka, I., Hiro, Y.: Automatic Hexahedral Mesh Generation with Feature Line Extraction. In: Proceedings of 15th International Meshing Roundtable, pp. 453–467 (2006)

[12] Maréchal, L.: A New Approach to Octree-Based Hexahedral Meshing. In: Proceedings of 10th International Meshing Roundtable, pp. 209–221 (2001)

[13] Yamakawa, S., Shimada, K.: HEXHOOP: Modular Templates for Converting a Hex-Dominant Mesh to an ALL-Hex Mesh. Engineering with Computers 18, 211–228 (2002)

[14] Mitchell, S.A.: The All-Hex Geode-Template for Conforming a Diced Tetrahedral Mesh to any Diced Hexahedral Mesh. In: Proceedings of 7th International Meshing Roundtable, pp. 295–305 (1998)

[15] White, D.R., Tautges, T.J.: Automatic Scheme Selection for Toolkit Hex Meshing. International Journal for Numerical Methods in Engineering 49, 127–144 (2000)

[16] Lai, M., Benzley, S., White, D.: Automated Hexahedral Mesh Generation by Generalized Multiple Source to Multiple Target Sweeping. International Journal for Numerical Methods in Engineering 49, 261–275 (2000)

[17] Shepherd, J., Mitchell, S.A., Knupp, P., White, D.: Methods for Multisweep Automation. In: Proceedings of 9th International Meshing Roundtable, pp. 77–87 (2000)

[18] Quadros, W.R., Shimada, K.: Hex-Layer: Layered All-Hex Mesh Generation on Thin Section Solids via Chordal Surface Transformation. In: Proceedings of 11th International Meshing Roundtable, pp. 169–180 (2002)

[19] Meshkat, S., Talmor, D.: Generating a Mixed Mesh of Hexahedra, Pentahedra and Tetrahedra from an Underlying Tetrahedral Mesh. International Journal for Numerical Methods in Engineering 49, 17–30 (2000)

[20] Meyers, R.J., Tautges, T.J., Tuchinsky, P.M.: The "Hex-Tet" Hex-Dominant Meshing Algorithm as Implemented in CUBIT. In: Proceedings of 7th International Meshing Roundtable, pp. 151–158 (1998)

[21] Owen, S.J., Saigal, S.: H-Morph: an Indirect Approach to Advancing Front Hex Meshing. International Journal for Numerical Methods in Engineering 49, 289–312 (2000)

[22] Yamakawa, S., Shimada, K.: Hex-Dominant Mesh Generation with Directionality Control via Packing Rectangular Solid Cells. In: Proceedings of Geometric Modeling and Processing (2002)

[23] Becker, E.B., Carey, G.F., Oden, J.T.: Finite Elements: an Introduction, vol. 1. Prentice-Hall, Englewood Cliffs (1981)

[24] Knupp, P.M.: Achieving Finite Element Mesh Quality via Optimization of the Jacobian Matrix Norm and Associated Quantities. Part II - A Framework for Volume Mesh Optimization and the Condition Number of the Jacobian Matrix. International Journal for Numerical Methods in Engineering 48, 1165–1185 (2000)

[25] Freitag, L.A., Plassmann, P.E.: Local Optimization-based Untangling Algorithms for Quadrilateral Meshes. In: Proceedings of 10th International Meshing Roundtable, pp. 397–406 (2001)

[26] Zhou, T., Shimada, K.: An Angle-Based Approach to Two-dimensional Mesh Smoothing. In: Proceedings of 9th International Meshing Roundtable, pp. 373–384 (2000)

[27] Calvo, N.A., Idelsohn, S.R.: All-Hexahedral Mesh Smoothing with a Node-Based Measure of Quality. International Journal for Numerical Methods in Engineering 50, 1957–1967 (2001)

[28] Klingner, B.M., Shewchuck, J.R.: Aggressive Tetrahedral Mesh Improvement. In: Proceedings of 16th International Meshing Roundtable, pp. 3–23 (2007)

[29] Joe, B.: Construction of Three-Dimensional Improved-Quality Triangulations Using Local Transformations. SIAM Journal on Scientific Computing 16, 1292–1307 (1995)

[30] Molinari, J.F., Ortiz, M.: Three-Dimensional Adaptive Meshing by Subdivision and Edge-Collapse in Finite-Deformation Dynamic-Plasticity Problems with Application to Adiabatic Shear Banding. International Journal for Numerical Methods in Engineering 53, 1101–1126 (2002)

[31] Yamakawa, S., Shimada, K.: Fully-Automated Hex-Dominant Mesh Generation with Directionality Control via Packing Rectangular Solid Cells. International Journal for Numerical Methods in Engineering 57, 2099–2129 (2003)

[32] Yamakawa, S., Shimada, K.: HEXHOOP: Modular Templates for Converting a Hex-dominant Mesh to an All-hex Mesh. In: Proceedings of 10th International Meshing Roundtable, pp. 235–246 (2001)

[33] Yamakawa, S., Shimada, K.: Converting a Tetrahedral Mesh to a Prism-Tetrahedral Hybrid Mesh for FEM Accuracy and Efficiency. In: Proceedings of ACM Symposium on Solid and Physical Modeling 2008, pp. 287–294 (2008)

[34] "Abaqus," Dassault Systemes, `http://www.simulia.com`

Local Dual Contributions on Simplices: A Tool for Block Meshing*

X. Roca and J. Sarrate

Laboratori de Càlcul Numèric (LaCàN)
Departament de Matemàtica Aplicada III
Universitat Politècnica de Catalunya
Jordi Girona 1-3, E-08034 Barcelona, Spain
{xevi.roca,jose.sarrate}@upc.edu
http://www-lacan.upc.edu

Summary. Our final goal is to automatically generate a block decomposition of a given domain without previously meshing its boundary. To this end, we propose to obtain directly a valid dual arrangement that leads to a block mesh. In particular, we introduce a tool based on the new concept of local dual contributions. That is, given a domain we first generate a reference mesh composed by simplices, triangles in 2D and tetrahedra in 3D. Then, we add local dual contributions in the elements of a reference mesh to describe a valid dual arrangement. These local dual contributions are added according to a set of hierarchical rules to ensure the correct matching of adjacent contributions. The first implementation of the tool has been successfully applied to the block decomposition of several geometries, ranging from convex and non-convex domains to geometries with holes. Further research is under way in order to extend the applicability of the presented tool to more complicated geometries.

Keywords: Mesh generation, dual, block decomposition, hexahedra, quadrilaterals.

1 Introduction

The finite element method and finite volume method are widely used to solve problems in applied sciences and engineering. Both techniques require a valid discretization of the problem geometry. There exist several mature techniques [1, 2] to obtain these discretizations. However, meshes composed by block elements, quadrilateral (2D) or hexahedra (3D), are preferred for specific problems [3, 4]. Therefore, the solution of these problems demands a tool that automatically generates block elements, see references [5, 6] for a survey of available techniques. On the one hand, several techniques attempt to generate this kind of mesh by directly generating hexahedral elements. On the other hand, the nature of the hexahedral meshes have induced to consider the dual problem [7, 8, 9, 10].

This work is devoted to obtaining a valid topological block decomposition of a given domain without a previous discretization of the boundary. We propose an algorithm that explicitly inserts descriptions of the dual curves (2D) or the dual

* This work was partially sponsored by the Spanish *Ministerio de Educación y Ciencia* under grants DPI2007-62395 and BIA2007-66965.

surfaces (3D) and handles their intersections. Finally, the primal mesh is obtained from this dual construction. Analogous algorithms have been previously sketched in the literature. In particular, Murdoch et al. [11] briefly introduce the *Twist Plane Insertion* algorithm, and Calvo [12] speculates about the *free stroking*[1] method. In both works such an algorithm is not detailed because there is not a procedure that manages the insertion of dual curves (2D) or surfaces (3D), and that properly describes the intersections. To fill this gap, we introduce here the new concept of local dual contributions which allow to obtain a discrete representation of the dual arrangement. In order to ensure the correct matching of adjacent contributions we present a set of hierarchical rules. The concept of local dual contribution and this set of hierarchical matching rules are the main contribution of our work.

2 Background

2.1 Block Meshes

A mesh in $\mathbb{R}^n$, for our applications n is either 2 or 3, is composed by d-dimensional entities where $d = 0, \ldots, n$. An entity of maximum dimension, i.e. a n-dimensional entity, is also referred as **element**. From 0-dimensional to 3-dimensional entities we use the following notation: **node**, **edge**, **face** and **cell**.

A mesh entity that is topologically equivalent (homeomorph) to a d-dimensional unit interval, i.e. $[0,1]^d$, is called a **block**. Therefore a mesh composed by blocks is called a **block mesh**. This notation allows to denote to quadrilateral and hexahedral entities with the same term. Hence, quadrilateral and hexahedral meshes are examples of block meshes.

An $(n-d)$-dimensional entity, i.e. an entity of co-dimension d, is referred as a **d-co-entity**, where d is an integer value such that $0 \le d \le n$. For instance, an element of a mesh is also a **co-node**, and a face in a 3-dimensional mesh is also a **co-edge**. Table 1 shows the correspondences between entities and their associated co-entities for the 2D and 3D cases.

Table 1. Correspondence between d-dimensional entities and their associated co-entities for the 2D (left) and 3D (right) cases

d-entity	d	co-entity	co-dimension
Node	0	co-face	2
Edge	1	co-edge	1
Face	2	co-node	0
–	–	–	–

d-entity	d	co-entity	co-dimension
Node	0	co-cell	3
Edge	1	co-face	2
Face	2	co-edge	1
Cell	3	co-node	0

2.2 Dual of a Block Mesh

The concept of mesh dual has been widely used in theoretical [13, 14] and practical [7, 8, 9, 10] works on block mesh generation. Specifically, Murdoch et al.

[1] Named "libre trazado" in the original work in Spanish.

Table 2. Primal entities for a block mesh and their associated dual entities for a quadrilateral and a hexahedral mesh, respectively

Primal	d	Dual	2D dual	3D dual
Node	0	$\mathcal{D}$-co-node	$\mathcal{D}$-face	$\mathcal{D}$-cell
Edge	1	$\mathcal{D}$-co-edge	$\mathcal{D}$-edge	$\mathcal{D}$-face
Quadrilateral	2	$\mathcal{D}$-co-face	$\mathcal{D}$-node	$\mathcal{D}$-edge
Hexahedron	3	$\mathcal{D}$-co-cell	–	$\mathcal{D}$-node

[11], Calvo [12], Thurston [13], and Mitchell [14] extend the concept of mesh dual and present it as an arrangement of intersecting co-curves that bisect blocks in each direction. In 2D, they consider the dual of a quadrilateral mesh defined by a set of curves that bisect the edges of the elements and that intersect in the center of the quadrilaterals. In 3D, the dual of a hexahedral mesh is presented as an arrangement of surfaces that bisect the faces of the hexahedra and that intersect in the center of each hexahedron. We denote this extended view of the dual, referred as Spatial Twist Continuum (STC) in [11], as the **$\mathcal{D}$-space** and one of its composing entities as a **$\mathcal{D}$-co-curve**. That is, in 2D the $\mathcal{D}$-space is an arrangement of **$\mathcal{D}$-curves**, and in 3D is an arrangement of **$\mathcal{D}$-surfaces**. Notice that not all the dual arrangements lead to a valid block mesh. Specifically, Mitchell presents in [14] the conditions that have to be satisfied by a dual arrangement in order to obtain a block mesh from it.

The remaining dual entities of a mesh are obtained by substituting each primal entity by its co-entity. That is, given a d-dimensional primal entity we have an associated dual entity of dimension $n - d$. We assume that dual entities are also in the $\mathcal{D}$-space and we denote them by **$\mathcal{D}$-entities**. Thus, a primal node is substituted by a **$\mathcal{D}$-element**, and a primal element by a **$\mathcal{D}$-node**. Table 2 presents the association between primal entities and their dual co-entities and entities for 2-dimensional and 3-dimensional meshes. For instance, the dual of an edge is always a **$\mathcal{D}$-co-edge** that corresponds to a **$\mathcal{D}$-edge** for the dual of a quadrilateral mesh and to a **$\mathcal{D}$-face** for the dual of a hexahedral mesh.

The $\mathcal{D}$-space is in fact a description of the inherent constraints of the block meshing problem. Note that each co-edge of a block element has an opposed co-edge, and both are connected by $(n-1)$ $\mathcal{D}$-co-curves. In 2D, each edge has an opposed edge and both are connected by a piece of a $\mathcal{D}$-curve. Along one of these $\mathcal{D}$-curves we have an associated row of quadrilateral elements. In 3D, each hexahedron face is bisected by two $\mathcal{D}$-surfaces that connect the face with an opposite face. The intersection of these $\mathcal{D}$-surfaces define a $\mathcal{D}$-curve associated to a stack of primal hexahedra. In addition, each $\mathcal{D}$-surface determines a layer of primal hexahedra. Note that each $\mathcal{D}$-co-curve is composed by several $\mathcal{D}$-co-edges separated by $\mathcal{D}$-co-faces. Therefore in 2D the $\mathcal{D}$-curves are composed by $\mathcal{D}$-edges separated by $\mathcal{D}$-nodes, and in 3D the $\mathcal{D}$-surfaces are composed by $\mathcal{D}$-faces separated by $\mathcal{D}$-edges.

2.3 Local Dual Contributions

The basic idea of the proposed algorithm is to generate a valid topological block decomposition of a given domain without a previous discretization of the boundary. Our algorithm explicitly inserts descriptions of the $\mathcal{D}$-co-curves in the $\mathcal{D}$-space. Finally, the primal mesh is obtained from this dual construction.

The crucial step is the development of an automatic tool that allows the construction, manipulation and management of $\mathcal{D}$-co-curves. Hence, we propose to create a discrete representation of the $\mathcal{D}$-space referred as $\overline{\mathcal{D}}$**-space**, see Figure 1. To this end, we first create a simple discretization of the domain called **reference mesh**. The proposed algorithm adds discretized versions of the $\mathcal{D}$-co-curves, the $\overline{\mathcal{D}}$**-co-curves**, in the elements of the reference mesh. Each one of the $\overline{\mathcal{D}}$-co-curves is composed by several discretized $\mathcal{D}$-co-edges, denoted by $\overline{\mathcal{D}}$**-co-edges**. We impose that the intersections between different $\overline{\mathcal{D}}$-co-curves are captured by the boundaries of the $\overline{\mathcal{D}}$-co-edges, the $\overline{\mathcal{D}}$**-co-faces**. That is, we consider that our discretized version of the $\mathcal{D}$-co-curves, the $\overline{\mathcal{D}}$-co-curves, define a non-manifold $(n-1)$-dimensional mesh. The intersections between $\overline{\mathcal{D}}$-co-curves and the regions bounded by this arrangement define the $\overline{\mathcal{D}}$-space.

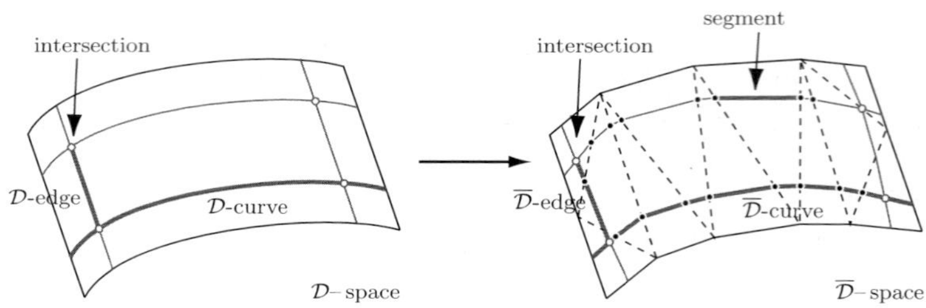

Fig. 1. $\mathcal{D}$-space for a 2D domain and its discretized version, the $\overline{\mathcal{D}}$-space

In our discrete approach the domain is substituted by a reference mesh of simplices. We use the simplices of the reference mesh as containers of linear entities of co-dimension one. We propose to describe the $\overline{\mathcal{D}}$-co-curves and their intersections in the $\overline{\mathcal{D}}$-space as the composition of several co-edges, the **local dual contributions**. Thus, for 2D applications we store in the triangles the segments that define the $\overline{\mathcal{D}}$-curves. In Figure 1 we present the $\overline{\mathcal{D}}$-space for a 2D domain. Note that a $\overline{\mathcal{D}}$-curve is composed by several $\overline{\mathcal{D}}$-edges (the piece-wise curves between two white circles). Moreover, each $\overline{\mathcal{D}}$-edge is composed by one or several segments (the straight lines between two black circles) that determine a local contribution to the dual of a triangle of the reference mesh. Analogously, for 3D problems we store in the tetrahedra the planar faces that compose the $\overline{\mathcal{D}}$-surfaces. We remark that each reference element can store none, one or several local dual contributions that can intersect or not between them.

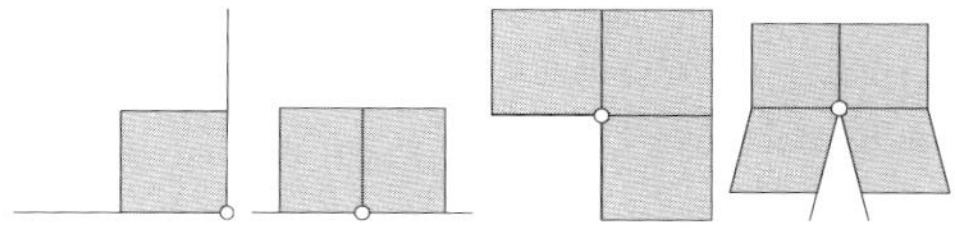

(a) (b) (c) (d)

Fig. 2. Vertex classification for 2D domains according to their ideal primal mesh close to the boundary: **(a)** *end*, **(b)** *side*, **(c)** *corner*, and **(d)** *reversal*

2.4 Features

We consider that a domain is described by the following types of entities: **vertex**, **curve**, **surface** and **volume**. Some of the special characteristics of the boundary of the domain, the features, have to be explicitly preserved by the final block mesh. To this end, in this section we classify the features according to the *ideal mesh* around each entity. Note that the name election is based on *Paving* [15] and *Submapping* [16] notation. For a 2D mesh the vertices are classified as, see Figure 2: *i)* ***side***, the ideal mesh is composed by two quadrilateral elements sharing the vertex. In addition, these quadrilaterals have only one edge on the boundary; *ii)* ***end***, the ideal mesh is composed by one quadrilateral element touching the vertex and with only two edges on the boundary; *iii)* ***corner***, the ideal mesh has three quadrilateral elements sharing the vertex. Two elements have only one edge on the boundary, and the middle element has four inner edges; and *iv)* ***reversal***, the ideal mesh has four quadrilateral elements touching the vertex. Two elements have only one edge on the boundary, and the two middle elements have four inner edges.

A vertex is of type ***feature*** if it is not of type ***side***. For 3D applications we have to classify curves and vertices. First, a piece of a curve can be classified as: *i)* ***side***, the ideal mesh is composed by two hexahedral elements sharing the curve and each one with one face on the boundary; *ii)* ***end***, its ideal mesh consists in one hexahedron touching the curve and with two faces on the boundary; *iii)* ***corner***, the ideal mesh has three hexahedral elements incident to the curve. Two elements have only one face on the boundary, and the middle element has six inner faces; and *iv)* ***reversal***, the ideal mesh has four hexahedra touching

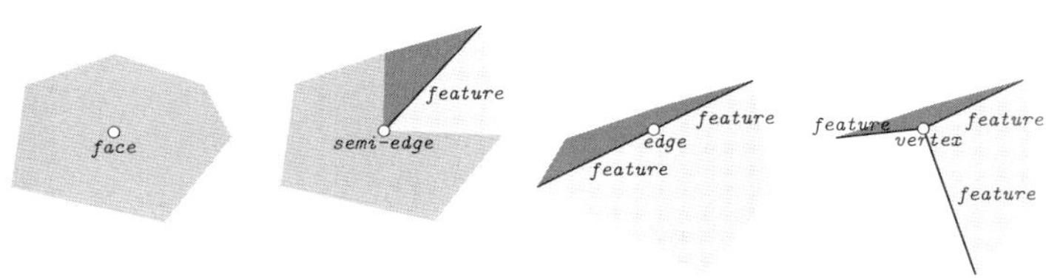

(a) (b) (c) (d)

Fig. 3. A sample for each vertex type: **(a)** *face* vertex; **(b)** *semi-edge* vertex; **(c)** *edge* vertex; and **(d)** *vertex* type the curve. Two elements have only one face on the boundary, and the two middle elements have six inner faces.

A curve is of type *feature* if it is not of type *side*. We determine the classification of the vertices according to the classification of their adjacent curves, see Figure 3: *i) face*, there is no adjacent curve of type *feature*; *ii) semi-edge*, one adjacent curve is of type *feature*; *iii) edge*, two adjacent curves are of type *feature*; and *iv) vertex*, more than two adjacent curves are of type *feature*.

3 Constructing the Dual

3.1 Requirements and Aim of the Algorithm

Any block mesh has to fullfill several topological, geometrical and qualitative requirements in order to be used in a numerical simulation. A detailed exposition of these conditions for block meshes can be found in [5, 6]. For the purposes of this work we highlight two of them:

- **Geometric matching.** The block mesh has to reproduce the geometric features of the domain. Moreover, the mesh has to be adapted to each kind of feature. Hence, in Section 2.4 we have defined the vertices (2D) and curves (3D) features according to the ideal mesh around them. For instance, Figure 2 and Figure 4 present the desired primal and dual configuration for each kind of vertex.
- **Boundary sensitivity.** The mesh has to present layers of blocks that follow the boundary of the domain. This condition can be expressed in the $\mathcal{D}$-space, where the layers of blocks are substituted by $\mathcal{D}$-co-curves. Specifically, we have to generate a series of $\mathcal{D}$-co-curves by shrinking the boundary of the domain inwards the domain.

The proposed algorithm is based on the following two paradigms:

- **Block decomposition oriented mesher.** One of the most reliable approaches to block meshing is to decompose the domain in several coarse blocks. The final mesh is obtained by meshing each block and verifying the mesh compatibilities between blocks. Several software packages implement this paradigm [17, 18] providing a set of tools in order to define, by hand, a block topology decomposition of the domain. The user is responsible for determining the topological

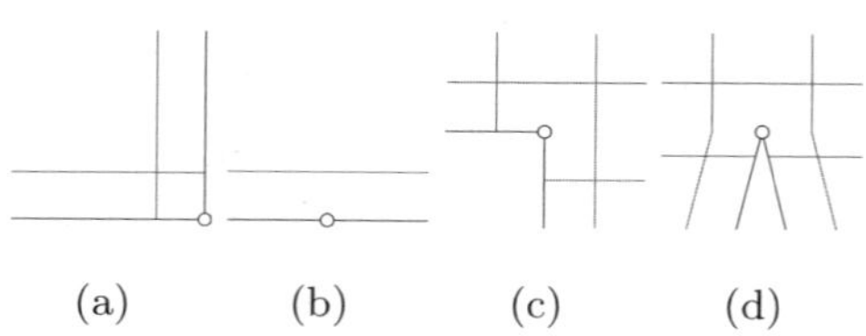

Fig. 4. Ideal mesh dual close to boundary features of a 2D domain: **(a)** *end*, **(b)** *side*, **(c)** *corner*, and **(d)** *reversal*

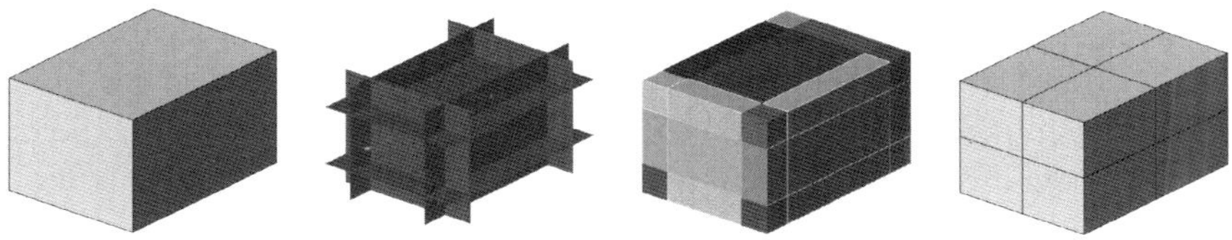

(a) (b) (c) (d)

Fig. 5. Block meshing steps in 3D: **(a)** obtain the domain; **(b)** add $\mathcal{D}$-surfaces; **(c)** create $\overline{\mathcal{D}}$-cells; and **(d)** block mesh

relation of the blocks but is allowed to loosely dispose the vertices of the blocks. Then the package is able to adapt the mesh to the domain and to untangle the edges of the blocks by relaxing node location. Therefore, our goal is to obtain a valid coarse block decomposition of the domain.

- **Unconstrained boundary mesher.** A previous discretization of the boundary adds additional constraints that are difficult to fulfill, in particular for hexahedral meshes. Several authors [11, 12] have suggested to develop algorithms in order to mesh domains without a prescribed boundary mesh. In particular, Staten et al. [19] fully develop this idea. In our work we also follow this approach and we propose an algorithm that directly generates the dual of a block mesh without a previous discretization of the boundary.

3.2 Algorithm Proposal

Taking into account the requirements of a valid dual [14] and the requirements of the block mesh, Section 3.1, we propose an algorithm that explicitly inserts descriptions of the $\mathcal{D}$-co-curves in the $\mathcal{D}$-space. Our algorithm is structured in four steps, see Figure 5 for a 3D graphical description.

The main task of the proposed algorithm is to build up a representation of the $\mathcal{D}$-co-curves. To this end, as we mentioned in Section 2.3, we propose to consider a discrete representation of the $\mathcal{D}$-space, the $\overline{\mathcal{D}}$-space. In the following we present an algorithm and we describe the second and the third step:

Algorithm 1. $\overline{\mathcal{D}}$-space block meshing algorithm

Reference mesh. We obtain a Constrained Delaunay Triangulation (**CDT**) of the domain. In our implementation we have used TetGen [20].
Add $\overline{\mathcal{D}}$-co-curves. Create an arrangement of intersecting $\overline{\mathcal{D}}$-co-curves composed by several $\overline{\mathcal{D}}$-co-edges.
Create $\overline{\mathcal{D}}$-co-nodes. Cap the $\overline{\mathcal{D}}$-co-edges with the boundary of the reference mesh. Note that the boundary of the reference mesh is composed by co-edges. We obtain a set of bounded regions, the $\overline{\mathcal{D}}$-co-nodes, sharing at most a common $\overline{\mathcal{D}}$-co-edge.
Block mesh. Once we have the $\overline{\mathcal{D}}$-space, a discretized representation of the $\mathcal{D}$-space, we obtain the final block mesh as the dual of the dual.

Add $\overline{\mathcal{D}}$-co-curves. Each $\overline{\mathcal{D}}$-co-curve is a mesh composed by co-edge elements. We modify the reference mesh in order to make it compatible with the added co-edges. Specifically, we split the elements of the reference mesh to match them with the inserted co-edges. In 2D, the discretized versions of $\mathcal{D}$-co-curves, the $\overline{\mathcal{D}}$-curves, are composed by segments that intersect in a set of shared nodes, the $\overline{\mathcal{D}}$-nodes. The $\overline{\mathcal{D}}$-edges are delimited by the intersections of the $\overline{\mathcal{D}}$-curves. We split the reference mesh elements to match these new segments, see Figure 1. In 3D, the discretized versions of the $\mathcal{D}$-co-curves, the $\overline{\mathcal{D}}$-surfaces, are represented by triangular meshes that can intersect in two possible configurations: *i)* two triangular meshes can intersect in a set of shared edges that compose the discretized versions of the $\mathcal{D}$-edges and $\mathcal{D}$-curves, denoted by $\overline{\mathcal{D}}$-edges and $\overline{\mathcal{D}}$-curves; and *ii)* three triangular meshes can intersect in a shared node which corresponds to a $\overline{\mathcal{D}}$-node. The intersections of the $\overline{\mathcal{D}}$-surfaces bound the discretized versions of the $\mathcal{D}$-faces, the $\overline{\mathcal{D}}$-faces. We split the tetrahedra of the reference mesh obtaining new entities that match properly with the additional triangles that conform $\mathcal{D}$-surfaces.

Create $\overline{\mathcal{D}}$-co-nodes. Taking into account that we have split the reference mesh to add the $\overline{\mathcal{D}}$-co-curves, we obtain a conformal n-dimensional mesh of the $\mathcal{D}$-space, the $\overline{\mathcal{D}}$-space. We obtain a set of regions, the **$\overline{\mathcal{D}}$-co-nodes**, bounded by the $\overline{\mathcal{D}}$-co-edges and the boundary co-edges. In 2D, we obtain a set of areas, the $\overline{\mathcal{D}}$-faces, composed by faces and delimited by the edges that compose the $\overline{\mathcal{D}}$-edges, and the split boundary of the reference mesh. In 3D, we obtain several volumes, the $\overline{\mathcal{D}}$-cells, composed by cells and delimited by the composing faces of the $\overline{\mathcal{D}}$-surfaces, and the split surface mesh of the boundary of the reference mesh.

4 Adding Local Dual Contributions

4.1 Reference Element Contributions

We have outlined an algorithm to obtain a block mesh by means of first generating a discretized version of the dual, see Section 3.2. In the second step of the algorithm, **add $\overline{\mathcal{D}}$-co-edges**, we need a tool for describing the $\overline{\mathcal{D}}$-co-edges and their intersections. To this end, we have introduced the concept of local dual contributions in Section 2.3. Recall that we use a reference mesh of simplices as a discrete representation of the geometry and that local dual contributions are added in the elements of the reference mesh. In this section we detail the geometrical definition for each type of local dual contribution.

In 2D, given a triangle of nodes $\{n_1, n_2, n_3\}$ in a reference mesh, we consider two types of local dual contributions, see Figure 6:

- `edge`. Given an edge e_i, we consider the opposite node n_i. Thus, e_i is determined by the nodes n_{i_1} and n_{i_2}, where $i_1 = i{+}1 (\mathrm{mod}\, 3)$ and $i_2 = i{+}2\, (\mathrm{mod}\, 3)$. This type of contribution is determined by the segment $\overline{r_{i_1} r_{i_2}}$, where $r_{i_1} := \frac{1}{4} n_i + \frac{3}{4} n_{i_1}$ and $r_{i_2} := \frac{1}{4} n_i + \frac{3}{4} n_{i_2}$.

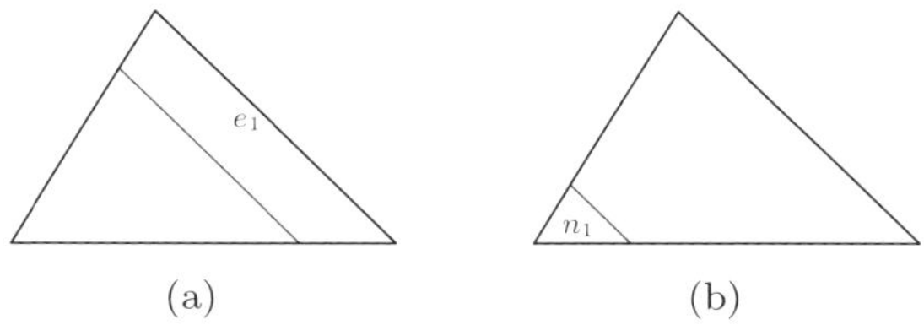

Fig. 6. Local dual contributions on a reference mesh triangle. **(a)** *edge* contribution; and **(b)** *node* contribution.

- *node*. Given the node n_i it is the contribution defined by the segment $\overline{q_{i_1}q_{i_2}}$, where $i_1 = i + 1(\text{mod}\,3)$ and $i_2 = i + 2\,(\text{mod}\,3)$, $q_{i_1} := \frac{3}{4}n_i + \frac{1}{4}n_{i_1}$ and $q_{i_2} := \frac{3}{4}n_i + \frac{1}{4}n_{i_2}$.

Note that the local dual contributions intersect with the edges of the reference mesh at distances $\frac{1}{4}$ and $\frac{3}{4}$ of the edge length. On the one hand, this selection ensures that the local dual contributions are compatible between adjacent elements. On the other hand, the intersections between local dual contributions have the proper multiplicity inside the reference elements.

In 3D, given a tetrahedron of nodes $\{n_1, n_2, n_3, n_4\}$ in a reference mesh, we consider three types of local dual contributions, see Figure 7:

- *face*. For a reference tetrahedron we consider a face f of nodes $\{n_{i_1}, n_{i_2}, n_{i_3}\}$ and opposite node n_i. The contribution of this face is defined by the triangle $\{q_{i_1}, q_{i_2}, q_{i_3}\}$ where $q_{i_j} := \frac{1}{5}n_i + \frac{4}{5}n_{i_j}$ for $j = 1, \ldots, 3$.
- *edge*. Consider an edge e delimited by the nodes n_{i_1} and n_{i_2}, and being n_{i_3} and n_{i_4} its opposite nodes. The contribution of e in the tetrahedron is defined by the quadrilateral $\{r_1, r_2, r_3, r_4\}$ with $r_1 := \frac{4}{5}n_{i_1} + \frac{1}{5}n_{i_4}$, $r_2 := \frac{4}{5}n_{i_1} + \frac{1}{5}n_{i_3}$, $r_3 := \frac{4}{5}n_{i_2} + \frac{1}{5}n_{i_3}$ and $r_4 := \frac{4}{5}n_{i_2} + \frac{1}{5}n_{i_4}$.
- *node*. The contribution of a node n_i with opposite nodes $\{n_{i_1}, n_{i_2}, n_{i_3}\}$ is defined by the triangle $\{s_{i_1}, s_{i_2}, s_{i_3}\}$, where $s_{i_j} := \frac{4}{5}n_i + \frac{1}{5}n_{i_j}$ for $j = 1, \ldots, 3$.

For 3D problems, the local dual contributions intersect with the edges of the reference mesh at distances $\frac{1}{5}$ and $\frac{4}{5}$ of the edge length. As in the 2D case, this selection ensures compatibility between neighboring elements and the correct multiplicity of the intersections inside the reference elements.

We propose to use local dual contributions to describe arrangements and intersections of $\overline{\mathcal{D}}$-co-curves. According to Section 3.2, local dual contributions have to define a non-manifold $(n-1)$-dimensional mesh with conformal elements. To capture all the possible intersections with other local dual contributions and obtain the conformal mesh, we decompose each type of local dual contribution in several simplices of co-dimension one. For instance, given a 3D reference mesh we consider the following decompositions for each local dual contribution type, see Figure 8:

- *face*. Given a face contribution defined by the nodes $\{q_1, q_2, q_3\}$, we consider the points $q_{121} := \frac{3}{4}q_1 + \frac{1}{4}q_2$, $q_{122} := \frac{1}{4}q_1 + \frac{3}{4}q_2$, $q_{231} := \frac{3}{4}q_2 + \frac{1}{4}q_3$, $q_{232} := \frac{1}{4}q_2 + \frac{3}{4}q_3$, $q_{311} := \frac{3}{4}q_3 + \frac{1}{4}q_1$, $q_{312} := \frac{1}{4}q_3 + \frac{3}{4}q_1$, $p_1 := \frac{2}{4}q_1 + \frac{1}{4}(q_2 + q_3)$,

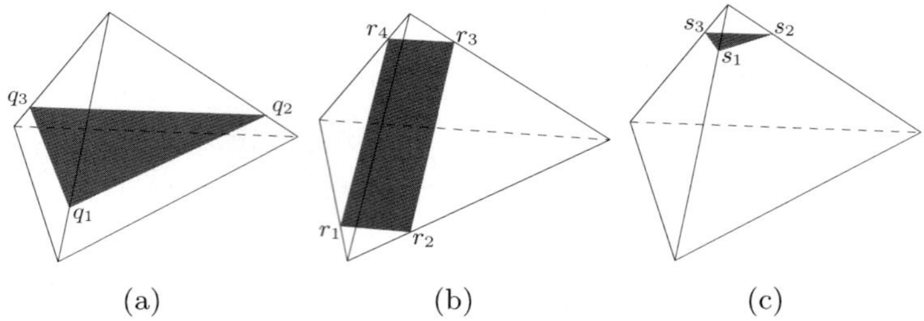

Fig. 7. A sample for each type of local dual contribution: **(a)** contribution of the base face; **(b)** contribution of the frontal edge; and **(c)** contribution of the top node

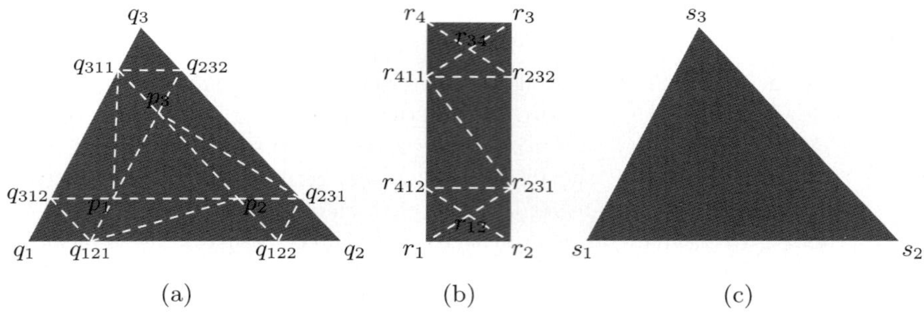

Fig. 8. Associated triangulation for each type of local contribution: **(a)** `face` contribution; **(b)** `edge` contribution; and **(c)** `node` contribution

$p_2 := \frac{2}{4}q_2 + \frac{1}{4}(q_1 + q_3)$, $p_3 := \frac{2}{4}q_3 + \frac{1}{4}(q_1 + q_2)$ and the decomposition of the contribution in the triangles $\{q_1, q_{121}, q_{312}\}$, $\{q_2, q_{231}, q_{122}\}$, $\{q_3, q_{311}, q_{232}\}$, $\{q_{121}, p_1, q_{312}\}$, $\{q_{231}, p_2, q_{122}\}$, $\{q_{311}, p_3, q_{232}\}$, $\{q_{121}, q_{122}, p_2\}$, $\{p_1, q_{121}, p_2\}$, $\{q_{231}, q_{232}, p_3\}$, $\{p_2, q_{231}, p_3\}$, $\{q_{311}, q_{312}, p_1\}$, $\{p_3, q_{311}, p_1\}$ and $\{p_1, p_2, p_3\}$.

- `edge`. Given an edge contribution defined by the nodes $\{r_1, r_2, r_3, r_4\}$, we consider the points $r_{231} := \frac{3}{4}r_2 + \frac{1}{4}r_3$, $r_{232} := \frac{1}{4}r_2 + \frac{3}{4}r_3$, $r_{411} := \frac{3}{4}r_4 + \frac{1}{4}r_1$, $r_{412} := \frac{1}{4}r_4 + \frac{3}{4}r_1$, $r_{12} := \frac{1}{4}(r_1 + r_2 + r_{231} + r_{412})$, $r_{34} := \frac{1}{4}(r_3 + r_4 + r_{411} + r_{232})$ and the decomposition of the contribution in the triangles $\{r_1, r_2, r_{12}\}$, $\{r_2, r_{231}, r_{12}\}$, $\{r_{231}, r_{412}, r_{12}\}$, $\{r_{412}, r_1, r_{12}\}$, $\{r_{411}, r_{231}, r_{232}\}$, $\{r_{412}, r_{232}, r_{411}\}$, $\{r_3, r_4, r_{34}\}$, $\{r_4, r_{411}, r_{34}\}$, $\{r_{411}, r_{232}, r_{34}\}$, and $\{r_{232}, r_3, r_{34}\}$.
- `node`. A node contribution defined by the nodes $\{s_1, s_2, s_3\}$ is not decomposed because it does not have inner intersections with other contributions.

4.2 Hard and Soft Contributions

In order to improve the robustness of the proposed algorithm we require that given two different reference meshes of a single domain the algorithm generates the same dual. For instance, given the reference meshes of the rectangular domain presented in Figures 9(a) and 9(b) we require that the algorithm generates the same dual, Figure 9(c), and the same primal, Figure 9(d).

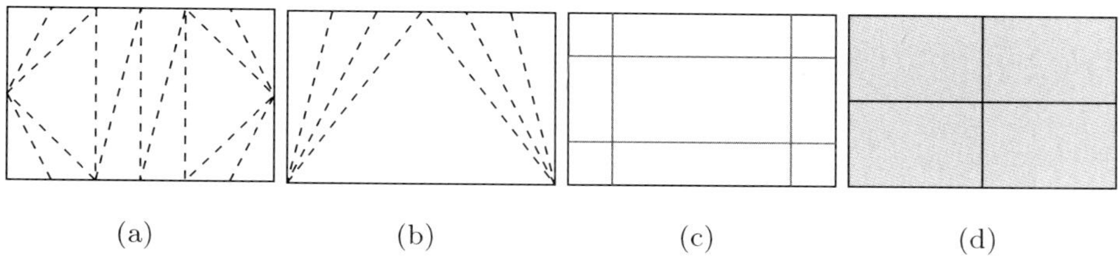

Fig. 9. Decomposition of a rectangular domain. **(a)** First reference mesh; **(b)** second reference mesh; **(c)** ideal mesh dual of the rectangular domain; and **(d)** ideal primal mesh.

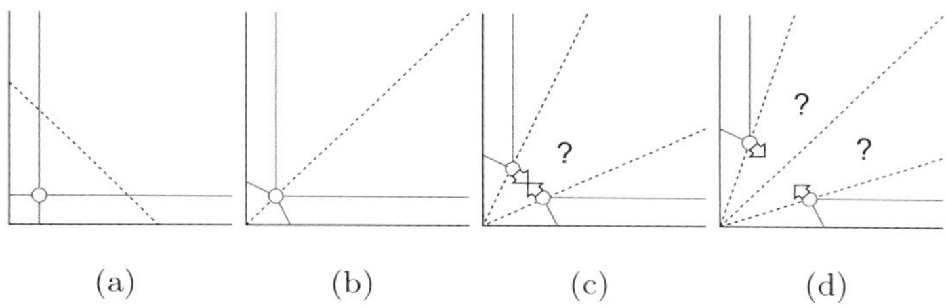

Fig. 10. Local dual contributions for several triangulations around an ***end*** node: **(a)** one adjacent triangle; **(b)** two adjacent triangles; **(c)** three adjacent triangles; and **(d)** four adjacent triangles. Solid black lines denote the boundary; slashed lines denote the edges of the triangular reference mesh; local dual contributions are denoted by red lines; and the red stroked circles represent the intersection of two local dual contributions.

To this end, consider an ***end*** vertex of a 2D domain. Depending on the reference mesh this ***end*** node may belong to different number of elements of the reference mesh, see Figure 10. If the ***end*** node belongs to one or two elements, see Figure 10(a) and Figure 10(b) respectively, we can add local dual contributions to obtain the ideal $\overline{\mathcal{D}}$-space curves around the ***end*** node. In the first case two ***edge*** local dual contributions are needed. In the second case one ***edge*** and one ***node*** local dual contributions are needed in each reference triangle adjacent to the vertex.

If the node belongs to more than two triangles, see Figures 10(c) and 10(d), the local dual contribution types presented in Section 4.1 are not sufficient to obtain a valid arrangement of the $\overline{\mathcal{D}}$-curves. The proposed solution for analogous configurations is to virtually collapse the two intersection points into a single virtual point. Then, we obtain the desired two $\overline{\mathcal{D}}$-curves that intersect in the virtual point. To generalize this behavior we classify the local dual contributions in two classes:

- ***hard.*** The contribution is respected during the construction of the primal mesh. That is, if at least one of the contributions that separate two $\overline{\mathcal{D}}$-co-nodes is ***hard*** we generate a primal edge connecting their associated primal nodes.

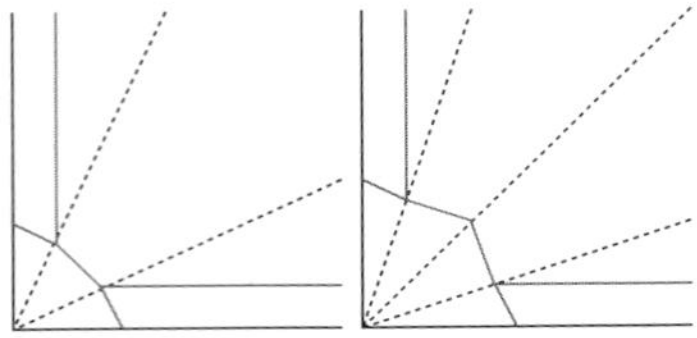

(a) (b)

Fig. 11. Local dual contributions for several possible triangulations around an ***end*** node: **(a)** three adjacent triangles; and **(b)** four adjacent triangles. Red segments represent ***hard*** contributions and blue segments represent ***soft*** contributions.

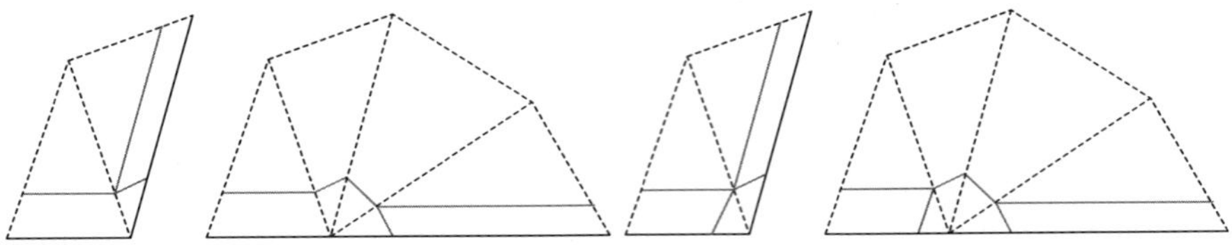

(a) (b) (c) (d)

Fig. 12. Examples of local dual contributions in reference triangles. Two non-valid duals: **(a)** three ***hard*** contributions intersecting at one point; and **(b)** a configuration composed by three ***hard*** contributions and two ***soft*** contributions. Two configurations defining a valid dual: **(c)** two ***hard*** contributions matching at one point; and **(d)** four ***hard*** contributions and two ***soft node*** contributions in the middle.

- ***soft***. The contribution is not respected during the construction of the primal mesh. That is, when all the contributions that separate two $\overline{\mathcal{D}}$-co-nodes are ***soft*** we do not generate a primal edge connecting their associated primal nodes.

The solution to the problem presented in Figures 10(c) and 10(d) can be interpreted in terms of ***hard*** and ***soft*** local dual contributions. On the one hand, the proposed solution is composed by ***soft node*** contributions of the ***end*** vertex in each inner triangle. On the other hand, it is composed by two ***hard node*** contributions of the ***end*** vertex, and one ***edge hard*** contribution on each triangle adjacent to the boundary of the reference mesh, see Figure 11. Then, the ***soft*** contributions will be ignored during the construction of the primal mesh, and this representation of the $\overline{\mathcal{D}}$-space is equivalent to virtually collapsing the ***soft*** contributions into one virtual point. Note that this virtual point is represented by blue segments in Figure 11.

4.3 Matching Rules

In order to obtain a valid dual we have to ensure that: *i)* local dual contributions compose $\overline{\mathcal{D}}$-co-edges without gaps, and *ii)* the intersections of the $\overline{\mathcal{D}}$-co-edges have the proper multiplicity. For instance, in 2D the multiplicity at the matching points of local dual contributions must be either two or four.

In this sense, the direct application of `hard` and `soft` local dual contributions does not necessarily lead to a valid dual. To illustrate this shortcoming, we consider two configurations that define a non-valid dual and that can be constructed by using `hard` and `soft` local dual contributions. First, Figure 12(a) shows three `hard` local dual contributions sharing an intersection point. Hence, we obtain three $\overline{\mathcal{D}}$-edges and not the required four to obtain a valid dual. Second, Figure 12(b) presents a configuration with `hard` and `soft` local dual contributions. Since `soft` local dual contributions are equivalent to virtually collapse the contributions into one single virtual point, we obtain again three $\overline{\mathcal{D}}$-edges sharing a point. These local dual contributions do not correspond to a valid dual.

To overcome these drawbacks we propose to match the number of contributions from the left and right of a reference mesh co-edge. For instance, in Figure 12(c) the `hard edge` and `node` contributions from the left match with the `hard edge` and `node` contributions from the right, and we obtain a correct piece of $\overline{\mathcal{D}}$-co-edge. Likewise, in Figure 12(d), if we virtually collapse the middle `soft node` contributions we have that `hard` contributions from the left match with the two `hard` contributions from the right. These contributions define the ending parts of four $\overline{\mathcal{D}}$-edges separated by a $\overline{\mathcal{D}}$-node, generated by virtually collapsing the `soft` contributions.

Note that by construction the local dual contributions do not have gaps nor intersections of wrong multiplicity inside the elements of the reference mesh. Thus, after considering that middle `soft` contributions are virtually collapsed, we have to ensure that the `hard` contributions match properly at the boundaries of reference elements. To this end, we propose the following strategies to ensure that the local dual contributions match properly.

- **2D.** The boundaries of the reference triangles are edges. Each edge has two possible intersection points of local dual contributions. We have to check local dual contributions around each one of these points:
  - **Boundary edge.** We allow either no contribution or one incident `hard` contribution from inside the element.
  - **Inner edge.** The number of `hard` contributions from the left have to be equal to the number of `hard` contributions from the right. Thus, the possibilities are: *i)* no incident contributions; *ii)* one from the left and one from the right defining a piece of a $\overline{\mathcal{D}}$-curve; and *iii)* two from the left and two from the right defining four $\overline{\mathcal{D}}$-edges ending at the intersection node of two $\overline{\mathcal{D}}$-curves.
- **3D.** The boundaries of a reference tetrahedron are determined by four triangular faces. Each face has six possible segments where local dual contributions can intersect. These segments are determined by all the possible `face`, `edge`, and `node` contributions inside the tetrahedron. Note that for each intersection segment there are only two possible contributions. Hence, we have to check local dual contributions around each one of these segments:
  - **Boundary face.** We allow either no contribution or one incident `hard` contribution from inside the element.

– **Inner face.** The number of ***hard*** contributions from the left has to be equal to the number of ***hard*** contributions from the right. Thus, the possibilities are: *i)* no incident contributions; *ii)* one from the left and one from the right defining a piece of a $\overline{\mathcal{D}}$-surface; and *iii)* two from the left and two from the right defining four $\overline{\mathcal{D}}$-faces ending at the intersection of two $\overline{\mathcal{D}}$-surfaces.

4.4 Implementation

We can add local dual contributions from the composing entities of the reference mesh. In addition, we can use ***hard*** and ***soft*** contributions in order to capture the desired dual. But we have to add these local dual contributions taking into account the set of matching rules. To this end, in our implementation we add the contributions using a hierarchical procedure: we start at the highest dimension entities (co-edge contributions) and finish at lowest dimensional entities (***node*** contributions). In this scheme, highest dimensional contributions are responsible of determining the main shape of the desired dual, and lower dimensional contributions are responsible of filling the gaps and matching local dual contributions properly. Algorithm 2, details a procedure for 3D applications.

We use this procedure to implement our local dual contributions tool. The last two steps are responsible of obtaining a valid dual once the ***face*** contributions have been added. These steps are done automatically according to the matching rules and the required ideal mesh around the features. The difficulties reside in the first step where we have to decide which ***face*** contributions to add to finally obtain the desire $\overline{\mathcal{D}}$-surfaces.

Algorithm 2. Add local dual contributions

Add *face* contributions. We add ***hard*** face contributions: *i)* to capture one layer of hexahedra that follows the boundary (adding a ***face*** contribution for each boundary face); *ii)* to split vertices with more than three incident feature curves; and *iii)* to capture ideal mesh close to curve and vertex features.

Add *edge* contributions. Taking into account previously added ***face*** contributions, we automatically add ***edge*** contributions according to the matching rules. Thus, we add ***edge*** contributions: *i)* to cover the gaps between ***face*** contributions with ***soft*** contributions; *ii)* to ensure that ***face*** contributions match with correct multiplicity using ***soft*** and ***hard*** contributions; and *iii)* to capture the ideal mesh close to curve and vertex features with ***hard*** contributions.

Add *node* contributions. Taking into account previously added ***face*** and ***edge*** contributions we can still have a non-valid arrangement of $\overline{\mathcal{D}}$-surfaces. To obtain the final set of intersecting $\overline{\mathcal{D}}$-surfaces we add ***node*** contributions: *i)* to fill all the remaining gaps between ***face*** and ***edge*** contributions with ***soft*** node contributions; *ii)* to match all the ***hard*** ***face*** and ***edge*** contributions and to define valid $\overline{\mathcal{D}}$-surfaces intersections using ***soft*** and ***hard*** contributions; and *iii)* to capture the ideal mesh at the vertex features with ***hard*** contributions.

5 Examples

The aim of the examples presented in this section is to show that the concept of local dual contributions has enough expressivity to capture the dual of the desired block mesh for a wide range of geometries: convex and non-convex domains, with 3 or 4-valenced vertices, and domains with holes.

We use our implementation to decompose these geometries in coarse blocks. It is important to point out that the proposed algorithm focuses on the generation of valid topological block decomposition. Then, the final node location can be improved using a relaxation procedure [17, 18]. In addition, a finer mesh can be obtained by meshing each block separately while keeping the mesh compatibility conditions between them.

For the three first examples, Figure 13 shows the five main steps involved in the block meshing process: *i)* to obtain hte reference mesh, the CDT of tetrahedra; *ii)* to add local dual contributions to define $\overline{\mathcal{D}}$-surfaces and their intersections, where red faces represent the ***hard*** local dual contributions and blue faces represent the *soft* local dual contributions; *iii)* to cap the $\overline{\mathcal{D}}$-surfaces and obtain a $\overline{\mathcal{D}}$-space where cells are colored according to the containing $\overline{\mathcal{D}}$-cell; finally, *iv)* and *v)* to obtain the primal nodes (colored as the associated $\overline{\mathcal{D}}$-cells), edges, faces and hexahedra of the block mesh.

In the first example, we decompose in eight blocks a brick object that is tapered in two orthogonal directions, see Figure 13(a). In this example all the vertices are 3-valenced. Thus, it can be meshed with submapping or midpoint subdivision. In fact, the obtained decomposition is equivalent to a coarse mesh that can be obtained by these methods. In this example all the $\overline{\mathcal{D}}$-surfaces follow the boundary of the domain and the possible gaps between contributions are filled with *soft* contributions.

For the second example, see Figure 13(b), we consider an extruded pentagon. The final block decomposition is equivalent to the mesh we would obtain using midpoint subdivision. Notice that in this example the submapping method will lead to a non-symmetrical block decomposition of lower quality. The $\overline{\mathcal{D}}$-surfaces are discretized versions of $\mathcal{D}$-surfaces that follow the boundaries of the domain. Close to the top front vertex the $\overline{\mathcal{D}}$-surfaces have a *soft* local dual contribution. When the algorithm caps the $\overline{\mathcal{D}}$-surfaces we obtain a region close to the top front vertex colored with cyan. This dual region, $\overline{\mathcal{D}}$-cell, is associated to the top front node of the primal, colored in cyan. Notice that since the local dual contribution is *soft*, the primal mesh does not show an edge connecting the top front node with the center node of the top of the volume. That is, during the generation of the primal mesh we ignore the *soft* contribution.

The third example presents the decomposition of a double pyramid, see Figure 13(c). Note that we can use neither submapping nor midpoint subdivision because all the vertices are 4-valenced. In this case our local dual contributions tool automatically splits the double pyramid in 4 tetrahedra. Moreover, the final mesh is composed by 16 hexahedra that correspond to the decomposition of each one of

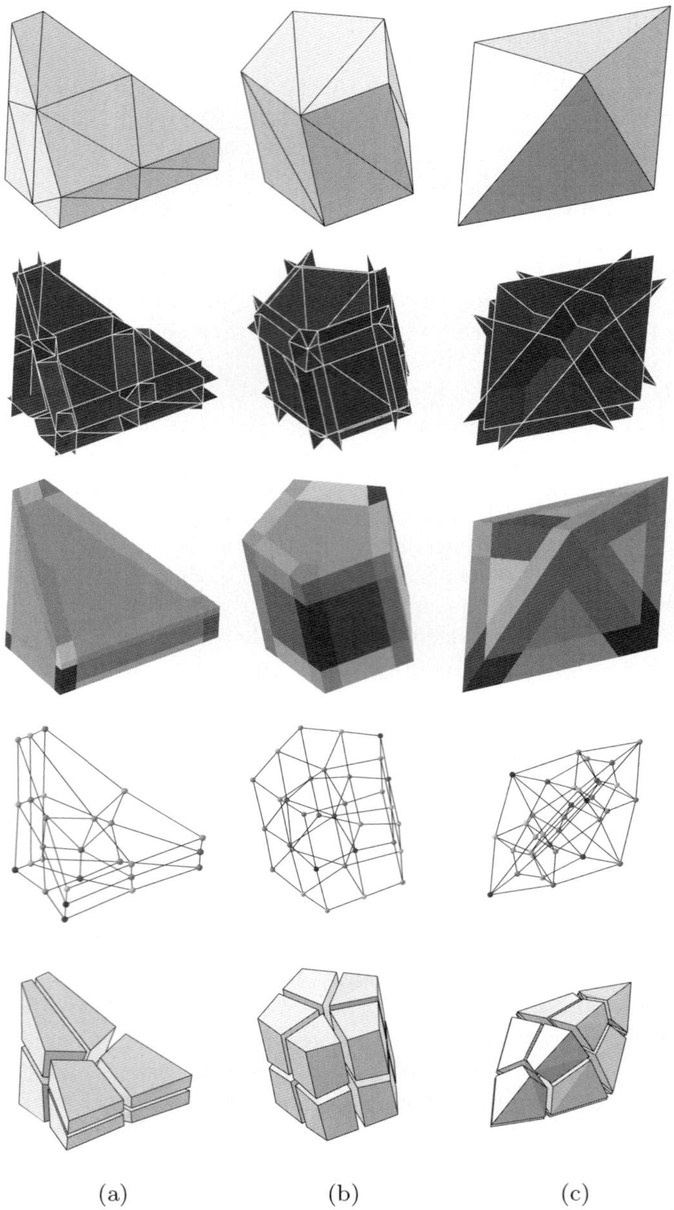

Fig. 13. Block meshing steps for three different domains: **(a)** a domain with 3-valenced vertices; **(b)** an extruded pentagon; and **(c)** a domain with 4-valenced vertices. Rows represent the algorithm steps: *i)* reference mesh; *ii)* $\overline{\mathcal{D}}$-surfaces composed by local dual contributions; *iii)* $\overline{\mathcal{D}}$-space where $\overline{\mathcal{D}}$-cells are colored in base of the containing $\overline{\mathcal{D}}$-cell; *iv)* primal nodes (colors associated to $\overline{\mathcal{D}}$-cells) and edges; and *v)* block mesh.

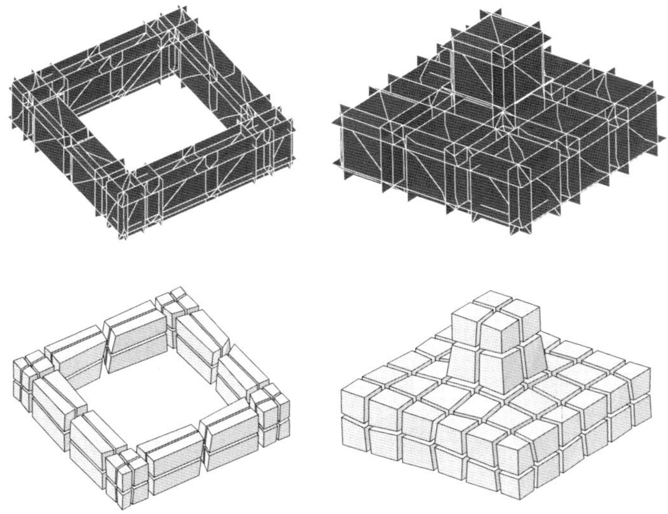

(a) (b)

Fig. 14. The $\overline{\mathcal{D}}$-surfaces and the block meshes for **(a)** a domain with a hole defined by four *corner* curves; and **(b)** a convex domain with a 3-valenced protrusion

the 4 tetrahedra in 4 hexahedra. In this case all the contributions are *hard*, and all the inner parallel $\overline{\mathcal{D}}$-surfaces are associated with an inner quadrilateral primal mesh that splits the domain in two parts.

The last two examples, two non-convex domains, are decomposed in a larger number of hexahedra than the three first examples. Therefore, we only show the local dual contributions and the obtained primal meshes, see Figure 14.

In the fourth example we discretize the extrusion of a square-shaped ring, Figure 14(a). This test domain has four *feature* curves of type *corner* (non-convex domain), a hole that cross the whole domain, and all the nodes are 3-valenced. Observe that we obtain ideal primal and dual configurations close to the *corner* curves. In fact, these configurations are the 3D generalization of the ideal configurations presented for *corner* vertices in Figure 4.

The last example is an extruded square with a protrusion on the top, Figure 14(b). All the nodes are 3-valenced and the curves that join the protrusion with the base are of type *corner*. We obtain a coarse decomposition of the domain that could also be obtained with a submapping algorithm. Since we consider the unconstrained boundary mesh approach, we do not have to previously compute a compatible interval assignment of the boundary. The four quadrilaterals that separate the protrusion from the rest of the mesh are associated to two parallel $\overline{\mathcal{D}}$-surfaces.

6 Concluding Remarks and Future Work

In this paper we present a new algorithm to generate a block decomposition of a given domain when there is not a prescribed boundary mesh. To this end, we propose an approach to directly construct a valid dual of the block mesh. The proposed algorithm is composed by two main steps. First, we create a reference mesh from the data of the boundary domain by means of a CDT. Second, we create the dual of the block mesh using the new concept of local dual contributions. These contributions are generated in the elements of the reference mesh. Specifically, we have implemented a tool that automatically adds local dual contributions following a set of hierarchical rules. The proposed implementation ensures that adjacent local contributions match properly. That is, the local dual contributions define a dual of the block mesh with intersections of the proper multiplicity and without gaps.

The main goal of the first implementation of the method is to obtain block meshes and their duals for some simple configurations: convex and non-convex domains defined by planar surfaces. Our current implementation allows to automatically mesh these simple domains. However, we can also mesh some domains with several vertices having a valence greater than three. In all theses cases the developed tool generates the expected mesh.

Practical results suggest that the proposed types of local dual contributions and the set of matching rules incorporate enough expressivity to describe the dual arrangement of a block mesh. However, it is expected that as the tool is applied to more complicated geometries (non-blocky geometries, domains with sharp angles, or assembly models) additional logic might be incorporated.

Note that the proposed tool is under continuous development. In this sense, the current version of this tool has several limitations that should be investigated and solved in the near future. First, we have to analyze the dependence of the algorithm from the reference mesh. For instance, it is not clear how the resolution of the reference mesh will affect the topology of the block decomposition. Second, it has to be investigated how the small perturbations of the boundary will affect the final primal mesh. Third, we have to apply the developed tool to domains described by curved boundaries that include smooth details. Finally, we have to check the robustness of the proposed tool in applications of practical interest.

References

1. Owen, S.J.: A survey fo unstructured mesh generation technology. In: 7th International Meshing Roundtable, pp. 239–267 (1998)
2. Baker, T.J.: Mesh generation: Art or science? Progress in Aerospace Sciences 41(1), 29–63 (2005)
3. Benzley, S., Perry, E., Merkley, K., Clark, B., Sjaardema, G.: A comparison of all-hexahedral and all-tetrahedral finite element meshes for elastic and elasto-plastic analysis. In: 4th International Meshing Roundtable, pp. 179–191 (1995)
4. Cifuentes, A.O., Kalbag, A.: A performance study of tetrahedral and hexahedral elements in 3-d finite element structural analysis. Finite Elements in Analysis and Design 12(3-4), 313–318 (1992)

5. Blacker, T.: Automated conformal hexahedral meshing constraints, challenges and opportunities. Engineering with Computers 17(3), 201–210 (2001)
6. Tautges, T.J.: The generation of hexahedral meshes for assembly geometry: survey and progress. International Journal for Numerical Methods in Engineering 50(12), 2617–2642 (2001)
7. Tautges, T.J., Blacker, T., Mitchell, S.A.: The whisker weaving algorithm: A connectivity-based method for constructing all-hexahedral finite element meshes. International Journal for Numerical Methods in Engineering 39(19), 3327–3350 (1996)
8. Folwell, N.T., Mitchell, S.A.: Reliable whisker weaving via curve contraction. Engineering with Computers 15(3), 292–302 (1999)
9. Mueller-Hannemann, M.: Hexahedral mesh generation by successive dual cycle elimination. Engineering with Computers 15(3), 269–279 (1999)
10. Calvo, N.A., Idelsohn, S.R.: All-hexahedral element meshing: Generation of the dual mesh by recurrent subdivision. Computer Methods in Applied Mechanics and Engineering 182(3-4), 371–378 (2000)
11. Murdoch, P., Benzley, S., Blacker, T., Mitchell, S.A.: The spatial twist continuum: A connectivity based method for representing all-hexahedral finite element meshes. Finite Elements in Analysis and Design 28(2), 137–149 (1997)
12. Calvo, N.A.: Generación de mallas tridimensionales por métodos duales. PhD thesis, Facultad de Ingeniería y Ciencias Hídricas (2005)
13. Thurston, W.: Hexahedral decomposition of polyhedra, `http://www.ics.uci.edu/~eppstein/gina/Thurston-hexahedra`
14. Mitchell, S.A.: A characterization of the quadrilateral meshes of a surface which admit a compatible hexahedral mesh of enclosed volume. In: Puech, C., Reischuk, R. (eds.) STACS 1996. LNCS, vol. 1046, pp. 465–478. Springer, Heidelberg (1996)
15. Blacker, T., Stephenson, M.B.: Paving: A new approach to automated quadrilateral mesh generation. International Journal for Numerical Methods in Engineering 32(4), 811–847 (1991)
16. Whiteley, M., White, D., Benzley, S., Blacker, T.: Two and three-quarter dimensional meshing facilitators. Engineering with Computers 12, 144–154 (1996)
17. Ansys. Ansys icem cfd, `http://www.ansys.com/products/icemcfd.asp`
18. Program Development Company. Gridpro, `http://www.gridpro.com/`
19. Staten, M.L., Owen, S.J., Blacker, T.: Unconstrained paving and plastering: A new idea for all hexahedral mesh generation. In: 14th International Meshing Roundtable (2005)
20. A quality tetrahedral mesh generator, `http://tetgen.berlios.de`

A Mesh Optimization Algorithm to Decrease the Maximum Error in Finite Element Computations

U. Hetmaniuk[1] and P. Knupp[2]

[1] Sandia National Laboratories*, P.O. Box 5800, MS 1320, Albuquerque, NM 87185-1320
ulhetma@sandia.gov

[2] Sandia National Laboratories, P.O. Box 5800, MS 1318, Albuquerque, NM 87185-1318
pknupp@sandia.gov

Abstract. We present a mesh optimization algorithm for adaptively improving the finite element interpolation of a function of interest. The algorithm minimizes an objective function by swapping edges and moving nodes. Numerical experiments are performed on model problems. The results illustrate that the mesh optimization algorithm can reduce the $W^{1,\infty}$ semi-norm of the interpolation error. For these examples, the L^2, L^∞, and H^1 norms decreased also.

1 Introduction

Engineering applications often involve solutions with different scales of variations along different directions. For these problems, recent works have illustrated the importance of anisotropic mesh adaptation, see [18, 21, 25] and the references therein. Courty et al. [12] show that meshes aligned with the solution characteristics allow to capture the solution details with fewer nodes than isotropic meshes. When using anisotropic meshes, they observe second-order convergence with respect to the number of degrees of freedom for representing a step function with continuous linear finite elements.

Several techniques exist for anisotropic mesh adaptation. Remeshing approaches define an anisotropic metric map that governs the generation of a new anisotropic mesh [1, 12, 18, 21]. Mesh modification techniques (refinement, coarsening, edge swapping, and node movement) operate locally to adapt the mesh [14, 25]. The approach followed in this paper is mesh optimization via edge swapping and node smoothing guided by the minimization of an appropriate objective function.

We can distinguish the objective functions available in the literature into two major categories. In the first category, the objective functions rely on a variational formulation or a governing partial differential equation. Examples of

* Sandia is a multiprogram laboratory operated by Sandia Corporation, a Lockheed Martin Company, for the United States Department of Energy under Contract DE-AC04-94AL85000.

such objective functions include the energy functional [15, 22, 28] and the least-squares norm of residuals [27, 29, 31]. For these examples, the exact solution minimizes the objective function. The functional provides a natural criterion for the design of computational grids adapted to this solution via node movement and edge swapping. In the second category, the objective functions rely on recovery of derivatives of the function of interest u. This second category is more function-centric and does not depend on the problem origin (variational formulation or partial differential equation). Typically, the objective function exploits an approximation to the Hessian matrix of u. Examples of such objective functions include geometric quality functionals in a transformed space [8, 14, 25] (where the metric map is built on the Hessian matrix of u) and functionals based on local interpolation errors [3, 6, 11, 23].

We choose here to focus on an Hessian-based objective function. Defining an optimization problem independent from a particular engineering application enables the development of a black box mesh optimization algorithm that can impact many different simulation groups and packages. Such a strategy has been implemented successfully in the Mesquite mesh optimization library [24]. Currently, though, Mesquite does not include an algorithm to adapt a mesh to the physical solution in an engineering simulation. The present work is a stepping-stone toward filling this gap.

Several mesh smoothing schemes have been designed to reduce the interpolation error. These works differ on how to measure the interpolation error. When measuring the interpolation error in the L^2 norm, the optimization [3, 8, 10, 11, 14, 25] aims to equidistribute the edge length under some metric map related to the Hessian matrix of the approximated function. Additional geometric criteria must be used in conjunction with the edge length criterion, otherwise inverted elements or sliver tetrahedra may appear. The approach is efficient at reducing the L^2 norm of the error. For many engineering applications, a good representation of the solution gradient is as important as a good representation of the solution itself. Unfortunately, a mesh optimized for reducing the L^2 norm of the interpolation error may be inappropriate for representing the gradient of the solution. Long thin elements are good for linear approximation if we measure the error in the L^2 norm [11, 26]. But such elements can be undesirable when we look for a good representation of the solution gradient (see [4]). Bank and Smith [6] (node movement) and Lagüe [23] (edge swapping and node movement) introduced two mesh optimization algorithms to minimize the H^1 norm of the interpolation error. Their objective functions incorporate naturally a "barrier" term, which helps prevent elements from becoming degenerate or tangled. To the best of our knowledge, these two works are the only ones optimizing for the gradient of the interpolation error.

Other choices of norms are possible. For example, one might consider constructing an objective function based on the L^∞-norm in order to reduce pointwise the interpolation error of some physical quantity of interest. Reducing the L^∞-norm of the error does not always guarantee a decrease in the error as measured by the L^p-norms (p finite), but it provides an upper bound for these error norms when the function of interest is smooth. A similar remark holds for the $W^{1,\infty}$ and $W^{1,p}$ norms. By reducing the $W^{1,\infty}$-norm, we control point values of the error and its gradient. This control can provide an upper bound for the $W^{1,p}$-norms, like the H^1-norm ($H^1 = W^{1,2}$). Since engineering problems are often interested in a good representation of the solution gradient at a point, we here construct an objective function that can lead to a reduction of the $W^{1,\infty}$ semi-norm of the interpolation error.

To the best of our knowledge, no work has studied the max-norm for the gradient of the interpolation error. The goal of this paper is to fill this gap and to assess numerically whether local mesh modification techniques can reduce the $W^{1,\infty}$ semi-norm of the linear interpolation error. In Section 2, we motivate and derive the adaptation algorithm. Starting from the objective function of Bank and Smith [6], we will present a new algorithm targeting the $W^{1,\infty}$ semi-norm of the interpolation error. In Section 3, we study on model problems the capability of this algorithm to reduce the interpolation error measured in the L^2, L^∞, H^1, and $W^{1,\infty}$ norms. Throughout the paper, by mesh optimization, we mean a combination of edge swapping and node movement.

2 Optimization-Based Adaptation Algorithm

In this section, we describe our approach. Section 2.1 reviews the algorithm of Bank and Smith [6]. Next, Section 2.2 motivates and derives the new objective function. Section 2.3 describes the edge swapping and node smoothing algorithms.

The focus of this paper is on two-dimensional triangular meshes because they are an important stepping stone toward a practical algorithm. For other meshes (quadrilateral, tetrahedral, ...), the method may apply but important details would need to be worked out, like swapping for tetrahedral elements. In our study, we do not incorporate h-adaptivity via refining or coarsening elements. Several works have illustrated the importance of combining h-adaptivity with mesh smoothing to generate adaptively anisotropic meshes (see [6, 12, 25]). In general, h-adaptivity and mesh smoothing are applied sequentially or in a black-box mode. We think that our algorithm could be plugged into such an approach. Most likely, some work would be required in order to get these approaches to play nicely together.

2.1 Review of the Bank and Smith Algorithm [6]

Given a domain Ω in $\mathbb{R}^2$ and a function $u \in H^1(\Omega)$, Bank and Smith [6] introduced an optimization of node locations to compute approximately

$$\min_{\mathcal{T} \in \mathcal{F}} \int_\Omega |\nabla u - \nabla u_L|^2 \, d\Omega, \tag{1}$$

where $\mathcal{F}$ is the family of conforming triangulations with fixed number of vertices. Members of $\mathcal{F}$ differ by the positions of the vertices in the mesh and, possibly, a finite number of edge flips. u_L denotes the continuous piecewise linear nodal interpolant of u defined on the triangulation $\mathcal{T}$.

Since problem (1) is expensive to solve, Bank and Smith [6] considered

$$\min_{\mathcal{T}\in\mathcal{F}} \int_{\Omega} |\nabla u_Q - \nabla u_L|^2 \, d\Omega, \tag{2}$$

where u_Q denotes the quadratic interpolant of u on K. They showed that

$$\int_K |\nabla u_Q - \nabla u_L|^2 \, dK \approx \frac{l_1^2 + l_2^2 + l_3^2}{192\,|K|} s(\mathbf{H}_K, K), \tag{3}$$

where the function $s(\mathbf{H}_K, K)$ is defined by

$$s(\mathbf{H}_K, K) = (l_1^2 \mathbf{t}_1^T \mathbf{H}_K \mathbf{t}_1)^2 + (l_2^2 \mathbf{t}_2^T \mathbf{H}_K \mathbf{t}_2)^2 + (l_3^2 \mathbf{t}_3^T \mathbf{H}_K \mathbf{t}_3)^2. \tag{4}$$

$\mathbf{t}_i$ is a unit tangent vector for edge i and l_i is the length of the same edge. $\mathbf{H}_K$ is the Hessian matrix for the quadratic interpolant u_Q. Their objective function approximates the squared H^1 semi-norm for the linear interpolation error in Ω,

$$\sum_{K\in\mathcal{T}} \frac{l_1^2 + l_2^2 + l_3^2}{192\,|K|} s(\mathbf{H}_K, K) \approx \int_{\Omega} |\nabla u_Q - \nabla u_L|^2 \, d\Omega \approx \int_{\Omega} |\nabla u - \nabla u_L|^2 \, d\Omega. \tag{5}$$

In objective function (5), the factor

$$\frac{l_1^2 + l_2^2 + l_3^2}{|K|}$$

is proportional to a well-known triangle quality metric [16]. It acts as a "barrier" function that guarantees untangled elements provided the initial mesh is untangled. Bank and Smith [6] combine a few sweeps of mesh smoothing with refinement and coarsening. They do not include edge swapping. Numerical experiments by Bank and Xu [7] illustrate reductions for the L^2 and H^1 interpolation error norms. However, they do not study the point-wise maximum norm. It is conceivable that the H^1-norm of the interpolation error decreases while the $W^{1,\infty}$-norm increases.

2.2 New Objective Function

To derive our objective function, we notice that

$$\max_{\mathbf{x}\in K} |\nabla u(\mathbf{x}) - \nabla u_L(\mathbf{x})|^2 \approx \frac{1}{|K|} \int_K |\nabla u - \nabla u_L|^2 \, d\Omega$$

when $u \in H^1(\Omega) \cap W^{1,\infty}(\Omega)$ and the mesh is fine enough. After exploiting the result of Bank and Smith (3), we define the local adaptive quality metric μ by

$$\mu(K) = \frac{l_1^2 + l_2^2 + l_3^2}{192\,|K|^2} s(\mathbf{H}_K, K) \approx \max_{\mathbf{x}\in K} |\nabla u_Q(\mathbf{x}) - \nabla u_L(\mathbf{x})|^2. \tag{6}$$

$\mu(K)$ differs from (3) only in that the power of $|K|$ is two instead of one. The first term in $\mu(K)$ still acts as a "barrier" function, but it is no longer a triangle shape-metric since it is not scale-invariant. A different but related approach to derive objective functions is used in [20], where we construct objective functions from upper bounds for the interpolation error.

Bank and Smith did not prove that objective function (5) is convex or has a unique minimum. To study the convexity of $\mu(K)$, we plot level curves for μ. In Figure 1, the triangle K is composed of two fixed vertices at $(0,0)$ and at $(1,0)$ and a free vertex at (x,y). Two different choices of matrix $\mathbf{H}_K$ are presented. When the free vertex gets closer to the horizontal axis, $\mu(K)$ increases highlighting the effect of the barrier function. The choice of matrix $\mathbf{H}_K$ modifies the level curves of μ. When $\mathbf{H}_K$ has both positive and negative eigenvalues, the contour lines are not convex and, consequently, the function μ is not a convex function. On the other hand, when the matrix $\mathbf{H}_K$ is positive definite, the contour lines are convex.

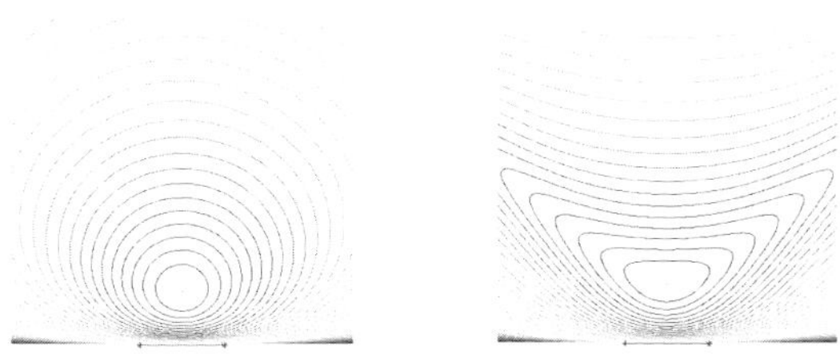

Fig. 1. Contour lines with $\mathbf{H}_K = diag(1,1)$ (left) and $\mathbf{H}_K = diag(1,-1)$ (right)

In Figure 2, showing contour lines for a local patch of elements, the interior node is free while the corner nodes are fixed. The plotted isolines are for

$$\max_{K_1,K_2,K_3,K_4} \mu(K), \tag{7}$$

when the interior node moves in the rectangle. The matrices vary per element such that

$$\mathbf{H}_{K_2} = \mathbf{H}_{K_4} = \begin{bmatrix} 1 & 0 \\ 0 & -3 \end{bmatrix} \quad \text{and} \quad \mathbf{H}_{K_1} = \mathbf{H}_{K_3} = 10^{-2}\mathbf{H}_{K_2}. \tag{8}$$

The isolines are not convex.

Convexity of the objective function is fundamental in optimization. If the optimization algorithm converges to a stationary point of a convex objective function, then the algorithm has converged to a global minimizer. When the matrix $\mathbf{H}_K$ has positive and negative eigenvalues, the function μ is not always convex. So we prefer to replace the matrix $\mathbf{H}_K$ with its absolute value and introduce the function

$$\tilde{\mu}(K) = \frac{l_1^2 + l_2^2 + l_3^2}{192\,|K|^2}[(l_1^2\mathbf{t}_1^T\,|\mathbf{H}_K|\,\mathbf{t}_1)^2 + (l_2^2\mathbf{t}_2^T\,|\mathbf{H}_K|\,\mathbf{t}_2)^2 + (l_3^2\mathbf{t}_3^T\,|\mathbf{H}_K|\,\mathbf{t}_3)^2]. \tag{9}$$

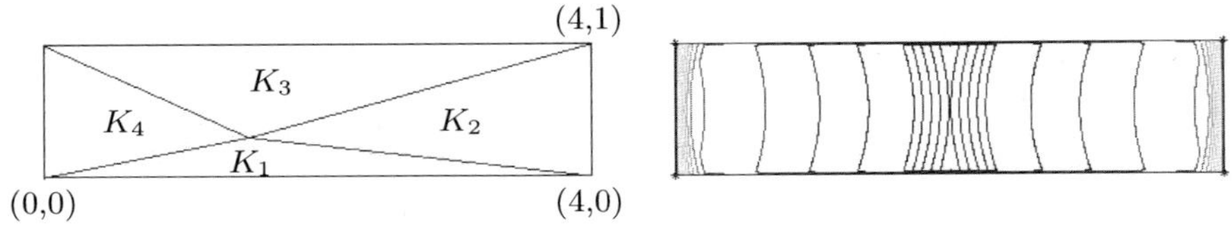

Fig. 2. Contour lines on a patch of four elements

$|\mathbf{H}_K|$ is the absolute value of the matrix $\mathbf{H}_K$, defined as $|\mathbf{H}_K| = \mathbf{S}\,|\boldsymbol{\Lambda}|\,\mathbf{S}^T$ where $(\mathbf{S}, \boldsymbol{\Lambda})$ are the eigenpairs for the matrix $\mathbf{H}_K$ and $|\boldsymbol{\Lambda}|$ is a diagonal matrix with the absolute values of eigenvalues as diagonal entries. For every vector $\mathbf{t}$, we have

$$\left(\mathbf{t}^T \mathbf{H}_K \mathbf{t}\right)^2 \leq \left(\mathbf{t}^T\, |\mathbf{H}_K|\, \mathbf{t}\right)^2 .$$

So the function $\tilde{\mu}$ is an upper bound for the local quality metric μ. Asymptotically, the function $\tilde{\mu}$ is also an upper bound for the max-norm of the gradient of the interpolation error. We were not able to prove that the function $\tilde{\mu}$ is convex. But our experiments gave convex level curves for this function. For example, with the matrices $\mathbf{H}_K$ defined in (8), the isolines for the functional

$$\max_{K_1,K_2,K_3,K_4} \tilde{\mu}(K), \tag{10}$$

are convex (see Figure 3).

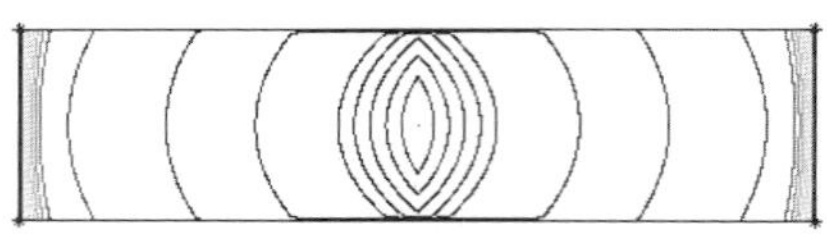

Fig. 3. Contour lines for functional (10) on a patch of four elements

Even though $\tilde{\mu}$ formally has a "barrier", the "barrier" term can become very weak when the matrix $|\mathbf{H}_K|$ approaches zero. The function $\tilde{\mu}$ satisfies

$$0 \leq \tilde{\mu}(K) \leq \|\mathbf{H}_K\|_2^2 \, \frac{l_1^2 + l_2^2 + l_3^2}{192\,|K|^2} \left(l_1^4 + l_2^4 + l_3^4\right) .$$

When the matrix $\mathbf{H}_K$ is exactly zero in a mesh region, the "barrier" term is canceled for the elements in that region. Then degenerate elements or elements of infinite size may appear where the function u is linear. When the matrix $\mathbf{H}_K$ has a small norm, *i.e.* when the function u is almost linear, the "barrier" term will prevent inverted elements. But a small norm for $\mathbf{H}_K$ can postpone the impact of the barrier function, *i.e.* the function $\tilde{\mu}$ can remain small even when the element K becomes flat.

To minimize the interpolation error, the aspect ratio of a triangle aligned with a function u must depend on the condition number of the Hessian matrix, *i.e.*

the ratio of largest over smallest eigenvalues (see [9, 13, 26]). A condition number of 1 will result in an equilateral triangle. A large condition number will result in a stretched triangle. To control the level of anisotropy in the optimized mesh, we modify $\tilde{\mu}$ and introduce the function $\tilde{\mu}_\varepsilon$,

$$\tilde{\mu}_\varepsilon(K) = \frac{l_1^2 + l_2^2 + l_3^2}{192\,|K|^2}\; s(|\mathbf{H}_K| + \varepsilon\mathbf{I}, K). \tag{11}$$

This regularization controls the level of anisotropy in the optimized mesh because it sets an upper bound on the condition number of $|\mathbf{H}_K| + \varepsilon\mathbf{I}$,

$$\frac{\lambda_{\max}(|\mathbf{H}_K| + \varepsilon\mathbf{I})}{\lambda_{\min}(|\mathbf{H}_K| + \varepsilon\mathbf{I})} = \frac{\lambda_{\max}(|\mathbf{H}_K|) + \varepsilon}{\lambda_{\min}(|\mathbf{H}_K|) + \varepsilon} \leq 1 + \frac{\lambda_{\max}(|\mathbf{H}_K|)}{\varepsilon}. \tag{12}$$

This upper bound on the condition number will limit the aspect ratios for triangles in the optimized mesh.

The goal of our mesh optimization algorithm is to compute approximately

$$\min_{\mathcal{T}\in\mathcal{F}}\; \max_{K\in\mathcal{T}} \tilde{\mu}_\varepsilon(K) \tag{13}$$

via edge swapping and node movement in order to reduce the interpolation error. The max-norm for the gradient of interpolation error and the function $\tilde{\mu}_\varepsilon$ satisfy

$$\max_{\mathbf{x}\in K}\; |\nabla u(\mathbf{x}) - \nabla u_L(\mathbf{x})|^2 \approx \mu(K) \leq \tilde{\mu}(K) \leq \tilde{\mu}_\varepsilon(K). \tag{14}$$

2.3 Edge Swapping and Node Movement Algorithms

In this section, we describe our algorithm to solve approximately (13). Problem (13) is a non-smooth optimization problem because the max-value function is non-differentiable. Such a problem requires special-purpose algorithms (see, for example, Fletcher [17]). Here we prefer to look for a cheap approximate solution to the optimization problem and we replace problem (13) by the following optimization problem

$$\min_{\mathcal{T}\in\mathcal{F}} \sum_{K\in\mathcal{T}} \tilde{\mu}_\varepsilon(K). \tag{15}$$

The goal of our algorithm is now to compute an approximate solution to (15).

First, we describe our edge swapping algorithm. Figure 4 illustrates the swapping of edge e into edge γ. Note that when edge e is flipped, the surrounding elements $K_{e,1}$ and $K_{e,2}$ are replaced with the elements $K_{\gamma,1}$ and $K_{\gamma,2}$. Let Ω_e denote the patch surrounding the edge e such that

$$\overline{\Omega_e} = \overline{K_{e,1}} \cup \overline{K_{e,2}} = \overline{K_{\gamma,1}} \cup \overline{K_{\gamma,2}}.$$

Algorithm 1 describes our edge swapping algorithm for problem (15). The first step is to check that the patch Ω_e is convex. When Ω_e is not convex, the flip would generate inverted elements, so the algorithm skips the edge. Then we

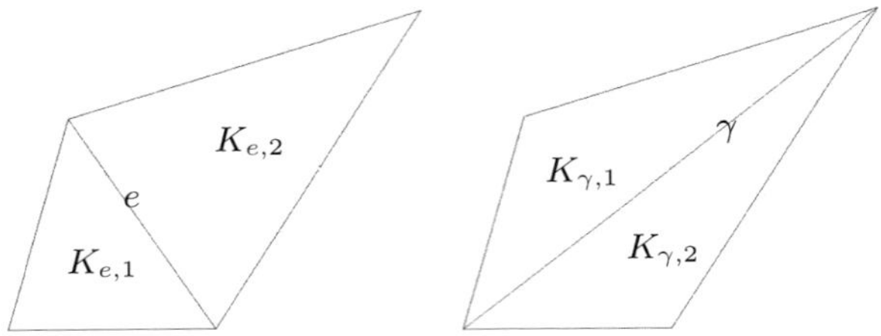

Fig. 4. Illustration of swapping the edge e into the edge γ

Algorithm 1. Edge Swapping Algorithm:

repeat
  for any edge e in the mesh **do**
    if Ω_e is not convex **then**
      Skip the edge
    end if
    Compute $\tilde{\mu}_\varepsilon(K_{e,1})$ and $\tilde{\mu}_\varepsilon(K_{e,2})$.
    Let γ denote the candidate flipped edge.
    Compute $\tilde{\mu}_\varepsilon(K_{\gamma,1})$ and $\tilde{\mu}_\varepsilon(K_{\gamma,2})$.
    if $\max(\tilde{\mu}_\varepsilon(K_{\gamma,1}), \tilde{\mu}_\varepsilon(K_{\gamma,1})) \geq \max(\tilde{\mu}_\varepsilon(K_{e,1}), \tilde{\mu}_\varepsilon(K_{e,1}))$ **then**
      Skip the edge.
    end if
    if $\tilde{\mu}_\varepsilon(K_{\gamma,1}) + \tilde{\mu}_\varepsilon(K_{\gamma,1}) \geq \tilde{\mu}_\varepsilon(K_{e,1}) + \tilde{\mu}_\varepsilon(K_{e,1})$ **then**
      Skip the edge.
    end if
    Perform the edge swapping.
  end for
until no edge is flipped.

compare the value of the function $\tilde{\mu}_\varepsilon$ on all the four elements. If the maximum and the sum of $\tilde{\mu}_\varepsilon$ over the patch both decrease, then we perform the flipping of edge e into edge γ. When swapping an edge, the value of the objective function, $\sum_{K\in\mathcal{T}} \tilde{\mu}_\varepsilon(K)$, varies only for the elements in Ω_e. So we perform checks on the maximum and the sum only over the patch Ω_e. These checks arise from the fact that the algorithm looks for an approximate solution to problem (15) and to the original problem (13).

A key component of the edge swapping algorithm is the evaluation of $\tilde{\mu}_\varepsilon$ on the four elements. This evaluation requires a matrix $|\mathbf{H}_K|$ on each element. The inputs for our mesh optimization algorithm are a non-inverted initial mesh $\mathcal{T}^0$ and, for every vertex V of $\mathcal{T}^0$, a matrix $\tilde{\mathbf{H}}_V$. The initial mesh will be stored as a background mesh for the piecewise linear interpolation of the absolute value of these nodal matrices, $\boldsymbol{\Pi}^0(\{|\tilde{\mathbf{H}}_V|\}_{V\in\mathcal{T}^0})$. For any point $\mathbf{x}$ in element K^0 of the background mesh $\mathcal{T}^0$, we choose

$$\boldsymbol{\Pi}^0(\{|\tilde{\mathbf{H}}_V|\}_{V\in\mathcal{T}^0})(\mathbf{x}) = |\tilde{\mathbf{H}}_{V_1}|\lambda^0_{V_1}(\mathbf{x}) + |\tilde{\mathbf{H}}_{V_2}|\lambda^0_{V_2}(\mathbf{x}) + |\tilde{\mathbf{H}}_{V_3}|\lambda^0_{V_3}(\mathbf{x}), \qquad (16)$$

where the vertices V_1, V_2, and V_3 form the element K^0 and $\lambda_V^0(\mathbf{x})$ denotes the barycentric coordinate of $\mathbf{x}$ with respect to vertex V in element K^0. To evaluate $\tilde{\mu}_\varepsilon$, we set the matrix $|\mathbf{H}_K|$ to be

$$|\mathbf{H}_K| = \boldsymbol{\Pi}^0(\{|\tilde{\mathbf{H}}_V|\}_{V\in\mathcal{T}^0})(\mathbf{G}_K), \tag{17}$$

where $\mathbf{G}_K$ is the centroid of element K. Requiring the matrices $\tilde{\mathbf{H}}_V$ as input of an adaptive scheme is reasonable. Indeed, $\mathbf{H}_K$ is the Hessian matrix for the quadratic interpolant, u_Q, of u in element K. As u_Q is usually unknown, the discrete Hessian matrix at vertex V is recovered from a discrete solution. Several nodal-based recovery schemes of Hessian matrix from a discrete solution are available. Discussing such recovery schemes is outside the scope of this paper. For further details, we refer to [30] and the references therein.

For the node movement algorithm, an iterative Gauss-Seidel-like method is used where we sweep through the vertices, locally optimizing the position of a single vertex while holding all others vertices fixed. Given a vertex V, let Ω_V be the patch of elements sharing V as a vertex. Algorithm 2 describes our node movement algorithm for problem (15). The first step is to compute the matrices $|\mathbf{H}_K|$ for all the elements K. To reduce computational expenses, these matrices are kept constant during one whole sweep through the mesh vertices. To optimize the position of vertex V, we use the nonlinear conjugate gradient algorithm CG_DESCENT [19] on the objective function, $\sum_{K\in\mathcal{T}} \tilde{\mu}_\varepsilon(K)$, viewed as a function of (x_V, y_V). Algorithm CG_DESCENT is stopped when the norm of the gradient of the objective function has been reduced by a factor 10^{-6}, when the relative change in objective function is smaller than 10^{-12}, or when 100 iterations have been performed. The barrier term in $\tilde{\mu}_\varepsilon$ will keep vertex V in Ω_V or in the subregion of Ω_V visible from all points on the boundary of Ω_V (when Ω_V is not convex). Moving a vertex V can affect the values of $\tilde{\mu}_\varepsilon$ only for the elements having V as vertex, *i.e.* the elements in Ω_V. Algorithm CG_DESCENT will reduce

Algorithm 2. Node Movement Algorithm:

for iter $= 1, \cdots, it_{\max}$ **do**
  Get all the matrices $|\mathbf{H}_K|$ with formula (17).
  Compute the objective function $\mu_{init} = \sum_{K\in\mathcal{T}} \tilde{\mu}_\varepsilon(K)$.
  for any vertex V in the mesh **do**
    Compute the maximum value of $\tilde{\mu}_\varepsilon$ on Ω_V, $\mu_{max,init}$.
    Apply the nonlinear conjugate gradient to move vertex V in Ω_V.
    Compute the maximum value of $\tilde{\mu}_\varepsilon$ on Ω_V, $\mu_{max,new}$.
    if $\mu_{max,new} \geq \mu_{max,init}$ **then**
      Reset vertex V to the latest position.
    end if
  end for
  Compute the objective function $\mu_{new} = \sum_{K\in\mathcal{T}} \tilde{\mu}_\varepsilon(K)$.
  if $|\mu_{new} - \mu_{init}| \leq tol \cdot \mu_{init}$ **then**
    Stop the iterations.
  end if
end for

the value $\sum_{K \in \Omega_V} \tilde{\mu}_\varepsilon(K)$. As the original optimization problem (13) looks at the maximum value of $\tilde{\mu}_\varepsilon$, we check also the maximum value of $\tilde{\mu}_\varepsilon$ on Ω_V. When the movement of vertex V creates an increase, we do not perform the movement and set vertex V to its latest position. Finally, we stop sweeping through vertices when the relative change in objective function is smaller than a tolerance, *tol*, or when a maximal number of iterations has been reached. For boundary vertices, we apply the same algorithm, but with appropriate constraints to keep the vertex on the boundary.

Several choices are available to combine the edge swapping and the node smoothing algorithms. In the following section, we will study on model problems how to combine these two algorithms.

3 Numerical Experiments

In this section, numerical experiments on model problems are presented to illustrate the mesh optimization algorithm. The goal is to assess whether the algorithm can reduce the interpolation error or not. The meshes are adapted to three given analytical functions with different anisotropic features, so that we can compute exactly the interpolation error. We do not use a realistic engineering problem because the exact solution would not be known.

The inputs for the mesh optimization algorithm are a non-inverted initial mesh and Hessian matrices at the vertices of the initial mesh. The input matrices at the vertices are computed with the analytical expression of the function to separate the potential error reduction, due to the mesh algorithm, from the effect of a Hessian recovery operator.

3.1 Experiment on the Regularization Parameter

For the first test case, only the node movement algorithm is used. Based on a scaling argument via (12), the regularization parameter ε is set to $\sigma \max_{V \in \mathcal{T}^0} \|\tilde{\mathbf{H}}_V\|_2$, where σ is a parameter in $[0, 1]$. We compare numerically the effect of σ on the optimization algorithm.

The first function u is given by

$$u(x, y) = \exp\left[-100(x - \frac{1}{2})^2 - 100(y - \frac{1}{2})^2\right] \qquad (18)$$

in the domain $\Omega = (0, 1) \times (0, 1)$. The Gaussian function u is symmetric, decays rapidly to zero, and exhibits a large region where it is flat and where the norm $\left\|D^2 u(\mathbf{x})\right\|_F$ is small. The function has no anisotropy in it. Figure 5 illustrates the initial uniform structured mesh with 32 elements per direction and the contour of function (18).

Table 1 lists the norms of interpolation error on the initial mesh and after optimization via node movement, when σ varies. The interpolation error norms decrease after optimization. Except for the L^2-norm, the decrease is almost monotone with σ. For the $W^{1,\infty}$ norm, the reduction of interpolation error

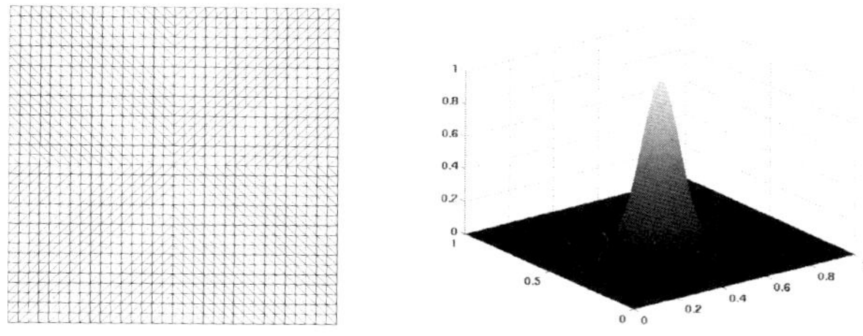

Fig. 5. Initial structured mesh (left) — Contour for function (18) (right)

Table 1. Evolution of interpolation error for function (18) as a function of σ

	$\|u-u_L\|_{L^2}$	$\|u-u_L\|_{H^1}$	$\|u-u_L\|_{L^\infty}$	$\|u-u_L\|_{W^{1,\infty}}$	$\|\mathbf{A}_K\|_F \|\mathbf{A}_K^{-1}\|_F$
Initial	3.28×10^{-3}	3.52×10^{-1}	3.80×10^{-2}	2.40	[2, 3]
$\sigma = 10^{-2}$	0.80×10^{-3}	1.57×10^{-1}	1.24×10^{-2}	1.67	[2, 10]
$\sigma = 10^{-4}$	1.63×10^{-3}	1.43×10^{-1}	1.01×10^{-2}	1.53	[2, 69]
$\sigma = 10^{-8}$	1.52×10^{-3}	1.42×10^{-1}	0.98×10^{-2}	1.51	[2, 560]
$\sigma = 0.0$	1.52×10^{-3}	1.43×10^{-1}	1.00×10^{-2}	1.52	[2, 567]

is between 30% and 37%. The other norms are reduced by at least 50%. Table 1 also lists the range for condition number of $\mathbf{A}_K$. We recall that $\mathbf{A}_K$ is the Jacobian matrix for mapping the unit right-angled triangle to the physical triangle K. The range of the condition numbers increase when σ gets smaller, indicating that some elements become more anisotropic.

Figure 6 describes the optimized meshes for $\sigma = 10^{-2}$ and $\sigma = 10^{-4}$. When $\|D^2u(\mathbf{x})\|_F$ is large, small isotropic elements are created. In the region where $\|D^2u(\mathbf{x})\|_F$ is small, the mesh optimization allows flat or stretched elements as these elements do not increase the interpolation error.

Function (18) has no anisotropic feature. Therefore, anisotropic elements are not needed to represent this function. The optimized meshes contain anisotropic elements because of the flatness of u, the boundary constraints, and the fixed

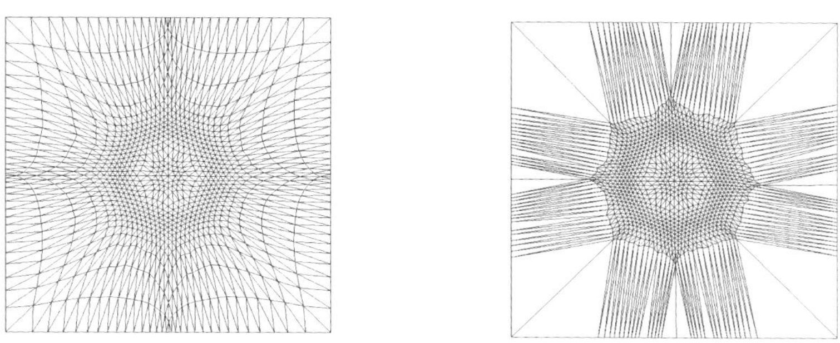

Fig. 6. Optimized meshes for (18) with $\sigma = 10^{-2}$ (left) and $\sigma = 10^{-4}$ (right)

topology. In order to reduce the error, the optimization gathers nodes in the regions that matter. In those regions, the mesh is notably isotropic. Elsewhere, the function is almost linear and anisotropic elements do not cause the error to increase. The regularization controls here the degree of anisotropy at the expense of a slight increase in the error.

3.2 Experiments on Combining Swapping and Smoothing

For the second series of tests, we combine edge swapping and node smoothing and compare numerically different combinations for the mesh optimization algorithm. In the following, the letter E denotes one call to Algorithm 1 and the letter N one call to Algorithm 2.

Table 2. Evolution of interpolation error for function (18) with $\sigma = 10^{-2}$

	$\|u-u_L\|_{L^2}$	$\|u-u_L\|_{H^1}$	$\|u-u_L\|_{L^\infty}$	$\|u-u_L\|_{W^{1,\infty}}$	$\|\mathbf{A}_K\|_F \|\mathbf{A}_K^{-1}\|_F$
Initial	3.28×10^{-3}	3.52×10^{-1}	3.80×10^{-2}	2.40	[2, 3]
N	0.804×10^{-3}	1.57×10^{-1}	1.24×10^{-2}	1.67	[2, 10]
(N,E,N)	0.702×10^{-4}	1.36×10^{-1}	0.986×10^{-2}	1.51	[2, 8.1]
E	2.71×10^{-3}	2.98×10^{-1}	3.79×10^{-2}	2.40	[2, 3]
(E,N)	1.57×10^{-3}	2.37×10^{-1}	2.23×10^{-2}	2.16	[2, 5]

σ is set at 10^{-2}. Table 2 lists the interpolation error norms with different combinations of edge swapping and node movement for function (18). All the interpolation error norms decrease after optimization. The combination (N, E, N) gives the smallest error norms. The $W^{1,\infty}$ norm is reduced by 37%, the H^1 norm by 55%, the L^2 norm by 78%, and the L^∞ norm by 67%. The $W^{1,\infty}$ error in Table 1 corresponding to Figure 6 (right) is nearly the same as the error in Table 2 corresponding to Figure 7 (left). Therefore, one can achieve the same error level with a lot less mesh anisotropy by adding the ability to swap edges to the adaptive algorithm.

Figure 7 plots the optimized meshes for the Gaussian function (18) with the combinations (N, E, N) and (E, N). The meshes are visually different from the optimized mesh with node movement only (see Figure 6). For this function and this initial mesh, starting with edge swapping creates a mesh topology that prevents outer vertices to gather in the middle of the mesh. This behavior limits the error reduction in comparison to combinations starting with node movement.

In the subsequent experiments, we use a heuristic formula to set the regularization parameter ε. We choose ε to be

$$\varepsilon = \sigma^0 \max_{V \in \mathcal{T}^0} \|\tilde{\mathbf{H}}_V\|_2 \text{ and } \sigma^0 = \frac{1}{100 \max_{K^0 \in \mathcal{T}^0} \|\mathbf{A}_{K^0}\|_F \|\mathbf{A}_{K^0}^{-1}\|_F}. \tag{19}$$

In this formula, the constant σ^0 depends only on the initial mesh $\mathcal{T}^0$.

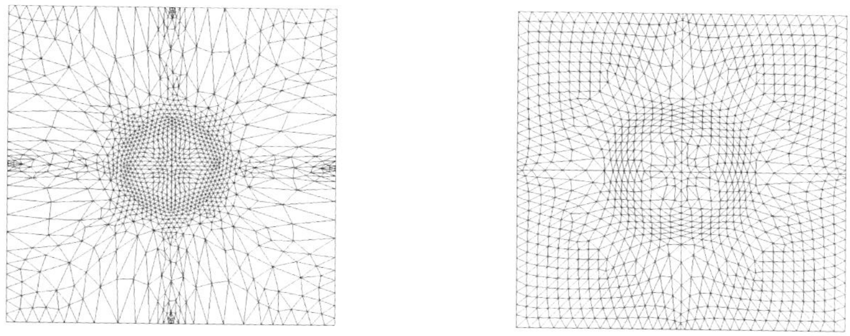

Fig. 7. Optimized meshes for (18) with (N, E, N) (left) and (E, N) (right)

The next function, from Huang [21], is given by

$$u(x, y) = \tanh(24y) - \tanh\left[24\left(x - y - \frac{1}{2}\right)\right] \tag{20}$$

in the domain $\Omega = (0, 1) \times (0, 1)$. This function simulates the interaction of a boundary layer (along the line $y = 0$) with an oblique shock wave (along the line $y = x - 1/2$). Figure 8 illustrates the initial unstructured mesh with 1002 vertices and 2006 elements and the surface defined by function (20). Based on (19), this initial mesh results in σ^0 to be 1/420.

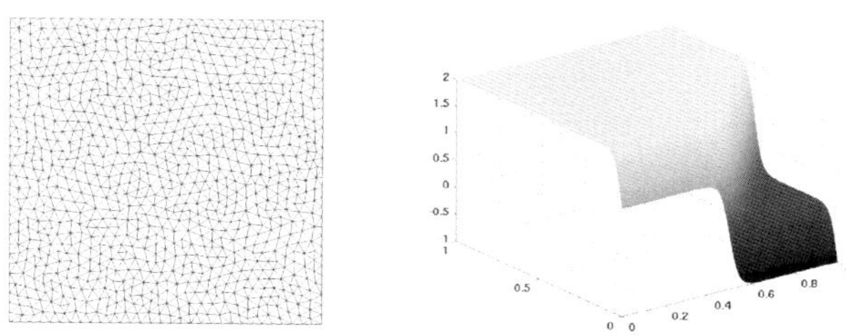

Fig. 8. Initial unstructured mesh (left) – Contour surface for function (20) (right)

Table 3 lists the interpolation error norms on the optimized mesh with different combinations of edge swapping and node smoothing. All the interpolation error norms decrease with more levels of optimization. Combining edge swapping and node movement results in smaller error norms than using only the edge swapping or only the node movement. Combinations starting with node movement produce larger reduction in error norms than the ones starting with edge swapping. After the combinations (N, E, N) and (E, N), adding more cycles of edge swapping and node smoothing does not reduce significantly the interpolation error. The combination (N, E, N) reduces the $W^{1,\infty}$ norm by 37% and the other norms by at least 63%. It reduces the error more than the combination (E, N) and exhibits a larger condition number range, indicating that anisotropic

Table 3. Evolution of interpolation error for function (20) for $\sigma^0 = 1/420$

	$\|u - u_L\|_{L^2}$	$\|u - u_L\|_{H^1}$	$\|u - u_L\|_{L^\infty}$	$\|u - u_L\|_{W^{1,\infty}}$	$\|\mathbf{A}_K\|_F \, \|\mathbf{A}_K^{-1}\|_F$
Initial	1.72×10^{-2}	1.74	1.75×10^{-1}	15.17	[2, 4.2]
N	0.655×10^{-2}	0.755	0.575×10^{-1}	9.56	[2, 28.0]
(N, E, N)	0.471×10^{-2}	0.628	0.575×10^{-1}	9.56	[2, 28.5]
$(N, (E, N)^2)$	0.467×10^{-2}	0.620	0.575×10^{-1}	9.56	[2, 28.4]
$(N, (E, N)^3)$	0.459×10^{-2}	0.619	0.575×10^{-1}	9.56	[2, 28.1]
E	0.933×10^{-2}	1.11	1.48×10^{-1}	15.17	[2, 23.4]
(E, N)	0.538×10^{-2}	0.78	0.902×10^{-1}	11.37	[2, 18.0]
$(E, N)^2$	0.525×10^{-2}	0.762	0.902×10^{-1}	11.37	[2, 23.2]
$(E, N)^3$	0.524×10^{-2}	0.759	0.902×10^{-1}	11.37	[2, 24.7]

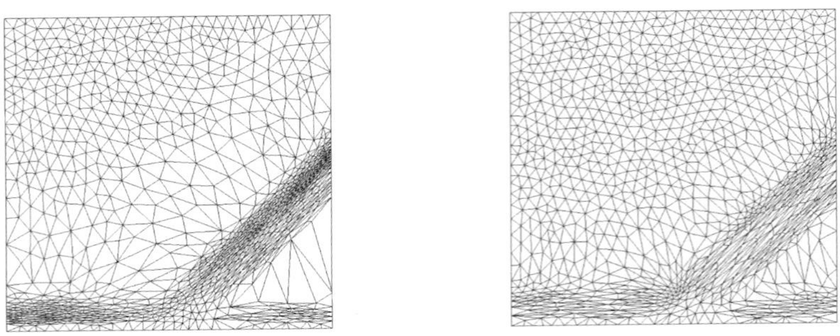

Fig. 9. Optimized meshes for (20) with (N, E, N) (left) and (E, N) (right)

elements help reducing the error. Figure 9 illustrates the optimized meshes for the combinations (N, E, N) and (E, N). Starting with edge swapping creates a mesh topology that prevents vertices to gather along the anisotropic features of function (20).

Finally, we study the function from [2]

$$u(x, y) = \sin\left[5(2x - 1.2)^3(4y^2 - 6y + 3)\right]. \tag{21}$$

The initial mesh was generated by the code BAMG of Hecht [5] and it is pre-adapted to function (21) so that the relative L^∞-norm of the interpolation error is smaller than 5 %. The mesh contains 3173 vertices and 5962 elements. Based on (19), this initial mesh results in σ^0 to be 1/73520. The value of σ^0 is smaller because the pre-adapted initial mesh contains stretched elements. Figure 10 illustrates the initial pre-adapted mesh and the contour for function (21). Table 4 lists the interpolation error norms on the optimized mesh with different combinations of edge swapping and node smoothing. All the interpolation error norms decrease after optimization. For this function and this initial mesh, the H^1 and $W^{1,\infty}$ norms are more reduced when the mesh optimization starts with node movement. For the L^2 and L^∞ norms, it is better to start with edge swapping. The combination (N, E, N) decreases the $W^{1,\infty}$ norm by 53% and the other

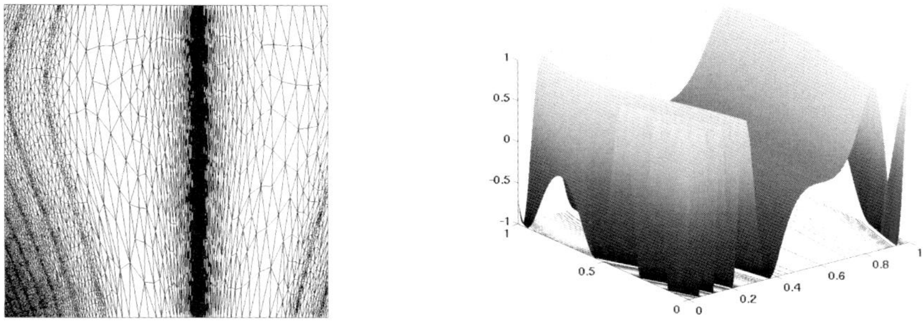

Fig. 10. Initial mesh pre-adapted for (21) (left) – Contour surface for (21) (right)

Table 4. Evolution of interpolation error for function (21) for $\sigma^0 = 1/73520$

	$\|u - u_L\|_{L^2}$	$\|u - u_L\|_{H^1}$	$\|u - u_L\|_{L^\infty}$	$\|u - u_L\|_{W^{1,\infty}}$	$\|\mathbf{A}_K\|_F \, \|\mathbf{A}_K^{-1}\|_F$
Initial	6.15×10^{-3}	1.67	3.85×10^{-2}	31.73	[2, 735.2]
N	5.19×10^{-3}	1.29	3.05×10^{-2}	17.49	[2, 430]
(N, E, N)	4.98×10^{-3}	1.04	3.05×10^{-2}	14.85	[2, 67.5]
$(N, (E, N)^2)$	4.94×10^{-3}	0.99	3.05×10^{-2}	14.24	[2, 54.65]
E	5.57×10^{-3}	1.31	2.92×10^{-2}	22.58	[2, 168]
(E, N)	4.83×10^{-3}	1.07	2.22×10^{-2}	17.24	[2, 52.8]
$(E, N)^2$	4.67×10^{-3}	1.01	2.20×10^{-2}	17.24	[2, 61.3]

norms by at least 19%. The combination (E, N) reduces the $W^{1,\infty}$ norm by 45% and the other norms by at least 21%. In comparison with the previous experiments, it is notable that even when the initial mesh is pre-adapted, our algorithm can still make worthwhile reductions in the error. Figure 11 plots the optimized meshes for the combinations (N, E, N) and (E, N).

Based on these experiments, we can draw the following conclusions. Combining edge swapping and node movement results in smaller error norms than using only the edge swapping or only the node movement. The edge swapping should

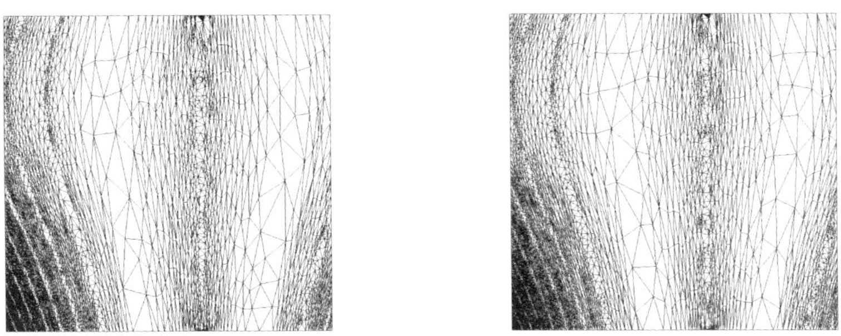

Fig. 11. Optimized meshes for (21) with (N, E, N) (left) and (E, N) (right)

be followed by node smoothing to benefit from the topology changes. When the initial mesh is not pre-adapted, it seems important to move the nodes before swapping any edge. The combination (N, E, N) is usually a good combination.

4 Conclusion

An optimization-based mesh adaptation algorithm has been presented. It combines edge swapping and node movement to minimize an objective function. The mesh adaptation algorithm exploits information from a discrete Hessian matrix. The objective function is based on the Bank and Smith [6] formula for the H^1 semi-norm of the linear interpolation error. In this formula, we replace the Hessian matrix for the quadratic interpolant of the fonction of interest u with a symmetric definite positive approximation. The objective function contains naturally a "barrier" term to ensure that starting from a mesh without inverted elements, the resulting mesh will not contain any inverted elements.

On model problems, we illustrated numerically that the algorithm can diminish the $W^{1,\infty}$ semi-norm of the interpolation error. This decrease is gratifying because our objective function is only asymptotically an upper bound to the $W^{1,\infty}$ semi-norm of the interpolation error. In our numerical experiments, the reduction in the $W^{1,\infty}$ semi-norm is accompanied by a decrease in the L^∞, L^2, and H^1 norms. The amount of decrease is problem-dependent. However, when the initial mesh is already appropriate to represent the function of interest, the decrease can be limited. When the initial mesh is not aligned with the function, the algorithm computes a better mesh topology and node distribution to represent the function. Then the reduction of the error norm can be significant.

A limitation of the algorithm is the lack of guarantee that all the error norms will always decrease. Report [20] contains a numerical experiment where some norms increase after optimization. Without more knowledge about the function of interest (like a governing partial differential equation for engineering applications), we are not aware of any algorithm with such a guarantee. The focus of this paper is on triangular two-dimensional meshes. For other situations (quadrilateral element, three-dimensional meshes), we expect that our approach will apply as well. However, important details remain to be worked out. Future directions for this work include the coupling with a recovery operator for the Hessian matrix, an efficient implementation into the Mesquite library [24], the extension to multiple dependent variables, and the use of quantities of interest for defining the objective function.

References

1. Agouzal, A., Lipnikov, K., Vassilevski, Y.: Adaptive generation of quasi-optimal tetrahedral meshes. East-West J. Numer. Math. 7(4), 223–244 (1999)
2. Alauzet, F., Frey, P.: Estimation d'erreur géométrique et métriques anisotropes pour l'adaptation de maillage. Partie II: exemples d'applications. Technical Report 4789, INRIA (2003)

3. Apel, T., Berzins, M., Jimack, P., Kunert, G., Plaks, A., Tsukerman, I., Walkley, M.: Mesh shape and anisotropic elements: Theory and practice. In: The Mathematics of Finite Elements and Applications X, pp. 367–376. Elsevier, Amsterdam (2000)
4. Babuška, I., Aziz, A.K.: On the angle condition in the finite element method. SIAM J. Numer. Anal. 13, 214–226 (1976)
5. BAMG, Bidimensional anisotropic mesh generator, `http://pauillac.inria.fr/cdrom/www/bamg/eng.htm`
6. Bank, R., Smith, R.: Mesh smoothing using a posteriori error estimates. SIAM J. Numer. Anal. 34, 979–997 (1997)
7. Bank, R., Xu, J.: Asymptotically exact a posteriori error estimators, part II: general unstructured grids. SIAM J. Numer. Anal. 41, 2313–2332 (2003)
8. Buscaglia, G., Dari, E.: Anisotropic mesh optimization and its application in adaptivity. Int. J. Numer. Meth. Engng. 40, 4119–4136 (1997)
9. Cao, W.: On the error of linear interpolation and the orientation, aspect ratio, and internal angles of a triangle. SIAM J. Numer. Anal. 43, 19–40 (2005)
10. Chen, L.: Mesh smoothing schemes based on optimal Delaunay triangulations. In: Proceedings of 13th International Meshing Roundtable (2004)
11. Chen, L., Sun, P., Xu, J.: Optimal anisotropic meshes for minimizing interpolation errors in L^p-norm. Math. Comp. 76, 179–204 (2007)
12. Courty, F., Leservoisier, D., George, P.L., Dervieux, A.: Continuous metric and mesh adaptation. Appl. Numer. Math. 56, 117–145 (2006)
13. D'Azevedo, E., Simpson, R.: On optimal interpolation triangle incidences. SIAM J. Sci. Comput. 10, 1063–1075 (1989)
14. Dompierre, J., Vallet, M.G., Bourgault, Y., Fortin, M., Habashi, W.: Anisotropic mesh adaptation: towards user-independent, mesh-independent and solver independent CFD. Part III: unstructured meshes 39, 675–702 (2002)
15. Felippa, C.: Optimization of finite element grids by direct energy search. Appl. Math. Modelling 1, 93–96 (1976)
16. Field, D.: Qualitative measures for initial meshes. Int. J. Numer. Meth. Engng. 47, 887–906 (2000)
17. Fletcher, R.: Practical methods of optimization. Wiley, Chichester (1987)
18. Frey, P., Alauzet, F.: Anisotropic mesh adaptation for CFD computations. Comput. Methods Appl. Mech. Engrg. 194, 5068–5082 (2005)
19. Hager, W., Zhang, H.: Algorithm 851: CG_DESCENT, a conjugate gradient method with guaranteed descent. ACM Trans. Math. Softw. 32, 113–137 (2006)
20. Hetmaniuk, U., Knupp, P.: Mesh optimization algorithms to decrease the linear interpolation error. SAND Report (2008)
21. Huang, W.: Mathematical principles of anisotropic mesh adaptation. Commun. Comput. Phys. 1, 276–310 (2006)
22. Jimack, P., Mahmood, R., Walkley, M., Berzins, M.: A multilevel approach for obtaining locally optimal finite element meshes. Advances in Engineering Software 33, 403–415 (2002)
23. Lagüe, J.F.: Optimisation de maillage basée sur une erreur locale d'interpolation. PhD thesis, Université de Paris VI (2006)
24. Mesquite, Mesh quality improvement toolkit, `http://www.cs.sandia.gov/optimization/knupp/Mesquite.html`
25. Remacle, J.F., Li, X., Shephard, M., Flaherty, J.: Anisotropic adaptive simulation of transient flows using discontinuous Galerkin methods. Int. J. Numer. Meth. Engng. 62, 899–923 (2005)

26. Rippa, S.: Long and thin triangles can be good for linear interpolation. SIAM J. Numer. Anal. 29, 257–270 (1992)
27. Roe, P., Nishikawa, H.: Adaptive grid generation by minimizing residuals. Int. J. Numer. Meth. Fluids 40, 121–136 (2002)
28. Tourigny, Y., Hülsemann, F.: A new moving mesh algorithm for the finite element solution of variational problems. SIAM J. Numer. Anal. 35, 1416–1438 (1998)
29. Tourigny, Y.: The optimisation of the mesh in first-order systems least-squares methods. J. Sci. Comput. 24, 219–245 (2005)
30. Vallet, M.G., Manole, C., Dompierre, J., Dufour, S., Guibault, F.: Numerical comparison of some Hessian recovery techniques. Int. J. Numer. Meth. Engng. 72, 987–1007 (2007)
31. Walkley, M., Jimack, P., Berzins, M.: Anisotropic adaptivity for the finite element solutions of three-dimensional convection-dominated problems. Int. J. Numer. Meth. Fluids 40, 551–559 (2002)

Grid Generation and Adaptation by Monge-Kantorovich Optimization in Two and Three Dimensions

John M. Finn, Gian Luca Delzanno, and Luis Chacón

T-15, Plasma Theory, Los Alamos National Laboratory, Mail stop: K717, Los Alamos, NM 87545

Summary. The derivation of the Monge-Ampère (MA) equation, as it results from a variational principle involving grid displacement, is outlined in two dimensions (2D). This equation, a major element of Monge-Kantorovich (MK) optimization, is discussed both in the context of grid generation and grid adaptation. It is shown that grids which are generated by the MA equation also satisfy equations of an alternate variational principle minimizing grid distortion. Numerical results are shown, indicating robustness to grid tangling. Comparison is made with the deformation method [G. Liao and D. Anderson, *Appl. Analysis* **44**, 285 (1992)], the existing method of equidistribution. A formulation is given for more general physical domains, including those with curved boundary segments. The Monge-Ampère equation is also derived in three dimensions (3D). Several numerical examples, both with more general 2D domains and in 3D, are given.

1 Introduction

In this paper we describe an approach to grid generation and adaptation in 2D and 3D based on *equidistribution*. This means that we take a monitor function or density $\rho(x,y)$ (in 2D) and find a mapping from the logical space $\boldsymbol{\xi} = (\xi,\eta) \in \Xi$ (the unit square in our case) to the physical domain $\mathbf{x} = (x,y) \in X$ satisfying $\rho(x,y)dxdy = d\xi d\eta$, or $\rho J_0 = 1$, where $J_0 = \partial(x,y)/\partial(\xi,\eta)$ is the Jacobian. Thus the rectangular grid cells in the logical space, all having equal area $d\xi d\eta = dA$, map to physical cells with equal measure $\rho(x,y)dxdy = dA$. In 3D this takes the form $\rho(x,y,z)dxdydz = dV$.

Equidistribution is an important general principle, and has potentially many applications. For example, the density ρ can be an estimate of the error in a numerical scheme. It has been shown that equidistribution of the local error leads to minimization of the global error [1]. Another example consists of compressible hydrodynamics when the sound speed c_s varies greatly over the domain, e. g. due to a large change of density. Then the CFL (Courant-Friedrichs-Lewy) condition for an explicit code $\Delta t \lesssim \Delta x/c_s$, where Δt is the time step and Δx is the grid cell size respectively, can in $1D$ be made uniform over the domain by equidistributing $\rho = 1/c_s$.

In one dimension, the equidistribution requirement $\rho(x)dx = d\xi$ determines $x(\xi)$. In higher dimensions, the equidistribution requirement cannot be satisfied uniquely. In Sec. 2.1 we introduce a variational principle to determine a unique solution. In this principle, the L_2 norm of the grid displacement is minimized. This leads to the Monge-Ampère (MA) equation. This is a key element of Monge-Kantorovich (MK) optimization, applied here to grid generation and adaptation. In Sec. 2.2 we discuss the relationship with another variational principle based on minimization of the grid distortion, defined as the trace of the covariant metric tensor. Minimization of grid displacement and particularly grid distortion suggest strongly that grids generated by such methods should be robust to tangling.

In Sec. 3 we discuss briefly the numerical methods used, namely multigrid preconditioned Newton-Krylov. In Sec. 4 we show numerical results on the grid produced by the MK method, in 2D with the physical domain X equal to the unit square. We compare these results with the grid produced by the deformation method of Ref. [2], the only other area/volume equidistribution method in the literature. The deformation method is based on finding a flow related to the required Jacobian of the map $\boldsymbol{\xi} \rightarrow \mathbf{x}$, and integrating an ODE with that flow. The flow is, however, not unique; in fact in Ref. [2] several examples of such a flow are given. We show that the grid obtained using one of these flows is considerably less smooth than the grid obtained by MK optimization, and more prone to tangling.

In Sec. 5 we formulate the problem in 2D with more general physical domains and show numerical examples. In Sec. 6 we outline the formulation in 3D, with a numerical example. In the Appendix we show some detail related to the analysis in Sec. 2, as a template for the analysis in Sec. 6.

2 Monge-Kantorovich Optimization in 2D

In this section we introduce a variational principle based on Monge-Kantorovich optimization as a method of grid generation and adaptation. We also discuss briefly the relationship with another variational principle based on minimizing the grid distortion, and discuss grid tangling. This material reviews Ref. [3].

2.1 Variational Principle with Local Constraint

Equidistribution of a density $\rho(x, y)$ is determined by the condition $\rho(x, y)dxdy = \rho_0(\xi, \eta)d\xi d\eta$, with $\rho_0(\xi, \eta) = 1$. Here, $\mathbf{x} = (x, y)$ are coordinates on the physical domain X, with bouondary ∂X. Also, $\boldsymbol{\xi} = (\xi, \eta)$ are logical variables on the logical domain Ξ, in our case the unit square $[0, 1] \times [0, 1]$. For grid *generation*, we consider $\psi_0 : \Xi \rightarrow X$, with $\rho(x, y)$ specified.

For grid *adaptation*, we consider the sequence $\Xi \rightarrow X \rightarrow X$, and the map $\psi : X \rightarrow X$ giving $\mathbf{x}' = \psi(\mathbf{x})$. See Fig. 1. For adaptation, both $\rho(x, y)$ and $\rho'(x', y')$ are specified. One application is for time-stepping with $\rho(x, y)$ an estimate of the local error of a PDE at time t, and $\rho'(x', y')$ the error estimate at time $t + \Delta t$. Both densities are normalized so that

$$\int \rho(x, y)dxdy = \int \rho'(x', y')dx'dy' = 1. \tag{1}$$

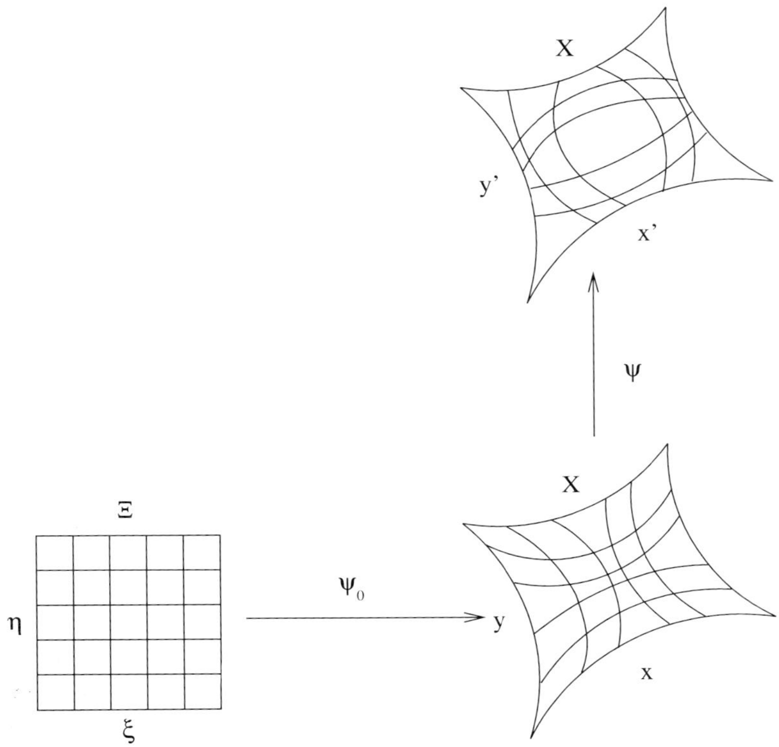

Fig. 1. Sketch of the map ψ_0 from the unit square Ξ to the physical domain X with curved boundary segments, and a second map $\psi : X \to X$

The new density ρ' is also equidistributed if $\rho(x,y)dxdy = \rho'(x',y')dx'dy' = d\xi d\eta$. In terms of the Jacobian J of the map $\psi : \mathbf{x} \to \mathbf{x}' = (x',y')$, this condition takes the form

$$J = \frac{\partial(x',y')}{\partial(x,y)} = \frac{\rho(x,y)}{\rho'(x',y')}. \tag{2}$$

In one dimension, the corresponding equation $\rho(x)dx = \rho'(x')dx'$ has the unique solution (up to an irrelevant integration constant) $R(x) = R'(x')$, where R, R' are cumulative distribution functions. In higher dimensions, Eq. (2) does not have a unique solution. In a time-stepping context, a grid $\mathbf{x}'(\boldsymbol{\xi})$ satisfying Eq. 2 with a large amount of rotation can be very different from $\mathbf{x}(\boldsymbol{\xi})$, even in the extreme case in which ρ and ρ' are equal.

In order to specify a map uniquely and optimally for grid generation and adaptation, we develop a variational principle with Eq. (2) as a constraint. Consider the first variation $\delta \int dxdy\mathcal{L}$ with [3]

$$\begin{aligned}\mathcal{L}(x,y,x',y') = \rho(x,y)\left[(x'-x)^2/2 + (y'-y)^2\right]/2 \\ -\lambda(x,y)\left[\rho'(x',y')\left(\partial_x x'\partial_y y' - \partial_x y'\partial_y x'\right) - \rho(x,y)\right].\end{aligned} \tag{3}$$

This is the standard L_2 form of Monge-Kantorovich optimization [4]. Here, $\lambda(x,y)$ is a local Lagrange multiplier, which ensures that the Jacobian condition (2) holds locally. In the time stepping context, this minimization of the L_2 norm of $\mathbf{x}' - \mathbf{x} = \psi(\mathbf{x}) - \mathbf{x}$ (weighted by ρ) minimizes the *grid velocity*. For the case for which X is the unit square, the boundary conditions on ∂X take a simple form: each side of X maps to itself under ψ. That is, $\hat{\mathbf{n}} \cdot (\mathbf{x}' - \mathbf{x}) = 0$, where $\hat{\mathbf{n}}$ is the vector normal to ∂X. These boundary conditions ensure that the boundary terms obtained by integrating $\delta \int dxdy\mathcal{L}$ by parts vanish [3].

The variation with respect to $\mathbf{x}'$ leads [3] after some analysis to

$$\mathbf{x}' = \mathbf{x} + \nabla\Phi(\mathbf{x}). \tag{4}$$

See the Appendix. That is, ψ is a gradient map $\mathbf{x}' = \nabla\Omega(\mathbf{x})$ with $\Omega(\mathbf{x}) = (x^2+y^2)/2+\Phi(\mathbf{x})$. This is a major conclusion in Monge-Kantorovich optimization theory [4]. Substituting into the Jacobian condition (2), we find

$$\nabla^2_{\mathbf{x}}\Phi + H_{\mathbf{x}}[\Phi] = \frac{\rho(x,y)}{\rho'(x',y')} - 1, \tag{5}$$

where $H_{\mathbf{x}}[\Phi]$ is the Hessian $\partial_{xx}\Phi\,\partial_{yy}\Phi - (\partial_{xy}\Phi)^2$. This is the 2D Monge-Ampère (MA) equation, a single nonlinear equation for $\Phi(x,y)$. (An approximate form of the MA equation was used for grid generation in Ref. [5].) There are two sources of nonlinearity: the Hessian and the dependence of the right side on $\mathbf{x}' = \mathbf{x}+\nabla\Phi$. The above boundary condition $\hat{\mathbf{n}} \cdot (\mathbf{x}' - \mathbf{x}) = 0$ leads to

$$\hat{\mathbf{n}} \cdot \nabla\Phi = 0 \tag{6}$$

on ∂X. It is known that a solution to the MA equation with these boundary conditions exists and is unique, and that the MA equation is elliptic [6].

2.2 Relation with Minimum Distortion

It is plausible that the variational principle using the Lagrangian $\mathcal{L}(x,y,x',y')$ of Eq. (3) should be helpful in preventing grid tangling. In a time-stepping context (for grid adaptation), if the cells are reasonably rectangular at one time, and the L_2 norm of the displacement is minimized, the cells at the next time step are expected to be reasonably rectangular for Δt small, i. e. $\rho(x,y)/\rho'(x',y')$ close to unity.

Based on these thoughts, consider a variational principle minimizing the cell distortion. We define $\mathcal{L}_2(x,y,x',y')$ by

$$\begin{aligned}\mathcal{L}_2(x,y,x',y') &= \rho(x,y)\,(g_{11}+g_{22})\,/2 \\ -\mu(x,y)\,&[\rho'(x',y')\,(\partial_x x'\partial_y y' - \partial_x y'\partial_y x') - \rho(x,y)]\,,\end{aligned} \tag{7}$$

where $T = g_{11} + g_{22} = (\partial_{x_i}x'_j)(\partial_{x_i}x'_j)$ (summation implied) is the trace of the covariant metric tensor $T = \mathrm{trace}(\mathrm{J}^{\mathrm{T}}\mathrm{J})$. This quantity measures the distortion of the $\mathbf{x}'$ cells relative to the $\mathbf{x}$ cells. Again, $\mu(x,y)$ is a local Lagrange multiplier, guaranteeing that Eq. (2) holds locally. In Ref. [3] we showed that for $\rho(x,y) = 1 + O(\epsilon)$, $\rho'(x',y') = 1 + O(\epsilon)$ with $\epsilon \ll 1$, solutions of the MA equation are solutions of the variational equations obtained from Eq. (7), i.e. are also minimum distortion solutions. This conclusion is further indication that MA solutions should be robust to tangling. We discuss numerical results related to tangling in Sec. 4.1.

3 Numerical Methods

We solve the MA equation by Jacobian-free Newton-Krylov methods [7, 8]. The particular solver is GMRES [9], preconditioned by multigrid. The fact that the MA equation is elliptic means that multigrid can be used effectively. The function Φ is defined at cell centers and the boundary conditions (6) are implemented using ghost cells in the logical domain Ξ. The efficiency and accuracy of these methods for solving the MA equation have been documented in Ref. [3]. Some of this material is reviewed in the next section.

4 Examples in a Square

Here, we consider X to be the unit square and $\rho(x,y) = 1$. For this case, we will describe the problem as grid adaptation.

4.1 Isotropic Example

The first case we show, from Ref. [3], has density

$$\rho'(x',y') = \frac{C}{2 + \cos(8\pi r')} \tag{8}$$

Table 1. Performance study for the Monge-Kantorovich approach with $\rho'(x',y')$ given by Eq. (8). Shown are the equidistribution error $\sim 1/N_x^2$, for grids with $N_x = N_y$; the CPU time; the grid quality measures $||\mathbf{p}||_2^{MK}$ and $||g_{11} + g_{22}||_1^{MK}$; and the number of linear and nonlinear iterations as functions of $N = N_x \times N_y$.

Number of Cells	**Error**	**CPU time [s]**	$\|\|\mathbf{p}\|\|_2^{MK}$	$\|\|g_{11}+g_{22}\|\|_1^{MK}$	Newton/ GMRES its.
16×16	9.64×10^{-2}	0.1	0.0173	1.449	3/3
32×32	2.28×10^{-2}	0.4	0.0173	1.466	4/4
64×64	5.78×10^{-3}	1.3	0.0173	1.470	4/4
128×128	1.46×10^{-3}	4.9	0.0174	1.470	4/4
256×256	3.67×10^{-4}	19	0.0174	1.471	4/4

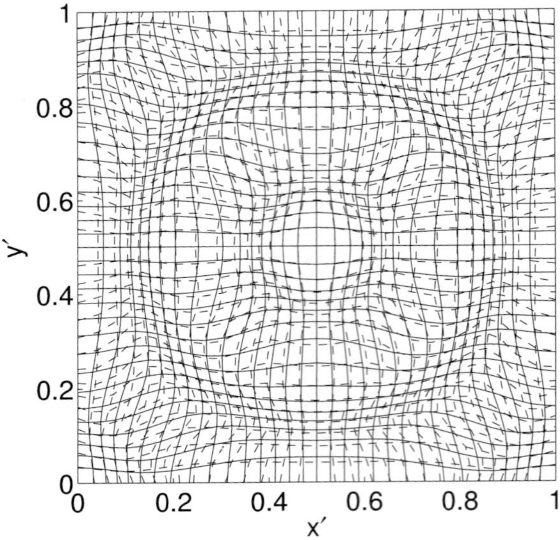

Fig. 2. The grid for Eq. (8) by MK (solid) and by deformation method (dashed)

Table 2. Performance study for the deformation method with time step $\Delta t = 0.01$ for Eq. (8). The grid quality measures $||\mathbf{p}||_2$ and $||g_{11} + g_{22}||_1$ are expressed in terms of variation with respect to the values in Table 1.

Number of Cells	Error	CPU time [s]	$\frac{\lVert\mathbf{p}\rVert_2}{\lVert\mathbf{p}\rVert_2^{MK}} - 1$	$\frac{\lVert g_{11}+g_{22}\rVert_1}{\lVert g_{11}+g_{22}\rVert_1^{MK}} - 1$
16×16	1.16×10^{-1}	0.2	+24%	+1%
32×32	3.53×10^{-2}	0.9	+28%	+2%
64×64	9.64×10^{-3}	3.4	+30%	+3%
128×128	2.46×10^{-3}	13.6	+30%	+3%
256×256	6.21×10^{-4}	55	+30%	+3%

with $r' = \sqrt{(x' - 1/2)^2 + (y' - 1/2)^2}$. The constant C is determined by normalization as in Eq. (1). The grid is shown in Fig. 2 (solid lines) for 32×32 cells. This isotropic example has $O(1)$ variations in density ($\rho_{max}/\rho_{min} = 3$) over scales $l \sim 0.1$ right up to the boundary. The performance data for this example are shown in Table 1. Note that the computational time required for convergence scales as the total number of grid points $N_x \times N_y$, i. e. it is optimal. This is traced to the fact that the number of Newton and GMRES iterations is nearly independent of the grid refinement.

In Fig. 2 (dashed lines) we have superimposed the grid determined by the deformation method using the proposed symmetric flow of Ref. [2]. It is clear that this grid is not as smooth as the MK grid. Table 2 shows performance for the deformation method for this example. In particular, note that the L_2 norm of $\mathbf{p} = \mathbf{x}' - \mathbf{x}$ is much larger for the deformation method.

4.2 Anisotropic Example

A second example has

$$\rho'(x', y') = C\left[1 + \frac{9}{1 + 100r'^2 \cos^2(\theta' - 20r'^2)}\right]. \tag{9}$$

Here, $r' = \sqrt{(x' - 0.7)^2 + (y' - 0.5)^2}$ and $\theta' = \tan^{-1}\left[(x' - 0.7)/(y' - 0.5)\right]$. The 64×64 grid obtained by MK optimization is shown in Fig. 3. This example mimics the spiral pattern that develops in the nonlinear Kelvin-Helmholtz instability. This example has $\rho'_{max}/\rho'_{min} \approx 9$ and very fine scales below $l = 0.02$. This is a very challenging example: the corresponding 64×64 grid obtained by the deformation method tangles [3], but the MK grid is quite smooth.

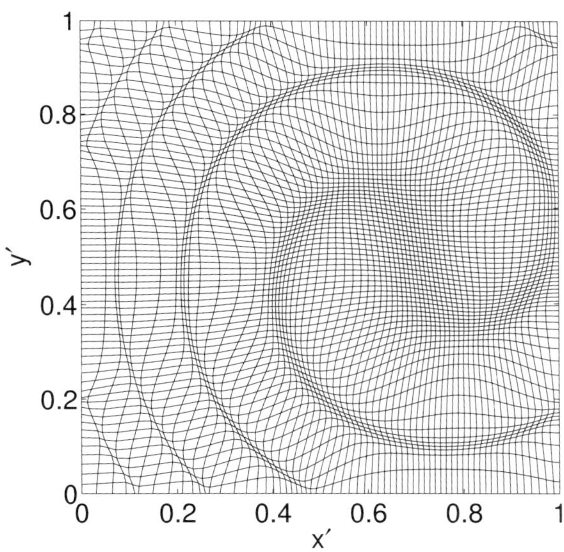

Fig. 3. 64×64 MK grid according to the anisotropic spiral density (9)

4.3 An Example with Fine Structures and Large Jumps in Density

To demonstrate the power of the MK method, we show an example of adapting to challenging images. This is the ubiquitous Lena image [10], commonly used as a standard image for testing image processing methods. We take as density $\rho'(x', y')$ the brightness of the monochrome Lena image. For 200×200 grid points as in Fig. 4, it shows much of the detail in the original image.

Fig. 4. The Lena image by the MK method, with a 200 × 200 grid

5 Formulation in More General 2D Domains; Examples

We present a formulation for more general physical domains, with examples.

5.1 Map from Logical to Physical Domains; Boundary Formulation

An important issue involves dealing with more general physical domains in 2D. Two questions arise. The first is: How are boundary conditions on an arbitrary shaped domain applied? The second is: Do the terms obtained by integrating by parts as in the variation of $\int dxdy\mathcal{L}$ in Eq. (3) again vanish in this more general setting?

To address the first issue, let us start by considering grid generation, i. e. the map $\psi_0 : \Xi \to X$, for a physical domain X whose boundary ∂X consists of four sides mapped from the four sides of the unit square in the logical space.

See Fig. 1. We shall specialize here to the case in which the four sides of the boundary are given respectively by

$$\begin{array}{rrl} \mathrm{L} - & \text{left side}: & x = x_1(y), \\ \mathrm{R} - & \text{right side}: & x = x_2(y), \\ \mathrm{B} - & \text{bottom}: & y = y_1(x), \\ \mathrm{T} - & \text{top}: & y = y_2(x). \end{array} \tag{10}$$

These boundary conditions on $\Phi(\xi, \eta)$ take the form

$$\begin{aligned} L\ (\xi = 0): &\quad \frac{\partial \Phi}{\partial \xi} = x_1\left(\eta + \frac{\partial \Phi}{\partial \eta}\right), \\ R\ (\xi = 1): &\quad 1 + \frac{\partial \Phi}{\partial \xi} = x_2\left(\eta + \frac{\partial \Phi}{\partial \eta}\right), \\ B\ (\eta = 0): &\quad \frac{\partial \Phi}{\partial \eta} = y_1\left(\xi + \frac{\partial \Phi}{\partial \xi}\right), \\ T\ (\eta = 1): &\quad 1 + \frac{\partial \Phi}{\partial \eta} = y_2\left(\xi + \frac{\partial \Phi}{\partial \xi}\right). \end{aligned} \tag{11}$$

As in Sec. 2, these boundary conditions require that each side of the square map to the corresponding side of the physical domain, but allow an arbitrary (but one-to-one) motion of points along each of the sides. These nonlinear conditions are applied by solving for the value of Φ in each ghost cell.

For grid adaptation, we deal with the map $\mathbf{x} \to \psi(\mathbf{x}) = \mathbf{x}'$ from X to X, and the corresponding boundary conditions, e. g. $x' = x_1(y')$, in terms of $\Phi(x, y)$, are: are

$$\begin{array}{rl} L\ (x = x_1(y)): & x_1(y) + \partial_x \Phi(x_1(y), y) = x_1\left(y + \partial_y \Phi(x_1(y), y)\right), \\ R\ (x = x_2(y)): & x_2(y) + \partial_x \Phi\left(x_2(y), y\right) = x_2\left(y + \partial_y \Phi(x_2(y), y)\right), \\ B\ (y = y_1(x)): & y_1(x) + \partial_y \Phi(x, y_1(x)) = y_1\left(x + \partial_x \Phi(x, y_1(x))\right), \\ T\ (y = y_2(x)): & y_2(x) + \partial_y \Phi(x, y_2(x)) = y_2\left(x + \partial_x \Phi(x, y_2(x))\right). \end{array} \tag{12}$$

Numerically, these conditions are implemented using ghost cells on the logical domain, by using $[J_0 \equiv \partial(x, y)/\partial(\xi, \eta)]$

$$\begin{aligned} \frac{\partial \Phi}{\partial x} &= \frac{\partial_\xi \Phi\, \partial_\eta y - \partial_\eta \Phi\, \partial_\xi y}{J_0}, \\ \frac{\partial \Phi}{\partial y} &= \frac{-\partial_\xi \Phi\, \partial_\eta x + \partial_\eta \Phi\, \partial_\xi x}{J_0}. \end{aligned} \tag{13}$$

The second issue relates to the terms obtained by integration by parts when taking the Euler-Lagrange equations from the variational principle in Eq. (3). Indeed, we find that the variation of $\int dxdy\mathcal{L}$ contains a term

$$\begin{aligned} -\int dxdy \lambda(x, y) \rho'(x', y') \left(\partial_x \delta x'\ \partial_y y' - \partial_y \delta x'\ \partial_x y'\right. \\ \left. + \partial_x x'\ \partial_y \delta y' - \partial_y x'\ \partial_x \delta y'\right) \end{aligned} \tag{14}$$

in addition to terms proportional to $\delta x'$ and $\delta y'$. Integrating by parts, this expression becomes

$$\int dxdy\,[\delta x'\,(\partial_x(\rho'\lambda\partial_y y') - \partial_y(\rho'\lambda\partial_x y')) + \delta y'\,(\partial_y(\rho'\lambda\partial_x x') - \partial_x(\rho'\lambda\partial_y x'))]$$

$$+ \int dy\lambda(x,y)\rho'(x',y')\,[\partial_y x'\;\delta y' - \partial_y y'\;\delta x']\,|_{x=x_1(y)}^{x=x_2(y)}$$

$$+ \int dx\lambda(x,y)\rho'(x',y')\,[\partial_x y'\;\delta x' - \partial_x x'\;\delta y']\,|_{y=y_1(x)}^{y=y_2(x)}.$$

The analysis involving the first terms (and the other terms proportional to $\delta x'$ and $\delta y'$) proceeds as in Ref. [3] and the Appendix. The terms in the bracket for $x = x_1(y)$, with $x' = x_1(y')$, are

$$\frac{dx_1(y')}{dy'}\frac{\partial y'}{\partial y}\delta y' - \frac{\partial y'}{\partial y}\frac{dx_1(y')}{dy'}\delta y',$$

which equals zero because $x' = x_1(y')$ implies $\delta x' = (dx_1(y')/dy')\delta y'$. The same is true for the boundary terms at $x = x_2(y)$, $y = y_1(x)$, and $y = y_2(x)$. Therefore the boundary terms obtained by the integration by parts vanish and the analysis proceeds as in Sec. 2.1, the Appendix, and Ref. [3]. In particular, we again conclude that ψ is a gradient map, and that the Monge-Ampère (MA) equation (5) applies.

5.2 Examples with Non-square Physical Domains

The first example, involving grid generation, has a physical domain X consisting of a parallelogram obtained by mapping the sides of the square according to

$$x = a\xi + b\eta, \quad y = b\xi + c\eta, \tag{15}$$

with $a = 1$, $b = 0.2$, and $c = (1 + b^2)/a$. The latter condition ensures that the area of the parallelogram equals unity. The symmetry of this linear map implies that it is a gradient map $\mathbf{x} = \nabla_\xi \Omega$ with $\Omega(\xi,\eta) = a\xi^2/2 + b\xi\eta + c\eta^2/2$. This map is therefore of the form $\mathbf{x} = \boldsymbol{\xi} + \nabla_\xi \Phi(\xi,\eta)$ with

$$\Phi = (a-1)\xi^2/2 + b\xi\eta + (c-1)\eta^2/2. \tag{16}$$

Clearly, $\Phi(\xi,\eta)$ is a solution of the MA equation $\nabla_\xi^2\Phi + H_\xi[\Phi] = \rho_0(\xi,\eta)/\rho(x,y) - 1$ with $\rho_0(\xi,\eta) = \rho(x,y) = 1$. [The determinant condition $c = (1 + b^2/a)$ is not essential: for determinant equal to J_0 (such that $c = (J_0 + b^2)/a$), the MA equation is satisfied with $\rho_0(\xi,\eta) = 1$ and $\rho(x,y) = 1/J_0$.]

From Eq. (15) we conclude that the boundary conditions on the four sides, according to the format of Eq. (10), are

$$\begin{array}{rr} \mathrm{L}(\xi = 0): & x = x_1(y) = by/c, \\ \mathrm{R}(\xi = 1): & x = x_2(y) = a + b(y-b)/c, \\ \mathrm{B}(\eta = 0): & y = y_1(x) = bx/a, \\ \mathrm{T}(\eta = 1): & y = y_2(x) = c + b(x-b)/a. \end{array} \tag{17}$$

We now formulate the problem for grid adaptation, using the MA equation to find the map $\psi : \mathbf{x} \to \mathbf{x}'$ with $\rho(x,y) = 1$ and $\rho'(x',y')$ specified. The boundary conditions are of the form (17) but with $x \to x'$, $y \to y'$. For example, the boundary condition on the left is $x' = by'/c$ or $x + \partial_x \Phi(x,y) = b[y + \partial_y \Phi(x,y)]/c$. From Eq. (17) we conclude $\partial_x \Phi(by/c, y) = b\partial_y \Phi(by/c, y)/c$. Summarizing for the four sides, we obtain

$$\begin{aligned} \text{L}: &\quad \partial_x \Phi(by/c, y) = b\partial_y \Phi(by/c, y)/c, \\ \text{R}: &\quad \partial_x \Phi(a + b(y-b)/c, y) = b\partial_y \Phi(a + b(y-b)/c, y)/c, \\ \text{B}: &\quad \partial_y \Phi(x, bx/a) = b\partial_x \Phi(x, bx/a)/a, \\ \text{T}: &\quad \partial_y \Phi(x, c + b(x-b)/a) = b\partial_x \Phi(x, c + b(x-b)/a)/a. \end{aligned} \tag{18}$$

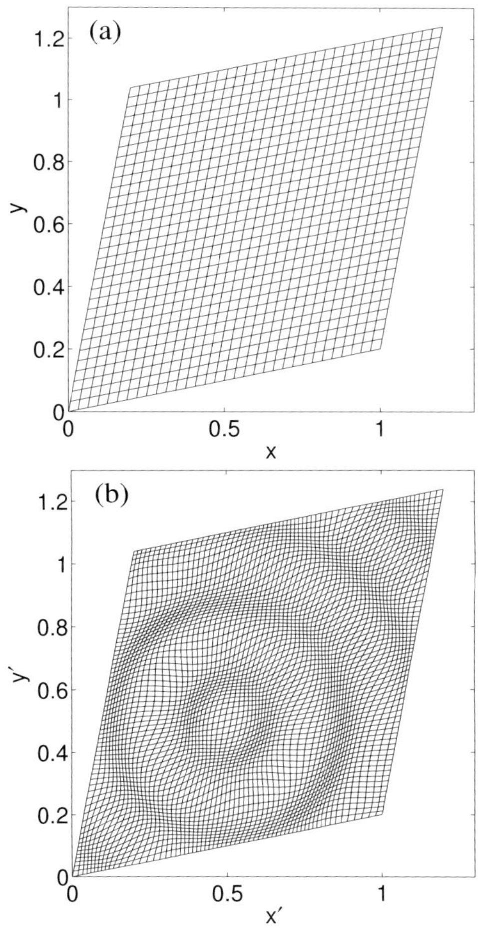

Fig. 5. (a) The linear map $\mathbf{x}(\boldsymbol{\xi})$ [Eq. (15)] from the unit square Ξ to a parallelogram X, with $\rho = 1$; (b) the composite map $\mathbf{x}'(\boldsymbol{\xi}) = \mathbf{x}'(\mathbf{x}(\boldsymbol{\xi}))$ with $\rho'(x',y')$ given by Eq. (8)

Again, these boundary conditions are implemented by means of ghost cells on the logical grid, using Eq. (13).

Results obtained with these boundary conditions for the isotropic case of Eq. (8) are shown in Fig. 5(a) for the (x, y) grid given by Eqs. (15), (16) and in Fig. 5(b) for the (x', y') grid solved numerically. Because the density ρ' is specified as a function of $\mathbf{x}'$, the density of grid lines in Fig. 5(b) is according to Eq. (8), without distortion. In this case, the (x, y) grid (i. e. ψ_0), is formed by a gradient map with $\rho = 1$. (For grid adaptation, it is not necessary that the initial map ψ_0 be a gradient map, i. e. be obtained by MK grid generation.)

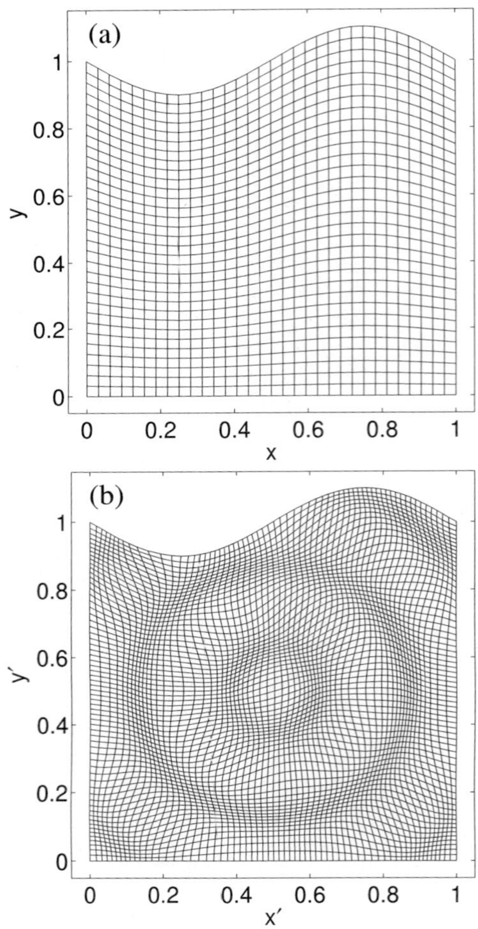

Fig. 6. (a) Sinusoidal map (19) with $\rho = 1$; (b) map obtained by MK grid adaptation with $\rho'(x', y')$ as in Eq. (8)

As a second example, consider X to be the set bounded by $x = 0$; $x = 1$; $y = 0$; $y = 1 - \epsilon \sin(2\pi x)$ for $\epsilon = 0.1$. We choose the 'sinusoidal' map ψ_0 from Ξ to X given by

$$\begin{aligned} x &= \xi, \\ y &= \eta[1 - \epsilon \sin(2\pi\xi)] \end{aligned} \tag{19}$$

and shown in Fig. 6(a). The Jacobian is $J_0 = \partial(x, y)/\partial(\xi, \eta) = 1 - \epsilon \sin(2\pi\xi) = 1 - \epsilon \sin(2\pi x)$. Since $\rho_0(\xi, \eta) = 1$, this implies $\rho(x, y) = [1 - \epsilon \sin(2\pi x)]^{-1}$. The map given by Eq. (19) is not a solution of the MA equation (and is in fact not even a gradient map). However, as we discussed above, this is not an essential requirement for the map ψ_0. For the map $\psi : X \to X$ we specify again $\rho'(x', y')$ according to Eq. (8), and solve the MA equation, Eq. (5). The boundary conditions are of the form (10), (12) with

$$\begin{aligned} \mathrm{L}: &\quad x_1(y) = 0, \\ \mathrm{R}: &\quad x_2(y) = 1, \\ \mathrm{B}: &\quad y_1(x) = 0, \\ \mathrm{T}: &\quad y_2(x) = 1 - \epsilon \sin(2\pi x). \end{aligned} \tag{20}$$

The results, showing the undistorted density of Eq. (8), are in Fig. 6(b).

6 Three Dimensions

In 3D we show that a similar minimization of the L_2 norm of $\mathbf{x}' - \mathbf{x}$ leads to a gradient map and to the 3D generalization of the MA equation. We also show a numerical example obtained by solving the 3D MA equation by methods discussed in Sec. 3.

6.1 Variational Principle and Gradient Map in 3D

In 3D, we will derive the variational principle corresponding to Eq. (3) for general $\rho(x, y, z)$ but for the special case $\rho' = 1$. We have derived the more general case [3], which is tedious and not particularly enlightening. Following the 2D derivation in the Appendix, we have

$$\mathcal{L} = \frac{\rho}{2} (x'_i - x_i)(x'_i - x_i)$$

$$-\lambda(x, y, z) \left[\epsilon_{ijk} \frac{\partial x'_i}{\partial x} \frac{\partial x'_j}{\partial y} \frac{\partial x'_k}{\partial z} - \rho(x, y, z) \right],$$

where ϵ_{ijk} are the components of the usual antisymmetric 3D Levi-Civita tensor, and repeated indices indicate summation. The Euler-Lagrange equations are

$$\rho (x'_i - x_i) + \epsilon_{ijk} \frac{\partial}{\partial x} \left[\lambda \frac{\partial x'_j}{\partial y} \frac{\partial x'_k}{\partial z} \right]$$

$$+\epsilon_{kij}\frac{\partial}{\partial y}\left[\lambda\frac{\partial x_k'}{\partial x}\frac{\partial x_j'}{\partial z}\right]+\epsilon_{jki}\frac{\partial}{\partial z}\left[\lambda\frac{\partial x_j'}{\partial x}\frac{\partial x_k'}{\partial y}\right]=0;$$

the boundary terms and the terms proportional to λ vanish as in the 2D case discussed in the Appendix. We find

$$\rho\left(x_i'-x_i\right)=-\frac{1}{2}\epsilon_{ijk}\left[\lambda,x_j',x_k'\right]_{\mathbf{x}},$$

where

$$[f,g,h]_{\mathbf{x}}=\epsilon_{pqr}\frac{\partial f}{\partial x_p}\frac{\partial g}{\partial x_q}\frac{\partial h}{\partial x_r}=\nabla f\cdot\nabla g\times\nabla h.$$

In the next step we show that

$$[f,g,h]_{\mathbf{x}}=J\,[f,g,h]_{\mathbf{x}'}\,. \tag{21}$$

We have

$$[f,g,h]_{\mathbf{x}}=\epsilon_{pqr}\frac{\partial x_s'}{\partial x_p}\frac{\partial x_t'}{\partial x_q}\frac{\partial x_u'}{\partial x_r}\,\frac{\partial f}{\partial x_s'}\frac{\partial g}{\partial x_t'}\frac{\partial h}{\partial x_u'}.$$

Now, since

$$\epsilon_{pqr}\frac{\partial x_s'}{\partial x_p}\frac{\partial x_t'}{\partial x_q}\frac{\partial x_u'}{\partial x_r}=\epsilon_{stu}J,$$

we obtain Eq. (21). Finally, using $\rho=J$ (for $\rho'=1$), this leads to

$$\left(x_i'-x_i\right)=-\frac{1}{2}\epsilon_{ijk}\left[\lambda,x_j',x_k'\right]_{\mathbf{x}'}=-\frac{\partial\lambda}{\partial x_i'}.$$

By following the reasoning at the end of the Appendix, we conclude that Monge-Kantorovich optimization in 3D also leads to a gradient map. This result also holds for general $\rho'(x',y',z')$.

6.2 3D Monge-Ampère Equation

Substituting $\mathbf{x}'=\mathbf{x}+\nabla\Phi(\mathbf{x})$ into the Jacobian equation

$$\frac{\partial(x',y',z')}{\partial(x,y,z)}=\frac{\rho(x,y,z)}{\rho'(x',y',z')},$$

we find

$$\det\begin{bmatrix}1+\partial_{xx}\Phi & \partial_{xy}\Phi & \partial_{xz}\Phi\\ \partial_{yx}\Phi & 1+\partial_{yy}\Phi & \partial_{yz}\Phi\\ \partial_{zx}\Phi & \partial_{zy}\Phi & 1+\partial_{zz}\Phi\end{bmatrix}=\frac{\rho(x,y,z)}{\rho'(x+\partial_x\Phi,y+\partial_y\Phi,z+\partial_z\Phi)}.$$

This is the 3D Monge-Ampère equation (generalizing to arbitrary $\rho'(x',y',z')\neq 1$). When the determinant is expanded out, it has $\nabla^2\Phi$ plus quadratic and cubic terms. The nonlinearities consist of the last two types of terms plus the denominator on the right.

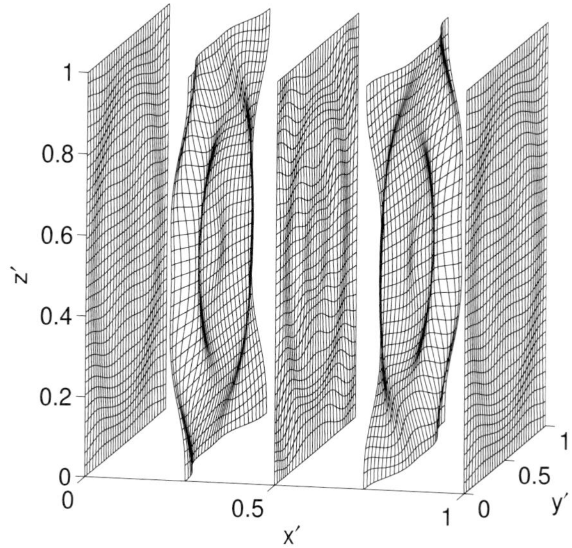

Fig. 7. Grid obtained by the 3D MA equation for the density $\rho'(x', y', z')$ prescribed according to Eq. (22): slices for $x' \approx 0$, 0.25, 0.5, 0.75, 1.0

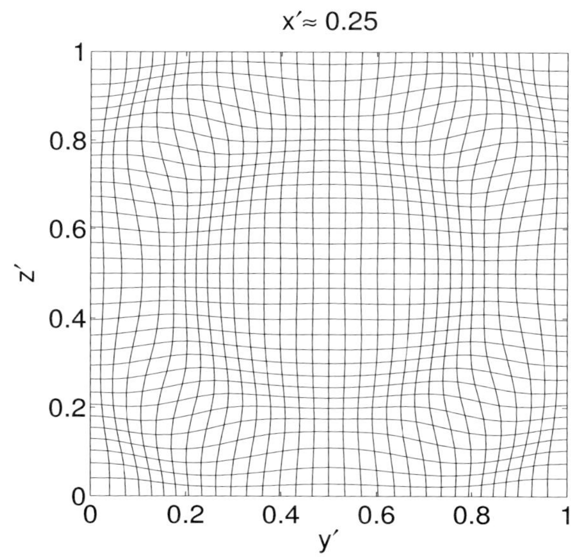

Fig. 8. Grid obtained by 3D MA equation for the density $\rho'(x', y', z')$ prescribed according to Eq. (22): a projection of the grid for $x' \approx 0.25$

6.3 Example in 3D

We treat a case with X the unit cube having $\rho(x, y, z) = 1$ and

$$\rho'(x', y', z') = \frac{C}{2 + \cos(8\pi r')}, \tag{22}$$

now with $r' = \sqrt{(x' - 1/2)^2 + (y' - 1/2)^2 + (z' - 1/2)^2}$. We show the mesh in (x', y', z') in Fig. 7, and Fig. 8 shows a projection of the mesh for $x' \approx 0.25$.

7 Conclusions

We have given a brief review of the main results of Ref. [3] in a square in 2D, showing how minimization of the L_2 norm of the grid displacement leads to equidistribution via the Monge-Ampère (MA) equation, a major element in Monge-Kantorovich (MK) optimization. (Detailed algorithmic issues were discussed in Ref. [3]). We have also shown the relation with minimum distortion and grid tangling. We have exhibited several examples showing how this method works, and comparing it with the deformation method of Ref. [2]. Table 1 and the performance tests in Ref. [3] show that the MK method uses a small amount of computer time and scales optimally with respect to grid size. The results of Ref. [3] also show that the method is robust in two important senses: (1) the increase in computational requirements with the complexity of the error measure $\rho'(x', y')$ is modest and (2) in very challenging densities $\rho'(x', y')$ the grid was not observed to fold.

We have shown new results in 2D on the formulation and application of the MK method in a more general class of physical domains X. The examples of applications include a parallelogram and an area with a sinusoidal boundary segment. These results suggest strongly that the method extends readily to more general physically realistic domains in 2D for grid generation and adaptation. Further work is underway to assess the difficulties involved with multiple block structured grids and unstructured grids.

We have shown a derivation of the Monge-Ampère equation in 3D. We have obtained results using this 3D Monge-Ampère equation in the unit cube.

Acknowledgements. We wish to thank P. Knupp, D. Knoll, G. Hansen, and X. Z. Tang for useful discussions.

References

1. Lapenta, G.: Variational grid adaptation based on the minimization of local truncation error: Time independent problems. J. Comput. Phys. 193, 159 (2004)
2. Liao, G., Anderson, D.: A new approach to grid generation. Appl. Anal. 44, 285–297 (1992)
3. Delzanno, G.L., Chacón, L., Finn, J.M., Chung, Y., Lapenta, G.: An optimal robust equidistribution method for two-dimensional grid generation based on Monge-Kantorovich optimization. J. Comp. Phys. (submitted, 2008)
4. Evans, L.C.: Partial differential equations and Monge-Kantorovich mass transfer. In: Yau, S.T. (ed.) Current Developments in Mathematics (1997)

5. Budd, C.J., Williams, J.F.: Parabolic Monge-Ampère methods for blow-up problems in several spatial dimensions. J. Phys. A 39, 5425 (2006)
6. Caffarelli, L., Nirenberg, L., Spruck, J.: The Dirichlet problem for nonlinear 2nd-order elliptic equations I. Monge-Ampère equation. Communications on Pure and Applied Mathematics 37(3), 369–402 (1984)
7. Kelley, C.T.: Iterative Methods for Linear and Nonlinear Equations. SIAM, Philadelphia (1995)
8. Dembo, R., Eisenstat, S., Steihaug, R.: Inexact Newton methods. J. Numer. Anal. 19, 400 (1982)
9. Saad, Y., Schultz, M.H.: GMRES: a generalized minimal residual algorithm for solving nonsymmetric linear systems. SIAM journal on scientific and statistical computing 7(3), 856–869 (1986)
10. `http://ndevilla.free.fr/lena/`

Appendix

In this appendix we show some technical details relating to minimization of $\int dxdy\mathcal{L}$ in 2D. This is a more compact version of the derivation in Ref. [3], and for our purposes here is a template for the 3D derivations in Sec. 6.

In 2D for the special case $\rho' = 1$, take (with summation notation)

$$\begin{aligned}\mathcal{L}(\mathbf{x},\mathbf{x}') = {} & \frac{1}{2}\rho(x,y)(x'_i - x_i)(x'_i - x_i) \\ & -\lambda(\mathbf{x})\left[\epsilon_{ij}\frac{\partial x'_i}{\partial x}\frac{\partial x'_j}{\partial y} - \rho(\mathbf{x})\right],\end{aligned} \tag{23}$$

where $\epsilon_{11} = \epsilon_{22} = 0$; $\epsilon_{12} = -\epsilon_{21} = 1$. [The Jacobian $J(\mathbf{x}) = \partial(x',y')/\partial(x,y)$ can be written in any of the alternate forms: $J = [x',y']_{\mathbf{x}}$, where $[\cdot,\cdot]$ is the Poisson bracket; $J = \epsilon_{ij}(\partial x'_i/\partial x)(\partial x'_j/\partial y)$; or $J = \epsilon_{ij}(\partial x'/\partial x_i)(\partial x'/\partial x_j)$.] The Euler-Lagrange equations give

$$\begin{aligned}&\frac{\partial\mathcal{L}}{\partial x'_i} - \frac{\partial}{\partial x_k}\frac{\partial\mathcal{L}}{\partial(\partial x'_i/\partial x_k)} = \\ &\rho(\mathbf{x})(x'_i - x_i) + \epsilon_{ij}\frac{\partial\lambda}{\partial x}\frac{\partial x'_j}{\partial y} - \epsilon_{ij}\frac{\partial\lambda}{\partial y}\frac{\partial x'_j}{\partial x} = 0.\end{aligned} \tag{24}$$

The terms proportional to λ cancel. The boundary terms take the form

$$-\int dy\left[\lambda\epsilon_{ij}\frac{\partial x'_j}{\partial y}\delta x'_i\right]_{x=0}^{x=1} - \int dx\left[\lambda\epsilon_{ij}\frac{\partial x'_i}{\partial x}\delta x'_j\right]_{y=0}^{y=1}, \tag{25}$$

each term of which equals zero. Note the Poisson bracket relation $[f,g]_{\mathbf{x}} = J[f,g]_{\mathbf{x}'}$, which follows from

$$\begin{aligned}[f,g]_{\mathbf{x}} = \epsilon_{kl}\frac{\partial f}{\partial x_k}\frac{\partial g}{\partial x_l} &= \epsilon_{kl}\frac{\partial x'_m}{\partial x_k}\frac{\partial f}{\partial x'_m}\frac{\partial x'_n}{\partial x_l}\frac{\partial g}{\partial x'_n} = [x'_m, x'_n]_{\mathbf{x}}\frac{\partial f}{\partial x'_m}\frac{\partial g}{\partial x'_n} \\ &= \epsilon_{mn}J\frac{\partial f}{\partial x'_m}\frac{\partial g}{\partial x'_n} = J[f,g]_{\mathbf{x}'}.\end{aligned}$$

This and the relation $\rho = J$ (for $\rho' = 1$) show that Eq. (24) becomes

$$x'_i - x_i = -\epsilon_{ij}[\lambda, x'_j]_{\mathbf{x}'}.$$

The whole point of writing the Poisson bracket in terms of $\mathbf{x}'$ is that this can be rewritten as

$$\begin{aligned} x'_i - x_i &= -\epsilon_{ij}\epsilon_{kl}(\partial\lambda/\partial x'_k)(\partial x'_j/\partial x'_l) \\ &= -\epsilon_{ij}\epsilon_{kj}(\partial\lambda/\partial x'_k) \\ &= -\partial\lambda/\partial x'_i. \end{aligned} \tag{26}$$

This equation shows that $x_i = \partial\left(x'_j x'_j/2 - \lambda\right)/\partial x'_i$. That is, the map ψ^{-1} is a gradient map (a Legendre transform). This implies that ψ is also a gradient map and can be written in the form used in Sec. 2,

$$\mathbf{x}' = \nabla\Omega(\mathbf{x}) = \mathbf{x} + \nabla\Phi(\mathbf{x}).$$

Metric Generation for a Given Error Estimation

Frédéric Hecht[1] and Raphaël Kuate[2]

[1] Laboratoire Jacques-Louis Lions, Université Pierre et Marie Curie
hecht@ann.jussieu.fr

[2] Laboratoire Jacques-Louis Lions, Université Pierre et Marie Curie
kuate@ann.jussieu.fr

Abstract. The mesh adaptation is a classical method for accelerating and improving the PDE finite element computation. Two tools are generally used: the metric [10] to define the mesh size and the error indicator to know if the solution is accurate enough.

A lot of algorithms used to generate adapted meshes suitable for a PDE numerical solution for instance [15, 1] for discrete metrics, or [15, 2] for the continuous one, use those tools for the local specification of the mesh size.

Lot of PDE softwares like *Freefem++* [14] use metrics to build meshes, the edges sizes of which are equal with respect to the metric field. The construction of metrics from the hessian matrix [16, 18, 13] is only justify for the piecewise linear Lagrange finite element.

So there is the problem of metric generation when we have another interpolation error estimator [6, 11] that could be used for instance when the Lagrange interpolation needed is a k degree polynomial, $k > 1$.

To answer that question, we propose in this paper an algorithm whose complexity is quasi-linear, in two spacial dimensions; assuming that the error is locally described by a closed curve representing the error level set. Some efficient numerical examples are given. This algorithm allows us to obtain the analytical metric [1], when the error indicator is based on the hessian matrix.

We have also done one comparison in the software *Freefem++* of mesh adaptation with metrics computed using this algorithm with respect to the interpolation error estimation described in [11], and the method with metrics based on the hessian. The results seem to be better for the maximal error.

1 Introduction

We suppose once a metric field on a mesh is known, we can built an adapted mesh for that metric field [8, chap 21.4], [12, chap 20.3]. If we know the metric value on each vertex of the mesh for example, the method described in [12] modify the mesh by evaluating each edge length in the metric field (control space) using an interpolation of the metrics defined on the two vertices of the edge. When an edge length exceeds the "unity", a subdivision of that edge is done in the control space. This method is used in the software *Freefem++* with the mesh generator Bamg [9]

Thanks to Cea lemma, the approximation error is upper bounded by the interpolation error. So our problem is posed for the interpolation error, we'll use the recent results described in [11] which aim to bound the interpolation error of a function on a simplex with the norm of a vector inside a metric given by the $D^{(l)}$ derivative of the function.

We suppose that we know an estimation of the local interpolation error of a function on a 2D mesh around a vertex P in each direction ξ given by a function, and we want to dispose the points M of the triangulation around P such that the local interpolation error should be satisfy.

Let $E_p(\xi)$ be an estimation of the interpolation error in the direction of the vector ξ. If the mesh size is less than or equal to $||\xi||$ in the direction of ξ then the interpolation error will be less than $E_p(\xi)$. We want to have locally

$$|E_p(\xi)| \leq \mathcal{E}; \xi \in \mathbb{R}^2. \tag{1}$$

The previous inequality represents an area bounded by the level set

$$|E_p(\xi)| = \mathcal{E}.$$

Dispose the points M of the triangulation around the vertex P as close as possible to the level set while being inside the area delimited by that level set is equivalent to say that the interpolation error evaluated at each vertex of the triangulation going from the vertex P is less than or equal to its value on the level set. If we want to have a mesh whose edges are equal with respect to a metric field, i.e all the points M are equi-distant from P in a local metric $\mathcal{M}$, so they belong to the unit ball of the metric $\mathcal{M}$; the biggest as possible that could be contained inside the closed line described by equation (1).

Let us find a metric prescribing the mesh size around that vertex P such as the triangles vertices should be close as possible to the level set while being inside the area delimited by that level set. How to build that metric?

In the following, after introducing shortly the results of [11] and the metric notion, [15, 10, 1], we formulate the optimization problem that goes with and it resolution algorithm. Then we'll explain the algebraic consequences of the obtained equations in the third section. The fourth section talk about our algorithm convergence and complexity and some illustrating figures. The last section before concluding explains shortly the analytic metric obtainability from the hessian matrix, that our method allows to approach. We also give in that section an example of mesh adaptation for a function with metrics generated by our algorithm with respect to the local interpolation error of [11], and we compare the results obtained with mesh adaptation using metrics based on the hessian matrix, both of them done with the software *Freefem++*.

2 Definitions

We use the notations [3, Chap. VII], which are introduced in the books [7] [5, chap II].

- $x = (x_1, \ldots, x_d) \in \mathbb{R}^d$, Sphere $S_d = \{x \in \mathbb{R}^d; ||x|| = 1\}$,
- $D^{(\ell)}u$ is the the derivative of u of order ℓ so if u is a real function, we have $D^{(0)}u = u$.

- $B^{\otimes \ell} : (\xi_1, ..., \xi_\ell) \mapsto (B\xi_1, ..., B\xi_\ell)$
- $W^{\ell,p}(K)$ is the set of functions of $L^p(K)$ such that all the partial derivatives of order $\leq \ell$ are in $L^p(K)$.
- $|u|_{W^{\ell,p}(K)} = (\int_K \sum_{\alpha \in [1..d]^\ell} |\partial_\alpha u|^p)^{\frac{1}{p}}$
- Let F be an affine application of $\mathbb{R}^d$ with values in $\mathbb{R}^d$,

$$||D^{(k)}(v \circ F)||_{L^p(K)} \leq ||D^{(1)}F||^k ||D^{(k)}v||_{L^p(F(K))}$$

d is the space dimension , K is a d-simplex (triangle if $d=2$, or tetrahedron if $d=3$), h_K the diameter of K and ρ_K the diameter of its inscribed ball, $|K|$ the measure of K and $|\partial K|$ the measure the border of K.

Definition 1 (d-simplex finite element of degree k)
A finite element (K, P_K, Σ_K) will be called d-simplex of degree k, if K is a d-simplex and if $P_k(K) \subset P_K$ where $P_k(K)$ is the set of polynomials $P[\mathbb{R}^d]$ restricted to K of degree less than or equal to k. Let $\mathcal{I}_K^k$ be the interpolation operator with values in P_K and which keep unchanged the polynomials of $P_k(K)$, i.e if $u \in P_k(K)$, then $u = \mathcal{I}_K^k u$.

Definition 2 (reference element)
The reference element is a regular d-simplex denoted $\hat{K}$, chosen with the size 1.

Let F_K be the affine transformation such that $K = F_K(\hat{K})$, we denote: B_K the linear part of F_K which corresponds to the derivate of F_K; and $\det(B_K)$ the determinant of B_K which is the Jacobian of F_K.

The size $h_{K,\xi}$ of an element K in the direction ξ, the size of the edges in the direction ξ of that edge. $||B_K^{-1}||^{-1}$ is the smallest size of the element denoted $h_K^- = \inf_{\xi \in S_d} h_{K,\xi}$. This element size can be represented by an ellipse associated to the "natural" metric of the element introduced in [17, 4].

Let u be a real function of K, we define its corresponding function $\hat{u} = u \circ F_K$ on $\hat{K}$. We have

$$x = F_K(\hat{x}), \qquad \hat{u}(\hat{x}) = u(x), \qquad u = \hat{u} \circ F_K^{-1}.$$

A finite element K is affine-equivalent if it has the following property

$$\widehat{\mathcal{I}_K^k u} = \mathcal{I}_{\hat{K}}^k \hat{u}.$$

Main Result
For any finite element (K, P_K, Σ_K) of definition 1 satisfying the previous property, with an associated metric $\mathcal{M}_K$; let u be a function. For all integer $\ell \in [0..k+1]$ and for all $(p,q) \in [1,\infty]^2$ such that $W^{\ell,p}(K)$ be included in $\mathcal{C}^0(K)$ and $u \in W^{\ell,p}(K)$. If it exists a real constant $\kappa \in \mathbb{R}$ such that

$$||(D^{(\ell)}u)(\xi, ..., \xi)||_{L^p(K)} \leq \kappa\, ||\xi||_{\mathcal{M}_K}^\ell (h_K^-)^\ell = \kappa\, (\xi, \mathcal{M}_K \xi)^{\frac{\ell}{2}} (h_K^-)^\ell, \tag{2}$$

then we have

$$||u - \mathcal{I}_K^k u||_{L^q(K)} \leq \kappa C\, |K|^{\frac{1}{q} - \frac{1}{p}} (h_K^-)^\ell,$$

where C is a real positive constant depending only on d, ℓ, p, q.

And if $W^{\ell,p}(K) \subset W^{1,q}(K)$ then we have

$$|u - \mathcal{I}_K^k u|_{W^{1,q}(K)} \leq \kappa C\,|K|^{\frac{1}{q}-\frac{1}{p}}(h_K^-)^{\ell-1},$$

with C positive constant depending only on d, ℓ, p, q.

Definition 3. *A constant metric $\mathcal{M}$ of $\mathbb{R}^2$ is a symmetric matrix positively definite that allows to define a distance in the euclidian metric space $\mathbb{R}^2$.*

In fact the scalar product of two vectors $\overrightarrow{X}$ and $\overrightarrow{Y}$ under the metric space $\mathbb{R}^2$ is defined as: [8]

$$\langle \overrightarrow{X}, \overrightarrow{Y} \rangle_{\mathcal{M}} = {}^t\overrightarrow{X}\mathcal{M}\overrightarrow{Y} \in \mathbb{R}.$$

The euclidian norm of a vector $\overrightarrow{AB}$ or the distance between the points A and B for the metric $\mathcal{M}$ is defined by:

$$d_{\mathcal{M}}(A, B) = \|\overrightarrow{AB}\|_{\mathcal{M}} = \sqrt{{}^t\overrightarrow{AB}\mathcal{M}\overrightarrow{AB}}.$$

There is the situation for which the metric is non constant but continuous. In this case we have a *metric field* $\mathcal{M}(X)$ defined in each point X of the space. In this case, the norm of a vector $\overrightarrow{AB}$ is given by

$$d_{\mathcal{M}}(A, B) = \|\overrightarrow{AB}\|_{\mathcal{M}} = \int_0^1 \sqrt{{}^t\overrightarrow{AB}\mathcal{M}(A + t\overrightarrow{AB})\overrightarrow{AB}}dt.$$

A metric $\mathcal{M}$ can be represented by it unit ball [1], The unit ball under the metric $\mathcal{M}$ defined at the vertex P is the set of points M that satisfies:

$$\|\overrightarrow{PM}\|_{\mathcal{M}} = \sqrt{{}^t\overrightarrow{PM}\mathcal{M}\overrightarrow{PM}} = 1.$$

Let us consider $\mathcal{M} = \begin{pmatrix} a & \frac{c}{2} \\ \frac{c}{2} & b \end{pmatrix}$ $\quad a > 0, b > 0,\ 4ab - c^2 > 0;\ P(x_p, y_p);\ M(x, y).$

$$\|\overrightarrow{PM}\|_{\mathcal{M}}^2 = 1 \qquad \Longleftrightarrow \qquad a(x - x_0)^2 + c(x - x0)(y - y_0) + b(y - y_0)^2 = 1.$$

By writing

$$a = \frac{\cos\theta^2}{\alpha^2} + \frac{\sin\theta^2}{\beta^2}; \quad b = \frac{\cos\theta^2}{\beta^2} + \frac{\sin\theta^2}{\alpha^2}; \quad c = \sin 2\theta\left(\frac{1}{\alpha^2} - \frac{1}{\beta^2}\right); \tag{3}$$

which are bounded by:

$$\frac{1}{\alpha^2} \leq a \leq \frac{1}{\beta^2}; \quad \frac{1}{\alpha^2} \leq b \leq \frac{1}{\beta^2}; \quad \frac{1}{\alpha^2} - \frac{1}{\beta^2} \leq c \leq \frac{1}{\beta^2} - \frac{1}{\alpha^2}; \tag{4}$$

we obtain:

$$\frac{((x - x_0)\cos\theta + (y - y_0)\sin\theta)^2}{\alpha^2} + \frac{((y - y_0)\cos\theta - (x - x0)\sin\theta)^2}{\beta^2} = 1,$$

which is an ellipse equation centered at P with semi-axes of length $\alpha > 0$ and $\beta > 0$; $\alpha \geq \beta$; where θ is the angle between the major axis and the abscises axis of the $\mathbb{R}^2$ canonic base.

3 The Problem Description

We do not take care of the different norms used and the different constants of equation (2). We consider that

$$|E_p(\xi)| = ||(D^{(\ell)}u)(\xi, ..., \xi)||_{(K)}. \tag{5}$$

We want to find a local metric $\mathcal{M}$ such that the right member of (2) should be closed to $\mathcal{E}$ in (1).

Let $F(x, y)$ be $E_p(\xi)$. Our problem becomes drawing an ellipse that have the maximum area centered at P contained inside that given error curve or finding the parameters α , β and θ of that ellipse.

Let us consider $\alpha_1 = \dfrac{1}{\alpha^2}$ and $\alpha_2 = \dfrac{1}{\beta^2}$, we obtain from (3)

$$\alpha_2 = \frac{\sqrt{(a-b)^2 + c^2} + (a+b)}{2}; \quad \alpha_1 = \frac{-\sqrt{(a-b)^2 + c^2} + (a+b)}{2}.$$

Maximizing the ellipse area, the parameters of which are α; β; θ is the same as maximizing $\alpha\beta$ or minimizing $\alpha_1\alpha_2$ which is equivalent to minimize $(4ab - c^2)$.

We look at the problem in a numerical point of view and suppose that the vertex P is at the origin.

Let us consider a discretization of our curve with n points M_i of position (x_i, y_i). We have to find three reals a, b and c that minimize $(4ab - c^2)$ under the constraints

$$\begin{cases} a > 0, b > 0, \\ ax_i^2 + by_i^2 + cx_iy_i \geq 1, \ 0 \leq i \leq n-1, \\ 4ab > c^2. \end{cases}$$

Without "good" properties of the minimization function (convexity, quasi-convexity), we didn't find any algorithm that guarantees a global minimum; although it exists because the linear constraints form a close set. The resolution of this problem in this form need the construction of the constraints polyhedron where we'll find the minimum. As $(a, b, c) \in \mathbb{R}^3$ the resolution algorithm will be $\mathcal{O}(n^3)$ complexity in the best case; but it is very cheap inside an a mesh adaptation process. To reduce that cost we propose to solve an approximated problem whose algorithm will be $\mathcal{O}(n^k)$; $k \leq 2$ complexity.

Let us consider two real non negative numbers ϵ_0 and ϵ close to zero, let us call M_0 the point of the error curve the most closed to the origin P of our ellipse, and M_i another point of that error curve. Then we find an ellipse that have the maximal area. Let ϵ_0 be the distance between that ellipse and M_0, and let ϵ be the distance between that ellipse and M_i (figure 1). The distance between the ellipse and a point M_j of the error curve will be consider as the distance between M_j and the intersection point of the vector $\overrightarrow{M_jP}$ and the ellipse. Let us consider a point X of the ellipse on the segment $[P, M_i]$ such that $\|\overrightarrow{M_iX}\| = \epsilon$, the value of X is $(1 - \frac{\epsilon}{\|X_i\|})X_i$. let us consider $r_i = \|X_i\|$.

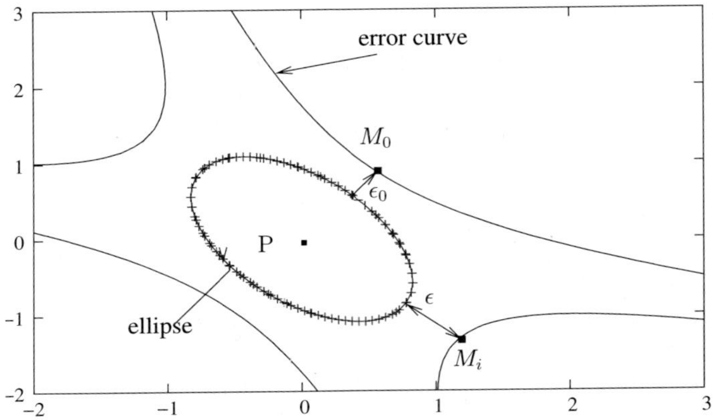

Fig. 1. Parameters

Our problem becomes:

Find three real numbers a, b and c that minimize $(4ab - c^2)$ under the constraints

$$\left\{\begin{array}{rcll} & a > 0, b > 0, & & \\ ax_0^2 + by_0^2 + cx_0y_0 & = & \left(\dfrac{r_0}{r_0 - \epsilon_0}\right)^2, & \\ ax_i^2 + by_i^2 + cx_iy_i & = & \left(\dfrac{r_i}{r_i - \epsilon}\right)^2, & 1 \le i \le n-1, \\ ax_k^2 + by_k^2 + cx_ky_k & \ge & 1, & 1 \le k \le n-1,\ k \ne i, \\ & 4ab > c^2. & & \end{array}\right. \tag{6}$$

Numerically, the constraint $4ab > c^2$ need a threshold $\mathcal{T} > 0$ for which a solution of $4ab - c^2 - \mathcal{T}$ will be consider as a minimizer of $4ab - c^2$.

Minimization of $4ab - c^2$.
Let us call R_{max} the euclidian norm of the error point of the curve the most far away from it center P; we obtain:

$$\mathcal{T} = \left(\frac{2}{R_{max}(r_0 - \epsilon_0)}\right)^2$$

4 Algebraic Resolution

Let us examine the two equations of system (6)
If $x_0 = 0$, then $b = \dfrac{1}{(|y_0| - \epsilon_0)^2}$.

- If $x_i = 0$, then $b = \dfrac{1}{(|y_i| - \epsilon)^2}$ so we must have

$$\epsilon = |y_i| - |y_0| + \epsilon_0. \tag{7}$$

- If $y_i = 0$ and $y_k = 0$, we must have

$$\epsilon \geq |x_i| - |x_k|. \tag{8}$$

- If $x_i y_0 - y_i x_0 = 0$ and $x_i \neq 0$, we must have in this case

$$\epsilon = r_i \left(1 - \frac{x_0(r_0 - \epsilon_0)}{x_i r_0}\right). \tag{9}$$

Remark. We observe that if one coefficient of system (6) equal zero or if the points M_i, M_0 and the origin P are aligned, the choice of the parameters ϵ (or ϵ_0 in some cases) is not arbitrary. Those parameters should satisfy (7), (8) or (9) for the previous cases, but it will add new constraints to the problem and will not allow the precision we want.

The points representing those coefficients, when they equal zero, are almost four (the four halve-axes). They will be turn out of the axes with a small angle and also the eventually point on the axe (P, M_0). Doing that will not sensitively modify the problem neither its solution. We will consider that $x_0, x_i, x_k, y_0, y_i, y_k$ are all not zero for all i and for all k, and that $x_i y_0 - y_i x_0$ is not zero for all i. From both two equations of system (6), we extract c and a as a function of b.

Let us write:

$$\Delta = \frac{x_i}{x_0(x_i y_0 - x_0 y_i)} \left(\frac{r_0}{r_0 - \epsilon_0}\right)^2 - \frac{x_0}{x_i(x_i y_0 - x_0 y_i)} \left(\frac{r_i}{r_i - \epsilon}\right)^2;$$

$$\gamma = -\frac{y_i}{x_0(x_i y_0 - x_0 y_i)} \left(\frac{r_0}{r_0 - \epsilon_0}\right)^2 + \frac{y_0}{x_i(x_i y_0 - x_0 y_i)} \left(\frac{r_i}{r_i - \epsilon}\right)^2;$$

$$\tau = \frac{y_i y_0}{x_i x_0}; \qquad \lambda = \frac{x_0 y_i + y_0 x_i}{x_i x_0}; \quad \beta_k = y_k^2 + x_k^2 \tau - x_k y_k \lambda;$$

$$C_k = 1 - \frac{x_k(y_k x_i - y_i x_k)}{x_0(x_i y_0 - x_0 y_i)} \left(\frac{r_0}{r_0 - \epsilon_0}\right)^2 + \frac{x_k(y_k x_0 - x_k y_0)}{x_i(x_i y_0 - x_0 y_i)} \left(\frac{r_i}{r_i - \epsilon}\right)^2;$$

then,

$$c = \Delta - \lambda b; \quad a = \gamma + \tau b.$$

The inequality before the last one of (6) becomes

$$\beta_k b \geq C_k, \qquad 1 \leq k \leq n-1; \ k \neq i. \tag{10}$$

So the function to minimize becomes

$$G(b) = (4\tau - \lambda^2) b^2 + 2(2\gamma + \lambda\Delta) b - \Delta^2 - \mathcal{T},$$

the roots of which are

$$b_1 = \left(\frac{1}{x_i y_0 - x_0 y_i}\right)^2 \left[\left(\frac{x_0 r_i}{r_i - \epsilon}\right)^2 + \left(\frac{x_i r_0}{r_0 - \epsilon_0}\right)^2 + \mathcal{D}\right];$$

$$b_2 = \left(\frac{1}{x_i y_0 - x_0 y_i}\right)^2 \left[\left(\frac{x_0 r_i}{r_i - \epsilon}\right)^2 + \left(\frac{x_i r_0}{r_0 - \epsilon_0}\right)^2 - \mathcal{D}\right];$$

where

$$\mathcal{D} = \frac{2|x_i x_0|}{(r_0 - \epsilon_0)} \sqrt{\left(\frac{r_i r_0}{r_i - \epsilon}\right)^2 + \left(\frac{x_0 y_i - y_0 x_i}{R_{max}}\right)^2}.$$

The inequalities (4), (10) and the following bounds

$$\frac{(r_0 - \epsilon_0)^2}{R_{max}} \leq \beta \leq r_0 - \epsilon_0 \leq \alpha \leq R_{max},$$

gives b_{min} and b_{max} such that $b \in [b_{min}, b_{max}]$. So, we will compare $G(b_{min})$ and $G(b_{max})$ to determine the value of b that minimize G.

5 Numerical Resolution

The algorithm consists in:

- Discretizing the error curve around the vertex P into n points, determining the point M_0 of the curve which is the most closed to the origin P and the point M_{max} of the curve which is the most far away from P;
- moving out the points on the axes and also the point which is eventually aligned with M_0 and the origin;
- for all fixed M_i, switching the constraints for all the M_k ($k \neq 0$; $k \neq i$) and determining the optimal values of a, b and c.

5.1 Algorithm Order

This algorithm is globally quasi-linear because for each fixed M_i only one (the first) unsatisfied constraint on the M_k implies an interruption of the M_k's loop and switch to the next M_{i+1}(when the distance between a point of the error curve and the ellipse is greater than ϵ, that point will not satisfy any constraints). The points that satisfy the constraints are only those in the regions where the ellipse is tangent to the error curve, and their number is few. The only case where for each fixed M_i we go trough all the M_k is the case where the error curve is already an ellipse and in this case the algorithm order is $\mathcal{O}(n^2)$. But if the error curve is already an ellipse, we don't need this algorithm.

5.2 Convergence of the Approximation

It is natural to think that M_0 will not be the point of the error curve that is the most far away from the ellipse. So we can find ϵ_0 for which the solution of our approach problem converges to the optimal solution.

We can have an upper bound of ϵ_0:

The disc area of radius r_0 is less than the ellipse area of parameters $r_0 - \epsilon_0$ and R_{max} so, $\epsilon_0 < r_0 \left(1 - \frac{r_0}{R_{max}}\right)$. Although we don't have any absolute criteria of

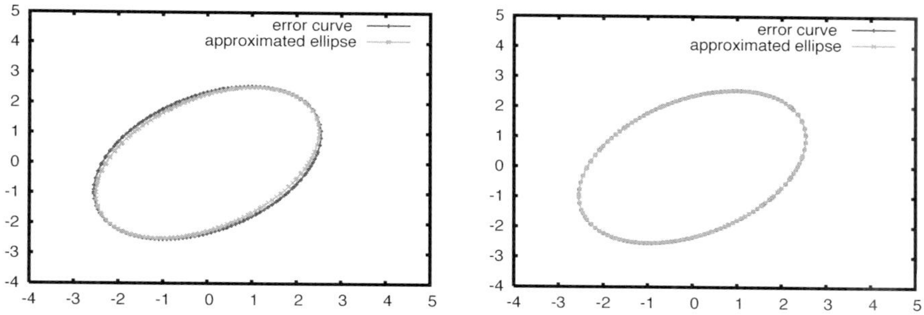

Fig. 2. An ellipse of parameters $\theta = \frac{\pi}{4}$, $\alpha = 3$, $\beta = 2$, $n = 159$; left: $\epsilon_0 = 10^{-1}$; right: $\epsilon_0 = 10^{-4}$

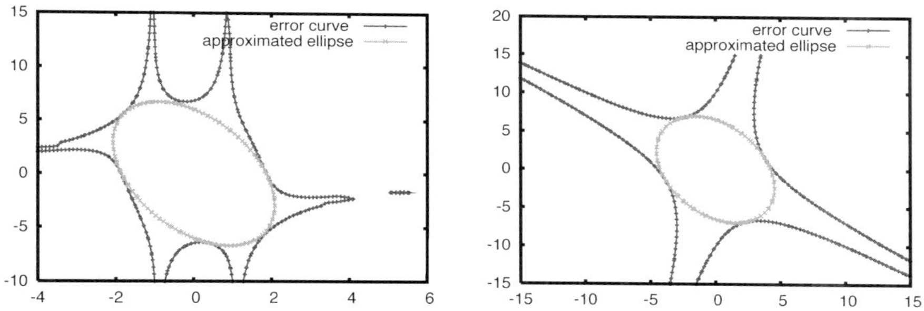

Fig. 3. left: $\mathcal{E} = 59$, ϵ_0 opt. $= 10^{-5}$, $n = 285$, $F(x, y) = -0.2y^3x^2 - yx^4 + 0.2y^3 - 9x^3 - 0.5xy^2 + 0.005x^2 - 0.1y^2$, right: $F(x, y) = 0.05x^2 - 0.01y^2 + 0.05xy$, $\mathcal{E} = 1$, ϵ_0 opt. $= 0.001$, $n = 304$

choosing ϵ_0, when changing its values within a few iterations (one dozen) inside the code between 10^{-5} and it upper bound, we optimize the approximation (see figures 2, 3).

6 Validation

When solving numerically a partial differential equation with piece wise linear Lagrange finite element, one can use the hessian matrix of the function to approach as error estimator to generate adapted meshes [1].

As the hessian matrix $\mathcal{H}$ is a symmetric matrix, it absolute value define bellow when invertible is positive semi definite so it can be used to define a metric.

$$|\mathcal{H}| = \mathcal{R}|\Lambda|\mathcal{R}^{-1},$$

where $|\Lambda|$ is the diagonal matrix formed by the absolute value of the eigenvalues and $\mathcal{R}$ the matrix of the eigenvectors of $\mathcal{H}$. When $\mathcal{H}$ is not invertible, the zero eigenvalue is replaced by $\frac{1}{h_{max}^2}$, where h_{max} is the maximal authorized edge size in the mesh.

6.1 Some Examples

When applying our algorithm in some cases with

$$F(x,y) = \langle (x,y), \mathcal{H} \cdot (x,y) \rangle,$$

we obtain comparative results with the metric given by the absolute value of $\mathcal{H}$ (analytical) figures 4 and 5.

In the example of figure 4 for the case $\mathcal{H}$ invertible, one can observe that the obtained ellipse is closed to the analytical one when the number n of discretization points is enough, which equation in that example is

$$x^2 + y^2 + xy = 1.$$

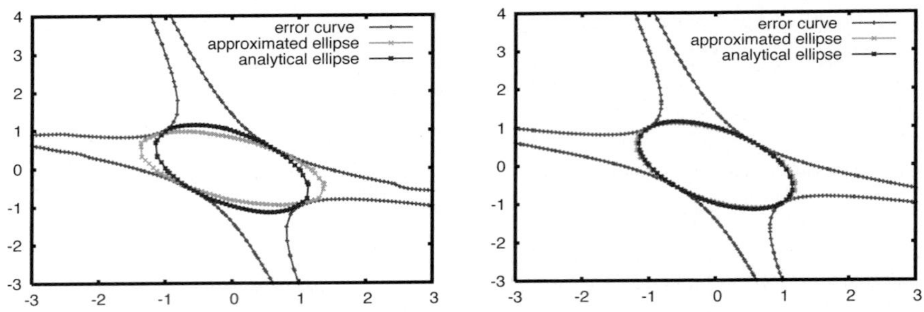

Fig. 4. $\mathcal{E} = 1$; $F(x,y) = 0.5x^2 + 0.5y^2 + 2xy$; left: $n = 228$; right: $n = 288$

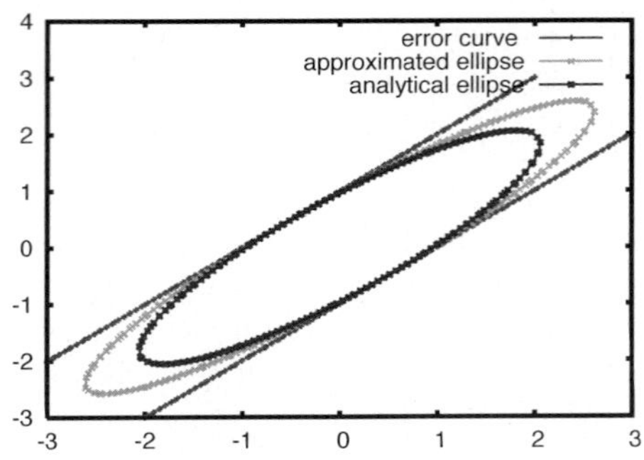

Fig. 5. $\mathcal{E} = 1$; $n = 234$ $F(x,y) = x^2 + y^2 - 2xy$

```
!"#$%&'()*+,-.#)(&
+/,!.$(&0"/,)+"/1($2&
3(45%65748.#*(9:;:;<9=!>1:?@A;9B!>1:?@AC?D
*(#!%=>;B>D
*(#!%,"(0%7E>D
0(42#,(%F5965;GE?D
0(42#,(%H5965;GA?D
0(42#,(%G5965;G>?D

0./,%0%7%)#/59,@94+/99:@B??!9A%@%=???I%9B@=@=?I%2"J9B;K?D
0./,%(**790===@=>@=>@=>IK@0==B@=>@=>@B>IK@0=BB@=>@B>@B>
        %%I0BBB@B>@B>@B>?@,"(0D
LL0===;0==B;0=BB;0BBB%$(*+M#)+M(4%"0%0%#/$%)5(%,"/4)#/)%,%
LL#*(%$(0+/($%+/%0+!(%0"/,)+"/1($2

0"*9+/)%+7ED+NE>D+II?
%%O
%%%%F5%3EE;3EA;3AAD%%%%
%%%%H5%0A70D
%%%%'()*+,-.#)(965;E:>;>1>>>E;K;(**;<3EE<C;3EA<C;3AA<CC;<=>;B>C?D
%%%%*(#!%,,7E>D
%%%%657#$#2)3(45965;,,@3EE;,,@3EA;,,@3AA;P4'()*+,7E;
                  %53+/7>1>>>>E;/QM=7E>>>>>>?D
%%%%,".)NN3EE<C13#=NN&%&NN%3EA<C13#=%NN&%&NN%3AA<C13#=%NN(/$!D
%%%%2!")965;J#+)7>;247&651(24&?D
%%%%2!")90A;J#+)7>?D
%%%%G5%(5%7%9#Q490A!0??D
%%%%G5%!(57%!"RE>9(5?D
%%%%*(#!<+/)C%M+4"7<!S;!T;!U;!V;!:;!W;!K;!A;!E;>;ECD
%%%%2!")9!(5;0+!!7E;J#+)7>;M+4"7M+4";M#!.(7E;247&!(51(24&?D
%%%%,".)NN+NN&1111&NN(5<C13+/NN&%&NN(5<C13#=NN&%&NN(5<C14.3L(5<C1/
        %NN&%&NN+/)A$965?9(5?L651#*(#NN&%&NN651/)NN&%&NN651/MNN(/$!D
%%X
65748.#*(9:;:;<9=!>1:?@A;9B!>1:?@AC?D

*(#!%,(**79>1>>:@T1>A:US(!>:?Y>1:%D
0"*9+/)%+7ED+NE>D+II?
%%O
%%%%F5%3EE;3EA;3AAD%%%%
%%%%H5%0A70D
%%%%657#$#2)3(45965;0;(**7,(**;53+/7>1>>>>E;/QM=7E>>>>>>?D
%%%%,".)NN%3EE<C13#=%NN%&%&%NN%3EA<C13#=%NN%&%&%NN%3AA<C13#=%NN(/$!D
%%%%2!")965;J#+)7>;247&65A1(24&?D
%%%%2!")90A;J#+)7>?D
%%%%G5%(5%7%9#Q490A!0??D
%%%%G5%!(57%!"RE>9(5?D
%%%%*(#!<+/)C%M+4"7<!S;!T;!U;!V;!:;!W;!K;!A;!E;>;ECD
%%%%2!")9!(5;0+!!7E;J#+)7>;M+4"7M+4";M#!.(7E;247&!(5A1(24&?D
%%%%,".)NN+NN&1111&NN(5<C13+/NN&%&NN(5<C13#=NN&%&NN(5<C14.3L(5<C1/
        %NN&%&NN+/)A$965?9(5?L651#*(#NN&%&NN651/)NN&%&NN651/MNN(/$!D
%%X
```

The example of figure 5 show the case $\mathcal{H}$ non invertible where h_{max} have been arbitrary chosen in the analytical case which equation in that example is

$$0.5\left((\frac{1}{h_{max}^2} + 2)(x^2+y^2) + 2(\frac{1}{h_{max}^2} - 2)xy \right) = 1.$$

We just need to variate the value of h_{max} to coincide with the approached solution.

6.2 Test with Freefem++

This algorithm have been implemented into Freefem++ and we present the obtained results when doing mesh adaptation for a function, in the following code.

In that code, the error function is given by the third derivates of the function f. That error function is written as an homogenous third degree polynomial of the coordinates of the local unit ball, as it's shown in [11]. We use our algorithm with $\mathcal{E} = 1$, to compute the metric on each vertex of the mesh, and the mesh is adapted with the obtained metric in the first loop. In the code, the parameters

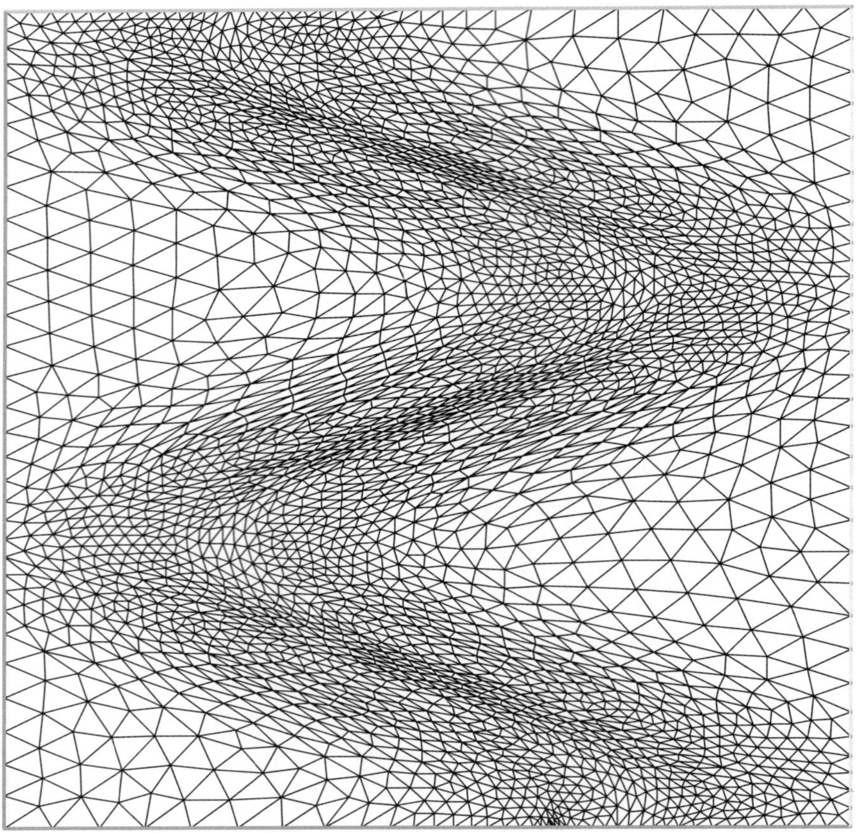

Fig. 6. Adapted meshes for the maximal error level around 10^{-5}, cc=10, coef=10, metrics computed with this algorithm for the (5) error estimator

cc and $coef$ are used to change the value of the error. Let err be the maximal error, and let Ct be a constant. err is given by

$$err = \frac{Ct}{coef \cdot cc^{l/2}}.$$

Using the second loop and the result of the first loop, we computed backward the real $cerr$ useful to have the same error level on each final mesh.

In the second loop, we made the usual mesh adaptation of Freefem++ for the same function f, based on metrics given by the hessian matrix.

Comments

In our example $l = 3$. According to the results of the second table of figure 9, if we take $Ct \simeq 2.6e{-}5$, we obtain $\frac{Ct}{10^{3/2}} = 8.22e{-}5$. It corresponds to the level of the maximal error in the first table.

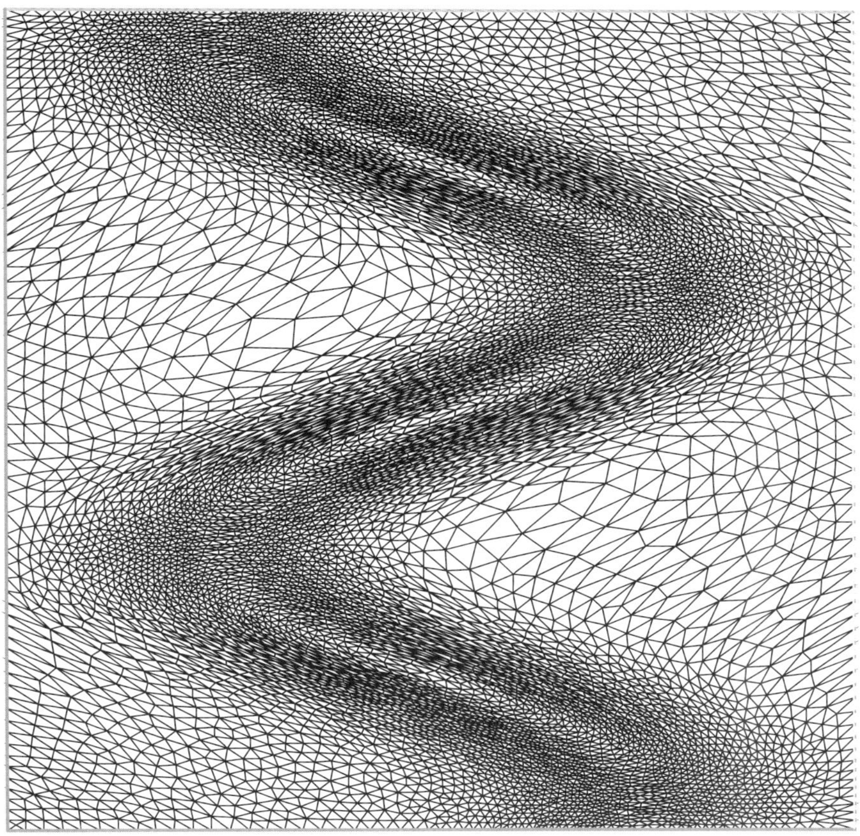

Fig. 7. Adapted meshes for the maximal error level around 10^{-5}, cc=10, coef=10, method based on the hessian

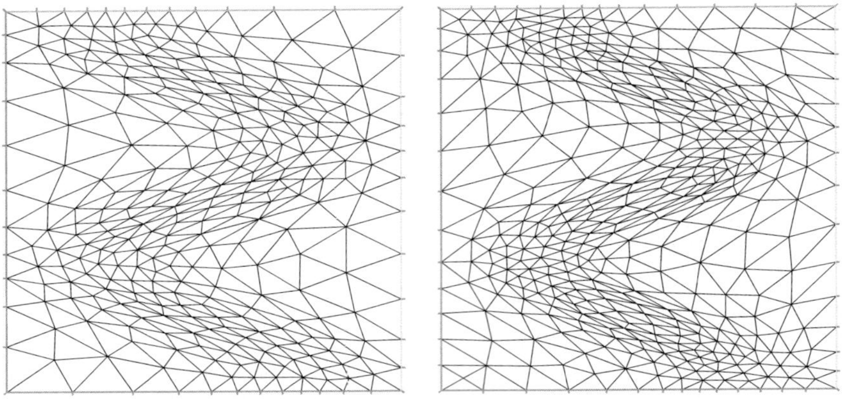

Fig. 8. Adapted meshes for the same maximal error level around 10^{-3}, cc=1, coef=10. Left: metrics computed with this algorithm for the (5) error estimator. Right: method based on the hessian.

iterations	error min 1	error min 2	error max 1	error max 2	vertices 1	vertices2
1	1.20059e-05	4.22866e-06	0.356521	0.358384	2050	3923
2	3.11655e-09	2.48466e-09	0.0043082	0.00437699	5194	17235
3	3.62071e-09	1.52523e-11	9.67516e-05	9.9182e-05	5198	15381
4	3.69877e-09	2.19181e-10	7.95773e-05	0.000103307	5194	15210
5	1.10912e-09	4.76638e-10	7.79361e-05	0.000106652	5264	15188
6	7.05385e-09	3.61698e-10	7.77929e-05	8.02886e-05	5200	15190
7	1.89551e-09	9.00541e-12	8.01738e-05	7.80091e-05	5200	15186
8	5.05435e-10	1.42422e-10	7.75686e-05	8.22038e-05	5200	15186
9	9.78865e-10	9.68075e-10	8.02579e-05	7.7833e-05	5218	15180
1	1.55839e-07	1.00832e-05	0.358347	0.35043	425	460
2	9.65766e-07	3.80745e-08	0.0134088	0.064436	536	743
3	3.58336e-06	2.93163e-08	0.00281739	0.00536942	536	733
4	2.34807e-07	1.97044e-07	0.0027611	0.00278402	540	733
5	1.10585e-06	9.35475e-08	0.00291751	0.0025183	540	732
6	7.26946e-08	3.9567e-07	0.00264979	0.00275217	540	733
7	1.49178e-06	7.63447e-08	0.00266049	0.00270342	540	733
8	8.80737e-08	5.55839e-08	0.00267491	0.00271847	540	733
9	1.04068e-06	2.8896e-07	0.00269423	0.00285419	540	733

Fig. 9.

We can observe on figures 8, 6, 7 and 9 that for the same maximal error level around 10^{-3} or 10^{-5}, the error estimator described in [11] and our algorithm uses much less elements than the classical mesh adaptation method with metrics based on the hessian matrix. We see that the convergence is obtained after five iterations, see figure 9. For the hessian based error built mesh, if we suppose that the function is a polynomial of degree 3, then the adaptation will be accurate. As our function is not a polynomial of degree three, we can observe that on figure 6, there is two thin mesh layers bounding the middle layer, and the corners of the domain are meshed with small size elements for the hessian based method. Those wrong effects are not observed with our new approach.

7 Conclusion

We have proposed a quasi-linear complexity algorithm in two spacial dimension useful to build numerically metrics for any given error estimation represented by a closed curve around a mesh vertex. As a metric can be represented by an ellipse, this algorithm is based on the optimization of an ellipse area. The main purpose of this work is to adapt mesh when doing numerical simulations where the interpolation error estimator is not based on the hessian matrix.

The obtained numerical results of our tests are encouraging. We have presented an example of mesh adaptation using metrics computed with this algorithm with respect to the new interpolation error estimation [11], done in Freefem++. That example seems to give better results than the usual adaptation procedure of Freefem++ with metrics based on the hessian, according the maximal error. We hope that we can expect "good" performances of the future new mesh adaptation routine of the software Freefem++ [14] where this algorithm could be used for the Lagrange finite element interpolation of higher order.

A future work on the local representation of other types of error estimators like residual estimators could be interesting. Then we will see how this algorithm could used or not on those cases. It development into dimension three is a future challenge in so far as the software Freefem++ is now moving from dimension two to dimension three.

We thank the reviewers for their comments which helped to improve the quality of this paper. We also thank Pascal Ventura who help us for the achievement of this work.

References

1. Alauzet, F.: Adaptation de maillage anisotrope en trois dimensions. Application aux simulations instationnaires en mecanique des fluides. PhD thesis, université de Montpellier II (October 2003)
2. Alauzet, F., Dervieux, A., Frey, P.: Multi-dimensional continuous metric for mesh adaptation. In: 15th International Meshing Roundtable, September 2006, pp. 191–213 (2006)
3. Bernardi, C., Maday, Y., Rapetti, F.: Discrétisations variationnelles de problèmes aux limites elliptiques. Mathématiques & Applications (Berlin) [Mathematics & Applications], vol. 45. Springer-Verlag, Berlin (2004)
4. Cao, W.: On the error of linear interpolation and the orientation, aspect ratio, and internal angles of a triangle. SIAM J. Numer. Anal. 43(1), 19–40 (2005) (electronic)

5. Ciarlet, P.G., Lions, J.-L. (eds.): Handbook of numerical analysis. Handbook of Numerical Analysis, vol. II, p. 928. North-Holland, Amsterdam (1991); Finite element methods. Part 1
6. Ciarlet, P.G., Wagschal, C.: Multipoint taylor formulas and applications to the finite element method. Numer. Math 17, 84–100 (1971)
7. Ciarlet, P.G.: The finite element method for elliptic problems. Classics in Applied Mathematics, vol. 40. Society for Industrial and Applied Mathematics (SIAM), Philadelphia (2002)
8. Frey, P., Georges, P.L.: Maillages: Applications aux Elements Finis, Hermés Science, Paris (1999)
9. Hecht, F.: The mesh adapting software: bamg. INRIA report (1998), `http://www-c.inria.fr/gamma/cdrom/www/bamg/eng.htm`
10. Hecht, F.: Métriques et indicateurs d'erreur. In: Cours maillage. Paris, INRIA, CEA, EDF, février (2003)
11. Hecht, F.: Quelques estimations d'erreurs d'interpolation anisotropes pour eléments finis de lagrange. Numerical Analysis and Scientific Computing for Partial Differential Equations and their Challenging Applications, février (2008)
12. Hecht, F., Georges, P.L.: Nonisotropic Grids, ch. 20. CRC Press, Boca Raton (1999)
13. Hecht, F., Kuate, R.: Reconstruction de la matrice hessienne d'une fonction discrète. Laboratoire Jacques-Louis Lions (preprint, 2008)
14. Hecht, F., Leharic, A., Pironneau, O.: Freefem++, language for finite element method and Partial Differential Equations (PDE). Université Pierre et Marie, Laboratoire Jacques-Louis Lions, `http://www.freefem.org/ff++/`
15. Leservoisier, D.: Stratégies D'adaptation et de raffinement de maillage en Mécanique des Fluides Numérique. PhD thesis, Université Pierre et Marie Curie (September 2001)
16. Manole, C.-M., Vallet, M.-G., Dompierre, J., Guibault, F.: Benchmarking second order derivatives recovery of a piecewise linear scalar field. Département de génie informatique, Ecole Polytechnique de Montréal Canada (2005)
17. Picasso, M.: An anisotropic error indicator based on Zienkiewicz-Zhu error estimator: application to elliptic and parabolic problems. SIAM J. Sci. Comput. 24(4), 1328–1355 (2003) (electronic)
18. Zienkiewicz, O.C., Zhu, J.Z.: The superconvergent patch recovery and posterior estimates, Part I: The recovery technique. Int. J. Numer. Meth. Engng 33(7), 1331–1364 (1992)

Tracking Adaptive Moving Mesh Refinements in 3D Curved Domains for Large-Scale Higher Order Finite Element Simulations

Xiaojuan Luo[1], Mark S. Shephard[1], Lie-Quan Lee[2], Cho Ng[2], and Lixin Ge[2]

[1] Rensselaer Polytechnic Institute, Troy, NY 12180
xluo@scorec.rpi.edu, shephard@scorec.rpi.edu
[2] Stanford linear Accelerator Center (SLAC), Menlo Park, CA 94025
liequan@slac.stanford.edu, cho@slac.stanford.edu, lge@slac.stanford.edu

Summary. When applying higher order finite elements to curved 3D domains in large-scale accelerator simulations, complexities that arise include needing valid curved finite elements and the capability to track the movement of mesh refinement in the critical domains. This paper presents a procedure which combines Bézier mesh curving and size driven mesh adaptation technologies to address those requirements. The intelligent selection of local mesh modifications to eliminate invalid curved elements and properly control the size distribution are the two key technical components. The procedure has been successfully applied by SLAC to generate 3D moving curved meshes in the large-scale electromagnetic modeling of next generation accelerator designs. The results demonstrated that valid curvilinear meshes not only make the time domain simulations more reliable but also improve the computational efficiency up to 30%.

1 Introduction

Higher order finite elements [1], which are well known for faster rates of convergence in terms of computational efficiency, can provide an effective approach for large-scale simulations. When applying higher order finite elements to three-dimensional curved domains, the elements must be properly curved to maintain the rate of convergence [2]. The common approach to the construction of such meshes is to apply a straight-sided mesh generation procedure [3, 4] and then curve the mesh edges and faces on the curved domain boundaries to the proper orders. This approach takes advantage of the conventional unstructured mesh generators to deal with the complexity of model geometry. However, the resulting meshes may become invalid because the curving of the mesh entities to model boundaries can lead to negative determinant of Jacobian in the closures of curved elements. Effective and efficient correction of those invalid elements is critical in curvilinear mesh construction and for its usage with higher order finite elements.

The electromagnetic simulation tools developed at Stanford Linear Accelerator Center (SLAC), supported by DOE SciDAC program [5, 6, 7] have successfully taken advantage of higher order finite elements to perform analysis needed for the design of the next generation linear accelerators, for example, short-range wakefield calculations. In those simulations, a small region near the beam requires sufficient refinement to resolve high frequency while the rest of the domain can have a large mesh size. This refinement region must move with the particles beam through the curved domains in the time dependent simulations to achieve acceptable computational efficiency. Since the domains are curved and higher order finite elements are used, the refined meshes must also be curved to provide a sufficiently higher order geometric approximation to ensure that the solution can effectively achieve the desired level of accuracy. The uniform refinement using smaller mesh size throughout the entire domain can produce over-refined meshes outside of the critical particle beam domains while larger mesh size can generate too coarse meshes which often become invalid during the curving procedure. These lead either to infeasibly large problem size, inaccurate results, or possible failure of the simulations.

To enable the higher order finite elements in large-scale simulations, a moving mesh adaptation in curvilinear domains that includes the combination of a general mesh curving tool [8, 9] and size driven mesh adaptation [10, 11] is presented. A full set of curved mesh modification procedures play an important role in ensuring the resulting curved meshes are valid and with minimal number of elements for the desired accuracy.

The outline of this paper is as follows. Section 2 describes a general mesh curving procedure to construct valid curvilinear meshes for three-dimensional curved domains. The procedure employs Bézier polynomial to represent the higher order geometric shapes for curved mesh entities. The extension of the size driven mesh adaptation procedure in curved domains is discussed in Section 3. Analyses results applied by SLAC for linear accelerator design are shown in Section 4. Conclusions and future works are given Section 5.

2 Mesh Curving

A flexible distributed mesh data structure [12] is employed in this paper to support the mesh adaptation in curved domains. The mesh data structure applies a general topology and classification of the entities with respect to the geometric model entity that the mesh entity is on [13]. M_i^d and G_i^d are used to describe the mesh and model topological entity of dimension d, $d = 0, 1, 2, 3$ represent mesh and model vertex, edge, face, and region respectively.

The approximation of the mesh to curved geometric domains is maintained by assigning appropriate Bézier higher order geometric shapes to mesh edges and faces on curved domain boundaries. The Bézier topology-based mesh geometry shape is constructed using Bernstein polynomials which possess a number of advantageous properties including [14]:

- The Convex Hull Property - A Bézier curve, surface, or volume is contained in the convex hull formed by its control points.

- Computationally efficient algorithms for degree elevation and subdivision are available which can be used to refine the shape's convex hull as well as adaptively refine the mesh's shape.

Those properties are useful to form the validity check algorithm for Bézier higher order curved elements and to determine local mesh modifications to correct invalid elements which have negative determinant Jacobian in the element closures due to the curving of mesh entities. The resulting curved meshes can always guarantee that each element has positive determinant of Jacobian in its closures.

2.1 Bézier Topology-Based Higher Order Shape Representation

Bernstein polynomials provide an effective means to define Bézier hierarchic higher order shapes for topological mesh entities in their parametric coordinates. A qth order Bézier mesh entity can be represented as [9],

$$\boldsymbol{x}(\boldsymbol{\xi}) = \sum_{|\boldsymbol{i}|=q} B_{|\boldsymbol{i}|}(\boldsymbol{\xi}) b_{|\boldsymbol{i}|} \boldsymbol{\xi}^{|\boldsymbol{i}|} \tag{1}$$

where $B_{|\boldsymbol{i}|}\boldsymbol{\xi}^{|\boldsymbol{i}|}$ are the Bernstein polynomials defined in the mesh entity parametric coordinate system as shown in Table 1. Note that the independent parametric coordinates for a topological mesh entity should be ξ_1, (ξ_1, ξ_2) and (ξ_1, ξ_2, ξ_3) respectively with the constraints that $\xi_2 = 1 - \xi_1$, $\xi_3 = 1 - \xi_1 - \xi_2$ and $\xi_4 = 1 - \xi_1 - \xi_2 - \xi_3$. $b_{|\boldsymbol{i}|}$ are the control points used to define the shapes of the Bèzier mesh edges, faces and regions. Figure 1 shows the control points for a quadratic curved mesh edge, triangle face and tetrahedral region.

Table 1. $\boldsymbol{\xi}$, $|\boldsymbol{i}|$, $B_{|\boldsymbol{i}|}$ and $\boldsymbol{\xi}^{|\boldsymbol{i}|}$ for topology mesh entity

	$\boldsymbol{\xi}$	$\lvert\boldsymbol{i}\rvert$	$B_{\lvert\boldsymbol{i}\rvert}$	$\boldsymbol{\xi}^{\lvert\boldsymbol{i}\rvert}$
Edge	(ξ_1, ξ_2), $\xi_1 + \xi_2 = 1$	$\lvert\boldsymbol{i}\rvert = i + j$	$\frac{q!}{i!j!}$	$\xi_1^i \xi_2^j$
Triangle	(ξ_1, ξ_2, ξ_3), $\xi_1 + \xi_2 + \xi_3 = 1$	$\lvert\boldsymbol{i}\rvert = i + j + k$	$\frac{q!}{i!j!k!}$	$\xi_1^i \xi_2^j \xi_3^k$
Tetrahedron	$(\xi_1, \xi_2, \xi_3, \xi_4)$, $\xi_1 + \xi_2 + \xi_3 + \xi_4 = 1$	$\lvert\boldsymbol{i}\rvert = i + j + k + l$	$\frac{q!}{i!j!k!l!}$	$\xi_1^i \xi_2^j \xi_3^k \xi_4^l$

In the case that a straight-sided mesh and its associated geometry CAD model are given, the control points for those mesh entities on the curved model edges/faces are determined based on Bézier curve and surface interpolation method to evaluate the model geometry at a set of discrete parametric locations. Common approaches often use the uniformly distributed parametric space points. However, alternative methods, such as chord length method or curvature-based procedure, will be used to improve the geometric approximation [14]. In the case that a curved mesh with different shape representation method is given, the control points are computed by converting the given shapes to Bézeir shapes. For example, a quadratic mesh edge using Lagrange interpolating method to represent its curved shape has three control points $\mathbf{l_1}$, $\mathbf{l_2}$ and $\mathbf{l_3}$, the control points

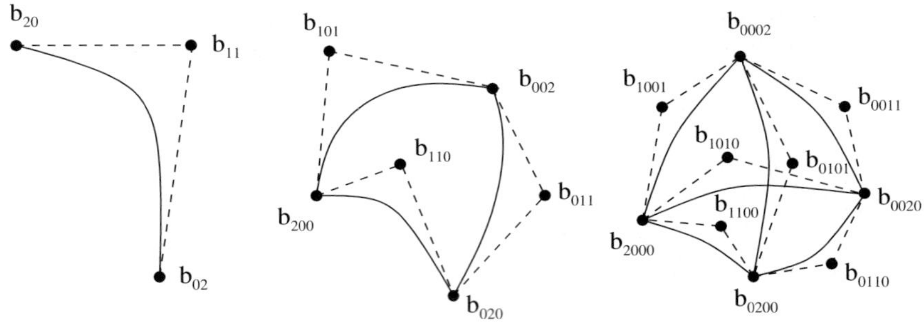

Fig. 1. Bézier control points for a curved mesh edge, face and region

for the Bézier shape of the edge, $\boldsymbol{x} = \mathbf{b}_{20}\xi_1^2 + 2\mathbf{b}_{11}\xi_1(1-\xi_1) + \mathbf{b}_{02}(1-\xi_1)^2$, can be computed by solving the following equations,

$$\begin{array}{lll} \boldsymbol{x}(\xi_1 = 0) & = \mathbf{b}_{02} & = \mathbf{l}_1 \\ \boldsymbol{x}(\xi_1 = 1/2) & = \frac{\mathbf{b}_{20}}{4} + \frac{\mathbf{b}_{11}}{2} + \frac{\mathbf{b}_{02}}{4} & = \mathbf{l}_2 \\ \boldsymbol{x}(\xi_1 = 1) & = \mathbf{b}_{20} & = \mathbf{l}_3 \end{array} \tag{2}$$

Therefore, $\mathbf{b}_{02} = \mathbf{l}_1$, $\mathbf{b}_{20} = \mathbf{l}_3$ and $\mathbf{b}_{11} = (4\mathbf{l}_2 - (\mathbf{l}_1 + \mathbf{l}_3))/2$.

2.2 Validity Check of Bézier Higher Order Curved Elements

When the finite element basis is increased in applying the higher order finite elements, the integration rules of the finite element approximation must be properly improved to ensure that the numerical integration error does not become the dominant error. The improvement of the integration rules requires evaluation of the geometric mapping of the curved element at new integration locations or knowledge that the Jacobian is positive throughout the element. To avoid the need to constantly recheck the integration points, a general element validity check is performed. This validity check for Bézier higher order curved elements takes advantage of its convex hull property to ensure that a valid curved element always has positive determinant of Jacobian in its closures, which is independent of the basis functions, the polynomial orders, or the applied integration rules [8].

Given a qth order Bézier tetrahedron mesh region described in Eq.1, the Jacobian matrix of the geometric mapping with respect to the independent parametric coordinates (ξ_1, ξ_2, ξ_3) is,

$$J = \left[\frac{\partial \mathbf{x}}{\partial \xi}\right] = \begin{bmatrix} \frac{\partial x_1}{\partial \xi_1} & \frac{\partial x_1}{\partial \xi_2} & \frac{\partial x_1}{\partial \xi_3} \\ \frac{\partial x_2}{\partial \xi_1} & \frac{\partial x_2}{\partial \xi_2} & \frac{\partial x_2}{\partial \xi_3} \\ \frac{\partial x_3}{\partial \xi_1} & \frac{\partial x_3}{\partial \xi_2} & \frac{\partial x_3}{\partial \xi_3} \end{bmatrix} \tag{3}$$

where $\mathbf{x} = (x_1, x_2, x_3)$ Therefore, the determinant of the Jacobian J is,

$$det(J) = (\frac{\partial \boldsymbol{x}}{\partial \xi_1} \times \frac{\partial \boldsymbol{x}}{\partial \xi_2}) \cdot (\frac{\partial \boldsymbol{x}}{\partial \xi_3}) \tag{4}$$

where $\frac{\partial \boldsymbol{x}}{\partial \xi_i}$ are the three partial derivatives of $\boldsymbol{x}^q$ which are $(q-1)th$ order Bézier functions. Therefore, the resulting determinant of Jacobian is a Bézier polynomial function with order $3(q-1)$,

$$det(J) = \sum_{|\boldsymbol{i}|=r} C_{|\boldsymbol{i}|} c_{|\boldsymbol{i}|} \boldsymbol{\xi}^{|\boldsymbol{i}|} \tag{5}$$

where $r = 3(q-1)$. $C_{|\boldsymbol{i}|}$ and $c_{|\boldsymbol{i}|}$ can be expressed using the coefficients $B_{|\boldsymbol{i}|}$ and $b_{|\boldsymbol{i}|}$ in Eq. 1. As an example, the quadratic tetrahedral region shown in Figure 1 can be expressed as,

$$\begin{aligned} \boldsymbol{x} = {} & B_{2000} b_{2000} \xi_1^2 + B_{0200} b_{0200} \xi_2^2 + B_{0020} b_{0020} \xi_3^2 + B_{0002} b_{0002} \xi_4^2 \\ & + B_{1100} b_{1100} \xi_1 \xi_2 + B_{1010} b_{1010} \xi_1 \xi_3 + B_{1001} b_{1001} \xi_1 \xi_4 \\ & + B_{0110} b_{0110} \xi_2 \xi_3 + B_{0101} b_{0101} \xi_2 \xi_4 + B_{0011} b_{0011} \xi_3 \xi_4 \end{aligned} \tag{6}$$

Considering that $\xi_4 = 1 - \xi_1 - \xi_2 - \xi_3$, $B_{2000} = B_{0200} = B_{0020} = B_{0002} = 1$ and the rest coefficients $B's$ equal to 2, therefore,

$$\begin{aligned} \frac{\partial \boldsymbol{x}}{\partial \xi_1} &= 2\{\underbrace{(b_{2000} - b_{1001})}_{a_1} \xi_1 + \underbrace{(b_{1100} - b_{0101})}_{b_1} \xi_2 \\ &\quad + \underbrace{(b_{1010} - b_{0111})}_{c_1} \xi_3 + \underbrace{(b_{1001} - b_{0002})}_{d_1} \xi_4\} \\ \frac{\partial \boldsymbol{x}}{\partial \xi_2} &= 2\{\underbrace{(b_{1100} - b_{1001})}_{a_2} \xi_1 + \underbrace{(b_{0200} - b_{0101})}_{b_2} \xi_2 \\ &\quad + \underbrace{(b_{0110} - b_{0011})}_{c_2} \xi_3 + \underbrace{(b_{0101} - b_{0002})}_{d_2} \xi_4\} \\ \frac{\partial \boldsymbol{x}}{\partial \xi_3} &= 2\{\underbrace{(b_{1000} - b_{1001})}_{a_3} \xi_1 + \underbrace{(b_{0110} - b_{0101})}_{b_3} \xi_2 \\ &\quad + \underbrace{(b_{0110} - b_{0020})}_{c_3} \xi_3 + \underbrace{(b_{0011} - b_{0002})}_{d_3} \xi_4\} \end{aligned} \tag{7}$$

The determinant of Jacobian is a cubic Bernstein polynomial and the coefficients $C_{|\boldsymbol{i}|}$ and $c_{|\boldsymbol{i}|}$ are listed in Table 2. a_i, b_i, c_i and d_i are the vectors defined by the corresponding control points shown in Eq. 6.

The convex hull property of Bézier polynomial indicated that the polynomial is bounded by its minimal and maximal control points [14]. So,

$$min(c_{|\boldsymbol{i}|}) \leq det(J) \leq max(c_{|\boldsymbol{i}|}) \tag{8}$$

Therefore, a curved tetrahedral region is valid in its closure as long as its $min(c_{|\boldsymbol{i}|}) > 0$.

Table 2. $C_{|i|}$ and $c_{|i|}$ for $det(J)$ of a quadratic tetrahedral region

$\lvert i\rvert$	$C_{\lvert i\rvert}$	$c_{\lvert i\rvert}$	$\lvert i\rvert$	$C_{\lvert i\rvert}$	$c_{\lvert i\rvert}$
3000	8	$(a_1 \times a_2) \cdot a_3$	0300	8	$(b_1 \times b_2) \cdot b_3$
0030	8	$(c_1 \times c_2) \cdot c_3$	0003	8	$(d_1 \times d_2) \cdot d_3$
2100	8	$(a_1 \times b_2 + a_2 \times b_1) \cdot a_3$	1200	8	$(a_1 \times b_2 + a_2 \times b_1) \cdot b_3$
...	...	...	...	...	...

2.3 Effective Procedure to Generate Valid Curvilinear Meshes

The mesh curving procedure can deal with a starting straight-sided mesh and an existing curved mesh with invalid elements. In the case of a straight-sided mesh, the procedure computes the Bézier control points for the mesh edges/faces on curved domain boundaries and curves them incrementally. In the case of a given curved mesh with different initial higher order shapes representation, for example, Lagrange interpolation, the shapes are converted to Bézier form and the invalid elements are detected and corrected incrementally. Central to both of the approaches is the selection of effective local mesh modification operations to eliminate the invalid elements till the resulting curvilinear meshes are valid.

The computation of the determinant of Jacobian to detect invalid elements can provide useful information to determine key mesh entities and appropriate operations to correct the invalidity. The invalid elements are defined as those curved elements having at least one negative coefficients, $c_{|i|} \leq 0$ as shown in Eq. 5. The key mesh entities are defined as those whose control points appear in the computation of the negative coefficients $c_{|i|}$. As an example, Figure 2 shows an invalid quadratic tetrahedral region and the computation of the determinant Jacobian shows that coefficient $c_{3000} < 0$. Based on Eq.8 and Table 2, the control points $b_{2000}, b_{1100}, b_{1010}$ and b_{1001} have been used to compute the c_{3000} which indicates that M_0^0, M_0^1, M_1^1 and M_2^1 are key mesh entities and applying local

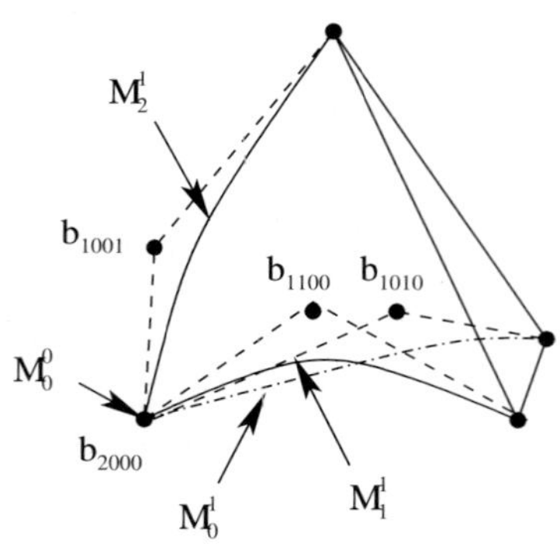

Fig. 2. The computation of det(J) indicates that the mesh entities M_0^0, M_0^1, M_1^1 and M_2^1 are key mesh entities

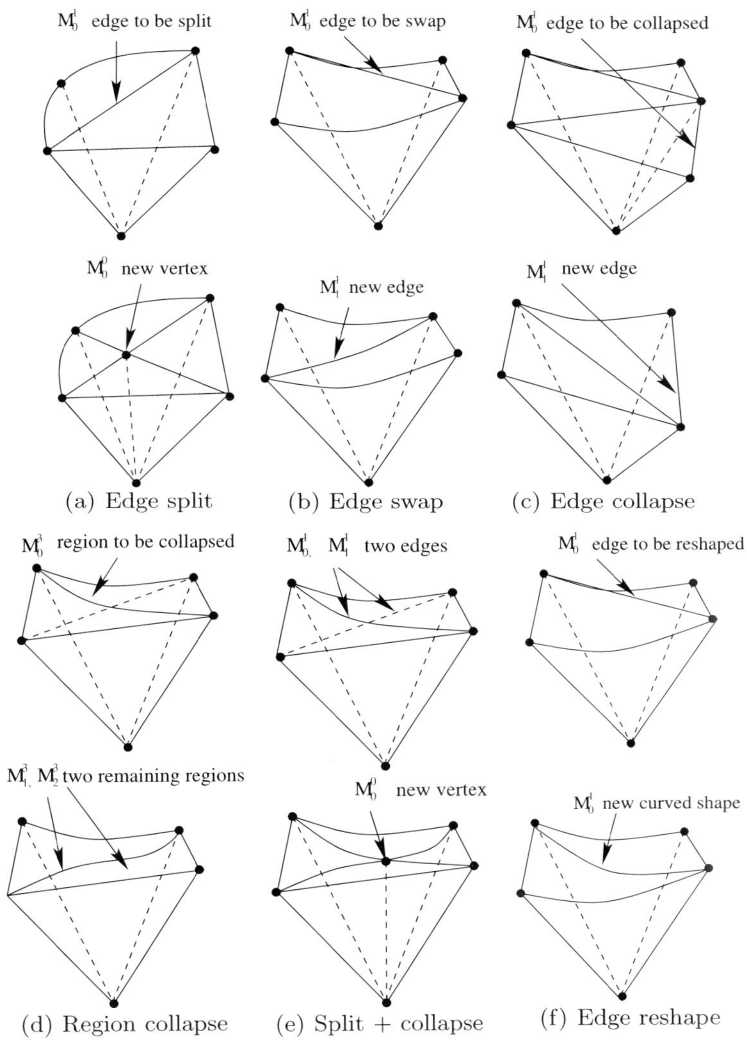

(a) Edge split (b) Edge swap (c) Edge collapse (d) Region collapse (e) Split + collapse (f) Edge reshape

Fig. 3. 3D curved local mesh modification operations

mesh modifications on any of them can effectively make c_{3000} positive and the curved element valid.

The set of curved local mesh modifications applied to create valid curvilinear meshes include edge split, edge swap, edge collapse, region collapse, double split+collapse, and edge re-shape [9] as shown in Figure 3. Comparing to the straight-sided mesh, the validity check algorithm discussed in Section 2.2 is used to determine whether a curved local mesh modification operation can be applied.

Those operations are essential to ensure the reliability of the mesh curving procedure to create valid curved elements.

The procedure processes one curved mesh entity at a time as follows [8]:

- Determine the key mesh entities to apply local mesh operations based on the negative coefficients, $c_{|\boldsymbol{i}|} \leq 0$, in computing the determinant of Jacobian.
- Determine if the invalidity is caused by pairs of neighboring mesh faces or edges classified on the boundary such that angles of 180^0 are created. In those cases, $c_{|\boldsymbol{i}|} \leq 0$ only happens at $i, j, k, l = r$. Apply either split (see Figure 3(a)) or swap operations (see Figure 3(b)) to introduce additional entities to subdivide those larger angles to correct the invalid curved element.
- Determine if the invalidity is caused by pairs of opposite mesh edges coming too close to each other in one curved region, where $c_{|\boldsymbol{i}|} \leq 0$ happens at $i, j, k, l \neq r$. Apply either region split (see Figure 3(d)) or split+collapse (see Figure 3(e)) to remove the invalid curved element.
- If neither of above two steps is successful, examine the applications of the remaining operations (see Figure 3(c), 3(f)) to correct the invalid curved elements.
- If the invalid curved element can not be corrected using those local mesh operations, local refinement is applied and all new mesh entities will be added to the list to be processed. Subdivision creates more options for applying operations later.

Figure 4 shows the straight-sided and curved mesh for a 3D curved model to demonstrate the effectiveness of the developed procedure. The mesh has 139 regions and 31 curved regions are invalid. 20 local mesh modifications are applied to correct those invalid elements. Curved meshes for more complex domains used by SLAC for electromagnetic linear accelerator analysis are shown in section 4.

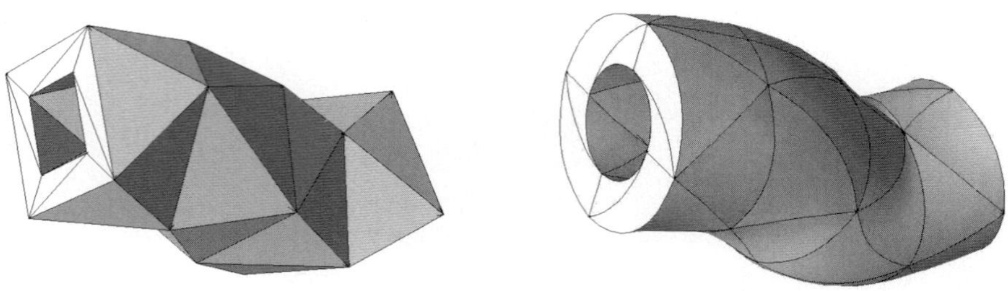

Fig. 4. Straight-sided mesh (left) and curved mesh (right) for a 3D curved domain

3 Moving Adaptive Mesh Refinements in 3D Curved Domains

The developed size driven mesh adaptation procedure [10] has been successfully applied in cardiovascular blood flow simulations [15], metal forming process [16], wave propagation simulations [17], etc. where the results demonstrated substantial computational efficiency can be improved using the isotropic or anisotropic adapted meshes to effectively resolve solution fields. The procedure has been extended to deal with curved meshes for higher order finite elements in large-scale simulations. The extended procedure maintains the existing functionalities developed for straight-sided meshes such as vertex-based size field specifications and selective local mesh modification applications [10]. In addition, the following two steps have been added in the case where the mesh is curved.

- The validity check algorithm described in Section 2.2 must be applied when the affecting cavities for a local mesh modification operation have curved mesh entities. This step ensures that resulting curved meshes are valid after applying the selected local mesh operation.
- Any newly created mesh entities on the curved domain boundaries must be properly curved to the model boundaries which ensures that the geometric approximation of the resulting adapted meshes is maintained. As an example, Figure 5 shows how the procedure to split a quadratic curved mesh edge M_0^1 which is classified on the curved model edge G_0^1. The two newly created mesh edges M_1^1 and M_2^1 are also curved to the model edge G_0^1.

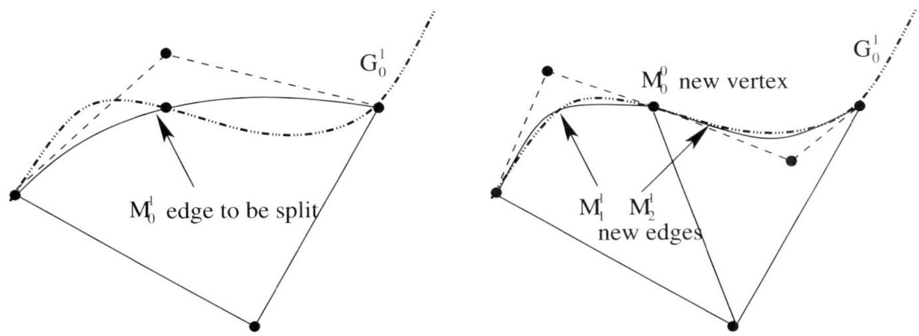

Fig. 5. Before (left) and after (right) refine a quadratic curved mesh edge M_0^1 on model edge G_0^1. New mesh edges M_1^1, M_2^1 have been appropriately curved to the model boundaries

In size driven mesh adaptation procedure, a mesh metric field, which can be either isotropic or anisotropic, is defined to specify the desired size of elements. The metric field is used to compute the edge length and directions of the current mesh with respect to this metric. A series of controlled mesh modification steps are applied to obtain a new mesh that satisfies the specified mesh metric field which consist of the following three steps [11]:

- Coarsening stage to eliminate the mesh edges that are shorter than the desired edge length in the metric field. This stage is accomplished by applying collapse operation on the identified shorter edges one at a time.

- Refinement stage to reduce the maximal mesh edge length to reach the desired edge length in the metric field. Edge-based refinement templates and application of local mesh modification to project the new created mesh vertices to the curved boundaries are iteratively applied until the adapted mesh satisfies the mesh size metric field requirements [10].
- Shape improvement stage to improve the quality of the resulting mesh using swap and/or vertex reposition operations.

For large-scale adaptive simulations, discretization error estimation is applied to construct size fields to control the mesh adaptation [15, 16, 17] in which the adapted meshes can conform to the size requirements. However, there are certain situations where other factors may also be applied to set the size field. For example, the size information being given for the short-range wakefield simulations performed by SLAC is supplemented to have a refined mesh in areas where particles currently reside. The specification of this refinement information is dictated by the initial locations of the particles beam and the desired mesh size around the beam which is often at least one order of magnitude smaller than the rest of the domains. The larger size difference between the finer particle domain and the coarse domain can lead to bad quality resulting meshes. The left mesh in Figure 6 shows an adapted curved mesh which uses size 1 and 10 to control the fine and coarse mesh in the model. The abrupt size field change causes meshes at the fine and coarse mesh interface not acceptable which clearly demonstrated that the control of the mesh gradation is needed.

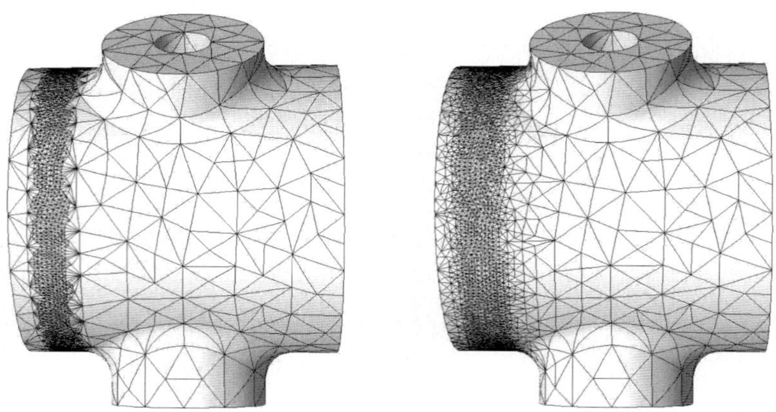

Fig. 6. Curved meshes without (left) and with (right) mesh size gradation control

The procedure described in [18] is adopted to control a smooth mesh size transition over the mesh. Central to the algorithm is that the ratio between the larger mesh size to the smaller mesh size at the two bounding mesh vertices of any mesh edge is under a prescribed factor β, where $\beta > 1$.

Let M_i^1 be a mesh edge, M_{j1}^0 and M_{j2}^0 are its two bounding mesh vertices, the given mesh sizes at each vertex are h_{j_1} and h_{j_2} which represent the desired edge lengths at those two vertices. We require that,

$$max\{\frac{h_{j_1}}{h_{j_2}}, \frac{h_{j2}}{h_{j_1}}\}^{\frac{1}{L(M_1^1)}} \leq \beta \tag{9}$$

where $L(M_1^1)$ represents the length of the mesh edge M_i^1 with respect to the mesh size field as defined as,

$$L(M_1^1) = ||M_i^1|| \int_0^1 \frac{1}{H(t)} dt \tag{10}$$

where $||M_i^1||$ denotes the length of the mesh edge and $H(t)$ is a monotonic interpolation function along the mesh edge such that $H(0) = h_{j_1}$ and $H(1) = h_{j_2}$. For the piecewise linear mesh size field used in this paper to track the moving mesh adaptation in curved domains, the function $H(t)$ is,

$$H(t) = h_{j_1} + (h_{j_2} - h_{j_1})t \tag{11}$$

Therefore, Eq.10 gives,

$$L(M_1^1) = ||M_i^1|| \frac{log(h_{j_1}/h_{j_2})}{h_{j_1} - h_{j_2}}, h_{j_1} \neq h_{j_2} \tag{12}$$

Therefore, for any mesh edge which is not satisfied in Eq.9, the larger mesh size of its bounding mesh vertices is decreased to $min(h_{j_1}, h_{j_2})\beta^{L(M_i^1)}$ to meet Eq.9. The process is iteratively performed over the mesh when all of the mesh edges satisfy Eq.9. For the mesh shown in Figure 6(b), β is adopted as 2.0. More moving adaptive meshes are shown in section 4.

4 Analysis Results

4.1 Curvilinear Meshes for FETD Electromagnetic Simulations

SLAC performs simulations for the wakefield effects of an 8-cavity cryomodule for the proposed International Linear Collider (ILC) using the FETD method, which applies a set of higher order hierarchical Nedelec basis functions [19] for the finite element spacial discretization that requires the meshes to be curved. A curvilinear mesh with 2.974 million quadratic isoparametric tetrahedral elements is used in this FETD simulation. The initial curvilinear mesh uses Lagrange interpolation to represent the higher order shapes for those curved mesh edges which have been converted to Bézeir representations using Eq.2. 515 invalid curved elements were detected and have been corrected using the procedure discussed in Section 2. The valid curved mesh was exported by converting the Bézier shapes back to Lagrange shapes to be suitable for the analysis simulation system. Figure 7 shows the curved mesh for one cavity of the model and the close-up mesh before and after curving.

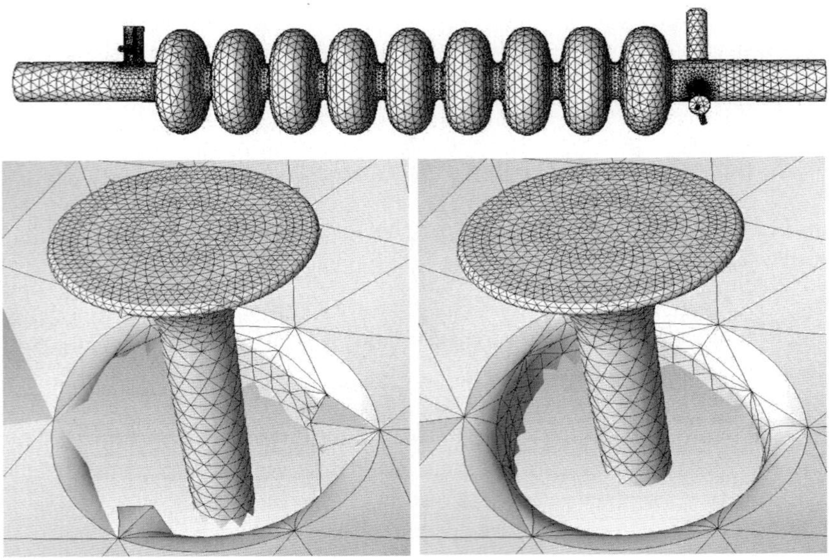

Fig. 7. The mesh for one cavity (top), close-up mesh before (bottom left) and after (bottom right) correcting the invalid curved elements marked as yellow

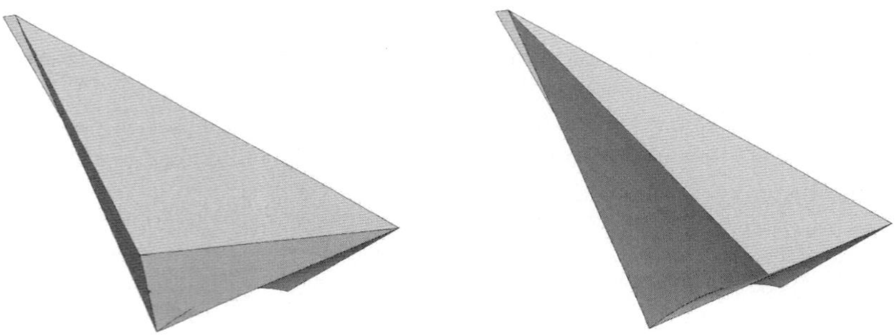

Fig. 8. Local mesh cavity before (left) and after (right) applying edge swap to correct the invalid element

Figure 8 shows how an edge collapse operation is applied to correct the invalid curved elements during curving process.

The mesh produced about 20 million degrees of freedom. The simulation used 256 Multi-stream processors on the Cray-X1E at Oak Ridge National Laboratory. It took a total runtime of 300 wall-hours through multiple jobs with checkpointing for a complete run and half terabtye of data was generated. Figure 9 shows a snapshot of the electric field distribution excited by a beam in the ILC cryomodule.

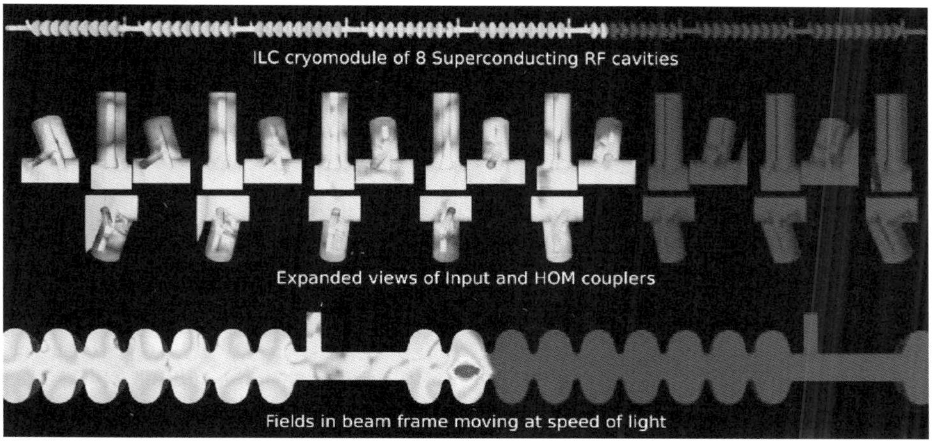

Fig. 9. A snapshot of the electric field distribution excited by a beam in an 8-cavity cryomodule for the proposed International Linear Collider

Table 3. Statistics for correcting the 2.97M mesh with 515 invalid curved regions

Time usage (sec)		Local mesh operations	
Import the mesh	381.162	Edge collapse	253
Create invalid region list	45.106	Region collapse	17
Correcting invalid regions	256.182	Edge swap	76
Export the mesh	64.911	Double edge split+collapse	13
		Recurving	32

The statistics for correcting the invalid curved regions is presented in Table 3. The data shows that the procedure used about 10 minutes to correct the invalid regions on a single processor linux workstation. The corrected curvilinear mesh not only leads to a stable time-domain simulation but also reduces the execution time per time-step by up to 30% due to better conditioned matrices, which is 90 wall-hours runtime efficiency improvement on the parallel computers.

4.2 Moving Adaptively Refined Meshes for Short-Range Wakefield Calculations

A series of moving adapted meshes in a curved domain were generated using the procedure described in Section 3 for short-range wakefield calculations by SLAC. Figure 10 shows the geometric model which has some complex components in the middle of domain. The initial location of the beam is at the left end of the domain, the desired mesh size inside the particle dense mesh is 1 and the size for the rest of the domains is 10. Figure 11 shows the moving adapted meshes up to step 5. Figure 12 shows the interior adapted mesh at step 5. Mesh size gradation control discussed in Section 3 is applied with $\beta = 2.0$. Figure 13 shows

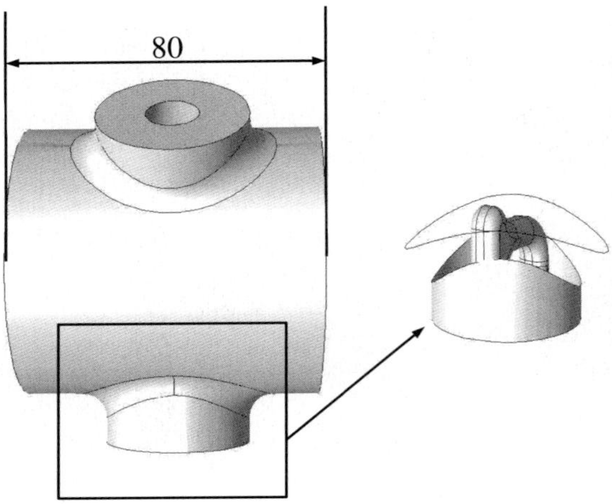

Fig. 10. Geometric model for the short-range wakefield simulation

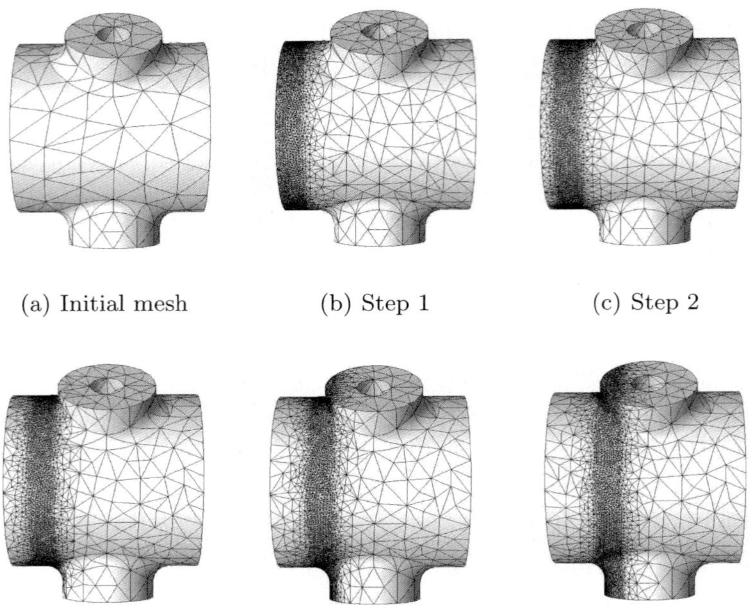

(a) Initial mesh (b) Step 1 (c) Step 2

(d) Step 3 (e) Step 4 (f) Step 5

Fig. 11. Moving adapted meshes in curved domain for short-range wakefield simulation

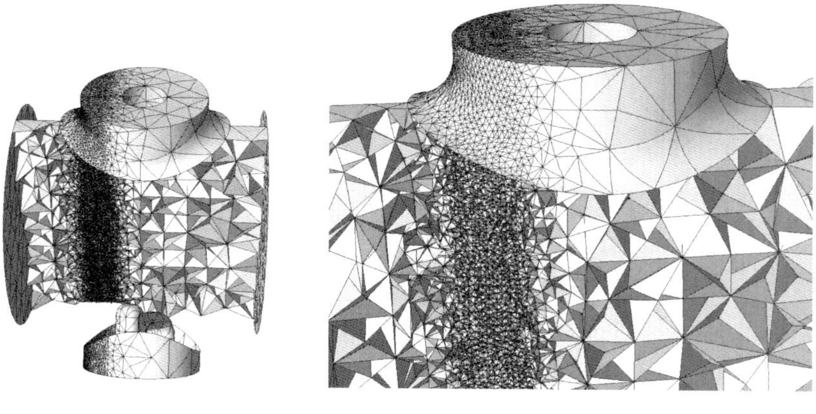

(a) Interior (b) Close-up

Fig. 12. Interior mesh for the adapted mesh at step 5

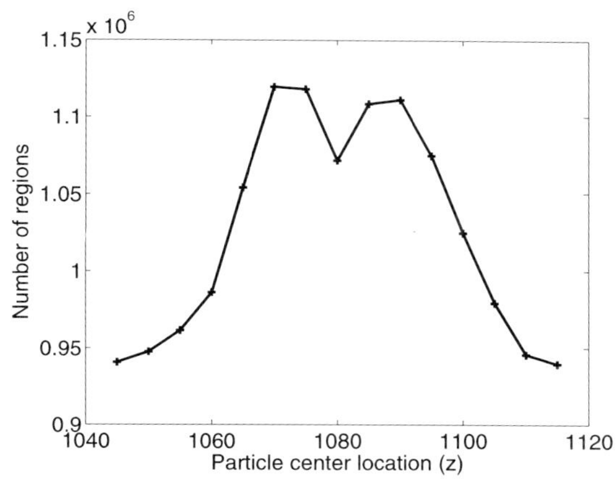

Fig. 13. Number of mesh regions at each step

the number of elements at each step which indicates that these adaptively refined meshes have around $1 \sim 1.15$ million elements compared to the uniform refined mesh with 6.5 million elements if the mesh size inside the particle domains is applied in the entire domain. The increase of the number of elements in the middle of the domain is due to the complex geometries as shown in Figure 10. The computation effort of short-range wakefield calculations using the moving adaptively refined meshes can be reduced by one order of magnitude compared to the uniformly-refined mesh.

5 Conclusion

This paper has presented a procedure to track moving adaptive mesh refinement in curved domains which is capable of generating suitable curvilinear meshes to enable large-scale accelerator simulations. The procedure combined a general mesh curving tool and size-driven mesh adaptation to produce valid curved meshes with substantially fewer elements and the analysis results demonstrated such meshes improved the computational efficiency and reliability. Future work will focus on the scalable parallelization of all steps for petascale simulations.

Acknowledgments

This work is supported by U.S. Department of Energy under DOE grant number DE-FC02-06ER25769 and DE-AC02-76SF00515.

References

1. Babuska, I., Szabo, B.A., Katz, I.N.: The p-version of the finite element method. Int. J. Numer. Meth. Engng. 18(3), 515–545 (1981)
2. Luo, X.J., Shephard, M.S., Remacle, J.F., O'Bara, R.M., Beall, M.W., Szabo, B.A., Actis, R.: p-version mesh generation issues. In: Proc.of 11th Meshing Roundtable, pp. 343–354 (2002)
3. CUBIT geometry and mesh generation toolkit, `http://cubit.sandia.gov`
4. Simmetrix Inc. Enabling simulation-based design, `http://www.simmetrix.com`
5. Akcelik, V., Ko, K., Lee, L.Q., Li, Z.H., Ng, C.K., Xiao, L.L.: Shape determination for deformed electromagnetic cavities. J. Comput. Physics 227(3), 1722–1738 (2008)
6. Xiao, L., Adolphsen, C., Akcelik, V., Kabel, A., Ko, K., Lee, L.Q., Li, Z., Ng, C.K.: Modeling imperfection effects on dipole modes in TESLA cavity. In: Proc. of 2007 Particle Accelerator Conference (2007)
7. Lee, L.Q., Akcelik, V., Chen, S., Gl, X., Prudencio, E., Schussman, G., Uplenchwar, R., Ng, C., Ko, K., Luo, X.J., Shephard, M.S.: Enabling technologies for petascale electromagnetic accelerator simulation. J. of Physics 78(Conference Series), 012040 (2007)
8. Luo, X.J., Shephard, M.S., Obara, R.M., Nastasia, R., Beall, M.W.: Automatic p-version mesh generation for curved domains. Engineering with Computers 20, 265–285 (2004)
9. Luo, X.J.: An Automatic Adaptive Directional Variable p-Version Method in 3D Curved Domains. PhD Thesis, Rensselaer Polytechnic Institute, New York (2005)
10. Li, X.R., Shephard, M.S., Beall, M.W.: Accounting for curved domains in mesh adaptation. Int. J. Numer. Meth. Engng. 150, 247–276 (2003)
11. Li, X.R., Shephard, M.S., Beall, M.W.: 3D anisotropic mesh adaptation by mesh modification. Comp. Meth. Appl. Mech. Engng. 194, 4915–4950 (2005)
12. Seegyoung, E., Shephard, M.S.: Efficient distributed mesh data structure for parallel automated adaptive analysis. Engineering with Computers 22, 197–213 (2006)
13. Beall, M.W., Shephard, M.S.: A general topology-based mesh data structure. Int. J. Numer. Meth. Engng. 40(9), 1573–1596 (1997)

14. Farin, G.: Curves and Surfaces for Computer Aided Geometric Design. Academic Press, London (1992)
15. Sahni, O., Muller, J., Jansen, K.E., Shephard, M.S., Taylor, C.: Efficient anisotropic adaptive discretization of the cardiovascular system. Comput. Methods Appl. Mech. Engrg. 195(41-43), 5634–5655 (2006)
16. Wan, J.: An Automated Adaptive Procedure for 3D Metal Forming Simulations. PhD Thesis, Rensselaer Polytechnic Institute, New York (2006)
17. Chevaugeon, N., Hillewaert, K., Gallez, X., Ploumhans, P., Remacle, J.F.: Optimal numerical parameterization of discontinuous Galerkin method applied to wave propagation problems. J. Comput. Physics 223(1), 188–207 (2007)
18. Borouchaki, H., Hecht, F., Frey, P.J.: Mesh gradation control. Int. J. Numer. Meth. Engng. 43(6), 1143–1165 (1998)
19. Sun, D.K., Lee, J.F., Cendes, Z.: Construction of nearly orthogonal Nedelec bases for rapid convergence with multilevel preconditioned solvers. SIAM J. on Sci Comput. 23(4), 1053–1076 (2001)

Localized Coarsening of Conforming All-Hexahedral Meshes

Adam C. Woodbury[1], Jason F. Shepherd[2], Matthew L. Staten[2,3], and Steven E. Benzley[1]

[1] Brigham Young University
aw18@byu.net, seb@byu.edu
[2] Sandia National Laboratories
jfsheph@sandia.gov, mlstate@sandia.gov
[3] Carnegie Mellon University
mstaten@cmu.edu

Abstract. Finite element mesh adaptation methods can be used to improve the efficiency and accuracy of solutions to computational modeling problems. In many applications involving hexahedral meshes, localized modifications which preserve a conforming all-hexahedral mesh are desired. Effective hexahedral refinement methods that satisfy these criteria have recently become available; however, due to hexahedral mesh topology constraints, little progress has been made in the area of hexahedral coarsening. This paper presents a new method to locally coarsen conforming all-hexahedral meshes. The method works on both structured and unstructured meshes and is not based on undoing previous refinement. Building upon recent developments in quadrilateral coarsening, the method utilizes hexahedral sheet and column operations, including pillowing, column collapsing, and sheet extraction. A general algorithm for automated coarsening is presented and examples of models that have been coarsened with this new algorithm are shown. While results are promising, further work is needed to improve the automated process.

Keywords: Hexahedral, Mesh, Coarsening, Simplification, Adaptivity, Refinement.

1 Introduction

The efficiency and accuracy of a finite element computational modeling solution are greatly influenced by the distribution of elements in the finite element mesh. In a given model, there are usually regions that require greater mesh density than others. A higher concentration of elements in these regions may be necessary to reduce error in the finite element approximation, increase resolution where there are high gradients, or more accurately represent the model geometry. Regions where high accuracy is not critical or where gradients are low can generally be modeled with lower mesh density. Since the computational time required in a finite element analysis is directly related to the number of elements in the model being analyzed, it is advantageous to produce a mesh that has as few elements as possible. Therefore, in an ideal analysis, each region in the model should have enough elements to produce a good solution, but no more.

For most real-world finite element models, current mesh generation algorithms are unable to create an initial mesh that optimizes both accuracy and efficiency in the finite element solution. Although some control over mesh density is possible, an initial mesh will almost always contain regions that have too few elements, regions that have too many elements, or both. In addition, some finite element applications require mesh density to evolve throughout an analysis as areas of high and low activity change with time [1, 2, 3, 4]. For these reasons, much research has been devoted to the development of mesh adaptation tools that make it possible to adjust element density in specific regions either before or during analysis.

Mesh adaptation consists of both refinement and coarsening. Refinement is the process of increasing mesh density by adding elements to a mesh, while coarsening is the process of decreasing mesh density by removing elements from a mesh. By refining areas that have too few elements and coarsening areas that have too many elements, a more accurate and efficient analysis can be performed.

To satisfy the requirements of some finite element solvers, a mesh must be topologically conforming and contain only one element type. In general, a conforming all-tetrahedral mesh can be locally modified with much greater ease than a conforming all-hexahedral mesh. However, in many modeling applications, hexahedral elements are preferred over tetrahedral elements because they provide greater efficiency and accuracy in the computational process [1, 5]. For this reason, work has been done to improve hexahedral mesh adaptation methods. As a result, robust hexahedral refinement algorithms are becoming available [6, 7, 8, 9]. However, few developments have been seen in the area of hexahedral coarsening. The lack of effective coarsening methods creates a major gap in the field of hexahedral mesh adaptation.

To effectively achieve the objectives of mesh adaptation, a truly general hexahedral coarsening algorithm should:

1. Preserve a conforming all-hexahedral mesh
2. Restrict mesh topology and density changes to defined regions
3. Work on both a structured and unstructured mesh
4. Not be limited to only undoing previous refinement

Although hexahedral coarsening has been utilized in some modeling applications, no single algorithm has been developed that satisfies all the criteria listed above. This is, in large part, due to the topology constraints that exist in a conforming all-hexahedral mesh. These constraints make it difficult to modify mesh density without causing topology changes to propagate beyond the boundaries of a defined region [10, 11].

Since current hexahedral coarsening methods are unable to satisfy all the requirements listed above, they have limited applications. For example, to prevent global topology changes, some algorithms introduce non-conforming or non-hexahedral elements into the mesh [1, 2, 12, 13, 14]. While this is a valid solution for some types of analysis, not all finite element solvers can accommodate hanging nodes or hybrid meshes. Other algorithms maintain a conforming all-hexahedral mesh, but they generally require either global topology changes beyond the defined coarsening region [11, 15, 16], structured mesh topology where predetermined transition templates can be used [8, 15], or prior refinement that can be undone [2, 12, 13]. These weaknesses severely limit the effectiveness of these algorithms on most real-world models.

This paper presents a new method to locally coarsen conforming all-hexahedral meshes. The method works on both structured and unstructured meshes and is not based on undoing previous refinement. The remainder of this paper is organized as follows. Section 2 provides an overview of some basic hexahedral mesh operations. Section 3 shows how these operations have been combined to produce localized hexahedral coarsening and how the coarsening process has been automated. In Section 4, some examples of models which have been coarsened are shown. Finally, in Section 5, some areas of future work are discussed.

2 Hexahedral Mesh Operations

In recent years, a greater understanding of hexahedral mesh topology has led to the development of many new hexahedral mesh operations [17, 18, 19]. In this section, three operations which are useful for hexahedral coarsening are presented. These operations are based on hexahedral sheets and columns, which are topology-based groups of hexahedra that always exist in a conforming hexahedral mesh.

2.1 Hexahedral Sheets and Columns

A hexahedral element contains three sets of four topologically parallel edges, as shown in Figure 1. Topologically parallel edges provide the basis for hexahedral sheets. The formation of a sheet begins with a single edge. Once an edge has been chosen, all elements which share that edge are identified. For each of these elements, the three edges which are topologically parallel to the original edge are also identified. These new edges are then used to find another layer of elements and topologically parallel edges. This process is repeated until no new adjacent elements can be found. The set of elements which are traversed during this process makes up a hexahedral sheet. Figure 2 shows a hexahedral mesh with one of the sheets in the mesh defined.

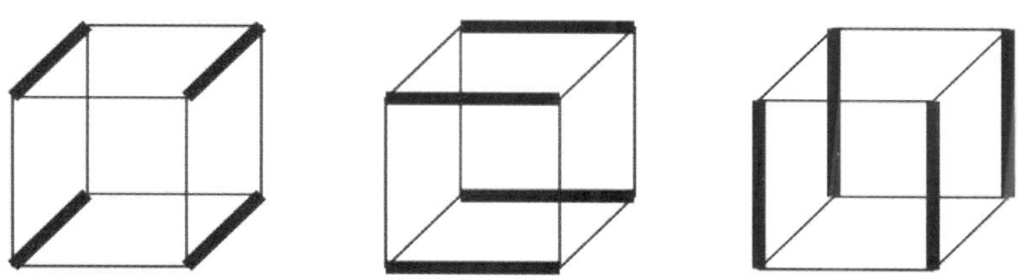

Fig. 1. A hexahedral element's three sets of topologically parallel edges

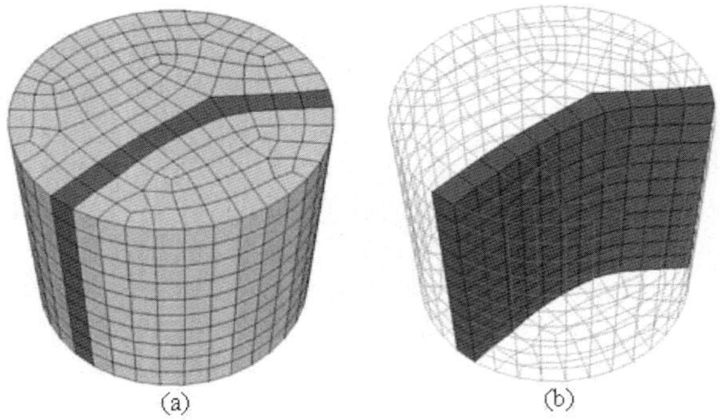

Fig. 2. A hexahedral sheet: (a) A hexahedral mesh with one sheet defined. (b) A view of the entire sheet

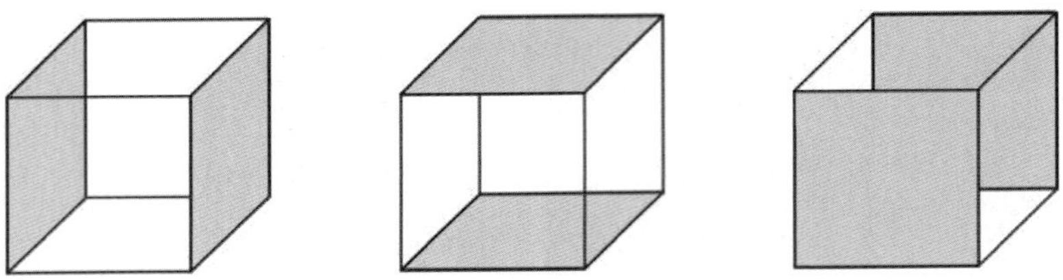

Fig. 3. A hexahedral element's three pairs of topologically opposite faces

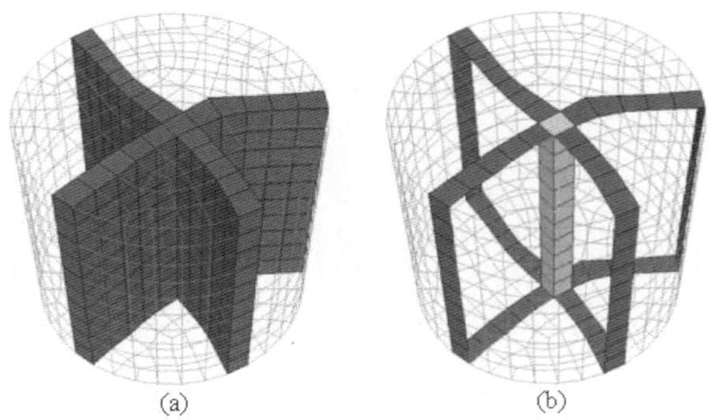

Fig. 4. A hexahedral column: (a) Two intersecting sheets. (b) The column that defines the intersection of the two sheets in (a).

A hexahedral element also contains three pairs of topologically opposite quadrilateral faces, as shown in Figure 3. Topologically opposite faces provide the basis for hexahedral columns. The formation of a column begins with a single face. Once a face has been chosen, the elements which share that face are identified. For each of these elements, the face which is topologically opposite of the original face is also identified. These new faces are then used to find another layer of elements and topologically opposite faces. This process is repeated until no new adjacent elements can be found. The set of elements which are traversed during this process makes up a hexahedral column. An important relationship between sheets and columns is that a column defines the intersection of two sheets. This relationship is illustrated in Figure 4.

2.2 Sheet and Column Operations

Hexahedral sheet and column operations can be used to modify a hexahedral mesh without introducing non-conforming elements. One such operation is known as sheet extraction [16]. Sheet extraction removes a sheet from a mesh by simply collapsing the edges that define the sheet and merging the two nodes on each edge, as shown in Figure 5. Merging nodes in this manner decreases

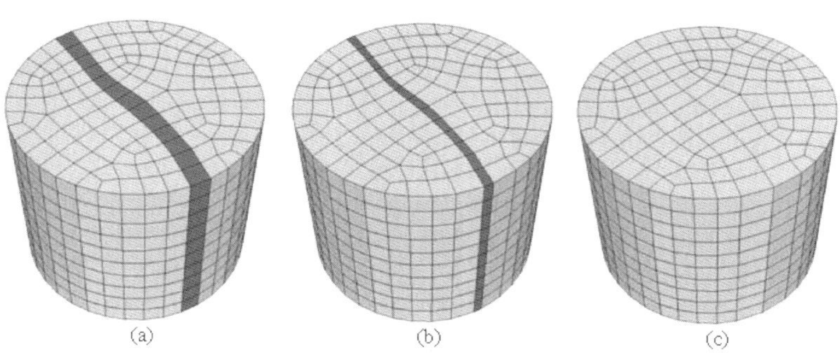

Fig. 5. Sheet extraction: (a) A sheet is selected for extraction. (b) The edges that define the sheet are collapsed. (c) The two nodes on each edge are merged, which eliminates the sheet and preserves a conforming hexahedral mesh.

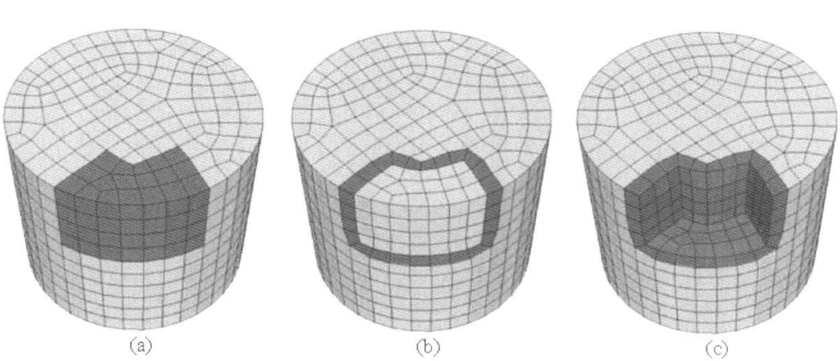

Fig. 6. Pillowing: (a) A shrink set is defined. (b) The elements in the shrink set are reduced in size and separated from the rest of the mesh. A sheet is inserted to fill in the gap and preserve a conforming hexahedral mesh. (c) The newly inserted sheet.

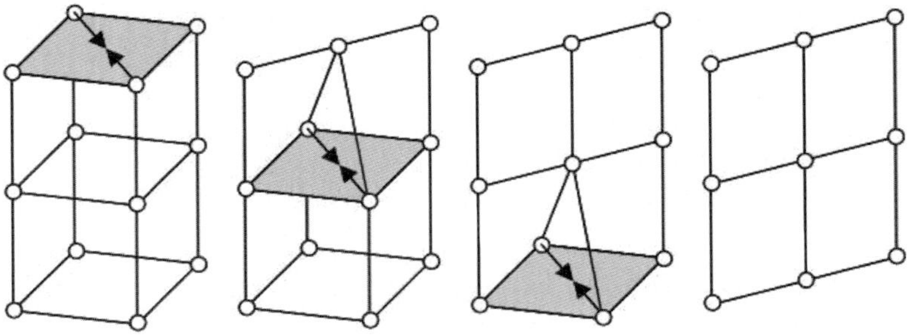

Fig. 7. Column collapse operation

element density in the vicinity of the extracted sheet and guarantees that the resulting mesh will be conforming.

Another hexahedral mesh operation that involves sheets is pillowing [19, 20]. Unlike sheet extraction, which removes an existing sheet from a mesh, pillowing inserts a new sheet into a mesh. As demonstrated in Figure 6, pillowing is performed on a contiguous group of hexahedral elements which make up a 'shrink' set. These elements are reduced in size and pulled away from the rest of the mesh, leaving a gap. A new sheet is then inserted into the gap by reconnecting each of the separated node pairs with a new edge. The new sheet increases element density in the vicinity of the shrink set and ensures the preservation of a conforming hexahedral mesh.

A third hexahedral mesh operation is known as column, or face, collapsing [18, 21]. A column is collapsed by merging diagonally opposite nodes in each quadrilateral face that defines the column, as shown in Figure 7. Since a quadrilateral face has two pairs of diagonally opposite nodes, a column can be collapsed in one of two different directions.

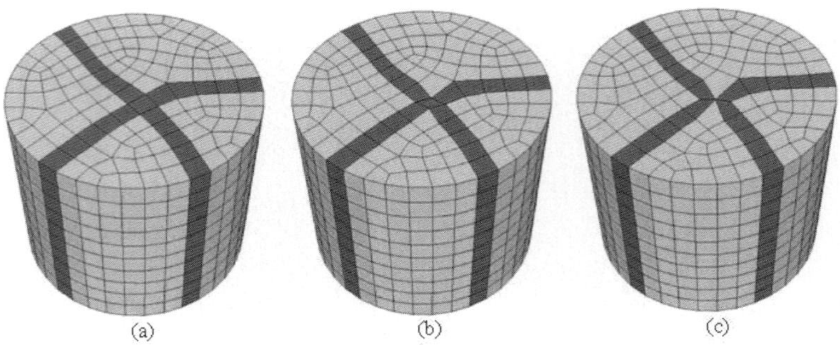

Fig. 8. Redirection of intersecting sheets through column collapsing: (a) Two intersecting sheets. (b) The column defining the intersection is collapsed. (c) The two sheets no longer intersect.

As previously mentioned, a column defines the intersection of two sheets. When a column is collapsed, two intersecting sheets are altered such that they no longer intersect, as illustrated in Figure 8. The paths of the new sheets are determined by the direction of the collapse. Just like sheet extraction and pillowing, the column collapse operation always preserves a conforming hexahedral mesh. In addition, similar to sheet extraction, the column collapse operation decreases element density in the vicinity of the collapsed column.

3 Hexahedral Mesh Coarsening

Utilizing the sheet and column operations described in Section 2, the hexahedral coarsening method presented in this section builds upon recent developments in quadrilateral coarsening [21]. While it is true that some quadrilateral coarsening operations can be directly extended to hexahedral coarsening, by themselves, these operations are not always able to prevent changes in element density from propogating beyond the boundaries of a defined hexahedral coarsening region.

3.1 Previously Developed Coarsening Techniques

As illustrated in Section 2.2, sheet extraction decreases mesh density by removing elements from a mesh. Therefore, sheet extraction is a very useful tool for hexahedral coarsening. However, sheet extraction by itself is generally not sufficient when localized coarsening is desired. This is due to the fact that sheets are rarely contained entirely within a region that has been selected for coarsening. As shown in Figure 9, extracting a sheet that extends beyond the boundaries of a defined region decreases mesh density in areas where coarsening is not desired. Therefore, before sheet extraction can occur, it is often necessary to modify the

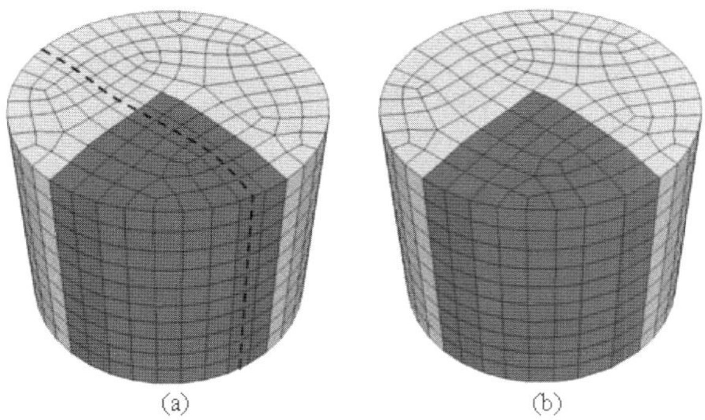

Fig. 9. Global coarsening: (a) A sheet passes through a region selected for coarsening. (b) When the sheet is extracted, mesh density is decreased both inside and outside the defined coarsening region.

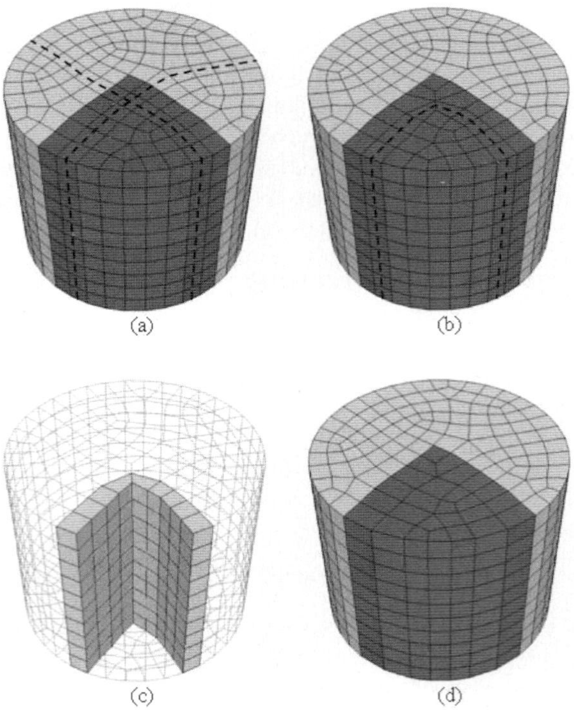

Fig. 10. Localized coarsening: (a) Two intersecting sheets pass through a region selected for coarsening. (b) The column defining the intersection of the two sheets in (a) is collapsed to produce a sheet contained entirely within the coarsening region. (c) The sheet that will be extracted. (d) When the sheet in (c) is extracted, mesh density is only decreased within the defined coarsening region.

mesh in such a way that produces sheets which are contained entirely within the boundaries of a defined coarsening region.

As described in Section 2.2, the paths of intersecting sheets can be altered using the column collapse operation. Figure 10, shows how this operation can be used to create a sheet that is contained entirely within a defined region. Such a sheet can then be extracted to coarsen the region without affecting any other part of the mesh.

The coarsening region shown in Figure 10 extends from the top to the bottom of the mesh. Suppose the coarsening region is modified so that it only extends a few layers from the top of the mesh, as shown in Figure 11. In this case, the column collapse operation can be used twice to produce a sheet that is contained entirely within the coarsening region. However, as seen in the figure, the first collapse operation is performed on a column which extends beyond the boundaries of the region. Collapsing this column modifies mesh topology and density in areas where coarsening is not desired. This shows that entirely localized coarsening cannot always be accomplished with the column collapse and sheet extraction operations alone.

3.2 Entirely Localized Coarsening

The previous examples demonstrate that entirely localized coarsening requires all operations to take place within the boundaries of the selected coarsening region. Referring to Figure 11, it can be seen that the second collapse operation was performed on a column contained within the coarsening region. Collapsing this column produced a sheet contained within the region without affecting any other part of the mesh. Of course, the formation of this column was accomplished through a previous collapse operation that did affect areas outside the coarsening region. Therefore, a critical aspect of entirely localized coarsening is the creation of local columns. Such columns must be formed in the coarsening region without affecting areas outside the region.

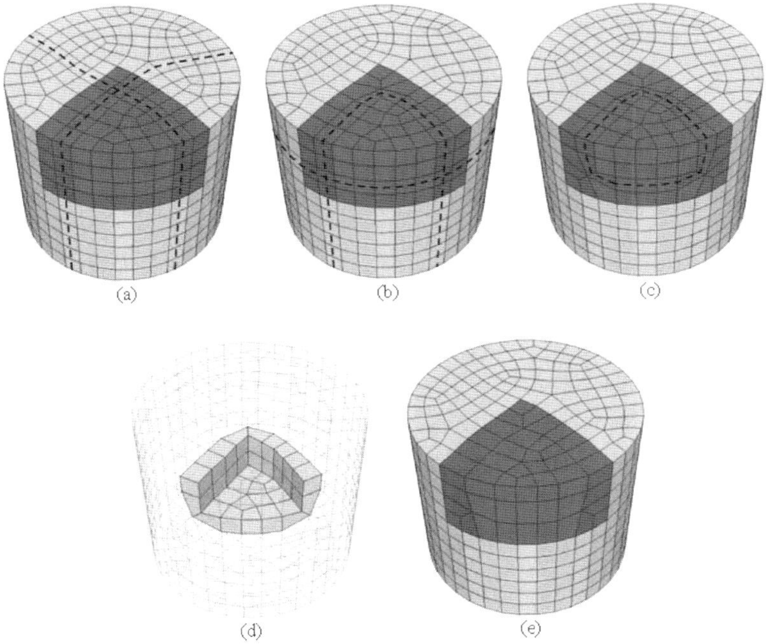

Fig. 11. Semi-localized coarsening: (a) Two intersecting sheets pass through a region selected for coarsening. (b) The column defining the intersection of the two sheets in (a) is collapsed. A sheet formed by the collapse and another intersecting sheet are shown. (c) The column defining the intersection of the two sheets in (b) is collapsed to produce a sheet contained entirely within the coarsening region. (d) The sheet that will be extracted. (e) When the sheet in (d) is extracted, mesh density is only decreased within the defined coarsening region.

One way to create local columns without affecting areas outside the coarsening region is through pillowing. As illustrated in Section 2.2, pillowing is a form of refinement because it increases mesh density in the vicinity of the shrink set. For this reason, pillowing is not an obvious solution for coarsening. However, due to the topology constraints that exist in a conforming all-hexahedral mesh, adding elements appears to be a necessary step when coarsening some regions.

Figure 12 shows how pillowing can be used to produce entirely localized coarsening. By pillowing the coarsening region, a sheet is inserted around the region. This sheet intersects other sheets that pass through the coarsening region and provides columns which follow the boundary of the region. Such columns can be collapsed to form sheets contained within the coarsening region without modifying mesh topology or density in areas where coarsening is not desired. These sheets can then be extracted to locally coarsen the region. It should be noted that many of the elements added through pillowing are removed through sheet extraction. Only those elements which are necessary to transition from higher to lower mesh density are left in the mesh. As long as the number of elements removed through sheet extraction is greater than the number of elements added through pillowing, the final mesh density in the coarsening region will be lower than the initial mesh density.

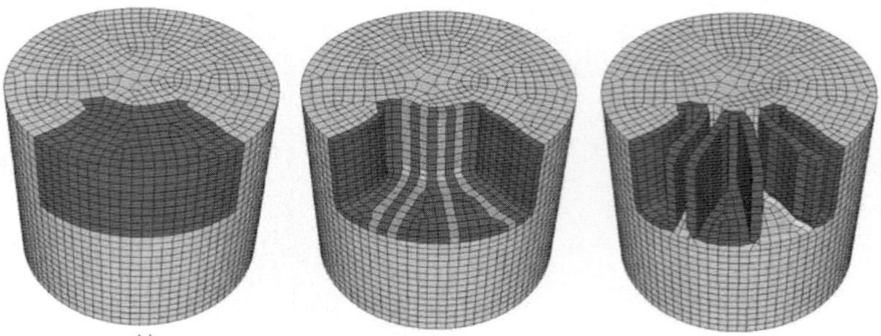

(a) (b) (c)

Fig. 12. Entirely localized coarsening: (a) A coarsening region is defined. (b) The sheet that forms when the coarsening region is pillowed. This sheet provides columns which follow the boundary of the region. (c) Collapsing the columns in (b) produces sheets contained entirely within the coarsening region.

3.3 Automated Coarsening Algorithm

For a given region, the process of pillowing, column collapsing, and sheet extraction can be repeated multiple times to achieve various levels of coarsening. A simple algorithm has been developed to automate this process for an arbitrary region and level of coarsening. The overall structure of the algorithm is briefly described by the following steps.

1. A coarsening region is defined and a target mesh density for that region is determined based on input given by a user.

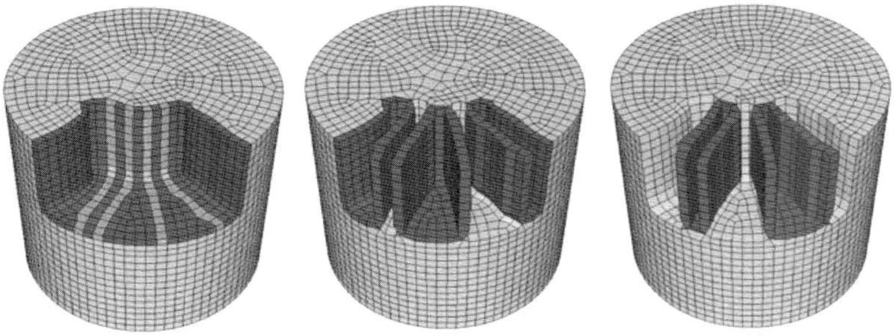

Fig. 13. Two coarsening options: (a) Columns selected for collapsing. (b) The sheets that will form if the columns are collapsed one way. (c) The sheets that will form if the columns are collapsed the other way.

2. Every sheet that passes through the coarsening region is found. Sheets contained entirely within the coarsening region are distinguished from those that extend beyond the region.
3. Due to a variety of geometry and mesh topology constraints, each sheet is examined to see if it will facilitate valid collapses and extractions during the coarsening process. Sheets that are unable to facilitate valid collapses and extractions are ignored from this point on.
4. For each acceptable sheet, a shape quality metric [22] is used to estimate how the quality of the mesh will be affected if that sheet, or the portion of that sheet contained in the coarsening region, is extracted. Sheets that will potentially produce a higher mesh quality are given higher priority.
5. If there are any sheets contained entirely within the coarsening region, then valid combinations of those sheets are analyzed. The combination that, when extracted, will produce a mesh density that is closest to the target mesh density without over-coarsening is saved. If no acceptable combination is found, the algorithm moves to step 6. Otherwise, steps 6 through 8 are skipped because no other operations are needed before sheet extraction.
6. If there are any sheets that extend beyond the coarsening region, then valid combinations of those sheets are analyzed. For each combination, two coarsening options are possible, as shown in Figure 13. These two coarsening options are distinguished by which direction the columns are collapsed. The combination that will produce a mesh density that is closest to the target mesh density without over-coarsening is saved. If no acceptable combination is found, steps 7 through 9 are skipped.
7. A sheet is inserted around the boundary of the coarsening region through pillowing.
8. Columns in the pillow sheet are collapsed in directions which were previously determined when the best sheet combination was saved. These collapses form sheets which are contained entirely within the coarsening region.
9. Sheets contained entirely within the coarsening region are extracted.

10. Steps 2 through 9 are repeated until the target mesh density is achieved (within a certain tolerance) or no more valid sheet combinations are found.
11. If coarsening took place, the remaining elements in the region are smoothed to improve mesh quality [23].

4 Examples

The following three examples show some results of the automated coarsening algorithm described in Section 3.3. In each example, the goal was to remove 25, 50, and 75 percent of the elements in the region selected for coarsening, while maintaining acceptable element quality. Quality was measured using the scaled Jacobian [24].

The first example was performed on a structured mesh of a cube, as shown in Figure 14. The second example was performed on an unstructured multiple-source to single-target swept mesh of a mechanical part, as shown in Figure 15.

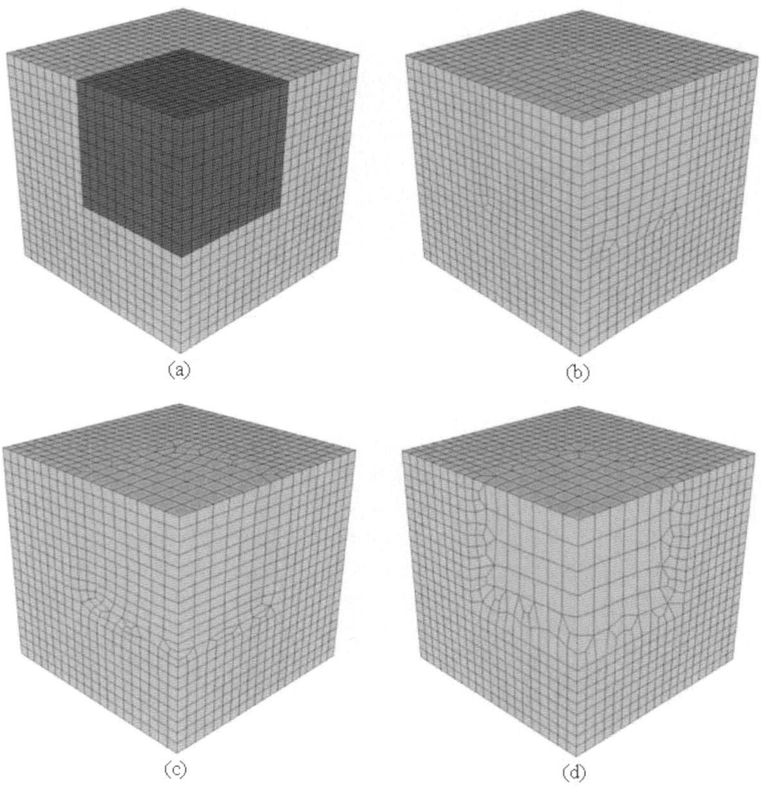

Fig. 14. Structured cube example: (a) Original mesh with coarsening region defined. (b) 25 percent coarsening. (c) 50 percent coarsening. (d) 75 percent coarsening.

Table 1. Coarsening Results for Cube Model

Target Percent Removal	Elements in Region	Actual Percent Removal	Min. Scaled Jacobian	Coarsening Time (sec)
0	1331	–	1.00	–
25	1056	20.7	0.47	0.7
50	684	48.6	0.41	0.9
75	355	73.3	0.34	1.1

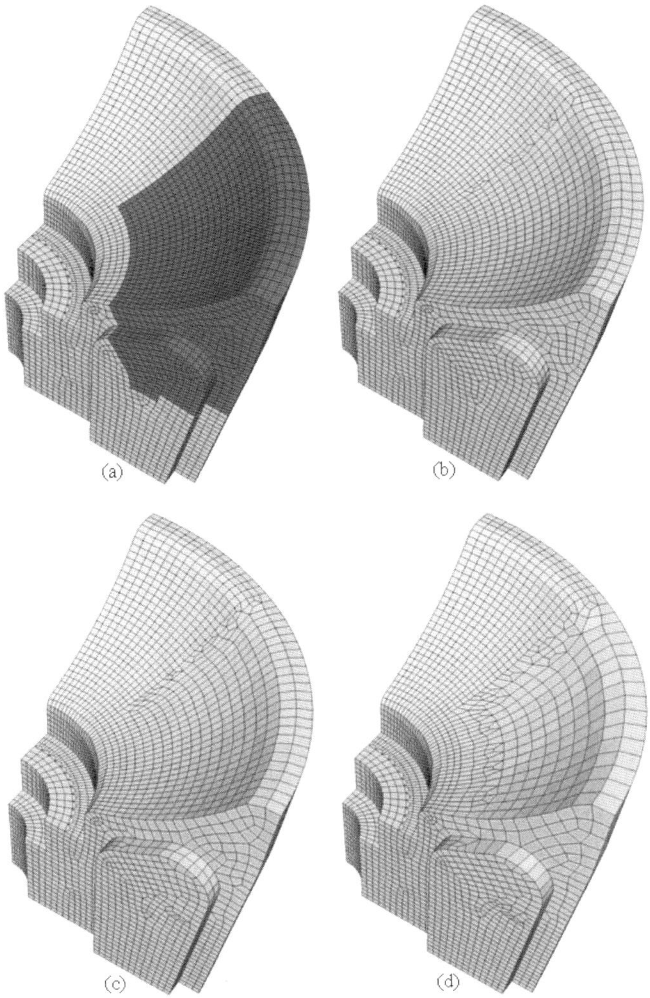

Fig. 15. Unstructured mechanical part example: (a) Original mesh with coarsening region defined. (b) 25 percent coarsening. (c) 50 percent coarsening. (d) 75 percent coarsening.

Table 2. Coarsening Results for Mechanical Part Model

Target Percent Removal	Elements in Region	Actual Percent Removal	Min. Scaled Jacobian	Coarsening Time (sec)
0	7641	–	0.77	–
25	5807	24.0	0.59	5.3
50	4057	46.9	0.32	9.6
75	2205	71.1	0.22	12.5

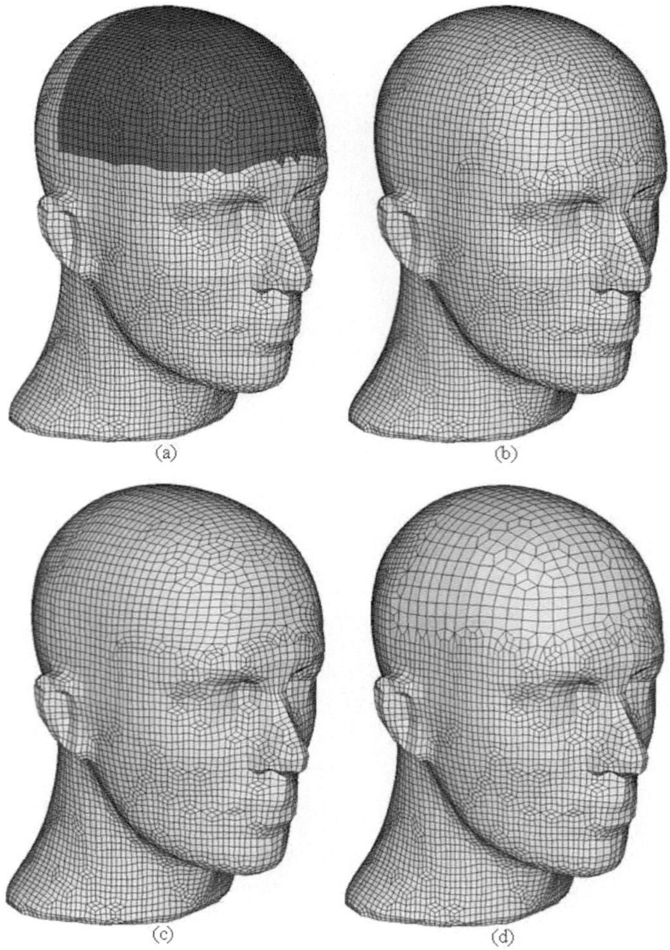

Fig. 16. Unstructured human head example: (a) Original mesh with coarsening region defined. (b) 25 percent coarsening. (c) 50 percent coarsening. (d) 75 percent coarsening.

Table 3. Coarsening Results for Human Head Model

Target Percent Removal	Elements in Region	Actual Percent Removal	Min. Scaled Jacobian	Coarsening Time (sec)
0	10080	–	0.48	–
25	7953	21.1	0.29	13.0
50	5129	49.1	0.17	17.9
75	2615	74.1	0.22	22.5

The final example was performed on an unstructured mesh of a human head generated with an octree based, sheet insertion algorithm [25], as shown in Figure 16. For both the mechanical part and human head models, refinement was performed prior to coarsening to create a higher starting mesh density.

Tables 1, 2, and 3 provide element removal, element quality, and coarsening time results for each model. In almost every case, acceptable element quality was maintained and a density that very nearly reflects the target mesh density was achieved.

5 Future Work

The automated coarsening algorithm described in Section 3.3 takes advantage of sheets already existing entirely within the coarsening region which can be extracted without any previous operations. However, it does not take advantage of columns already existing entirely within the coarsening region which can be collapsed without any previous operations. Modifying the algorithm to take advantage of such columns would improve the efficiency and effectiveness of the coarsening process in certain situations.

While the automated coarsening algorithm guarantees a topologically conforming mesh, it does not guarantee that the final quality of the mesh will be acceptable. Further research is needed to ensure that hexahedral coarsening does not degrade mesh quality below an acceptable threshold. This might be accomplished through more sophisticated methods which prevent poor quality elements from forming, or through cleanup operations which fix bad elements without significantly affecting mesh density. Many effective methods to cleanup a quadrilateral mesh have recently been developed. It is hoped that further research will lead to similar methods for a hexahedral mesh.

The coarsening method presented in this paper has been shown to work on unstructured meshes. However, even though these meshes are considered to be unstructured, they are usually structured in one dimension. Little work has been done to test this method on completely unstructured meshes. In theory, the method should work for any hexahedral mesh. However, it is likely that some meshes cannot be coarsened without degrading element quality below an acceptable level. Further work is needed to determine the limits of this coarsening method.

6 Conclusion

By utilizing sheet and column operations such as pillowing, column collapsing, and sheet extraction, entirely localized coarsening can be achieved in conforming all-hexahedral meshes. This method of coarsening works on both structured and unstructured meshes and is not based on undoing previous refinement. Although not fully developed, automation of this hexahedral coarsening method has already shown promising results. However, further work is needed to ensure acceptable element quality and to improve the efficiency of the overall process.

References

1. Rupak, B., Strawn, R.C.: Tetrahedral and hexahedral mesh adaptation for CFD problems. Applied Numerical Mathematics 26, 135–151 (1998)
2. Morton, D., Tyler, J.M.: A new 3D adaptive finite element scheme with 1-irregular hexahedral element meshes. In: Proceedings of the 2000 ACM Symposium on Applied Computing, pp. 99–104 (2000)
3. Tam, A., Ait-Ali-Yahia, D., Robichaud, M.P., Moore, M., Kozel, V., Habashi, W.G.: Anisotropic mesh adaptation for 3D flows on structured and unstructured grids. Computer Methods in Applied Mechanics and Engineering 189, 1205–1230 (2000)
4. Cavallo, P.A., Sinha, N., Feldman, G.M.: Peter, A., Cavallo, N.: Parallel unstructured mesh adaptation for transient moving body and aeropropulsive applications. In: 42nd AIAA Aerospace Sciences Meeting and Exhibit, pp. 6555–6565 (2004)
5. Benzley, S.E., Perry, E., Merkley, K., Clark, B.: A comparison of all hexagonal and all tetrahedral finite element meshes for elastic and elasto-plastic analysis. In: Proceedings, 4th International Meshing Roundtable, October 1995, pp. 179–191. Sandia National Laboratories (1995)
6. Tchon, K.-F., Dompierre, J., Camarero, R.: Conformal refinement of all-quadrilateral and all-hexahedral meshes according to an anisotropic metric. In: Proceedings, 11th International Meshing Roundtable, September 2002, pp. 231–242. Sandia National Laboratories (2002)
7. Harris, N., Benzley, S.E., Owen, S.J.: Conformal refinement of all-hexahedral meshes based on multiple twist plane insertion. In: Proceedings, 13th International Meshing Roundtable, September 2004, pp. 157–168. Sandia National Laboratories (2004)
8. Benzley, S.E., Harris, N.J., Scott, M.A., Borden, M.J., Owen, S.J.: Conformal refinement and coarsening of unstructured hexahedral meshes. Journal of Computing and Information Science in Engineering 5, 330–337 (2005)
9. Parrish, M., Borden, M.J., Staten, M.L., Benzley, S.E.: A selective approach to conformal refinement of unstructured hexahedral finite element meshes. In: Proceedings, 16th International Meshing Roundtable, September 2007, pp. 251–268. Sandia National Laboratories (2007)
10. Tautges, T.J., Knoop, S.: Topology modification of hexahedral meshes using atomic dual-based operations. In: Proceedings, 12th International Meshing Roundtable, September 2003, pp. 415–423. Sandia National Laboratories (2003)
11. Sheffer, A., Ungor, A.: Efficient adaptive meshing of parametric models. Journal of Computing and Information Science in Engineering 123, 366–375 (2001)

12. Choi, C.K., Lee, E.J., Yu, W.J.: Adaptive mesh refinement/recovery strategy for FEA. Structural Engineering and Mechanics 17, 379–391 (2004)
13. Kallinderis, Y., Kavouklis, C.: A dynamic adaptation scheme for general 3D hybrid meshes. Computer Methods in Applied Mechanics and Engineering 194, 5019–5050 (2005)
14. Kirk, B.S., Peterson, J.W., Stogner, R.H., Carey, G.F.: libmesh: A c++ library for parallel adaptive mesh refinement/coarsening simulations. Engineering With Computers 22, 237–254 (2006)
15. Taghavi, R.: Automatic, parallel and fault tolerant mesh generation from CAD. Engineering With Computers 12, 178–185 (1996)
16. Borden, M.J., Benzley, S.E., Shepherd, J.F.: Coarsening and sheet extraction for all-hexahedral meshes. In: Proceedings, 11th International Meshing Roundtable, September 2002, pp. 147–152. Sandia National Laboratories (2002)
17. Murdoch, P.J., Benzley, S.E.: The spatial twist continuum. In: Proceedings, 4th International Meshing Roundtable, October 1995, pp. 243–251. Sandia National Laboratories (1995)
18. Shepherd, J.F.: Topologic and Geometric Constraint-Based Hexahedral Mesh Generation, May 2007. Published Doctoral Dissertation, University of Utah (2007)
19. Merkley, K., Ernst, C.D., Shepherd, J.F., Borden, M.J.: Methods and applications of generalized sheet insertion for hexahedral meshing. In: Proceedings, 16th International Meshing Roundtable, September 2007, pp. 233–250. Sandia National Laboratories (2007)
20. Mitchell, S.A., Tautges, T.J.: Pillowing doublets: Refining a mesh to ensure that faces share at most one edge. In: Proceedings, 4th International Meshing Roundtable, October 1995, pp. 231–240. Sandia National Laboratories (1995)
21. Staten, M.L., Benzley, S.E., Scott, M.A.: A methodology for quadrilateral finite element mesh coarsening. Engineering With Computers 24 (2008)
22. Knupp, P.M.: Algebraic mesh quality metrics for unstructured initial meshes. Finite Elements in Analysis and Design 39, 217–241 (2003)
23. Knupp, P.M.: Hexahedral and tetrahedral mesh shape optimization. International Journal for Numerical Methods in Engineering 58(1), 319–332 (2003)
24. Knupp, P.M.: Algebraic mesh quality metrics. SIAM J. Sci. Comput. 23(1), 193–218 (2001)
25. Shepherd, J.F., Zhang, Y., Tuttle, C., Silva, C.T.: Quality improvement and boolean-like cutting operations in hexahedral meshes. In: Proceedings, 10th Conference of the International Society of Grid Generation, Crete, Greece (September 2007)

Author Index

Printing: Krips bv, Meppel, The Netherlands
Binding: Stürtz, Würzburg, Germany